Lecture Notes in Computer Sci

Edited by G. Goos, J. Hartmanis and J. van

T0237740

Springer
*Berlin
Heidelberg
New York
Barcelona
Hong Kong
London
Milan
Paris
Singapore
Tokyo*

Afonso Ferreira   Horst Reichel (Eds.)

STACS 2001

18th Annual Symposium
on Theoretical Aspects of Computer Science
Dresden, Germany, February 15-17, 2001
Proceedings

Springer

Series Editors

Gerhard Goos, Karlsruhe University, Germany
Juris Hartmanis, Cornell University, NY, USA
Jan van Leeuwen, Utrecht University, The Netherlands

Volume Editors

Afonso Ferreira
CNRS, I3S & INRIA Sophia Antipolis
INRIA, 2004 Route des Lucioles, 06902 Sophia-Antipolis, France
E-mail: ferreira@sophia.inria.fr

Horst Reichel
TU Dresden
Institut für Theoretische Informatik, Fakultät Informatik
01062 Dresden, Germany
E-mail: reichel@tcs.inf.tu-dresden.de

Cataloging-in-Publication Data applied for

Die Deutsche Bibliothek - CIP-Einheitsaufnahme

STACS <18, 2001, Dresden>:
Proceedings / STACS 2001 / 18th Annual Symposium on Theoretical
Aspects of Computer Science, Dresden, Germany, February 15 - 17, 2001.
Afonso Ferreira ; Horst Reichel (ed.). - Berlin ; Heidelberg ; New
York ; Barcelona ; Hong Kong ; London ; Milan ; Paris ; Singapore ;
Tokyo : Springer, 2001
    (Lecture notes in computer science ; Vol. 2010)
    ISBN 3-540-41695-1

CR Subject Classification (1998): F, E.1, I.3.5, G.2

ISSN 0302-9743
ISBN 3-540-41695-1 Springer-Verlag Berlin Heidelberg New York

Springer-Verlag Berlin Heidelberg New York
a member of BertelsmannSpringer Science+Business Media GmbH
© Springer-Verlag Berlin Heidelberg 2001
Printed in Germany

Typesetting: Camera-ready by author, data coversion by Christian Grosche
Printed on acid-free paper   SPIN 10782078   06/3142   5 4 3 2 1 0

Preface

The Symposium on Theoretical Aspects of Computer Science (STACS) is held annually, alternating between France and Germany. The STACS meetings are organized jointly by the Special Interest Group for Theoretical Computer Science of the Gesellschaft für Informatik (GI) in Germany and the Maison de l'Informatique et des Mathématiques Discrètes (MIMD) in France.

STACS 2001 was the 18th in this series, held in Dresden, February 15-17, 2001. Previous STACS symposia took place in Paris (1984), Saarbrücken (1985), Orsay (1986), Passau (1987), Bordeaux (1988), Paderborn (1989), Rouen (1990), Hamburg (1991), Cachan (1992), Würzburg (1993), Caen (1994), München (1995), Grenoble (1996), Lübeck (1997), Paris (1998), Trier (1999), and Lille (2000). It may be worth noting that in 2001 the symposium was held in one of the new states of reunited Germany for the first time. The proceedings of all of these symposia have been published in the Lecture Notes in Computer Science series of Springer–Verlag.

STACS has become one of the most important annual meetings in Europe for the theoretical computer science community. It covers a wide range of topics in the area of foundations of computer science: algorithms and data structures, automata and formal languages, computational and structural complexity, logic, verification, and current challenges. This year, 153 submissions were received, mostly in electronic form, from more than 30 countries, with a fair portion from non–European countries. We would like to thank Jochen Bern who designed the electronic submission procedure which performed marvelously and was of great help to the program committee. The program committee met for two days in Dresden and selected 46 out of the 153 submissions. Most of the papers were evaluated by four members of the program committee, partly with the assistance of subreferees. We thank the program committee for the thorough and careful work. Our gratitude extends to the numerous subreferees. The program committee was impressed by the high scientific quality of the submissions as well as the broad spectrum they covered. Because of the constraints imposed by the limited period of the symposium, a number of good papers could not be accepted.

We thank the three invited speakers at this symposium, Julien Cassaigne (Marseille), Martin Grohe (Chicago), and Dexter Kozen (Ithaca) for accepting our invitation to share their insights on new developments in their research areas.

We would like to express our sincere gratitude to all the members of the *Institut für Theoretische Informatik* and to the local organizing committee who invested their time and energy to organize this conference.

We would like to acknowledge the various sources of financial support for STACS 2001, especially the Deutsche Forschungsgemeinschaft (DFG), the Ministerium für Wissenschaft und Kunst des Landes Sachsen, and Freunde und Förderer der TU Dresden.

December 2000

Afonso Ferreira
Horst Reichel

Program Committee

D. Barrington
A. Bertone
J. Blömer
V. Blondel
J. D. Boissonnat
M. Bousquet-Mélou
M. Dietzfelbinger
A. Ferreira
L. Fribourg
G. Gastin
A. Marchetti-Spacamella
H. Reichel (Chair)
G. Rote
N. Vereshchagin
Th. Wilke

Local Arrangements Committee

H. Vogler
P. Buchholz

Referees[1]

Parosh Abdulla	Danièle Beauquier	Yegor Bryukhov
Luca Aceto	Luca Becchetti	Peter Buergisser
Micah Adler	Béatrice Berard	Jan van den Bussche
Marc Aiguier	Anna Bernasconi	Christian Cachin
Susanne Albers	Gilles Bernot	Benoit Caillaud
Paola Alimonti	J. Berstel	T. Calamoneri
Eric Allender	Andre Berthiaume	Cristian Calude
Jean-Paul Allouche	A. Bertoni	Olivier Carton
Helmut Alt	Ingrid Biehl	Julien Cassaigne
Karine Altisen	Norbert Blum	Franck Cassez
Andris Ambainis	Luc Boasson	N. Cesa-Bianchi
Klaus Ambos–Spies	Bernard Boigelot	Aleksey Chernov
Christoph Ambuehl	Paolo Boldi	Alexey Chernov
Eric Angel	M. A. Bonuccelli	Christian Choffrut
A. Arnold	Ahmed Bouajjani	Francesc Comellas
Giorgio Ausiello	Amar Bouali	Hubert Comon
Philippe Balbiani	Olivier Bournez	David Coudert
E. Bampis	Mireille Bousquet-Melou	Jean-Michel Couvreur
Klaus Barthelmann	Julian Bradfield	Yves Crama
Reuven Bar-Yehuda	Andreas Brandstädt	Pierluigi Crescenzi
Christina Bazgan	Peter Brass	M. Crochemore
Cristina Bazgan	Roberto Bruni	Carsten Damm
Marie-Pierre Beal	Véronique Bruyère	Fabrizio d'Amore

---

[1] The list of referees was compiled automatically from the database that was used by the program committee. We apologize for any omissions and inaccuracies.

Guy Melançon
C. Mereghetti
Yves Metivier
Friedhelm Meyer auf der Heide
C. Moore
Nicole Morawe
Michel Morvan
Larry Moss
Till Mossakowski
Laurent Mounier
Andrei Muchnik
Madhavan Mukund
Markus Müller-Olm
Anca Muscholl
Kedar Namjoshi
Giri Narasimhan
Mark-Jan Nederhof
Sotiris Nikoletseas
Damian Niwinski
Vincnet van Oostrom
Friedrich Otto
Martin Otto
Christophe Paul
Gheorghe Paun
W. Penczek
Paolo Penna
Mati Pentus
Stephane Perennes
Pino Persiano
Antoine Petit
Rossella Petreschi
Ulrich Pferschy
Birgit Pfitzmann
Claudine Picaronny
Giovanni Pighizzini
Jean-Eric Pin
Axel Poigne
Lorant Porkolab
Alex Rabinovich
R. Ramanujam
Klaus Reinhardt

Steffen Reith
Eric Rémila
Antonio Restivo
Mark Reynolds
Herve Rivano
J. M. Robson
Andrei E. Romashchenko
Gianluca Rossi
Jörg Rothe
Michel de Rougemont
Jan Rutten
Mike Saks
Massimo Santini
Martin Sauerhoff
P. Savicky
Christian Schindelhauer
Michael Schmitt
Philippe Schnoebelen
Uwe Schöning
Rainer Schuler
Nicole Schweikardt
Thomas Schwentick
Camilla Schwind
Geraud Senizergues
Alexander Shen
Amin Shokrollahi
Mihaela Sighireanu
David Simplot
Vaclav Snasel
Roberto Solis-Oba
Paul Spirakis
Ludwig Staiger
Yannis Stamatiou
Kathleen Steinhöfel
Andrea Sterbini
Iain Stewart
Colin Stirling
Howard Straubing
Gregoire Sutre
Till Tantau
Sergey Tarasov
Gábor Tardos

Hendrik Tews
Bernhard Thalheim
Denis Therien
Peter Thiemann
Martin Thimm
Wolfgang Thomas
Sophie Tison
Jacobo Toran
Mauro Torelli
Francoise Tort
Denis Trystram
Savio Tse
Max Ushakov
Erich Valkema
Brigitte Vallee
Oleg Verbitsky
Yann Verhoeven
Adrian Vetta
M. Vidyasagar
Sebastiano Vigna
Berthold Vöcking
Heribert Vollmer
M. Vyalyi
Michael Vyugin
Vladimir V'yugin
Stephan Waack
Klaus Wagner
RFC Walters
Rolf Wanka
Osamu Watanabe
Klaus Weihrauch
Pascal Weil
Emo Welzl
Rolf Wiehagen
Gerhard Wöginger
Haiseung Yoo
Stanislav Zak
Thomas Zeugmann
Martin Ziegler
Wieslaw Zielonka
Alexander Zvonkin
Uri Zwick

Sponsoring Institutions

- Deutsche Forschungsgemeinschaft (DFG)
- Ministerium für Wissenschaft und Kunst des Landes Sachsen
- Freunde und Förderer der TU Dresden.

Table of Contents

Invited Presentations

Contributions

Recurrence in Infinite Words

(Extended Abstract)

Julien Cassaigne

Institut de Mathématiques de Luminy
Case 907, F-13288 Marseille Cedex 9, France
cassaigne@iml.univ-mrs.fr

Abstract. We survey some results and problems related to the notion of recurrence for infinite words.

1 Introduction

The notion of recurrence comes from the theory of dynamical systems. A system $T : X \to X$ is *recurrent* when any trajectory eventually returns arbitrarily near its starting point, or in more formal terms, when for any open subset U of X and any $x \in U$, there exists an integer $n \geq 1$ such that $T^n(x) \in U$. And if this n — the *return time* — can be chosen independently of x for a given U, the system is said to be *uniformly recurrent*.

When $X = \overline{O(\mathbf{u})}$ is the subshift generated by an infinite word $\mathbf{u}$, the recurrence of X can be expressed as a combinatorial property of the word $\mathbf{u}$. Moreover, it is possible to compute return times and this allows to quantify the speed of recurrence in the system, via the *recurrence function* of the word $\mathbf{u}$. This point of view was initiated by Morse and Hedlund in their 1938 article on symbolic dynamics [13].

In this article, we survey some results and problems concerning recurrence in infinite words.

2 Preliminaries

Let A be a finite alphabet, with at least two elements. We denote by A^* the set of finite words over A (i.e., the free monoid generated by A), including the empty word ε, and by $A^{\mathbb{N}}$ the set of one-way infinite words over A. Given an infinite word $\mathbf{u} \in A^{\mathbb{N}}$, we denote by $F(\mathbf{u})$ the set of factors (or subwords) of $\mathbf{u}$, and, for any $n \in \mathbb{N}$, by $F_n(\mathbf{u})$ the set of factors of length n of $\mathbf{u}$.

The *shift* is the operator T on $A^{\mathbb{N}}$ defined by $T(u_0 u_1 u_2 u_3 \ldots) = u_1 u_2 u_3 \ldots$. The set $A^{\mathbb{N}}$ equipped with the product topology is a compact topological space, and under the action of T it becomes a discrete dynamical system named the *full shift*. A closed subset of $A^{\mathbb{N}}$ invariant under T is a *subshift*, and in particular any infinite word $\mathbf{u} \in A^{\mathbb{N}}$ generates a subshift $\overline{O(\mathbf{u})}$, the adherence of $O(\mathbf{u}) = \{T^n(\mathbf{u}): n \in \mathbb{N}\}$.

A. Ferreira and H. Reichel (Eds.): STACS 2001, LNCS 2010, pp. 1–11, 2001.

3   Recurrence

3.1   Recurrent and Uniformly Recurrent Words

An infinite word **u** is said to be *recurrent* if any factor of **u** occurs infinitely often in **u**. Recurrence can be characterized using an apparently much weaker property:

Proposition 1. *An infinite word* $\mathbf{u} \in A^{\mathbb{N}}$ *is recurrent if and only if any prefix of* **u** *occurs at least twice in* **u**.

An infinite word is said to be *uniformly recurrent* if it is recurrent and additionally, for any factor w of **u**, the distance between two consecutive occurrences of w in **u** is bounded by a constant that depends only on w.

For instance, any purely periodic infinite word is uniformly recurrent. Many classical infinite words like the Thue-Morse and Fibonacci words are uniformly recurrent. An eventually periodic word which is not purely periodic is not recurrent. The word

01011011101111011111011111101111111101111111110111111111...

(where the number of ones between consecutive zeros increases each time by one) is not recurrent and cannot be made recurrent by removing a prefix.

There are infinite words which are recurrent but not uniformly recurrent. Examples are easily constructed as fixed points of substitutions on A^*. For instance, the word

01011101011111111010111010111111111111111111111111110101110101...

is a fixed point of the substitution $0 \mapsto 010$, $1 \mapsto 111$. It is recurrent but not uniformly recurrent (since it is a fixed point of a substitution, it is sufficient for this to show that both symbols 0 and 1 occur infinitely often, but with unbounded intervals in the case of 0). Note that this word can also be defined as the characteristic word of the set of nonnegative integers that have at least one 1 in their ternary expansion.

3.2   The Recurrence Function

The *recurrence function* of an infinite word **u** is the function $R_{\mathbf{u}}\colon \mathbb{N} \to \mathbb{N} \cup \{+\infty\}$ defined by

$$R_{\mathbf{u}}(n) = \inf\left(\{N \in \mathbb{N}\colon \forall v \in F_N(\mathbf{u}), F_n(v) = F_n(\mathbf{u})\} \cup \{+\infty\}\right) .$$

In other words, $R_{\mathbf{u}}(n)$ is the size of the smallest window such that, whatever the position of the window on **u**, all factors of length n that occur in **u** occur at least once inside the window. We shall write $R(n) = R_{\mathbf{u}}(n)$ when there is no ambiguity on the relevant infinite word (this convention also applies to other notations with an infinite word as an index).

For instance, in the Thue-Morse word

0110100110010110100101100110100110010110011010010110100110010110...

(the fixed point of the substitution θ with $\theta(0) = 01$ and $\theta(1) = 10$), we have $R(0) = 0$, $R(1) = 3$ (every factor of length 3 contains at least one 0 and one 1), $R(2) = 9$ (the factor 01011010 of length 8 does not contain 00), $R(3) = 11$, etc. The recurrence function of this word is studied in detail in [13]. We shall present in Sect. 5 a method to compute $R(n)$ in general for words similar to this one.

An infinite word $\mathbf{u}$ is uniformly recurrent if and only if $R_{\mathbf{u}}$ takes only finite values.

3.3 Return Times and Return Words

A closely related notion is that of *return time*. Given a factor w of a recurrent infinite word $\mathbf{u} = u_0 u_1 u_2 \ldots$, an integer i is the position of an occurrence of w in $\mathbf{u}$ if $u_i u_{i+1} \ldots u_{i+|w|-1} = w$. Let us denote by i_0 the smallest such position, by i_1 the next one, etc., so that $(i_0, i_1, i_2, i_3 \ldots)$ is the increasing sequence of all positions of occurrences of w in $\mathbf{u}$. Then define the set of words

$$r_{\mathbf{u}}(w) = \{u_{i_j} u_{i_j+1} \ldots u_{i_{j+1}-1} : j \in \mathbb{N}\} \ .$$

Elements of $r_{\mathbf{u}}(w)$ are called *return words* of w in $\mathbf{u}$, and the (possibly infinite) number $\ell_{\mathbf{u}}(w) = \sup\{|v| : v \in r_{\mathbf{u}}(w)\}$ is called the *(maximal) return time* of w in $\mathbf{u}$. Note that return words of w either have w as a prefix, or are prefixes of w. The latter case happens when two occurrences of w in $\mathbf{u}$ overlap.

Finally, for all $n \in \mathbb{N}$ define $\ell_{\mathbf{u}}(n) = \max\{\ell_{\mathbf{u}}(w) : w \in F_n(\mathbf{u})\}$. Then

Proposition 2. *For any recurrent infinite word $\mathbf{u} \in A^{\mathbb{N}}$ and for any $n \in \mathbb{N}$, one has $R_{\mathbf{u}}(n) = \ell_{\mathbf{u}}(n) + n - 1$ (with the convention that $+\infty + n - 1 = +\infty$).*

For instance, consider the Fibonacci word

0100101001001010010100100101001001010010100100101001010010010100...

(the fixed point of the substitution $0 \mapsto 01$, $1 \mapsto 0$). The factor 0010 occurs at positions 2, 7, 10, 15, 20, 23, etc. and two return words can be observed in the prefix of $\mathbf{u}$ shown here, 001 and 00101. In fact, $r(0010) = \{001, 00101\}$ and $\ell(0010) = 5$, but other factors of the same length have longer return times and $\ell(4) = \ell(0101) = 8$, hence $R(4) = 11$

Return times have a direct dynamic interpretation. In the subshift $X = \overline{O(\mathbf{u})}$ generated by $\mathbf{u}$, given a finite word $w \in F(\mathbf{u})$, the set $[w] = \{\mathbf{v} \in X : w$ is a prefix of $\mathbf{v}\}$ is both open and closed and is called a *cylinder*. Then the definition of $\ell_{\mathbf{u}}(w)$ can be rephrased as

$$\ell_{\mathbf{u}}(w) = \inf\{N \in \mathbb{N} : \forall \mathbf{v} \in [w], \exists n \in \mathbb{N}, 1 \le n \le N \text{ and } T^n(\mathbf{v}) \in [w]\} \ ,$$

i.e., $\ell_{\mathbf{u}}(w)$ is the maximum time before which the system returns to the cylinder $[w]$.

The set $r_{\mathbf{u}}(w)$ of return words of a factor w is always a *circular code*, and if i_0 is the position of the first occurrence of w in $\mathbf{u}$, then $T^{i_0}(\mathbf{u})$ can be factored over this code. In particular, if w is a prefix of $\mathbf{u}$ and $r_{\mathbf{u}}(w)$ is finite, then $\mathbf{u}$ can be *recoded* as $\mathbf{u} = f(\mathbf{v})$, where $\mathbf{v} \in B^{\mathbb{N}}$ is an infinite word on a new alphabet B, and f is a one-to-one map from B to $r_{\mathbf{u}}(w)$, extended as a substitution. Such a word $\mathbf{v} = \Delta_w(\mathbf{u})$ is said to be *derived* from $\mathbf{u}$. The following characterization is due to F. Durand (a substitution $f\colon A^* \to A^*$ is *primitive* if there exists an integer $n \geq 1$ such that, for all $a \in A$, $f^n(a)$ contains every letter of A at least once):

Theorem 1 (Durand [9]). *An infinite word* $\mathbf{u} \in A^{\mathbb{N}}$ *is a fixed point of a primitive substitution on a subset of A if and only if $\mathbf{u}$ is uniformly recurrent and the number of distinct (up to letter renaming) infinite words derived from* $\mathbf{u}$ *is finite.*

For instance, the Thue-Morse word, which is a fixed point of a primitive substitution, has three distinct derived words, the Thue-Morse word $\mathbf{t}$ itself:

0110100110010110100101100110100110010110011010010110100110010110...

the derivated word associated with the prefix 0, $\Delta_0(\mathbf{t})$, with $\mathbf{t} = f_1(\Delta_0(\mathbf{t}))$ where $f_1(0) = 011$, $f_1(1) = 01$, $f_1(2) = 0$:

012021012102012021020121012021012102012102021020120210121020120...

and the derivated word associated with all prefixes of length 2 or more, $\mathbf{v} = \Delta_{01}(\mathbf{t}) = \Delta_{011}(\mathbf{t}) = \cdots$, with $\mathbf{t} = f_2(\mathbf{v}) = f_3(\mathbf{v}) = \cdots$, where $f_2(0) = 011$, $f_2(1) = 010$, $f_2(2) = 0110$, $f_2(3) = 01$, and $f_{2+k} = \theta^k \circ f_2$ for all $k \in \mathbb{N}$:

012301320123201301230132013012320123013201232013012320123013201320 13...

3.4   Recurrence and Subword Complexity

Another numerical function associated with an infinite word $\mathbf{u}$ is the (subword) complexity function $p_{\mathbf{u}}\colon \mathbb{N} \to \mathbb{N}$ defined by $p_{\mathbf{u}}(n) = \#F_n(\mathbf{u})$, the number of factors of length n in $\mathbf{u}$. There is no direct relation between the functions $p_{\mathbf{u}}$ and $R_{\mathbf{u}}$, but only an inequality.

Proposition 3 (Morse and Hedlund [13]). *For any infinite word* $\mathbf{u} \in A^{\mathbb{N}}$ *and for any $n \in \mathbb{N}$, one has $\ell_{\mathbf{u}}(n) \geq p_{\mathbf{u}}(n)$ and $R_{\mathbf{u}}(n) \geq p_{\mathbf{u}}(n) + n - 1$.*

For non-periodic words, this inequality is not optimal and Morse and Hedlund show that it can be improved to $R_{\mathbf{u}}(n) \geq p_{\mathbf{u}}(n) + n$.

In the other direction, no such inequality holds since it is possible to construct infinite words with $p(n) = n + 1$ (Sturmian words) for which $R(n)$ grows as fast as desired, while remaining finite.

4   Linear Recurrence

4.1   Linearly Recurrent Words

When the recurrence function grows slowly, it means that all factors have to occur rather often and this gives much structure to the infinite word. Of particular interest are words for which $R(n)/n$ is bounded. An infinite word is said to be *linearly recurrent* with constant K if $\ell(n) \leq Kn$ for all $n \geq 1$ [11].

Proposition 4 (Durand, Host, and Skau [11]). *Let* $\mathbf{u} \in A^{\mathbb{N}}$ *be a linearly recurrent infinite word with constant* K. *Then*

(i) *For all* $n \geq 1$, $R(n) \leq (K + 1)n - 1$.
(ii) *For all* $n \geq 1$, $p(n) \leq Kn$.
(iii) $\mathbf{u}$ *is* $(K + 1)$-*power free (i.e, it does not contain any factor of the form* w^{K+1} *with* $w \in A^* \setminus \{\varepsilon\}$).
(iv) *For all* $w \in F(\mathbf{u})$ *and* $v \in r_{\mathbf{u}}(w)$, $|w|/K < |v| \leq K|w|$.
(v) *For all* $w \in F(\mathbf{u})$, $\#r_{\mathbf{u}}(w) \leq K(K + 1)^2$.

Property (iv) shows that in a linearly recurrent word, return times can be neither too long nor too short. By property (ii), linearly recurrent words are a special case of words with linear subword complexity. In particular, this implies that $p_{\mathbf{u}}(n + 1) - p_{\mathbf{u}}(n)$ is bounded by a constant that depends only on K and $\#A$ [3].

The structure of linearly recurrent words can be characterized using *primitive S-adic infinite words*, words obtained by applying in an appropriate order substitutions taken from a finite set (see [10] for a precise definition):

Theorem 2 (Durand [10]). *An infinite word* $\mathbf{u}$ *is linearly recurrent if and only if it is an element of the subshift generated by some primitive S-adic infinite word.*

4.2   Recurrence Quotient

Another way to define linearly recurrent words is to use the *recurrence quotient* $\rho_{\mathbf{u}}$. For any infinite word $\mathbf{u}$, let

$$\rho_{\mathbf{u}} = \limsup_{n \to +\infty} \frac{R_{\mathbf{u}}(n)}{n} \in \mathbb{R} \cup \{+\infty\} .$$

Then $\rho_{\mathbf{u}}$ is a finite real number if and only if $\mathbf{u}$ is linearly recurrent. Moreover, if $\mathbf{u}$ is linearly recurrent with constant K, $\rho_{\mathbf{u}} \leq K + 1$.

If $\mathbf{u}$ is a purely periodic word, it is clear that $\rho_{\mathbf{u}} = 1$. For non-periodic words, Hedlund and Morse [13] asked as an open problem to find the best lower bound for $\rho_{\mathbf{u}}$. Proposition 3 together with the fact that $p_{\mathbf{u}}(n) \geq n + 1$ ([13]) implies that $\rho_{\mathbf{u}} \geq 2$. Using graph representations, we improve this result to

Theorem 3. *Let* $\mathbf{u} \in A^{\mathbb{N}}$ *be an infinite word which is not purely periodic. Then* $\rho_{\mathbf{u}} \geq 3$.

Rauzy [16] conjectured that the minimal value of $\rho_{\mathbf{u}}$ for non-periodic word is still larger:

Conjecture 1 (Rauzy [16]). Let $\mathbf{u} \in A^{\mathbb{N}}$ be an infinite word which is not purely periodic. Then $\rho_{\mathbf{u}} \geq \frac{5+\sqrt{5}}{2} \simeq 3.618$.

This value $(5 + \sqrt{5})/2$ is exactly the recurrence quotient of the Fibonacci word (see below), so if the conjecture holds then it is optimal. We believe that the techniques used to prove Theorem 3 (see Sect. 7), and in particular the extensive study of possible Rauzy graphs, will lead to a proof of this conjecture.

5   Computing the Recurrence Function

5.1   Singular Factors

Wen and Wen [18] defined *singular words* as particular factors of the Fibonacci word (the factors 0, 1, 00, 101, 00100, 10100101, etc., of length the successive Fibonacci numbers, that when concatenated in this order yield the infinite word itself). Here we define singular factors for any infinite word, generalizing one of the properties of Wen and Wen's singular words.

Let $\mathbf{u}$ be an infinite word. A factor w of $\mathbf{u}$ is said to be *singular* for $\mathbf{u}$ if $|w| = 1$ or if there exist a word $v \in A^*$ and letters $x, x', y, y' \in A$ such that $w = xvy$, $x \neq x'$, $y \neq y'$ and $\{xvy, x'vy, xvy'\} \subset F(\mathbf{u})$. In other words, a factor w is singular if there is a way to alter its first letter and still have a factor of $\mathbf{u}$, and symmetrically with the last letter.

When $w = xvy$ is singular, then v is *bispecial*, i.e., v can be extended in at least two different ways both to the right and to the left (see [4]).

Proposition 5 ([7]). *Let $\mathbf{u}$ be an infinite word and $n \geq 1$. If $\ell(n-1) < \ell(n)$, then there exists a singular factor w of $\mathbf{u}$ such that $\ell(n) = \ell(w)$.*

A singular factor w is said to be an *essential singular factor* if $\ell(w) = \ell(|w|) > \ell(|w| - 1)$. We denote by $S(\mathbf{u})$ the set of singular factors of $\mathbf{u}$, and by $S'(\mathbf{u})$ the set of essential singular factors of $\mathbf{u}$.

Theorem 4 ([7]). *Let $\mathbf{u}$ be an infinite word and $n \geq 1$. Then*

$$\ell(n) = \sup\{\ell(w) : w \in S(\mathbf{u}) \text{ and } |w| \leq n\} = \sup\{\ell(w) : w \in S'(\mathbf{u}) \text{ and } |w| \leq n\} .$$

5.2   Computation Method

Theorem 4 allows to explicitly compute the recurrence function $R(n)$ as long as one is able to describe singular factors (or at least essential singular factors) and their return time. Since singular factors are extensions of bispecial factors, techniques presented in [4] can be used to describe them when the infinite word is a fixed point of a substitution, or more generally when it is defined using a finite number of substitutions (S-adic words). This results in the following procedure:

1. Determine bispecial factors. Usually a small number of bispecial factors of small length generate all other bispecial factors through recurrence relations.
2. Deduce the form of singular words, and compute their length.
3. For a given singular words, determine the associated return words and compute their length. Singular words with shorter return time can be left out since they are not essential.
4. Deduce the function $\ell(n)$, which will be typically staircase-like, the position and height of each step being expressed with a (usually linear) recurrence relation.

As an example, let us apply this procedure to the Thue-Morse word.

1. Apart from the empty word and letters, there are four families of bispecial factors: 01, 10, 010 and 101 are bispecial, and if w is bispecial then $\theta(w)$ is also bispecial.
2. Bispecial factors in the families generated by 01 and 10 each give rise to four singular factors, which can be summarized as $x\theta^k(y)z$ with $x, y, z \in \{0, 1\}$ and $k \geq 1$, of length $2^k + 2$. The two other families do not produce singular factors (because they are *weak bispecial factors*), and the remaining singular factors are all words of length 1 and 2, as well as 010 and 101.
3. Observation yields

$$r(0) = \{0, 01, 001\} \ ,$$
$$r(00) = \{0011, 001101, 001011, 00101101\} \ ,$$
$$r(01) = \{01, 010, 011, 0110\} \ ,$$
$$r(010) = \{010, 01011, 0100110, 010110011\} \ ,$$

the case of 1, 11, 10, and 101 being symmetric. The word 0010 always occurs in the form $0\theta(00)1^{-1}$, hence

$$r(0010) = \{0\theta(v)0^{-1} : v \in r(00)\}$$
$$= \{00101101, 001011010011, 001011001101, 0010110011010011\} \ .$$

Similarly,

$$r(0011) = \{1^{-1}\theta(v)1 : v \in r(101)\} \ ,$$
$$r(1010) = \{\theta(v) : v \in r(11)\} \ , \text{ and}$$
$$r(1011) = \{0^{-1}\theta(v)0 : v \in r(00)\} \ .$$

Then the return words of $x\theta^{k+1}(y)z$ can be deduced from those of $\bar{x}\theta^k(y)z$ by applying θ and conjugating by x. One has $\ell(0010) = \ell(1010) = \ell(1011) = 16$ and $\ell(0011) = 18$, so obviously only 0011 and the family it generates are essential. Essential singular factors of length 4 or more therefore have length $2^k + 2$ and return time 9.2^k (actually, this also holds for $k = 0$).
4. The function $\ell(n)$ is defined by $\ell(0) = 1$, $\ell(1) = 3$, $\ell(2) = 8$, and $\ell(n) = 9.2^k$ for $2^k + 2 \leq n \leq 2^{k+1} + 1$. Consequently $\rho = 1 + \limsup \ell(n)/n = 10$.

6   The Recurrence Quotient of Sturmian Words

6.1   Computing ρ Using Continued Fractions

Sturmian words are infinite words for which $p(n) = n + 1$; see [12] for equivalent definitions, properties and references. They are all uniformly recurrent.

In this particular case, the method given in the previous section to compute the recurrence function amounts to the method described by Morse and Hedlund [14] using continued fraction expansions. As far as the recurrence function is concerned, it is sufficient to study standard Sturmian words: given an irrational number $\alpha \in [0, 1] \setminus \mathbb{Q}$, the standard Sturmian word of density α is the word $\mathbf{u} = u_0 u_1 u_2 \ldots$ where $u_n = 1$ if the fractional part of $(n + 2)\alpha$ is less than α, $u_n = 0$ otherwise.

Proposition 6 (Morse and Hedlund [14]). *The essential singular factors of the standard Sturmian word $\mathbf{u}$ of density $\alpha \in [0, 1] \setminus \mathbb{Q}$ constitute a sequence (w_i) with $|w_i| = q_i$ and $\ell(w_i) = q_i + q_{i+1}$, where p_i/q_i are the convergents associated with the continued fraction expansion of the density, $\alpha = [0; a_1, a_2, a_3, \ldots]$. The recurrence quotient of $\mathbf{u}$ is $\rho = 2 + \limsup [a_i; a_{i-1}, \ldots, a_1]$.*

For instance, the Fibonacci word is the standard Sturmian word of density $\alpha = (3 - \sqrt{5})/2 = [0; 2, 1, 1, \ldots]$. Its recurrence quotient is therefore $\rho = 2 + \limsup [1; 1, 1, 1, \ldots, 1] = [3; 1, 1, 1, \ldots] = (5 + \sqrt{5})/2$. The denominators of the convergents are the classical Fibonacci numbers, $q_0 = F_1 = 1$, $q_1 = F_2 = 2$, $q_2 = F_3 = 3$, $q_3 = F_4 = 5$, $q_4 = F_5 = 8$, etc. and they correspond to the lengths of the essential singular factors, $w_0 = 1$, $w_1 = 00$, $w_2 = 101$, $w_3 = 00100$, $w_4 = 10100101$, etc. The associated return times are $\ell(w_i) = q_i + q_{i+1} = F_{i+3}$. Finally, the recurrence function satisfies $R(n) = F_{i+2} + n - 1$ if $F_i \leq n < F_{i+1}$.

A consequence of this proposition is that a Sturmian word is linearly recurrent if and only if the continued fraction expansion of its density is bounded. Then, if $a = \limsup a_i$, one has $a + 2 < \rho < a + 3$.

6.2   The Spectrum of Values of ρ

Let $S \subset \mathbb{R} \cup \{+\infty\}$ be the set of values taken by ρ for Sturmian words. The set S has an interesting topological structure (we treat sequences of integers $\mathbf{b} = (b_i)_{i \in \mathbb{N}}$ as infinite words on the infinite alphabet $\mathbb{N}^*$, so that the notation $[\mathbf{b}]$ means $[b_0; b_1, b_2, b_3, \ldots]$ and $[T^k(\mathbf{b})] = [b_k; b_{k+1}, b_{k+2}, b_{k+3}, \ldots]$):

Theorem 5 ([7]). *The set S is given by*

$$S = \{2 + [\mathbf{b}] : \mathbf{b} \in (\mathbb{N}^*)^{\mathbb{N}} \text{ and } \forall k \in \mathbb{N}, [\mathbf{b}] \geq [T^k(\mathbf{b})]\} \cup \{+\infty\} .$$

It is a compact subset of $[0, +\infty]$, with empty interior. It has the power of the continuum. Its smallest accumulation point is the transcendental number $\rho_0 = 2 + [\mathbf{v}] \simeq 4.58565$, where $\mathbf{v} = v_0 v_1 v_2 \ldots \in \{1, 2\}^{\mathbb{N}}$ is the fixed point of the substitution $1 \mapsto 2$, $2 \mapsto 211$. The intersection of S with the set of quadratic numbers is dense in S. Every non-countable interval of S contains a sub-interval which is isomorphic to S as an ordered set.

The transcendence of ρ_0 was proved by Allouche et al. in [2]. Some questions remain open about the structure of S, for instance its Hausdorff dimension.

7 Main Ideas for the Proof of Theorem 3

Assume that $\mathbf{u} \in A^{\mathbb{N}}$ is a non-periodic infinite word with $\rho < 3$. Assume also that $\mathbf{u}$ is a binary word (i.e., $\#A = 2$), since the general case can easily be reduced to the binary case by projection.

Let $s(n) = p(n+1) - p(n)$: since $\mathbf{u}$ is not eventually periodic, $s(n) \geq 1$ for all $n \in \mathbb{N}$. By Proposition 3, $\limsup p(n)/n < 2$, which implies that $s(n) = 1$ for infinitely many values of n. There are now two cases: either there is some n_0 such that $s(n) = 1$ for all $n \geq n_0$, or there are infinitely many n such that $s(n) = 1$ and $s(n+1) > 1$.

The first case is essentially the case of Sturmian words, and it is not difficult to adapt the method of [14] to prove that $\rho \geq (5 + \sqrt{5})/2 > 3$ in this case.

In the second case, we have infinitely many n for which the Rauzy graph (see [16,4]) is "eight-shaped" and contains a strong bispecial factor. For subsequent values of n, the Rauzy graphs get more complicated, and it is possible to express return times of certain words as lengths of paths in these graphs, for which lower bounds can be given. Combining these bounds, we get a contradiction with the assumption $\rho < 3$.

To prove a lower bound for ρ larger than 3, one would have to consider also infinite words for which $\liminf s(n) = 2$. Rauzy graphs for these words can have ten different shapes, which where first classified by Rote [17], and the study of their evolutions would involve a large number of subcases.

8 Two Other Functions

Two functions associated with an infinite word $\mathbf{u}$ and similar to the recurrence function have also been considered. The first one is $R'(n)$, the size of the smallest *prefix* w of $\mathbf{u}$ such that $F_n(w) = F_n(\mathbf{u})$, defined by Allouche and Bousquet-Mélou [1] to study a conjecture of Pomerance, Robson, and Shallit [15]. The second one is $R''(n)$, the size of the smallest *factor* w of $\mathbf{u}$ such that $F_n(w) = F_n(\mathbf{u})$, studied in [6]. These functions compare with each other as follows:

Proposition 7 ([6]). *For any infinite word $\mathbf{u} \in A^{\mathbb{N}}$ and any $n \in \mathbb{N}$, the functions $p_{\mathbf{u}}$, $R_{\mathbf{u}}$, $R'_{\mathbf{u}}$, and $R''_{\mathbf{u}}$ satisfy the inequality $p_{\mathbf{u}}(n) + n - 1 \leq R''_{\mathbf{u}}(n) \leq R'_{\mathbf{u}}(n) \leq R_{\mathbf{u}}(n)$.*

It should be noted that, whereas the functions R and R'' depend only on the set of factors of $\mathbf{u}$, or equivalently on the subshift generated by $\mathbf{u}$, the function R' depends on the specific word $\mathbf{u}$. The conjecture by Shallit et al., rephrased using the function R', was very similar to Conjecture 1: if $\mathbf{u}$ is an infinite word which is not eventually periodic, then $\limsup R'(n)/n \geq (3 + \sqrt{5})/2$. This is indeed true for standard Sturmian words, but considering non-standard Sturmian words (which have the same factors, but not the same prefixes) we were able to construct a counter-example, and to prove its optimality:

Theorem 6 ([5]). *Let* $\mathbf{u} \in A^{\mathbb{N}}$ *be an infinite word that is not eventually periodic. Then*

$$\limsup_{n \to +\infty} \frac{R'_{\mathbf{u}}(n)}{n} \geq \frac{29 - 2\sqrt{10}}{9} \simeq 2.51949 \;,$$

and this value is optimal since it is attained by the Sturmian word

$$\mathbf{z}_3 = 01001010010010010100100101001001001010010010010010010\ldots$$

fixed point of the substitution $0 \mapsto 01001010$, $1 \mapsto 010$.

The function $R''(n)$ seems to have less interesting properties. It is not difficult to see that the minimal value of $\limsup R''(n)/n$ for an infinite word that is not eventually periodic is 2. This minimal value is attained, among others, by all Sturmian words, which can in fact be characterized using R'':

Proposition 8 ([6]). *An infinite word* $\mathbf{u} \in A^{\mathbb{N}}$ *is Sturmian if and only if* $R''_{\mathbf{u}}(n) = 2n$ *for every* $n \geq 0$.

9   Concluding Remarks

Some of the properties that we have presented deal with the connections between recurrence and other properties of infinite words: subword complexity, frequencies, repetitions, special factors, etc. These connections have not been completely explored: for instance, the inequality between $p(n)$ and $R(n)$ in Proposition 3 can certainly be improved. We mainly focused on linearly recurrent words, and did not say much about recurrence of infinite words with very high subword complexity: while *complete* words (i.e., infinite words with maximal complexity $p(n) = \#A^n$) are not uniformly recurrent, how fast can the complexity of a uniformly recurrent word grow?

Not much is known about the spectrum of values taken by ρ for all infinite words, not just Sturmian ones. A proof of Conjecture 1 would provide the minimum of this spectrum, but there are many other questions on its structure. Is it similar to that of S in Theorem 5, or does it contain full intervals of real numbers? It seems that at least the minimum is an isolated point, what is the smallest accumulation point, is it different from ρ_0?

Another problem from [13] is still open: is it true in general that $R(n)/n$ does not converge to a limit? Some progress in this direction has been recently made by N. Chekhova [8].

References

1. J.-P. ALLOUCHE AND M. BOUSQUET-MÉLOU, On the conjectures of Rauzy and Shallit for infinite words, *Comment. Math. Univ. Carolinae* **36** (1995), 705–711.

2. J.-P. ALLOUCHE, J. L. DAVISON, M. QUEFFÉLEC, AND L. Q. ZAMBONI, Transcendence of Sturmian or morphic continued fractions. Preprint.

3. J. CASSAIGNE, Special factors of sequences with linear subword complexity, in *Developments in Language Theory II*, pp. 25–34, World Scientific, 1996.

4. J. CASSAIGNE, Facteurs spéciaux et complexité, *Bull. Belg. Math. Soc.* **4** (1997), 67–88.

5. J. CASSAIGNE, On a conjecture of J. Shallit, in *ICALP'97*, pp. 693–704, *Lect. Notes Comput. Sci.* **1256**, Springer-Verlag, 1997.

6. J. CASSAIGNE, Sequences with grouped factors, in *Developments in Language Theory III*, pp. 211–222, Aristotle University of Thessaloniki, 1998.

7. J. CASSAIGNE, Limit values of the recurrence quotient of Sturmian sequences, *Theoret. Comput. Sci.* **218** (1999), 3–12.

8. N. CHEKHOVA, Fonctions de récurrence des suites d'Arnoux-Rauzy et réponse à une question de Morse et Hedlund. Preprint.

9. F. DURAND, A characterization of substitutive sequences using return words, *Discrete Math.* **179** (1998), 89–101.

10. F. DURAND, Linearly recurrent subshifts, Research report 98-02, Institut de Mathématiques de Luminy, Marseille, France, 1998.

11. F. DURAND, B. HOST, AND C. SKAU, Substitutions, Bratteli diagrams and dimension groups, *Ergod. Th. Dyn. Sys.* **19** (1999), 952–993.

12. M. LOTHAIRE, Algebraic combinatorics on words. To appear. Available online at `http://www-igm.univ-mlv.fr/~berstel/Lothaire/`.

13. M. MORSE AND G. A. HEDLUND, Symbolic dynamics, *Amer. J. Math.* **60** (1938), 815–866.

14. M. MORSE AND G. A. HEDLUND, Symbolic dynamics II: Sturmian trajectories, *Amer. J. Math.* **61** (1940), 1–42.

15. C. POMERANCE, J. M. ROBSON, AND J. SHALLIT, Automaticity II: Descriptional complexity in the unary case, *Theoret. Comput. Sci.* **180** (1997), 181–201.

16. G. RAUZY, Suites à termes dans un alphabet fini, *Sém. Théor. Nombres Bordeaux*, 1982–1983, 25.01–25.16.

17. G. ROTE, Sequences with subword complexity $2n$, *J. Number Theory* **46** (1994), 196–213.

18. Z.-X. WEN AND Z.-Y. WEN, Some properties of the singular words of the Fibonacci word, *European J. Combin* **15** (1994), 587–598.

Generalized Model-Checking Problems
for First-Order Logic

Martin Grohe

Department of Mathematics, Statistics, and Computer Science, UIC
851 S. Morgan St. (M/C 249), Chicago, IL 60607-7045, USA
grohe@uic.edu

1 Introduction

Descriptive complexity theory provides a convenient and intuitive way to model a large variety of computational problems. The basic problem studied in descriptive complexity is of the following form:

> *Given a finite relational structure A and a sentence φ of some logic L, decide if φ is satisfied by A.*

We call this problem the *model-checking problem* for L. The model-checking problem and variants of it appear in different areas of computer science:

The name model-checking is usually associated with automated verification. Here state spaces of finite state systems are modeled by Kripke structures and specifications are formulated in modal or temporal logics. Then checking whether the system has the specified property means checking whether the Kripke structure satisfies the specification, that is, model-checking. In recent years, this approach has very successfully been applied to the verification circuits and protocols.

There has always been a close connection between descriptive complexity theory and database theory. As a matter of fact, some of the roots of descriptive complexity can be found in research on the expressiveness and complexity of database query languages (e.g. [3,31]). The basic link between the two areas is that relational databases and finite relational structures are just the same. Therefore, the problem of evaluating a Boolean query of some query language L against a relational database is the model-checking problem for L. (A Boolean query is a query with a yes/no answer.) More generally, evaluating a k-ary query $\varphi(x_1, \ldots, x_k)$ in a structure (or database) A amounts to computing all tuples $(a_1, \ldots, a_k)$ such that A satisfies $\varphi(a_1, \ldots, a_k)$. We call this variant of the model-checking problem the *evaluation problem* for L. The most important logic to be considered in this database context is first-order logic, which closely resembles the commercial standard query language SQL.

A third application area for model-checking problems is artificial intelligence. Constraint satisfaction problems can easily be formulated in terms of the model-checking problem for monadic second-order logic. This observation goes back to Feder and Vardi [11]. Research of the last few years has shown that there is an intimate connection between constraint satisfaction problems and database theory [25,32,19].

Model-checking problems can also be used as a framework for reasoning about standard problems considered in algorithms and complexity theory. To model, for example,

A. Ferreira and H. Reichel (Eds.): STACS 2001, LNCS 2010, pp. 12–26, 2001.
© Springer-Verlag Berlin Heidelberg 2001

the clique-problem by a model-checking problem, for every $k \geq 1$ we write a sentence $\varphi^k_{\text{clique}}$ of first-order logic saying that a graph G has a clique of size k. This shows that the clique problem can be seen as a special case of the model-checking problem for first-order logic. Similarly, we can consider the graph coloring problem as a special case of the model-checking problem for monadic second-order logic. Thus algorithms for model-checking problems can also be seen as "meta-algorithms" for more concrete problems. While the algorithms for concrete problems obtained this way will usually not be the most efficient, they often highlight the structural reasons making the problems tractable. Moreover, these meta-algorithms can be taken as a starting point for the development of more refined algorithms taking into account the special properties of the particular problem at hand.

Often, we are not only interested in a model-checking problem itself, but also in certain variants. One example is the evaluation of database queries - model-checking in the strict sense only corresponds to the evaluation of Boolean queries, that is, queries with a yes/no answer, but queries whose output is a set of tuples of database entries also need to be evaluated. Constraint satisfaction problems provide another example - usually, we are not only interested in the question of whether a constraint satisfaction problem has a solution, but we actually want to construct a solution. Sometimes, we want to count the number of solutions, or generate a random solution, or construct a solution that is optimal in some sense. We refer to such problems as *generalized model-checking problems*. An abstract setting for studying the complexity of such problems is given in [21].

As the title suggests, we focus on first-order logic here. Examples of problems that can be described as generalized model-checking problems for first-order logic are given in Section 2.4, among them such well-known problems as CLIQUE, DOMINATING SET, SUBGRAPH ISOMORPHISM, HOMOMORPHISM, and CONJUNCTIVE QUERY EVALUATION. The main purpose of the paper is to give a survey of known results and explain the basic techniques applied to prove them. One new result, which nicely illustrates the use of locality in model-checking algorithms, is concerned with first-order model-checking on graphs of low degree and on sparse random graphs.

2 Generalized Model-Checking Problems

2.1 Structures and Queries

A *vocabulary* is a finite set τ of *relation symbols*. Associated with every relation symbol R is a positive integer $\text{ar}(R)$, the *arity* of R. A τ-structure $\mathcal{A}$ consists of a set A called the *universe* of $\mathcal{A}$ and, for every $R \in \tau$, an $\text{ar}(R)$-ary relation $R^{\mathcal{A}} \subseteq A^{\text{ar}(R)}$. *In this paper, we only consider structures whose universe is finite.* STR denotes the class of all (finite) structures. If C is a class of structures, $C[\tau]$ denotes the subclass of all τ-structures in C.

For example, we can consider *graphs* as $\{E\}$-structures $\mathcal{G}$, where E is a binary relation symbol and $E^{\mathcal{G}}$ is symmetric and anti-reflexive.

Hypergraphs can be modeled as $\{V, I\}$-structures $\mathcal{H}$, where V is unary and I is binary and $I^{\mathcal{H}} \subseteq V^{\mathcal{H}} \times H \setminus V^{\mathcal{H}}$. (The *hyperedges* of $\mathcal{H}$ are the sets $\{v \in V^{\mathcal{H}} \mid (v, e) \in I^{\mathcal{H}}\}$, for $e \in H \setminus V^{\mathcal{H}}$.)

Boolean circuits can be modeled as $\{E, A, N\}$-structures $\mathcal{B}$, where E is binary, A, N are unary, and $(B, E^{\mathcal{B}})$ is a directed acyclic graph that has precisely one vertex of out-degree 0 (the *output node*), $A^{\mathcal{B}}$ is a subset of all vertices with in-degree at least 2 (the *and nodes*; all other vertices of in-degree at least 2 are considered as *or-nodes*), and $N^{\mathcal{B}}$ is a subset of the nodes of in-degree 1 (the *negation nodes*).

2.2 First-Order Logic

Atomic formulas, or *atoms*, are expressions of the form $x = y$ or $Rx_1 \ldots x_r$, where R is an r-ary relation symbol and $x, y, x_1, \ldots, x_r$ are *variables*. The formulas of *first-order logic* are build up in the usual way from the atomic formulas using the connectives $\land, \lor, \neg, \rightarrow$, and the quantifiers $\forall, \exists$.

The class of all first-order formulas is denoted by FO. The *vocabulary* of a first-order formula φ, denoted by $\mathrm{voc}(\varphi)$, is the set of all relation symbols occurring in φ. If $\Phi \subseteq$ FO is a class of first-order formulas, then $\Phi[\tau]$ denotes the class of all $\varphi \in \Phi$ with $\mathrm{voc}(\varphi) \subseteq \tau$. A *free variable* of a first-order formula is a variable x not in the scope of a quantifier $\exists x$ or $\forall x$. The set of all free variables of a formula φ is denoted by $\mathrm{free}(\varphi)$. The notation $\varphi(x_1, \ldots, x_k)$ indicates that $\mathrm{free}(\varphi) = \{x_1, \ldots, x_k\}$. A *sentence* is a formula without free variables.

For a formula $\varphi(x_1, \ldots, x_k) \in \mathrm{FO}[\tau]$, a τ-structure $\mathcal{A}$, and $a_1, \ldots, a_k \in A$, we write $\mathcal{A} \models \varphi(a_1, \ldots, a_k)$ to say that $\mathcal{A}$ satisfies φ if the variables $x_1, \ldots, x_k$ are interpreted by the elements $a_1, \ldots, a_k$, respectively. We let

$$\varphi(\mathcal{A}) := \big\{(a_1, \ldots, a_k) \in A^k \mid \mathcal{A} \models \varphi(a_1, \ldots, a_k)\big\}.$$

To extend this definition to sentences in a reasonable way, we identify the set consisting of the empty tuple with TRUE and the empty set with FALSE.

2.3 Generalized Model-Checking Problems

The basic model-checking problem asks whether a given structure satisfies a given sentence. In our more general setting, we shall consider formulas with free variables. For every class $\Phi \subseteq$ FO of formulas and every class C $\subseteq$ STR of structures we consider the following four basic problems:

The input is always a structure $\mathcal{A} \in$ C and a formula $\varphi \in \Phi$.

The decision problem. Decide if $\varphi(\mathcal{A})$ is non-empty. Essentially, this problem is the same as the *model-checking problem*. Therfore, we refer to this problem as Φ-MODEL-CHECKING on C.

The construction problem. Find a tuple $\bar{a} \in \varphi(\mathcal{A})$ if such a tuple exists. We refer to this problem as Φ-CONSTRUCTION on C.

The listing problem. Compute the set $\varphi(\mathcal{A})$. Because of its database application, we refer to this problem as Φ-EVALUATION on C.

The counting problem. Compute the size of $\varphi(\mathcal{A})$. We refer to this problem as Φ-COUNTING on C.

If C is the class STR of all structures, we usually do not mention it explicitly and speak of Φ-MODEL-CHECKING, Φ-CONSTRUCTION, et cetera.

Another interesting problem is the *sampling problem*, that is, the problem of generating a *random* element of $\varphi(\mathcal{A})$. To model combinatorial optimization problems, one may consider structures $\mathcal{A}$ on which a suitable cost-function is defined and then search for *optimal* solutions in $\varphi(\mathcal{A})$. But in this paper, we focus on the four basic problems.

This uniform view on combinatorial problems associated with a binary relation that relates *instances* (in our case pairs $(\mathcal{A}, \varphi) \in C \times \Phi$ with *solutions* ($\bar{a} \in \varphi(\mathcal{A})$ for us) is well-studied in complexity theory [17,29,23]. An abstract model-theoretic framework for considering such problems as *generalized model-checking problems* is presented in [21].

2.4 Examples

Before we proceed, let us consider a few examples of problems that can be described as generalized model-checking problems for first-order logic.

Example 1. Let $\varphi_{\text{clique}}^k(x_1, \ldots, x_k)$ be the formula $\bigwedge_{1 \leq i < j \leq k} E x_i x_j$, and let $\Phi_{\text{clique}} := \{\varphi_{\text{clique}}^k \mid k \geq 1\}$. Observe that for every graph $\mathcal{G}$, $\varphi_{\text{clique}}^k(\mathcal{G})$ is the set of all k-cliques of $\mathcal{G}$. Thus Φ_{CLIQUE}-MODEL-CHECKING on the class of all graphs is just the well-known CLIQUE problem. Φ_{CLIQUE}-CONSTRUCTION is the problem of finding a clique of given size in a given graph. Φ_{CLIQUE}-EVALUATION is the problem of finding all cliques of given size in a given graph, and Φ_{CLIQUE}-COUNTING is the problem of counting the number of such cliques.

Two well-known generalizations of the CLIQUE problem are the GRAPH HOMOMORPHISM problem and the SUBGRAPH ISOMORPHISM problem:

Example 2. For every τ-structure $\mathcal{A}$ with universe $A = \{a_1, \ldots, a_k\}$, let

$$\varphi_{\text{hom}}^{\mathcal{A}}(x_1, \ldots, x_k) := \bigwedge_{\substack{R \in \tau \\ r\text{-ary}}} \bigwedge_{(a_{i_1}, \ldots, a_{i_r}) \in R^{\mathcal{A}}} R x_{i_1} \ldots x_{i_r},$$

and let $\Phi_{\text{hom}} := \{\varphi_{\text{hom}}^{\mathcal{A}} \mid \mathcal{A} \in \text{STR}\}$. Then for every τ-structure $\mathcal{B}$, $\varphi_{\text{hom}}^{\mathcal{A}}(\mathcal{B})$ is the set of homomorphic images of $\mathcal{A}$ in $\mathcal{B}$. Thus Φ_{HOM}-MODEL-CHECKING is the same as HOMOMORPHISM (on relational structures).

Note that $\varphi_{\text{clique}}^k = \varphi_{\text{hom}}^{\mathcal{K}}$ for the complete graph $\mathcal{K}$ with k vertices.

Adding clauses stating that the variables are pairwise distinct to the formulas in Φ_{hom}, we obtain a class Φ_{sub} of formulas whose MODEL-CHECKING problem is the SUBSTRUCTURE ISOMORPHISM problem.

Another slight modification yields a description of INDUCED SUBSTRUCTURE ISOMORPHISM.

HOMOMORPHISM is closely related to the following problem playing a fundamental role in database theory:

Example 3. *Conjunctive queries* are relational database queries that can be described by first-order formulas of the form $\exists y_1 \ldots \exists y_l(\alpha_1 \wedge \ldots \wedge \alpha_n)$, where $\alpha_1, \ldots, \alpha_n$ are atomic formulas. Let CQ denote the class of all such formulas. Note that $\Phi_{\text{hom}} \subseteq \text{CQ}$.

CQ-EVALUATION is the problem of evaluating a conjunctive query against a finite relational database.

From a logical point of view, the formulas considered so far are all very simple; they are all *quantifier-free* or *existential* first-order formulas. The following examples require more complicated formulas.

Example 4. For every $k \geq 1$, let $\varphi_{\text{DS}}^k(x_1, \ldots, x_k) := \forall y \bigvee_{i=1}^k (x_i = y \vee Ex_iy)$, and let $\Phi_{\text{DS}} := \{\varphi_{\text{DS}}^k \mid k \geq 1\}$. Then for every graph $\mathcal{G}$, φ_{DS}^k is the set of all dominating sets of $\mathcal{G}$ of size k. Thus Φ_{DS}-MODEL-CHECKING on the class of all graphs is the DOMINATING SET problem.

Example 5. For all τ-structures $\mathcal{A} \subseteq \mathcal{A}^+$ with universes $A = \{a_1, \ldots, a_k\}$ and $A^+ = A \cup \{a_{k+1}, \ldots, a_l\}$, let

$$\varphi_{\text{ext}}^{\mathcal{A},\mathcal{A}^+} := \forall x_1 \ldots x_k \left(\varphi_{\text{sub}}^{\mathcal{A}}(x_1, \ldots, x_k) \rightarrow \exists x_{k+1} \ldots x_l \varphi_{\text{sub}}^{\mathcal{A}^+}(x_1, \ldots, x_l)\right),$$

and let $\Phi_{\text{ext}} := \{\varphi_{\text{ext}}^{\mathcal{A},\mathcal{A}^+} \mid \mathcal{A} \subseteq \mathcal{A}^+ \in \text{STR}\}$. Then for every structure $\mathcal{B}$ we have $\varphi_{\text{ext}}^{\mathcal{A},\mathcal{A}^+}(\mathcal{B}) = \text{TRUE}$ if, and only if, every substructure of $\mathcal{B}$ that is isomorphic to $\mathcal{A}$ can be extended to a substructure isomorphic to $\mathcal{A}^+$.

Φ_{ext}-MODEL-CHECKING is the EXTENSION problem for relational structures.

Example 6. Recall that Boolean circuits can be viewed as $\{E, A, N\}$-structures. The *depth* of a Boolean circuit is the length of the longest path from an input node to the output node. It is easy to find, for all $d, k \geq 1$, a first-order formula $\varphi_{\text{sat}}^{d,k}(x_1, \ldots, x_k)$ such that for every Boolean circuit $\mathcal{C}$, a tuple $(c_1, \ldots, c_k) \in C^k$ is contained in $\varphi_{\text{sat}}^{d,k}(\mathcal{C})$ if, and only if, the assignment that sets $c_1, \ldots, c_k$ to TRUE and all other input nodes to FALSE satisfies $\mathcal{C}$. Note that the formula $\varphi_{\text{sat}}^{d,k}$ may have up to d quantifier alternations. Let $\Phi_{\text{sat}} := \{\varphi_{\text{sat}}^{d,k} \mid d, k \geq 1\}$ Then for Φ_{sat}-MODEL-CHECKING is essentially the CIRCUIT SATISFIABILITY problem.

Example 7. Using the well-known equivalence of the *relational calculus* and first-order logic, we observe that FO-EVALUATION is equivalent to the problem of evaluating relational calculus queries against finite relational databases.

2.5 Complexity

Our underlying model of computation is the standard RAM-model with addition and subtraction as arithmetic operations (cf. [1,30]). In our complexity analysis we use the uniform cost measure. Structures are coded in a straightforward way by first describing their vocabulary, then listing all elements of the universe, listing all tuples of all relations, and then listing the constants. For details, we refer the reader to [12]. The length of the encoding of a structure $\mathcal{A}$ is denoted by $||\mathcal{A}||$. For a fixed vocabulary τ we have $||\mathcal{A}|| \in \Theta(|A| + \sum_{R \in \tau} |R^{\mathcal{A}}|)$ for all $\mathcal{A} \in \text{STR}[\tau]$. For instance, for graphs $\mathcal{G}$ with n vertices and m edges this means $||\mathcal{G}|| \in \Theta(n + m)$. We fix some reasonable way to encode formulas; the length of the encoding of a formula φ is denoted by $||\varphi||$.

The most straightforward way to measure the complexity of a model-checking problem is to measure it just in the length of the input, that is, in $||\mathcal{A}|| + ||\varphi||$. This is usually referred to as the *combined complexity* of the problem [31]. In particular, we say that the combined complexity of Φ-MODEL-CHECKING on C is in PTIME if there is a polynomial $p(X)$ and an algorithm solving the problem in time at most $p(||\mathcal{A}|| + ||\varphi||)$. The same definition applies to the other generalized model-checking problem.

However, since Φ may contain formulas with arbitrarily many free-variables, in general there is no polynomial bound on the size of $\varphi(\mathcal{A})$ (in terms of $||\mathcal{A}|| + ||\varphi||$). There are different complexity theoretic notions to handle listing problems with potentially large output (see, for example, [18,24]). The simplest is to measure the complexity both in terms of the size of the input and the size of the output. We say that the combined complexity of Φ-EVALUATION on C is in PTT (*polynomial total time*) if there is a polynomial $p(X)$ and an algorithm solving the problem in time at most $p(||\mathcal{A}|| + ||\varphi|| + ||\varphi(\mathcal{A})||)$. A stricter notion is that of *polynomial delay*. An $f(n)$-*delay algorithm* for a listing problem is an algorithm that generates its first output in at most $f(n)$ steps and thereafter never takes more than $f(n)$ steps between generating two outputs. We say that Φ-EVALUATION on C is in PD if there is a polynomial $p(X)$ and a $p(||\mathcal{A}|| + ||\varphi||)$-delay algorithm solving the problem.

Theorem 8 (Stockmeyer [28], Vardi [31]). *The combined complexity of* FO-MODEL-CHECKING *is* PSPACE *complete.*

Of course under the assumption that PTIME $\neq$ PSPACE, this implies that the combined complexity of none of the generalized model-checking problems for FO is in PTIME, or PTT for the listing problem.

The proof of the hardness part of Theorem 8 is by a reduction from the QUANTIFIED BOOLEAN FORMULA problem — essentially, QUANTIFIED BOOLEAN FORMULA is the same as FO-model-checking on a structure with just two distinguishable elements representing the Boolean values TRUE and FALSE.

Thus the high complexity of FO-MODEL-CHECKING is caused by large and complicated input formulas. The input structure only plays a very limited role, it can actually be a fixed two-element structure whose vocabulary only consists of one unary relation symbol. However, instances of practical problems modeled by model-checking problems usually consist of large input structures representing some "real-world" *data* such as a network or a relational database, and much smaller formulas representing a *specification* of the information we want to gather from these data. Thus the relevance of Theorem 8 is somewhat limited.

The simplest way to take into account that usually the input structure is much larger than the input formula is to completely disregard the formula and measure the complexity just in terms of the size of the input structure. This complexity measure is known as the *data complexity* [31]. In particular, we say that the data complexity of Φ-MODEL-CHECKING on C is in PTIME if for every formula $\varphi \in \Phi$ there is a polynomial $p_\varphi(X)$ and an algorithm that solves the model-checking problem in at most $p_\varphi(||\mathcal{A}||)$ steps.[*] It is easy to see that the data complexity of FO-MODEL-CHECKING,

---

[*] Here we may consider both a *uniform* version where there is just one algorithm solving the model-checking problem, or the *non-uniform* version where for every formula φ there is an algorithm solving $\{\varphi\}$-MODEL-CHECKING on C in time $p_\varphi(X)$. Usually, data complexity refers to the non-uniform version.

FO-CONSTRUCTION, FO-EVALUATION, and FO-COUNTING is in PTIME. More precisely, there is an $||\mathcal{A}||^{O(||\varphi||)}$-algorithm for each of the problems.

However, this is still not completely satisfying because even for very small input formulas φ, $||\mathcal{A}||^{O(||\varphi||)}$ cannot be seen as a feasible complexity. On the other hand, a complexity such as $2^{||\varphi||} \cdot ||\mathcal{A}||$ is acceptable. *Parameterized complexity* is a refined measure taking these considerations into account. It may often be the most appropriate way to measure the complexity of generalized model-checking problems. We say that Φ-MODEL-CHECKING on C is *fixed-parameter tractable*, or equivalently that the parameterized complexity of Φ-MODEL-CHECKING on C is in FPT, if there is a constant $c > 0$, a computable function f, and an algorithm solving the problem in time at most $f(||\varphi||) \cdot ||\mathcal{A}||^c$.** The parameterized complexity of the other generalized model-checking problems can be defined analogously. Similarly as for the combined complexity, we can usually not expect the parameterized complexity of Φ-EVALUATION to be in FPT simply because the output may get too large. It is straightforward to define parameterized analogues of the classes PTT and PD, which we refer to as FPTTT and FPTD. For further background in parameterized complexity theory, we refer the reader to [9].

The following theorem can be seen as the parameterized analogue of Theorem 8. Under the complexity theoretic assumption that the parameterized complexity classes AW[∗] and FPT are distinct, it implies that the generalized model-checking problems for FO are not fixed-parameter tractable.

Theorem 9 (Downey, Fellows, Taylor [10]). FO-MODEL-CHECKING *is complete for* AW[∗] *under parameterized reductions.*

So we are facing the situation that both the combined complexity and the parameterized complexity of the generalized model-checking problems for first-order logic are very high. In the next section, we shall study restrictions on the class Φ of formulas that make the problems tractable, and in Section 4 we shall study restrictions on the class C of structures.

Remark 10. It is interesting to see what the relation among the complexities of the different generalized model-checking problems is. It can be shown [21] that under mild closure conditions on the class Φ of formulas and the class C of structures, the following four statements are equivalent: (i) The combined complexity of Φ-MODEL-CHECKING on C is in PTIME. (ii) The combined complexity of Φ-CONSTRUCTION on C is in PTIME. (iii) The combined complexity of Φ-EVALUATION on C is in PD. (iv) The combined complexity of Φ-EVALUATION on C is in PTT.

An analogous statement holds with respect to parameterized complexity.

In general, Φ-COUNTING on C is a harder problem, as a suitable formalization of the MAXIMUM MATCHING problem on bipartite graphs shows.

3   Simple Formulas

In this section we look for restrictions on the class $\Phi \subseteq$ FO of formulas making a generalized model-checking problem tractable. A first idea is to restrict quantifier alternations

---

** As for the data complexity, there is also a *non-uniform* version of this definition, but for parameterized complexity the uniform version is more common.

and just look at existential formulas. Conjunctive queries (see Example 3) are a good starting point because they are particularly simple existential formulas that are nevertheless very important. As expected, model-checking problems for conjunctive queries are of lower complexity than those for full first-order logic, but unfortunately they are still not tractable:

Theorem 11 (Chandra and Merlin [4], Papadimitriou and Yannakakis [26]).
The combined complexity of CQ-MODEL-CHECKING *is* NP-*complete, and the parameterized complexity is* W[1]-*complete.*

W[1] is a parameterized complexity class that plays a similar role in parameterized complexity theory as NP does in classical complexity theory. In particular, it is believed that FPT $\neq$ W[1].

To get tractable model-checking problems, we need to consider even simpler formulas, and it is not clear how they might look. A fruitful idea is to study the graph $\mathcal{G}_\varphi$ of a conjunctive query φ: The vertex set of $\mathcal{G}_\varphi$ is the set of all variables of φ, and there is an edge between variables x and y if, and only if, there is an atom α of φ that contains both x and y. The hope is that model-checking is easy for queries with a "simple" graph, and indeed this is true for the right notion of "simplicity", which turns out to be "tree-likeness", or more precisely, bounded *tree-width.**** For a class C of graphs, we let CQ(C)-denote the class of all conjunctive queries φ with $\mathcal{G}_\varphi \in$ C.

Example 12. Let TREE denote the class of all trees. In this example, we consider the class CQ(TREE) of all conjunctive queries whose underlying graph is a tree. We shall prove the following:

CQ(TREE)-MODEL-CHECKING *can be solved in time* $O(||\mathcal{A}|| \cdot ||\varphi||)$.

Let $\varphi(x_1, \ldots, x_k) := \exists x_{k+1} \ldots \exists x_l (\alpha_1 \wedge \ldots \wedge \alpha_m) \in$ CQ(TREE)[τ], and let $\mathcal{A}$ be a τ-structure. Without loss of generality we assume that φ contains no equalities. If $x_i = x_j$ is an atom of φ, we can just delete that atom and replace x_j by x_i everywhere. The resulting formula is equivalent to φ and still in CQ(TREE).

The graph $\mathcal{G}_\varphi$ is a tree with universe $T := \{x_1, \ldots, x_l\}$. We declare x_1 to be the *root* of this tree. We define the *parent* and the *children* of a node x_i in the usual way. Let T denote the directed tree with universe T and edge relation $E^T := \{xy \mid x \text{ parent of } y\}$. We define the tree-order $\leq^T$ to be the reflexive transitive closure of E^T.

For every node $x \in T$, we let δ_x be the conjunction of all atoms α_i of φ with free(α_i) = $\{x\}$. For every edge $xy \in E^T$, we let ε_{xy} be the conjunction of atoms α_i of φ with free(α_i) = $\{x, y\}$. Then every atom of φ occurs in precisely one δ_x or ε_{xy}. Thus φ is equivalent to the formula $\exists x_{k+1} \ldots x_l \left(\bigwedge_{x \in T} \delta_x \wedge \bigwedge_{xy \in E^T} \varepsilon_{xy} \right)$.

Our algorithm is a straightforward dynamic programming algorithm. It starts by doing some pre-computations setting up the data structures needed in the actual dynamic programming phase. For every vertex $x \in T$ it computes $\delta_x(\mathcal{A})$ and stores it in a Boolean array with one entry for every $a \in A$. This requires time $O(|A| \cdot ||\delta_x||)$. Similarly, for every edge $xy \in E^T$ it computes the set $\varepsilon_{xy}(\mathcal{A})$. Then for every $a \in A$ it produces a linked list that contains all b such that $ab \in \varepsilon_{xy}(\mathcal{A})$. This can be done in time $O(||\mathcal{A}|| \cdot ||\varepsilon_{xy}||)$ (see [12] for details).

***For graph theoretic notions such as *tree-width* or *minors* that are left unexplained here, we refer the reader to [8].

Thus the overall time required for these pre-computations is $O(||\mathcal{A}|| \cdot ||\varphi||)$.

Let $y_0 \in T$ and let $y_1, \ldots, y_p$ all descendants of y_0 in T, that is, all nodes $x \in T \setminus \{y_0\}$ such that $y_0 \leq^T x$. The *subtree-formula of y_0* is the formula

$$\sigma_{y_0} := \exists y_1 \ldots \exists y_p \Big(\bigwedge_{0 \leq i \leq p} \delta_{y_i} \wedge \bigwedge_{\substack{0 \leq i,j \leq p \\ y_i y_j \in E^T}} \varepsilon_{y_i y_j} \Big).$$

Note that free$(\sigma_{y_0}) = \{y_0\}$.

Now the dynamic programming phase starts. Inductively from the leaves to the root of T, for every $x \in T$ the algorithm computes the set $\sigma_x(\mathcal{A})$ and stores it in a Boolean array. For the leaves x of T, we have $\sigma_x = \delta_x$. Since the algorithm has already computed $\delta_x(\mathcal{A})$, there is nothing to do. For a node x with children $y_1, \ldots, y_q$ we let $S_0 := \delta_x(\mathcal{A})$ and, for $i \geq 1$, $S_i := \{a \in S_{i-1} \mid \exists b \in A : b \in \sigma_{y_i}(\mathcal{A})$ and $ab \in \varepsilon_{xy}(\mathcal{A})\}$. Using the arrays for $\delta_x(\mathcal{A})$ and $\sigma_{y_i}(\mathcal{A})$ and, for every $a \in \sigma_x(\mathcal{A})$, the list of all b such that $ab \in \varepsilon_x(\mathcal{A})$, it is easy to see that S_i can be computed from S_{i-1} in time $O(||\mathcal{A}||)$, and thus $S_q = \sigma_x(\mathcal{A})$ can be computed in time $O(q \cdot ||\mathcal{A}||)$.

Hence the overall time required in the dynamic-programming phase is $O(||\mathcal{A}|| \cdot |T|) \subseteq O(||\mathcal{A}|| \cdot ||\varphi||)$.

Remember that x_1 is the root of T and observe that $\sigma_{x_1}(\mathcal{A})$ is the set of all $a_1 \in A$ for which there exist $a_2, \ldots, a_l \in A$ such that $(a_1, \ldots, a_l) \in \big(\bigwedge_{i=1}^m \alpha_i \big)(\mathcal{A})$ and thus $(a_1, \ldots, a_k) \in \varphi(\mathcal{A})$. This implies $\varphi(\mathcal{A}) \neq \emptyset$ if, and only if, $\sigma_{x_1}(\mathcal{A}) \neq \emptyset$.

Therefore, our algorithm returns TRUE if $\sigma_{x_1}(\mathcal{A}) \neq \emptyset$ and FALSE otherwise.

So model-checking for formulas whose underlying graph is a tree is tractable. The algorithm of the previous example can easily be extended to formulas whose underlying graph has bounded tree-width. Theorem 14 shows that this is essentially all we can do; for conjunctive queries whose underlying graph is more complicated, model-checking gets intractable.

Theorem 13 (Chekuri and Rajaraman [5]). *Let C be a class of graphs of bounded tree-width. Then the combined complexity of* CQ(C)-MODEL-CHECKING *is in* PTIME, *and the combined complexity* CQ(C)-EVALUATION *is in* PTT.

It is not hard to see that the combined complexity of CQ(C)-CONSTRUCTION and CQ(C)-COUNTING for classes C of graphs of bounded tree-width is also in PTIME and that the combined complexity CQ(C)-EVALUATION is actually in PD [21]. Of course these result imply that the parameterized complexity of the respective problems is in FPT or FPTD.

Theorem 14 (Grohe, Schwentick, and Segoufin [22]).

(1) Let C be a class of graphs of unbounded tree-width that is closed under taking minors. Then CQ(C)-MODEL-CHECKING *is* NP-*complete.*

(2) Let C be a class of graphs of unbounded tree-width. Then CQ(C)-MODEL-CHECK-ING *is* W[1]-*complete.*

Instead of a graph, we may also associate a *hypergraph* with a conjunctive query in a natural way. It turns out that the generalized model-checking problems also become tractable for conjunctive queries with tree-like hypergraphs. In a fundamental paper that is underlying all the work described in this section, Yannakakis [33] proved that the combined complexity of MODEL-CHECKING for conjunctive queries with an *acyclic* hypergraph is in PTIME, and the combined complexity of the EVALUATION problem for such queries is in PTT. As a matter of fact, the algorithm described in Example 12 is essentially the one suggested by Yannakakis. As for graphs, the acyclicity restriction can also be relaxed for hypergraphs. Gottlob, Leone, and Scarcello [20] introduce the notion of *bounded hypertree-width* and show that conjunctive queries whose hypergraph has bounded hypertree-width have tractable model-checking problems. As a matter of fact, they show that MODEL-CHECKING for all the tree-like classes of conjunctive queries considered here is in the parallel complexity class LOGCFL and actually complete for this class.

If we look beyond conjunctive queries, there are two well-known classes of first-order formulas whose model-checking problems have a polynomial time combined complexity: The *finite variable fragments* of first-order logic and the *guarded fragment*. Surprisingly, these two fragments are closely related to the tree-like classes of conjunctive queries. A straightforward generalization of the class of conjunctive queries whose underlying graph has tree-width at most $(k + 1)$ to full first-order logic yields the k variable fragment. Similarly, a generalization of the class of conjunctive queries whose underlying hypergraph is acyclic yields the guarded fragment [12].

4   Simple Structures

In this section, we restrict the class C of input structures of first-order generalized model-checking problems. We first note that there is not much we can do about the combined complexity: The PSPACE-completeness of the QUANTIFIED BOOLEAN FORMULA problem implies that FO-MODEL-CHECKING is already PSPACE complete on the class $\{\mathcal{B}\}$ consisting of just one structure $\mathcal{B}$ with universe $\{0, 1\}$ and one unary relation $P^{\mathcal{B}} = \{1\}$. So we concentrate on parameterized complexity here.

4.1   Gaifman's Locality Theorem

An important property that distinguishes first-order logic from most stronger logics is that it can only express *local* properties. The model-checking algorithms that we shall consider in this section crucially depend on locality.

The *Gaifman graph* of a τ-structure $\mathcal{A}$ is the graph with universe A that has an edge between distinct elements $a, b \in A$ if there is a relation symbol $R \in \tau$ and tuple $\bar{a}$ of elements of A such that $\bar{a} \in R^{\mathcal{A}}$, and both a and b appear in $\bar{a}$. The distance $d^{\mathcal{A}}(a, b)$ between two elements $a, b \in A$ in $\mathcal{A}$ is the length of the shortest path from a to b in the Gaifman graph of $\mathcal{A}$. For every $r \geq 0$ and $a \in A$, the r-*neighborhood of a in* $\mathcal{A}$ is the set $N_r^{\mathcal{A}}(a) := \{b \in A \mid d^{\mathcal{A}}(a, b) \leq r\}$. For a set $B \subseteq A$ we let $N_r^{\mathcal{A}}(B) := \bigcup_{b \in B} N_r^{\mathcal{A}}(b)$.

It is easy to see that for every vocabulary τ and every $r \geq 0$ there is a formula $\delta_r(x, y) \in \text{FO}[\tau]$ such that for every τ-structure $\mathcal{A}$ we have $\delta_r(\mathcal{A}) = \{(a, b) \in A^2 \mid d^{\mathcal{A}}(a, b) \leq r\}$. We write $d(x, y) \leq r$ instead of $\delta_r(x, y)$. For a sentence $\varphi \in \text{FO}[\tau]$ we let $\varphi^{N_r(x)}(x)$ denote the *relativization* of φ to $N_r(x)$, that is, the formula obtained from φ by replacing every subformula of the form $\exists y \psi$ by $\exists y (d(x, y) \leq r \wedge \psi)$ and every subformula of the form $\forall y \psi$ by $\forall y (d(x, y) \leq r \rightarrow \psi)$. Here we assume, without loss of generality, that x does not occur in φ.

Observe that the formula $\varphi^{N_r(x)}(x)$ is *r-local* in the following sense: For every τ-structure $\mathcal{A}$ and for every $a \in A$ we have

$$\mathcal{A} \models \varphi^{N_r(x)}(a) \iff \langle N_r^{\mathcal{A}}(a) \rangle \models \varphi^{N_r(x)}(a). \qquad (\star)$$

Here $\langle N_r^{\mathcal{A}}(a) \rangle$ denotes the substructure induced by $\mathcal{A}$ on $N_r^{\mathcal{A}}(a)$.

Theorem 15 (Gaifman [16]). *Every first-order sentence φ is equivalent to a Boolean combination of sentences of the form*

$$\exists x_1 \dots \exists x_k \Big(\bigwedge_{1 \leq i < j \leq k} d(x_i, x_j) > 2r \wedge \bigwedge_{1 \leq i \leq k} \psi^{N_r(x_i)}(x_i) \Big), \qquad (\star\star)$$

where $k, r \geq 0$ and ψ is a first-order sentence.
    Furthermore, such a Boolean combination can be effectively computed from φ.

4.2   Structures of Low Degree

In this subsection we generalize a theorem due to Seese [27] stating that first-order definable properties of structures of bounded degree can be checked in linear time. Our main purpose is to illustrate how Gaifman's theorem can be used in model-checking algorithms. Seese's proof is different, it uses Hanf's locality theorem instead of Gaifman's theorem. This technique cannot be used in order to prove Theorem 16.

The *degree* of a structure $\mathcal{A}$, denoted by $\deg(\mathcal{A})$, is the maximal degree of a vertex in the Gaifman graph of $\mathcal{A}$. We say that a class C of structures has *low degree* if for every $\varepsilon > 0$ there is an integer N_ε such that for all $\mathcal{A} \in$ C with $|A| \geq N_\varepsilon$ we have $\deg(\mathcal{A}) \leq |A|^\varepsilon$. For example, the class of all structures whose degree is at most logarithmic in their size has low degree.

Theorem 16. *There is an algorithm $\mathfrak{A}$ for FO-MODEL-CHECKING and a function f such that for every class C of structures that has low degree and for every $\varepsilon > 0$ the runtime of $\mathfrak{A}$ on an input $(\mathcal{A}, \varphi) \in$ C $\times$ FO is in $O(f(||\varphi||) \cdot |A|^{1+\varepsilon})$.*

Proof: By Gaifman's Theorem, it suffices to find an algorithm that model-checks formulas φ of the form $(\star\star)$. So let

$$\varphi = \exists x_1 \dots \exists x_k \Big(\bigwedge_{1 \leq i < j \leq k} d(x_i, x_j) > 2r \wedge \bigwedge_{1 \leq i \leq k} \psi^{N_r(x_i)}(x_i) \Big),$$

and let $\mathcal{A}$ be a structure.

Our model-checking algorithm first computes the set $\psi^{N_r(x)}(\mathcal{A})$, proceeding as follows: For every $a \in A$, it computes $N_r^{\mathcal{A}}(a)$ and then it checks whether $\langle N_r^{\mathcal{A}}(a) \rangle \models \psi^{N_r(x)}(a)$. By $(\star)$, this is equivalent to $a \in \psi^{N_r(x)}(\mathcal{A})$.

$\langle N_r^{\mathcal{A}}(a) \rangle \models \psi^{N_r(x)}(a)$ is checked in a straightforward way; this requires time $||\langle N_r^{\mathcal{A}}(a) \rangle||^{O(||\psi||)}$. Hence the overall time needed to compute $\psi^{N_r(x)}(\mathcal{A})$ is at most $\sum_{a \in A} ||\langle N_r^{\mathcal{A}}(a) \rangle||^{c||\psi||}$ for some constant $c > 0$.

Let $S := \psi^{N_r(x)}(\mathcal{A})$. To decide whether $\mathcal{A} \models \varphi$, it remains to check whether there are $a_1, \dots, a_k \in S$ such that $d^{\mathcal{A}}(a_i, a_j) > 2r$ for $1 \leq i < j \leq r$. A simple algorithm doing this is described in [15]: It starts by picking an arbitrary $a_1 \in S$ (if S is empty, the algorithm immediately rejects). Having picked $a_1, \ldots, a_l \in S$ of pairwise distance greater than $2r$, the algorithm tries to find an $a_{l+1} \in S$ of distance greater than $2r$ from $a_1, \ldots, a_l$. Either it will eventually find $a_1, \ldots, a_k \in S$ and accept, or after having found $a_1, \ldots, a_l$ for some $l < k$ it will get stuck. This means that $S \subseteq N_{2r}^{\mathcal{A}}(\{a_1, \ldots, a_l\})$. Noting that for all sets $B \subseteq A$ and $a, b \in B$ we have $d^{\mathcal{A}}(a, b) > 2r \iff d^{\langle N_r^{\mathcal{A}}(B) \rangle}(a, b) > 2r$, it now suffices to find out if there are $a_1', \ldots, a_k' \in S$ such that $d^{\langle N_{3r}^{\mathcal{A}}(\{a_1, \ldots, a_l\}) \rangle}(a_i', a_j') > 2r$ for $1 \le i \le k$. Our algorithm does this by first computing a distance matrix for $N_{3r}^{\mathcal{A}}(\{a_1, \ldots, a_l\})$ and then exhaustively searching all k-tuples. This requires time $O(k^2 || \langle N_{3r}^{\mathcal{A}}(\{a_1, \ldots, a_l\}) \rangle ||^{\max\{3, k\}})$. Finding $a_1, \ldots, a_l$ requires time $O(k \cdot ||\mathcal{A}||)$.

A few straightforward computations show that on classes of input structures of low degree, our algorithm satisfies the requirements posed on its runtime. □

We note that on classes C of bounded degree, the algorithm $\mathfrak{A}$ of Theorem 16 actually runs in time linear in $|A|$, which implies Seese's result mentioned above.

Corollary 17. *For every class* C *of structures of low degree, the parameterized complexity of* FO-MODEL-CHECKING *on* C *is in* FPT.

For every $n \ge 1$ and $p \in [0, 1]$, we let $G(n, p)$ denote the probability space of all graphs with vertex set $\{1, \ldots, n\}$ and edge-probability p. We call a function $p : \mathbb{N} \to [0, 1]$ *sparse* if $p(n) \in O(n^{-1+\varepsilon})$ for all $\varepsilon > 0$. For example, the function p defined by $p(n) = \log(n)/n$ is sparse.

Corollary 18. *For every first-order sentence* φ *there is an algorithm* $\mathfrak{A}$ *that, given a graph* $\mathcal{G}$, *decides if* $\mathcal{G} \models \varphi$ *such that for every sparse* $p : \mathbb{N} \to [0, 1]$ *we have:*
For every $\varepsilon > 0$, *the average runtime of* $\mathfrak{A}$ *on input* $\mathcal{G} \in G(n, p(n))$ *is in* $O(n^{1+\varepsilon})$.

Proof: We use the algorithm of Theorem 16. Some simple computations show that the probability that a graph $\mathcal{G} \in G(n, p(n))$ has degree greater than n^δ is exponentially low (for every $\delta > 0$). Even on the few high-degree graphs in $G(n, p(n))$, the runtime is in $n^{O(||\varphi||)}$, so we obtain a low average runtime. □

4.3 Tree-Width, Local Tree-Width, and Excluded Minors

Model-checking does not only become tractable if the input formulas have a tree-like structure, but also if the input structures are tree-like. This is even true for monadic second-order logic MSO, which is much more powerful than FO. The underlying reason for this is that MSO-sentences on trees can be translated to tree-automata, and it is easy to check whether a given tree-automaton accepts a given tree. A well-known result due to Courcelle [7] states that MSO-MODEL-CHECKING on classes of structures of bounded tree-width is possible in time $O(f(||\varphi||)|A|)$ (for a suitable function f) and therefore fixed-parameter tractable. Unfortunately, the function f is extremely fast-growing. It is non-elementary; essentially, it is a tower of 2s whose height is the number of quantifier-alternations of φ. Arnborg, Lagergren, and Seese [2] show that

MSO-COUNTING on classes of structures of bounded tree-width is possible in time $O(f(||\varphi||)|A|)$, and Flum, Frick, and Grohe [12] show that MSO-CONSTRUCTION on such classes is in time $O(f(||\varphi||)|A|)$ and that MSO-EVALUATION is in (total) time $O(f(||\varphi||)(|A| + ||\varphi(A)||)$. Since FO $\subseteq$ MSO, the corresponding results for generalized model-checking problems for FO on classes of structures of bounded tree-width follow.

Remembering that first-order logic is local, we observe that actually we do not need the whole input structures to have bounded tree-width in order to make model-checking tractable. It suffices to have structures that locally have bounded tree-width. Then we can apply Gaifman's Theorem to the input sentence, evaluate the local formulas $\psi^{N_r(x_i)}(x_i)$ (see ($\star\star$)) using Courcelle's approach, and put everything together as in the proof of Theorem 16. Let us make this precise: A class C of structures has *bounded local tree-width* if there is a function $\lambda : \mathbb{N} \to \mathbb{N}$ such that for all structures $A \in C$, all $a \in A$, and all $r \in \mathbb{N}$, the substructure $\langle N_r^A(a) \rangle$ of A has tree-width at most $\lambda(r)$. Many interesting classes have bounded local tree-width, among them the class of planar graphs and more generally all classes of graphs of bounded genus, all classes of structures of bounded degree (but not all classes of low degree), and, trivially, all classes of structures of bounded tree-width.

Theorem 19 (Frick and Grohe [15]). *Let C be a class of structures of bounded local tree-width. Then the parameterized complexity of* FO-MODEL-CHECKING *on C is in* FPT. *More precisely, there is a function f and, for every $\varepsilon > 0$, an algorithm that solves the problem in time $O(f(||\varphi||) \cdot |A|^{1+\varepsilon})$.*

Requiring the class C to be *locally tree-decomposable*, which is slightly more restrictive than just requiring it to have bounded local tree-width, we can actually find a model-checking algorithm that is linear in $|A|$ [15]. All examples of classes of bounded local tree-width that we have seen above are actually examples of locally tree-decomposable classes. In his forthcoming dissertation, Frick [14] is able to extend this result to the other generalized model-checking problem: He gives algorithms solving FO-CONSTRUCTION and FO-COUNTING in time linear in $|A|$ and FO-EVALUATION in total time linear in $(|A| + ||\varphi(A)||)$.

Building on the ideas of using locality and tree-decompositions to evaluate first-order formulas more efficiently, Flum and Grohe [13] showed that for every class C of graphs with an excluded minor, the parameterized complexity of FO-MODEL-CHECKING on C is in FPT. Here we say that a class C of graphs has an excluded minor if there is a graph H such that H is not a minor of any graph in C.

5   Conclusions

We have seen different approaches towards finding tractable instances for the generalized model-checking problems for first-order logic, which are known to be hard in general.

All known fragments of FO that have a tractable model-checking problem can be characterized as being tree-like in some sense. Efficient model-checking for such classes can be done by a relatively simple dynamic programming algorithm that goes back to Yannakakis [33]. Theorem 14 indicates that in some sense, these results are optimal.

Model-checking problems for full first-order logic become fixed-parameter tractable on several interesting classes of input structures, among them classes of low degree and the class of planar graphs. The fixed-parameter tractable algorithms make crucial use of the locality of first-order logic.

The results discussed in Section 4, in particular Theorem 19, imply a number of known results on the fixed-parameter tractability of more concrete problems on certain classes of graphs (for instance, the result that DOMINATING SET is fixed-parameter tractable on planar graphs [9]). The original proofs of these results are often very ad-hoc and vary a lot from problem to problem and for the different classes of structures. The results on FO-MODEL-CHECKING give a nice uniform explanation for all of these results. Moreover, they give us a simple way to see that a particular problem is fixed-parameter tractable, say, on the class of planar graphs, or, to give a fancier example, on the class of all graphs that have a knot-free embedding into $\mathbb{R}^3$:[†] Just show that the problem is first-order definable.

The price we pay for this generality is that we obtain algorithms with huge hidden constants that are only of theoretical interest. But often it is a good starting point to have at least some fixed-parameter tractable algorithm for a particular problem (or just to know that such an algorithm exists) when designing one that is more practical.

References

1. A.V. Aho, J.E. Hopcroft, and J.D. Ullman. *The Design and Analysis of Computer Algorithms*. Addison-Wesley, 1974.
2. S. Arnborg, J. Lagergren, and D. Seese. Easy problems for tree-decomposable graphs. *Journal of Algorithms*, 12:308–340, 1991.
3. A. Chandra and D. Harel. Structure and complexity of relational queries. *Journal of Computer and System Sciences*, 25:99–128, 1982.
4. A.K. Chandra and P.M. Merlin. Optimal implementation of conjunctive queries in relational data bases. In *Proceedings of the 9th ACM Symposium on Theory of Computing*, pages 77–90, 1977.
5. Ch. Chekuri and A. Rajaraman. Conjunctive query containment revisited. In Ph. Kolaitis and F. Afrati, editors, *Proceedings of the 5th International Conference on Database Theory*, volume 1186 of *Lecture Notes in Computer Science*, pages 56–70. Springer-Verlag, 1997.
6. J.H. Conway and C.McA. Gordon. Knots and links in spatial graphs. *Journal of Graph Theory*, 7:445–453, 1983.
7. B. Courcelle. Graph rewriting: An algebraic and logic approach. In J. van Leeuwen, editor, *Handbook of Theoretical Computer Science*, volume 2, pages 194–242. Elsevier Science Publishers, 1990.
8. R. Diestel. *Graph Theory*. Springer-Verlag, second edition, 2000.
9. R.G. Downey and M.R. Fellows. *Parameterized Complexity*. Springer-Verlag, 1999.
10. R.G. Downey, M.R. Fellows, and U. Taylor. The parameterized complexity of relational database queries and an improved characterization of $W[1]$. In Bridges, Calude, Gibbons, Reeves, and Witten, editors, *Combinatorics, Complexity, and Logic – Proceedings of DMTCS '96*, pages 194–213. Springer-Verlag, 1996.

[†] K_7 is an excluded minor for this class [6].

11. T. Feder and M.Y. Vardi. Monotone monadic SNP and constraint satisfaction. In *Proceedings of the 25th ACM Symposium on Theory of Computing*, pages 612–622, 1993.
12. J. Flum, M. Frick, and M. Grohe. Query evaluation via tree-decompositions. In Jan van den Bussche and Victor Vianu, editors, *Proceedings of the 8th International Conference on Database Theory*, Lecture Notes in Computer Science. Springer Verlag, 2001. To appear.
13. J. Flum and M. Grohe. Fixed-parameter tractability and logic. Submitted for publication.
14. M. Frick. *Easy Instances for Model Checking*. PhD thesis, Albert-Ludwigs-Universität Freiburg. To appear.
15. M. Frick and M. Grohe. Deciding first-order properties of locally tree-decomposable structures. Submitted for publication. A preliminary version of the paper appeared in *Proceedings of the 26th International Colloquium on Automata, Languages and Programming*, LNCS 1644, Springer-Verlag, 1999.
16. H. Gaifman. On local and non-local properties. In *Proceedings of the Herbrand Symposium, Logic Colloquium '81*. North Holland, 1982.
17. M.R. Garey and D.S. Johnson. *Computers and Intractability: A Guide to the Theory of NP-Completeness* Freeman, 1979.
18. L.A. Goldberg. *Efficient Algorithms for Listing Combinatorial Structures*. Cambridge University Press, 1993.
19. G. Gottlob, N. Leone, and F. Scarcello. A comparison of structural CSP decomposition methods. In Thomas Dean, editor, *Proceedings of the Sixteenth International Joint Conference on Artificial Intelligence*, pages 394–399. Morgan Kaufmann, 1999.
20. G. Gottlob, N. Leone, and F. Scarcello. Hypertree decompositions and tractable queries. In *Proceedings of the 18th ACM Symposium on Principles of Database Systems*, pages 21–32, 1999.
21. M. Grohe. The complexity of generalized model-checking problems. In preparation.
22. M. Grohe, T. Schwentick, and L. Segoufin. When is the evaluation of conjunctive queries tractable, 2000. Submitted for publication.
23. M.R. Jerrum, L.G. Valiant, and V.V. Vazirani. Random generation of combinatorial structures from a uniform distribution. *Theoretical Computer Science*, 43:169–188, 1986.
24. D.S. Johnson, C.H. Papadimitriou, and M. Yannakakis. On generating all maximal independent sets. *Information Processing Letters*, 27:119–123, 1988.
25. Ph.G. Kolaitis and M.Y. Vardi. Conjunctive-query containment and constraint satisfaction. In *Proceedings of the 17th ACM Symposium on Principles of Database Systems*, pages 205–213, 1998.
26. C.H. Papadimitriou and M. Yannakakis. On the complexity of database queries. In *Proceedings of the 17th ACM Symposium on Principles of Database Systems*, pages 12–19, 1997.
27. D. Seese. Linear time computable problems and first-order descriptions. *Mathematical Structures in Computer Science*, 6:505–526, 1996.
28. L.J. Stockmeyer. *The Complexity of Decision Problems in Automata Theory*. PhD thesis, Department of Electrical Engineering, MIT, 1974.
29. L.G. Valiant. The complexity of combinatorial computations: An introduction. In *GI 8. Jahrestagung Informatik, Fachberichte 18*, pages 326–337, 1978.
30. P. van Emde Boas. Machine models and simulations. In J. van Leeuwen, editor, *Handbook of Theoretical Computer Science*, volume 1, pages 1–66. Elsevier Science Publishers, 1990.
31. M.Y. Vardi. The complexity of relational query languages. In *Proceedings of the 14th ACM Symposium on Theory of Computing*, pages 137–146, 1982.
32. M.Y. Vardi. Constraint satisfaction and database theory: A tutorial. In *Proceedings of the 19th ACM Symposium on Principles of Database Systems*, pages 76–85, 2000.
33. M. Yannakakis. Algorithms for acyclic database schemes. In *7th International Conference on Very Large Data Bases*, pages 82–94, 1981.

Myhill–Nerode Relations on Automatic Systems and the Completeness of Kleene Algebra

Dexter Kozen

Department of Computer Science
Cornell University, Ithaca, NY 14853-7501, USA
kozen@cs.cornell.edu

Abstract. It is well known that finite square matrices over a Kleene algebra again form a Kleene algebra. This is also true for infinite matrices under suitable restrictions. One can use this fact to solve certain infinite systems of inequalities over a Kleene algebra. *Automatic systems* are a special class of infinite systems that can be viewed as infinite-state automata. Automatic systems can be collapsed using Myhill–Nerode relations in much the same way that finite automata can. The Brzozowski derivative on an algebra of polynomials over a Kleene algebra gives rise to a triangular automatic system that can be solved using these methods. This provides an alternative method for proving the completeness of Kleene algebra.

1 Introduction

Kleene algebra (KA) is the algebra of regular expressions. It dates to a 1956 paper of Kleene [7] and was further developed in the 1971 monograph of Conway [4]. Kleene algebra has appeared in one form or another in relational algebra [16,20], semantics and logics of programs [8,17], automata and formal language theory [14,15], and the design and analysis of algorithms [1,6,9]. Many authors have contributed over the years to the development of the algebraic theory; see [11] and references therein. There are many competing definitions and axiomatizations, and in fact there is no universal agreement on the definition of Kleene algebra.

In [10], a Kleene algebra was defined to be an idempotent semiring such that a^*b is the least solution to $b + ax \leq x$ and ba^* the least solution to $b + xa \leq x$. This is a finitary universal Horn axiomatization (universally quantified equations and equational implications). These axioms were shown in [10] to be sound and complete for the equational theory of the regular sets, improving a 1966 result of Salomaa [19]. Salomaa's axiomatization is sound and complete for the regular sets, but his axiom for * involves a nonalgebraic side condition that renders it unsound over other interpretations of importance, such as relational models. In contrast, the axiomatization of [10] is sound over a wide variety of models that arise in computer science, including relational models. No finitary axiomatization consisting solely of equations exists [18].

Matrices over a Kleene algebra, under the proper definition of the matrix operators, again form a Kleene algebra. This fundamental construction has many applications: the solution of systems of linear inequalities, construction of regular expressions equivalent to a given finite automaton, an algebraic treatment of finite automata in terms of their transition matrices, shortest path algorithms in directed graphs. In [10] it is used to encode algebraically various combinatorial constructions in the theory of finite automata,

A. Ferreira and H. Reichel (Eds.): STACS 2001, LNCS 2010, pp. 27–38, 2001.

including determinization via the subset construction and state minimization via the formation of a quotient modulo a Myhill–Nerode relation (see [5,12]). A key theorem of Kleene algebra used in both these constructions is

$$ax = xb \rightarrow a^*x = xb^*. \tag{1}$$

Intuitively, x represents a transformation between two state spaces, and a and b are transition relations of automata on those respective state spaces. The theorem represents a kind of bisimulation relationship. The completeness proof depends on the uniqueness of minimal deterministic automata: given two regular expressions representing the same regular set, it is shown how the construction of the unique minimal deterministic automaton can be carried out purely algebraically and the equivalence deduced from the axioms of Kleene algebra.

In this paper we give a new proof of completeness that does not depend on the uniqueness of minimal automata. Our approach is via a generalization of Myhill–Nerode relations. We introduce *automatic systems*, a special class of infinite systems that can be viewed as infinite-state automata. Automatic systems can be collapsed using Myhill–Nerode relations in much the same way that finite automata can. Again, the chief property describing the relationship between the collapsed and uncollapsed systems is (1). The Brzozowski derivative [3] on an algebra of polynomials over a Kleene algebra gives rise to a triangular automatic system that can be solved using these methods. Completeness is proved essentially by showing that two equivalent systems have a common Myhill–Nerode unwinding.

2   Kleene Algebra

Kleene algebra was introduced by S. C. Kleene (see [4]). We define a Kleene algebra to be an idempotent semiring such that a^*b is the least solution to $b + ax \le x$ and ba^* the least solution to $b + xa \le x$. This axiomatization is from [10], to which we refer the reader for further definitions and basic results.

The free Kleene algebra $\mathcal{F}_\Sigma$ on a finite set of generators Σ is normally constructed as the set of regular expressions over Σ modulo the Kleene algebra axioms. This is the same as $\mathbf{2}[\Sigma]$, the algebra of Kleene polynomials over indeterminates Σ, where $\mathbf{2}$ is the two-element Kleene algebra. As shown in [10], $\mathcal{F}_\Sigma$ is isomorphic to $\mathbf{Reg}_\Sigma$, the Kleene algebra of regular sets of strings over Σ.

The evaluation morphism $\varepsilon : \mathbf{2}[\Sigma] \rightarrow \mathbf{2}$, where $\varepsilon(a) = 0$ for $a \in \Sigma$, corresponds to the *empty word property* (EWP) discussed by Salomaa [2,19]. This map satisfies the property that $\varepsilon(\beta) = 1$ if $1 \le \beta$, 0 otherwise.

3   Generalized Triangular Matrices

Let A be a set and $\le$ a preorder (reflexive and transitive) on A. The preordered set A is *finitary* if all principal upward-closed sets $A_\alpha \stackrel{\text{def}}{=} \{\beta \in A \mid \alpha \le \beta\}$ are finite.

A (*generalized*) *triangular matrix* on a finitary preordered set A over a Kleene algebra K is a map $e : A^2 \rightarrow K$ such that $e_{\alpha,\beta} = 0$ whenever $\alpha \not\le \beta$. The family of generalized triangular matrices on A over K is denoted $\mathrm{Mat}(A, K)$.

There are several ways this definition generalizes the usual notion of triangular matrix. Ordinarily, the index set is finite and totally ordered, usually $\{1, \ldots, n\}$ with its natural order, and *triangular* is defined with respect to this order. In the present development, the index set A can be infinite and the order can be any finitary preorder. There can be pairwise incomparable elements, as well as "loops" with distinct elements α, β such that $\alpha \leq \beta$ and $\beta \leq \alpha$.

Nevertheless, the restrictions we have imposed are sufficient to allow the definition of the usual matrix operations on $\mathrm{Mat}(A, K)$. For $e, f \in \mathrm{Mat}(A, K)$, let

$$(e + f)_{\alpha,\beta} \stackrel{\mathrm{def}}{=} e_{\alpha,\beta} + f_{\alpha,\beta} \qquad 1_{\alpha,\beta} \stackrel{\mathrm{def}}{=} \begin{cases} 1, \text{ if } \alpha = \beta \\ 0, \text{ otherwise} \end{cases}$$

$$(ef)_{\alpha,\beta} \stackrel{\mathrm{def}}{=} \sum_{\gamma} e_{\alpha,\gamma} f_{\gamma,\beta} \qquad 0_{\alpha,\beta} \stackrel{\mathrm{def}}{=} 0.$$

Because A is finitary, the sum in the definition of matrix product is finite. It is not difficult to verify that the structure $\mathrm{Mat}(A, K)$ forms an idempotent semiring under these definitions.

Now we wish to define the operator $*$ on $\mathrm{Mat}(A, K)$ so as to make it a Kleene algebra. That A is finitary is elemental here. We define $e_{\alpha,\beta}$ to be $(e \restriction A_\alpha)^*_{\alpha,\beta}$, where $e \restriction A_\alpha$ is the restriction of e to domain A_α^2. Since A_α is finite, $e \restriction A_\alpha$ is a finite square submatrix of e, so $(e \restriction A_\alpha)^*$ exists. Actually, we could have restricted e to any finite upward-closed subset $B \subseteq A$ containing α and gotten the same result.

Formally, let 1_B denote the restriction of 1 to domain $A \times B$, where $B \subseteq A$ is upward-closed. The restriction of e to domain B^2 can be represented matricially by $1_B^T e \, 1_B$. If B is finite, then $1_B^T e \, 1_B$ is a finite square matrix, therefore the $*$ operator can be applied to obtain the matrix $(1_B^T e \, 1_B)^*$. We define

$$e^* \stackrel{\mathrm{def}}{=} \sup_B \, 1_B \, (1_B^T e \, 1_B)^* \, 1_B^T, \tag{2}$$

where the supremum is taken over all finite upward-closed subsets $B \subseteq A$. It can be shown by elementary arguments that the value of the right-hand side of (2) at α, β is a constant independent of B if $\alpha \in B$ and 0 if $\alpha \notin B$. Since there is at least one finite upward-closed subset of A containing α (namely A_α), the supremum exists.

4   Infinite Systems of Linear Inequalities

We can exploit the Kleene algebra structure of $\mathrm{Mat}(A, K)$ to solve triangular systems of linear inequalities indexed by the infinite set A. Such a system is represented by a triangular matrix $e \in \mathrm{Mat}(A, K)$ and vector $c : A \to K$ as

$$\sum_\beta e_{\alpha,\beta} X_\beta + c_\alpha \leq X_\alpha, \quad \alpha \in A,$$

where X is a vector of indeterminates. This is equivalent to the infinite matrix-vector inequality $eX + c \leq X$.

A solution of the system (A, e, c) over K is a map $\sigma : A \to K$ such that

$$\sum_{\beta} e_{\alpha,\beta}\sigma_\beta + c_\alpha \leq \sigma_\alpha, \quad \alpha \in A,$$

or in other words $e\sigma + c \leq \sigma$. As in the finite case, the unique least solution to this system is e^*c.

5 Automatic Systems

We now focus on index sets A of a special form. Let Σ be a finite set of functions acting on A. The value of the function $a \in \Sigma$ on $\alpha \in A$ is denoted αa. Each finite-length string $x \in \Sigma^*$ induces a function $x : A \to A$ defined inductively by

$$\alpha\varepsilon \overset{\text{def}}{=} \alpha \qquad \alpha(xa) \overset{\text{def}}{=} (\alpha x)a.$$

Define $\alpha \leq \beta$ if $\beta = \alpha x$ for some $x \in \Sigma^*$. This is a preorder on A, and it is finitary iff for all $\alpha \in A$, the set $A_\alpha = \{\alpha x \mid x \in \Sigma^*\}$ is finite. Since Σ is assumed to be finite, it follows from König's lemma that A is finitary iff every $\leq$-chain $\alpha_0 \leq \alpha_1 \leq \cdots$ has only finitely many distinct elements; equivalently, for every α, every sufficiently long string $x \in \Sigma^*$ has two distinct prefixes y and z such that $\alpha y = \alpha z$.

Now let $e \in \text{Mat}(A, K)$ be a triangular matrix and $c : A \to K$ a vector over A representing a triangular system of linear inequalities as described in the last section. Assume further that if $\beta \neq \alpha a$ for any $a \in \Sigma$, then $e_{\alpha,\beta} = 0$. The system of inequalities represented by e and c is thus

$$\sum_{a \in \Sigma} e_{\alpha,\alpha a} X_{\alpha a} + c_\alpha \leq X_\alpha, \quad \alpha \in A.$$

A linear system of this form is called *automatic*. This name is meant to suggest a generalization of finite-state automata over $\mathbf{Reg}_\Sigma$ to infinite-state systems over arbitrary Kleene algebras. One can regard A as a set of states and elements of Σ as input symbols. An ordinary finite-state automaton is essentially a finite automatic system over the Kleene algebra $\mathbf{Reg}_\Sigma$.

6 Myhill–Nerode Relations

Myhill–Nerode relations are fundamental in the theory of finite-state automata. Among other applications, they allow an automaton to be collapsed to a unique equivalent minimal automaton. Myhill–Nerode relations can also be defined on finitary automatic systems.

Given a finitary automatic system $S = (A, e, c)$, an equivalence relation $\equiv$ on A is called *Myhill–Nerode* if the following conditions are satisfied: for all $\alpha, \beta \in A$ and $a \in \Sigma$,

   (i) if $\alpha \equiv \beta$, then $\alpha a \equiv \beta a$;
   (ii) if $\alpha \equiv \beta$, then $\sum_{ab \equiv \alpha a} e_{\alpha,\alpha b} = \sum_{\beta b \equiv \beta a} e_{\beta,\beta b}$;
   (iii) if $\alpha \equiv \beta$, then $c_\alpha = c_\beta$.

For any Myhill–Nerode relation $\equiv$ on $S = (A, e, c)$, we can construct a quotient system $S/\equiv$ as follows:

$$[\alpha] \stackrel{\text{def}}{=} \{\beta \in A \mid \beta \equiv \alpha\}$$
$$[\alpha]a \stackrel{\text{def}}{=} [\alpha a]$$
$$A/\equiv \stackrel{\text{def}}{=} \{[\alpha] \mid \alpha \in A\}$$

$$(e/\equiv)_{[\alpha],[\alpha]a} \stackrel{\text{def}}{=} \sum_{\alpha b \equiv \alpha a} e_{\alpha, \alpha b}$$
$$(c/\equiv)_{[\alpha]} \stackrel{\text{def}}{=} c_\alpha$$
$$S/\equiv \stackrel{\text{def}}{=} (A/\equiv, e/\equiv, c/\equiv).$$

The matrix $e/\equiv$ and vector $c/\equiv$ are well defined by the restrictions in the definition of Myhill–Nerode relation. The original system S can be thought of as an "unfolding" of the collapsed system $S/\equiv$.

The set Σ acts on $A/\equiv$ by $[\alpha]a \stackrel{\text{def}}{=} [\alpha a]$. This is well defined by clause (i) in the definition of Myhill–Nerode relation. The preorder $\leq$ on $A/\equiv$ is defined as in Section 5. This relation is easily checked to be reflexive, transitive, and finitary on $A/\equiv$. Moreover, the matrix $e/\equiv$ is triangular. Thus $S/\equiv$ is an automatic system.

We now describe the relationship between the solutions of the systems S and $S/\equiv$. First, any solution of the collapsed system $S/\equiv$ can be lifted to a solution of the original system S. If $\sigma : A/\equiv \to K$ is a solution of $S/\equiv$, define $\hat{\sigma} : A \to K$ by $\hat{\sigma}_\alpha \stackrel{\text{def}}{=} \sigma_{[\alpha]}$. It is easily verified that $\hat{\sigma}$ is a solution of S:

$$\sum_{a \in \Sigma} e_{\alpha, \alpha a} \hat{\sigma}_{\alpha a} + c_\alpha = \sum_{a \in \Sigma} e_{\alpha, \alpha a} \sigma_{[\alpha a]} + c_\alpha$$
$$= \sum_{a \in \Sigma} (e/\equiv)_{[\alpha],[\alpha a]} \sigma_{[\alpha a]} + (c/\equiv)_{[\alpha]}$$
$$\leq \sigma_{[\alpha]} = \hat{\sigma}_\alpha.$$

It is more difficult to argue that $\hat{\sigma}$ is the least solution to S. The unfolded system S is less constrained than $S/\equiv$, and it is conceivable that a smaller solution could be found in which different but $\equiv$-equivalent α, β are assigned different values, whereas in the collapsed system $S/\equiv$, α and β are unified and must have the same value. We show that this cannot happen.

Example 1. Consider the 2×2 system

$$aY + c \leq X$$
$$aX + c \leq Y.$$

This is represented by the matrix-vector equation

$$\begin{bmatrix} 0 & a \\ a & 0 \end{bmatrix} \cdot \begin{bmatrix} X \\ Y \end{bmatrix} + \begin{bmatrix} c \\ c \end{bmatrix} \leq \begin{bmatrix} X \\ Y \end{bmatrix}.$$

We can collapse this system by a Myhill–Nerode relation to the single inequality $aX + c \leq X$. The least solution of the 2×2 system is given by

$$
\begin{bmatrix} X \\ Y \end{bmatrix} = \begin{bmatrix} 0 & a \\ a & 0 \end{bmatrix}^* \cdot \begin{bmatrix} c \\ c \end{bmatrix}
$$
$$
= \begin{bmatrix} (aa)^* & (aa)^*a \\ (aa)^*a & (aa)^* \end{bmatrix} \cdot \begin{bmatrix} c \\ c \end{bmatrix}
$$
$$
= \begin{bmatrix} (aa)^*c + (aa)^*ac \\ (aa)^*ac + (aa)^*c \end{bmatrix}
$$
$$
= \begin{bmatrix} a^*c \\ a^*c \end{bmatrix},
$$

which is the same as that obtained by lifting the least solution a^*c of the collapsed system $aX + c \leq X$.

We show that in general, the least solution of S is obtained by lifting the least solution of $S/\equiv$. Define $\chi : A \times A/\equiv \; \rightarrow K$ by

$$
\chi_{\alpha,[\beta]} \stackrel{\text{def}}{=} \begin{cases} 1, & \text{if } \alpha \equiv \beta \\ 0, & \text{otherwise.} \end{cases}
$$

The matrix χ is called the *characteristic matrix* of $\equiv$. To lift a solution from $S/\equiv$ to S, we multiply it on the left by χ; thus in the above example, $\hat{\sigma} = \chi\sigma$.

Now for any α, γ,

$$
(e\chi)_{\alpha,[\gamma]} = \sum_{\alpha a} e_{\alpha,\alpha a} \chi_{\alpha a,[\gamma]}
$$
$$
= \sum_{\alpha a \equiv \gamma} e_{\alpha,\alpha a}
$$
$$
= (e/\equiv)_{[\alpha],[\gamma]}
$$
$$
= \sum_{[\beta]} \chi_{\alpha,[\beta]} (e/\equiv)_{[\beta],[\gamma]}
$$
$$
= (\chi(e/\equiv))_{\alpha,[\gamma]},
$$

therefore $e\chi = \chi(e/\equiv)$. By (1) (see [13]), $e^*\chi = \chi(e/\equiv)^*$. Since $c = \chi(c/\equiv)$, we have

$$
e^*c = e^*\chi(c/\equiv) = \chi(e/\equiv)^*(c/\equiv),
$$

which shows that the least solution e^*c of S is obtained by lifting the least solution $(e/\equiv)^*(c/\equiv)$ of $S/\equiv$.

7 Brzozowski Derivatives

For $x \in \Sigma^*$, the *Brzozowski derivative* was originally defined by Brzozowski [3,4] as a map $2^{\Sigma^*} \rightarrow 2^{\Sigma^*}$ such that

$$
D_x(A) \stackrel{\text{def}}{=} \{y \in \Sigma^* \mid xy \in A\};
$$

that is, the set of strings obtained by removing x from the front of a string in A. It follows from elementary arguments that $D_x(A)$ is a regular set if A is.

Here we wish to consider D_x as an operator on $\mathcal{F}_\Sigma$. Without knowing that $\mathcal{F}_\Sigma \cong \mathbf{Reg}_\Sigma$, we could have defined D_x on $\mathcal{F}_\Sigma$ inductively as follows. For $a \in \Sigma$,

$$D_a(0) = D_a(1) = D_a(b) \stackrel{\text{def}}{=} 0, \quad b \neq a$$
$$D_a(a) \stackrel{\text{def}}{=} 1$$
$$D_a(\alpha + \beta) \stackrel{\text{def}}{=} D_a(\alpha) + D_a(\beta)$$
$$D_a(\alpha\beta) \stackrel{\text{def}}{=} D_a(\alpha)\beta + \varepsilon(\alpha)D_a(\beta) \tag{3}$$
$$D_a(\alpha^*) \stackrel{\text{def}}{=} D_a(\alpha)\alpha^*,$$

where $\varepsilon : \mathcal{F}_\Sigma \to \mathbf{2}$ is the evaluation morphism $\varepsilon(a) = 0$, $a \in \Sigma$. We then define inductively

$$D_\varepsilon(\alpha) \stackrel{\text{def}}{=} \alpha \qquad D_{xa}(\alpha) \stackrel{\text{def}}{=} D_a(D_x(\alpha)).$$

This definition agrees with Brzozowski's on $\mathbf{Reg}_\Sigma$ [3]. However, we must argue axiomatically that it is well defined on elements of $\mathcal{F}_\Sigma$; that is, if $\alpha = \beta$ is a theorem of Kleene algebra, then $D_a(\alpha) = D_a(\beta)$. This can be done by induction on the lengths of proofs. We argue the case of the Horn axiom $\alpha\gamma + \beta \leq \gamma \to \alpha^*\beta \leq \gamma$ explicitly. Suppose we have derived $\alpha^*\beta \leq \gamma$ by this rule, having previously proved $\alpha\gamma + \beta \leq \gamma$. By the induction hypothesis, we have $D_a(\alpha\gamma + \beta) \leq D_a(\gamma)$ and we wish to prove that $D_a(\alpha^*\beta) \leq D_a(\gamma)$.

$$D_a(\alpha^*\beta) = D_a(\alpha^*)\beta + \varepsilon(\alpha^*)D_a(\beta)$$
$$= D_a(\alpha)\alpha^*\beta + D_a(\beta)$$
$$\leq D_a(\alpha)\gamma + D_a(\beta)$$
$$\leq D_a(\alpha)\gamma + \varepsilon(\alpha)D_a(\gamma) + D_a(\beta)$$
$$= D_a(\alpha\gamma + \beta)$$
$$\leq D_a(\gamma).$$

The following lemmas list some basic properties of Brzozowski derivatives. All of these properties are well known and are easily derived by elementary inductive arguments using the laws of Kleene algebra.

Lemma 1. *Let* $R : \mathcal{F}_\Sigma \to \mathbf{Reg}_\Sigma$ *be the canonical interpretation* $R(a) = \{a\}$.

(i) *For* $a \in \Sigma$, $aD_a(\beta) \leq \beta$;
(ii) *If* $1 \leq \beta$, *then for* $m \geq n = |x|$, $D_x(\beta^m) = D_x(\beta^n)\beta^{m-n}$;
(iii) *For* $n = |x|$, $D_x(\alpha^*) = D_x((1+\alpha)^n)\alpha^*$;
(iv) $D_x(\alpha\beta) = D_x(\alpha)\beta + \sum_{x=yz} \varepsilon(D_y(\alpha))D_z(\beta)$;
(v) $\varepsilon(D_x(\alpha\beta)) = \sum_{x=yz} \varepsilon(D_y(\alpha)D_z(\beta))$;
(vi) $D_x(\alpha^*) = D_x(1) + D_x(\alpha)\alpha^* + \sum_{\substack{x=yz \\ z \neq x}} \varepsilon(D_y(\alpha))D_z(\alpha^*)$;

(vii) $x \in R(\alpha)$ *iff* $\varepsilon(D_x(\alpha)) = 1$.

Proof. All follow by elementary inductive arguments from the definition of D_x and the laws of Kleene algebra. We prove (vii) explicitly. Proceeding by induction on α, the base cases $\alpha = 0, 1$, or $a \in \Sigma$ are immediate. For expressions of the form $\alpha + \beta$, the result follows from the linearity of R, ε, and D_x. For the other compound expressions,

$$
\begin{aligned}
x \in R(\alpha\beta) &\Longleftrightarrow \exists y, z \; x = yz \wedge y \in R(\alpha) \wedge z \in R(\beta) \\
&\Longleftrightarrow \exists y, z \; x = yz \wedge \varepsilon(D_y(\alpha)) = 1 \wedge \varepsilon(D_z(\beta)) = 1 \\
&\Longleftrightarrow \sum_{x=yz} \varepsilon(D_y(\alpha)D_z(\beta)) = 1 \\
&\Longleftrightarrow \varepsilon(D_x(\alpha\beta)) = 1 \qquad \text{by (v);}
\end{aligned}
$$

$$
\begin{aligned}
x \in R(\alpha^*) &\Longleftrightarrow x \in R((1+\alpha)^n), \quad \text{where } n = |x| \\
&\Longleftrightarrow \varepsilon(D_x((1+\alpha)^n)) = 1 \\
&\Longleftrightarrow \varepsilon(D_x((1+\alpha)^n))\varepsilon(\alpha^*) = 1 \\
&\Longleftrightarrow \varepsilon(D_x((1+\alpha)^n)\alpha^*) = 1 \\
&\Longleftrightarrow \varepsilon(D_x(\alpha^*)) = 1 \qquad \text{by (iii).}
\end{aligned}
$$

8   Brzozowski Systems

A class of automatic systems can be defined in terms of Brzozowski derivatives. We take the set A in Section 5 to be $\mathcal{F}_\Sigma$ and define the action of $a \in \Sigma$ on $\mathcal{F}_\Sigma$ as D_a. That is, for all $\alpha \in \mathcal{F}_\Sigma$, $\alpha a \overset{\text{def}}{=} D_a(\alpha)$. We must argue that the induced preorder is finitary. The proof of Brzozowski (see [4]) depends on the interpretation $\mathbf{Reg}_\Sigma$, but we must argue axiomatically.

Lemma 2. *For any α, the set $\{\alpha x \mid x \in \Sigma^*\} = \{D_x(\alpha) \mid x \in \Sigma^*\}$ is finite.*

Proof. The proof proceeds by induction on α. For α of the form 0, 1, or $a \in \Sigma$, the result is easy. For $\alpha + \beta$, the result follows from the linearity of D_x and the induction hypothesis. For $\alpha\beta$, the result follows from Lemma 1(iv) and the induction hypothesis. Finally, for α^*, the result follows from Lemma 1(vi) and the induction hypothesis.

Now consider the system $S = (\mathcal{F}_\Sigma, e, c)$, where

$$
e_{\alpha,\alpha a} \overset{\text{def}}{=} \sum_{\alpha b = \alpha a} b \qquad c_\alpha \overset{\text{def}}{=} \varepsilon(\alpha).
$$

We call this system the *Brzozowski system* on Σ. The least solution of this system over $\mathcal{F}_\Sigma$ is $\ell = e^* c$. The key property that we need is that ℓ, considered as a map $\ell : \mathcal{F}_\Sigma \to \mathcal{F}_\Sigma$, is a homomorphism. We show in fact that ℓ is ι, the identity on $\mathcal{F}_\Sigma$.

Lemma 3. *The identity map $\iota : \alpha \mapsto \alpha$ is the least solution to the Brzozowski system.*

Proof. First we show that $\ell \leq \iota$. It suffices to show that ι is a solution to S. We must argue that for all $\alpha \in \mathcal{F}_\Sigma$,

$$
\sum_{a \in \Sigma} a D_a(\alpha) + \varepsilon(\alpha) \leq \alpha.
$$

But this is immediate from Lemma 1(i) and the property $\varepsilon(\beta) \leq \beta$ noted in Section 2.

Now we show that ι is the least solution to S. The major portion of the work is involved in showing that if $\alpha \leq \beta$, then $\ell_\alpha \leq \ell_\beta$. We use the Myhill–Nerode theory developed in Section 6 to find a common unwinding of the Brzozowski system S, allowing us to compare ℓ_α and ℓ_β.

First, lift the system S to the product $\mathcal{F}_\Sigma \times \mathcal{F}_\Sigma$ under each of the two projection maps to obtain two systems $U = (\mathcal{F}_\Sigma \times \mathcal{F}_\Sigma, e, c)$ and $V = (\mathcal{F}_\Sigma \times \mathcal{F}_\Sigma, e, d)$, where

$$e_{(\gamma,\delta),(\gamma,\delta)a} \overset{\mathrm{def}}{=} \sum_{\substack{\gamma b = \gamma a \\ \delta b = \delta a}} b \qquad c_{\gamma,\delta} \overset{\mathrm{def}}{=} \varepsilon(\gamma) \qquad d_{\gamma,\delta} \overset{\mathrm{def}}{=} \varepsilon(\delta).$$

The relations defined by the two projections,

$$(\gamma, \delta) \equiv_1 (\gamma', \delta') \overset{\mathrm{def}}{\Longleftrightarrow} \gamma = \gamma' \qquad (\gamma, \delta) \equiv_2 (\gamma', \delta') \overset{\mathrm{def}}{\Longleftrightarrow} \delta = \delta',$$

are Myhill–Nerode.

Now restrict these systems to the finite induced subsystems on

$$(\mathcal{F}_\Sigma \times \mathcal{F}_\Sigma)_{(\alpha,\beta)} = \{(\alpha x, \beta x) \mid x \in \Sigma^*\}$$

to obtain $U' = ((\mathcal{F}_\Sigma \times \mathcal{F}_\Sigma)_{(\alpha,\beta)}, e', c')$ and $V' = ((\mathcal{F}_\Sigma \times \mathcal{F}_\Sigma)_{(\alpha,\beta)}, e', d')$, where e', c', and d' are e, c, and d, respectively, restricted to $(\mathcal{F}_\Sigma \times \mathcal{F}_\Sigma)_{(\alpha,\beta)}$. The least solution of U' is $e'^* c'$ and the least solution of V' is $e'^* d'$. Moreover, by linearity, $\varepsilon(D_x(\alpha)) \leq \varepsilon(D_x(\beta))$ for all $x \in \Sigma^*$, therefore $c' \leq d'$ and

$$\ell_\alpha = (e'^* c')_{\alpha,\beta} \leq (e'^* d')_{\alpha,\beta} = \ell_\beta.$$

We have shown that $\alpha \leq \beta$ implies $\ell_\alpha \leq \ell_\beta$. It follows that

$$\ell_\alpha + \ell_\beta \leq \ell_{\alpha+\beta}. \tag{4}$$

Now we show that $\alpha \leq \ell_\alpha$ for all α by induction on α. We actually show by induction that $\alpha \ell_\beta \leq \ell_{\alpha\beta}$ for all α and β by induction on α.

For atomic expressions, we have

$$\ell_{0\beta} = \ell_0 = 0 = 0\ell_\beta;$$
$$\ell_{1\beta} = \ell_\beta = 1\ell_\beta;$$
$$\ell_{b\beta} = \sum_{a \in \Sigma} a\ell_{D_a(b\beta)} + \varepsilon(b\beta)$$
$$= \sum_{a \in \Sigma} a\ell_{D_a(b)\beta}$$
$$= b\ell_{D_b(b)\beta}$$
$$= b\ell_\beta, \quad b \in \Sigma.$$

For compound expressions,

$$(\alpha + \gamma)\ell_\beta = \alpha\ell_\beta + \gamma\ell_\beta$$
$$\leq \ell_{\alpha\beta} + \ell_{\gamma\beta} \qquad \text{by the induction hypothesis}$$
$$\leq \ell_{(\alpha+\gamma)\beta} \qquad \text{by (4);}$$

$$\alpha\gamma\ell_\beta \leq \alpha\ell_{\gamma\beta} \qquad \text{by the induction hypothesis on } \gamma$$
$$\leq \ell_{\alpha\gamma\beta} \qquad \text{by the induction hypothesis on } \alpha.$$

Finally, to show $\alpha^*\ell_\beta \leq \ell_{\alpha^*\beta}$, by an axiom of Kleene algebra it is enough to show $\ell_\beta + \alpha\ell_{\alpha^*\beta} \leq \ell_{\alpha^*\beta}$. We have

$$\ell_\beta + \alpha\ell_{\alpha^*\beta} \leq \ell_\beta + \ell_{\alpha\alpha^*\beta} \qquad \text{by the induction hypothesis}$$
$$\leq \ell_{\beta+\alpha\alpha^*\beta} \qquad \text{by (4)}$$
$$= \ell_{\alpha^*\beta}.$$

Thus $\ell_\alpha \leq \alpha$ since ι is a solution and ℓ is the least solution, and $\alpha \leq \ell_\alpha$ by taking $\beta = 1$ in the argument above, therefore $\ell_\alpha = \alpha$.

9 Completeness

The completeness result of [10], which states that the free Kleene algebra $\mathcal{F}_\Sigma$ and the Kleene algebra of regular sets $\mathbf{Reg}_\Sigma$ are isomorphic, follows from the considerations of the previous sections. Let $R : \mathcal{F}_\Sigma \to \mathbf{Reg}_\Sigma$ be the canonical interpretation in which $R(a) = \{a\}$. If $R(\alpha) = R(\beta)$, then for all $x \in \Sigma^*$, $x \in R(\alpha)$ iff $x \in R(\beta)$, therefore by Lemma 1(vii), $\varepsilon(D_x(\alpha)) = \varepsilon(D_x(\beta))$. This says that the common unwinding of the Brzozowski system S on $\mathcal{F}_\Sigma \times \mathcal{F}_\Sigma$ restricted to $(\mathcal{F}_\Sigma \times \mathcal{F}_\Sigma)_{(\alpha,\beta)}$ gives identical systems, therefore their solutions are equal. In particular, $\ell_\alpha = \ell_\beta$. By Lemma 3, $\alpha = \beta$.

10 The Commutative Case

A similar completeness result holds for *commutative* Kleene algebra, in which we postulate the commutativity axiom $\alpha\beta = \beta\alpha$. The free commutative Kleene algebra on n generators is the Kleene algebra $\mathbf{Par}_n$ of regular subsets of $\mathbb{N}^n$. Elements of $\mathbb{N}^n$ are often called *Parikh vectors*. We interpret regular expressions over $\Sigma = \{a_1, \ldots, a_n\}$ as follows:

$$L(a_i) \stackrel{\text{def}}{=} \{(\underbrace{0, \ldots, 0}_{i-1}, 1, \underbrace{0, \ldots, 0}_{n-i})\}$$

$$L(\alpha + \beta) \stackrel{\text{def}}{=} L(\alpha) \cup L(\beta)$$

$$L(\alpha\beta) \stackrel{\text{def}}{=} \{u + v \mid u \in L(\alpha),\ v \in L(\beta)\}$$

$$L(\alpha^*) \stackrel{\text{def}}{=} \bigcup_m L(\alpha)^m$$

$$L(0) \stackrel{\text{def}}{=} \varnothing$$

$$L(1) \stackrel{\text{def}}{=} \{(0, \ldots, 0)\}.$$

A set of Parikh vectors is *regular* if it is $L(\alpha)$ for some α. The family of all regular sets of Parikh vectors forms a commutative Kleene algebra under the above operations. We denote this algebra by $\mathbf{Par}_n$.

The completeness result follows from a characterization due to Redko (see [4]) of the equational theory of $\mathbf{Par}_n$ as the consequences of a certain infinite but easily-described set of equations, namely the equational axioms for commutative idempotent semirings plus the equations

$$(x + y)^* = (x^*y)^*x^* \qquad x^{**} = x^*$$
$$(xy)^*x = x(yx)^* \qquad x^*y^* = (xy)^*(x^* + y^*)$$
$$x^* = 1 + xx^* \qquad x^* = (x^m)^*(1 + x)^{m-1}, \quad m \geq 1.$$

All these are theorems of commutative Kleene algebra.

The proof of Redko, as given in [4], is quite involved and depends heavily on commutativity. We began this investigation in a attempt to give a uniform completeness proof for both the noncommutative and commutative case. Our hope was to give a simpler algebraic proof along the lines of [10] for commutative Kleene algebra, although the technique of [10] does not apply directly, since minimal automata are not unique. For example, the three-state deterministic automata corresponding to the expressions $(ab)^*$ and $(ba)^*$ are both minimal and represent the same set of Parikh vectors $\{(m, m) \mid m \geq 0\}$. The usual construction of the canonical deterministic automaton directly from the set itself (see [12, Lemma 16.2]) yields infinitely many states.

Nevertheless, one can define the free commutative Kleene algebra $\mathcal{C}_\Sigma$ on generators Σ and attempt to show that L, factored through $\mathcal{C}_\Sigma$, gives an isomorphism $\mathcal{C}_\Sigma \to \mathbf{Par}_n$. The Brzozowski derivatives $D_a : \mathcal{C}_\Sigma \to \mathcal{C}_\Sigma$ are defined differently on products in the commutative case:

$$D_a(\alpha\beta) \overset{\text{def}}{=} D_a(\alpha)\beta + \alpha D_a(\beta).$$

The action of D_a on other expressions is as defined in Section 7. As in that section, we can argue that D_a respects the axioms of Kleene algebra. Here we must also show that it respects the commutativity axiom; in other words, $D_a(\alpha\beta) = D_a(\beta\alpha)$. Also, for any $x, y \in \Sigma^*$, $D_{xy}(\alpha) = D_{yx}(\alpha)$. Unfortunately, the principal upward closed sets $\{D_x(\alpha) \mid x \in \Sigma^*\}$ are not necessarily finite, and it is not clear how to define a Kleene algebra structure of infinite matrices as in Section 3. Nevertheless, the set $\{D_x(\alpha) \mid x \in \Sigma^*\}$ does exhibit a regular $(n - 1)$-dimensional linear geometric structure which is respected by the action of the Brzozowski derivatives. It remains a topic for future investigation to see how this structure can be exploited.

Acknowledgments

The support of the National Science Foundation under grant CCR-9708915 is gratefully acknowledged.

References

1. Alfred V. Aho, John E. Hopcroft, and Jeffrey D. Ullman. *The Design and Analysis of Computer Algorithms*. Addison-Wesley, Reading, Mass., 1975.
2. Roland Carl Backhouse. *Closure Algorithms and the Star-Height Problem of Regular Languages*. PhD thesis, Imperial College, London, U.K., 1975.
3. Janusz A. Brzozowski. Derivatives of regular expressions. *J. Assoc. Comput. Mach.*, 11:481–494, 1964.
4. John Horton Conway. *Regular Algebra and Finite Machines*. Chapman and Hall, London, 1971.
5. J. E. Hopcroft and J. D. Ullman. *Introduction to Automata Theory, Languages, and Computation*. Addison-Wesley, 1979.
6. Kazuo Iwano and Kenneth Steiglitz. A semiring on convex polygons and zero-sum cycle problems. *SIAM J. Comput.*, 19(5):883–901, 1990.
7. Stephen C. Kleene. Representation of events in nerve nets and finite automata. In C. E. Shannon and J. McCarthy, editors, *Automata Studies*, pages 3–41. Princeton University Press, Princeton, N.J., 1956.
8. Dexter Kozen. On induction vs. *-continuity. In Kozen, editor, *Proc. Workshop on Logic of Programs*, volume 131 of *Lecture Notes in Computer Science*, pages 167–176, New York, 1981. Springer-Verlag.
9. Dexter Kozen. *The Design and Analysis of Algorithms*. Springer-Verlag, New York, 1991.
10. Dexter Kozen. A completeness theorem for Kleene algebras and the algebra of regular events. *Infor. and Comput.*, 110(2):366–390, May 1994.
11. Dexter Kozen. Kleene algebra with tests and commutativity conditions. In T. Margaria and B. Steffen, editors, *Proc. Second Int. Workshop Tools and Algorithms for the Construction and Analysis of Systems (TACAS'96)*, volume 1055 of *Lecture Notes in Computer Science*, pages 14–33, Passau, Germany, March 1996. Springer-Verlag.
12. Dexter Kozen. *Automata and Computability*. Springer-Verlag, New York, 1997.
13. Dexter Kozen. Typed Kleene algebra. Technical Report 98-1669, Computer Science Department, Cornell University, March 1998.
14. Werner Kuich. The Kleene and Parikh theorem in complete semirings. In T. Ottmann, editor, *Proc. 14th Colloq. Automata, Languages, and Programming*, volume 267 of *Lecture Notes in Computer Science*, pages 212–225, New York, 1987. EATCS, Springer-Verlag.
15. Werner Kuich and Arto Salomaa. *Semirings, Automata, and Languages*. Springer-Verlag, Berlin, 1986.
16. K. C. Ng. *Relation Algebras with Transitive Closure*. PhD thesis, University of California, Berkeley, 1984.
17. Vaughan Pratt. Dynamic algebras as a well-behaved fragment of relation algebras. In D. Pigozzi, editor, *Proc. Conf. on Algebra and Computer Science*, volume 425 of *Lecture Notes in Computer Science*, pages 77–110, Ames, Iowa, June 1988. Springer-Verlag.
18. V. N. Redko. On defining relations for the algebra of regular events. *Ukrain. Mat. Z.*, 16:120–126, 1964. In Russian.
19. Arto Salomaa. Two complete axiom systems for the algebra of regular events. *J. Assoc. Comput. Mach.*, 13(1):158–169, January 1966.
20. Alfred Tarski. On the calculus of relations. *J. Symb. Logic*, 6(3):65–106, 1941.

2-Nested Simulation Is Not Finitely Equationally Axiomatizable

Luca Aceto[1], Wan Fokkink[2], and Anna Ingólfsdóttir[1]

[1] **BRICS** (**B**asic **R**esearch **i**n **C**omputer **S**cience)
Centre of the Danish National Research Foundation
Department of Computer Science, Aalborg University
Fr. Bajersvej 7E, 9220 Aalborg Ø, Denmark
luca@cs.auc.dk, annai@cs.auc.dk
[2] CWI, Department of Software Engineering
Kruislaan 413, 1098 SJ Amsterdam, The Netherlands
wan@cwi.nl

Abstract. 2-nested simulation was introduced by Groote and Vaandrager [10] as the coarsest equivalence included in completed trace equivalence for which the tyft/tyxt format is a congruence format. In the linear time-branching time spectrum of van Glabbeek [8], 2-nested simulation is one of the few equivalences for which no finite equational axiomatization is presented. In this paper we prove that such an axiomatization does not exist for 2-nested simulation.

1 Introduction

Labelled transition systems (LTSs) [11] are a fundamental model of concurrent computation, which is widely used in light of its flexibility and applicability. In particular, they are the prime model underlying Plotkin's Structural Operational Semantics [18] and, following Milner's pioneering work on CCS [14], are by now the standard semantic model for various process description languages.

LTSs model processes by explicitly describing their states and their transitions from state to state, together with the actions that produced them. Since this view of process behaviours is very detailed, several notions of behavioural equivalence and preorder have been proposed for LTSs. The aim of such behavioural semantics is to identify those (states of) LTSs that afford the same "observations", in some appropriate technical sense. The lack of consensus on what constitutes an appropriate notion of observable behaviour for reactive systems has led to a large number of proposals for behavioural equivalences for concurrent processes. (Cf. the encyclopaedic study [8], where van Glabbeek presents the linear time-branching time spectrum—a lattice that contains all the known behavioural equivalences and preorders over LTSs, ordered by inclusion.)

One of the criteria that has been put forward for studying the mathematical tractability of the behavioural equivalences in the linear time-branching time spectrum is that they afford elegant, finite equational axiomatizations over fragments of process algebraic languages. Equationally based proof systems play an

A. Ferreira and H. Reichel (Eds.): STACS 2001, LNCS 2010, pp. 39–50, 2001.

important role in both the practice and the theory of process algebras. From the point of view of practice, these proof systems can be used to perform system verifications in a purely syntactic way, and form the basis of axiomatic verification tools like, e.g., PAM [12]. From the theoretical point of view, complete axiomatizations of behavioural equivalences capture the essence of different notions of semantics for processes in terms of a basic collection of identities, and this often allows one to compare semantics which may have been defined in very different styles and frameworks. A review of existing complete equational axiomatizations for many of the behavioral semantics in van Glabbeek's spectrum is offered in [8]. The equational axiomatizations offered *ibidem* are over Milner's Basic CCS (abbreviated to BCCS in what follows), a fragment of CCS suitable for describing finite synchronization trees, and characterize the differences between behavioural semantics in terms of a few revealing axioms.

The main omission in this menagerie of equational axiomatizations for the behavioural semantics in van Glabbeek's spectrum is an axiomatization for 2-nested simulation semantics. 2-nested simulation was introduced by Groote and Vaandrager [10] as the coarsest equivalence included in completed trace equivalence for which the tyft/tyxt format is a congruence format. It thus characterizes the distinctions amongst processes that can be made by observing their termination behaviour in program contexts that can be built using a wide array of operators. (The interested reader is referred to *op. cit.* for motivation and the basic theory of 2-nested simulation.) 2-nested simulation can be decided over finite LTSs in time that is quadratic in their number of transitions [21], and can be characterized by a single parameterised modal logic formula [15]. However, as previously mentioned, no equational axiomatization for it has ever been proposed, even for the language BCCS.

In this paper, we offer a possible mathematical justification for the lack of an equational axiomatization for the 2-nested simulation equivalence and preorder even for the language of finite synchronization trees. More precisely, we show that neither of these two behavioural relations has a finite equational axiomatization over the language of BCCS. These results hold in a very strong form. Indeed, we prove that no finite collection of inequations that are sound with respect to the 2-nested simulation preorder can prove all of the inequalities of the form

$$a^{2m} \lesssim a^{2m} + a^m \qquad (m \geq 0) \ ,$$

which are sound with respect to the 2-nested simulation preorder. Similarly, we establish a result to the effect that no finite collection of equations that are sound with respect to 2-nested simulation equivalence can be used to derive all of the sound equalities of the form

$$a(a^{2m} + a^m) \approx a(a^{2m} + a^m) + a^{2m+1} \qquad (m \geq 0) \ .$$

The import of these two results is that not only the equational theory of 2-nested simulation is not finitely equationally axiomatizable, but neither is the collection of (in)equivalences that hold between BCCS terms over one action and without occurrences of variables. This state of affairs should be contrasted with the elegant equational axiomatizations over BCCS for most of the other behavioural equivalences in the linear time–branching time spectrum that are reviewed by van Glabbeek in [8]. Only in the case of additional, more complex operators, such as iteration, are these equivalences known to lack a finite equational axiomatization; see, e.g., [3,6,7,19,20]. Of special relevance for concurrency theory are Moller's results to the effect that the process algebras ACP and CCS (without the auxiliary left merge operator from [5]) do not have a finite equational axiomatization modulo bisimulation equivalence [16,17]. Aceto, Ésik and Ingólfsdóttir [2] proved that there is no finite equational axiomatization that is ω-complete for the max-plus algebra of the natural numbers, a result whose process algebraic implications are discussed in [1].

The paper is organized as follows. We begin by presenting preliminaries on the language BCCS and (in)equational logic (Sect. 2). We then proceed to define 2-nested simulation, and study some of its basic properties that play a major role in the proof of our main results (Sect. 3). The definition of 2-nested simulation suggests a natural conditional inference system for it. This is presented in Sect. 4. Our main results on the non-existence of finite (in)equational axiomatizations for 2-nested equivalence and preorder are the topic of Sects. 5 and 6. The paper concludes with a result to the effect that the 3-nested simulation preorder has no finite inequational axiomatization, and some open problems (Sect. 7).

2   Preliminaries

The Language BCCS. The process algebra BCCS [14] is a basic formalism to express finite process behaviour. Its syntax consists of (process) terms that are constructed from a countably infinite set of variables (with typical elements x, y, z), a constant $\mathbf{0}$, a binary operator $+$ called *alternative composition*, and unary *prefixing* operators a, where a ranges over some nonempty set Act of *atomic actions*. We shall use the meta-variables t, u, v to range over process terms, and write $var(t)$ for the collection of variables occurring in the term t.

A process term is *closed* if it does not contain any variables. Closed terms will be typically denoted by p, q, r. Intuitively, closed terms represent completely specified finite process behaviours, where $\mathbf{0}$ does not exhibit any behaviour, $p+q$ combines the behaviours of p and q, and ap can execute action a to transform into p. This intuition for the operators of BCCS is captured, in the style of Plotkin [18], by the transition rules in Table 1. These transition rules give rise to transitions between process terms. The operational semantics for BCCS is thus given by the labelled transition system [11] whose states are terms, and whose Act-labelled transitions are those that are provable using the rules in Table 1.

A (closed) substitution is a mapping from process variables to (closed) BCCS terms. For every term t and (closed) substitution σ, the (closed) term obtained by replacing every occurrence of a variable x in t with the (closed) term $\sigma(x)$ will be written $\sigma(t)$.

Table 1. Transition Rules for BCCS

$$\frac{x \xrightarrow{a} x'}{x + y \xrightarrow{a} x'} \qquad \frac{y \xrightarrow{a} y'}{x + y \xrightarrow{a} y'} \qquad ax \xrightarrow{a} x$$

Table 2. Axioms for BCCS

A1	$x + y \approx y + x$
A2	$(x + y) + z \approx x + (y + z)$
A3	$x + x \approx x$
A4	$x + \mathbf{0} \approx x$

In the remainder of this paper, process terms are considered modulo associativity and commutativity of $+$, and modulo absorption of $\mathbf{0}$ summands. In other words, we do not distinguish $t+u$ and $u+t$, nor $(t+u)+v$ and $t+(u+v)$, nor $t+\mathbf{0}$ and t. This is justified because all of the behavioural equivalences we consider satisfy axioms A1, A2 and A4 in Table 2. In what follows, the symbol $=$ will denote syntactic equality modulo axioms A1, A2 and A4. We use a *summation* $\sum_{i \in \{1,\dots,k\}} t_i$ to denote $t_1 + \cdots + t_k$, where the empty sum represents $\mathbf{0}$. It is easy to see that, modulo the equations A1, A2 and A4, every BCCS term t has the form $\sum_{i \in I} x_i + \sum_{j \in J} a_j t_j$, for some finite index sets I, J, terms t_j $(j \in J)$ and variables x_i $(i \in I)$.

Equational Logic. An *axiom system* is a collection of (in)equations over the language BCCS. We say that an equation $t \approx u$ (resp. an inequation $t \lesssim u$) is derivable from an axiom system E if it can be proven from the axioms in E using the standard rules of equational (resp. inequational) logic. It is well-known (cf., e.g., Sect. 2 in [9]) that if an (in)equation relating two closed terms can be proven from an axiom system E, then there is a closed proof for it.

In the proofs of our main results (cf. Thms. 3 and 4), it will be convenient to use a different formulation of the notion of provability of an (in)equation from a set of axioms. This we now proceed to define for the sake of clarity.

A *context* $C[]$ is a closed BCCS term with exactly one occurrence of a hole $[]$ in it. For every context $C[]$ and closed term p, we write $C[p]$ for the closed term that results by placing p in the hole in $C[]$. It is not hard to see that an equation $p \approx q$ is provable from an equational axiom system E iff there is a sequence $p_1 \approx \cdots \approx p_k$ $(k \geq 1)$ such that

- $p = p_1$, $q = p_k$ and
- $p_i = C[\sigma(t)] \approx C[\sigma(u)] = p_{i+1}$ for some closed substitution σ, context $C[]$ and pair of terms t, u with either $t \approx u$ or $u \approx t$ an axiom in E $(1 \leq i < k)$.

The obvious modification of the above observation applies to proofs of inequations from inequational axiom systems. In what follows, we shall refer to sequences of the form $p_1 \approx \cdots \approx p_k$ (resp. $p_1 \lesssim \cdots \lesssim p_k$) as *equational* (resp. *in-equational*) *derivations*.

For later use, note that, using axioms A1, A2 and A4 in Table 2, every context can be proven equal to either one of the form $C[b([] + p)]$ or to one of the form $[] + p$, for some action b and closed BCCS term p.

3   2-Nested Simulation

In this paper, we shall study the (in)equational theory of 2-nested simulation semantics over BCCS. This is a behavioural semantics for processes that stems from [10], where it was characterized as the largest congruence with respect to the tyft/tyxt format of transition rules which is included in completed trace semantics.

Definition 1. *A binary relation R between closed terms is a* simulation *iff $p\ R\ q$ together with $p \xrightarrow{a} p'$ implies that there is a transition $q \xrightarrow{a} q'$ with $p'\ R\ q'$.*

For closed terms p, q, we write $p \subseteq^1 q$ iff $p\ R\ q$ with R a simulation. The kernel of $\subseteq^1$ (i.e., the equivalence $\subseteq^1 \cap(\subseteq^1)^{-1}$) is denoted by $\leftrightarrows^1$.

The relation $\subseteq^1$ is the well-known *simulation preorder* [13].

Definition 2. *For closed terms p, q, we write $p \subseteq^2 q$ iff $p\ R\ q$ with R a simulation and R^{-1} included in $\subseteq^1$. The kernel of $\subseteq^2$ (i.e., the equivalence $\subseteq^2 \cap(\subseteq^2)^{-1}$) is denoted by $\leftrightarrows^2$.*

The relations $\subseteq^2$ and $\leftrightarrows^2$ are the *2-nested simulation preorder* and the *2-nested simulation equivalence*, respectively. It is easy to see that $\subseteq^2$ is included in $\leftrightarrows^1$. In the remainder of this paper we will use, instead of Definition 2, the following more descriptive, fixed-point characterization of 2-nested simulation. To the best of our knowledge, this characterization is new.

Theorem 1. *Let p, q be closed BCCS terms. Then $p \subseteq^2 q$ iff*

(1) for all $p \xrightarrow{a} p'$ there is a $q \xrightarrow{a} q'$ with $p' \subseteq^2 q'$, and
(2) $q \subseteq^1 p$.

The transition rules in Table 1 are in tyft/tyxt format, that is a (pre)congruence format for $\subseteq^2$ and $\leftrightarrows^2$ [10]. Hence, we immediately have that:

Lemma 1. *The relations $\subseteq^2$ and $\leftrightarrows^2$ are preserved by the operators of BCCS.*

The relations $\subseteq^2$ and $\leftrightarrows^2$ are extended to arbitrary BCCS terms thus:

Definition 3. *Let t, u be BCCS terms. The inequation $t \lesssim u$ is* sound *with respect to $\subseteq^2$ iff $\sigma(t) \subseteq^2 \sigma(u)$ holds for every closed substitution σ. Similarly, the equation $t \approx u$ is* sound *with respect to $\leftrightarrows^2$ iff $\sigma(t) \leftrightarrows^2 \sigma(u)$ holds for every closed substitution σ.*

Examples of (in)equations that are sound with respect to $\subseteq^2$ are those in Table 2 and $a(x + y) \lesssim a(x + y) + ax$.

Table 3. Axiom for Simulation

$$\text{S} \quad x \lesssim_1 x + y$$

Table 4. Axiom for 2-Nested Simulation

$$\text{2S} \quad y \lesssim_1 x \;\Rightarrow\; x \lesssim_2 x + y$$

Norm and Depth. We now present some results on the depth and the norm of BCCS terms that are related in 2-nested simulation semantics. These will find important applications in the proofs of our main results, and shed light on the nature of the identifications made by 2-nested simulation semantics.

Definition 4. *A sequence $a_1 \cdots a_k \in \mathsf{Act}^*$ ($k \geq 0$) is a* termination trace *of a term t iff there exists a sequence of transitions $t = t_0 \xrightarrow{a_1} t_1 \xrightarrow{a_2} \cdots \xrightarrow{a_k} t_k$ with t_k a term without outgoing transitions.*

Definition 5. *The* depth *and the* norm *of a BCCS term t, denoted by $depth(t)$ and $norm(t)$, are the lengths of the longest and of the shortest termination trace of t, respectively.*

Lemma 2. *If $p \subseteq^2_\lesssim q$, then*

1. *each termination trace of p is a termination trace of q;*
2. *$depth(p) = depth(q)$; and*
3. *$norm(p) \geq norm(q)$.*

4   A Conditional Axiomatization

The definition of 2-nested simulation immediately suggests an implicational proof system for the 2-nested simulation preorder. It is folklore that the axioms in Tables 2 and 3 give a complete axiomatization of the simulation preorder over the language BCCS [8]. To obtain a complete inference system for the 2-nested simulation preorder, it is sufficient to add the conditional axiom in Table 4 to the axiom system in Table 2. In axioms S and 2S, the relation symbol $\lesssim_1$ refers to inequations that are provable using the proof system for the simulation preorder, while the relation symbol $\lesssim_2$ refers to inequations that are provable using the proof system for the 2-nested simulation preorder. Not too surprisingly, we have that:

Theorem 2. *A1-4+2S is sound and complete for BCCS modulo the 2-nested simulation preorder.*

Proof. The soundness proof is left to the reader. We prove that A1-4+2S is complete modulo the 2-nested simulation preorder. Suppose $p \subseteq^2 q$. We prove, by induction on the depth of p, that $p \lesssim q$ can be derived from A1-4+2S.

Let $p = \sum_{i \in I} a_i p_i$ and $q = \sum_{j \in J} b_j q_j$. Since $p \subseteq^2 q$, for every $i \in I$ there is a $j_i \in J$ such that $a_i = b_{j_i}$ and $p_i \subseteq^2 q_{j_i}$. By the induction hypothesis, $p_i \lesssim q_{j_i}$ can be derived from A1-4+2S. Hence, $\sum_{i \in I} a_i p_i \lesssim \sum_{i \in I} a_i q_{j_i}$ can be proven from A1-4+2S.

Vice versa, since $q \subseteq^1 p$, for each $l \in J$ there is an $i_l \in I$ such that $b_l = a_{i_l} = b_{j_{i_l}}$ and $q_l \subseteq^1 p_{i_l} \subseteq^2 q_{j_{i_l}}$. By completeness of A1-4+S for the simulation preorder, $b_l q_l \lesssim a_{i_l} q_{j_{i_l}}$ can be derived from A1-4+S. So $a_{i_l} q_{j_{i_l}} \lesssim a_{i_l} q_{j_{i_l}} + b_l q_l$ can be derived using 2S. Hence, $\sum_{l \in J} a_{i_l} q_{j_{i_l}} \lesssim \sum_{j \in J} b_j q_j$ can be proven from A1-4+2S. As the index set $\{j_{i_l} \mid l \in J\}$ is included in the set $\{j_i \mid i \in I\}$, we can derive from A1-4+2S that

$$\sum_{i \in I} a_i q_{j_i} \approx \sum_{i \in I} a_i q_{j_i} + \sum_{l \in J} a_{i_l} q_{j_{i_l}} \lesssim \sum_{i \in I} a_i q_{j_i} + \sum_{j \in J} b_j q_j \approx \sum_{j \in J} b_j q_j = q \ .$$

By transitivity we conclude that $p \lesssim q$ can be derived from A1-4+2S. □

The aforementioned proof system for the 2-nested simulation preorder, albeit very natural, includes the conditional axiom 2S; moreover, the condition of this axiom contains an auxiliary relation symbol that is not defined inductively on the syntax of BCCS. This raises the question of whether there exists a finite purely (in)equational axiomatization of 2-nested simulation preorder and/or equivalence at least over the language BCCS. The remainder of this study is devoted to showing that no finite (in)equational axiomatization of 2-nested simulation exists over BCCS.

5   Inaxiomatizability of the 2-Nested Simulation Preorder

In this section we prove that the 2-nested simulation preorder is not finitely inequationally axiomatizable. The following lemma will play a key role in the proof of this statement. In the lemma, and in the remainder of this paper, we let a^0 denote $\mathbf{0}$, and a^{m+1} denote $a(a^m)$.

Lemma 3. *If $p \subseteq^2 a^{2m} + a^m$, then either $p \leftrightarrows^2 a^{2m}$ or $p \leftrightarrows^2 a^{2m} + a^m$.*

The idea behind the proof that the 2-nested simulation preorder is not finitely inequationally axiomatizable is as follows. Assume a finite inequational axiomatization E for BCCS that is sound modulo $\subseteq^2$. We show that, if m is sufficiently large, then, for all closed inequational derivations $a^{2m} \lesssim p_1 \lesssim \cdots \lesssim p_k$ from E with $p_k \subseteq^2 a^{2m} + a^m$, we have that $p_k \leftrightarrows^2 a^{2m}$. So $a^{2m} \lesssim a^{2m} + a^m$ cannot be derived from E. Note that $a^{2m} \subseteq^2 a^{2m} + a^m$.

Lemma 4. *Let $t \lesssim u$ be sound modulo $\subseteq^2$. Let m be greater than the depth of t. Assume that $C[\sigma(u)] \subseteq^2 a^{2m} + a^m$. Then $C[\sigma(t)] \leftrightarrows^2 a^{2m}$ implies $C[\sigma(u)] \leftrightarrows^2 a^{2m}$.*

Proof. Let $C[\sigma(t)] \leftrightarrows^2 a^{2m}$; we prove $C[\sigma(u)] \leftrightarrows^2 a^{2m}$. Since $C[\sigma(u)] \subseteq^2 a^{2m} + a^m$, it is sufficient to show that $a^{2m} + a^m \not\subseteq^2 C[\sigma(u)]$. In fact, if $C[\sigma(u)] \subseteq^2 a^{2m} + a^m$ and $a^{2m} + a^m \not\subseteq^2 C[\sigma(u)]$, by Lemma 3 it follows that $C[\sigma(u)] \leftrightarrows^2 a^{2m}$, which is to be shown. We prove that $a^{2m} + a^m \not\subseteq^2 C[\sigma(u)]$ holds by distinguishing two cases, depending on the form of the context $C[]$.

- *Case 1*: Suppose $C[]$ is of the form $C'[b([] + r)]$.
  Consider a transition $C[\sigma(u)] \xrightarrow{a} q'$. Since $C[]$ is of the form $C'[b([] + r)]$, clearly there is a transition $C[\sigma(t)] \xrightarrow{a} p'$ where p' can be obtained by replacing at most one subterm $\sigma(u)$ of q' by $\sigma(t)$. Since $\sigma(t) \subseteq^2 \sigma(u)$, by Lemma 2(2), $\sigma(t)$ and $\sigma(u)$ have the same depth; so p' and q' have the same depth as well. Since $C[\sigma(t)] \leftrightarrows^2 a^{2m}$, it follows that $p' \subseteq^2 a^{2m-1}$. So by Lemma 2(2), $depth(p') = depth(q') = 2m - 1$. As $depth(a^{m-1}) \neq 2m - 1$, by Lemma 2(2) $a^{m-1} \not\subseteq^2 q'$. This holds for all transitions $C[\sigma(u)] \xrightarrow{a} q'$, and $a^{2m} + a^m \xrightarrow{a} a^{m-1}$, so $a^{2m} + a^m \not\subseteq^2 C[\sigma(u)]$.
- *Case 2*: Suppose $C[]$ is of the form $[] + r$.
  As $\rho(t) \subseteq^2 \rho(u)$ for all closed substitutions ρ, by Lemma 2(2) $\rho(t)$ and $\rho(u)$ have the same depth for all ρ. Clearly this implies that $depth(t) = depth(u)$, and moreover that t and u contain exactly the same variables.
  Since $\sigma(t) + r \leftrightarrows^2 a^{2m}$, by Lemma 2(3) $norm(\sigma(t)) \geq 2m$ and $norm(r) \geq 2m$. As $\sigma(u) + r \subseteq^2 a^{2m} + a^m$, again by Lemma 2(3), we have that $norm(\sigma(u)) \geq m$.
  Since $depth(t) < m$ and $norm(\sigma(t)) \geq 2m$, for each variable $x \in var(t) = var(u)$ we have $norm(\sigma(x)) > m$.
  By the fact that $depth(u) = depth(t) < m$ and $norm(\sigma(u)) \geq m$, each termination trace of $\sigma(u)$ must become, after less than m transitions, a termination trace of a $\sigma(x)$ with $x \in var(u)$. Since for all $x \in var(u) = var(t)$ we have $norm(\sigma(x)) > m$, it follows that $norm(\sigma(u)) > m$. Since moreover $norm(r) \geq 2m$, we have $norm(\sigma(u) + r) > m$. As $a^{2m} + a^m$ has norm m, by Lemma 2(3) we may conclude that $a^{2m} + a^m \not\subseteq^2 \sigma(u) + r$. □

Remark 1. The inequation $ax \lesssim ax + a^1$ is sound modulo $\subseteq^2$. However, $a^4 \not\leftrightarrows^2 a^4 + a^1$. So the side condition in the statement of Lemma 4 that $C[\sigma(u)] \subseteq^2 a^{2m} + a^m$ cannot be omitted. (Note that $a^4 + a^1 \not\subseteq^2 a^4 + a^2$.)

Theorem 3. *BCCS modulo the 2-nested simulation preorder is not finitely inequationally axiomatizable.*

Proof. Let E be a finite, non-empty inequational axiomatization for BCCS that is sound modulo $\subseteq^2$. Let $m > \max\{depth(t) \mid t \lesssim u \in E\}$.

By Lemma 4, and using induction on the length of derivations, it follows that if the closed inequation $a^{2m} \lesssim r$ can be derived from E and $r \subseteq^2 a^{2m} + a^m$, then $r \leftrightarrows^2 a^{2m}$. As $a^{2m} + a^m \not\subseteq^2 a^{2m}$ (Lemma 2(3)), it follows that $a^{2m} \lesssim a^{2m} + a^m$ cannot be derived from E. Since $a^{2m} \subseteq^2 a^{2m} + a^m$, we may conclude that E is not complete modulo $\subseteq^2$. □

6 Inaxiomatizability of 2-Nested Simulation Equivalence

We now proceed to prove that the 2-nested simulation equivalence is not finitely equationally axiomatizable. The following lemma will play a key role in the proof of this statement.

Lemma 5. *Let the inequational axiom $u \lesssim t$ be sound modulo $\subseteq^2$. If t is of the form $\sum_{i \in I} x_i + \sum_{j \in J} a_j t_j$ and u is of the form $\sum_{k \in K} y_k + \sum_{\ell \in L} b_\ell u_\ell$, then*

- *$\{y_k \mid k \in K\} \subseteq \{x_i \mid i \in I\}$, and*
- *for each $\ell \in L$ there is a $j \in J$ such that $var(t_j) \subseteq var(u_\ell)$.*

Proof. Let m be greater than the depth of u.

Assume, towards a contradiction, that $y_k \notin \{x_i \mid i \in I\}$ for some $k \in K$. Let $\sigma(y_k) = a^m$ and let $\sigma(z) = \mathbf{0}$ for $z \neq y_k$. As $\sigma(y_k) \xrightarrow{a} a^{m-1}$, it follows that $\sigma(u) \xrightarrow{a} a^{m-1}$; so $\sigma(u)$ has a termination trace of length m. On the other hand, $\sigma(x_i) \leftrightarroweq^2 \mathbf{0}$ for $i \in I$, and it is easy to see that no $\sigma(a_j t_j)$ for $j \in J$ has a termination trace of length m; so $\sigma(t)$ does not have a termination trace of length m. As $\sigma(u) \subseteq^2 \sigma(t)$ by the soundness of $u \lesssim t$, this contradicts Lemma 2(1).

Assume, towards a contradiction, that there is an $\ell \in L$ such that $var(t_j) \not\subseteq var(u_\ell)$ for all $j \in J$. Let $\rho(z) = \mathbf{0}$ for $z \in var(u_\ell)$ and let $\rho(z) = a^m$ for $z \notin var(u_\ell)$. Since $\rho(z) = \mathbf{0}$ for $z \in var(u_\ell)$, clearly $depth(\rho(u_\ell)) \leq depth(u) - 1 < m - 1$. On the other hand, for all transitions $\rho(t) \xrightarrow{c} p'$ we have $depth(p') \geq m - 1$. Namely, each transition of $\rho(t)$ is of the form $\rho(t) \xrightarrow{a} a^{m-1}$ or $\rho(t) \xrightarrow{a_j} \rho(t_j)$; by assumption, for every $j \in J$, the term t_j contains a variable $z \notin var(u_\ell)$, implying that $depth(\rho(t_j)) \geq m$. Since $\rho(u) \subseteq^2 \rho(t)$ and $\rho(u) \xrightarrow{b_\ell} \rho(u_\ell)$, it follows that there is a transition $\rho(t) \xrightarrow{b_\ell} q'$ with $\rho(u_\ell) \subseteq^2 q'$. Since $depth(\rho(u_\ell)) < m - 1$ and $depth(q') \geq m - 1$, this contradicts Lemma 2(2). □

Assume a finite equational axiomatization E for BCCS that is sound modulo $\leftrightarroweq^2$. The idea behind the proof that E cannot be complete modulo $\leftrightarroweq^2$ is as follows. We show that, if m is sufficiently large, then, for all closed derivations $a(a^{2m} + a^m) \approx p_1 \approx \cdots \approx p_k$ from E, $p_k \xrightarrow{a} p'_k$ implies $norm(p'_k) = m$. Clearly, $a(a^{2m} + a^m) + a^{2m+1}$ does not satisfy the latter property, so $a(a^{2m} + a^m) \approx a(a^{2m} + a^m) + a^{2m+1}$ cannot be derived from E. Note that $a(a^{2m} + a^m) \leftrightarroweq^2 a(a^{2m} + a^m) + a^{2m+1}$.

Theorem 4. *BCCS modulo 2-nested simulation equivalence is not finitely equationally axiomatizable.*

Proof. Let E be a finite, non-empty equational axiomatization for BCCS that is sound modulo $\leftrightarroweq^2$. Let $m > \max\{depth(t) \mid t \approx u \in E\}$.

First we prove the following fact:

Claim: Let $t \approx u \in E$ and let σ be a closed substitution such that $C[\sigma(t)]$ only has termination traces of lengths $m + 1$ and $2m + 1$. Suppose moreover that for every transition $C[\sigma(t)] \xrightarrow{b} p'$ we have $norm(p') = m$. Then, for every transition $C[\sigma(u)] \xrightarrow{c} q'$ we have $norm(q') = m$.

Proof of the claim. First of all, note that, as $C[\sigma(t)] \leftrightarrows^2 C[\sigma(u)]$, by Lemma 2(1) we know that $C[\sigma(u)]$ only has termination traces of lengths $m+1$ and $2m+1$. We now proceed with the proof by distinguishing two cases, depending on the form of the context $C[]$.

- *Case 1*: Suppose $C[]$ is of the form $C'[d([] + r)]$.

  Consider a transition $C[\sigma(u)] \xrightarrow{c} q'$. Since $C[]$ is of the form $C'[d([] + r)]$, clearly there is a transition $C[\sigma(t)] \xrightarrow{c} p'$ where p' can be obtained by replacing at most one subterm $\sigma(u)$ of q' by $\sigma(t)$. Since $\sigma(t) \leftrightarrows^2 \sigma(u)$, by Lemma 2(3) $\sigma(t)$ and $\sigma(u)$ have the same norm; so p' and q' have the same norm as well. By assumption $norm(p') = m$, so $norm(q') = m$.

- *Case 2*: Suppose $C[]$ is of the form $[] + r$.

  Let t be of the form $\sum_{i \in I} x_i + \sum_{j \in J} a_j t_j$ and let u be of the form $\sum_{k \in K} y_k + \sum_{\ell \in L} b_\ell u_\ell$. Consider a transition $\sigma(u) + r \xrightarrow{c} q'$. We distinguish three possible cases.

- *Case 2.1*: Let $r \xrightarrow{c} q'$. Then $\sigma(t) + r \xrightarrow{c} q'$, which implies $norm(q') = m$.

- *Case 2.2*: Let $\sigma(y_k) \xrightarrow{c} q'$ for some $k \in K$. By Lemma 5, $y_k = x_i$ for some $i \in I$, so $\sigma(x_i) \xrightarrow{c} q'$. Then $\sigma(t) + r \xrightarrow{c} q'$, which implies $norm(q') = m$.

- *Case 2.3*: Let $q' = \sigma(u_\ell)$ for some $\ell \in L$. By Lemma 5, $var(t_j) \subseteq var(u_\ell)$ for some $j \in J$. Since $depth(t) < m$, we have $depth(t_j) < m$. On the other hand, $\sigma(t) + r \xrightarrow{a_j} \sigma(t_j)$ implies $norm(\sigma(t_j)) = m$. Hence, each termination trace of $\sigma(t_j)$ (so in particular its shortest one) must become, after less than m transitions, a termination trace of a $\sigma(x)$ with $x \in var(t_j)$. So $norm(\sigma(t_j)) = m$ implies $norm(\sigma(x)) \leq m$ for some $x \in var(t_j)$. Since $x \in var(u_\ell)$ and $depth(u_\ell) < m$, we have $norm(\sigma(u_\ell)) < 2m$. Since $\sigma(u)$ only has termination traces of lengths $m+1$ and $2m+1$, and moreover $\sigma(u) \xrightarrow{b_\ell} \sigma(u_\ell)$, it follows that $\sigma(u_\ell)$ can only have termination traces of lengths m and $2m$. Hence, $norm(\sigma(u_\ell)) = m$. (*End of the proof of the claim*)

Suppose now that p only has termination traces of lengths $m+1$ and $2m+1$. Suppose moreover that for every transition $p \xrightarrow{b} p'$ we have $norm(p') = m$. By induction on the length of equational derivations from E, using the claim that we have just proven, it is easy to show that if $p \approx q$ can be derived from E, then for every transition $q \xrightarrow{c} q'$ we have $norm(q') = m$.

Concluding, $a(a^{2m} + a^m)$ only has termination traces of lengths $m+1$ and $2m+1$. Moreover, its only transition is $a(a^{2m} + a^m) \xrightarrow{a} a^{2m} + a^m$, and $a^{2m} + a^m$ has norm m. Finally, $a(a^{2m} + a^m) + a^{2m+1} \xrightarrow{a} a^{2m}$, and a^{2m} does not have norm m. So $a(a^{2m} + a^m) \approx a(a^{2m} + a^m) + a^{2m+1}$ cannot be derived from E. Since $a(a^{2m} + a^m) \leftrightarrows^2 a(a^{2m} + a^m) + a^{2m+1}$, we may conclude that E is not complete modulo $\leftrightarrows^2$. $\qquad\square$

7   The 3-Nested Simulation Preorder and Beyond

Groote and Vaandrager [10] actually introduced a hierarchy of n-nested simulation preorders for $n \geq 2$. The following definition generalizes Definition 2.

Definition 6. *For $n \geq 1$, $p \subseteq^{n+1} q$ iff $p \ R \ q$ with R a simulation and R^{-1} included in $\subseteq^n$. The kernel of $\subseteq^{n+1}$ is denoted by $\leftrightarrows^{n+1}$.*

It is easy to see that $\subseteq^{n+1}$ is included in $\leftrightarrows^n$, for $n \geq 1$. The characterization of the 2-nested simulation preorder in Theorem 1 generalizes to the n-nested simulation preorders for $n \geq 3$. Also, the idea behind the conditional axiomatization for the 2-nested preorder (see Theorem 2) generalizes to the n-nested simulation preorders for $n \geq 3$. The proofs of these results are omitted.

Theorem 5. *For $n \geq 1$, and for closed process terms p and q over BCCS, $p \subseteq^{n+1} q$ iff*

(1) for all $p \xrightarrow{a} p'$ there is a $q \xrightarrow{a} q'$ with $p' \subseteq^{n+1} q'$, and
(2) $q \subseteq^n p$.

Definition 7. *For $n \geq 1$, let $\precsim_{n+1}$ be the preorder generated by the equational axioms A1-4 together with $y \precsim_n x \Rightarrow x \precsim_{n+1} x + y$.*

Theorem 6. *For $n \geq 1$, and for closed process terms p and q over BCCS, $p \precsim_n q$ iff $p \subseteq^n q$.*

It follows from the proof of Theorem 4 that there does not exist a finite inequational axiomatization for the 3-nested simulation preorder.

Theorem 7. *BCCS modulo the 3-nested simulation preorder is not finitely inequationally axiomatizable.*

Proof. Let E be a finite inequational axiomatization for BCCS that is sound modulo $\subseteq^3$. Since $\subseteq^3$ is included in $\leftrightarrows^2$, clearly the equational axiomatization $E' = \{t \approx u \mid t \precsim u \in E\}$ is sound modulo $\leftrightarrows^2$. Let $m > \max\{depth(t) \mid t \approx u \in E'\}$. In the proof of Theorem 4 it was shown that $a(a^{2m} + a^m) \approx a(a^{2m} + a^m) + a^{2m+1}$ cannot be derived from E'. Hence, $a(a^{2m} + a^m) \precsim a(a^{2m} + a^m) + a^{2m+1}$ cannot be derived from E. Since $a(a^{2m} + a^m) \subseteq^3 a(a^{2m} + a^m) + a^{2m+1}$, it follows that E is not complete modulo $\subseteq^3$. □

We leave it as an open question whether there exist finite equational axiomatizations for n-nested simulation equivalence if $n \geq 3$, and finite inequational axiomatizations for the n-nested simulation preorder if $n \geq 4$.

References

1. L. ACETO, Z. ÉSIK, AND A. INGÓLFSDÓTTIR, *On the two-variable fragment of the equational theory of the max-sum algebra of the natural numbers*, in Proceedings of the 17th STACS, H. Reichel and S. Tison, eds., vol. 1770 of Lecture Notes in Computer Science, Springer-Verlag, Feb. 2000, pp. 267–278.

2. L. ACETO, Z. ÉSIK, AND A. INGÓLFSDÓTTIR, *The max-plus algebra of the natural numbers has no finite equational basis*, research report, BRICS, Department of Computer Science, Aalborg University, October 1999. Pp. 25. To appear in *Theoretical Computer Science*.

3. L. ACETO, W. FOKKINK, AND A. INGÓLFSDÓTTIR, *A menagerie of non-finitely based process semantics over BPA*—from ready simulation to completed traces*, Mathematical Structures in Computer Science, 8 (1998), pp. 193–230.

4. J. BAETEN AND J. KLOP, eds., *Proceedings CONCUR 90*, Amsterdam, vol. 458 of Lecture Notes in Computer Science, Springer-Verlag, 1990.

5. J. BERGSTRA AND J. W. KLOP, *Fixed point semantics in process algebras*, Report IW 206, Mathematisch Centrum, Amsterdam, 1982.

6. J. H. CONWAY, *Regular Algebra and Finite Machines*, Mathematics Series (R. Brown and J. De Wet eds.), Chapman and Hall, London, United Kingdom, 1971.

7. J. L. GISCHER, *The equational theory of pomsets*, Theoretical Comput. Sci., 61 (1988), pp. 199–224.

8. R. VAN GLABBEEK, *The linear time – branching time spectrum*, in Baeten and Klop [4], pp. 278–297.

9. J. F. GROOTE, *A new strategy for proving ω–completeness with applications in process algebra*, in Baeten and Klop [4], pp. 314–331.

10. J. F. GROOTE AND F. VAANDRAGER, *Structured operational semantics and bisimulation as a congruence*, Information and Computation, 100 (1992), pp. 202–260.

11. R. KELLER, *Formal verification of parallel programs*, Comm. ACM, 19 (1976), pp. 371–384.

12. H. LIN, *An interactive proof tool for process algebras*, in 9th Annual Symposium on Theoretical Aspects of Computer Science, vol. 577 of Lecture Notes in Computer Science, Cachan, France, 13–15 Feb. 1992, Springer, pp. 617–618.

13. R. MILNER, *An algebraic definition of simulation between programs*, in Proceedings 2nd Joint Conference on Artificial Intelligence, William Kaufmann, 1971, pp. 481–489.

14. ——, *Communication and Concurrency*, Prentice-Hall International, Englewood Cliffs, 1989.

15. W. MITCHELL AND D. CARLISLE, *Modal observation equivalence of processes*, Technical Report UMCS-96-1-1, Manchester University, Computer Science, 1996.

16. F. MOLLER, *The importance of the left merge operator in process algebras*, in Proceedings 17th ICALP, Warwick, M. Paterson, ed., vol. 443 of Lecture Notes in Computer Science, Springer-Verlag, July 1990, pp. 752–764.

17. ——, *The nonexistence of finite axiomatisations for CCS congruences*, in Proceedings 5th Annual Symposium on Logic in Computer Science, Philadelphia, USA, IEEE Computer Society Press, 1990, pp. 142–153.

18. G. PLOTKIN, *A structural approach to operational semantics*, Report DAIMI FN-19, Computer Science Department, Aarhus University, 1981.

19. V. REDKO, *On defining relations for the algebra of regular events*, Ukrainskii Matematicheskii Zhurnal, 16 (1964), pp. 120–126. In Russian.

20. P. SEWELL, *Nonaxiomatisability of equivalences over finite state processes*, Annals of Pure and Applied Logic, 90 (1997), pp. 163–191.

21. S. K. SHUKLA, D. J. ROSENKRANTZ, H. B. HUNT III, AND R. E. STEARNS, *A HORNSAT based approach to the polynomial time decidability of simulation relations for finite state processes*, in DIMACS Workshop on Satisfiability Problem: Theory and Applications, D. Du, J. Gu, and P. M. Pardalos, eds., vol. 35 of DIMACS Series in Discrete Mathematics and Computer Science, 1996, pp. 603–642.

On the Difference between Polynomial-Time Many-One and Truth-Table Reducibilities on Distributional Problems*

Shin Aida[1], Rainer Schuler[2], Tatsuie Tsukiji[1], and Osamu Watanabe[2]

[1] School of Informatics and Sciences
Nagoya University, Nagoya 464-8601
[2] Dept. of Mathematical and Computing Sciences
Tokyo Institute of Technology Tokyo 152-8552.

Abstract. In this paper we separate many-one reducibility from truth-table reducibility for distributional problems in $\text{Dist}\mathcal{NP}$ under the hypothesis that $\mathcal{P} \neq \mathcal{NP}$. As a first example we consider the 3-Satisfiability problem (3SAT) with two different distributions on 3CNF formulas. We show that 3SAT using a version of the standard distribution is truth-table reducible but not many-one reducible to 3SAT using a less redundant distribution unless $\mathcal{P} = \mathcal{NP}$.

We extend this separation result and define a distributional complexity class $\mathcal{C}$ with the following properties:
(1) $\mathcal{C}$ is a subclass of $\text{Dist}\mathcal{NP}$, this relation is proper unless $\mathcal{P} = \mathcal{NP}$.
(2) $\mathcal{C}$ contains $\text{Dist}\mathcal{P}$, but it is not contained in $\text{Ave}\mathcal{P}$ unless $\text{Dist}\mathcal{NP} \subseteq \text{Ave}\mathcal{ZPP}$.
(3) $\mathcal{C}$ has a $\leq_m^p$-complete set.
(4) $\mathcal{C}$ has a $\leq_{tt}^p$-complete set that is not $\leq_m^p$-complete unless $\mathcal{P} = \mathcal{NP}$.
This shows that under the assumption that $\mathcal{P} \neq \mathcal{NP}$, the two completeness notions differ on some non-trivial subclass of $\text{Dist}\mathcal{NP}$.

1 Introduction

Since the discovery of $\mathcal{NP}$-complete problems by Cook and Levin [Coo71,Lev73], a considerable number of $\mathcal{NP}$-complete problems have been reported from various areas in computer science. It is quite interesting and even surprising that most of these $\mathcal{NP}$-completeness results, except only few cases [VV83], have been proven by showing a polynomial-time *many-one* reduction from some other known $\mathcal{NP}$-complete problems. Recall that there are various reducibility types (among polynomial-time deterministic reducibilities) and that polynomial-time many-one reducibility is of the most restrictive type. For example, polynomial-time truth-table reducibility is, by definition, more general than polynomial-time many-one reducibility, and in fact, it has been shown [LLS75] that these two reducibilities differ on some problem. Nevertheless, no $\mathcal{NP}$-complete problem is known that requires (even seems to require) polynomial-time truth-table reducibility for proving its $\mathcal{NP}$-completeness.

* Supported in part by JSPS/NSF cooperative research: Complexity Theory for Strategic Goals, 1998–2001.

A. Ferreira and H. Reichel (Eds.): STACS 2001, LNCS 2010, pp. 51–62, 2001.

Many researchers have studied the difference between these polynomial-time reducibility types; see, e.g., [LY90,Hom97]. Notice first that showing the difference between many-one and more stronger reducibilities on $\mathcal{NP}$ implies that $\mathcal{P} \neq \mathcal{NP}$ (because if $\mathcal{P} = \mathcal{NP}$, then any nontrivial set in $\mathcal{NP}$ is $\mathcal{NP}$-complete under many-one reducibility). Thus, it is more reasonable to assume (at least) $\mathcal{P} \neq \mathcal{NP}$ and to ask about the difference between, e.g., many-one and truth-table reducibilities on $\mathcal{NP}$ under this assumption. Unfortunately, however, the question is still open even assuming that $\mathcal{P} \neq \mathcal{NP}$. Maybe the difference is too subtle to see it in $\mathcal{NP}$ by only assuming $\mathcal{P} \neq \mathcal{NP}$. In this paper we show that this subtle difference appears when we use reducibility for analyzing distributional $\mathcal{NP}$ problems.

The notion of "distributional problem" has been introduced by Levin [Lev86] in his framework for studying average-case complexity of $\mathcal{NP}$ problems. A distributional problem is a pair (A, μ) of a decision problem A (as usual, A is a set of positive instances of the problem) and an input distribution μ. Intuitively (see below for the formal definition), by the complexity of (A, μ), we mean the complexity of A when inputs are given under the distribution μ. Analog to the class $\mathcal{NP}$, Levin proposed to study a class Dist$\mathcal{NP}$, the class of all distributional problems (A, μ) such that $A \in \mathcal{NP}$ and μ can be computed in polynomial-time. Also he introduced a class Ave$\mathcal{P}$, the class of distributional problems solvable in polynomial-time on average. Then the question analog to the $\mathcal{P}$ versus $\mathcal{NP}$ question is whether Dist$\mathcal{NP} \subseteq$ Ave$\mathcal{P}$. Levin also extended the notion of reducibility for distributional problems, and somewhat surprisingly, he proved that distributional problem (BH, μ_{st}), where BH is a canonical $\mathcal{NP}$-complete set and μ_{st} is a standard uniform distribution, is complete in Dist$\mathcal{NP}$ by using many-one reducibility. (See, e.g., [Gur91,Wang97] for detail explanation and basic results on Levin's average-case complexity theory.)

Unlike the worst-case complexity, only a small number of "natural" distributional problems have been shown as complete for Dist$\mathcal{NP}$. Intuitively, it seems that most $\mathcal{NP}$ problems are not hard enough to become complete under natural distributions. More technically, the condition required for the reducibility (in the average-case framework) is strong, it is affected by even some small change of distribution. Aida and Tsukiji [AT00] pointed out that this sensitivity could be used to show the subtle difference between many-one and more general reducibilities. They showed two problems (A, μ_A) and (B, μ_B) in Dist$\mathcal{NP}$ such that $(A, \mu_A) \leq_{tt}^{p} (B, \mu_B)$ but $(A, \mu_A) \not\leq_{m}^{p} (B, \mu_B)$ unless $\mathcal{P} = \mathcal{NP}$. Unfortunately, though, these distributions μ_A and μ_B are so small that these two problems are trivially in Ave$\mathcal{P}$. In fact, e.g. any problem in $\mathcal{EXP}$ is in Ave$\mathcal{P}$ for some artificial distribution and hence any separation result in some larger class still holds within Ave$\mathcal{P}$. It has been left open to show such difference on nontrivial distributional $\mathcal{NP}$ problems.

We solve this open question in this paper. We separate many-one reducibility from truth-table reducibility for nontrivial problems in Dist$\mathcal{NP}$ under the hypothesis that $\mathcal{P} \neq \mathcal{NP}$. Furthermore, we show some nontrivial subclass of Dist$\mathcal{NP}$ in which many-one and truth-table completeness notions differ unless $\mathcal{P} = \mathcal{NP}$.

First we define two versions of the distributional 3-Satisfiability problem (3SAT) by considering different distributions on 3CNF formulas. The first distribution μ is defined by modifying a standard uniform distribution on 3CNF formulas. Here the standard distribution gives each formula the probability that it is generated by a random process, where every literal is chosen randomly from the set of variables and their complements. For the second distribution ν, we consider less redundant 3CNF representation. Note that a 3CNF formula F usually has many trivially equivalent formulas; for example, permuting the order of clauses in F, we can easily get a different but equivalent formula. We consider some restriction on the form of formulas to reduce this redundancy, and define the second distribution ν so that non-zero probability is given only on such formulas that satisfy our restriction. By this way, the probability of each formula (of the required form) gets increased considerably (compared with ν). By using this increase, we prove that $(3SAT, \nu)$ is not many-one reducible to $(3SAT, \mu)$ unless $\mathcal{P} = \mathcal{NP}$. On the other hand, by using the self-reducibility of 3SAT, we prove that even $(3SAT, \nu)$ is truth-table reducible $(3SAT, \mu)$.

Next we extend this separation technique and define a subclass $\mathcal{C}$ of Dist$\mathcal{NP}$ in which many-one and truth-table completeness notions differ unless $\mathcal{P} = \mathcal{NP}$. Furthermore, we can show that $\mathcal{C}$ is not contained in Ave$\mathcal{P}$ (thus it is not trivial) unless all Dist$\mathcal{NP}$ are solvable in polynomial-time on average by randomized zero-error computation.

2   Preliminaries

We use standard notations and definitions from computability theory, see, e.g., [BDG88]. We briefly recall the definitions of the average-case complexity classes used in the following. For definitions and discussion, see [Gur91].

A distributional problem consists of a set L and a distribution on strings defined by the distribution function μ, i.e., a (real) valued function such that $\sum_x \mu(x) = 1$. A distribution μ is called *polynomial-time computable* if the binary expansion of the distribution function μ^*, defined by $\mu^*(x) = \sum_{y \le x} \mu(x)$ for all x, is polynomial-time computable in the sense that for any x and n, the first n bits of $\mu^*(x)$ is computable within polynomial time w.r.t. $|x|$ and n.

Let Dist$\mathcal{NP}$ denote the class of all distributional problems (L, μ) such that $L \in \mathcal{NP}$ and μ is polynomial-time computable. Similarly, Dist$\mathcal{P}$ denotes the class of distributional problems $(L, \mu) \in$ Dist$\mathcal{NP}$ such that L is in $\mathcal{P}$.

The average-case analog of $\mathcal{P}$ is denoted by Ave$\mathcal{P}$ and defined as follows. A distributional problem (L, μ) is decidable in polynomial-time on average, if L is decidable by some t-time bounded Turning machine, and t is *polynomial on μ-average*, which means that t, a function from $\Sigma^* \to \mathbf{N}$, satisfies the following for some constant $\epsilon > 0$ [Lev86,Gur91].

$$\sum_x \frac{t^\epsilon(x)}{|x|} \mu(x) \ < \ \infty.$$

Let Ave$\mathcal{P}$ denote the class of all distributional problems that are decidable in polynomial-time on average. Similarly let Ave$\mathcal{ZPP}$ denote the class of all distributional problems that are decidable in polynomial-time on average by randomized Turing machines (without error), see, e.g., [Imp95]. Here we have to be a little careful defining average polynomial-time for randomized computation [Gur91]. Let $t(x, r)$ denote the running time of M on input x using random bits r. We say that M is *polynomial-time* on average if

$$\sum_x \sum_r 2^{-|r|} \frac{t^\epsilon(x, r)}{|x|} \mu(x) \; < \; \infty,$$

where r ranges over all binary strings such that M on input x halts using r but it does not halt using any prefix r' of r.

Finally, we define "reducibility" between distributional problems. A distributional problem (A, μ) is *polynomial-time reducible* to (B, ν), if there exists an oracle Turing machine M and a polynomial p such that the following three conditions hold.

(1) The running time of M (with oracle B) is polynomially bounded.
(2) For every x, we have $x \in A \Leftrightarrow x \in L(M, B)$, where $L(M, B)$ is the set of strings accepted by M with oracle B.
(3) For any x, let $Q(M, B, x)$ denote the set of oracle queries made by M with oracle B and input x. The following condition holds for every y.

$$\nu(y) \; \geq \; \sum_{x \, : \, y \in Q(M, B, x)} \frac{\mu(x)}{p(|x|)}.$$

From these three conditions, any problem (A, μ) that is polynomial-time reducible to some problem in Ave$\mathcal{P}$ also belongs to Ave$\mathcal{P}$ [Lev86,Gur91]. The above condition (3) is called a *dominance condition*.

By restricting the type of queries, we can define finer reducibilities. A reduction M is called a *truth-table reduction* if for every x, the oracle queries of M on input x are made non-adaptively, i.e., they are independent of the oracle set. M is a *many-one reduction* if for every x, M on input x makes exactly one query, and it accepts x iff the query is in the oracle set. We can define more general reduction types by considering randomized computation. That is, a reduction is called a *randomized reduction* if the oracle Turing machine is randomized. In this paper, we consider the most restrictive randomized reduction type that requires "zero error" to the oracle Turing machine M, i.e., M is correct and polynomial-time bounded for all inputs and all possible random bits. The dominance condition needs to be revised for randomized reductions. For any x and any r, let $Q(M, B, x, r)$ denote the set of oracle queries made by $M^B(x; r)$, i.e., the execution of M with oracle B on input x using random bits r. Here we assume that $M^B(x; r)$ halts consuming all bits of r and $M^B(x; r')$ does not halt for any prefix r' of r. (If r does not satisfy this condition, then we simply define $Q(M, B, x, r)$ to be empty.) Then our dominance condition is stated as follows.

(3') For every y, we have $$\nu(y) \; \geq \; \sum_{x, r \, : \, y \in Q(M, B, x, r)} \frac{\mu(x) \cdot 2^{-|r|}}{p(|x|)}.$$

3   Separation on 3SAT

Our first separation is on 3SAT, i.e., the set of all *satisfiable* 3CNF formulas
F. We recall some basic definitions on 3SAT. A formula F is in *3CNF* if F is
a conjunction of *clauses* which contain at most 3 *literals*, i.e., F is of the form
$C_1 \wedge C_2 \wedge \cdots \wedge C_m$, where $C_i = l_{j_1} \vee l_{j_2} \vee l_{j_3}$ and l_{j_k} is either the variable v_{j_k}
or its negation. (We use the index of j_k of each literal l_{j_k} to denote that of its
variable.) We use $\mathcal{F}^{(n,m)}$ to denote the set of 3CNF formulas with n variables
and m clauses. (We assume that $m \leq 8n^3$.)

The standard distribution μ_{st} assigns to any formula F in $\mathcal{F}^{(n,m)}$ the probability

$$\frac{1}{n(n+1)} \cdot \frac{1}{8n^3} 2^{-3m(1+\lceil \log n \rceil)}.$$

That is, we have the following random experiment in mind.

Choose n (number of variables) randomly. Choose $m \in \{1, \cdots, 8n^3\}$ (number
of clauses) randomly. Choose each of the $3m$ literals l randomly from the set of
variables and negated variables of size $2n$. Let F denote the resulting formula.
Output F.

In order to simplify our discussion, we restrict the form of formulas so that
$m = f_0(n)$, where $f_0(n) = \lceil n \log n \rceil$. Since m is determined from n, the standard
distribution is modified as follows.

$$\mu_{st_{f0}}(F) = \begin{cases} 8n^3 \cdot \mu_{st}(F), & \text{if } F \in \mathcal{F}^{(n,f_0(n))}, \text{ and} \\ 0, & \text{otherwise.} \end{cases}$$

We should note here that the same result holds by considering any "smooth"
function for f such that $n \leq f(n) \leq n \log n$ for all n. Here a function f is
smooth if there is no big jump from $f(n-1)$ to $f(n)$; more precisely, there
exists constants $c_f > 1$ and $d_f > 0$ such that for any sufficiently large n and
for some $k < d_f \log n$, we have $f(n) - c_f \log n < f(n-k) < f(n) - \log n$. For
example, consider $f(n) = n\lceil \log n \rceil$. While this function satisfies our smoothness
condition for most n, we have $f(n) \geq f(n-k) + \log n$ for any $k = O(\log n)$ if n is
sufficiently large and $\lceil \log n \rceil = 1 + \lceil \log(n-1) \rceil$. On the other hand, a function
like $f(n) = \lceil n \log n \rceil$ satisfies this smoothness condition for $k = 1$ and $c_f = 2$.

Note that it is still open whether (3SAT, $\mu_{st_{f0}}$) is in Ave$\mathcal{P}$, i.e., polynomial-
time solvable on average. (Though using $\mu_{st_{f0}}$ most formulas are in fact unsatisfiable, and standard algorithms perform well an average [KiSe94]). On the other
hand, it has been shown that (3SAT, μ_{st_f}) defined using $f(n) \geq dn^2$ for some
$d > 0$ is indeed in Ave$\mathcal{P}$ [KP92].

Now define the first distribution. ¿From some technical reason, we consider
3CNF formulas with some additional clauses. For any $n > 0$, let $d(n) = f_0(n) -
f_0(n-1)$ (where $f_0(0) = 0$). A 3CNF $P = C_1 \wedge \cdots C_{d(n)}$ is called a *type-I prefix*
for n if each C_i is of the form $C_i = (v_{j_i} \vee v_{j_i} \vee v_{j_i})$ for some $j_i \in \{3i-2, 3i-1, 3i\}$.
Note that there are $3^{d(n)} \leq n^4$ type-I prefixes for n. We consider only formulas G
that are of the form $P \wedge F$ for some type-I prefix P for n and $F \in \mathcal{F}^{(n,f_0(n-1))}$. We
use $\mathcal{G}^{(n)}$ to denote the set of such formulas. This somewhat artificial requirement
is just to simplify our analysis of a truth-table reduction defined in Lemma 3.

Our first distribution is defined as follows.

$$\mu(G) = \begin{cases} \frac{1}{n(n+1)} 3^{-d(n)} \cdot 2^{-3f_0(n-1)(1+\lceil \log n \rceil)}, & \text{if } G \text{ is in } \mathcal{G}^{(n)}, \text{ and} \\ 0, & \text{otherwise.} \end{cases}$$

Next we define the second distribution. As mentioned in the Introduction, the 3CNF representation has redundancy; i.e., a 3CNF formula (usually) has many trivially equivalent formulas. Here we introduce one restriction on the form of formulas for reducing some redundancy, which is not essential for the hardness of the satisfiability problem.

For any n, a 3CNF $P = C_1 \wedge \cdots \wedge C_{d(n)}$ is called the *type-II prefix for n* if each C_i is of the form $C_i = (v_{3i-2} \vee v_{3i-1} \vee v_{3i})$. Note that for each n, the type-II prefix for n is uniquely determined. We consider only formulas F in $\mathcal{F}^{(n,f_0(n))}$ such that the first $d(n)$ clauses of F are the type-II prefix for n. Let $\mathcal{F}^{(n)}$ denote the set of such formulas. Note that $\mathcal{F}^{(n)}$ and $\mathcal{G}^{(n)}$ are subsets of $\mathcal{F}^{(n,f_0(n))}$. As shown in the next Lemma, the restriction to formulas of type $\mathcal{F}^{(n)}$ is not essential for the hardness of the satisfiability problem.

Lemma 1. *For any 3CNF formula $F \in \mathcal{F}^{(n,f_0(n))}$, we can either convert it to an equivalent formula $F' \in \mathcal{F}^{(n)}$ (by (i) reordering clauses and (ii) renaming and/or changing the signs of variables) or determine the satisfiability of F in polynomial-time.*

Now our distribution is defined as follows.

$$\nu(F) = \begin{cases} \frac{1}{n(n+1)} 2^{-3(f_0(n)-d(n))(1+\lceil \log n \rceil)}, & \text{if } F \in \mathcal{F}^{(n)}, \\ 0, & \text{otherwise.} \end{cases}$$

Intuitively, ν corresponds to the following random generation.

Choose n (number of variables) randomly. Fix first $d(n)$ clauses as required for the type-II prefix. Then choose the remaining $f_0(n) - d(n)$ clauses as in the standard distribution. Output F.

We observe that the distributions μ and ν defined above are polynomial time computable. Thus, both distributional problems $(3\text{SAT}, \mu)$ and $(3\text{SAT}, \nu)$ belong to Dist$\mathcal{NP}$.

For our separation result, we first show that $(3\text{SAT}, \nu)$ is not $\leq_m^p$ to $(3\text{SAT}, \mu)$ unless $\mathcal{P} = \mathcal{NP}$.

Lemma 2. *If $(3\text{SAT}, \nu) \leq_m^p (3\text{SAT}, \mu)$, then we have $3\text{SAT} \in \mathcal{P}$ and hence $\mathcal{P} = \mathcal{NP}$.*

Proof. Assume there exists a many-one reduction R from $(3\text{SAT}, \nu)$ to $(3\text{SAT}, \mu)$. Consider the 3SAT solver defined in the Figure 1.

The correctness is clear by the definition of the many-one reducibility. The polynomial-time bound of this algorithm is guaranteed as follows. The reduction R reduces (in each iteration) a formula of F in $\mathcal{F}^{(n,f_0(n))}$ to a formula F' in $\mathcal{F}^{(n'-1,f_0(n'-1))}$ with $n' \leq n$. That is, the number of variables is reduced by at least one in each while-iteration.

Algorithm 3SAT Solver
  input F in $\mathcal{F}^{(n,f_0(n))}$
  $F' \leftarrow F$; $n' \leftarrow n$;
  while $n' > \log n$ **do**
      modify F' to an equivalent formula F in $\mathcal{F}^{(n')}$;
      % The procedure mentioned in Lemma 1 is used.
      % If this fails, then the satisfiability of F' can be determined directly.
      $G \leftarrow R(F)$; $n' \leftarrow$ the number of variables in G;
      % G is in $\mathcal{G}^{(n')}$; i.e.,
      % $G = P \wedge F'$ with some type-I prefix P for n' and $F' \in \mathcal{F}^{(n',f_0(n'-1))}$.
      remove each clause $(v_{k_i} \vee v_{k_i} \vee v_{k_i})$ of P by assinging $v_{k_i} = 1$ in F';
      % F' may be reduced to a simpler formula.
      (if necessary) add redundant variables or clauses so that $F' \in \mathcal{F}^{(n'-1,f_0(n'-1))}$;
  end-while
  output 1 if the final F' is satisfiable, and output 0 otherwise;
end-algorithm.

Fig. 1. SAT Solver

This claim is proved by using the dominance condition. ¿From the dominance condition, for some constant $c > 0$ and for any sufficiently large n, we have

$$\frac{1}{n(n+1)} 2^{-3(f_0(n)-d(n))(1+\lceil \log n \rceil)}$$
$$= \nu(F) \leq n^c \cdot \mu(F') = \frac{n^c}{n'(n'+1)} 3^{-d(n)} \cdot 2^{-3f_0(n'-1)(1+\lceil \log n' \rceil)}.$$

Since $d(n) \geq \lceil \log n \rceil$, this implies

$$\frac{1}{n(n+1)} 2^{-3(f_0(n)-\lceil \log n \rceil)(1+\lceil \log n \rceil)} \leq \frac{n^c}{(n')^5(n'+1)} 2^{-3f_0(n'-1)(1+\lceil \log n' \rceil)}.$$

Now suppose that $n' > n$. Then from the above, it should hold that $c \log n > 3 \log^2 n$, which is impossible for sufficiently large n. Therefore, we have $n' \leq n$.

On the other hand, some $\leq_{tt}^p$-reduction exists from $(3SAT, \nu)$ to $(3SAT, \mu)$.

Lemma 3. $(3SAT, \nu) \leq_{tt}^p (3SAT, \mu)$.

Proof. We define a truth-table reduction from $(3SAT, \nu)$ to $3SAT, \mu$. For our discussion, consider any formula F in $\mathcal{F}^{(n)}$. Recall that $F = C_1 \wedge \cdots \wedge C_{d(n)} \wedge E$, where each C_i, $1 \leq i \leq 2\lceil \log n \rceil$, is of the form $(v_{3i-2} \vee v_{3i-1} \vee v_{3i})$. We would like to solve the satisfiability of F by asking polynomially many non-adaptive queries to 3SAT. Note that all queried formulas have to be of some appropriate form, more precisely, they should belong to $\mathcal{G}^{(n')}$ for some n'. Furthermore, since $\nu(F)$ (for $F \in \mathcal{F}^{(n)}$) is much bigger than $\mu(G)$ (for $G \in \mathcal{G}^{(n+1)}$), we cannot increase the size of queried formulas. Our idea is simple. We delete the first $d(n)$ clauses $C_1, ..., C_{d(n)}$ by considering all possible partial assignments satisfying all these clauses. Since each C_i is $(v_{3i-2} \vee v_{3i-1} \vee v_{3i})$, we only have to assign 1 to one of three variables $v_{3i-2}, v_{3i-1}, v_{3i}$ for satisfying C_i. That is, for every partial assignment, which assigns 1 to one of three variables $v_{3i-2}, v_{3i-1}, v_{3i}$ for each i, $1 \leq i \leq d(n)$, we can substitute the first $d(n)$ clauses by a type-I prefix for n. The resulting formula G is in $\mathcal{G}^{(n)}$ (i.e., has (at most) n variables and consists of a type-I prefix for n followed by $f_0(n) - d(n) = f_0(n-1)$ clauses). Note that there are $3^{d(n)} \leq n^4$ such partial assignments and that F is satisfiable if and only if one of the obtained formula G is satisfiable. Therefore, the above procedure is indeed a disjunctive truth-table reduction that asks a polynomial number of formulas (of the same size).

The dominance condition, is satisfied since (i) $\nu(F) \leq n^c \cdot \mu(G)$ and (ii) any query formula G is asked for only one formula F. The condition (ii) is satisfied since the type-II prefix of F is unique, and G is identical to F on all other clauses.

The fact that $\nu(F) \leq n^c \cdot \mu(G)$ for some $c > 0$. is immediate by comparing $\nu(F)$ and $\mu(G)$ as follows.

$$\nu(F) = \frac{1}{n(n+1)} 2^{-3(f_0(n)-d(n))(1+\log n)} = \frac{1}{n(n+1)} 2^{-3f_0(n-1)(1+\log n)}$$
$$= 3^{d(n)} \cdot \frac{1}{n(n+1)} 3^{-d(n)} \cdot 2^{-3f_0(n-1)(1+\log n)} \leq n^c \cdot \mu(G)$$

From above two lemmas, we have the following separation result.

Theorem 1. *There exist polynomial time computable distributions ν and μ such that $(3SAT, \nu) \leq_{tt}^p (3SAT, \mu)$, but $(3SAT, \nu) \not\leq_m^p (3SAT, \mu)$ unless $\mathcal{P} = \mathcal{NP}$.*

4   Separating Completeness Notions

In this section we define some subclass of $\text{Dist}\mathcal{NP}$ in which we can show the difference between many-one and truth-table completeness notions. More specifically, we will define a distributional complexity class $\mathcal{C}$ with the following properties:

(1) $\mathcal{C}$ is a subclass of $\text{Dist}\mathcal{NP}$, and furthermore, the relation is proper unless $\mathcal{P} = \mathcal{NP}$.
(2) $\mathcal{C}$ contains $\text{Dist}\mathcal{P}$, but $\mathcal{C}$ is not contained in $\text{Ave}\mathcal{P}$ unless $\text{Dist}\mathcal{NP} \subseteq \text{Ave}\mathcal{ZPP}$.
(3) $\mathcal{C}$ has a $\leq_m^p$-complete set.
(4) There exists a problem $C \in \mathcal{C}$ that is $\leq_{tt}^p$-complete in $\mathcal{C}$ but that is not $\leq_m^p$-complete in $\mathcal{C}$ unless $\mathcal{P} = \mathcal{NP}$.

That is, if $\mathcal{P} \neq \mathcal{NP}$, then two completeness notions differ on some subclass of $\text{Dist}\mathcal{NP}$. Recall that it is not known whether the assumption that $(3SAT, \mu_{st_{f0}}) \in \text{Ave}\mathcal{P}$ has some unlikely consequence such as $\text{Dist}\mathcal{NP} \subseteq \text{Ave}\mathcal{ZPP}$ above. Hence we cannot simply define $\mathcal{C}$ as the set of distributional problems that are many one reducible to $(3SAT, \mu_{st_{f0}})$.

First we define the complexity class $\mathcal{C}$. For this purpose, we consider the following version of bounded halting problem, which we call *Bounded Halting problem with Padding*. Here for some technical reason, we consider only Turing machines M using one tape as both an input and a work tape. We also assume that M's tape alphabet is $\{0, 1, B\}$ and that M cannot go beyond the cells containing 0 or 1. Note that this is not an essential restriction if we assume that M's reachable tape cells are initially filled by 0. On the other hand, with this assumption, we can represent the content of the whole tape of M by a string in $\{0,1\}^*$ of fixed length.

Below we use ϕ to denote any fixed function on $\mathbf{N}$ such that $n \le \phi(n) \le p(n)$ for some polynomial p and $\phi(n)$ is computable within polynomial-time in n.

$\mathrm{BHP}_\phi = \{\langle M, q, i, w, y\rangle :$

(i) M is NDTM, q is a state, i, $1 \le i \le |w|$, is a head position, and
         $w, y \in \{0,1\}^*$, where w is M's tape and y is padding,

(ii) $|y| = \phi(|M| + |w| + t)$ for some $t \in \mathbf{N}$, and

(iii) M has an accepting path of length t from configuration (q, i, w).$\}$

Notice here that w represents the content of the whole M's tape. We assume that M's tape head does not go outside of w. We assume some reasonable encoding of M and its state q, and $|M|$ and $|q|$ are the length of the descriptions of M and q under this encoding. Again for simplifying our discussion below, we assume that for each M and w, the length of $|q|$ and $|i|$ is fixed.

In the literature, the following versions of the halting problem BH and its padded version BH' have been studied [Gur91]. Our BHP_ϕ is regarded a variation of of BH' when ϕ is defined as $\phi(n) = n$.

$$\mathrm{BH} = \{ \langle M, x, 0^t\rangle : M \text{ accepts } x \text{ in } t \text{ steps. } \}, \text{ and}$$
$$\mathrm{BH}' = \{ \langle M, x, y\rangle : M \text{ accepts } x \text{ in } |y| \text{ steps. } \}.$$

As a distribution we consider the standard distribution extended on tuples, e.g., every instance $\langle M, q, i, x, y\rangle$ of BHP_ϕ has the following probability.

$$\mu_{\mathrm{st}}(\langle M, q, i, w, y\rangle) = \frac{1}{\alpha(|M|, |q|, |i|, |w|, |y|)} \cdot 2^{-(|M|+|q|+|i|+|w|+|y|)},$$

where $\alpha(n_1, n_2, \ldots, n_k) = \prod_{i=1}^k n_i(n_i + 1)$. Note however that a unary padding string has probability inverse polynomial to its length; for example, for any instance $\langle M, x, 0^t\rangle$ for BH, we have $\mu_{\mathrm{st}}(\langle M, x, 0^t\rangle) = \frac{1}{\alpha(|M|,|x|,t)} \cdot 2^{-(|M|+|x|)}$.

First it should be mentioned that $(\mathrm{BH}, \mu_{\mathrm{st}})$ is reducible to $(\mathrm{BHP}_\phi, \mu_{\mathrm{st}})$ via a randomized reduction of the strongest type, i.e., the one with no error.

Proposition 1. *For any polynomially bounded $\phi(n)$ that is polynomial-time computable w.r.t. n, there is a polynomial-time randomized reduction (with no error) from $(\mathrm{BH}, \mu_{\mathrm{st}})$ to $(\mathrm{BHP}_\phi, \mu_{\mathrm{st}})$.*

Since $(\mathrm{BH}, \mu_{\mathrm{st}})$ a complete problem in $\mathrm{Dist}\mathcal{NP}$ [Gur91], this proposition shows that $(\mathrm{BHP}_\phi, \mu_{\mathrm{st}})$ is complete in $\mathrm{Dist}\mathcal{NP}$ under the zero-error randomized reducibility. On the other hand, since $(\mathrm{BHP}_\phi, \mu_{\mathrm{st}})$ is a distributional problem with a flat distribution, as we will see below, $(\mathrm{BH}, \mu_{\mathrm{st}})$ is not $\le_{\mathrm{m}}^p$-reducible to $(\mathrm{BHP}_\phi, \mu_{\mathrm{st}})$ unless $\mathcal{P} = \mathcal{NP}$.

We may use any reasonable function for ϕ. Here for the following discussion, we fix $\phi(n) = n \log n$, by which we formally mean that $\phi(n) = \lceil n \log n \rceil$ (see the smoothness discussion in the previous section). Let BHP denote the class BHP_ϕ

with this ϕ. Now our class $\mathcal{C}$ is defined as a class of distributional problems (L, μ) such that (i) μ is polynomial-time computable, and (ii) (L, μ) is $\leq_{\mathrm{m}}^{p}$-reducible to $(\mathrm{BHP}, \mu_{\mathrm{st}})$.

Note first that if (L, μ) is $\leq_{\mathrm{m}}^{p}$-reducible to $(\mathrm{BHP}, \mu_{\mathrm{st}})$, then L must be in $\mathcal{NP}$. Thus, $\mathcal{C}$ is contained in Dist$\mathcal{NP}$. But $(\mathrm{BH}, \mu_{\mathrm{st}})$ is not $\leq_{\mathrm{m}}^{p}$-reducible to $(\mathrm{BHP}, \mu_{\mathrm{st}})$ unless $\mathcal{P} = \mathcal{NP}$. Thus, if $\mathcal{P} \neq \mathcal{NP}$, then $\mathcal{C}$ is a proper subclass of Dist$\mathcal{NP}$ because $(\mathrm{BH}, \mu_{\mathrm{st}})$ does not belong to $\mathcal{C}$. On the other hand, since $(\mathrm{BHP}, \mu_{\mathrm{st}})$ is complete in Dist$\mathcal{NP}$ under the zero-error randomized reducibility, it cannot be in Ave$\mathcal{P}$ unless Dist$\mathcal{NP} \subseteq \mathrm{Ave}\mathcal{ZPP}$; that is, $\mathcal{C} \not\subseteq \mathrm{Ave}\mathcal{P}$ unless Dist$\mathcal{NP} \subseteq \mathrm{Ave}\mathcal{ZPP}$.

Proposition 2. *The class $\mathcal{C}$ defined above has the following complexity.*

(1) *It is a subclass of* Dist$\mathcal{NP}$, *and the relation is proper unless* $\mathcal{P} = \mathcal{NP}$.
(2) *It contains* Dist$\mathcal{P}$, *but is not contained in* Ave$\mathcal{P}$ *unless* Dist$\mathcal{NP} \subseteq \mathrm{Ave}\mathcal{ZPP}$.

Clearly, the class $\mathcal{C}$ has $\leq_{\mathrm{m}}^{p}$-complete sets, e.g., $(\mathrm{BHP}, \mu_{\mathrm{st}})$ is one of them. On the other hand, we can define some $\leq_{\mathrm{tt}}^{p}$-complete problem in $\mathcal{C}$ that is not $\leq_{\mathrm{m}}^{p}$-complete unless $\mathcal{P} = \mathcal{NP}$.

Theorem 2. *Define* BHP$'$ *as follows with* $\phi'(n) = n \log n + \log^2 n$ (*or, more formally,* $\phi'(n) = \lceil n \log n + \log^2 n \rceil$). *Then we have* $(\mathrm{BHP}, \mu_{\mathrm{st}}) \leq_{\mathrm{tt}}^{p} (\mathrm{BHP}', \mu_{\mathrm{st}})$, *but* $(\mathrm{BHP}, \mu_{\mathrm{st}}) \not\leq_{\mathrm{m}}^{p} (\mathrm{BHP}', \mu_{\mathrm{st}})$ *unless* $\mathcal{P} = \mathcal{NP}$. *That is,* $(\mathrm{BHP}', \mu_{\mathrm{st}})$ *is* $\leq_{\mathrm{tt}}^{p}$-*complete in $\mathcal{C}$ but it is not* $\leq_{\mathrm{m}}^{p}$-*complete unless* $\mathcal{P} = \mathcal{NP}$.

$\mathrm{BHP}' = \{\langle M, q, i, w, u, v\rangle$:
   (i) M *is NDTM,* q *is a state,* i *is a head position,* $w, u, v \in \{0, 1\}^*$,
   (ii) $|v| = \phi'(|M| + |w| + t - |u|)$ *for some* t,
   (iii) $|u| = \log(|M| + |w| + t)$, *and*
   (iv)   *starting from configuration* (q, i, w),
            M *has an accepting path of length* $t - |u|$ *whose prefix is* u. $\}$

Proof. First we show that $(\mathrm{BHP}', \mu_{\mathrm{st}})$ is $\leq_{\mathrm{m}}^{p}$-reducible to $(\mathrm{BHP}, \mu_{\mathrm{st}})$. This implies that $(\mathrm{BHP}', \mu_{\mathrm{st}})$ is indeed contained in the class $\mathcal{C}$. Let $\langle M, q, i, w, u, v\rangle$ be any instance of BHP$'$ satisfying the syntactic conditions, i.e., the conditions (i) $\sim$ (iii), of BHP$'$ for some number t. Let $m = |M| + |w| + t - |u|$. We map this instance to $\langle M, q', i', w', y'\rangle$, where q', i', w' are respectively M's state, head position, and tape content after executing $|u|$ steps on the path u starting from configuration (q, i, w). In order to satisfy the syntactic conditions of BHP (and keep the consistency as a reduction), y' should be a string of length $\phi(m)$. But since $\phi(m) = \phi'(m) - \log^2 m$ (recall that $|w| = |w'|$), we have $|y'| \leq |v| - \log^2(m)$; hence, we can simply use the prefix of v of appropriate length for y'. Notice that this mapping may not be one-to-one. But first note that

$$\mu_{\mathrm{st}}(\langle M, q', i', w', y'\rangle) = \sum_{\tilde{v} \in V(y')} \mu_{\mathrm{st}}(\langle M, q, i, w, u, \tilde{v}\rangle),$$

where $V(y')$ is the set of $\tilde{v}$ of length $\phi'(m)$ whose prefix is y'. Also for considering all configurations reachable to (q', i', w'), let $C(q', i', w')$ be the set of pairs of M's configurations $(\tilde{q}, \tilde{i}, \tilde{w})$ and $\tilde{u}$ of length $\log(|M| + |w| + t)$ such that the configuration (q', i', w') is reached after executing $|\tilde{u}| = \log(|M| + |w| + t)$ steps from $(\tilde{q}, \tilde{i}, \tilde{w})$ following $\tilde{u}$. Since $|\tilde{u}| = \log(|M| + |\tilde{w}| + t) = \log(|M| + |w| + t)$,

$C(q', i', w')$ has at most $|M|(|M| + |w| + t)^2 \times (|M| + |w| + t)$ elements. Thus, we have
$$\frac{\sum_{(\tilde{q}, \tilde{i}, \tilde{w}), \tilde{u} \in C(q', i', w')} \sum_{\tilde{v} \in V(y')} \mu_{\mathrm{st}}(\langle M, q, i, w, u, \tilde{v}\rangle)}{|M|(|M| + |w| + t)^3} \leq \mu_{\mathrm{st}}(\langle M, q', i', w', y'\rangle).$$

Therefore the dominance condition is satisfied.

We observe here that the many-one reduction decreases the length of the instance by order $(\log)^2$. Let $\ell = |M| + |q| + |i| + |w| + |u| + |v|$ and $\ell' = |M| + |q'| + |i'| + |w'| + |y'|$, then if ℓ is sufficiently large, we have $\ell' \leq \ell - \log^2(m) \leq \ell - \log^2(l^{1/2}) = \ell - \frac{1}{4}\log^2 \ell$,

since we may assume that $m^2 \geq m \log m + \log^2 m + (|M| + |w| + |q| + |i| + |u|) = |M| + |q| + |i| + |w| + |u| + |v| = \ell$, for sufficiently large ℓ.

Next suppose that there is a $\leq_{\mathrm{m}}^p$-reduction from $(\mathrm{BHP}, \mu_{\mathrm{st}})$ to $(\mathrm{BHP}', \mu_{\mathrm{st}})$. We will show that this assumption implies $\mathcal{P} = \mathcal{NP}$. Consider any $\langle M, q, i, w, y\rangle$ satisfying the syntax of BHP, and let $\langle M', q', i', w', u', v'\rangle$ be the instance of BHP$'$ obtained by the assumed reduction. We may assume that $\langle M', q', i', w', u', v'\rangle$ satisfies the syntax of BHP$'$ for some t'; i.e., $|v| = \phi'(|M'| + |w'| + t' - |u'|)$. Let $\ell = |M| + |q| + |i| + |w| + |y|$. By using the reduction from BHP$'$ to BHP explained above, we reduce further the instance $\langle M', q', i', w', u', v'\rangle$ to some instance $\langle M', q'', i'', w'', y''\rangle$ of BHP. Note that $|y''| = \phi(|M'| + |w''| + t'')$ where $t'' = t' - |u'|$.

We estimate $\ell' = |M'| + |q'| + |i'| + |w'| + |u'| + |v'|$ and $\ell'' = |M'| + |q''| + |i''| + |w''| + |y''|$, and prove that $\ell'' < \ell$, i.e., $\langle M', q'', i'', w'', y''\rangle$ is shorter than $\langle M, q, i, w, y\rangle$. First from the above analysis, we have $\ell'' \leq \ell' - \frac{1}{4}\log^2 \ell'$ Now consider the case that $\ell' < \ell/2$. Then from the above bound, we immediately have $\ell'' < \ell$ for sufficiently large ℓ. Thus, consider the other case, i.e., $\ell' \geq \ell/2$. Even in this case, ℓ' cannot be so large. This is because from the dominance condition, we have $\ell' \leq \ell + d \log \ell$ for some constant $d > 0$, and hence,
$$\ell'' \leq (\ell + d \log \ell) - \frac{1}{4}\log^2(\ell + d \log \ell) \leq (\ell + d \log \ell) - \frac{1}{4}\log^2 \ell,$$
which, by using the assumption that $\ell' \geq \ell/2$, implies $\ell'' < \ell$ if ℓ is large enough.

Therefore, the obtained instance $\langle M', q'', i'', w'', y''\rangle$ is at least one bit shorter than the original instance $\langle M, q, i, w, y\rangle$. Thus, applying this process for enough number of times, which is still polynomially bounded, we can obtain a trivial instance for BHP. Thus BHP is in $\mathcal{P}$, which implies that $\mathcal{P} = \mathcal{NP}$.

Finally, we show a $\leq_{\mathrm{tt}}^p$-reduction from $(\mathrm{BHP}, \mu_{\mathrm{st}})$ to $(\mathrm{BHP}', \mu_{\mathrm{st}})$. For a given instance $\langle M, i, q, w, y\rangle$ of BHP with $|y| = \phi(|M| + |w| + t)$ for some t, we only have to ask queries of the form $\langle M, i, q, w, u, v\rangle$ for all $u \in \{0, 1\}^{\log m}$, where $m = |M| + |w| + t$, and v is the prefix of y of length $\phi'(|M| + |w| + t - \log m)$. (We will see below that $\phi'(|M| + |w| + t - \log m)$ is smaller than $\phi(|M| + |w| + t)$; hence, this choice of v is possible.)

Clearly, this reduction works as a disjunctive truth-table reduction from BHP to BHP$'$. To check the dominance condition, consider any $\langle M, i, q, w, u, v\rangle$ satisfying the syntax of BHP$'$, we estimate the probability of instances in BHP that ask $\langle M, i, q, w, u, v\rangle$ in our $\leq_{\mathrm{tt}}^p$-reduction. First note that
$$
\begin{aligned}
|v| &= \phi'(|M| + |w| + t - \log m) \\
&= (m - \log m)\log(m - \log m) + (\log(m - \log m))^2 \\
&\leq m \log m = \phi(|M| + |w| + t) = |y|.
\end{aligned}
$$

Let I be the set of instances in BHP that ask $\langle M, i, q, w, u, v \rangle$. Then I consists of strings $\langle M, i, q, w, vy' \rangle$ for some y'. Thus, $\mu_{\text{st}}(I)$, the total probability of instances in BHP that ask $\langle M, i, q, w, u, v \rangle$ is estimated as follows.

$$\mu_{\text{st}}(I) = \sum_{\langle M, i, q, w, vy' \rangle \in I} (1/\alpha) \cdot 2^{-(|M|+|i|+|q|+|w|+|v|+|y'|)}$$

$$= 2^{|y'|} \times (1/\alpha) \cdot 2^{-(|M|+|i|+|q|+|w|+|v|+|y'|)}$$

$$= (1/\alpha) \cdot 2^{-(|M|+|i|+|q|+|w|+|v|)} \le |u|^2 2^{|u|} \cdot (1/\alpha') \cdot 2^{-(|M|+|i|+|q|+|w|+|u|+|v|)}$$

$$= (\log m)^2 2^{\log m} \cdot \mu_{\text{st}}(\langle M, w, u, v \rangle).$$

Here $\alpha = \alpha(|M|, |q|, |i|, |w|, |v|, |y'|)$ and $\alpha' = \alpha(|M|, |q|, |i|, |w|, |u|, |v|)$. Note that $1/\alpha \le |u|^2/\alpha'$. Since $(\log m)^2 2^{\log m}$ is bounded by $p(|\langle M, w, u, v \rangle|)$ with some polynomial p, the dominance condition is satisfied.

References

[AT00]   S. Aida and T. Tsukiji, On the difference among polynomial-time reducibilities for distributional problems (*Japanese*), in *Proc. of the LA Symposium, Winter*, RIMS publication, 2000.

[BDG88]   J. Balcázar, J. Díaz, and J. Gabarró, *Structural Complexity I*, EATCS Monographs on Theoretical Computer Science, Springer-Verlag, 1988.

[Betal92] S. Ben-David, B. Chor, O. Goldreich, and M. Ludy, On the theory of average case complexity, *Journal of Comput. and Syst. Sci.*, 44:193-219, 1992.

[Coo71]   S.A. Cook, The complexity of theorem proving procedures, in *the Proc. of the third ACM Sympos. on Theory of Comput.*, ACM, 151-158, 1971.

[Gur91]   Y. Gurevich, Average case completeness, *Journal of Comput. and Syst. Sci.*, 42:346–398, 1991.

[Hom97]   S. Homer, Structural properties of complete problems for exponential time, in *Complexity Theory Retrospective 2* (A.L. Selman Ed.), Springer-Verlag, 135–154, 1997.

[Imp95]   R. Impagliazzo, A personal view of average-case complexity, in *Proc. 10th Conference Structure in Complexity Theory*, IEEE, 134–147, 1995.

[KiSe94]   S. Kirkpatrick and B. Selman, Critical Behauviour in Satisfiablility of Random Boolean Expressions, *Science*. 264, 1297–1301, 1994.

[KP92]   E. Koutsoupias and C. Papadimitriou, On the greedy algorithm for satisfiability, *Infom. Process. Lett.* 43, 53–55, 1992.

[LLS75]   R. Ladner, N. Lynch, and A. Selman, A Comparison of polynomial time reducibilities, *Theoretical Computer Science*, 1:103–123, 1975.

[Lev73]   L.A. Levin, Universal sequential search problem, *Problems of Information Transmission*, 9:265–266, 1973.

[Lev86]   L.A. Levin, Average case completeness classes, *SIAM J. Comput.*, 15:285–286, 1986.

[LY90]   L. Longpré and P. Young, Cook reducibility is faster than Karp reducibility, *J. Comput. Syst. Sci.*, 41, 389–401, 1990.

[VV83]   U. Vazirani and V. Vazirani, A natural encoding scheme proved probabilistic polynomial complete, *Theoret. Comput. Sci.*, 24, 291–300, 1983.

[Wang97]   J. Wang, Average-case computational complexity theory, in *Complexity Theory Retrospective 2* (A.L. Selman Ed.), Springer-Verlag, 295–328, 1997.

Matching Polygonal Curves with Respect to the Fréchet Distance

Helmut Alt, Christian Knauer, and Carola Wenk[*]

Institut für Informatik, Freie Universität Berlin
Takustraße 9, D–14195 Berlin, Germany
{alt,knauer,wenk}@inf.fu-berlin.de

Abstract. We provide the first algorithm for matching two polygonal curves P and Q under translations with respect to the Fréchet distance. If P and Q consist of m and n segments, respectively, the algorithm has runtime $\mathcal{O}((mn)^3(m+n)^2\log(m+n))$. We also present an algorithm giving an approximate solution as an alternative. To this end, we generalize the notion of a reference point and observe that all reference points for the Hausdorff distance are also reference points for the Fréchet distance. Furthermore we give a new reference point that is substantially better than all known reference points for the Hausdorff distance. These results yield a $(1 + \epsilon)$-approximation algorithm for the matching problem that has runtime $\mathcal{O}(\epsilon^{-2}mn)$.

Keywords: Computational geometry, Shape matching, Fréchet distance, Parametric search, Approximation algorithm, Reference point, Steiner point.

1 Introduction

The task of comparing two two-dimensional shapes arises naturally in many applications, e.g., in computer graphics, computer vision and computer aided design. Often two-dimensional shapes are given by the planar curves forming their boundaries which directly leads to the problem of comparing two planar curves. There are several possible distance measures to assess the 'resemblance' of the shapes, and there are also different kinds of transformations that are allowed to match them, see [5] for a survey. We will focus here on the *Fréchet distance* δ_F for polygonal curves, and we will search for a *translation* which, when applied to the first curve, minimizes the Fréchet distance to the second one. In [4] it is shown how to compute the Fréchet distance for two polygonal curves.

The only algorithm we know of that decides whether there is a transformation that, when applied to the first curve, results in a Fréchet distance less or equal than some given parameter ϵ (this is called the *decision problem*, see Problem 2 below) is presented in [10], where the admissible transformations are translations

---

[*] This research was supported by the Deutsche Forschungsgemeinschaft under Grant No. AL 253/4-3.

A. Ferreira and H. Reichel (Eds.): STACS 2001, LNCS 2010, pp. 63–74, 2001.

in a fixed direction. But to our knowledge there is no algorithm which actually computes the Fréchet distance under a non-trivial class of transformations[1].

In the following we will adopt some basic definitions and results from [4] on which we will subsequently build up.

Definition 1 (Polygonal Curve) *A continuous mapping* $f \colon [a, b] \to \mathbb{R}^2$ *with* $a, b \in \mathbb{R}$ *and* $a < b$ *is called a* curve. *A polygonal curve is a curve* $P \colon [0, n] \to \mathbb{R}^2$ *with* $n \in \mathbb{N}$, *such that for all* $i \in \{0, 1, ..., n - 1\}$ *each* $P_i := P|_{[i,i+1]}$ *is affine, i.e.,* $P(i + \lambda) = (1 - \lambda)P(i) + \lambda P(i + 1)$ *for all* $\lambda \in [0, 1]$.

Definition 2 (Fréchet Distance) *Let* $f \colon [a, a'] \to \mathbb{R}^2$ *and* $g \colon [b, b'] \to \mathbb{R}^2$ *be curves. Then* $\delta_F(f, g)$ *denotes their* Fréchet distance, *defined as*

$$\delta_F(f, g) := \inf_{\substack{\alpha[0,1] \to [a,a'] \\ \beta[0,1] \to [b,b']}} \max_{t \in [0,1]} ||f(\alpha(t)) - g(\beta(t))||.$$

where $||.||$ *denotes the* L_2 *norm, and* α, β *range over continuous and increasing functions with* $\alpha(0) = a$, $\alpha(1) = a'$, $\beta(0) = b$ *and* $\beta(1) = b'$ *only.*

As a popular illustration of the Fréchet-metric suppose a man is walking his dog, he is walking on the one curve the dog on the other. Both are allowed to control their speed but are not allowed to go backwards. Then the Fréchet distance of the curves is the minimal length of a leash that is necessary.

In the rest of the paper we will develop algorithms for the following two problems:

Problem 1 (δ_F – Optimization Problem)
    Given *two polygonal curves* P, Q, *and a class of transformations* $\mathcal{T}$.
    Find *a* $\tau \in \mathcal{T}$ *such that* $\delta_F(\tau(P), Q)$ *is as small as possible.*

Similar to [4] we will first consider the decision problem which we will afterwards optimize applying Megiddo's parametric search technique, c.f. [8]. The decision problem in our setting is the following:

Problem 2 (δ_F – Decision Problem)
    Given   *two polygonal curves* P, Q, *a class of transformations* $\mathcal{T}$, *and* $\epsilon \geq 0$.
    Decide, *whether there exists a* $\tau \in \mathcal{T}$ *such that* $\delta_F(\tau(P), Q) \leq \epsilon$.

We will show that in the case of translations we can solve the decision problem in $\mathcal{O}((mn)^3(m+n)^2)$ time. The parametric search adds only a logarithmic overhead, since we can apply Cole's trick for parametric search based on sorting, so we can solve the optimization problem in $\mathcal{O}((mn)^3(m + n)^2 \log(m + n))$ time.

---

[1] We recently learned that Efrat et al. [7] have independently developed an algorithm for the decision problem under translations. However, the runtime they achieve is by a quadratic factor slower than ours, and their result is rather complicated and relies on complex data structures.

2   Computing the Fréchet Distance

Throughout the rest of the paper let $P : [0, m] \rightarrow \mathbb{R}^2$ and $Q : [0, n] \rightarrow \mathbb{R}^2$ be polygonal curves. Unless stated otherwise $\epsilon \geq 0$ is a fixed real parameter. In the sequel we will use the notion of a *free space* which was introduced in [4]:

Definition 3 (Free Space, [4]) *The set $F_\epsilon(P, Q) := \{(s, t) \in [0, m] \times [0, n] \mid \|P(s) - Q(t)\| \leq \epsilon\}$, or F_ϵ for short, denotes the* free space *of P and Q.*

Sometimes we refer to $[0, m] \times [0, n]$ as the *free space diagram*; the *feasible* points $p \in F_\epsilon$ will be called 'white' and the *infeasible* points $p \in [0, m] \times [0, n] - F_\epsilon$ will be called 'black' (for obvious reasons, c.f. Figure 1). Consider $[0, m] \times [0, n]$ as composed of the mn cells $C_{i,j} := [i - 1, i] \times [j - 1, j]$ $1 \leq i \leq n$, $1 \leq j \leq m$. Then $F_\epsilon(P, Q)$ is composed of the mn free spaces for each pair of edges $F_\epsilon(P_{i-1}, Q_{j-1}) = F_\epsilon(P, Q) \cap C_{i,j}$.

The following results from [4] describe the structure of the free space and link it to the problem of computing δ_F.

Lemma 4 (Alt/Godau, [4]) *The free space of two line segments is the intersection of the unit square with an affine image of the unit disk, i.e., with an ellipse, possibly degenerated to the space between two parallel lines.*

Lemma 5 (Alt/Godau, [4]) *For polygonal curves P and Q we have $\delta_F(P, Q) \leq \epsilon$, exactly if there exists a curve within $F_\epsilon(P, Q)$ from $(0, 0)$ to (m, n) which is monotone in both coordinates.*

For proofs of the above two Lemmas see [4]. Figure 1 shows polygonal curves P, Q, a distance ϵ, and the corresponding diagram of cells $C_{i,j}$ with the free space F_ϵ. Observe that the curve as a continuous mapping from $[0, 1]$ to $[0, m] \times [0, n]$ directly gives feasible reparametrizations, i.e., two reparametrizations α and β, such that $\max_{t \in [0,1]} \|f(\alpha(t)) - g(\beta(t))\| \leq \epsilon$.

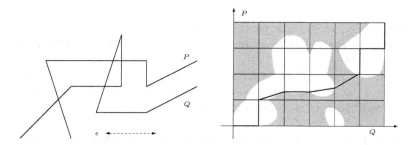

Fig. 1. Two polygonal curves P and Q and their free space diagram for a given ϵ. An example monotone curve in the free space (c.f. Lemma 5) is drawn bold.

For $(i, j) \in \{1, \ldots, m\} \times \{1, \ldots, n\}$ let $L_{i,j}^F := \{i - 1\} \times [a_{i,j}, b_{i,j}]$ (or $B_{i,j}^F := [c_{i,j}, d_{i,j}] \times \{j - 1\}$) be the left (or bottom) line segment bounding $C_{i,j} \cap F_\epsilon$ (see Figure 2).

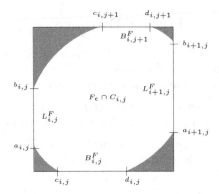

Fig. 2. Intervals of the free space on the boundary of a cell.

By induction it can easily be seen that those parts of the segments $L_{i,j}^F$ and $B_{i,j}^F$ which are reachable from $(0,0)$ by a monotone path in F_ϵ are also line segments. Using a dynamic programming approach one can compute them, and thus decide if $\delta_F(P,Q) \leq \epsilon$. For details we refer the reader to the proof of the following theorem in [4]:

Theorem 6 (Alt/Godau, [4]) *For given polygonal curves P, Q and $\epsilon \geq 0$ one can decide in $\mathcal{O}(mn)$ time, whether $\delta_F(P,Q) \leq \epsilon$.*

Now let us observe a continuity property of F_ϵ: As we have already mentioned, each (possibly clipped ellipse) in F_ϵ is the affine image of a unit disk. Thus each ellipse in F_ϵ varies continuously in ϵ. This implies the following observation:

Observation 7 (See [4]) *If $\epsilon = \delta_F(P,Q)$, then F_ϵ contains at least one monotone path from $(0,0)$ to (m,n) and for each such path π one of the following cases occurs:*

a) *$L_{i,j}^F$ or $B_{i,j}^F$ is a single point on π for some pair (i,j). (The path passes through a passage between two neighboring cells that consists of a single point.)*

b) *$a_{i,j} = b_{k,j}$ (or $c_{i,j} = d_{i,k}$) for some i,j,k and π passes through $(i,a_{i,j})$ and $(k,b_{k,j})$ (or π passes through $(c_{i,j},j)$ and $(d_{i,k},k)$). (The path contains a 'clamped' horizontal or vertical passage, see Figure 3.)*

Figure 4 shows the geometric situations that correspond to these two cases. In case a) the reparametrization maps the point $P(i-1)$ to the only point on the edge Q_j that has distance ϵ from $P(i-1)$. In case b) it maps the part of P between $P(i-1)$ and $P(k-1)$ to the only point on the edge Q_j that has distance ϵ from $P(i-1)$ and $P(k-1)$. This situation covers the case of horizontally clamped paths. The geometric situations that involve a vertically clamped passage are similar, with the roles of P and Q interchanged. Note that we can actually view case a) as a special case of case b) with $i=k$.

Fig. 3. The path contains a 'clamped' horizontal passage in the j-th row between the spikes A and B.

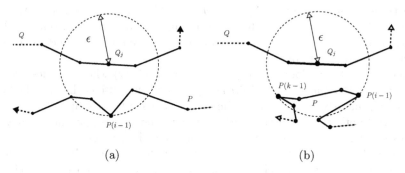

(a)   (b)

Fig. 4. The geometric situations corresponding to a horizontally clamped path.

3 Minimizing the Fréchet Distance

First we give a rough sketch of the basic idea of our algorithm: Assume that there is at least one translation that moves P to a Fréchet distance at most ϵ to Q. Then we can move P to a position $\tau_=$ where the Fréchet distance to Q is exactly ϵ. According to Observation 7 the free space diagram $F_\epsilon(\tau_=(P), Q)$ then contains at least one clamped path. As a consequence, one of the geometric situations from Figure 4 must occur. Therefore the set of translations that attain a Fréchet distance of exactly ϵ is a subset of the set of translations that realize at least one of those geometric situations. The set of translations that create a geometric situation involving the two different vertices $P(i-1)$ and $P(k-1)$ from P and the edge Q_j from Q consist of two segments in transformation space, i.e., it can be described geometrically.

Now assume that the geometric situation from above is specified by the two vertices $P(i-1)$ and $P(k-1)$ and the edge Q_j. When we move P in such a way that $P(i-1)$ and $P(k-1)$ remain at distance ϵ from a common point on an edge of Q (i.e., we shift P 'along' Q), we will preserve one geometric situation (namely the one involving $P(i-1)$ and $P(k-1)$ and some edge of Q). At some point however, we will reach a placement where the Fréchet distance becomes larger than ϵ. This means that immediately before that point it was exactly ϵ, so there is a second placement $\tau'_=$ such that $F_\epsilon(\tau'_=(P), Q)$ contains at least one clamped path and consequently another geometric situation must occur. So the set of translations that attain a Fréchet distance of exactly ϵ is a subset of the set of translations that realize at least *two* such geometric situations.

After this informal description of the basic ideas let us go into more detail now:

Convention: *In this section T_2 denotes the group of planar translations, P and Q are polygonal curves with m and n vertices, respectively and $\epsilon \geq 0$ is a real parameter.*

A translation $\tau = \langle (x, y) \mapsto (x + \delta_x, y + \delta_y) \rangle \in T_2$ can be specified by the pair $(\delta_x, \delta_y) \in \mathbb{R}^2$ of *parameters*. The set of parameters of all translations in T_2 is called the parameter space of T_2, or *translation space* for short, and we identify T_2 with its parameter space.

Let us now take a look at the free space $F_\epsilon(\tau(P), Q)$ when τ varies over T_2. Now we show that each of the $\mathcal{O}(mn)$ ellipses (and thus also each clipped ellipse) varies continuously in $\tau \in T_2$. In fact we consider all mn ellipses, even those that have an empty intersection with their corresponding square in the diagram (let us call these 'invisible'). Note that an ellipse is generated by two linearly independent line segments; one from P and one from Q. Parallel line segments generate only a 'degenerate ellipse', namely the space between two parallel lines. So if we fix a translation $\tau \in T_2$ it is easy to see that each (possibly invisible) ellipse in $F_\epsilon(\tau(P), Q)$ is a translation of the corresponding ellipse in $F_\epsilon(P, Q)$. In fact, the translation is $(-\lambda, \mu)$ where λ and μ are the coefficients which are obtained by representing τ as a linear combination of the direction unit vectors of the line segments. Thus each ellipse varies continuously in $\tau \in T_2$. Note that this is still true if an ellipse is visible but its translate is invisible or vice versa. A similar argument holds for degenerate ellipses.

Definition 8 (Configuration) *A triple (p, p', s) that consists of two (not necessarily different) vertices p and p' of P and an edge s of Q is called an h-configuration. v-configurations are defined analogously with the roles of P and Q exchanged. A configuration is an h- or v-configuration.*

Definition 9 (Critical Translations) *Let $c = (x, y, s)$ be an h-configuration and $c' = (x', y', s')$ be a v-configuration. The sets*

$$T_{crit}(c) := \{\tau \in T_2 \mid \exists z \in s : ||\tau(x) - z|| = ||\tau(y) - z|| = \epsilon\} \quad \text{and}$$
$$T_{crit}(c') := \{\tau \in T_2 \mid \exists z' \in s' : ||x' - \tau(z')|| = ||y' - \tau(z')|| = \epsilon\}$$

are called the sets of critical translations *for c and c'. A translation is called* critical *if it is critical for some configuration.*

Lemma 10 *If $\delta_F(\tau(P), Q) = \epsilon$, then τ is critical.*

Proof. By Observation 7 there is a path π in $F_\epsilon(\tau(P), Q)$ for which case a) or b) occurs. If the corresponding geometric situation (c.f. Figure 4) involves the vertices $\tau(P(i-1))$ and $\tau(P(k-1))$ on $\tau(P)$ and a point on the edge Q_j then the translation τ is critical for the h-configuration $(P(i-1), P(k-1), Q_j)$. If the geometric situation involves vertices from Q and a segment of $\tau(P)$, the same argument yields a v-configuration.   □

Note that the condition in Lemma 10 is only necessary but not sufficient, i.e., there are indeed critical translations τ with $\delta_F(\tau(P), Q) \neq \epsilon$. This is because a critical translation for a configuration (x, y, s) does not even have to map the part of the curve between x and y within distance ϵ to the corresponding point on s.

Let us now take a closer look at the critical translations in $\mathcal{T}_2$: For a given configuration (x, y, s) with two different vertices (which corresponds to case (b) in Figure 4) the set of critical translations is described by two parallel line segments in translation space, where each line segment is a translate of s. If the two vertices in the configuration are the same (which is case (a) in Figure 4) the set of critical translations is described by a 'racetrack' in translation space, which is the locus of points having distance ϵ to a translate of s. Note that a 'racetrack' consists of line segments and circular arcs.

We call the arrangement in translation space consisting of the curves describing all critical translations of all configurations the *arrangement of critical translations*. There are $\mathcal{O}(mn(m+n))$ different configurations, so the combinatorial complexity of the arrangement of critical translations (i.e., the number of vertices line segments and circular arcs) is $\mathcal{O}((mn(m+n))^2)$.

Lemma 11 *If there is a translation $\tau_\leq \in \mathcal{T}_2$ such that $\delta_F(\tau_\leq(P), Q) \leq \epsilon$ then there is a translation $\tau_= \in \mathcal{T}_2$ that is critical such that $\delta_F(\tau_=(P), Q) = \epsilon$.*

Proof. Pick any translation $\tau_> \in \mathcal{T}_2$ such that $\delta_F(\tau_>(P), Q) > \epsilon$. By continuity, there exists a translation $\tau_=$ on any curve between $\tau_\leq$ and $\tau_>$ in translation space such that $\delta_F(\tau_=(P), Q) = \epsilon$. By Lemma 10 the translation $\tau_=$ is critical.   □

This result states that, whenever there is some translation $\tau_\leq$ that moves P into Fréchet distance at most ϵ to Q, there is also a 'canonical' translation $\tau_=$ that results in a Fréchet distance exactly ϵ and that lies on the arrangement of critical translations. So in order to check if there is a translation that moves P into Fréchet distance at most ϵ to Q, it is sufficient to check all translations on the arrangement of critical translations.

However, since the translation space has more than one degree of freedom, the arrangement of critical translations contains an infinite number of translations. So our observation does not help from an algorithmic point of view. Lemma 14 shows that we can restrict our attention to the zero-dimensional parts of the arrangement, i.e., intersection points and endpoints of the curves describing the critical translations. First we need the following two observations:

Observation 12 *Let $c = (P(i-1), P(k-1), Q_j)$, with $i \neq k$, be an h-configuration. Then $a_{i,j} = b_{k,j}$ in $F_\epsilon(\tau(P), Q)$ for all $\tau \in T_{crit}(c)$, i.e., the relative position of the two spikes stays the same for all $\tau \in T_{crit}(c)$ (c.f. Figures 3 and 4).*

Observation 13 *Let $c = (P(i-1), P(k-1), Q_j)$, with $i \neq k$, be an h-configuration. Now we consider a feasible reparametrization for some $\tau_= \in T_{crit}(c)$, that maps the part of $\tau_=(P)$ between $\tau_=(P(i-1))$ and $\tau_=(P(k-1))$ to a point on Q_j. This corresponds to a path $\pi_=$ in $F_\epsilon(\tau_=(P), Q)$ that is clamped between the two corresponding vertical spikes in cell (i,j) and (k,j) of $F_\epsilon(\tau_=(P), Q)$. Now from Observation 12 it follows that for each $\tau \in T_{crit}(c)$ the relative position of the spikes does not change, i.e., we cannot 'destroy' $\pi_=$ locally by moving along $T_{crit}(c)$.*

Of course both observations remain true if we consider h-configurations that correspond to case a) of Observation 7 (where $i = k$) or v-configurations (where the roles of P and Q are interchanged).

Lemma 14 *If there is a translation $\tau_\leq \in T_2$ such that $\delta_F(\tau_\leq(P), Q) \leq \epsilon$ then there is a translation $\tau_= \in T_2$ that lies on a vertex of the arrangement of critical translations such that $\delta_F(\tau_=(P), Q) = \epsilon$.*

Proof. Suppose all vertices of the arrangement of critical translations yield a Fréchet distance greater than ϵ. By Lemma 11 there is a critical translation $\tau_= \in T_2$ with $\delta_F(\tau_=(P), Q) = \epsilon$, and by definitions 8 and 9 there is a configuration c such that $\tau_= \in T_{crit}(c)$. Now pick any translation $\tau_> \in T_2$ such that $\delta_F(\tau_>(P), Q) > \epsilon$. We can assume without loss of generality that $\tau_=$ lies in an 'extreme' position on $T_{crit}(c)$, which means that $\delta_F(\tau(P), Q) > \epsilon$ for every $\tau \in T_{crit}(c)$ that lies 'between' $\tau_=$ and $\tau_>$. Considering the free space diagram this means that $F_\epsilon(\tau_=(P), Q)$ contains a monotone path, but $F_\epsilon(\tau(P), Q)$ does not contain this or any other monotone path anymore. By continuity this can only happen, if each monotone path in $F_\epsilon(\tau_=(P), Q)$ is 'clamped' between two 'spikes' which close the narrow passage in the free space when moving from $\tau_=$ to $\tau_>$ on $T_{crit}(c)$.

But according to Observation 12 this cannot be true for the spikes corresponding to the critical translations $T_{crit}(c)$. Thus there must be another configuration c' such that $\tau_= \in T_{crit}(c')$, and close to $\tau_=$ the curve describing $T_{crit}(c')$ differs from $T_{crit}(c)$. Since both curves are algebraic, $T_{crit}(c) \cap T_{crit}(c')$ is zero-dimensional, and thus a vertex of the arrangement.   □

So in order to solve the decision problem for a given ϵ it is sufficient to check for all translations τ that correspond to vertices of the arrangement of critical translations whether $\delta_F(\tau(P), Q) \leq \epsilon$. We thus have altogether $\mathcal{O}((mn)^2(m + n)^2)$ translations for each of which we check in $\mathcal{O}(mn)$ time if it brings P into distance at most ϵ to Q, which solves Problem 4 for the case of translations and yields the following theorem:

Theorem 15 *For given polygonal curves P, Q and $\epsilon \geq 0$ one can decide in $\mathcal{O}((mn)^3(m + n)^2)$ time whether there is a translation $\tau \in T_2$ such that $\delta_F(\tau(P), Q) \leq \epsilon$.*

In order to find a translation that minimizes the Fréchet distance between the two polygonal curves we apply the parametric search paradigm. For this we generalize the approach of [4]. Remember that for a given configuration $c = (x, y, s)$ the set of critical translations $T_{crit}(c)$ is described by two parallel line segments or by a 'racetrack' in translation space. Now when we let ϵ vary $T_{crit}(c)$ changes accordingly, namely the distance between the parallel line segments or the radius of the 'racetrack' varies depending on ϵ. Note that for small ϵ, $T_{crit}(c)$ might even be empty, which happens for example when $||x - y|| < \epsilon$.

For a given ϵ let $S(\epsilon)$ be the set of $\mathcal{O}\big((mn(m+n))^2\big)$ vertices of the arrangement of critical translations. In fact, one can track each vertex in $S(\epsilon)$ for varying ϵ, i.e., one can interpret each vertex in $S(\epsilon)$ as a function of ϵ. Let S be the set of these vertex-functions. Note that the vertex-functions in S might not be defined for small ϵ. For each of the $\mathcal{O}\big((mn(m+n))^2\big)$ translation functions $\tau(\epsilon)$ in S we compute the free space $F_\epsilon(\tau(\epsilon)(P), Q)$ depending on ϵ. In fact we only compute all $a_{i,j}(\tau, \epsilon)$, $b_{i,j}(\tau, \epsilon)$, $c_{i,j}(\tau, \epsilon)$, and $d_{i,j}(\tau, \epsilon)$ which depend on ϵ and τ, and of which there are $\mathcal{O}\big((mn)^3(m+n)^2\big)$. For the parametric search an ϵ is critical if two of these functions have the same value (for the same translation function τ). A parametric search over all $\mathcal{O}\big((mn)^3(m+n)^2\big)$ values of $a_{i,j}(\tau, \epsilon)$, $b_{i,j}(\tau, \epsilon)$, $c_{i,j}(\tau, \epsilon)$, and $d_{i,j}(\tau, \epsilon)$ thus yields an optimum ϵ together with an optimum translation. As in [4] we apply a parallel sorting algorithm which generates a superset of the critical values of ϵ we need. By utilizing Cole's trick [6] for parametric search based on sorting, which in general yields a runtime of $\mathcal{O}\big((k + T_{seq}) \log k\big)$ where T_{seq} is the sequential runtime for the decision problem and k is the number of values to be sorted, we obtain a runtime of $\mathcal{O}\big((mn)^3(m+n)^2 log(m+n)\big)$. This solves Problem 3 for the case of translations and proves the following theorem:

Theorem 16 *For given polygonal curves P, Q one can compute a translation τ_{min} in $\mathcal{O}\big((mn)^3(m+n)^2 log(m+n)\big)$ time, such that $\delta_F(\tau_{min}(P), Q) = \min_{\tau \in T_2} \delta_F(\tau(P), Q)$.*

3.1   Other Transformation Classes

We are currently investigating the application of the techniques from above to other classes of transformations, such as translations in a fixed direction, rotations around a fixed center, rigid motions, and arbitrary affine maps, for matching curves in two and higher dimensions.

In the parameter space of the transformation class under consideration the set of critical transformations for a configuration is a semi-algebraic set in general, which is defined by a constant number of polynomials of bounded degree. Therefore we can define the arrangement of critical transformations in the same way as before.

A suitable generalization of Lemma 14 should imply that only the zero-dimensional pieces of this arrangement have to be considered as candidates for a successful match. This immediately yields an algorithm with a runtime that depends on the complexity of the arrangement of critical transformations, which in turn depends on the dimension of the parameter space as well as on the dimension of the underlying Euclidean space.

4   Approximately Minimizing the Fréchet Distance

The algorithms we described so far cannot be considered to be efficient. To remedy this situation, we present approximation algorithms which do not necessarily compute the optimal transformation, but one that yields a Fréchet distance which differs from the optimum value by a constant factor only. To this end, we generalize the notion of a *reference point*, c.f. [2] and [1], to the Fréchet metric and observe that all reference points for the Hausdorff distance are also reference points for the Fréchet distance.

We first need the concept of a *reference point* that was introduced in [1]. A reference point of a figure is a characteristic point with the property that similar figures have reference points that are close to each other. Therefore we get a reasonable matching of two figures if we simply align their reference points.

Definition 17 (Reference Point, [1]) *Let $\mathcal{K}$ be a set of compact subsets of $\mathbb{R}^2$ and δ be a metric on $\mathcal{K}$. A mapping $\mathbf{r} : \mathcal{K} \to \mathbb{R}^2$ is called a δ–reference point for $\mathcal{K}$ of quality $c > 0$ with respect to a set of transformations $\mathcal{T}$ on $\mathcal{K}$, if the following holds for any two sets $P, Q \in \mathcal{K}$ and each transformation $\tau \in \mathcal{T}$:*

$$\text{(Equivariance)} \quad \mathbf{r}(\tau(P)) = \tau(\mathbf{r}(P)) \tag{1}$$

$$\text{(Lipschitz continuity)} \quad ||\mathbf{r}(P) - \mathbf{r}(Q)|| \leq c \cdot \delta(P, Q). \tag{2}$$

In other words a reference point is a Lipschitz-continuous mapping between the metric spaces $(\mathcal{K}, \delta)$ and $(\mathbb{R}^2, || \cdot ||)$ with Lipschitz constant c, which is equivariant under $\mathcal{T}$. Various reference points are known for a variety of distance measures and classes of transformations, like, e.g., the centroid of a convex polygon which is a reference point of quality $11/3$ for translations, using the area of the symmetric difference as a distance measure, see [3]. However, most work on reference points has focused on the Hausdorff distance, see [1].

Definition 18 (Hausdorff Distance) *Let P and Q be curves. Then $\delta_H(P, Q)$ denotes their Hausdorff distance, defined as*

$$\delta_H(P, Q) := \max(\tilde{\delta}_H(P, Q), \tilde{\delta}_H(Q, P)), \quad with$$

$$\tilde{\delta}_H(X, Y) := \sup_{x \in X} \inf_{y \in Y} ||x - y||, \quad the \text{ one-sided Hausdorff distance } from X \text{ to } Y.$$

We will only mention the following result that provides a δ_H–reference point for polygonal curves with respect to similarities, the so called *Steiner point*. The Steiner point of a polygonal curve is the weighted average of the vertices of the convex hull of the curve, where each vertex is weighted by its exterior angle divided by 2π.

Theorem 19 (Aichholzer et al., [1]) *The Steiner point is a δ_H–reference point with respect to similarities of quality $4/\pi$. It can be computed in linear time.*

Note that the Steiner point is an *optimal* δ_H-reference point with respect to similarities, i.e., the quality of any δ_H-reference point for that transformation class is at least $4/\pi$, see [1].

Two feasible reparametrizations α and β of P and Q demonstrate, that for each point $P(\alpha(t))$ there is a point $Q(\beta(t))$ with $||P(\alpha(t)) - Q(\beta(t))|| \leq \epsilon$ (and vice versa), thus $\delta_H(P, Q) \leq \delta_F(P, Q)$. This shows the following observation:

Observation 20 *Let $c > 0$ be a constant and $\mathcal{T}$ be a set of transformations on $\mathcal{K}$. Then each δ_H-reference point with respect to $\mathcal{T}$ is also a δ_F-reference point with respect to $\mathcal{T}$ of the same quality.*

This shows that we can use the known δ_H-reference points to obtain δ_F-reference points. However, since each reparametrization has to map $P(0)$ to $Q(0)$, the distance $||P(0) - Q(0)||$ is a lower bound for $\delta_F(P, Q)$. So we get a new reference point that is substantially better than all known reference points for the Hausdorff distance.

Observation 21 *Let $\mathcal{C}_o$ be the set of all planar curves. The mapping*

$$\mathbf{r}_o : \begin{cases} \mathcal{C}_o \to \mathbb{R}^2 \\ P \mapsto P(0) \end{cases}$$

is a δ_F-reference point for curves of quality 1 with respect to translations.

The quality of this reference point, i.e., 1, is better than the quality of the Steiner point, which is $4/\pi$. Since the latter is an optimal reference point for the Hausdorff distance, this shows that for the Fréchet distance substantially better reference points exist. For closed curves however $\mathbf{r}_o$ is not defined at all.

Based on the existence of a δ_F-reference point for $\mathcal{T}_2$ we obtain the following algorithm for approximate matchings with respect to the Fréchet distance under the group of translations, which is the same procedure as already used in [1] for the Hausdorff distance.

Algorithm T: Compute $\mathbf{r}(P)$ and $\mathbf{r}(Q)$, translate P by $\tau := \mathbf{r}(Q) - \mathbf{r}(P)$, and output this matching as the approximate solution, together with $\delta_F(\tau(P), Q)$.

Theorem 22 *Suppose that $\mathbf{r}$ is a δ_F-reference point of quality c with respect to translations that can be computed in $\mathcal{O}(T_\mathbf{r}(n))$ time. Then algorithm T produces a $(c + 1)$-approximation to Problem 1 in $\mathcal{O}(mn + T_\mathbf{r}(m) + T_\mathbf{r}(n))$ time.*

Proof. Let τ_{opt} be a translation, such that $\min_\tau \delta_F(\tau(P), Q) = \delta_F(\tau_{opt}(P), Q)$. Then

$$||\mathbf{r}(\tau_{opt}(P)) - \mathbf{r}(Q)|| \leq c \cdot \delta_F(\tau_{opt}(P), Q).$$

Let $\tau_{diff} := \mathbf{r}(\tau_{opt}(P)) - \mathbf{r}(Q) \in \mathcal{T}_2$; then

$$\tau_{approx} := \tau_{diff} \circ \tau_{opt}$$

maps $\mathbf{r}(P)$ onto $\mathbf{r}(Q)$ and

$$\delta_F(\tau_{approx}(P), Q) \leq \delta_F(\tau_{opt}(P), Q) + ||\tau_{diff}|| \leq (c + 1) \cdot \delta_F(\tau_{opt}(P), Q).$$

The proof of the claimed time bound is obvious. $\qquad\square$

Note that with an idea from [9] it is possible to reduce the approximation constant for reference point based matching to $(1 + \epsilon)$ for any $\epsilon > 0$; the idea places a sufficiently small grid of size $\mathcal{O}(1/\epsilon^2)$ around the reference point of Q and checks each grid point as a potential image point for the reference point of P. The runtime increases by a factor proportional to the grid size.

Acknowledgements

We would like to thank Günter Rote and an anonymous referee for helpful comments on an earlier version of this paper.

References

1. O. Aichholzer, H. Alt, and G. Rote. Matching shapes with a reference point. *Internat. J. Comput. Geom. Appl.*, 7:349–363, 1997.
2. H. Alt, B. Behrends, and J. Blömer. Approximate matching of polygonal shapes. *Ann. Math. Artif. Intell.*, 13:251–266, 1995.
3. H. Alt, U. Fuchs, G. Rote, and G. Weber. Matching convex shapes with respect to the symmetric difference. *Algorithmica*, 21:89–103, 1998.
4. H. Alt and M. Godau. Computing the Fréchet distance between two polygonal curves. *Internat. J. Comput. Geom. Appl.*, 5:75–91, 1995.
5. H. Alt and L. J. Guibas. Discrete geometric shapes: Matching, interpolation, and approximation. In J.-R. Sack and J. Urrutia, editors, *Handbook of Computational Geometry*, pages 121–153. Elsevier Science Publishers B.V. North-Holland, Amsterdam, 2000.
6. R. Cole. Slowing down sorting networks to obtain faster sorting algorithms. *Journal of the ACM*, 34(1):200–208, 1987.
7. A. Efrat, P. Indyk, and S. Venkatasubramanian. Pattern matching for sets of segments, September 2000. Manuscript, accepted to the 12th Symposium on Discrete Algorithms, 2001.
8. N. Megiddo. Applying parallel computation algorithms in the design of serial algorithms. *J. ACM*, 30(4):852–865, 1983.
9. S. Schirra. Über die Bitkomplexität der ϵ-Kongruenz. Master's thesis, Fachbereich Informatik, Universität des Saarlandes, 1988.
10. S. Venkatasubramanian. *Geometric Shape Matching and Drug Design*. PhD thesis, Department of Computer Science, Stanford University, August 1999.

On the Class of Languages Recognizable by 1-Way Quantum Finite Automata

Andris Ambainis[1*], Arnolds Ķikusts[2**], and Māris Valdats[2**]

[1] Computer Science Division, University of California
Berkeley, CA94720, USA
ambainis@cs.berkeley.edu
[2] Institute of Mathematics and Computer Science
University of Latvia, Raiņa bulv. 29, Rīga, Latvia
sd70053@lanet.lv, sd70066@lanet.lv

Abstract. It is an open problem to characterize the class of languages recognized by quantum finite automata (QFA). We examine some necessary and some sufficient conditions for a (regular) language to be recognizable by a QFA. For a subclass of regular languages we get a condition which is necessary and sufficient.

Also, we prove that the class of languages recognizable by a QFA is not closed under union or any other binary Boolean operation where both arguments are significant.

1 Introduction

A 1-way quantum finite automaton (QFA)[1] is a theoretical model for a quantum computer with a finite memory.

Compared to classical (non-quantum) automata, QFAs have both strengths and weaknesses. The strength of QFAs is shown by the fact that quantum automata can be exponentially more space efficient than deterministic or probabilistic automata [AF 98]. The weakness of QFAs is caused by the fact that any quantum process has to be reversible (unitary). This makes QFAs unable to recognize some regular languages.

The first result of this type was obtained by Kondacs and Watrous [KW 97] who showed that there is a language that can be recognized by a deterministic finite automaton (DFA) but cannot be recognized by QFA. Later, Brodsky and Pippenger [BP 99] generalized the construction of [KW 97] and showed that any regular language that does not satisfy the partial order condition cannot be

---

* Research supported by Berkeley Fellowship for Graduate Studies and, in part, NSF Grant CCR-9800024.

** Research supported by Grant No.96.0282 from the Latvian Council of Science and European Commission, contract IST-1999-11234.

[1] For the rest of the paper, we will omit "1-way" because this is the only model of QFAs that we consider in this paper. For other models of QFAs, see [KW 97] and [AW 99].

A. Ferreira and H. Reichel (Eds.): STACS 2001, LNCS 2010, pp. 75–86, 2001.
© Springer-Verlag Berlin Heidelberg 2001

recognized by a QFA. They also conjectured that all regular languages satisfying the partial order condition can be recognized by a QFA.

In this paper, we disprove their conjecture. We show that, for a language to be recognizable by a QFA, its minimal deterministic automaton must not contain several "forbidden fragments". One of fragments is equivalent to the automaton not satisfying the partial order condition. The other fragments are new.

A somewhat surprising feature of our "forbidden fragments" is that they consist of several parts (corresponding to different beginnings of the word) and the language corresponding to every one of them can be recognized but one cannot simultaneously recognize the whole language without violating unitarity.

Our result implies that the set of languages recognizable by QFAs is not closed under union. In particular, the language consisting of all words in the alphabet $\{a, b\}$ that have an even number of a's after the first b is not recognizable by a QFA, although it is a union of two recognizable languages. (The first language consists of all words with an even number of a's before the first b and an even number of a's after the first b, the second language consists of all words with an odd number of a's before the first b and an even number of a's after it.) This answers a question of Brodsky and Pippenger [BP 99].

For a subclass of regular languages (languages that do not contain "two cycles in a row" construction shown in Fig. 3), we show that our conditions are necessary and sufficient for a language to be recognizable by a QFA. For arbitrary regular languages, we only know that these conditions are necessary but we do not know if all languages satisfying them can be recognized by a QFA.

Due to space constraints of these proceedings, most of proofs are omitted.

1.1 Definitions

Quantum finite automata (QFA) were introduced independently by Moore and Crutchfield [MC 97] and Kondacs and Watrous [KW 97]. In this paper, we consider the more general definition of QFAs [KW 97] (which includes the definition of [MC 97] as a special case).

Definition 1.1. *A QFA is a tuple* $M = (Q; \Sigma; V; q_0; Q_{acc}; Q_{rej})$ *where Q is a finite set of states, Σ is an input alphabet, V is a transition function (explained below), $q_0 \in Q$ is a starting state, and $Q_{acc} \subseteq Q$ and $Q_{rej} \subseteq Q$ are sets of accepting and rejecting states ($Q_{acc} \cap Q_{rej} = \emptyset$). The states in Q_{acc} and Q_{rej}, are called* halting states *and the states in $Q_{non} = Q - (Q_{acc} \cup Q_{rej})$ are called* non halting states.

States of M. The state of M can be any superposition of states in Q (i. e., any linear combination of them with complex coefficients). We use $|q\rangle$ to denote the superposition consisting of state q only. $l_2(Q)$ denotes the linear space consisting of all superpositions, with l_2-distance on this linear space.

Endmarkers. Let κ and \$ be symbols that do not belong to Σ. We use κ and \$ as the left and the right endmarker, respectively. We call $\Gamma = \Sigma \cup \{\kappa; \$\}$ the *working alphabet* of M.

Transition Function. The transition function V is a mapping from $\Gamma \times l_2(Q)$ to $l_2(Q)$ such that, for every $a \in \Gamma$, the function $V_a : l_2(Q) \to l_2(Q)$ defined by $V_a(x) = V(a, x)$ is a unitary transformation (a linear transformation on $l_2(Q)$ that preserves l_2 norm).

Computation. The computation of a QFA starts in the superposition $|q_0\rangle$. Then transformations corresponding to the left endmarker κ, the letters of the input word x and the right endmarker \$ are applied. The transformation corresponding to $a \in \Gamma$ consists of two steps.

1. First, V_a is applied. The new superposition ψ' is $V_a(\psi)$ where ψ is the superposition before this step.

2. Then, ψ' is observed with respect to $E_{acc}, E_{rej}, E_{non}$ where $E_{acc} = span\{|q\rangle : q \in Q_{acc}\}$, $E_{rej} = span\{|q\rangle : q \in Q_{rej}\}$, $E_{non} = span\{|q\rangle : q \in Q_{non}\}$. If the state before the measurement was

$$\psi' = \sum_{q_i \in Q_{acc}} \alpha_i |q_i\rangle + \sum_{q_j \in Q_{rej}} \beta_j |q_j\rangle + \sum_{q_k \in Q_{non}} \gamma_k |q_k\rangle$$

then the measurement accepts ψ' with probability $p_a = \Sigma \alpha_i^2$, rejects with probability $p_r = \Sigma \beta_j^2$ and continues the computation (applies transformations corresponding to next letters) with probability $p_c = \Sigma \gamma_k^2$ with the system having the (normalized) state $\frac{\psi}{\|\psi\|}$ where $\psi = \Sigma \gamma_k |q_k\rangle$.

We regard these two transformations as reading a letter a.

Unnormalized States. Normalization (replacing ψ by $\frac{\psi}{\|\psi\|}$) is needed to make the probabilities of accepting, rejecting and non-halting after the next letter sum up to 1. However, normalizing the state after every letter can make the notation quite messy. (For the state after k letters, there would be k normalization factors $\frac{1}{\|\psi_1\|}, \ldots, \frac{1}{\|\psi_k\|}$ - one for each letter!)

For this reason, we do not normalize the states in our proofs. That is, we apply the next transformations to the unnormalized state ψ instead of $\frac{\psi}{\|\psi\|}$.

There is a simple correspondence between unnormalized and normalized states. If, at some point, the unnormalized state is ψ, then the normalized state is $\frac{\psi}{\|\psi\|}$ and the probability that the computation has not stopped is $\|\psi\|^2$. The sums $p_a = \Sigma \alpha_i^2$ and $p_r = \Sigma \beta_i^2$ are the probabilities that the computation has not halted before this moment but accepts (rejects) at this step.

Notation. We use V_a' to denote the transformation consisting of V_a followed by projection to E_{non}. This is the transformation mapping ψ to the non-halting part of $V_a(\psi)$. We use V_w' to denote the product of transformations $V_w' = V_{a_n}' V_{a_{n-1}}' \cdots V_{a_2}' V_{a_1}'$, where a_i is the i-th letter of the word w.

We also use ψ_w to denote the (unnormalized) non-halting part of QFA's state after reading the left endmarker κ and the word $w \in \Sigma^*$. From the notation it follows that $\psi_w = V_{\kappa w}'(|q_0\rangle)$.

Recognition of Languages. A QFA M recognizes a language L with probability p ($p > \frac{1}{2}$) if accepts any word $x \in L$ with probability $\geq p$ and rejects any word $x \notin L$ with probability $\geq p$. If we say that a QFA M recognizes a language L (without specifying the accepting probability), this means that M recognizes L with probability $\frac{1}{2} + \epsilon$ for some $\epsilon > 0$.

1.2   Previous Work

The previous work on quantum automata has mainly considered 3 questions:

1. What is the class of languages recognized by QFAs?
2. What accepting probabilities can be achieved?
3. How does the size of QFAs (the number of states) compare to the size of deterministic (probabilistic) automata?

In this paper, we consider the first question. The first results in this direction were obtained by Kondacs and Watrous [KW 97].

Theorem 1.1. *[KW 97]*

1. *All languages recognized by QFAs are regular.*
2. *There is a regular language that cannot be recognized by a QFA.*

Brodsky and Pippenger [BP 99] generalized the second part of Theorem 1.1 by showing that any language satisfying a certain property is not recognizable by a QFA.

Theorem 1.2. *[BP 99] Let L be a language and M be its minimal automaton (the smallest DFA recognizing L). Assume that there are words x and y such that M contains states q_1, q_2 satisfying:*

1. *$q_1 \neq q_2$,*
2. *If M starts in the state q_1 and reads x, it passes to q_2,*
3. *If M starts in the state q_2 and reads x, it passes to q_2, and*
4. *If M starts in q_2 and reads y, it passes to q_1,*

then L cannot be recognized by a quantum finite automaton(Fig.1).

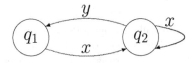

Fig. 1. Conditions of Theorem 1.2

A language L with the minimal automaton not containing a fragment of Theorem 1.2 is called *satisfying the partial order condition* [MT 69]. [BP 99] conjectured that any language satisfying the partial order condition is recognizable by a QFA. In this paper, we disprove this conjecture.

Another direction of research studies the accepting probabilities of QFAs. First, Ambainis and Freivalds [AF 98] proved that the language a^*b^* is recognizable by a QFA with probability 0.68... but not with probability $7/9 + \epsilon$ for any $\epsilon > 0$. Thus, the classes of languages recognizable with different probabilities are different. Next results in this direction were obtained by [ABFK 99] who studied the probability with which the languages $a_1^* \ldots a_n^*$ can be recognized.

There is also a lot of results about the number of states needed for QFA to recognize different languages. In some cases, it can be exponentially less than for deterministic or even for probabilistic automata [AF 98,K 98]. In other cases, it can be exponentially bigger than for deterministic automata [ANTV 98,N 99].

A good survey about quantum automata is Gruska [G 00].

2 Main Results

2.1 Necessary Condition

First, we give the new condition which implies that the language is not recognizable by a QFA. Similarly to the previous condition (Theorems 1.2), it can be formulated as a condition about the minimal deterministic automaton of a language. In Section 3, we will give an example of a language that satisfies the condition of Theorem 2.1 but not the previously known condition of Theorem 1.2 (the language L_1).

Theorem 2.1. *Let L be a language. Assume that there are words x, y, z_1, z_2 such that its minimal automaton M contains states q_1, q_2, q_3 satisfying:*
  1. $q_2 \neq q_3$,
  2. if M starts in the state q_1 and reads x, it passes to q_2,
  3. if M starts in the state q_2 and reads x, it passes to q_2,
  4. if M starts in the state q_1 and reads y, it passes to q_3,
  5. if M starts in the state q_3 and reads y, it passes to q_3,
  6. for any word $t \in (x|y)^*$ there exists a word $t_1 \in (x|y)^*$ such that if M starts in the state q_2 and reads tt_1, it passes to q_2,
  7. for any word $t \in (x|y)^*$ there exists a word $t_1 \in (x|y)^*$ such that if M starts in the state q_3 and reads tt_1, it passes to q_3,
  8. if M starts in the state q_2 and reads z_1, it passes to an accepting state,
  9. if M starts in the state q_2 and reads z_2, it passes to a rejecting state,
  10. if M starts in the state q_3 and reads z_1, it passes to a rejecting state,
  11. if M starts in the state q_3 and reads z_2, it passes to an accepting state.
Then L cannot be recognized by a QFA.

Proof. We use lemmas from [BV 97] and [AF 98].

Lemma 2.1. *[BV 97] If ψ and ϕ are two quantum states and $\|\psi - \phi\| < \varepsilon$ then the total variational distance between probability distributions generated by the same measurement on ψ and ϕ is at most[2] 2ε.*

---

[2] The lemma in [BV 97] has 4ε but it can be improved to 2ε.

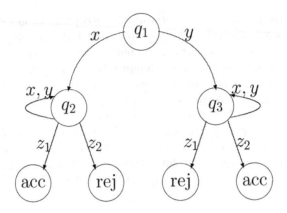

Fig. 2. Conditions of Theorem 2.1, Conditions 6 and 7 Are Shown Symbolically

Lemma 2.2. *[AF 98] Let $x \in \Sigma^+$. There are subspaces E_1, E_2 such that $E_{non} = E_1 \oplus E_2$ and*

(i) If $\psi \in E_1$, then $V'_x(\psi) \in E_1$ and $\|V'_x(\psi)\| = \|\psi\|$,
(ii) If $\psi \in E_2$, then $\|V'_{x^k}(\psi)\| \to 0$ when $k \to \infty$.

Lemma 2.2 can be viewed as a quantum counterpart of the *classification of states for Markov chains* [KS 76]. The classification of states divides the states of a Markov chain into *ergodic* sets and *transient* sets. If the Markov chain is in an ergodic set, it never leaves it. If it is in a transient set, it leaves it with probability $1 - \epsilon$ for an arbitrary $\epsilon > 0$ after sufficiently many steps.

In the quantum case, E_1 is the counterpart of an ergodic set: if the quantum random process defined by repeated reading of x is in a state $\psi \in E_1$, it stays in E_1. E_2 is a counterpart of a transient set: if the state is $\psi \in E_2$, E_2 is left (for an accepting or rejecting state) with probability arbitrarily close to 1 after sufficiently many x's.

The next Lemma is our generalization of Lemma 2.2 for the case of two different words x and y.

Lemma 2.3. *Let $x, y \in \Sigma^+$. There are subspaces E_1, E_2 such that $E_{non} = E_1 \oplus E_2$ and*

(i) If $\psi \in E_1$, then $V'_x(\psi) \in E_1$ and $V'_y(\psi) \in E_1$ and $\|V'_x(\psi)\| = \|\psi\|$ and $\|V'_y(\psi)\| = \|\psi\|$,
(ii) If $\psi \in E_2$, then for any $\epsilon > 0$, there exists $t \in (x|y)^*$ such that $\|V'_t(\psi)\| < \epsilon$.

Proof. Omitted. □

Let L be a language with its minimal automaton M containing the "forbidden construction" and M_q be a QFA. We show that M_q cannot recognize L.

For a word w, let $\psi_w = \psi_w^1 + \psi_w^2$, $\psi_w^1 \in E_1$, $\psi_w^2 \in E_2$.

Fix a word w after reading which M is in the state q_1. We find a word $a \in (x|y)^*$ such that after reading xa M is in the state q_2 and the norm of $\psi^2_{wxa} = V'_a(\psi^2_{wx})$ is at most some fixed $\epsilon > 0$. (Such word exists due to Lemma 2.3 and conditions 6 and 7.) We also find a word b such that $\|\psi^2_{wyb}\| \leq \epsilon$.

Because of unitarity of V'_x and V'_y on E_1 (part (i) of Lemma 2.3), there exist integers i and j such that $\|\psi^1_{w(xa)^i} - \psi^1_w\| \leq \epsilon$ and $\|\psi^1_{w(yb)^j} - \psi^1_w\| \leq \epsilon$.

Let p be the probability of M_q accepting while reading κw. Let p_1 be the probability of accepting while reading $(xa)^i$ with a starting state ψ_w, p_2 be the probability of accepting while reading $(yb)^j$ with a starting state ψ_w and p_3, p_4 be the probabilities of accepting while reading $z_1\$$ and $z_2\$$ starting at ψ^1_w.

Let us consider four words $\kappa w(xa)^i z_1\$$, $\kappa w(xa)^i z_2\$$, $\kappa w(yb)^j z_1\$$, $\kappa w(yb)^j z_2\$$.

Lemma 2.4. M_q accepts $\kappa w(xa)^i z_1\$$ with probability at least $p + p_1 + p_3 - 4\epsilon$ and at most $p + p_1 + p_3 + 4\epsilon$.

Proof. The probability of accepting while reading κw is p. After that, M_q is in the state ψ_w and reading $(xa)^i$ from ψ_w causes it to accept with probability p_1.

The remaining state is $\psi_{w(xa)^i} = \psi^1_{w(xa)^i} + \psi^2_{w(xa)^i}$. If it was ψ^1_w, the probability of accepting while reading the rest of the word ($z_1\$$) would be exactly p_3. It is not quite ψ^1_w but it is close to ψ^1_w. Namely, we have

$$\|\psi_{w(xa)^i} - \psi^1_w\| \leq \|\psi^2_{w(xa)^i}\| + \|\psi^1_{w(xa)^i} - \psi^1_w\| \leq \epsilon + \epsilon = 2\epsilon.$$

By Lemma 2.1, the probability of accepting during $z_1\$$ is between $p_3 - 4\epsilon$ and $p_3 + 4\epsilon$. □

Similarly, on the second word M_q accepts with probability between $p + p_1 + p_4 - 4\epsilon$ and $p + p_1 + p_4 + 4\epsilon$. On the third word M_q accepts with probability between $p + p_2 + p_3 - 4\epsilon$ and $p + p_2 + p_3 + 4\epsilon$. On the fourth word M_q accepts with probability $p + p_2 + p_4 - 4\epsilon$ and $p + p_2 + p_4 + 4\epsilon$.

This means that the sum of accepting probabilities of two words that belong to L (the first and the fourth) differs from the sum of accepting probabilities of two words that do not belong to L (the second and the third) by at most 16ϵ. Hence, the probability of correct answer of M_q on one of these words is at most $\frac{1}{2} + 4\epsilon$. Since such 4 words can be constructed for arbitrarily small ϵ, M_q does not recognize L. □

2.2 Necessary and Sufficient Condition

For languages whose minimal automaton does not contain the construction of Figure 3, this condition (together with Theorem 1.2) is necessary and sufficient.

Theorem 2.2. *Let U be the class of languages whose minimal automaton does not contain "two cycles in a row" (Fig. 3). A language that belongs to U can be recognized by a QFA if and only if its minimal deterministic automaton does not contain the "forbidden construction" from Theorem 1.2 and the "forbidden construction" from Theorem 2.1.*

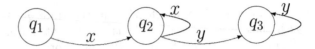

Fig. 3. Conditions of Theorem 2.2

3   Non-closure under Union

In particular, Theorem 2.1 implies that the class of languages recognized by QFAs is not closed under union.

Let L_1 be the language consisting of all words that start with <u>any</u> number of letters a and after first letter b (if there is one) there is an odd number of letters a. Its minimal automaton G_1 is shown in Fig.4.

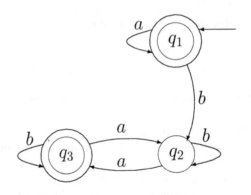

Fig. 4. Automaton G_1

This language satisfies the conditions of Theorem 2.1. (q_1, q_2 and q_3 of Theorem 2.1 are just q_1, q_2 and q_3 of G_1. x, y, z_1 and z_2 are b, aba, a and b.) Hence, it cannot be recognized by a QFA.

Consider 2 other languages L_2 and L_3 defined as follows.

L_2 consists of all words which start with an <u>even</u> number of letters a and after first letter b (if there is one) there is an odd number of letters a.

L_3 consists of all words which start with an <u>odd</u> number of letters a and after first letter b (if there is one) there is an odd number of letters a.

It is easy to see that $L_1 = L_2 \bigcup L_3$.

The minimal automata G_2 and G_3 are shown in Fig.5 and Fig.6. They do not contain any of the "forbidden constructions" of Theorem 2.2. Therefore, L_2 and L_3 can be recognized by a QFA and we get

Theorem 3.1. *There are two languages L_2 and L_3 which are recognizable by a QFA but the union of them $L_1 = L_2 \bigcup L_3$ is not recognizable by a QFA.*

Corollary 3.1. *The class of languages recognizable by a QFA is not closed under union.*

This answers a question of Brodsky and Pippenger [BP 99].

As $L_2 \bigcap L_3 = \emptyset$ then also $L_1 = L_2 \Delta L_3$. So the class of languages recognizable by QFA is not closed under symmetric difference. From this and from the fact that this class is closed under complement, it follows:

Corollary 3.2. *The class of languages recognizable by a QFA is not closed under any binary boolean operation where both arguments are significant.*

Instead of using the general construction of Theorem 2.2, we can also use a construction specific to languages L_2 and L_3. This gives simpler QFAs and achieves a better probability of correct answer. (Theorem 2.2 gives QFAs for L_2 and L_3 with the probability of correct answer $3/5$. Our construction below achieves the probability of correct answer $2/3$.)

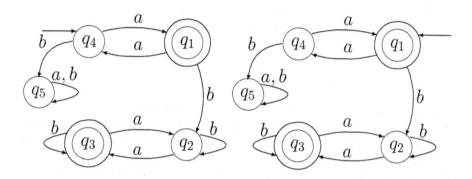

Fig. 5. Automaton G_2 **Fig. 6.** Automaton G_3

Theorem 3.2. *There are two languages L_2 and L_3 which are recognizable by a QFA with probability $\frac{2}{3}$ but the union of them $L_1 = L_2 \bigcup L_3$ is not recognizable with a QFA (with any probability $1/2 + \epsilon$, $\epsilon > 0$).*

This is the best possible, as shown by the following theorem.

Theorem 3.3. *If 2 languages L_1 and L_2 are recognizable by a QFA with probabilities p_1 and p_2 and $\frac{1}{p_1} + \frac{1}{p_2} < 3$ then $L = L_1 \bigcup L_2$ is also recognizable by QFA with probability $\frac{2p_1p_2}{p_1+p_2+p_1p_2}$.*

Corollary 3.3. *If 2 languages L_1 and L_2 are recognizable by a QFA with probabilities p_1 and p_2 and $p_1 > 2/3$ and $p_2 > 2/3$, then $L = L_1 \bigcup L_2$ is recognizable by QFA with probability $p_3 > 1/2$.*

4   More "Forbidden" Constructions

If we allow the "two cycles in a row" construction, Theorem 2.2 is not longer true. More and more complicated "forbidden fragments" that imply non-recognizability by a QFA are possible.

Theorem 4.1. *Let L be a language and M be its minimal automaton. If M contains a fragment of the form shown in Figure 7 where $a, b, c, d, e, f, g, h, i \in \Sigma^*$ are words and $q_0,\ q_a,\ q_b,\ q_c,\ q_{ad},\ q_{ae},\ q_{bd},\ q_{bf},\ q_{ce},\ q_{cf}$ are states of M and*

1. *If M reads $x \in \{a, b, c\}$ in the state q_0, its state changes to q_x.*
2. *If M reads $x \in \{a, b, c\}$ in the state q_x, its state again becomes q_x.*
3. *If M reads any string consisting of a, b and c in the state q_x ($x \in \{a, b, c\}$), it moves to a state from which it can return to the same q_x by reading some (possibly, different) string consisting of a, b and c.*
4. *If M reads $y \in \{d, e, f\}$ in the state q_x ($x \in \{a, b, c\}$), it moves to q_{xy}.[3]*
5. *If M reads $y \in \{d, e, f\}$ in the state q_{xy}, its state again becomes q_{xy}.*
6. *If M reads any string consisting of d, e and f in the state q_{xy} it moves to a state from which it can return to the same state q_{xy} by reading some (possibly, different) string consisting of d, e and f.*
7. *Reading h in the state q_{ad}, i in the state q_{be} and g in the state q_{cf} lead to accepting states. Reading g in q_{ae}, h in q_{bf} and i in q_{cd} lead to rejecting states.*

then L is not recognizable by a QFA.

The existence of the "forbidden construction" of Theorem 4.1 does not imply the existence of any of previously shown "forbidden constructions". To show this, consider the alphabet $\Sigma = \{a, b, c, d, e, f, g, h, i\}$ and languages of the form $L_{x,y,z} = x(a|b|c)^* y(d|e|f)^* z$ where $x \in \{a, b, c\}$, $y \in \{d, e, f\}$, $z \in \{g, h, i\}$. Let L be the union of languages $L_{x,y,z}$ corresponding to black squares in Figure 8.

Theorem 4.2. *The minimal automaton of L does not contain the "forbidden constructions" of Theorems 1.2 and 2.1.*

However, one can easily see that the minimal automaton of L contains the "forbidden construction" of Theorem 4.1. (Just take q_0 to be the starting state and make a, b, ..., i of Theorem 4.1 equal to corresponding letters in the alphabet Σ.) This means that the existence of "forbidden construction" of Theorem 4.1 does not imply the existence of previous "forbidden constructions".

Theorem 4.1 can be generalized to any number of levels (cycles following one another) and any number of branchings at one level as long as every arc from one vertex to other is traversed the same number of times in paths leading to accepting states and in paths leading to rejecting states.

A general "forbidden construction" is as follows.

---

[3] Note: we do not have this constraint (and the next two constraints) for pairs $x = a, y = f$, $x = b$, $y = e$ and $x = c$, $y = d$ for which the state q_{xy} is not defined.

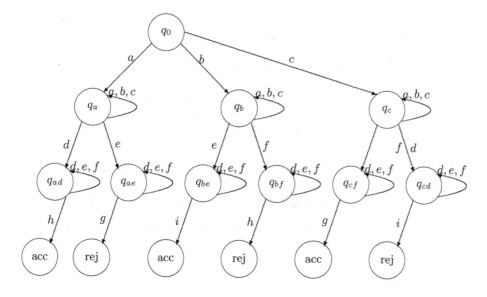

Fig. 7. Conditions of Theorem 4.1

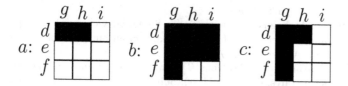

Fig. 8. The Language L

Level 1 of a construction consists of a state q_1 and some words a_{11}, a_{12}, ...

Level 2 consists of the states q_{21}, q_{22}, ... where the automaton goes if it reads one of words of Level 1 in a state in Level 1. We require that, if the automaton starts in one of states of Level 2 and reads any string consisting of words of Level 1 it can return to the same state reading some string consisting of these words. Level 2 also has some words a_{21}, a_{22}, ...

Level 3 consists of the states q_{31}, q_{32}, ... where the automaton goes if it reads one of words of Level 2 in a state in Level 2. We require that, if the automaton starts in one of states of Level 3 and reads any string consisting of words of Level 2 it can return to the same state reading some string consisting of these words. Again, Level 3 also has some words a_{31}, a_{32}, ...

Level n consists of the states q_{n1}, q_{n2}, $\ldots$ where the automaton goes if it reads one of words of Level $n - 1$ in a state in Level $n - 1$.

Let us denote all different words in this construction as $a_1, a_2, a_3, \ldots, a_m$.

For a word a_i and a level j we construct sets of states B_{ij} and D_{ij}. A state q in level $j + 1$ belongs to B_{ij} if the word a_i belongs to level j and M moves to q after reading a_i in some state in level j. A state belongs to D_{ij} if this state belongs to the Level n and it is reachable from B_{ij}.

Theorem 4.3. *Assume that the minimal automaton M of a language L contains the "forbidden construction" of the general form described above and, in this construction, for each D_{ij} the number of accepting states is equal to the number of rejecting states. Then, L cannot be recognized by a QFA.*

Theorems 2.1 and 4.1 are special cases of this theorem (with 3 and 4 levels, respectively).

References

[ABFK 99] A. Ambainis, R. Bonner, R. Freivalds, A. Ķikusts. Probabilities to accept languages by quantum finite automata. Proc. COCOON'99, *Lecture Notes in Computer Science*, 1627:174-183. Also `quant-ph/9904066`[4].

[AF 98] A. Ambainis, R. Freivalds. 1-way quantum finite automata: strengths, weaknesses and generalizations. *Proc. FOCS'98*, pp. 332–341. Also `quant-ph/9802062`.

[ANTV 98] A. Ambainis, A. Nayak, A. Ta-Shma, U. Vazirani. Dense quantum coding and a lower bound for 1-way quantum automata. *Proc. STOC'99*, pp. 376-383. Also `quant-ph/9804043`.

[AW 99] A. Ambainis, J. Watrous. Two-way finite automata with quantum and classical states. cs.CC/9911009. Submitted to *Theoretical Computer Science*.

[BV 97] E. Bernstein, U. Vazirani, Quantum complexity theory. *SIAM Journal on Computing,* 26:1411-1473, 1997.

[BP 99] A. Brodsky, N. Pippenger. Characterizations of 1-way quantum finite automata. `quant-ph/9903014`.

[G 00] J. Gruska. Descriptional complexity issues in quantum computing. *Journal of Automata, Languages and Combinatorics,* 5:191-218, 2000.

[KS 76] J. Kemeny, J. Laurie Snell. *Finite Markov Chains.* Springer-Verlag, 1976.

[K 98] A. Ķikusts. A small 1-way quantum finite automaton. `quant-ph/9810065`.

[KW 97] A. Kondacs, J. Watrous. On the power of quantum finite state automata. *Proc. FOCS'97*, pp. 66–75.

[MT 69] A. Meyer, C. Thompson. Remarks on algebraic decomposition of automata. *Mathematical Systems Theory*, 3:110–118, 1969.

[MC 97] C. Moore, J. Crutchfield. Quantum automata and quantum grammars. *Theoretical Computer Science*, 237:275–306, 2000. Also `quant-ph/9707031`.

[N 99] A. Nayak. Optimal lower bounds for quantum automata and random access codes. *Proc. FOCS'99*, pp. 369-376. Also `quant-ph/9904093`.

---

[4] quant-ph preprints are available at `http://www.arxiv.org/abs/quant-ph/preprint-number`

Star-Free Open Languages and Aperiodic Loops*

Martin Beaudry[1], François Lemieux[2], and Denis Thérien[3]

[1] Département de mathématiques et d'informatique
Université de Sherbrooke, Sherbrooke (Qc) Canada, J1K 2R1
beaudry@dmi.usherb.ca
[2] Département d'informatique et de mathématique
Université du Québec à Chicoutimi
Chicoutimi (Qc), Canada, G7H 2B1
flemieux@uqac.uquebec.ca
[3] School of Computer Science, McGill University
3480 rue University, Montréal (Qc), Canada, H3A 2A7
denis@cs.mcgill.ca

Abstract. It is known that recognition of regular languages by finite monoids can be generalized to context-free languages and finite groupoids, which are finite sets closed under a binary operation. A loop is a groupoid with a neutral element and in which each element has a left and a right inverse. It has been shown that finite loops recognize exactly those regular languages that are open in the group topology. In this paper, we study the class of aperiodic loops, which are those loops that contain no nontrivial group. We show that this class is stable under various definitions, and we prove some closure properties. We also prove that aperiodic loops recognize only star-free open languages and give some examples. Finally, we show that the wreath product principle can be applied to groupoids, and we use it to prove a decomposition theorem for recognizers of regular open languages.

1 Introduction

A *monoid* M is a set closed under a binary associative operation and that contains a two-sided identity element. The *free monoid* over an alphabet A is denoted by A^* and is defined as the set of all finite sequences of letters in A, with concatenation being the operation and the empty sequence ϵ playing the role of the identity.

The cornerstone of the algebraic theory of machines is the observation that any finite monoid can be seen as a finite state machine and that recognition of regular languages reduces to multiplication in a monoid. More formally, let $L \subseteq A^*$ be a language, let M be a monoid, and let $\phi : A^* \to M$ be a morphism. We say that M recognizes L if there exists $F \subseteq M$ such that $L = \phi^{-1}(F)$. Kleene's Theorem can then be stated as follows: a language is regular if and only if it is recognized by a finite monoid (e.g. see [16,23]).

---

* Work supported by NSERC (Canada) and FCAR (Québec). The last author is also supported by the von Humboldt foundation

A. Ferreira and H. Reichel (Eds.): STACS 2001, LNCS 2010, pp. 87–98, 2001.

Using this definition, it becomes possible to classify regular languages according to the algebraic properties of the monoids that recognize them. Two famous examples are the following.

A language $L \subseteq A^*$ is *piecewise-testable* if it is in the Boolean closure of languages of the form $A^* a_1 A^* \cdots A^* a_n A^*$ where $n \geq 0$ and $a_i \in A$, for all $0 \leq i \leq n$. A monoid is J-trivial iff any two distinct elements generate distinct two-sided ideals. A theorem of Simon ([28]) says that a language is piecewise-testable if and only if it is recognized by a J-trivial monoid.

A language is *star-free* if it is in the closure of $\{\{a\} : a \in A\} \cup \{\epsilon\}$ under Boolean operations and concatenation. A monoid M is *aperiodic* if there exists an integer n such that for every $a \in M$ we have $a^n = a^{n+1}$. Then, a deep result due to Schützenberger ([27]) states that star-free languages are precisely those languages that can be recognized by an aperiodic monoid.

This algebraic approach can also be used in the context of parallel complexity. By replacing homomorphisms by *polynomial-length programs*, it is possible to characterize algebraically well-known classes of Boolean circuits such as NC^1, AC^0 and ACC^0. Important open questions about the computing power of these models of parallel computation can be phrased in purely algebraic terms (see [4,5]).

Hence, the study of regular languages has become a rich theory with many deep results and applications, and it remains an active field that continues to challenge researchers. This makes more striking the observation that no such theory exists for context-free languages. Nevertheless, this topic has been the subject of recent investigations (e.g. [18,21,10,13,19,7,8,20]) that we briefly describe here.

A *groupoid* G is a set with a binary operation that can be non-associative. All groupoids considered in this paper are finite. Groupoids can be used as language recognizers as follows. For any $w \in G^*$, denote with $G(w)$ the set of all elements $g \in G$ such that w can be evaluated to g using some parenthesization. Let $L \subseteq A^*$ be a language, let G be a groupoid, and let $\phi : A^* \to G^*$ be a morphism induced by a function $\phi : A \to G$. We say that G recognizes L if there exists a subset $F \subseteq G$ such that for any $w \in A^*$ we have that $w \in L$ if and only if $G(\phi(w)) \cap F \neq \emptyset$. When G is associative, this definition corresponds to the recognition by monoid defined above. Our interest in groupoids comes from the fact that a language is context-free if and only if it is recognized by a finite groupoid (e.g. see [18,10]).

In the absence of a general theory of groupoids, a classification of the context-free languages based on the algebraic properties of the groupoids that recognize them is still a major research project. This approach could also have implications in complexity theory since context-free languages are related to the class SAC^1 of polynomial size, logarithmic depth circuits constructed with NOT gates, binary AND gates and OR gates of unbounded fan-in. Hence, a better understanding of the algebraic nature of these languages could be an important tool in the study of small circuit classes.

A subclass of groupoids that has been studied intensively (see [2,3,11,12,14], [22]) is the family of finite loops. A *loop* is a groupoid that possesses a unique identity element and that satisfies the *cancellation law*: every element has a right and a left inverse (not necessarily identical). We observe that the multiplication table of a finite loop is such that every row and column is a permutation. Hence, a group is an associative loop but not all loops are associative (the smallest example is the loop B_5 of Section 4).

In [13], it has been shown that any language recognized by a finite loop is regular. Despite this *lack of power* of finite loops, their investigation is essential in order to better understand the non-associativity of general groupoids and the languages they recognize. This result has been refined in [9], where an exact characterization of the languages recognized by finite loops is given. A language is recognized by a loop if and only if it is an open regular language in the group topology (see [25]). A simpler way to express this result is that a language $L \subseteq A^*$ is recognized by a loop if and only if it is a finite union of languages of the form $L_0 a_1 L_1 \cdots L_{n-1} a_n L_n$, where $n \geq 0$, $a_i \in A$, and each L_i is a group language.

One of the main tool used in [9] is an operation, called the *wreath product*, that takes a loop and a group to get another loop. The wreath product plays an important role in the algebraic theory of machines, as it is the algebraic formalization of the notion of *series connection*. A fundamental decomposition result states that any monoid can be decomposed as a wreath product of components, each of which is either a group or an aperiodic monoid (see [16]). In this paper we will prove a similar decomposition theorem for loops.

Our presentation is divided as follows. In Section 2, we define aperiodic and group-free loops, and we prove that these two classes of loops are equivalent. In Section 3, we show that aperiodic loops recognize only star-free languages. Some closure properties of aperiodic loops are demonstrated in Section 4. In Section 5, we introduce the notion of *algebraic transduction* which is a sequential function performed by a pushdown machine. We then relate this kind of transduction with the wreath product of loops, generalizing the *wreath product principle* of monoids (e.g. see [23]). Using this relation, we prove in Section 6 that aperiodic loops can recognize languages of the form $A^* a_1 A^* \cdots A^* a_n A^*$. As a consequence, any regular open language is the finite union of languages S_i that can be recognized by the wreath product of a loop B_i and a group G_i, where B_i is a loop that recognizes a language of the form $A^* a_1 A^* \cdots A^* a_n A^*$.

2 Aperiodic and Group-Free Loops

A loop B is said to *divide* another loop L if B is a morphic image of a subloop of L. A loop B is said to be *group-free* if there exists no nontrivial group that divides B. Given a word $w \in B^*$, we use $B(w)$ to denote the set of all elements that can be obtained from the evaluation of w in B using any parenthesization. When there is no ambiguity, we simply write w instead of $B(w)$. As an example, if $b \in B$, we write $b \in w$ (or $w \ni b$) whenever there exists a way to evaluate w in order to get b. When B is associative, then the set w is a singleton and we simply write $w = b$.

An element b of a loop B is said to be *aperiodic* if there exists an integer n such that for any $m \geq n$, the set b^m is equal to b^n. A loop is aperiodic if all its elements are aperiodic. We observe that for any element g of any loop B there exists $n > 0$ such that g^n contains the identity. It follows that if g is aperiodic then for a sufficiently large m we have $g^m = \langle g \rangle$ where $\langle g \rangle$ is the subloop of B generated by g.

The above definitions can be extended to any groupoid G in a straightforward way. In particular, when G is a monoid, it is known that G is group-free if and only if it is aperiodic [27]. We will prove in this section that this is also true for loops.

Given any loop B we define the set $I(B) = \{g \in B : \exists n_0 > 0, \forall n \geq n_0, e \in g^n\}$ where e is the identity of B. The following lemma shows that $I(B)$ corresponds precisely to the set of all aperiodic elements of B.

Lemma 1. *Let g be an element of a loop B. If $g \in I(B)$ then there exists n_1 such that for all $n \geq n_1$ we have $g^n = \langle g \rangle$.*

Proof. Let e be the identity of B. Since $g \in I(B)$, there exists n_0 such that $e \in g^n$ for all $n \geq n_0$. Let k_0 be such that for any $b \in \langle g \rangle$ there exists $k < k_0$ such that $b \in g^k$. Hence, for all $b \in \langle g \rangle$ and for all $m \geq n_0 + k_0$ we have $b \in g^m$.

Proposition 1. *If a loop B is aperiodic then it is group-free.*

Proof. Let $B = I(B)$ and suppose that a nontrivial group G divides B. Then, there exists a subloop $S \subseteq B$ and a surjective morphism $\phi : S \to G$. Since G is nontrivial, there must exist $g \in G$ such that $g^n \neq g^{n+1}$ for all $n > 0$. This means that $\phi^{-1}(g^n) \cap \phi^{-1}(g^{n+1}) = \emptyset$ for all $n > 0$. Hence $\phi^{-1}(g) \cap I(B) = \emptyset$ contradicting the aperiodicity of B.

To prove the other direction, we need the following classical lemma from number theory.

Lemma 2. *Let p and q be two coprime integers. There exists an integer n_0 such that for all $n > n_0$ there exist $a, b > 1$ such that $n = ap + bq$.*

Lemma 3. *Let $b \in B$ et let $n, m \geq 1$ be two coprime integers such that $e \in b^n$ and $e \in b^m$, where e is the identity of B. Then, there exists k_0 such that for all $k \geq k_0$ we have $e \in b^k$.*

Proof. Using Lemma 2 we have $b^k = b^{an+bm} \ni e$ when k is large enough.

Lemma 4. *Let B be a loop with identity e and let $g \in B$. Suppose there exists an integer $n \geq 2$ such that for any integer m, if $e \in g^m$ then $n \mid m$. Then, B is divided by a nontrivial group.*

Proof. Define for each $0 \leq i < n$ the set $S_i = \{b \in B : \exists c > 0, b \in g^{cn+i}\}$. Hence, S_0 is a subloop of B since if $a \in g^{c_1 n}$ and $b \in g^{c_2 n}$ then $ab \in g^{(c_1+c_2)n}$.

Also, if $0 \leq i < j \leq n$ then $S_i \cap S_j = \emptyset$. Indeed, if $a \in S_i \cap S_j$ then there exists $b \in B$ such that $b \in S_0 \cap S_k$ where $k = i + n - j$. Let $d > 0$ be such that $e \in bg^d$. Since $b \in S_0$, n must divide d and since $b \in S_k$ we must have $e \in S_{k+d}$. But, this contradicts the condition on B given in the statement of the lemma.

Hence, S_0 is a normal subloop of B and $B/S_0 \simeq Z_n$ (see [22] for a discussion on normal subloops).

Proposition 2. *If B is a group-free loop then it is also aperiodic.*

Proof. Lemmas 3 and 4 imply that $B = I(B)$. The conclusion follows from Lemma 1.

We have thus proved

Theorem 1. *A loop is group-free if and only if it is aperiodic.*

3 Aperiodic Loops Recognize only Star-Free Languages

Lemma 5 (Schützenberger). *A language is star-free if and only if it can be recognized by an aperiodic monoid.*

Lemma 6. *A regular language $L \subseteq A^*$ is star-free is and only if there exists $n_0 > 0$ such that for all $x, y, w \in A^*$ and for all $n \geq n_0$ we have $xw^n y \in L$ iff $xw^{n+1}y \in L$.*

Lemma 7. *Let B be an aperiodic loop. There exists $p_0 > 0$ such that for all $x, y, w \in B^*$ and all $n \geq p_0$ we have $xw^n y = xw^{n+1}y$.*

Proof. Let p be such that $|xw^p y|$ is maximal, where $|S|$ denote the cardinality of the set S. Let $n_0 > 0$ such that $e \in w^n$ for all $n \geq n_0$, where e is the identity of B. Hence, $xw^p y \subseteq xw^{p+n}y$ for all $n \geq n_0$ and since $|xw^p y|$ is maximal, $xw^p y = xw^{p+n}y = xw^{p+n+1}y$ for all $n \geq n_0$. Thus, it is sufficient to take $p_0 = p + n_0$.

As a consequence of the above two lemmas we have

Proposition 3. *Aperiodic loops can recognize only star-free languages.*

We have shown the following theorem.

Theorem 2. *Let B be a finite loop. The following conditions are equivalent.*

1. *B is group-free*
2. *B is aperiodic*
3. *$B = I(B)$*
4. *B recognizes only star-free languages*
5. *There exists $n_0 > 0$ such that $ab^n c = ab^{n+1}c$ for all $a, b, c \in B$ and all $n \geq n_0$.*

4   Other Properties of Aperiodic Loops

We begin this section with two examples of aperiodic (group-free) loops. In the sequel we will refer to these loops as B_5 and B_7.

$$
\begin{array}{c}
\begin{array}{ccccc}
0 & 1 & 2 & 3 & 4 \\
1 & 2 & 0 & 4 & 3 \\
2 & 3 & 4 & 0 & 1 \\
3 & 4 & 1 & 2 & 0 \\
4 & 0 & 3 & 1 & 2 \\
\end{array}
\qquad
\begin{array}{ccccccc}
0 & 1 & 2 & 3 & 4 & 5 & 6 \\
1 & 2 & 0 & 4 & 3 & 6 & 5 \\
2 & 0 & 3 & 5 & 6 & 4 & 1 \\
3 & 4 & 5 & 6 & 1 & 2 & 0 \\
4 & 3 & 6 & 1 & 5 & 0 & 2 \\
5 & 6 & 4 & 2 & 0 & 1 & 3 \\
6 & 5 & 1 & 0 & 2 & 3 & 4 \\
\end{array}
\end{array}
$$

The reader will verify (hopefully using some software!) that all elements in B_5 and B_7 are aperiodic. Loop B_7 is commutative but no aperiodic loop of even order can be commutative. Actually, we can prove something stronger.

Let a be an element of a loop B that is different from the identity e. The left inverse a^λ and the right inverse a^ρ of a are defined as the unique solution to the equation $a^\lambda a = aa^\rho = e$.

Lemma 8. *If B is an aperiodic loop of even order, then there exists $a \in B$ such that $a^\lambda \neq a^\rho$. In particular, there exists no commutative aperiodic loop of even order.*

Proof. Suppose that B has even order and that $a^\lambda = a^\rho$ for all $a \in B$. Then, for each $a \in B$ there exists a unique $b \in B$ such that $ab = ba = e$. Since $|B|$ is even and $ee = e$ there must exist an element $c \in B$ which is different from e and such that $cc = e$. This implies that B is not group-free.

Lemma 9. *Let B be a loop that does not recognize the language $OR = A^*aA^*$. Then, for every $b \in B$ different from the identity e, there exists $k \geq 2$ such that $b^k = \{e\}$.*

Proof. Let $b \in B$ be different from the identity e. If $|b^i| = 1$ for all $i > 0$ then b generates a group and $b^k = e$ for some $k \geq 2$. Otherwise, let $j \geq 3$ be the smallest integer such that $|b^j| \geq 2$. If for every $i < j$ we have $b^i \neq \{e\}$, then B can recognize the language $OR = A^*aA^*$. It suffices to map the letter a to the element b and all other letters in A to the identity e. The accepting set is $B - \{e\}$.

Corollary 1. *If B is an aperiodic loop of even order then B recognizes the language $OR = A^*aA^*$.*

Proof. Suppose that B cannot recognize OR. By the above lemma, we have that for all $b \in B$ different from the identity e, there exists $k \geq 3$ such that $b^k = \{e\}$. Observe that we cannot have $k = 2$ since B is aperiodic. By the cancellation law, there exists $c \in B$ such that $b^{k-1} = \{c\}$. Thus, we must have $bc = cb = e$ contradicting Lemma 8

Indeed, loop B_7 is an example of an aperiodic loop such that $a^\lambda = a^\rho$ for all elements a. However, this loop as all aperiodic loops known by the authors can recognize the language OR (we will see in Section 6 the importance of this fact). Actually, the only languages known to be recognizable by a finite aperiodic loop are unions of

- languages of the form $A^*\{aa + bb\}A^*$,
- languages of the form $A^*a_1A^* \cdots A^*a_nA^*$,
- cofinite languages

We show in Section 6 how to recognize languages of the second form. A construction for cofinite languages is given in the full version of this paper. The following aperiodic loop can recognize languages of the first form.

$$
\begin{array}{cccccccccc}
0 & 1 & 2 & 3 & 4 & 5 & 6 & 7 & 8 & 9 \\
1 & 3 & 0 & 6 & 5 & 7 & 8 & 9 & 2 & 4 \\
2 & 0 & 4 & 5 & 6 & 9 & 7 & 8 & 3 & 1 \\
3 & 5 & 6 & 8 & 9 & 1 & 4 & 2 & 7 & 0 \\
4 & 6 & 5 & 7 & 8 & 3 & 1 & 0 & 9 & 2 \\
5 & 9 & 7 & 4 & 0 & 8 & 2 & 3 & 1 & 6 \\
6 & 8 & 9 & 2 & 7 & 4 & 3 & 1 & 0 & 5 \\
7 & 4 & 8 & 1 & 2 & 0 & 9 & 6 & 5 & 3 \\
8 & 2 & 3 & 9 & 1 & 6 & 0 & 5 & 4 & 7 \\
9 & 7 & 1 & 0 & 3 & 2 & 5 & 4 & 6 & 8 \\
\end{array}
$$

It suffices to use the morphism induced by $a \rightarrow 1$ and $b \rightarrow 2$. The accepting set is $\{3, 4, 5, 6, 7, 8, 9\}$. One can verify that all words not in the language evaluate only to $0, 1$ or 2. All words of length 7 evaluate to at least 4 distinct values (the only exceptions are words of the form $22121212\ldots$ that evaluate to only three distinct values). Finally all other words can be checked exhaustively.

The fact that languages recognized by aperiodic loops are closed under union is a direct consequence of the following proposition that is proved in the full version of this paper.

Proposition 4. *The direct product of two (aperiodic) loops is a (aperiodic) loop.*

The above result can be generalized to the wreath product. The *wreath product* of two loops is defined as in the associative case. Let S and T be two loops with identity e_S and e_T, respectively. The product $S \circ T$ is the set $S^T \times T$ together with the binary operation

$$(f_1(u), t_1)(f_2(u), t_2) = (f_1(u)f_2(t_1u), t_1t_2)$$

Proposition 5. *The wreath product of two (aperiodic) loops is a (aperiodic) loop.*

5   The Wreath Product Principle

We consider in this section a finite groupoid as an algebraic structure that can be used to recognize both word languages and tree languages. The *free groupoid* over an alphabet A is denoted with $A^{(*)}$ and defined as the set of all well parenthesized expressions over A. A *tree* over an alphabet A is an element of $A^{(*)}$. We define the function Yield : $A^{(*)} \to A^*$ as follows. For all $a \in A \cup \{\epsilon\}$ we have Yield$(a) = a$ and if $t, t_1, t_2 \in A^{(*)}$ with $t = (t_1 t_2)$ then Yield$(t) = $ Yield(t_1)Yield(t_2). We say that a word $w \in A^*$ is the *yield* of a tree $t \in A^{(*)}$ whenever $w = $ Yield(t).

We recursively define the set of *(right) combs* in $A^{(*)}$ as follows. Each element of $A \cup \{\epsilon\}$ is a comb; if $a \in A$ and c is a comb then (ac) is a comb; nothing else is a comb. Hence, a comb in $A^{(*)}$ is simply a word in A^* that is parenthesized from right to left.

One can define an (associative) operation on combs such that the product of two combs c_1 and c_2 gives another comb $c = c_1 \cdot c_2$ which is the concatenation of the yields of c_1 and c_2 parenthesized from right to left.

We define a stack automaton as a tuple $M = (A, G, F, \phi)$, where

- A is an alphabet
- G is a finite groupoid whose elements form the set of states
- $F \subseteq G$ is the set of final states
- $\phi : A \to G$ is a function

We admit only two type of transitions.

1. Read the next input character $a \in A$
   Push the current state $h \in G$
   Go to the next state $\phi(a)$
2. Pop $g \in G$
   Go to the state gh, where h is the current state.

Initially, the stack is empty. The machine accepts its input if and only if there is no more character to be read, the stack is empty, and the current state is in F.

Remark. Recognition with a stack automaton is essentially identical to recognition with a finite groupoid. In particular, the input of a stack automaton can be a word in A^* or a tree in $A^{(*)}$. In the first case, the machine is seen as a non-deterministic device while it is deterministic in the second case.

Stack automata will be used to define a special kind of transductions that we call *algebraic*. Let $M = (A, G, F, \phi)$ be a stack automaton. For each $a \in G$, let $L(a)$ be the function $L(a) : x \to ax$ and let $\mathcal{M}_L(G)$ be the monoid generated by the set $\{L(g) : g \in G\}$. An *algebraic transducer* is a tuple $T = (M, B, \tau)$ where

- $M = (A, G, F, \phi)$ is a stack automaton
- B is an alphabet
- $\tau : \mathcal{M}_L(G) \times A \to B^{(*)}$ is the output function

Such a transducer behaves according to the two following rules.

1. After each transition of type 1, the transducer writes the expression $\tau(h(x), a)$ $\in B^{(*)}$, where $x \in G^*$ is the content of the stack and $h : G^* \to \mathcal{M}_L(G)$ is the morphism induced by $h(a) = L(a)$.
2. After each transition of type 2, the transducer replaces the two last written expressions, e_1 and e_2, with the expression $(e_1 e_2)$.

We will be particularly interested by the case where

- $A \subseteq G$
- $B = \mathcal{M}_L(G) \times A$
- τ is the identity function.

We then say that T is the *natural transducer* of G over the alphabet A.

We also denote with $T : A^{(*)} \to B^{(*)}$, the function computed by the algebraic transducer T. Such a function is called *algebraic transduction*. When $A \subseteq G$, the natural transduction is defined in an evident way from the natural transducer. One easily verifies the following proposition.

Proposition 6. *Any algebraic transduction is a homomorphic image of some natural transduction.*

Let G and H be two groupoids et let A and $B = \mathcal{M}_L(G) \times A$ be two alphabets. Let $M = (A, H, F, \phi)$ be a stack automaton and $T = (M, B, \tau)$ be the natural transducer of M. Moreover, let $F \subseteq G \times H$ and $h : (B \times A)^{(*)} \to G \times H$, a morphism. Recall that we also use T to denote the natural transduction $T : A^{(*)} \to B^{(*)}$.

We define the tree language $L_{h,F} \subseteq A^{(*)}$ as the set of all trees u such that $h(T(u), u) \in F$. We can now define the class of languages $T(G, H) = \{L_{h,F} : F \in G \times H \text{ and } h : (B \times A)^{(*)} \to G \times H \text{ is a morphism}\}$.

The wreath product principle is stated with the following two theorems.

Theorem 3. *If $L \in T(G, H)$ then L is recognized by $G \circ H$.*

Theorem 4. *If $L \subseteq A^{(*)}$ is recognized by $G \circ H$ then L is in the positive Boolean closure of $T(G, H)$ (more precisely, L is a finite union of finite intersections of languages in $T(G, H)$).*

We close this section with an observation that will be useful in the next section. Let $F : A^{(*)} \to B^{(*)}$ be an algebraic function and let $t_1, t_2, t_3 \in A^{(*)}$. Since algebraic transductions preserve the parenthesization, the transduction of $(t_1 t_2)$ is $(s_1 s_2)$ for some $s_1, s_2 \in B^{(*)}$. Similarly, $F((t_1 t_3)) = (s_3 s_4)$ for some $s_3, s_4 \in B^{(*)}$. Moreover, we must have $s_1 = s_3$ since an algebraic transducer would output the transduction of t_1 before reading the rest of the input. In this sense, algebraic transductions are sequential functions.

The above observation remains true in a slightly more complex situation.

Lemma 10. *Let $F : A^{(*)} \to B^{(*)}$ be an algebraic transduction. Let $t \in A^{(*)}$ be a tree and $c = c_1 \cdot c_2$ be a comb in $A^{(*)}$. Let $F((tc)) = (s_1 d)$ and $F(((tc_1)c_2)) = ((s_2 e)g)$ for some $s_1, s_2, d, e, g \in B^{(*)}$. Then $s_1 = s_2$ and d, e, g are combs. Moreover, there exists two combs d_1, d_2 in $B^{(*)}$ such that $d = d_1 \cdot d_2$ and such that the yields of e and d_1 are identical.*

Proof. Given a tree $t \in A^{(*)}$ a stack automaton behaves as follows. It ignores open parentheses, it uses transitions of type 1 on elements of A and it uses transitions of type 2 on close parentheses. On another hand, a natural transducer writes a symbol in B after a transition of type 1 and modifies the parenthesisation of the current output after a transition of type 2. Let k be the number of leaves in (tc_1).

The above description of the behaviour of a natural transducer implies that after doing k transitions of type 1, there is no difference between the output produced on inputs (tc) and $((tc_1)c_2)$. This is because in both cases, up to that point, the transducer has seen the same input t (recall that open parentheses are ignored). The result follows from the fact that future transitions of the transducer cannot modified the yield of this part of the output.

6   Application of the Wreath Product

We show in this section that if $L \subseteq A^*$ is a language recognized by a finite loop P and if $a \in A$ then the language LaA^* is recognized by $B \circ P$, where B is any group-free loop that possesses a certain property. As a consequence, we have that languages of the form $A^* a_1 A^* \cdots A^* a_n A^*$ are recognized by a group-free loop. We conclude with a decomposition theorem for finite loops.

Let L be a loop and $g \in L$. Denote with $g^{(n)}$ the comb with n leaves in $\{g\}^{(*)}$. Let $\langle g \rangle$ be the subloop generated by g and let $C(g)$ be the subset of $\langle g \rangle$ defined by $C(g) = \{g^{(n)} : n > 0\}$. Let $\mathcal{C}$ be the class of loops for which there exist $x, g \in L$ such that $g \in C(x)$ but $g^\rho \neq C(x)$ (recall that g^ρ is the right inverse of g).

As we will see, loops in $\mathcal{C}$ have some nice properties. In particular, we have the following lemma.

Lemma 11. *Any loop in $\mathcal{C}$ can recognize $A^* a A^*$*

Proof. Let B be a loop that cannot recognize $A^* a A^*$. Hence, by Lemma 9, for all $g \in B$, there exists $k > 0$ such that $g^k = \{e\}$, where e is the identity of B. This means that $C(g) = \{g^{(n)} : 0 < n \le k\}$. And since $gg^i = g^i g$ for all $i < k$, we have that the right (and left) inverse of any element in $C(g)$ is also in $C(g)$. Hence $B \notin \mathcal{C}$.

Let P be a loop. For each $a \in P$, define the permutations $L(a) : x \mapsto xa$ and $R(a) : \mapsto xa$. We define the *multiplication group* $\mathcal{M}(P)$ as the group generated by all $L(a)$ and $R(a)$, $a \in P$.

Proposition 7. *Let $L \subseteq A^*$ be a language recognized by a loop P and let G be a loop in $\mathcal{C}$. Then, LaA^* is recognized by $G \circ P$.*

Proof. Let $1 \in G$ be an element such that $1^\rho \notin C(1)$ and denote with 0 the identity of G. Assume that the accepting set of P is F. Let $B = \mathcal{M}(P) \times A$ and let $h : B^{(*)} \to G$ be the morphism induced by $h(D, x) = 1$ whenever $D(1) \in F$ and $x = a$, and $h(D, x) = 0$ otherwise. Let T be the natural transduction of P.

By Proposition 3, it suffices to show that a word $w \in A^*$ is in L if and only if there exists a tree $t \in A^{(*)}$ such that w is the yield of t and such that $h(T(t))$ evaluates in G to an element different from 0.

Let $w \in LaA^*$ and let $w = uav$ where $u \in L$ such that u is as small as possible (this means that u is not in LaA^*). We can parenthesize u into a tree $s \in A^{(*)}$ that evaluates to an element in F, and we can parenthesize v into a comb $c \in A^{(*)}$ to get a tree $t = (s(ac)) \in A^{(*)}$ with yield w.

Let $q = (z(xy)) \in B^{(*)}$ be the transduction of t. It is clear that $x = [L(u), a] \in B$ is such that $h(x) = 1$. Moreover, for all leaves g of z we have $h(g) = 0$, since u is not in LaA^*. If $h(q) \neq 0$ then q is accepted by G and everything is fine.

Consider the case where $h(q) = 0$. This can happen only if the number k of leaves in the comb $h(xy)$ that are labeled with 1 is congruent to zero modulo the cardinality of $C(1)$. Since $h(x) = 1$, we have that $k > 0$. Let $d \in C(1)$ be such that $d^p \notin C(g)$. There exists $i < k$ such that $g^{(i)} = d$ and we can write $xy = \alpha \cdot \beta$, where $h(\alpha) = d$.

By Lemma 10, we can parenthesize w as $w' = ((sa')c')$ such that the transduction of w' is $((zx')y')$, where $x' = \alpha$ and y' is some comb in $B^{(*)}$. This means that $h(zx')y'$ is different from 0 since $h(xy') = d$ and $d^p \notin C(g)$.

Loop B_5 of Section 4 is an example of a loop in $\mathcal{C}$. Not all aperiodic loops are in $\mathcal{C}$, however. This is the case of loop B_7 of Section 4. The question as to whether any aperiodic loop can be used in Proposition 7 is open.

Given a loop B and an integer $n > 0$, we define the loop B^n recursively on n as follows. $B^1 = B$ and, for $i > 1$, $B^i = B \circ B^{i-1}$. By Proposition 5, if B is group-free then so is B^n.

Corollary 2. *Let A be a finite alphabet. For any loop $B \in \mathcal{C}$, for any $n > 0$ and for any $a_1, \ldots, a_n \in A$, the language $A^* a_1 A^* \cdots A^* a_n A^*$ is recognized by B^n.*

Theorem 5. *A regular language is open if and only if it is recognized by a finite direct product of loops of the form $B^n \circ G$, where B is a group-free loop and G is a group.*

Proof. In [9], it is proved that a regular language L is open if and only if it is the finite union of languages that can be recognized by a loop of the form $P \circ G$ where G is a group and P any loop that can recognize the language $A^* a_1 A^* \cdots A^* a_n A^*$ for a large enough n. The theorem follows from Corollary 2.

References

1. M Ajtai, Σ_1^1-formulae on finite structures, Annals of Pure and Applied Logic, **24** pp.1-48, 1983.
2. A.A. Albert, Quasigroups. I, *Trans. Amer. Math. Soc.*, Vol. 54 (1943) pp.507-519.
3. A.A. Albert, Quasigroups. II, *Trans. Amer. Math. Soc.*, Vol. 55 (1944) pp.401-419.
4. D.A. Barrington, *Bounded-Width Polynomial-Size Branching Programs Recognize Exactly those Languages in* NC^1, *JCSS* **38**, 1 (1989), pp. 150-164.
5. D. Barrington and D. Thérien, *Finite Monoids and the Fine Structure of* NC^1, *JACM* **35**, 4 (1988), pp. 941-952.

6. P.W. Beam, S.A. Cook, and H.J. Hoover, *Log Depth Circuits for Division and Related Problems*, in *Proc. of the 25th IEEE Symp. on the Foundations of Computer Science* (1984), pp. 1-6.

7. M. Beaudry, *Languages recognized by finite aperiodic groupoids* , Theoretical Computer Science, vol. 209, 1998, pp. 299-317.

8. M. Beaudry, *Finite idempotent groupoids and regular languages* , Theoretical Informatics and Applications, vol. 32, 1998, pp. 127-140.

9. M. Beaudry, F. Lemieux, and D. Thérien, *Finite loops recognize exactly the regular open languages*, in Proc. of the 24th International Colloquium on Automata, Languages and Programming, Springer Lecture Notes in Comp. Sci. 1256 (1997), pp. 110-120.

10. F. Bédard, F. Lemieux and P.McKenzie, Extensions to Barrington's M-program model, *TCS* **107** (1993), pp. 31-61.

11. R.H. Bruck, Contributions to the Theory of Loops, *Trans. AMS*, (60) 1946 pp.245-354.

12. R.H. Bruck, *A Survey of Binary Systems*, Springer-Verlag, 1966.

13. H. Caussinus and F. Lemieux, *The Complexity of Computing over Quasigroups*, In the Proceedings of the 14th annual FST&TCS Conference, LNCS 1256, Springer-Verlag 1994, pp.36-47.

14. O. Chein, H.O. Pfugfelder, and J.D.H. Smith, *Quasigroups and Loops: Theory and Applications*, Helderman Verlag Berlin, 1990.

15. S.A. Cook, *A Taxonomy of Problems with Fast Parallel Algorithms*, Information and Computation **64** (1985), pp. 2-22.

16. S. Eilenberg, *Automata, Languages and Machines*, Academic Press, Vol. B, (1976).

17. M.L. Furst, J.B. Saxe and M. Sipser, *Parity, Circuits, and the Polynomial-Time Hierarchy*, Proc. of the 22nd IEEE Symp. on the Foundations of Computer Science (1981), pp. 260-270. Journal version Math. Systems Theory **17** (1984), pp. 13-27.

18. F. Lemieux, *Complexité, langages hors-contextes et structures algebriques non-associatives*, Masters Thesis, Université de Montréal, 1990.

19. F. Lemieux, *Finite Groupoids and their Applications to Computational Complexity*, Ph.D. Thesis, McGill University, May 1996.

20. C. Moore, F. Lemieux, D. Thérien, J. Berman, and A. Drisko, *Circuits and Expressions with Nonassociative Gates*, JCSS **60** (2000) pp.368-394.

21. A. Muscholl, Characterizations of LOG, LOGDCFL and NP based on groupoid programs, Manuscript, 1992.

22. H.O. Pfugfelder, *Quasigroups and Loops: Introduction*, Heldermann Verlag, 1990.

23. J.-E. Pin, *Variétés de langages formels*, Masson (1984). Also *Varieties of Formal Languages*, Plenum Press, New York, 1986.

24. J.-E. Pin, *On Reversible Automata*, in Proceedings of the first LATIN Conference, Sao-Paulo, Notes in Computer Science 583, Springer Verlag, 1992, 401-416

25. J.-E. Pin, *Polynomial closure of group languages and open set of the Hall topology*, Theoretical Computer Science 169 (1996) 185-200

26. J.E. Savage, *The complexity of computing*, Wiley, 1976.

27. M.P. Schützenberger, On Finite Monoids having only trivial subgroups, *Information and Control* **8** (1965), pp. 190-194.

28. I. Simon, *Piecewise testable Events*, Proc. 2nd GI Conf., LNCS 33, Springer, pp. 214-222, 1975.

29. H. Venkateswaran, *Circuit definitions of nondeterministic complexity classes*, Proceedings of the 8th annual FST&TCS Conference, 1988.

A $\frac{5}{2}n^2$–Lower Bound for the

Multiplicative Complexity of

$n \times n$–Matrix Multiplication

Markus Bläser

Institut für Theoretische Informatik, Med. Universität zu Lübeck
Wallstr. 40, 23560 Lübeck, Germany
blaeser@tcs.mu-luebeck.de

Abstract. We prove a lower bound of $\frac{5}{2}n^2 - 3n$ for the multiplicative complexity of $n \times n$–matrix multiplication over arbitrary fields. More general, we show that for any finite dimensional semisimple algebra A with unity, the multiplicative complexity of the multiplication in A is bounded from below by $\frac{5}{2} \dim A - 3(n_1 + \cdots + n_t)$ if the decomposition of $A \cong A_1 \times \cdots \times A_t$ into simple algebras $A_\tau \cong D_\tau^{n_\tau \times n_\tau}$ contains only noncommutative factors, that is, the division algebra D_τ is noncommutative or $n_\tau \geq 2$.

1 Introduction

One of the leading problems in algebraic complexity theory is the determination of good (lower as well as upper) bounds for the multiplicative complexity of $n \times n$–matrix multiplication. Loosely speaking, the problem is the following: given $n \times n$–matrices $X = (X_{i,j})$ and $Y = (Y_{i,j})$ with indeterminates $X_{i,j}$ and $Y_{i,j}$ over some ground field k, how many *essential* multiplications and divisions are needed to compute the entries of the product XY? Here, "essential" means that additions, subtractions, and scalar multiplications are free of costs. According to Strassen [20], we may reformulate the problem over infinite fields as follows: the multiplicative complexity of $n \times n$–matrix multiplication is the smallest number ℓ of products

$$p_\lambda = u_\lambda(X_{i,j}, Y_{i,j}) \cdot v_\lambda(X_{i,j}, Y_{i,j})$$

with linear forms u_λ and v_λ in the $X_{i,j}$ and $Y_{i,j}$ such that each entry of XY is contained in the linear span of $p_1, \ldots, p_\ell$, i.e.,

$$\sum_{\nu=1}^{n} X_{i,\nu} Y_{\nu,j} \in \lim\{p_1, \ldots, p_\ell\} \qquad \text{for } 1 \leq i, j \leq n.$$

In other words, we may restrict our attention to computations that contain only "normalized" multiplications and no divisions. (Since we are considering lower bounds in this work, the above restriction to infinite fields does not impose any

A. Ferreira and H. Reichel (Eds.): STACS 2001, LNCS 2010, pp. 99–109, 2001.

problems: lower bounds over a field K also hold over any subfield $k \subset K$.) For a modern and comprehensive introduction to algebraic complexity theory, we refer to [9].

A related quantity is the bilinear complexity (or rank) of $n \times n$–matrix multiplication. Here the products $p_\lambda = u_\lambda(X_{i,j}) \cdot v_\lambda(Y_{i,j})$ are bilinear products, that is, products of linear forms u_λ in the $X_{i,j}$ and linear forms v_λ in the $Y_{i,j}$. (Note that the entries of XY are bilinear forms.) The concept of bilinear complexity has been utilized with great success in the design of asymptotically fast matrix multiplication algorithms, see for example [19,2,18,22,10]. Obviously, the multiplicative complexity is a lower bound for the bilinear complexity and it is not hard to see that twice the multiplicative complexity is an upper bound for the bilinear complexity (see e.g. [9, Eq. 14.8]). Therefore, we usually want to have upper bounds for the bilinear complexity and lower bounds for the multiplicative complexity. While the difference between multiplicative and bilinear complexity seems to be minor at a first glance, it is much harder to cope with the multiplicative complexity when dealing with lower bounds. One reason among others is the fact that the bilinear complexity of a tensor of a bilinear map (see below for a definition) is invariant under permutations whereas the multiplicative complexity might not, see also [9, Chap. 14.2] for a further discussion.

1.1   Model of Computation

Of course, we can define multiplicative complexity not only for the multiplication of $n \times n$–matrices (which is a bilinear map $k^{n \times n} \times k^{n \times n} \rightarrow k^{n \times n}$) but also for arbitrary bilinear maps. When considering lower bounds, it is often more convenient to use a coordinate-free definition of multiplicative complexity, see e.g. [9, Chap. 14.1]. In the following, if V is a vector space, let V^* denote the dual space of V, i.e., the vector space of all linear forms on V.

Definition 1. *Let k be a field, U, V, and W finite dimensional vector spaces over k, and $\phi : U \times V \rightarrow W$ a bilinear map.*

1. *A sequence $\beta = (f_1, g_1, w_1, \ldots, f_\ell, g_\ell, w_\ell)$ with $f_\lambda, g_\lambda \in (U \times V)^*$ and $w_\lambda \in W$ is called a quadratic computation for ϕ over k of length ℓ if*

$$\phi(u, v) = \sum_{\lambda=1}^{\ell} f_\lambda(u, v) g_\lambda(u, v) w_\lambda \qquad \text{for all } u \in U, v \in V.$$

2. *The length of a shortest quadratic computation for ϕ is called the multiplicative complexity of ϕ and is denoted by $C(\phi)$.*
3. *If A is a finite dimensional associative k-algebra with unity, then the multiplicative complexity of A is defined as the multiplicative complexity of the multiplication map of A, which is a bilinear map $A \times A \rightarrow A$, and is denoted by $C(A)$.*

If we want to emphasize the underlying ground field k, we will sometimes write $C_k(\phi)$ and $C_k(A)$ instead of $C(\phi)$ and $C(A)$, respectively. Using the language from above, the multiplicative complexity of $n \times n$–matrix multiplication is denoted by $C(k^{n \times n})$.

If we require that $f_\lambda \in U^*$ and $g_\lambda \in V^*$ in the above Definition 1, we get *bilinear computations* and *bilinear complexity* (also called *rank*). We denote the bilinear complexity of a bilinear map ϕ by $R(\phi)$ or $R_k(\phi)$ and the bilinear complexity of an associative algebra A by $R(A)$ or $R_k(A)$. For any bilinear map ϕ, we have

$$C(\phi) \leq R(\phi) \leq 2 \cdot C(\phi). \tag{1}$$

Except for trivial cases, the second inequality is always strict, see [14].

1.2   Previous Results

In 1978, Brockett and Dobkin [7] proved the bound $R(k^{n \times n}) \geq 2n^2 - 1$ and in the same year, Lafon and Winograd [15] extended this result to the multiplicative complexity. Three years later, Alder and Strassen [1] unified most of the lower bounds known at that time, including the last one, in a single theorem: for any finite dimensional associative k-algebra A

$$C(A) \geq 2 \dim A - t, \tag{2}$$

where t is the number of maximal twosided ideals in A. Recently, some progress has been made in the case of matrix multiplication, namely $C(k^{n \times n}) \geq 2n^2 + n - 3$ (see [4]). Bshouty [8] obtained the bound $R_{GF(2)}(GF(2)^{n \times n}) \geq \frac{5}{2}n^2 - o(n^2)$ for the special case $k = GF(2)$ using methods from coding theory. He claims that this bound also holds for the multiplicative complexity (over $GF(2)$) but does not give a proof. Finally, in [3] we proved the lower bound

$$R(k^{n \times n}) \geq \frac{5}{2}n^2 - 3n. \tag{3}$$

for arbitrary fields k.

1.3   New Results

As our first main result, we show that (3) also holds for the multiplicative complexity. The following theorem is proven in Section 4.

Theorem 1. *For any field k, $C(k^{n \times n}) \geq \frac{5}{2}n^2 - 3n$.*

One of the main ingredients of its proof is Lemma 5 which is essentially a novel combination of a result by Ja'Ja' [14] on the relation between multiplicative and bilinear complexity, Strassen 3-slice tensor technique [21], and the substitution method [16]. We prove Lemma 5 in Section 3. Before doing so, we introduce the so-called "tensorial notion" in the next Section 2 and present a (well-known) alternative characterization of multiplicative complexity, which is more suited for our purposes.

The bound of Theorem 1 is a special case of the following lower bound (which we show in Section 5).

Theorem 2. *Let $A \cong A_1 \times \cdots \times A_t$ be a semisimple algebra over an arbitrary field k with $A_\tau = D_\tau{}^{n_\tau \times n_\tau}$ for all τ, where D_τ is a k-division algebra. Assume that each factor A_τ is noncommutative, that is, $n_\tau \geq 2$ or D_τ is noncommutative. Moreover, let $n = n_1 + \cdots + n_t$. Then $C(A) \geq \frac{5}{2} \dim A - 3n$.*

Our new bound of Theorem 2 is the first lower bound over arbitrary fields for the multiplicative complexity of a semisimple algebra—in particular of the important algebra $k^{n \times n}$—significantly above the Alder–Strassen bound (2). In [5] (see also the forthcoming [6]), we obtained the same bound for the easier case of the bilinear complexity. While in this case, our bounds can also be extended to arbitrary finite dimensional algebras A provided that the (semisimple) quotient algebra $A/\operatorname{rad} A$ satisfies the assumptions of Theorem 2, we provide an example in Section 6 which shows that our methods cannot yield this generalization for the multiplicative complexity (at least without any extra considerations). This again gives evidence for the intricate nature of multiplicative complexity.

2   Characterizations of Multiplicative Complexity

In the previous section, we have introduced the multiplicative complexity of a bilinear map in terms of computations. A second useful characterization of multiplicative complexity is the so-called "tensorial notion" (see [9, Chap. 14.4] for the bilinear complexity). With a bilinear map $\phi : U \times V \to W$, we may associate a *coordinate tensor* (or tensor for short) which is basically a "three-dimensional matrix": we fix bases $u_1, \ldots, u_m$ of U, $v_1, \ldots, v_n$ of V, and $w_1, \ldots, w_p$ of W. There are unique scalars $t_{\mu,\nu,\rho} \in k$ such that

$$\phi(u_\mu, v_\nu) = \sum_{\rho=1}^{p} t_{\mu,\nu,\rho} w_\rho \qquad \text{for all } 1 \leq \mu \leq m, \, 1 \leq \nu \leq n. \qquad (4)$$

Then $t = (t_{\mu,\nu,\rho}) \in k^{m \times n \times p}$ is the tensor of ϕ (with respect to the chosen bases). On the other hand, any given tensor also defines a bilinear map after choosing bases. We define the multiplicative complexity of the tensor t by $C(t) := C(\phi)$. In the same way, the bilinear complexity of t is $R(t) := R(\phi)$. (This is in both cases well-defined, since the multiplicative resp. bilinear complexity is robust with respect to invertible linear transformations, i.e., with respect to changes of bases.)

If ϕ is the multiplication in an algebra A, then we may instantiate the above three bases with one and the same basis. In this case, the tensor consists of the structural constants (see [11] for a definition) of the algebra A (with respect to the chosen basis).

With each tensor $t = (t_{\mu,\nu,\rho})$, we may associate three sets of matrices, the *slices* of t. The matrices $Q_\mu = (t_{\mu,\nu,\rho})_{1 \leq \nu \leq n, 1 \leq \rho \leq p} \in k^{n \times p}$ with $1 \leq \mu \leq m$ are called the 1-slices of t, the matrices $S_\nu = (t_{\mu,\nu,\rho})_{1 \leq \mu \leq m, 1 \leq \rho \leq p} \in k^{m \times p}$ with $1 \leq \nu \leq n$ the 2-slices, and finally $T_\rho = (t_{\mu,\nu,\rho})_{1 \leq \mu \leq m, 1 \leq \nu \leq n} \in k^{m \times n}$ with $1 \leq \rho \leq p$ are called the 3-slices of t. When dealing with bilinear complexity, it makes no difference which of the three sets of slices we consider. In the case of multiplicative complexity, however, the 3-slices play a distinguished role. (This is one reason why proving lower bounds for the multiplicative complexity is hard.)

Lemma 1. *Let k be a field and t be a tensor with 3-slices $T_1, \ldots, T_p \in k^{m \times n}$. Then $C(t) \leq \ell$ if and only if there are (column) vectors $u_\lambda, v_\lambda \in k^{m+n}$ for $1 \leq \lambda \leq \ell$ such that with $P_\lambda := u_\lambda \cdot v_\lambda^\top \in k^{(m+n) \times (m+n)}$*

$$\begin{pmatrix} 0 & T_1 \\ T_1^\top & 0 \end{pmatrix}, \ldots, \begin{pmatrix} 0 & T_p \\ T_p^\top & 0 \end{pmatrix} \in \mathrm{lin}\{P_1 + P_1^\top, \ldots, P_\ell + P_\ell^\top\}. \tag{5}$$

Here, $T^\top$ denotes the transpose of a matrix T and $\mathrm{lin}\{\ldots\}$ denotes the linear span. A proof of this lemma is straight forward. (One possibility is to follow the lines of the proof of Theorem 3.2 in [14].) The rank one matrices P_λ correspond to the products of a quadratic computation. By transposing, we identify the product xy of two indeterminates with yx.

If $T_1, \ldots, T_p$ are the 3-slices of a tensor t, we will occasionally also write $C(T_1, \ldots, T_p)$ instead of $C(t)$ and $R(T_1, \ldots, T_p)$ instead of $R(t)$. By multiplying (5) with

$$\begin{pmatrix} X & 0 \\ 0 & Y^\top \end{pmatrix} \quad \text{and} \quad \begin{pmatrix} X^\top & 0 \\ 0 & Y \end{pmatrix}$$

from the left and right, respectively, it follows from Lemma 1 that if $X \in k^{m \times m}$ and $Y \in k^{n \times n}$ are invertible matrices, then

$$C(T_1, \ldots, T_p) = C(X \cdot T_1 \cdot Y, \ldots, X \cdot T_p \cdot Y). \tag{6}$$

This multiplication of the slices with X and Y corresponds to a change of the bases $u_1, \ldots, u_m$ and $v_1, \ldots, v_n$ in (4).

3   Lower Bounds

In the present section, we prove our main lemma (Lemma 5). Its proof is done by a combination of the so-called substitution method (first used for proving lower bounds in algebraic complexity theory by Pan [16]), Strassen's lower bound for the (border) rank of a 3-slice tensor [21], and a result by Ja'Ja' [14] which relates multiplicative and bilinear complexity in a more sophisticated way than (1).

We first state the results of Strassen and Ja'Ja'. For two elements b, c of an associative algebra, let $[b, c] := bc - cb$ denote their *Lie product* (or *commutator*). For a matrix $M \in k^{N \times N}$, let $\mathrm{rk}\, M$ denote its (usual) rank. We denote the identity matrix of $k^{N \times N}$ by I_N.

Lemma 2 (Strassen). *Let t be a tensor with 3-slices $I_N, B, C \in k^{N \times N}$ over some field k. Then $R(t) \geq N + \frac{1}{2} \mathrm{rk}[B, C]$.*

For a proof, see [21, Thm. 4.1] or [9, Thm. 19.12]. Strassen actually proves the above lemma for the so-called border rank of t (see e.g. [9, Chap. 15.4] for a definition) which is a lower bound for the bilinear complexity.

Lemma 3 (Ja'Ja'). *Let t be a tensor with 3-slices $T_1, \ldots, T_p \in k^{N \times N}$ over some field k. Then*

$$C(t) \geq \tfrac{1}{2} R\!\left(\begin{pmatrix} T_1 & 0 \\ 0 & T_1^\top \end{pmatrix}, \ldots, \begin{pmatrix} T_p & 0 \\ 0 & T_p^\top \end{pmatrix} \right).$$

For a proof, see [14, Thm. 3.4].

Combining these two lemmata, we obtain a lower bound for the multiplicative complexity of a 3-slice tensor.

Lemma 4. *Let k be a field. Let t be a tensor with 3-slices $I_N, B, C \in k^{N \times N}$. Then $C(t) \geq N + \tfrac{1}{2} \operatorname{rk}[B, C]$.*

Proof. By the previous Lemmata 3 and 2

$$C(t) \geq \tfrac{1}{2} R\!\left(I_{2N}, \begin{pmatrix} B & 0 \\ 0 & B^\top \end{pmatrix}, \begin{pmatrix} C & 0 \\ 0 & C^\top \end{pmatrix} \right)$$

$$\geq \tfrac{1}{2}\!\left(2N + \tfrac{1}{2} \operatorname{rk}\left[\begin{pmatrix} B & 0 \\ 0 & B^\top \end{pmatrix}, \begin{pmatrix} C & 0 \\ 0 & C^\top \end{pmatrix} \right] \right). \tag{7}$$

Since $B^\top C^\top - C^\top B^\top = -[B, C]^\top$,

$$\operatorname{rk}\left[\begin{pmatrix} B & 0 \\ 0 & B^\top \end{pmatrix}, \begin{pmatrix} C & 0 \\ 0 & C^\top \end{pmatrix} \right] = \operatorname{rk}\begin{pmatrix} BC - CB & 0 \\ 0 & B^\top C^\top - C^\top B^\top \end{pmatrix} = 2 \operatorname{rk}[B, C].$$

Thus, the right-hand side of (7) equals $N + \tfrac{1}{2} \operatorname{rk}[B, C]$. $\square$

We now come to the proof of our main lemma. We combine the substitution method (here manifesting itself in the Steinitz exchange) with the previous Lemma 4.

Lemma 5. *Let t be a tensor with linearly independent 3-slices $T_1, \ldots, T_p \in k^{N \times N}$ over some field k. Assume there are integers s and q such that for each basis $U_1, \ldots, U_p$ of $\operatorname{lin}\{T_1, \ldots, T_p\}$ there are indices $i_1, \ldots, i_s$ and $j_1, \ldots, j_q$ with the following properties: the linear span of $U_{i_1}, \ldots, U_{i_s}$ contains an invertible matrix E and the linear span of $U_{j_1}, \ldots, U_{j_q}$ contains matrices B and C with $\operatorname{rk}[B, C] = N$. Then $C(t) \geq p - s - q + \tfrac{3}{2} N$.*

Proof. Let $\ell := C(t)$. By Lemma 1, there are $2N \times 2N$–matrices $P_1, \ldots, P_\ell$ of rank one and $S_\rho \in \operatorname{lin}\{P_1, \ldots, P_\ell\}$ such that

$$\begin{pmatrix} 0 & T_\rho \\ T_\rho^\top & 0 \end{pmatrix} = S_\rho + S_\rho^\top \qquad \text{for } 1 \leq \rho \leq p. \tag{8}$$

The matrices $S_1, \ldots, S_p$ are linearly independent, since otherwise we would obtain a linear dependence of the 3-slices $T_1, \ldots, T_p$ from (8), a contradiction. We now exploit the Steinitz exchange in a rather explicit way: write $S_\rho = \sum_{\lambda=1}^{\ell} \xi_{\rho,\lambda} P_\lambda$ with scalars $\xi_{\rho,\lambda} \in k$ for $1 \leq \rho \leq p$. Let $X = (\xi_{\rho,\lambda}) \in k^{p \times \ell}$.

The matrix X has full rank p, since the S_ρ are linearly independent. By permuting the P_λ, we may assume that the matrix X' consisting of the first p columns of X is invertible. Let Y' be its inverse. We may augment Y' to a matrix $Y = (\eta_{\rho,\lambda}) \in k^{p \times \ell}$ such that

$$\underbrace{\sum_{i=1}^{p} \eta_{\rho,i} S_i}_{=: M_\rho} = P_\rho + \sum_{\lambda=p+1}^{\ell} \eta_{\rho,\lambda} P_\lambda \qquad \text{for } 1 \le \rho \le p. \tag{9}$$

The above defined matrices $M_1, \ldots, M_p$ are linearly independent and their linear span equals $\lin\{S_1, \ldots, S_p\}$. By virtue of (8),

$$\begin{pmatrix} 0 & T_\rho \\ T_\rho^\top & 0 \end{pmatrix} \in \lin\{M_1 + M_1^\top, \ldots, M_p + M_p^\top\} \qquad \text{for } 1 \le \rho \le p. \tag{10}$$

Let L_ρ be the $N \times N$–matrix defined by

$$\begin{pmatrix} 0 & L_\rho \\ L_\rho^\top & 0 \end{pmatrix} = M_\rho + M_\rho^\top .$$

($M_\rho + M_\rho^\top$ is of the above form by (8) and the linear independence of $T_1, \ldots, T_p$.) By (10), $\lin\{L_1, \ldots, L_p\}$ equals $\lin\{T_1, \ldots, T_p\}$. By the assumption of the lemma, there are indices $i_1, \ldots, i_s$ such that $\lin\{L_{i_1}, \ldots, L_{i_s}\}$ contains an invertible matrix E. Exploiting (6), we may replace L_ρ with $E^{-1} L_\rho$ for $1 \le \rho \le p$. Thereafter, $I_N \in \lin\{L_{i_1}, \ldots, L_{i_s}\}$. Again by assumption, there are indices $j_1, \ldots, j_q$ such that $\lin\{L_{j_1}, \ldots, L_{j_q}\}$ contains matrices B and C with $\rk[B, C] = N$.

By (9), it follows that

$$\begin{pmatrix} 0 & I_N \\ I_N^\top & 0 \end{pmatrix}, \begin{pmatrix} 0 & B \\ B^\top & 0 \end{pmatrix}, \begin{pmatrix} 0 & C \\ C^\top & 0 \end{pmatrix} \in \lin\{P_{i_1} + P_{i_1}^\top, \ldots, P_{i_s} + P_{i_s}^\top,$$
$$P_{j_1} + P_{j_1}^\top, \ldots, P_{j_q} + P_{j_q}^\top,$$
$$P_{p+1} + P_{p+1}^\top, \ldots, P_\ell + P_\ell^\top\}.$$

Thus, $C(I_N, B, C) \le \ell - p + s + q$ yielding $C(t) \ge p - s - q + C(I_N, B, C)$. By Lemma 4, $C(I_N, B, C) \ge \frac{3}{2}N$. $\qquad\square$

4   Matrix Multiplication

As the first and most important example, we apply Lemma 5 to the algebra $k^{n \times n}$. Our aim is to utilize the following two lemmata which are proven in [3, Sect. 4].

Lemma 6. *Let k be an infinite field and let V be a subspace of $k^{n \times n}$ that contains an invertible matrix. Then for any basis $v_1, \ldots, v_d$ of V there are $s \le n$ and indices $i_1, \ldots, i_s$ such that already the linear span of $v_{i_1}, \ldots, v_{i_s}$ contains an invertible matrix.*

Lemma 7. *Let k be an infinite field, let $n \geq 2$, and let $v_1, \ldots, v_p$ be a basis of $k^{n \times n}$, where $p = n^2$. Then there are $q \leq 2n$, indices $j_1, \ldots, j_q$, and $b, c \in \lin\{v_{j_1}, \ldots, v_{j_q}\}$ such that $[b, c]$ is invertible.*

To exploit the above two lemmata, we have to relate the structure of the algebra $k^{n \times n}$ to the structure of the 3-slices of the coordinate tensor of $k^{n \times n}$.

For the moment, consider an arbitrary associative algebra A of dimension p. For an element $x \in A$, let ℓ_x and r_x denote the vector space endomorphisms defined by $y \mapsto xy$ and $y \mapsto yx$ for all $y \in A$, respectively. Let $a_1, \ldots, a_p$ be a basis of A and t be the corresponding coordinate tensor. ¿From (4) (where each of the three bases is instantiated with $a_1, \ldots, a_p$) it is clear that the ρth 1-slice of t is the matrix of the left multiplication ℓ_{a_ρ} (with respect to $a_1, \ldots, a_p$). Thus, the homomorphism that maps a_ρ to the ρth 1-slice of t for each ρ is a faithful representation of A. Therefore, the subalgebra of $k^{p \times p}$ generated by the 1-slices of t is isomorphic to A. In the same way, the subalgebra of $k^{p \times p}$ generated by the 2-slices is isomorphic to the opposite algebra A^o of A (see [11] for a definition). The question how the 3-slices of t are related to the structure of A is a more subtle question.

In the case of the algebra $k^{n \times n}$, we are lucky. It is well known that the structure of the coordinate tensor t of $k^{n \times n}$ is invariant under permutations, see for example [9, Eq. 14.21]. It follows that if $Q_1, \ldots, Q_p$ denote the 1-slices of t and $T_1, \ldots, T_p$ denote the 3-slices of t (where $p = n^2$), then there are invertible matrices $X \in k^{p \times p}$ and $Y \in k^{p \times p}$ (even permutation matrices) such that

$$\lin\{Q_1, \ldots, Q_p\} = \lin\{X \cdot T_1 \cdot Y, \ldots, X \cdot T_p \cdot Y\}.$$

By (6), we may replace T_ρ by $X \cdot T_\rho \cdot Y$ for $1 \leq \rho \leq p$. Thereafter, the subalgebra of $k^{p \times p}$ generated by $T_1, \ldots, T_p$ is isomorphic to the algebra $k^{n \times n}$. By the above Lemmata 6 and 7, $T_1, \ldots, T_p$ fulfill the assumptions of Lemma 5 with $s = n$ and $q = 2n$. (The restriction that k is infinite imposes no problem here, since $C_k(k^{n \times n}) \geq C_K(K^{n \times n})$ for any extension field $K \supset k$.) Now Lemma 5 yields $C(T_1, \ldots, T_p) \geq n^2 - n - 2n + \frac{3}{2}n^2$. This completes the proof of Theorem 1.

5   Further Applications: Semisimple Algebras

In the present section, we generalize the results of the preceding section to semisimple algebras.

As seen above, we have to relate the 3-slices of the coordinate tensor of an algebra with the structure of that algebra. We start with the case of a simple k-algebra A. By Wedderburn's Structure Theorem (see [11]), A is isomorphic to an algebra $D^{n \times n} \cong D \otimes k^{n \times n}$ for some positive integer n and some k-division algebra D. Let $p := \dim A = n^2 \cdot \dim D$. Let $a_1, \ldots, a_p$ be a basis of A and let $a_1^*, \ldots, a_p^*$ denote its dual basis. Let $\ell_{a_i}^*$ denote the dual of the left multiplication with a_i, that is, the linear map $A^* \to A^*$ defined by $b \mapsto b \circ \ell_{a_i}$. Applying a_ρ^* to (4) yields

$$a_\rho^*(\ell_{a_\mu}(a_\nu)) = t_{\mu, \nu, \rho},$$

hence

$$\ell^*_{a_\mu}(a^*_\rho) = \sum_{\nu=1}^{p} t_{\mu,\nu,\rho} a^*_\nu \qquad \text{for all } 1 \le \mu, \rho \le p. \tag{11}$$

De Groote [12, Prop. 1.1] shows that for any simple algebra A there is a vector space isomorphism $S : A \to A^*$ fulfilling

$$\ell^*_x = S \circ r_x \circ S^{-1} \qquad \text{for all } x \in A.$$

Substituting this into (11), we obtain

$$S^{-1}(a^*_\rho) \cdot a_\mu = \sum_{\nu=1}^{p} t_{\mu,\nu,\rho} S^{-1}(a^*_\nu) \qquad \text{for all } 1 \le \mu, \rho \le p.$$

Thus, the ρth 3-slice of the tensor of A with respect to the basis $a_1, \ldots, a_p$ is the matrix of the homomorphism $\ell_{S^{-1}(a^*_\rho)} : A \to A$ with respect to the two bases $a_1, \ldots, a_p$ and $S^{-1}(a^*_1), \ldots, S^{-1}(a^*_p)$. Hence, there are invertible matrices $X, Y \in k^{p \times p}$ such that the subalgebra of $k^{p \times p}$ generated by the matrices $X \cdot T_1 \cdot Y, \ldots, X \cdot T_p \cdot Y$ is isomorphic to A.

Next, we consider the case of a semisimple algebra. By Wedderburn's Structure Theorem, any semisimple algebra A is isomorphic to a direct product of simple algebras $A_1 \times \cdots \times A_t$ where each $A_\tau = D_\tau{}^{n_\tau \times n_\tau}$ for some k-division algebra D_τ. If we choose a basis with respect to this decomposition of A, then the corresponding coordinate tensor is a direct sum of the tensors of $A_1, \ldots, A_t$. By applying the above considerations for the simple case separately to each A_τ, we conclude that there are invertible matrices $X, Y \in k^{p \times p}$ such that the subalgebra of $k^{p \times p}$ spanned by $X \cdot T_1 \cdot Y, \ldots, X \cdot T_p \cdot Y$ is isomorphic to A.

The following analogue of Lemma 6 and Lemma 7 is proven in [5]:

Lemma 8. *Let $A \cong A_1 \times \cdots \times A_t$ be a semisimple algebra over an infinite field k with $A_\tau = D_\tau{}^{n_\tau \times n_\tau}$ for all τ, where D_τ is a k-division algebra. Assume that each factor A_τ is noncommutative, that is, $n_\tau \ge 2$ or D_τ is noncommutative. Moreover, let $n = n_1 + \cdots + n_t$ and $v_1, \ldots, v_p$ be a basis of A.*

1. *There are $s \le n$ and indices $i_1, \ldots, i_s$ such that $\mathrm{lin}\{v_{i_1}, \ldots, v_{i_s}\}$ contains an invertible element.*
2. *There are $q \le 2n$, indices $j_1, \ldots, j_q$, and $b, c \in \mathrm{lin}\{v_{j_1}, \ldots, v_{j_q}\}$ such that $[b, c]$ is invertible.*

By the above Lemma 8, the 3-slices $T_1, \ldots, T_p$ fulfill the assertion of Lemma 5 with $s = n$ and $q = 2n$ where $n = n_1 + \cdots + n_t$. (Again, the restriction that k is infinite does not impose any problems here, we can switch over from k to $k(x)$ with some extra indeterminate x. This does not have any relevant impact on the structure of A.) Altogether, this proves Theorem 2.

6   A Limiting Example

While for the bilinear complexity, our bounds can also be extended to arbitrary finite dimensional algebras A provided that the (semisimple) quotient algebra $A/\operatorname{rad} A$ fulfill the assumptions of Theorem 2 (see [5] and the forthcoming [6]), we here construct an example that satisfies these assumptions but for which our method fails in the case of multiplicative complexity. Of course, this does not mean that our method cannot be applied to arbitrary associative algebras, we just have to examine the 3-slices of the algebra explicitly.

Let $X_1, \ldots, X_n$ be indeterminates over some field k. Furthermore, let I denote the ideal generated by all monomials of total degree two. Consider the algebra $A = k[X_1, \ldots, X_n]/I$. We have $X_i \cdot X_j = 0$ in A for all i, j. With respect to the basis $1, X_1, \ldots, X_n$, the tensor t_A of A looks as follows:

$$
t_A = \begin{pmatrix}
 & 1 & 2 & 3 & \cdots & n+1 \\
1 & & & & & \\
2 & & & & & \\
3 & & & & & \\
\vdots & & & & & \\
n+1 & & & & &
\end{pmatrix}
$$

Above, a ρ in position (μ, ν) means that the ρth 3-slice has the entry one in position (μ, ν). Unspecified entries are zero.

The algebra A is commutative, so $A/\operatorname{rad} A$ does not fulfill the assumptions of Theorem 2. (In fact, A is of minimal rank.) Instead, consider $A' = D \otimes A$ for some noncommutative central division algebra D. We have $A'/\operatorname{rad} A' = D$, thus $A'/\operatorname{rad} A'$ satisfies the assumptions of Theorem 2. However, any matrix in the linear span of the 3-slices of the tensor of A' (which we obtain from t_A by substituting each one in t_A by the tensor of D) has rank at most $2 \dim D$. Consequently, the Lie product of any two such matrices has rank at most $4 \dim D$. Therefore, if n is large, we are not able to obtain the additional $\frac{1}{2} \dim A'$ that we achieved in the bound of Theorem 2.

References

1. A. Alder and V. Strassen. On the algorithmic complexity of associative algebras. *Theoret. Comput. Sci.*, 15:201–211, 1981.
2. Dario Bini, Milvio Capovani, Grazia Lotti, and Francesco Romani. $O(n^{2.7799})$ complexity for matrix multiplication. *Inf. Proc. Letters*, 8:234–235, 1979.
3. Markus Bläser. A $\frac{5}{2}n^2$–lower bound for the rank of $n \times n$–matrix multiplication over arbitrary fields. In *Proc. 40th Ann. IEEE Symp. on Found. Comput. Sci. (FOCS)*, pages 45–50, 1999.
4. Markus Bläser. Lower bounds for the multiplicative complexity of matrix multiplication. *Comput. Complexity*, 8:203–226, 1999.
5. Markus Bläser. *Untere Schranken für den Rang assoziativer Algebren*. Dissertation, Universität Bonn, 1999.
6. Markus Bläser. Lower bounds for the bilinear complexity of associative algebras. *Comput. Complexity*, to appear.

7. Roger W. Brockett and David Dobkin. On the optimal evaluation of a set of bilinear forms. *Lin. Alg. Appl.*, 19:207–235, 1978.

8. Nader H. Bshouty. A lower bound for matrix multiplication. *SIAM J. Comput.*, 18:759–765, 1989.

9. Peter Bürgisser, Michael Clausen, and M. Amin Shokrollahi. *Algebraic Complexity Theory*. Springer, 1997.

10. Don Coppersmith and Shmuel Winograd. Matrix multiplication via arithmetic progression. *J. Symbolic Comput.*, 9:251–280, 1990.

11. Yurij A. Drozd and Vladimir V. Kirichenko. *Finite Dimensional Algebras*. Springer, 1994.

12. Hans F. de Groote. Characterization of division algebras of minimal rank and the structure of their algorithm varieties. *SIAM J. Comput.*, 12:101–117, 1983.

13. Hans F. de Groote. *Lectures on the Complexity of Bilinear Problems*. LNCS 245. Springer, 1986.

14. Joseph Ja'Ja'. On the complexity of bilinear forms with commutativity. *SIAM J. Comput.*, 9:717–738, 1980.

15. Jean-Claude Lafon and Shmuel Winograd. A lower bound for the multiplicative complexity of the product of two matrices. Technical report, Centre de Calcul de L'Esplanade, U.E.R. de Mathematique, Univ. Louis Pasteur, Strasbourg, France, 1978.

16. Victor Yu. Pan. Methods for computing values of polynomials. *Russ. Math. Surv.*, 21:105–136, 1966.

17. Richard S. Pierce. *Associative Algebras*. Springer, 1982.

18. Arnold Schönhage. Partial and total matrix multiplication. *SIAM J. Comput.*, 10:434–455, 1981.

19. Volker Strassen. Gaussian elimination is not optimal. *Num. Math.*, 13:354–356, 1969.

20. Volker Strassen. Vermeidung von Divisionen. *J. Reine Angew. Math.*, 264:184–202, 1973.

21. Volker Strassen. Rank and optimal computation of generic tensors. *Lin. Alg. Appl.*, 52/53:645–685, 1983.

22. Volker Strassen. Relative bilinear complexity and matrix multiplication. *J. Reine Angew. Math.*, 375/376:406–443, 1987.

23. Volker Strassen. Algebraic complexity theory. In J. van Leeuven, editor, *Handbook of Theoretical Computer Science Vol. A*, pages 634–672. Elsevier Science Publishers B.V., 1990.

Evasiveness of Subgraph Containment and Related Properties[*]

Amit Chakrabarti, Subhash Khot, and Yaoyun Shi

Department of Computer Science
Princeton University, Princeton NJ 08544, USA
{amitc,khot,shiyy}@cs.princeton.edu

Abstract. We prove new results on evasiveness of monotone graph properties by extending the techniques of Kahn, Saks and Sturtevant [4]. For the property of containing a subgraph isomorphic to a fixed graph, and a fairly large class of related n-vertex graph properties, we show evasiveness for an arithmetic progression of values of n. This implies a $\frac{1}{2}n^2 - O(n)$ lower bound on the decision tree complexity of these properties.

We prove that properties that are preserved under taking graph minors are evasive for all sufficiently large n. This greatly generalizes the evasiveness result for planarity [1]. We prove a similar result for bipartite subgraph containment.

Keywords: Decision Tree Complexity, Monotone Graph Properties, Evasiveness, Graph Property Testing.

1 Introduction

Suppose we have an input graph G and are required to decide whether or not it has a certain (isomorphism invariant) property P. The graph is given by an oracle which answers queries of the form "is (x, y) an edge of G?" A *decision tree algorithm* for P is a strategy that specifies a sequence of such queries to the oracle, where each query may depend upon the outcomes of the previous ones, terminating when sufficient information about G has been obtained to decide whether or not P holds for G. The *cost* of such a decision tree algorithm is the worst case number of queries that it makes. The *decision tree complexity* of P is the minimum cost of any decision tree algorithm for P.

Since an n-vertex graph has $\frac{1}{2}n(n-1)$ vertex pairs each of which could either be an edge or not, it is clear that any property of n-vertex graphs has complexity at most $\frac{1}{2}n(n-1)$. If a property happens to have complexity *exactly* $\frac{1}{2}n(n-1)$ then it is said to be *evasive*.[1]

A property of n-vertex graphs is said to be *monotone* if, starting with a graph which has the property, the addition of edges does not destroy the property. It is said to be *nontrivial* if there exists an n-vertex graph which has the property and one which does not. Connectedness, non-planarity, non-k-colorability and

---

[*] This work was supported in part by NSF Grant CCR-96-23768, NSF Grant CCR-98-20855 and ARO Grant DAAH04-96-1-0181.

[1] Some authors call such properties "elusive" instead of evasive.

the property of containing a perfect matching are all examples of nontrivial monotone properties (for sufficiently large n). Rosenberg [7] attributes to Karp the following conjecture which, remarkably, remains open even today.

Karp Conjecture: Every nontrivial monotone graph property is evasive.

As a first step towards a resolution of this conjecture, Rivest and Vuillemin [6] proved that such properties have complexity at least $n^2/16$, thereby settling the Aanderaa-Rosenberg conjecture [7] of an $\Omega(n^2)$ complexity lower bound. The next big advance was the work of Kahn, Saks and Sturtevant [4] where an interesting *topological approach* was used to prove that the Karp Conjecture holds whenever n is a prime power. Triesch [8] used this approach, together with a complicated construction, to prove the evasiveness of some special classes of properties. Similar topological ideas were used by Yao [9] to prove a related result: namely, that nontrivial monotone *bipartite* graph properties are always evasive. Prior to the work of Kahn et al., adversarial strategies had been devised to prove the evasiveness of certain specific graph properties for all n in [5], [1] and [3]. These strategies worked for the properties of acyclicity, connectedness, 2-connectedness, planarity and simple variants on these. The most sophisticated of these adversarial strategies was one used by Bollobás [2] to prove the evasiveness of the property of containing a k-clique, for any k, $2 \leq k \leq n$.

Let H be any fixed graph. For n-vertex graphs, let Q_n^H denote the property of containing H as a subgraph (not necessarily as an induced subgraph). From the work of Bollobás [2] we know that Q_n^H is evasive for all n in the special case when H is a complete graph. This raises the natural question: what can we say about general H?

In this paper, we study this question and some related ones, extending, for the first time, the topological approach of [4] to a fairly general class of graph properties. For each of these properties, we draw stronger inferences than [4]. Our Main Theorem is stated below.

Theorem 1.1 (Main Theorem). *For any fixed graph H there exists an integer r_0 with the following property. Suppose $n = \sum_{i=1}^{r} q^{\alpha_i}$ where q is a prime power, $q \geq |H|$, each $\alpha_i \geq 1$ and $r \equiv 1 \pmod{r_0}$. Then Q_n^H is evasive.*

In order to understand the significance and strength of this theorem, consider the following statements (proven in this paper). Each of these statements follows either from the Main Theorem or from the techniques used in proving it.

- For any graph H, there is an *arithmetic progression* such that Q_n^H is evasive for all n in the progression. Note that this is a much stronger inference than can be drawn by applying the results of [4].
- The decision tree complexity of Q_n^H is $\frac{1}{2}n^2 - O(n)$. This bound does not follow from the results of [4].
- If the graph H is *bipartite*, then Q_n^H is evasive for large enough n.
- Any n-vertex nontrivial graph property that is preserved under taking graph minors is evasive for large enough n. This includes lots of very natural graph properties such as embeddability on any surface, outerplanarity, linkless embeddability in $\mathbf{R}^3$, the property of being a series-parallel graph, etc. Thus,

our result generalizes a result of Best et al. [1] who show that planarity is
evasive.

- Any monotone boolean combination of the properties Q_n^H for several different
graphs H still satisfies our Main Theorem. Thus, for example, if H_1, H_2 and
H_3 are fixed graphs, then the property of containing as subgraph either H_1
or both of H_2 and H_3 is still evasive for those n which satisfy the conditions
of the Main Theorem.

The remainder of the paper is organized as follows. In Section 2 we review
the basics of the topological approach of Kahn et al.[4], establishing a connection
between proving evasiveness of monotone properties and computing Euler characteristics of abstract complexes. Then in Section 3 we define a certain auxiliary
property of graphs and prove a technical result (called the Main Lemma) about
this property. This result is then used in Section 4 to prove our main theorem.
In Section 5, we provide proofs for the additional results itemized above. We end
with some concluding remarks in Section 6.

Notations, Terminology, and Conventions: We call a graph *trivial* if it has no
edges. Throughout this paper, all graphs will be assumed to be nontrivial, have no
loops and no parallel edges. For a graph G, $|G|$ will denote the number of vertices in
G, also called the *size* of G, $V(G)$ will denote its vertex set, $E(G)$ its edge set, chr(G)
its chromatic number and clq(G) the size of its largest clique. Graphs which occur
as "input graphs" on which boolean properties are to be tested are assumed to be
always vertex-labeled. All other graphs are assumed to be unlabeled, unless otherwise
specified. When we speak of an "edge" in an input graph, we really mean an unordered
vertex pair which may or may not be an edge.

2   Review of the Topological Approach

A property of m boolean variables $x_1, \ldots, x_m$ is a function $P : \{0,1\}^m \rightarrow \{0,1\}$; we say that the m-tuple $(x_1, \ldots, x_m)$ has (or satisfies) property P if
$P(x_1, \ldots, x_m) = 1$. We say that P is *monotone* if for every m-tuple $(x_1, \ldots, x_m)$
that satisfies P, increasing any x_i from 0 to 1 yields an m-tuple that also satisfies
P. We say that P is *evasive* if any decision tree algorithm for P has cost m. In
our study of graph properties, the variables will be unordered pairs of vertices
(i.e., potential edges of the graph) and P will be required to be invariant under
relabelings of the graph.

Let $[m]$ denote the set $\{1, 2, \ldots, m\}$ and consider the collection of subsets
$S \subseteq [m]$ with the following property: setting the variables indexed by S to 1
and those indexed by $[m] \setminus S$ to 0 yields an m-tuple which *does not satisfy*
P. Since P is monotone, this collection of sets is downward closed under set
inclusion. Recall that such a downward closed collection of sets is called an
abstract complex, and that the sets in this collection are called the *faces* of the
complex. This observation motivates

Definition 2.1. *If P is monotone, then the abstract complex associated with P,
denoted $\Delta(P)$, is defined as follows:*

$$\Delta(P) = \{S \subseteq [m] \ : \ \text{If } x_i = 1 \Longleftrightarrow i \in S, \text{ then } (x_1, \ldots, x_m) \text{ does not satisfy } P\}.$$

Associated with an abstract complex Δ is a topologically important number called its *Euler characteristic* which is denoted $\chi(\Delta)$ and is defined as follows:

$$\chi(\Delta) = \sum_{\emptyset \neq F \in \Delta} (-1)^{|F|-1}. \tag{1}$$

Kahn et al.[4] showed that non-evasiveness of P has topological consequences for $\Delta(P)$. The following theorem is implicit in their work:

Theorem 2.2 (Kahn et al. [4]). *If the monotone property P is not evasive, then $\chi(\Delta(P)) = 1$.* □

For our result, we shall need to use a stronger theorem which can also be found in [4]. Let Δ be an abstract complex defined on $[m]$ and let Γ be a finite group which acts on the set $[m]$, preserving the faces of Δ. The action partitions $[m]$ into orbits, say $A_1, \ldots, A_k$. We use the action of Γ to define another abstract complex Δ_Γ on $[k]$ as follows:

$$\Delta_\Gamma = \{S \subseteq [k] : \bigcup_{i \in S} A_i \in \Delta\} \tag{2}$$

Sometimes, as is the case with our work, it is not easy to say much about $\Delta(P)$ for a monotone property P. But it is possible to find some group Γ such that its action produces a more understandable abstract complex $(\Delta(P))_\Gamma$. The next theorem, the most important tool in [4], says that if Γ has certain rather restrictive properties, then non-evasiveness of P has a topological consequence on this new complex.

Theorem 2.3 (Kahn et al. [4]). *Suppose Γ has a normal subgroup Γ_1 which is such that $|\Gamma_1|$ is a prime power and the quotient group Γ/Γ_1 is cyclic. Then if P is not evasive, we have $\chi((\Delta(P))_\Gamma) = 1$.* □

An application of this result leads to the following theorem which is the main result of [4].

Theorem 2.4 (Kahn et al. [4]). *Let P_n be a nontrivial monotone property of n-vertex graphs. If n is a prime power, then P_n is evasive.* □

In order to derive Theorem 2.4 from Theorem 2.3, Kahn et al. construct a group which acts on the vertices of the input graph and thus, indirectly, on the edges. The number theoretic constraint on n is a consequence of the fact that this action depends crucially on being able to view the vertices of the graph as elements of a finite field. Our approach to proving evasiveness for more general n will be to devise a more sophisticated group action. Before we do so, we will need an auxiliary result which we shall establish in the next section.

3   The Main Lemma

Consider the following operation on a graph G. Let the vertices of G be colored, using *all* the colors in some set C, so that no two adjacent vertices get the same color. Let G' be a graph with vertex set C where two distinct vertices $c_1, c_2 \in C$ are adjacent iff the coloring assigns colors c_1 and c_2 to the end-points of some edge in G. We shall call G' a *compression of graph G induced by coloring C.* If there exists a C which induces a compression G' of G, we shall write $G' \lhd G$.

Definition 3.1. *A family $\mathcal{F}$ of graphs is said to be closed under compression if for graphs G, H such that $G \in \mathcal{F}$ and $H \lhd G$ we have $H \in \mathcal{F}$.*

Let $\mathcal{F}$ be a nonempty finite family of (nontrivial) graphs that is closed under compression. The property $P_n^{\mathcal{F}}$ that an input graph G on n vertices contains some member of $\mathcal{F}$ as a subgraph is clearly nontrivial, for n large enough, and monotone. Let $\Delta_n^{\mathcal{F}}$ be the abstract complex associated with this property and let $\chi_n = \chi(\Delta_n^{\mathcal{F}})$ be the Euler characteristic of this complex.

The purpose of this section is to establish that for any such family $\mathcal{F}$, we have $\chi_n \neq 1$ infinitely often. Let us set

$$T = 2^{2^t}, \text{ where } t \text{ is the smallest integer such that } T \geq \min_{F \in \mathcal{F}} |F|. \qquad (3)$$

We shall prove

Lemma 3.2 (Main Lemma). *If $n \equiv 1 \pmod{T-1}$ then $\chi_n \equiv 0 \pmod 2$.*

Since we only care about χ_n mod 2, we can use the fact that addition and subtraction are equivalent mod 2 in (1) to get[2]

$$\chi_n \equiv \#\{G : G \text{ is nontrivial and does not satisfy } P_n^{\mathcal{F}}\} \pmod 2. \qquad (4)$$

Consider n-vertex input graphs with vertices labeled with integers from 0 to $n-1$. For $n > T$, let us define a group action on such graphs as follows. For $a, b \in \{0, 1, 2, \ldots, T-1\}$ and a odd, let permutation $\phi_{a,b}$ be defined by mapping vertex i to vertex $(ai + b) \bmod T$ for $i \in \{0, 1, \ldots, T-1\}$. The other $n - T$ vertices are left fixed. It is routine to check that the set of all these permutations forms a group Φ under composition, thereby defining a group action on the labeled vertices. This action induces an action on graphs in the obvious manner, thereby partitioning the set of all labeled n-vertex graphs into orbits. Since the order $|\Phi|$ of the group is $T^2/2$, a power of 2, each orbit has size a power of 2. Therefore, (4) can be modified to

$$\chi_n \equiv \#\left\{G : \begin{array}{c} G \text{ is nontrivial, invariant under} \\ \Phi \text{ and does not satisfy } P_n^{\mathcal{F}} \end{array}\right\} \pmod 2. \qquad (5)$$

The action of Φ on the vertices also induces an action on edges (or rather, on unordered pairs of distinct vertices, each of which may or may not be an edge), not to be confused with the action on labeled graphs mentioned above. Therefore the set of edges amongst vertices $0, 1, \ldots, T-1$ is partitioned into orbits. Since any odd integer is invertible mod T, we get 2^t orbits $E_0, E_1, \ldots, E_{2^t-1}$, where

$$E_i = \{(x, y) : 0 \leq x < y < T, \ y - x = 2^i k \text{ for some odd number } k\}. \qquad (6)$$

---

[2] Note that we are counting not graphs, but labeled graphs.

Let G be an invariant graph. From now on, let us refer to the vertices $0, 1, \ldots, T-1$ as *left vertices* and the rest as *right vertices*. Let G_{left} and G_{right} denote the subgraphs of G induced by the left and right vertices, respectively. By invariance of G, the set of right vertices adjacent to any left vertex is the same for each left vertex; let $\mathcal{R}(G)$ denote this set. Also, the set of edges $E(G_{\text{left}})$ is the union of a certain number of the orbits E_i; let $\mathrm{orb}(G)$ denote this number. We shall show that whether or not G has the property $P_n^{\mathcal{F}}$ is completely determined once $G_{\text{right}}, \mathcal{R}(G)$ and this number $\mathrm{orb}(G)$ are fixed; the specific G_{left} does not matter.

Lemma 3.3. *For any invariant G, we have* $\mathrm{chr}(G_{\text{left}}) = \mathrm{clq}(G_{\text{left}}) = 2^{\mathrm{orb}(G)}$.

Proof. Let $I \subseteq \{0, 1, \ldots, 2^t - 1\}$ be such that $E(G_{\text{left}}) = \bigcup_{i \in I} E_i$; then we have $|I| = \mathrm{orb}(G)$. Consider two vertices x, y of G_{left}. If their binary representations agree on the bit positions indexed by I, then $x - y = \sum_{i \in I'} \pm 2^i$ for some set I' disjoint from I. By (6), this implies $(x, y) \notin E(G_{\text{left}})$. Therefore, the vertices of G_{left} can be partitioned into $2^{|I|}$ independent sets; thus $\mathrm{chr}(G_{\text{left}}) \le 2^{\mathrm{orb}(G)}$. On the other hand, if x, y are such that the bits in positions outside I are all zero, then $x - y = \sum_{i \in I''} \pm 2^i$ for some $I'' \subseteq I$, which by (6) implies that $(x, y) \in E(G_{\text{left}})$. Therefore, G_{left} has a clique of size $2^{|I|} = 2^{\mathrm{orb}(G)}$. The lemma follows. $\square$

Lemma 3.4. *Let G_1, G_2 be two invariant n-vertex labeled graphs with $G_{1,\text{right}} = G_{2,\text{right}}$, $\mathcal{R}(G_1) = \mathcal{R}(G_2)$ and $\mathrm{orb}(G_1) = \mathrm{orb}(G_2)$. Then G_1 has property $P_n^{\mathcal{F}}$ if and only if G_2 does.*

Proof. Suppose G_1 has property $P_n^{\mathcal{F}}$; we shall show that G_2 does too. Suppose G_1 contains $F \in \mathcal{F}$ as a subgraph. We fix a particular occurrence of F within G_1 so that we can talk about $F_{\text{left}}, F_{\text{right}}$ and $\mathcal{R}(F) := \mathcal{R}(G_1) \cap V(F)$.

Using Lemma 3.3 and the hypothesis, we obtain $\mathrm{chr}(F_{\text{left}}) \le \mathrm{chr}(G_{1,\text{left}}) = \mathrm{clq}(G_{2,\text{left}})$. Let $h = \mathrm{chr}(F_{\text{left}})$; from the above inequality it is clear that $G_{2,\text{left}}$ contains K_h as a subgraph. Fix a particular occurrence of K_h and, starting with the graph F_{right}, connect each of the h left vertices in this occurrence to each vertex in $\mathcal{R}(F)$. Let F' be the resulting graph. Since $\mathcal{R}(F) \subseteq \mathcal{R}(G_1) = \mathcal{R}(G_2)$ and since F_{right} is a subgraph of $G_{1,\text{right}} = G_{2,\text{right}}$, it follows that F' is a subgraph of G_2.

Consider the following coloring of the graph F: we use h colors for its left vertices and color each right vertex with a distinct color, never using any of these h colors. Let $F'' \triangleleft F$ be the compression of F induced by this coloring. It is not hard to see that F'' is a subgraph of F' and therefore of G_2. Since $\mathcal{F}$ is closed under compression, $F'' \in \mathcal{F}$. Therefore G_2 has property $P_n^{\mathcal{F}}$. $\square$

Lemma 3.5. *For $n \ge T = 2^{2^t}$, we have $\chi_n \equiv \chi_{n-T+1} \pmod 2$.*

Proof. Let k be a fixed integer with $0 \le k \le 2^t$. Recall that the group action induced on the edges creates 2^t orbits. Consider the family of all n-vertex invariant graphs G with $\mathcal{R}(G)$ and G_{right} fixed, and $\mathrm{orb}(G) = k$. By Lemma 3.4, either all graphs in this family have property $P_n^{\mathcal{F}}$ or none of them does. The size of this family is $\binom{2^t}{k}$ which is even if $k \neq 0$ and $k \neq 2^t$. If $k = 2^t$, G_{left} is a complete graph, and so G contains a clique of size T. From (3), we see that G has property $P_n^{\mathcal{F}}$. Therefore, by (5),

$$\chi_n \equiv \# \left\{ G : \begin{array}{l} \text{orb}(G) = 0 \text{ and } G \text{ is nontrivial,} \\ \text{invariant and doesn't satisfy } P_n^{\mathcal{F}} \end{array} \right\} \pmod{2} . \qquad (7)$$

Suppose we take such a G with $\text{orb}(G) = 0$ and collapse all its left vertices into one vertex which we connect to every vertex in $\mathcal{R}(G)$ and to no others, thereby yielding a graph $\hat{G}$. This gives a bijection from n-vertex invariant graphs G with $\text{orb}(G) = 0$ to $(n - T + 1)$-vertex graphs.

It is clear that if $\hat{G}$ has property $P_{n-T+1}^{\mathcal{F}}$, then G has property $P_n^{\mathcal{F}}$. Now suppose G has property $P_n^{\mathcal{F}}$ and let $F \in \mathcal{F}$ be a subgraph of G. Since $\text{orb}(G) = 0$, the vertices in F_{left} form an independent set; thus we may color them all with one color and then color each remaining vertex of F with a distinct color different from the one just used. This coloring produces a compression $\hat{F} \triangleleft F$ which clearly is a subgraph of $\hat{G}$. Since $\mathcal{F}$ is closed under compression, we have $\hat{F} \in \mathcal{F}$ and so $\hat{G}$ has property $P_{n-T+1}^{\mathcal{F}}$. Thus our bijection respects the relevant property and this completes the proof. $\qquad \square$

We now have all the pieces needed for the

Proof. **(of Lemma 3.2)** Set $n = T = 2^{2^t}$. The only way for an n-vertex graph to have $\text{orb}(G) = 0$ is for it to have no edges. Using (7), this implies $\chi_T \equiv 0 \pmod{2}$. Invoking Lemma 3.5 completes the proof. $\qquad \square$

4   Proof of the Main Theorem

We now return to proving Theorem 1.1. According to the theorem's hypotheses

$$n = \sum_{i=1}^{r} q^{\alpha_i} \qquad (8)$$

where q is a prime power, $q \geq |H|$, each $\alpha_i \geq 1$ and $r \equiv 1 \pmod{r_0}$. Our goal is to show that Q_n^H is evasive under these hypotheses for some choice of r_0.

The chief difficulty in applying the topological approach outlined in Section 2 lies in having to construct a group action natural enough for the property under consideration and satisfying the stringent conditions on the underlying group necessary for Theorem 2.3 to apply. In this section we shall come up with a group action that allows us to "merge together" big clusters of vertices in our graph, in the process changing the property under consideration from Q_n^H to $P_r^{\mathcal{F}}$ for some family $\mathcal{F}$ of graphs, r being as in (8).

We partition vertex set of our n-vertex graph into clusters $V_1, \ldots, V_r$, with $|V_i| = q^{\alpha_i}$ and identify vertices in V_i with elements of the finite field $\mathbb{F}_{q^{\alpha_i}}$. Define a permutation group Γ on the vertices as follows:

$$\Gamma = \{\langle a, b_1, b_2, \ldots, b_r \rangle : a \in \mathbb{F}_q^*, \, b_i \in \mathbb{F}_{q^{\alpha_i}}\} , \qquad (9)$$

where $\langle a, b_1, b_2, \ldots, b_r \rangle$ denotes a permutation which sends $x \in V_i = \mathbb{F}_{q^{\alpha_i}}$ to $ax + b_i \in V_i$. Let $\Gamma_1 = \{\langle 1, b_1, \ldots, b_r \rangle \, : \, b_i \in \mathbb{F}_{q^{\alpha_i}}\}$. It is easy to check that Γ_1 is a normal subgroup of Γ, $|\Gamma_1| = q^{\alpha_1 + \cdots + \alpha_r}$, a prime power, and $\Gamma / \Gamma_1 \cong \mathbb{F}_q^*$, a cyclic group. Thus Γ satisfies the hypotheses of Theorem 2.3.

As in Section 3, the action of Γ induces a group action on the edges and thus partitions the edges into orbits. Let $\mathcal{A}$ denote the set of these orbits and let $\Delta = \Delta(Q_n^H)$ denote the abstract complex associated with property Q_n^H. Define a complex Δ_Γ on $\mathcal{A}$ as in (2):

$$\Delta_\Gamma = \{\mathcal{D} \subseteq \mathcal{A} \, : \, \bigcup_{A \in \mathcal{D}} A \in \Delta\} \, . \tag{10}$$

Our intention is to show that the Euler characteristic $\chi(\Delta_\Gamma) \neq 1$. By Theorem 2.3, evasiveness of Q_n^H will follow. To this end, let us investigate what the faces of Δ_Γ look like. Call an edge an *intracluster edge* if both its end points lie in the same V_i for some i; call the edge an *intercluster edge* otherwise.

Lemma 4.1. *An orbit containing an intracluster edge is not contained in any face of Δ_Γ.*

Proof. Let $A \in \mathcal{A}$ be the orbit of the intracluster edge (u, v), $u, v \in V_i$. Then $A = \{(au + b, av + b) \, : \, b \in \mathbb{F}_{q^{\alpha_i}}, a \in \mathbb{F}_q^*\}$. Set $w = v - u$. Then $(0, w) \in A$. Consider the set of vertices $X = \{wz \, : \, z \in \mathbb{F}_q\}$. For $0 \neq x \in X$ we clearly have $(0, x) \in A$. Thus for any pair of distinct vertices $x_1, x_2 \in X$, we have $(0, x_2 - x_1) \in A$, whence $(x_1, x_2) \in A$. So A contains all edges among vertices in X. Since $|X| = q \geq |H|$, the orbit A contains H as a subgraph. By definition, Δ cannot contain a face that includes A and so no face of Δ_Γ can contain A. $\square$

If $u \in V_i, v \in V_j, i < j$, then the orbit of the intercluster edge (u, v) is the set E_{ij} of all edges between V_i and V_j. Let $\mathcal{E} = \{E_{ij} | \, i < j\} \subseteq \mathcal{A}$. From the preceding lemma and (10) it is clear that

$$\Delta_\Gamma = \{\mathcal{D} \subseteq \mathcal{E} \, : \, \bigcup_{A \in \mathcal{D}} A \in \Delta\} \, . \tag{11}$$

Let $\mathcal{D}$ be any subset of $\mathcal{E}$. Then $G_\mathcal{D} = \bigcup_{A \in \mathcal{D}} A$ is a graph on n vertices with no intracluster edges and such that if $i \neq j$, the edges between V_i and V_j are either all present or all absent. Define a graph $\hat{G}_\mathcal{D}$ on r vertices $v_1, \ldots, v_r$ such that (v_i, v_j) is an edge iff all edges between V_i, V_j are present in $G_\mathcal{D}$.

Let $\mathcal{T}_H$ denote the family of all graphs $\hat{H}$ such that $\hat{H} \lhd H$. It is easy to check that $\mathcal{T}_H$ is closed under compression (refer to Definition 3.1). The following lemma is simple to prove and connects this section with Section 3.

Lemma 4.2. *H is a subgraph of $G_\mathcal{D}$ if and only if there is a $\hat{H} \in \mathcal{T}_H$ such that $\hat{H}$ is a subgraph of $\hat{G}_\mathcal{D}$. In other words, $G_\mathcal{D}$ satisfies Q_n^H iff $\hat{G}_\mathcal{D}$ satisfies $P_r^{\mathcal{T}_H}$.*

Proof. Suppose H is a subgraph of $G_\mathcal{D}$. Consider the following coloring of $G_\mathcal{D}$: all vertices in a cluster are colored the same and no two clusters use the same color. This is a valid coloring since each cluster of vertices is an independent set. This coloring induces a coloring of H which in turn induces a compression $\hat{H} \lhd H$. Clearly this $\hat{H}$ is a subgraph of $\hat{G}_{\mathcal{D}}$.

Now suppose $\hat{H} \lhd H$ is a subgraph of $\hat{G}_{\mathcal{D}}$. Consider the graph H_1 with vertices in $\cup_{i=1}^r V_i$ formed by taking all edges in E_{ij} whenever v_i and v_j are adjacent in $\hat{H}$. Since each $|V_i| \geq q \geq |H|$, it follows that H is a subgraph of H_1, and therefore of $G_{\mathcal{D}}$. □

We are ready to prove our Main Theorem.

Proof. **(of Theorem 1.1)** Suppose Q_n^H is not evasive. From Theorem 2.3, we have $\chi(\Delta_\Gamma) = 1$. If $r = 1$, there is only one cluster, so by Lemma 4.1 we have $\Delta_\Gamma = \{\emptyset\}$, whence $\chi(\Delta_\Gamma) = 0$, a contradiction. Therefore $r > 1$. Equation (11) and Lemma 4.2 imply that there is a one-to-one correspondence between faces of Δ_Γ and nontrivial r-vertex graphs not satisfying property $P_r^{T_H}$. Hence the abstract complex Δ_Γ is same as the abstract complex $\Delta_r^{T_H}$ defined in Section 3. It follows from the definition of compression that T_H contains the complete graph on $\mathrm{chr}(H)$ vertices and contains no smaller graph. Therefore, (3) yields $t = \lceil \lg\lg \mathrm{chr}(H) \rceil$. Setting $r_0 = 2^{2^t} - 1$ and applying Lemma 3.2 we have $\chi(\Delta_r^{T_H}) \neq 1$ and so $\chi(\Delta_\Gamma) \neq 1$, a contradiction. □

5   Consequences and Extensions

Our techniques enable us to prove certain results with "cleaner" statements than our Main Theorem 1.1; we prove four such results below. The first two are simple corollaries of Theorem 1.1 while the other two can be easily proved using the machinery of its proof. Finally, we present an interesting generalization of our Main Theorem.

Theorem 5.1. *For any graph H there exist infinitely many primes p with the following property: for all sufficiently large n divisible by p, the property Q_n^H is evasive.*

Remark: Note that this establishes the evasiveness of Q_n^H for an arithmetic progression of values of n.

Proof. Choose an integer t such that $T = 2^{2^t}$ is at least $|H|$. By Dirichlet's Theorem there exist infinitely many primes p such that $p \equiv 2 \pmod{T - 1}$. Fix one such $p \geq T$ and pick any $n \geq p^2(T - 1)$ divisible by p. Now $p - 1$ is relatively prime to $T - 1$, therefore there is an integer x such that $x(p - 1) \equiv n/p - 1 \pmod{T - 1}$ and $0 \leq x < T - 1$. From the lower bound on n we have $n/p - px > 0$. Therefore we can write

$$n = \sum_{i=1}^{x} p^2 + \sum_{i=1}^{n/p-px} p$$

which is an expression of n as a sum of powers of p. The number of summands in this expression is $x + n/p - px \equiv 1 \pmod{T - 1}$. Since $p \geq T \geq |H|$, we can apply Theorem 1.1 to conclude that Q_n^H is evasive. □

Corollary 5.2. *For any graph H there exists a constant $c = c(H)$ such that for all sufficiently large n, the decision tree complexity of Q_n^H is at least $\frac{1}{2}n^2 - cn$.*

□

Theorem 5.3. *If the graph H is bipartite, then Q_n^H is evasive for all sufficiently large n.*

Proof. Since $\mathrm{chr}(H) = 2$, in the proof of Theorem 1.1, using the notation of that proof, we may take $t = 0$ which gives $r_0 = 1$. The condition $r \equiv 1 \pmod{r_0}$ is now trivially satisfied. The condition on n becomes a simple requirement that n be divisible by a prime power $q \geq |H|$. But if n is sufficiently large then it clearly satisfies this condition.

□

Theorem 5.4. *Let $\mathcal{M}$ be an infinite minor-closed family of graphs that does not include all graphs. For n-vertex graphs, let $R_n^{\mathcal{M}}$ be the property of being in $\mathcal{M}$. Then $R_n^{\mathcal{M}}$ is evasive for all sufficiently large n.*

Remark: Planarity was already known to be evasive. This result is a major generalization.

Proof. Let H be a graph not in $\mathcal{M}$ with minimum size, and let $h = |H|$. Then H is a minor of both the complete graph K_h and the complete bipartite graph $K_{h,h}$; therefore no graph in $\mathcal{M}$ can contain either K_h or $K_{h,h}$ as a subgraph.

Suppose n is divisible by a prime power $q \geq h$, a condition that always holds if n is sufficiently large. Following the argument of Section 4 we divide the labeled vertices of the candidate graph G into clusters of size q and consider the orbits of the edges created by the action of the group Γ described there. Let Δ be the abstract complex associated *with the negation*[3] of $R_n^{\mathcal{M}}$. An orbit containing an intracluster edge cannot be included in a face of Δ_Γ because its edges, if present, would create a K_q subgraph. An orbit containing an intercluster edge cannot be included either because its edges, if present, would create a $K_{q,q}$ subgraph. Thus, $\Delta_\Gamma = \{\emptyset\}$ and so $\chi(\Delta_\Gamma) = 0 \neq 1$. By Theorem 2.3, the negation of $R_n^{\mathcal{M}}$ is evasive and therefore so is $R_n^{\mathcal{M}}$.

□

The next theorem generalizes our Main Theorem and can be proved essentially using the same argument as that for the Main Theorem.

Theorem 5.5. *Let $f : \{0,1\}^k \to \{0,1\}$ be a nontrivial monotone boolean function and let $H_1, \ldots, H_k$ be arbitrary graphs. Define the composite property $Q_n = f(Q_n^{H_1}, \ldots, Q_n^{H_k})$. Then there exists an integer r_0 with the following property. Suppose $n = \sum_{i=1}^{r} q^{\alpha_i}$ where q is a prime power, $q \geq \max_{1 \leq i \leq k} |H_i|$, each $\alpha_i \geq 1$ and $r \equiv 1 \pmod{r_0}$. Then Q_n is evasive.*

□

Remark: This theorem shows, for instance, that properties like "G either contains H_1 as a subgraph or else contains both H_2 and H_3 as subgraphs" are evasive for several values of n. This theorem has corollaries similar to Theorem 5.1 and Corollary 5.2.

---

[3] Notice that the property $R_n^{\mathcal{M}}$ is not monotone. However, its negation is monotone. Clearly a property is evasive if its negation is.

6 Concluding Remarks

The major open question in the area of decision tree complexity of graph properties is to settle the Karp Conjecture. The pioneering work of Kahn et al. [4] has given us a possible direction to follow in attempting to settle this conjecture. Since the publication of that work, our work is the first which extends their topological approach for a fairly general class of graph properties.

An obvious open question raised by our work is: how far can one enlarge the set of values of n for which our results hold? We conjecture that in the notation of Section 3, we have $\chi_n \neq 1$ for large enough n. If proved true, this conjecture would remove all number theoretic restrictions in the Main Theorem.

Acknowledgments

We would like to express our sincere thanks to Professors Sanjeev Arora, Bernard Chazelle and Andrew Yao for their valuable comments and suggestions. We are grateful to the referees for their comments and for pointing out a simplification in the proof of Lemma 3.3.

References

1. Best, M.R., van Emde Boas, P., Lenstra, H.W., Jr. *A sharpened version of the Aanderaa-Rosenberg Conjecture*, Report ZW 30/74, Mathematisch Centrum Amsterdam, 1974.
2. Bollobás, B. *Complete subgraphs are elusive*, J. Combinatorial Th. (B), **21** (1976), 1–7.
3. Bollobás, B. *Extremal Graph Theory*, Academic Press, 1978, Chapter 8.
4. Kahn, J., Saks, M., Sturtevant, D. *A topological approach to evasiveness*, Combinatorica, **4** (1984), 297–306.
5. Milner, E.C., Welsh, D.J.A. *On the computational complexity of graph theoretical properties*, Proc. 5th British Columbia Conf. on Combinatorics (C.St.J.A. Nash-Williams and J. Sheehan, Eds.), 1975, 471–487.
6. Rivest, R.L., Vuillemin, J. *On recognizing graph properties from adjacency matrices*, Theoret. Comput. Sci., **3** (1976), 371–384.
7. Rosenberg, A.L. *On the time required to recognize properties of graphs: A problem*, SIGACT News, **5** (1973), 15–16.
8. Triesch, E. *Some results on elusive graph properties*, SIAM J. Comput., **23** (1994), 247–254.
9. Yao, A.C. *Monotone bipartite graph properties are evasive*, SIAM J. Comput., **17** (1988), 517–520.

On the Complexity of Computing Minimum Energy Consumption Broadcast Subgraphs

Andrea E.F. Clementi[1], Pilu Crescenzi[2,*], Paolo Penna[1,**],
Gianluca Rossi[2,*], and Paola Vocca[1]

[1] Dipartimento di Matematica, Università di Roma "Tor Vergata"
Via della Ricerca Scientifica, I-00133 Roma, Italy
{clementi,penna,vocca}@mat.uniroma2.it
[2] Dipartimento di Sistemi e Informatica, Università di Firenze
Via C. Lombroso 6/17, I-50134 Firenze, Italy
{piluc,rossig}@dsi.unifi.it

Abstract. We consider the problem of computing an optimal range assignment in a wireless network which allows a specified source station to perform a broadcast operation. In particular, we consider this problem as a special case of the following more general combinatorial optimization problem, called Minimum Energy Consumption Broadcast Subgraph (in short, MECBS): Given a weighted directed graph and a specified source node, find a minimum cost range assignment to the nodes, whose corresponding transmission graph contains a spanning tree rooted at the source node. We first prove that MECBS is not approximable within a sub-logarithmic factor (unless P=NP). We then consider the restriction of MECBS to wireless networks and we prove several positive and negative results, depending on the geometric space dimension and on the distance-power gradient. The main result is a polynomial-time approximation algorithm for the NP-hard case in which both the dimension and the gradient are equal to 2: This algorithm can be generalized to the case in which the gradient is greater than or equal to the dimension.

1 Introduction

Wireless networking technology will play a key role in future communications and the choice of the network architecture model will strongly impact the effectiveness of the applications proposed for the mobile networks of the future. Broadly speaking, there are two major models for wireless networking: *single-hop* and *multi-hop*. The single-hop model [22], based on the cellular network model, provides one-hop wireless connectivity between mobile hosts and static nodes known as *base stations*. This type of networks relies on a fixed backbone infrastructure that interconnects all base stations by high-speed wired links. On the other hand, the multi-hop model [15] requires neither fixed, wired infrastructure nor predetermined interconnectivity. *Ad hoc* networking [12] is the most popular type of multi-hop wireless networks because of its simplicity:

---

* Research partially supported by Italian MURST project "Algoritmi per Grandi Insiemi di Dati: Scienza ed Ingegneria".
** Part of this work has been done while visiting INRIA Sophia Antipolis (MASCOTTE Project).

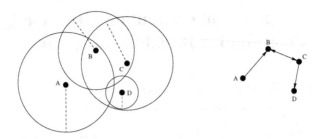

Fig. 1. A Range Assignment and Its Corresponding Directed Transmission Graph.

Indeed, an *ad hoc* wireless network is constituted by a homogeneous system of *mobile* stations connected by wireless links. In ad hoc networks, to every station is assigned a transmission range: The overall range assignment determines a transmission (directed) graph since one station s can transmit to another station t if and only if t is within the transmission range of s (see Fig. 1).

The range transmission of a station depends, in turn, on the energy power supplied to the station: In particular, the power P_s required by a station s to correctly transmit data to another station t must satisfy the inequality

$$\frac{P_s}{d(s,t)^\alpha} > \gamma \tag{1}$$

where $d(s,t)$ is the Euclidean distance between s and t, $\alpha \geq 1$ is the *distance-power gradient*, and $\gamma \geq 1$ is the *transmission-quality* parameter. In an ideal environment (i.e. in the empty space) it holds that $\alpha = 2$ but it may vary from 1 to more than 6 depending on the environment conditions of the place the network is located (see [19]). The fundamental problem underlying any phase of a dynamic resource allocation algorithm in ad-hoc wireless networks is the following: Find a transmission range assignment such that (1) the corresponding transmission graph satisfies a given property π, and (2) the overall energy power required to deploy the assignment (according to Eq. 1) is minimized.

A well-studied case of the above problem consists in choosing π as follows: The transmission graph has to be strongly connected. In this case, it is known that: (a) the problem is not solvable in polynomial time (unless P=NP) [6,14], (b) it is possible to compute a range assignment which is at most twice the optimal one (that is, the problem is 2-approximable), for multi-dimensional wireless networks [14], (c) there exists a constant $r > 1$ such that the problem is not r-approximable (unless P=NP), for d-dimensional networks with $d \geq 3$ [6], and (d) the problem can be solved in polynomial time for one-dimensional networks [14]. Another analyzed case consists in choosing π as follows: The diameter of the transmission graph has to be at most a fixed value h. In this case, while non-trivial negative results are not known, some tight bounds (depending on h) on the minimum energy power have been proved in [7], and an approximation algorithm for the one-dimensional case has been given in [5]. Other trade-offs between connectivity and energy consumption have been obtained in [16,21,24].

In this paper we address the case in which π is defined as follows: *Given a source station s, the transmission graph has to contain a directed spanning tree rooted at s.* This case has been posed as an open question by Ephremides in [10]: Its relevance is due to the fact that any transmission graph satisfying the above property allows the source station to perform a *broadcast* operation. Broadcast is a task initiated by the source station which transmits a message to all stations in the wireless network: This task constitutes a major part of real life multi-hop radio network [2,3].

The Optimization Problem. The broadcast range assignment problem described above is a special case of the following combinatorial optimization problem, called MINI-MUM ENERGY CONSUMPTION BROADCAST SUBGRAPH (in short, MECBS). Given a weighted directed graph $G = (V, E)$ with edge weight function $w : E \rightarrow \mathcal{R}^+$, a *range assignment* for G is a function $r : V \rightarrow \mathcal{R}^+$: The *transmission graph* induced by G and r is defined as $G_r = (V, E')$ where

$$E' = \bigcup_{v \in V} \{(v, u) : (v, u) \in E \wedge w(v, u) \leq r(v)\}.$$

The MECBS problem is then defined as follows: Given a *source node* $s \in V$, find a range assignment r for G such that G_r contains a spanning tree of G rooted at s and $\text{cost}(r) = \sum_{v \in V} r(v)$ is minimized.

Let us consider, for any $d \geq 1$ and for any $\alpha \geq 1$, the family of graphs N_d^α, called *(d-dimensional) wireless networks*, defined as follows: A complete (undirected) graph G belongs to N_d^α if it can be embedded on a d-dimensional Euclidean space such that the weight of an edge is equal to the αth power of the Euclidian distance between the two endpoints of the edge itself. The restriction of MECBS to graphs in N_d^α is denoted by MECBS[N_d^α]: It is then clear that the previously described broadcast range assignment problem in the ideal 2-dimensional environment is MECBS[N_2^2].

Our Results. In this paper, we analyze the complexity of the MINIMUM ENERGY CONSUMPTION BROADCAST SUBGRAPH problem both in the general case and in the more realistic case in which the instances are wireless networks. In particular, we first prove that MECBS *is not approximable within a sub-logarithmic factor, unless* P=NP (see Sect. 2). Subsequently, we consider MECBS[N_d^α], for any $d \geq 1$ and for any $\alpha \geq 1$, and we prove the following results (see Sect. 3):

- *For any $d \geq 1$, MECBS[N_d^1] is solvable in polynomial time:* This result is based on a simple observation.
- MECBS[N_d^α] *is not solvable in polynomial time (unless* P=NP), *for any $d \geq 2$ and for any $\alpha > 1$:* This negative result uses the same arguments of [6].
- *For any $\alpha \geq 2$, MECBS[N_2^α] is approximable within a constant factor:* This is the main result of the paper. A major positive aspect of the approximation algorithm lies on the fact that it is just based on the computation of a standard minimum spanning tree (shortly, MST). In a network with dynamic power control, the range assigned to the stations can be modified at any time: Our algorithm can thus take advantage of all known techniques to dynamically maintain MSTs (see, for example,

[9,11,18]). MSTs have already been used in order to develop approximation algorithms for range assignment problems in wireless networks: However, we believe that the analysis of the performance of our algorithm (which is based on computational geometry techniques) is rather interesting by itself.

Finally, in Sect. 4 we first observe that our approximation algorithm can be generalized in order to deal with MECBS[N_d^α], for any $d \geq 2$ and for any $\alpha \geq d$: However, we also prove that the approximation ratio grows at least exponentially with respect to d. We then briefly consider the behavior of our approximation algorithm when applied to MECBS[N_d^α] with $\alpha < d$ and we summarize some questions left open by this paper.

Prerequisites. We assume the reader to be familiar with the basic concepts of computational complexity theory (see, for example, [4,20]) and with the basic concepts of the theory of approximation algorithms (see, for example, [1]).

2   The Complexity of MECBS

In this section, we prove that the MINIMUM ENERGY CONSUMPTION BROADCAST SUBGRAPH problem is not approximable within a sub-logarithmic factor (unless P= NP). To this aim, we provide a reduction from MIN SET COVER to MECBS. Recall that MIN SET COVER is defined as follows: given a collection C of subsets of a finite set S, find a minimum cardinality subset $C' \subseteq C$ such that every element in S belongs to at least one member of C'. It is known that, unless P=NP, MIN SET COVER is not approximable within $c \log n$, for some $c > 0$, where n denotes the cardinality of S [23] (see, also, the list of optimization problems contained in [1]).

Theorem 1. *If* P $\neq$ NP, *then* MECBS *is not approximable within a sub-logarithmic factor.*

Proof (Sketch). Let x be an instance of the MIN SET COVER problem. In the full version of the paper, we show how to construct an instance y of MECBS such that there exists a feasible solution for x whose cardinality is equal to k if and only if there exists a feasible solution for y whose cost is equal to $k + 1$. This clearly implies that if MECBS is approximable within a sub-logarithmic factor, then MIN SET COVER is approximable within a sub-logarithmic factor: The theorem hence follows from the non-approximability of MIN SET COVER.   □

One interesting feature of the reduction used in the previous proof is that it also allows us to show that MECBS is not approximable within a constant factor (unless P=NP), when the problem is restricted to undirected graphs.

3   The Restriction to Wireless Networks

In this section we analyze the complexity of the MINIMUM ENERGY CONSUMPTION BROADCAST SUBGRAPH problem restricted to wireless networks, that is, MECBS[N_d^α] with $d, \alpha \geq 1$. First of all, observe that if $\alpha = 1$ (that is, the edge weights coincide with the Euclidian distances), then the optimal range assignment is simply obtained by assigning to s the distance from its farthest node and by assigning 0 to all other nodes. We then have that the following result holds.

Theorem 2. *For any $d \geq 1$, then MECBS[N_d^1] is solvable in polynomial time.*

It is, instead, possible to prove the following result, whose proof is an adaptation of the one given in [6] to prove the NP-hardness of computing a minimum range assignment that guarantees the strong connectivity of the corresponding transmission graph (the proof will be given in the full version of the paper).

Theorem 3. *For any $d \geq 2$ and for any $\alpha > 1$, MECBS[N_d^α] is not solvable in polynomial time (unless P= NP).*

Because of the above negative result, it is reasonable to look for polynomial-time algorithms that compute approximate solutions for MECBS restricted to wireless networks. We now present and analyze an efficient approximation algorithm for MECBS[N_2^α], for any $\alpha \geq 2$. In what follows, given a graph $G \in N_2^\alpha$, we denote by $G^{1/\alpha}$ the graph obtained from G by setting the weight of each edge to the αth root of the weight of the corresponding edge in G: Hence, $G^{1/\alpha} \in N_2^1$, that is, there exists an embedding of $G^{1/\alpha}$ on the plane such that the Euclidean distance $d(u, v)$ between two nodes u and v coincides with the weight of the edge (u, v) in $G^{1/\alpha}$.

> **The Approximation Algorithm** MST-ALG. Given a graph $G \in N_2^\alpha$ and a specified source node s, the algorithm first computes a MST T of G (observe that this computation does not depend on the value of α). Subsequently, it makes T downward oriented by rooting it at s. Finally, the algorithm assigns to each vertex v the maximum among the weights of all edges of T outgoing from v. Clearly, the algorithm runs in polynomial time and computes a feasible solution.

3.1 The Performance Analysis of the Approximation Algorithm

The goal of this section is to prove that, for any instance $x = \langle G = (V, E), w, s \rangle$ of MECBS[N_2^α] with $\alpha \geq 2$, the range assignment r computed by MST-ALG satisfies the following inequality:

$$\mathsf{cost}(r) \leq 10^{\alpha/2} \cdot 2^\alpha \mathsf{opt}(x), \tag{2}$$

where $\mathsf{opt}(x)$ denotes the cost of an optimal range assignment. First notice that

$$\mathsf{cost}(r) \leq w(T),$$

where, for any subgraph G' of G, $w(G')$ denotes the sum of the weights of the edges in G'. As a consequence of the above inequality, it now suffices to show that there exists a spanning subgraph G' of G such that $w(G') \leq 10^{\alpha/2} \cdot 2^\alpha \mathsf{opt}(x)$. Indeed, since the weight of T is bounded by the weight of G', we have that Eq. 2 holds.

In order to prove the existence of G', we make use of the following theorem whose proof is given in Sect. 3.2.

Theorem 4. *Let $G \in N_2^\alpha$ with $\alpha \geq 2$ and let R be the diameter of $G^{1/\alpha}$, that is, the maximum distance between two nodes in $G^{1/\alpha}$. Then, for any MST T of G,*

$$w(T) \leq 10^{\alpha/2} R^\alpha.$$

Let r_{opt} be an optimal assignment for x. For any $v \in V$, let

$$S(v) = \{u \in V : w(v, u) \leq r_{opt}(v)\}$$

and let $T(v)$ be a MST of the subgraph of G induced by $S(v)$. From Theorem 4, it follows that $w(T(v)) \leq 10^{\alpha/2} \cdot 2^\alpha r_{opt}(v)$. Consider the spanning subgraph $G' = (V, E')$ of G such that

$$E' = \bigcup_{v \in V} \{e \in E : e \in T(v)\}.$$

It then follows that

$$w(G') \leq \sum_{v \in V} w(T(v)) \leq 10^{\alpha/2} \cdot 2^\alpha \sum_{v \in V} r_{opt}(v) = 10^{\alpha/2} \cdot 2^\alpha \text{opt}(x).$$

We have thus proved the following result.

Theorem 5. *For any $\alpha \geq 2$, MECBS[N_2^α] is approximable within $10^{\alpha/2} \cdot 2^\alpha$.*

3.2 Proof of Theorem 4

Given a graph $G \in N_2^\alpha$ with $\alpha \geq 2$, we identify the nodes of G with the points corresponding to an embedding of $G^{1/\alpha}$ on the plane: Recall that the Euclidean distance $d(u, v)$ between two points u, v coincides with the weight of the edge (u, v) in $G^{1/\alpha}$.

Let us first consider the case $\alpha = 2$ and let $e_i = (u_i, v_i)$ be the ith edge in T, for $i = 1, \ldots, |V| - 1$ (any fixed ordering of the edges is fine). We denote by D_i the *diametral open circle* of e_i, that is, the open disk whose center c_i is on the midpoint of e_i and whose diameter is $d(u_i, v_i)$. From Lemma 6.2 of [17], it follows that D_i contains no point from the set $V - \{u_i, v_i\}$. The following lemma, instead, states that, for any two diametral circles, the center of one circle is not contained in the other circle.

Lemma 1. *For any $i, j \in \{1, \ldots, |V| - 1\}$ with $i \neq j$, c_i is not contained in D_j.*

Proof. Suppose by contradiction that there exist two diametral circles D_i and D_j such that c_i is contained in D_j. We will show that the longest edge between e_i and e_j can be replaced by a strictly shorter one, still maintaining the connectivity of T: Since T is a MST the lemma will follow. Let us assume, without loss of generality, that $d(u_j, v_j) \geq d(u_i, v_i)$. We first prove that

$$\max\{d(u_i, u_j), d(v_i, v_j)\} < d(u_j, v_j) \tag{3}$$

Let Y^+ and Y^- be the half-planes determined by the line identified by c_i and c_j: Without loss of generality, we may assume that v_i and v_j (respectively, u_i and u_j) are both contained in Y^+ (respectively, Y^-), as shown in Fig. 2. Assume also that

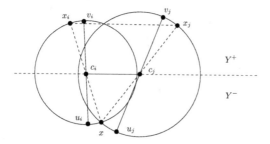

Fig. 2. The Proof of Lemma 1.

$d(v_i, v_j) \geq d(u_i, u_j)$ (the other case can be proved in a similar way). Let x be the intersection point in Y^- between the two circumferences determined by D_i and D_j (notice that, since D_i and D_j are open disks, neither D_i nor D_j contains x) and let x_i and x_j be the points diametrically opposite to x with respect to c_i and c_j, respectively. Clearly, $d(v_i, v_j) \leq d(x_i, x_j)$. Eq. 3 easily follows from the following

Fact 1. $d(x_i, x_j) < d(u_j, v_j)$.

Proof (of Fact 1). By definition, c_i (respectively, c_j) is the median of the segment $\overline{xx_i}$ (respectively, $\overline{xx_j}$). Thus, the triangles $\triangle(xx_ix_j)$ and $\triangle(xc_ic_j)$ are similar. From the hypothesis that $c_i \in D_j$, it follows that $d(c_i, c_j) < d(x, c_j)$. Thus, by similarity, it must hold that

$$d(x_i, x_j) < d(x, x_j) = d(u_j, v_j)$$

and the fact follows. □

As a consequence of Eq. 3, we can replace in T, $e_j = (u_j, v_j)$ by either (u_i, u_j) or (v_i, v_j) (the choice depends on the topology of T), thus obtaining a better spanning tree. □

We now use the above lemma in order to bound the number of diametral circles any point on the plane belongs to.

Lemma 2. *For any point p on the plane, p is contained in at most five diametral circles.*

Proof. Suppose by contradiction that there exist a point p covered by (at least) six diametral circles. Then, there must exist two circles D_1 and D_2 such that their respective centers c_1 and c_2 form with p an angle $\beta \leq \pi/3$ (see Fig. 3(a)). Let R_1 and R_2 denote the diameters of D_1 and D_2, respectively. Since $\beta \leq \pi/3$, we have that

$$d(c_1, c_2) \leq \max\{d(c_1, p), d(c_2, p)\} < \max\{R_1, R_2\}$$

where the strict inequality is due to the fact that $p \in D_1 \cap D_2$ and that both D_1 and D_2 are open disks. Hence, either $c_1 \in D_2$ or $c_2 \in D_1$, thus contradicting Lemma 1. □

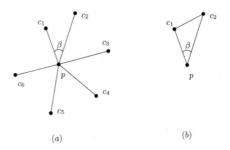

Fig. 3. The Proof of Lemma 2

For any i with $1 \leq i \leq |V| - 1$, let $\overline{D_i}$ denote the smallest closed disk that contains D_i. The last lemma of this section states that the union of all $\overline{D_i}$s is contained in a closed disk whose diameter is comparable to the diameter of $G^{1/\alpha}$.

Lemma 3. *Let* $D = \bigcup_{e_i \in T} \overline{D_i}$. *Then, D is contained into the closed disk whose diameter is equal to $\sqrt{2}R$ and whose center coincides with the center of D.*

Proof. Consider any two points x and y within D. It is easy to see that the worst case corresponds to the case in which both x and y are on the boundary of D. Consider the closed disk whose diameter is equal to $d(x, y)$ and whose center c' is on the midpoint of the segment $\overline{xy}$, and let z be any point on its boundary (see Fig. 4). It holds that $d(c, z) \leq \sqrt{2}R/2$, where c is the center of D. Indeed, from the triangular inequality we have that

$$d(c, z) \leq d(c, c') + d(c', z) = d(c, c') + d(x, y)/2.$$

Moreover, since the angle $cc'y$ is equal to $\pi/2$,

$$d(c, c')^2 + d(c', y)^2 = d(c, y)^2 = R^2/4.$$

Thus,

$$d(c, z) \leq \sqrt{\frac{R^2 - d(x, y)^2}{4}} + d(x, y)/2.$$

The right end of this equation reaches its maximum when $d(x, y) = \sqrt{2}R/2$, which implies $d(c, z) \leq \sqrt{2}R/2$. Hence the lemma follows. $\square$

We are now able to prove Theorem 4. In particular, we have to prove that

$$\sum_{i=1}^{|V|-1} d(u_i, v_i)^2 \leq 10R^2, \tag{4}$$

where (u_i, v_i) is the ith edge in T, for $i = 1, \ldots, |V| - 1$. Indeed, let $\text{Area}(D_i)$ denote the area of $\overline{D_i}$. It then holds that

$$\sum_{i=1}^{|V|-1} d(u_i, v_i)^2 = \frac{4}{\pi} \sum_{i=1}^{|V|-1} \text{Area}(D_i). \tag{5}$$

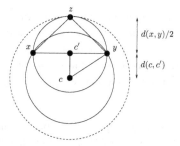

Fig. 4. The Proof of Lemma 3.

By combining Lemma 3 and 2, we have that

$$\sum_{i=1}^{|V|-1} \mathsf{Area}(D_i) \le 5 \cdot \left[\pi \left(\frac{\sqrt{2}R}{2} \right)^2 \right] = \frac{5}{2}\pi R^2. \tag{6}$$

By combining Eq. 5 and 6 we obtain Eq. 4, which proves the lemma for $\alpha = 2$.
   Finally, we consider the case $\alpha > 2$. By using simple computations, we get

$$\mathsf{cost}(r) = \sum_{i=1}^{|V|-1} d(u_i, v_i)^\alpha = \sum_{i=1}^{|V|-1} \left(d(u_i, v_i)^2 \right)^{\alpha/2}$$

$$\le \left(\sum_{i=1}^{|V|-1} d(u_i, v_i)^2 \right)^{\alpha/2} \le 10^{\alpha/2} R^\alpha,$$

where the last inequality follows from Eq. 4. This completes the proof of Theorem 4.

4 Further Results and Open Questions

Algorithm MST-ALG can be generalized to higher dimensions. In particular, it is possible to prove the following result.

Theorem 6. *There exists a function* $f : \mathcal{N} \times \mathcal{R} \to \mathcal{R}$ *such that, for any* $d \ge 2$ *and for any* $\alpha \ge d$, MECBS[N_d^α] *is approximable within factor* $f(d, \alpha)$.

The proof of the above theorem is again based on the computation of a MST of the input graph: Indeed, the algorithm is exactly the same. Unfortunately, the following result (whose proof is based on results in [8,13,25] and will be given in the full version of the paper) shows that the function f in the statement of the theorem grows exponentially with respect to d.

Theorem 7. *There exists a positive constant* γ *such that, for any* d *and for any* k, *an instance* $x_{k,d}$ *of* MECBS[N_d^α] *exists such that* $\mathsf{opt}(x_{k,d}) = k^d$ *while the cost of the range assignment computed by* MST-ALG *is at least* $k^d \cdot 2^{\gamma d}$.

One could also ask whether our algorithm approximates $\text{MECBS}[N_d^\alpha]$ in the case in which $d \geq 2$ and $\alpha < d$. Unfortunately, it is not difficult to produce an instance x such that $\text{opt}(x) = O(n^{\alpha/d})$ while the cost of the range assignment computed by MST-ALG is $\Omega(n)$, where n denotes the number of vertices: For example, in the case $d = 2$, we can just consider the two dimensional grid of side $\sqrt{n}$ and the source node positioned on its center.

Open Problems. Three main problems are left open by this paper. The first one is to improve the analysis of MST-ALG (or to develop a different algorithm with a better performance ratio). Actually, we have performed several experiments and it turns out that the practical value of the performance ratio of MST-ALG (in the case in which $d = 2$ and $\alpha = 2$) is between 2 and 3. The second open problem is to analyze the approximability properties of $\text{MECBS}[N_d^\alpha]$ when $\alpha < d$: In particular, it would be very interesting to study the three-dimensional case. As previously observed, the MST-based algorithm does not guarantee any approximation, and it seems thus necessary to develop approximation algorithms based on different techniques. The last open problem is to consider $\text{MECBS}[N_1^\alpha]$, for any $\alpha \geq 1$: In particular, we conjecture that this problem is solvable in polynomial time.

References

1. G. Ausiello, P. Crescenzi, G. Gambosi, V. Kann, A. Marchetti-Spaccamela, and M. Protasi. *Complexity and Approximation – Combinatorial optimization problems and their approximability properties.* Springer Verlag, 1999.
2. R. Bar-Yehuda, O. Goldreich, and A. Itai. On the time complexity of broadcast operations in multi-hop radio networks: an exponential gap between determinism and randomization. *J. Computer and Systems Science*, 45:104–126, 1992.
3. R. Bar-Yehuda, A. Israeli, and A. Itai. Multiple communication in multi-hop radio networks. *SIAM J. on Computing*, 22:875–887, 1993.
4. D. P. Bovet and P. Crescenzi. *Introduction to the Theory of Complexity.* Prentice Hall, 1994.
5. A.E.F. Clementi, A. Ferreira, P. Penna, S. Perennes, and R. Silvestri. The minimum range assignment problem on linear radio networks. In *Proc. 8th Annual European Symposium on Algorithms*, volume 1879 of *LNCS*, pages 143–154, 2000.
6. A.E.F. Clementi, P. Penna, and R. Silvestri. Hardness results for the power range assignment problem in packet radio networks. In *Proc. of Randomization, Approximation and Combinatorial Optimization*, volume 1671 of *LNCS*, pages 197–208, 1999. Full version available as ECCC Report TR00-54.
7. A.E.F. Clementi, P. Penna, and R. Silvestri. The power range assignment problem in radio networks on the plane. In *Proc. 17th Annual Symposium on Theoretical Aspects of Computer Science*, volume 1770 of *LNCS*, pages 651–660, 2000. Full version available as ECCC Report TR00-54.
8. J.H. Conway and N.J.A. Sloane. *Sphere Packings, Lattices and Groups.* Springer-Verlag, 1988.
9. B. Dixon, M. Rauch, and R.E. Tarjan. Verification and sensitivity analysis of minimum spanning trees in linear time. *SIAM J. Comput.*, 21:1184–1192, 1992.
10. A. Ephremides. Complicating factors for the use of distributed algorithms in wireless networks. In *1st Int. Workshop on Approximation and Randomized Algorithms in Communication Networks*, invited talk, 2000.

11. D. Eppstein. Offline algorithms for dynamic minimum spanning tree problem. *J. of Algorithms*, 17:237–250, 1994.

12. Z. Haas and S. Tabrizi. On some challenges and design choices in ad hoc communications. In *Proc. IEEE MILCOM'98*, 1998.

13. G.A. Kabatiansky and V.I. Levenshtein. Bounds for packings on a sphere and in space (in russian). *Problemy Peredachi Informatsii*, 14(1):3–25, 1978. English translation: *Problems of Information Theory*, 14(1):1–17, 1978.

14. L.M. Kirousis, E. Kranakis, D. Krizanc, and A. Pelc. Power consumption in packet radio networks. *Theoretical Computer Science*, 243:289–306, 2000.

15. G.S. Lauer. *Packet radio routing*, chapter 11. Prentice-Hall, 1995.

16. R. Mathar and J. Mattfeldt. Optimal transmission ranges for mobile communication in linear multi-hop packet radio networks. *Wireless Networks*, 2:329–342, 1996.

17. C. Monma and S. Suri. Transitions in geometric minimum spanning tree. *Discrete and Computational Geometry*, 8:265–293, 1992.

18. E. Nardelli, G. Proietti, and P. Widmayer. Maintainig a minimum spanning tree under transient node failures. In *Proc. 8th Annual European Symposium on Algorithms*, to appear, 2000.

19. K. Pahlavan and A. Levesque. *Wireless information networks*. Wiley-Interscience, 1995.

20. C. H. Papadimitriou. *Computational Complexity*. Addison Wesley, 1994.

21. P. Piret. On the connectivity of radio networks. *IEEE Trans. on Inform. Theory*, 37:1490–1492, 1991.

22. D. Raychaudhuri and N.D. Wilson. ATM-based transport architecture for multiservices wireless personal communication networks. *IEEE J. Selected Areas in Communications*, 12:1401–1414, 1994.

23. R. Raz and S. Safra. A sub-constant error-probability low-degree test, and sub-constant error-probability pcp characterization of np. In *Proc. 29th Ann. ACM Symp. on Theory of Comp.*, pages 784–798, 1997.

24. S. Ulukus and R.D. Yates. Stocastic power control for cellular radio systems. *IEEE Trans. Comm.*, 46:784–798, 1996.

25. A.D. Wyner. Capabilities of bounded discrepancy decoding. *BSTJ*, 44:1061–1122, 1965.

On Presburger Liveness of Discrete Timed Automata

Zhe Dang[1], Pierluigi San Pietro[2], and Richard A. Kemmerer[3]

[1] School of Electrical Engineering and Computer Science
Washington State University, Pullman, WA 99164, USA
[2] Dipartimento di Elettronica e Informazione
Politecnico di Milano, Italia
[3] Department of Computer Science
University of California at Santa Barbara, CA 93106, USA

Abstract. Using an automata-theoretic approach, we investigate the decidability of liveness properties (called *Presburger liveness properties*) for timed automata when Presburger formulas on configurations are allowed. While the general problem of checking a temporal logic such as TPTL augmented with Presburger clock constraints is undecidable, we show that there are various classes of Presburger liveness properties which are decidable for *discrete* timed automata. For instance, it is decidable, given a discrete timed automaton $\mathcal{A}$ and a Presburger property P, whether there exists an ω-path of $\mathcal{A}$ where P holds infinitely often. We also show that other classes of Presburger liveness properties are indeed undecidable for discrete timed automata, e.g., whether P holds infinitely often *for each* ω-path of $\mathcal{A}$. These results might give insights into the corresponding problems for timed automata over dense domains, and help in the definition of a fragment of linear temporal logic, augmented with Presburger conditions on configurations, which is decidable for model checking timed automata.

1   Introduction

Timed automata [3] are widely regarded as a standard model for real-time systems, because of their ability to express quantitative time requirements in the form of clock *regions*: a clock or the difference of two clocks is compared against an integer constant, e.g., $x - y > 5$, where x and y are clocks. A fundamental result in the theory of timed automata is that region reachability is decidable. This has been proved by using the region technique [3]. This result is very useful since in principle it allows some forms of automatic verification of timed automata. In particular, it helps in developing a number of temporal logics [2,6,13,15,4,16], in investigating the model-checking problem and in building model-checking tools [12,17,14] (see [1,18] for surveys).

In real-world applications [7], clock constraints represented as clock regions are useful but often not powerful enough. For instance, we might want to argue whether a non-region property such as $x_1 - x_2 > x_3 - x_4$ (i.e., the difference of clocks x_1 and x_2 is larger than that of x_3 and x_4) always holds when a

A. Ferreira and H. Reichel (Eds.): STACS 2001, LNCS 2010, pp. 132–143, 2001.
© Springer-Verlag Berlin Heidelberg 2001

timed automaton starts from clock values satisfying another non-region property. Hence, it would be useful to consider Presburger formulas as clock constraints, were it not for the fact that a temporal logic like TPTL [6] is undecidable when augmented with Presburger clock constraints [6]. However, recent work [9,10] has found decidable characterizations of the binary reachability of timed automata, giving hope that *some* important classes of non-region properties are still decidable for timed automata.

In this paper, we look at *discrete* timed automata (*dta*), i.e., timed automata where clocks are integer-valued. Discrete time makes it possible to apply, as underlying theoretical tools, a good number of automata-theoretic techniques and results. Besides the facts that discrete clocks are usually easier to handle than dense clocks also for practitioners, and that *dta*s are useful by themselves as a model of real-time systems [5], results on *dta*s may give insights into corresponding properties of dense timed automata [11].

The study of *safety* properties and *liveness* properties is of course of the utmost importance for real-life applications. In [10] (as well as in [9]), it has been shown that the Presburger safety analysis problem is decidable for discrete timed automata. That is, it is decidable whether, given a discrete timed automaton $\mathcal{A}$ and two sets I and P of configurations of $\mathcal{A}$ (tuples of control state and clock values) definable by Presburger formulas, $\mathcal{A}$ always reaches a configuration in P when starting from a configuration in I.

In this paper we concentrate on the Presburger liveness problem, by systematically formulating a number of Presburger liveness properties and investigating their decidability. For instance, we consider the ∃-Presburger-i.o. problem: whether there exists an ω-path p for $\mathcal{A}$ such that p starts from I and P is satisfied on p infinitely often. Another example is the ∀-Presburger-eventual problem: whether for all ω-paths p that start from I, P is eventually satisfied on p.

The main results of this paper show that (using an obvious notation, once it is clear that ∃ and ∀ are path quantifiers):

  - The ∃-Presburger-i.o. problem and the ∃-Presburger-eventual problem are both decidable. So are their duals, the ∀-Presburger-almost-always problem and the ∀-Presburger-always problem.
  - The ∀-Presburger-i.o. problem and the ∀-Presburger-eventual problem are both undecidable. So are their duals, the ∃-Presburger-almost-always problem and the ∃-Presburger-always problem.

These results can be helpful in formulating a weak form of a Presburger linear temporal logic and in defining a fragment thereof that is decidable for model-checking *dta*. The proofs are based on the definition of a version of *dta*, called *static dta*, which does not have enabling conditions on transitions. The decidability of the previous Presburger liveness problems is the same for *dta* and static *dta*. Hence, proofs can be easier, since static *dta* are much simpler to deal with than *dta*.

The paper is organized as follows. Section 2 introduces the main definitions, such as discrete timed automata and the Presburger liveness properties. Section 3 shows the decidability of the ∃-Presburger-i.o. and of the ∃-Presburger-eventual problems, by introducing static *dta*. Section 4 shows the undecidability of the

∀-Presburger-i.o. and of the ∀-Presburger-eventual problems. Section 5 discusses some aspects related to the introduction of Presburger conditions in temporal logic, and to the extension of our results to dense time domains.

The proofs of some lemmas and theorems can be found in the full version of the paper available at http://www.eecs.wsu.edu/~zdang.

2 Preliminaries

A timed automaton [3] is a finite state machine augmented with a number of real-valued clocks. All the clocks progress synchronously with rate 1, except when a clock is reset to 0 at some transition. In this paper, we consider *integer-valued* clocks. A *clock constraint* (or a *region*) is a Boolean combination of *atomic clock constraints* in the following form: $x \# c, x - y \# c$ where $\#$ denotes $\leq, \geq, <, >$, or $=$, c is an integer, x, y are integer-valued clocks. Let $\mathcal{L}_X$ be the set of all clock constraints on clocks X. Let $\mathbf{N}$ be the set of nonnegative integers.

Definition 1. *A discrete timed automaton (dta) is a tuple $\mathcal{A} = \langle S, X, E \rangle$ where S is a finite set of (control) states, X is a finite set of clocks with values in $\mathbf{N}$, and $E \subseteq S \times 2^X \times \mathcal{L}_X \times S$ is a finite set of edges or transitions.*

Each edge $\langle s, \lambda, l, s' \rangle$ denotes a transition from state s to state s' with *enabling condition* $l \in \mathcal{L}_X$ and a set of clock resets $\lambda \subseteq X$. Note that λ may be empty: in this case, the edge is called a *clock progress transition*. Since each pair of states may have more than one edge between them, in general $\mathcal{A}$ is nondeterministic.

The semantics of *dtas* is defined as follows. We use $\mathbf{A}, \mathbf{B}, \mathbf{V}, \mathbf{W}, \mathbf{X}, \mathbf{Y}$ to denote clock vectors (i.e., vectors of clock values) with $\mathbf{V}_x$ being the value of clock x in $\mathbf{V}$. $\mathbf{I}$ denotes the identity vector in $\mathbf{N}^{|X|}$; i.e., $\mathbf{I}_x = 1$ for each $x \in X$.

Definition 2. (Configuration, One-Step Transition Relation $\rightarrow^{\mathcal{A}}$) *A configuration $\langle s, \mathbf{V} \rangle \in S \times (\mathbf{N})^{|X|}$ is a tuple of a control state s and a clock vector $\mathbf{V}$. $\langle s, \mathbf{V} \rangle \rightarrow^{\mathcal{A}} \langle s', \mathbf{V}' \rangle$ denotes a one-step transition from configuration $\langle s, \mathbf{V} \rangle$ to configuration $\langle s', \mathbf{V}' \rangle$ satisfying all the following conditions:*

- *There is an edge $\langle s, \lambda, l, s' \rangle$ in $\mathcal{A}$ connecting state s to state s',*
- *The enabling condition of the edge is satisfied, that is, $l(\mathbf{V})$ is true,*
- *Each clock changes according to the edge. If there are no clock resets on the edge, i.e., $\lambda = \emptyset$, then clocks progress by one time unit, i.e., $\mathbf{V}' = \mathbf{V} + \mathbf{I}$. If $\lambda \neq \emptyset$, then for each $x \in \lambda$, $\mathbf{V}'_x = 0$ while for each $x \notin \lambda$, $\mathbf{V}'_x = \mathbf{V}_x$.*

A configuration $\langle s, \mathbf{V} \rangle$ is a *deadlock configuration* if there is no configuration $\langle s', \mathbf{V}' \rangle$ such that $\langle s, \mathbf{V} \rangle \rightarrow^{\mathcal{A}} \langle s', \mathbf{V}' \rangle$. $\mathcal{A}$ is *total* if every configuration is not a deadlock configuration. A *path* is a finite sequence $\langle s_0, \mathbf{V}^0 \rangle \cdots \langle s_k, \mathbf{V}^k \rangle$ such that $\langle s_i, \mathbf{V}^i \rangle \rightarrow^{\mathcal{A}} \langle s_{i+1}, \mathbf{V}^{i+1} \rangle$ for each $0 \leq i \leq k - 1$. A path is a *progress* path if there is at least one clock progress transition on the path. An ω-*path* is an infinite sequence $\langle s_0, \mathbf{V}^0 \rangle \cdots \langle s_k, \mathbf{V}^k \rangle \cdots$ such that each prefix $\langle s_0, \mathbf{V}^0 \rangle \cdots \langle s_k, \mathbf{V}^k \rangle$ is a path. An ω-path is *divergent* if there is an infinite number of clock progress transitions on the ω-path. Without loss of generality, in this paper we consider timed automata without event labels [3], since they can be built into the control states.

Let Y be a finite set of variables over integers. For all integers a_y with $y \in Y$, b and c (with $c > 0$), $\sum_{y \in Y} a_y y < b$ is an *atomic linear relation* on Y and $\sum_{y \in Y} a_y y \equiv_b c$ is a *linear congruence* on Y. A *linear relation* on Y is a Boolean combination (using $\neg$ and $\wedge$) of atomic linear relations on Y. A *Presburger formula* on Y is the Boolean combination of atomic linear relations on Y and of linear congruences on Y. A set P is *Presburger-definable* if there exists a Presburger formula $\mathcal{F}$ on Y such that P is exactly the set of the solutions for Y that make $\mathcal{F}$ true. Since Presburger formulas are closed under quantifications, we will allow quantifiers over integer variables.

Write $\langle s, \boldsymbol{V} \rangle \leadsto^{\mathcal{A}} \langle s', \boldsymbol{V}' \rangle$ if $\langle s, \boldsymbol{V} \rangle$ reaches $\langle s', \boldsymbol{V}' \rangle$ through a path in $\mathcal{A}$. The binary relation $\leadsto^{\mathcal{A}}$ can be considered as a subset of configuration tuples and called *binary reachability*. It has been shown recently that,

Theorem 1. *The binary reachability $\leadsto^{\mathcal{A}}$ is Presburger-definable [9,10].*

The *Presburger safety analysis problem* is to consider whether $\mathcal{A}$ can only reach configurations in P starting from any configuration in I, given two Presburger-definable sets I and P of configurations. Because of Theorem 1, the Presburger safety analysis problem is decidable [10] for *dtas*.

In this paper, we consider *Presburger liveness analysis problems* for *dtas*, obtained by combining a path-quantifier with various modalities of satisfaction on an ω-path. Let I and P be two Presburger-definable sets of configurations, and let p be an ω-path $\langle s_0, \boldsymbol{V}^0 \rangle, \langle s_1, \boldsymbol{V}^1 \rangle \ldots$. Define the following modalities of satisfactions of P and I over p:

- p is *P-i.o.* if P is satisfied infinitely often on the ω-path, i.e., there are infinitely many k such that $\langle s_k, \boldsymbol{V}^k \rangle \in P$.
- p is *P-always* if for each k, $\langle s_k, \boldsymbol{V}^k \rangle \in P$.
- p is *P-eventual* if there exists k such that $\langle s_k, \boldsymbol{V}^k \rangle \in P$.
- p is *P-almost-always* if there exists k such that for all $k' > k$, $\langle s_{k'}, \boldsymbol{V}^{k'} \rangle \in P$.
- p *starts* from I if $\langle s_0, \boldsymbol{V}^0 \rangle \in I$.

Definition 3. (Presburger Liveness Analysis Problems) *Let $\mathcal{A}$ be a dta and let I and P be two Presburger-definable sets of configurations of $\mathcal{A}$. The $\exists$-Presburger-i.o. (resp. always, eventual and almost-always) problem is to decide whether the following statement holds: there is an ω-path p starting from I that is P-i.o. (resp. P-always, P-eventual and P-almost-always). The $\forall$-Presburger-i.o. (resp. always, eventual and almost-always) problem is to decide whether the following statement holds: for every ω-path p, if p starts from I, then p is P-i.o. (resp. always, eventual and almost-always).*

3   Decidability Results

In this section, we show that the $\exists$-Presburger-i.o. problem is decidable for *dtas*. Proofs of an infinitely-often property usually involve analysis of cycles in the transition system. However, for *dtas*, this is difficult for the following reasons. A discrete timed automaton $\mathcal{A}$ can be treated as a transition graph on control states with clock reset sets properly assigned to each edge, and augmented with tests (i.e., clock constraints) on edges. The tests are dynamic – the results of the tests depend upon the current values of each clock and obviously determine which edges may be taken. This is an obstacle to applying cyclic analysis techniques on the transition graph of $\mathcal{A}$.

A solution to these difficulties is to introduce *static* discrete timed automata, i.e., *dtas* with all the enabling conditions being simply *true*. The lack of enabling conditions simplifies the proof that the $\exists$-Presburger-i.o. problem is decidable for static *dtas*. Then, we show that each $\exists$-Presburger-i.o. problem for a *dta* can be translated into an $\exists$-Presburger-i.o. problem for a static *dta*, and hence it is decidable as well.

3.1   The $\exists$-Presburger-i.o. Problem for Static *dtas*

Let $\mathcal{A}$ be a static *dta*. We show that the $\exists$-Presburger-i.o. problem for static *dtas* is decidable. Given two sets I and P of configurations definable by Presburger formulas, an ω-path $p = \langle s_0, \boldsymbol{V}^0 \rangle \cdots \langle s_k, \boldsymbol{V}^k \rangle \cdots$ is a *witness* if it is a solution of the $\exists$-Presburger-i.o. problem, i.e., p is P-i.o. and p starts from I ($\langle s_0, \boldsymbol{V}^0 \rangle \in I$). There are two cases to a witness p: (1) p is not divergent; (2) p is divergent. For Case (1), we can establish the following lemma by expressing the existence of p into a Presburger formula obtained from the binary reachability of $\mathcal{A}$.

Lemma 1. *The existence of a non-divergent witness is decidable.*

The difficult case, however, is when the witness p is divergent. The remainder of this subsection is devoted to the proof that the existence of a divergent witness is decidable. For now, we fix a choice of a control state s and a set $X_r \subseteq X$ of clocks (there are only finitely many of them). To ensure that p is divergent, each path from $\langle s_{k_i} = s, \boldsymbol{V}^{k_i} \rangle$ to $\langle s_{k_{i+1}} = s, \boldsymbol{V}^{k_{i+1}} \rangle$ is picked so that it contain at least one clock progress transition, i.e., a *progress cycle*, as follows.

Definition 4. *For all clock vectors $\boldsymbol{V}, \boldsymbol{V}'$, we write $\langle s, \boldsymbol{V} \rangle \leadsto^{\mathcal{A}}_{X_r} \langle s, \boldsymbol{V}' \rangle$ if*

1. *there exists $\langle s_0, \boldsymbol{V}^0 \rangle \in I$ such that $\langle s_0, \boldsymbol{V}^0 \rangle \leadsto^{\mathcal{A}} \langle s, \boldsymbol{V} \rangle$, i.e., $\langle s, \boldsymbol{V} \rangle$ is reachable from a configuration in I,*
2. *$\langle s, \boldsymbol{V}' \rangle \in P$,*
3. *$\langle s, \boldsymbol{V} \rangle \leadsto^{\mathcal{A}} \langle s, \boldsymbol{V}' \rangle$ through a progress path on which all the clocks in X_r are reset at least once and all the clocks not in X_r are never reset.*

The proof proceeds as follows. First, we show (Lemma 2) that the relation $\leadsto^{\mathcal{A}}_{X_r}$ is Presburger-definable. Then, since $\mathcal{A}$ is finite state, there exists a P-i.o. ω-path p iff there is a state s such that P holds infinitely often on p at state s. This is equivalent to saying (Lemma 3) that there exist clock vectors $\boldsymbol{V}^1, \boldsymbol{V}^2, \ldots$ such that $\langle s, \boldsymbol{V}^i \rangle \leadsto^{\mathcal{A}}_{X_r} \langle s, \boldsymbol{V}^{i+1} \rangle$ for each $i > 0$. Since the actual values of the clocks in X_r may be abstracted away (Lemma 4 and Definition 5) and the clocks in $X - X_r$ progress synchronously, this is equivalent to saying that there exist $\boldsymbol{V}, d^1 > 0, d^2 > 0, \ldots$ such that $\boldsymbol{V}^i_x = \boldsymbol{V}_x + d^i$ for all $x \in X - X_r$ (Lemma 5). The set $\{d^i\}$ may be defined with a Presburger formula, as shown in Lemma 7, since each d^i may always be selected to be of the form $c^i + f(c^i)$, where the set $\{c^i\}$ is a periodic set (hence, Presburger definable) and f is a Presburger-definable function. This is based on the fact that static automata have no edge conditions, allowing us to increase the length d of a progress cycle to a length nd (Lemma 6), for every $n > 0$. The decidability result on the existence of a divergent witness follows directly from Lemma 7.

Lemma 2. $\leadsto^{\mathcal{A}}_{X_r}$ *is Presburger-definable. That is, given* $s \in S$, $\langle s, \boldsymbol{V} \rangle \leadsto^{\mathcal{A}}_{X_r}$ $\langle s, \boldsymbol{V}' \rangle$ *is a Presburger formula, when the clock vectors* $\boldsymbol{V}, \boldsymbol{V}'$ *are regarded as integer variables.*

Based upon the above analysis, the following lemma is immediate:

Lemma 3. *There is a divergent witness p iff there are* s, X_r *and clock vectors* $\boldsymbol{V}^1, \boldsymbol{V}^2, \ldots$ *such that* $\langle s, \boldsymbol{V}^i \rangle \leadsto^{\mathcal{A}}_{X_r} \langle s, \boldsymbol{V}^{i+1} \rangle$ *for each* $i > 0$.

$\langle s, \boldsymbol{V} \rangle \leadsto^{\mathcal{A}}_{X_r} \langle s, \boldsymbol{W} \rangle$ denotes the following scenario. Starting from some configuration in I, $\mathcal{A}$ can reach $\langle s, \boldsymbol{V} \rangle$ and return to s again with clock values $\boldsymbol{W}$. The cycle at s is a progress one such that each clock in X_r resets at least once and all clocks not in X_r do not reset. Since $\mathcal{A}$ is static, the cycle can be represented by a sequence $s_0 s_1 \cdots s_t$ of control states, with $s_0 = s_t = s$, and such that, for each $0 \le i < t$, there is an edge in $\mathcal{A}$ connecting s_i and s_{i+1}. Observe that, since each $x \in X_r$ is reset in the cycle, the starting clock values $\boldsymbol{V}_x$ for $x \in X_r$ at $s_0 = s$ are insensitive to the ending clock values $\boldsymbol{W}_x$ with $x \in X_r$ at $s_t = s$ (those values of $\boldsymbol{W}_x$ only depend on the sequence of control states). We write $\boldsymbol{V} =_{X-X_r} \boldsymbol{U}$ if $\boldsymbol{V}$ and $\boldsymbol{U}$ agree on the values of the clocks not in X_r, i.e., $\boldsymbol{V}_x = \boldsymbol{U}_x$, for each $x \in X - X_r$. The insensitivity property is stated in the following lemma.

Lemma 4. *For all clock vectors* $\boldsymbol{U}, \boldsymbol{V}, \boldsymbol{W}$, *if* $\langle s, \boldsymbol{V} \rangle \leadsto^{\mathcal{A}}_{X_r} \langle s, \boldsymbol{W} \rangle$ *and* $\langle s, \boldsymbol{U} \rangle$ *is reachable from some configuration in* I *with* $\boldsymbol{V} =_{X-X_r} \boldsymbol{U}$, *then* $\langle s, \boldsymbol{U} \rangle \leadsto^{\mathcal{A}}_{X_r}$ $\langle s, \boldsymbol{W} \rangle$.

Also note that, since all clocks not in X_r do not reset on the cycle, the differences $\boldsymbol{W}_x - \boldsymbol{V}_x$ for each $x \in X - X_r$ are equal to the *duration* of the cycle (i.e., the number of progress transitions in the cycle). The following technical definition allows us to "abstract" clock values for X_r away in $\langle s, \boldsymbol{V} \rangle \leadsto^{\mathcal{A}}_{X_r} \langle s, \boldsymbol{W} \rangle$.

Definition 5. *For all clock vectors* $\boldsymbol{Y}$ *and for all positive integers d, we write* $\boldsymbol{Y} \leadsto^{\mathcal{A}}_{\langle s, X_r \rangle} \boldsymbol{Y} + d\boldsymbol{I}$ *if there exist two clock vectors* $\boldsymbol{V}$ *and* $\boldsymbol{W}$ *such that* $\langle s, \boldsymbol{V} \rangle \leadsto^{\mathcal{A}}_{X_r}$ $\langle s, \boldsymbol{W} \rangle$ *with* $\boldsymbol{Y} =_{X-X_r} \boldsymbol{V}$ *and* $\boldsymbol{Y} + d\boldsymbol{I} =_{X-X_r} \boldsymbol{W}$.

Obviously, in the previous definition, the cycle from $\langle s, \boldsymbol{V} \rangle$ to $\langle s, \boldsymbol{W} \rangle$ has duration d. Also, the relation $\leadsto^{\mathcal{A}}_{\langle s, X_r \rangle}$ is Presburger-definable (over $\boldsymbol{Y}$ and d).

Lemma 5. *There exists a divergent witness for* $\mathcal{A}$ *if, and only if, there are* s, X_r, $\boldsymbol{Y}$, $d^1, d^2, \ldots$ *such that* $0 \le d^1 < d^2 < \ldots$ *and* $\boldsymbol{Y} + d^i \boldsymbol{I} \leadsto^{\mathcal{A}}_{\langle s, X_r \rangle} \boldsymbol{Y} + d^{i+1}\boldsymbol{I}$, *for each* $i \ge 1$.

The following technical lemma, based on Definition 5 and Lemma 5, will be used in the proof of Lemma 7.

Lemma 6. *For all* $\boldsymbol{Y}, \boldsymbol{Y}'$, *and for all* $n > 1, d > 0$, *if* $\boldsymbol{Y} \leadsto^{\mathcal{A}}_{\langle s, X_r \rangle} \boldsymbol{Y} + d\boldsymbol{I}$ *and* $\boldsymbol{Y} + nd\boldsymbol{I} \leadsto^{\mathcal{A}}_{\langle s, X_r \rangle} \boldsymbol{Y}'$, *then* $\boldsymbol{Y} + d\boldsymbol{I} \leadsto^{\mathcal{A}}_{\langle s, X_r \rangle} \boldsymbol{Y}'$.

Lemma 7. *It is decidable whether there exists a divergent witness for a static dta $\mathcal{A}$.*

Proof. We claim that, there are s, X_r such that the Presburger formula

$$(*) \quad \exists Y \forall m > 0 \exists d_1 \geq m \exists d_2 > 0 \ (\ Y + d_1 I \leadsto^{\mathcal{A}}_{\langle s, X_r \rangle} Y + (d_1 + d_2)I \ )$$

holds if and only if there is a divergent witness for $\mathcal{A}$. The statement of the lemma then follows immediately.

Assume there is a divergent witness. Hence, by Lemma 3, there exist $V^1, V^2,$ $\ldots$ and $d^1, d^2, \ldots$ such that, for each $i \geq 1$, $\langle s, V^i \rangle \leadsto^{\mathcal{A}}_{X_r} \langle s, V^{i+1} \rangle$ with a progress cycle of duration $d^i > 0$. Let Y be such that $Y =_{X - X_r} V^1$. By Definition 5, $Y + \left(\sum_{j=1}^{i-1} d^j \right) I \leadsto^{\mathcal{A}}_{\langle s, X_r \rangle} Y + \left(\sum_{j=1}^{i} d^j \right) I$ for each $i \geq 1$. For each $m > 0$, let $d_1 = \sum_{j=1}^{m} d^j$, $d_2 = d^{m+1}$. It is immediate that $(*)$ holds.

Conversely, let Y_0 be one of the vectors Y such that $(*)$ holds. Apply skolemization to the formula $\exists d_2 > 0 \left(Y_0 + d_1 I \leadsto^{\mathcal{A}}_{\langle s, X_r \rangle} Y_0 + (d_1 + d_2)I \right)$, by introducing a function $f(d_1)$ to replace the variable d_2. Since $(*)$ holds, then the formula $H(d_1)$, defined as $Y_0 + d_1 I \leadsto^{\mathcal{A}}_{\langle s, X_r \rangle} Y_0 + (d_1 + f(d_1))I$, holds for infinitely many values of d_1. Combining the fact that $H(d_1)$ is Presburger-definable (because Y_0 is fixed), there is a periodic set included in the infinite domain of H, i.e., there exist $n > 1, k \geq 0$ such that for all $d \geq 0$ if $d \equiv_n k$ then $H(d)$ holds. Let c^0 be any value in the periodic set, and let $c^i = c^{i-1} + nf(c^{i-1})$, for every $i \geq 1$. Obviously, every c^i satisfies the periodic condition: $c^i \equiv_n k$, and therefore $H(c^i)$ holds. Hence, for every $i \geq 1$, $Y_0 + c^i I \leadsto^{\mathcal{A}}_{\langle s, X_r \rangle} Y_0 + (c^i + f(c^i))I$.

Since $Y_0 + c^{i+1}I = Y_0 + c^i I + nf(c^i)I \leadsto^{\mathcal{A}}_{\langle s, X_r \rangle} Y_0 + (c^{i+1} + f(c^{i+1}))I$, we may apply Lemma 6, with: $Y = Y_0 + c^i I$, $d = f(c^i)$, $Y' = Y + (c^{i+1} + f(c^{i+1}))I$. Lemma 6 then gives $Y + dI \leadsto^{\mathcal{A}}_{\langle s, X_r \rangle} Y'$, i.e., $Y_0 + (c^i + f(c^i))I \leadsto^{\mathcal{A}}_{\langle s, X_r \rangle} Y_0 + (c^{i+1} + f(c^{i+1}))I$, for every $i \geq 1$. By Lemma 5, with $d^i = c^i + f(c^i)$, there is a divergent witness. ∎

By Lemmas 1 and 7, we have:

Theorem 2. *The $\exists$-Presburger-i.o. problem is decidable for static dtas.*

3.2   The $\exists$-Presburger-i.o. Problem for *dtas*

In the full paper, we use a technique modified from [10] to show that the tests in $\mathcal{A}$ can be eliminated. That is, $\mathcal{A}$ can be effectively transformed into $\mathcal{A}''$ where all the tests are simply *true* and $\mathcal{A}''$ has (almost) the same static transition graph as $\mathcal{A}$. This is based on an encoding of the tests of $\mathcal{A}$ into the finite state control of $\mathcal{A}''$. Now we look at the $\exists$-Presburger-i.o. problem for $\mathcal{A}$. Recall that the problem is to determine, given two Presburger-definable sets I and P of configurations of $\mathcal{A}$, whether there exists a P-i.o. ω-path p starting from I. We relate the instance of the $\exists$-Presburger-i.o. problem for $\mathcal{A}$ to an instance of the $\exists$-Presburger-i.o. problem for $\mathcal{A}''$:

Lemma 8. *Given a dta $\mathcal{A}$, and two Presburger-definable sets I and P of configurations of $\mathcal{A}$, there exist a static dta $\mathcal{A}''$ and two Presburger definable sets I'' and P'' of configurations of $\mathcal{A}''$ such that: the existence of a witness to the $\exists$-Presburger-i.o. for $\mathcal{A}$, given I and P, is equivalent to the existence of a witness to the $\exists$-Presburger-i.o. for $\mathcal{A}''$, given I'' and P''.*

Since $\mathcal{A}''$ is a static *dta*, the decidability of the $\exists$-Presburger-i.o. for $\mathcal{A}$ follows from Theorem 2 and Lemma 8.

Theorem 3. *The $\exists$-Presburger-i.o. problem and the $\forall$-Presburger-almost-always problem are decidable for dtas.*

3.3   Decidability of the $\exists$-Presburger-Eventual Problem

Given a *dta* $\mathcal{A}$, and two Presburger-definable sets I and P of configurations, the $\exists$-Presburger-eventual problem is to decide whether there exists a P-eventual ω-path p starting from I. Define I' to be the set of all configurations in P that can be reached from a configuration in I. From Theorem 1, I' is Presburger-definable. Let P' be simply *true*. It can be shown that the existence of a witness for the $\exists$-Presburger-eventual problem (given I and P) is equivalent to the existence of a witness for the $\exists$-Presburger-i.o. problem (given I' and P'). From Theorem 3,

Theorem 4. *The $\exists$-Presburger-eventual problem and the $\forall$-Presburger-always problem are decidable for dtas.*

It should be noted that there is a slight difference between the $\forall$-Presburger-always problem and the Presburger safety analysis problem mentioned before. The difference is that the Presburger safety analysis problem considers (finite) paths while the $\forall$-Presburger-always problem considers ω-paths.

4   Undecidability Results

The next three subsections show that the undecidability of the $\forall$-Presburger-eventual problem and of the $\forall$-Presburger-i.o. problem. We start by demonstrating the fact that a two-counter machine can be implemented by a generalized version of a *dta*. This fact is then used in the following two subsections to show the undecidability results.

4.1   Counter Machines and Generalized Discrete Timed Automata

Consider a counter machine M with counters $x_1, \cdots, x_k$ over nonnegative integers and with a finite set of locations $\{l_1, \cdots, l_n\}$. M can increment, decrement and test against 0 the values of the counters. It is well-known that a two-counter machine can simulate a Turing machine.

We now define *generalized* discrete timed automata. They are defined similarly to *dtas* but for each edge $\langle s, \lambda, l, s' \rangle$ the formula l is of the form $\sum_i a_i x_i \# c$, where a_i and c are integers. Generalized *dtas* are Turing-complete, since they can simulate any counter machine:

Lemma 9. *Given a deterministic counter machine M, there exists a deterministic generalized dta that can simulate M.*

From now on, let M be a deterministic counter machine and let $\mathcal{A}$ be a deterministic generalized *dta* that implements M. We may assume that $\mathcal{A}$ is total (i.e., there are no deadlock configurations), since $\mathcal{A}$ can be made total by adding a new self-looped state s_f, and directing every a deadlock configuration to this new state. Now we define the *static version* $\mathcal{A}^-$, to which $\mathcal{A}$ can be modified as follows. $\mathcal{A}^-$ is a discrete timed automaton with the enabling condition on each edge being simply *true*. Each state in $\mathcal{A}^-$ is a pair of states in $\mathcal{A}$. $\langle\langle s_1, s_1'\rangle, \lambda_1, true, \langle s_2, s_2'\rangle\rangle$ is an edge of $\mathcal{A}^-$ iff there are edges $\langle s_1, \lambda_1, l_1, s_1'\rangle$ and $\langle s_2, \lambda_2, l_2, s_2'\rangle$ in $\mathcal{A}$ with $s_1' = s_2$. We define a set P, called the *path restriction of* $\mathcal{A}$, of configurations of $\mathcal{A}^-$ as follows. For each configuration $\langle\langle s, s'\rangle, \mathbf{V}\rangle$ of $\mathcal{A}^-$, $\langle\langle s, s'\rangle, \mathbf{V}\rangle \in P$ iff there exists an edge $e = \langle s, \lambda, l, s'\rangle$ such that the clock values $\mathbf{V}$ satisfy the linear relation l in e. Clearly, P is Presburger-definable. Since $\mathcal{A}$ is total and deterministic, the above edge e always exists and is unique for each configuration $\langle s, \mathbf{V}\rangle$ of $\mathcal{A}$. Using this fact, we have,

Theorem 5. *Let $\mathcal{A}$ be a total and deterministic generalized dta with path restriction P, and let $\mathcal{A}^-$ be the static version of $\mathcal{A}$. An ω-sequence $\langle s_0, \mathbf{V}^0\rangle \cdots \langle s_k, \mathbf{V}^k\rangle \cdots$ is an ω-path of $\mathcal{A}$ iff $\langle\langle s_0, s_1\rangle, \mathbf{V}^0\rangle \cdots \langle\langle s_k, s_{k+1}\rangle, \mathbf{V}^k\rangle \cdots$ is an ω-path of $\mathcal{A}^-$ with $\langle\langle s_k, s_{k+1}\rangle, \mathbf{V}^k\rangle \in P$ for each k.*

4.2   Undecidability of the $\forall$-Presburger-Eventual Problem

We consider the negation of the $\forall$-Presburger-eventual problem, i.e., the $\exists$-Presburger-always problem, which can be formulated as follows: given a discrete timed automaton $\mathcal{A}$ and two Presburger-definable sets I and P of configurations, decide whether there exists a $\neg P$-always ω-path of $\mathcal{A}$ starting from I.

Consider a deterministic counter machine M with the initial values of the counters being 0 and the first instruction labeled l_0. Let $\mathcal{A}$ be the deterministic generalized *dta* implementing M, as defined by Lemma 9, with P being the path restriction of $\mathcal{A}$. As before, $\mathcal{A}$ is total. Let $\mathcal{A}^-$ be the static version of $\mathcal{A}$. It is well known that the halting problem for (deterministic) counter machines is undecidable. That is, it is undecidable, given M and an instruction label l, whether M executes the instruction l. Define P' to be the set of configurations $\langle\langle s, s'\rangle, \mathbf{V}\rangle \in P$ with $s \neq l$. Let I be the set of initial configurations of $\mathcal{A}^-$ with all the clocks being 0 and the first component of the state (note that each state in $\mathcal{A}^-$ is a state pair of $\mathcal{A}$) being l_0. I is finite, thus Presburger-definable. From Theorem 5 and the fact that $\mathcal{A}$ implements M, we have: M does not halt at l iff $\mathcal{A}^-$ has a P'-always ω-path starting from a configuration in I. Thus, we reduce the negation of the halting problem to the $\exists$-Presburger-always problem for *dtas* with configuration sets P' and I. Therefore,

Theorem 6. *The $\exists$-Presburger-always problem and the $\forall$-Presburger-eventual problem are undecidable for discrete timed automata.*

4.3   Undecidability of the $\forall$-Presburger-i.o. Problem

In this subsection, we show that the $\exists$-Presburger-almost-always problem is undecidable. Therefore, the $\forall$-Presburger-i.o. problem is also undecidable. In the previous subsection, we have shown that the existence of a P-always ω-path of $\mathcal{A}$ is undecidable. But this result does not directly imply that the existence of a P-almost-always ω-path is also undecidable.

In fact, let $\mathcal{A}^-$ be the static version of a generalized discrete timed automaton $\mathcal{A}$ that implements a deterministic counter machine M, let P be the path restriction of $\mathcal{A}$, and let p be an ω-path of $\mathcal{A}^-$. In the previous subsection, we argued that the existence of a P'-always ω-path p is undecidable where P' is $P \cap \{\langle\langle s, s'\rangle, \boldsymbol{V}\rangle : s \neq l\}$ with l being a given instruction label in M. But when considering a P'-almost-always path p, the situation is different: p may have a prefix that does not necessarily satisfy P' (i.e., it does not obey the exact enabling conditions on the edges in $\mathcal{A}$).

Consider a deterministic two-counter machine M with an input tape, and denote with $M(i)$ the result of the computation of M when given $i \in \mathbf{N}$ in input. It is known that the finiteness problem for deterministic two-counter machines (i.e., finitely many i such that $M(i)$ halts) is undecidable. Now we reduce the finiteness problem to the $\exists$-almost-always problem for $dtas$.

We can always assume that M halts when and only when it executes an operation labeled $halt$. Let M' be a counter machine (without input tape) that enumerates all the computations of M on every $i \in \mathbf{N}$. M' works as follows. We use $M_j(i)$ to denote the j-th step of the computation of $M(i)$. If $M(i)$ halts in less than j steps, then we assume that $M_j(i)$ is a special null operation that does nothing. Thus, the entire computation of $M(i)$ is an ω-sequence $M_1(i), \cdots, M_j(i), \cdots$ (when $M(i)$ halts, the sequence is composed of a finite prefix, the halt operation and then infinitely many occurrences of the special null operation). Each step of the computation may or may not execute the instruction labeled $halt$, but of course an halt may be executed only at most once for each input value i. M' implements the following program:

$k := 0; z := 0;$
while true do
    $k := k + 1;$
    for $i := 0$ *to* $k - 1$ *do*
        $z := 1;$
        simulate $M(i)$ *for the first* k *steps* $M_1(i), M_2(i), \ldots, M_k(i);$
        if $M_k(i)$ *executes the instruction labeled* halt, *then* $z := 0;$

M' is still a deterministic counter machine (with various additional counters to be able to simulate M and keep track of k, i, z). In the enumeration, whenever $M_k(i)$ executes the instruction labeled $halt$ (at most once for each i, by the definition of M' as above), M' sets the counter z to 0, bringing it back to 1 immediately afterwards – M' resets z to 0 for only finitely many times iff the domain of M (i.e., the set of i such that $M(i)$ halts) is finite. Let $\mathcal{A}^-$ be the static version of a generalized discrete timed automaton $\mathcal{A}$ that implements M'. Let P be the path restriction of $\mathcal{A}$. P' is $P \cap \{\langle\langle s, s'\rangle, \boldsymbol{V}\rangle : \boldsymbol{V}_z \neq 0\}$. It can be established, by using Lemma 9 and Theorem 5, that there are only finitely many i such that $M(i)$ halts iff $\mathcal{A}^-$ is $\exists$-Presburger-almost-always for P' and I where I contains only the initial configuration. Therefore,

Theorem 7. *The $\exists$-Presburger-almost-always problem and the $\forall$-Presburger-i.o. problem are undecidable for discrete timed automata.*

5   Discussions and Future Work

It is important to provide a uniform framework to clarify what kind of temporal Presburger properties can be automatically checked for timed automata. Given a *dta* $\mathcal{A}$, the set of linear temporal logic formulas $\mathcal{L}_\mathcal{A}$ with respect to $\mathcal{A}$ is defined by the following grammar: $\phi := P | \neg\phi | \phi \wedge \phi | \bigcirc \phi | \phi U \phi$, where P is a Presburger-definable set of configurations of $\mathcal{A}$, $\bigcirc$ denotes "next", and U denotes "until". Formulas in $\mathcal{L}_\mathcal{A}$ are interpreted on ω-sequences p of configurations of $\mathcal{A}$ in a usual way. We use p^i to denote the ω-sequence resulting from the deletion of the first i configurations from p. We use p_i to indicate the i-th element in p. The satisfiability relation $\models$ is recursively defined as follows, for each ω-sequence p and for each formula $\phi \in \mathcal{L}_\mathcal{A}$ (written $p \models \phi$):

$p \models P$ if $p_1 \in P$,

$p \models \neg\phi$ if not $p \models \phi$,

$p \models \phi_1 \wedge \phi_2$ if $p \models \phi_1$ and $p \models \phi_2$,

$p \models \bigcirc\phi$ if $p^1 \models \phi$,

$p \models \phi_1 U \phi_2$ if $\exists j (p^j \models \phi_2$ and $\forall k < j (p^k \models \phi_1))$.

where the variables i, j, k range over $\mathbf{N}$. We adopt the convention that $\Diamond\phi$ (eventual) abbreviates $(true U \phi)$ and $\Box\phi$ (always) abbreviates $(\neg\Diamond\neg\phi)$.

Given $\mathcal{A}$ and a formula $\phi \in \mathcal{L}_\mathcal{A}$, the model-checking problem is to check whether each ω-path p of $\mathcal{A}$ satisfies $p \models \phi$. The satisfiability-checking problem, which is the dual of the model-checking problem, is to check whether there is an ω-path p of $\mathcal{A}$ satisfying $p \models \phi$. The results of this paper show that:

- The satisfiability-checking problem is decidable for formulas in $\mathcal{L}_\mathcal{A}$ in the form $I \wedge \Box\Diamond P$ and $I \wedge \Diamond P$, where I and P are Presburger.
- The model-checking problem is undecidable for formulas in $\mathcal{L}_\mathcal{A}$, even when the formulas are in the form $\Box\Diamond P$ and $\Diamond P$.
- Hence, both the satisfiability-checking problem and the model-checking problem are undecidable for the entire $\mathcal{L}_\mathcal{A}$, even when the "next" operator $\bigcirc$ is excluded from the logic $\mathcal{L}_\mathcal{A}$.

Future work may include investigating a fragment of $\mathcal{L}_\mathcal{A}$ that has a decidable satisfiability-checking/model-checking problem. For instance, we don't know whether the satisfiability-checking problem is decidable for $I \wedge \Box\Diamond P \wedge \Box\Diamond Q$ (i.e., find an ω-path that is both P-i.o. and Q-i.o.). A decidable subset of $\mathcal{L}_\mathcal{A}$ may be worked out along the recent work of Comon and Cortier [8] on model-checking a decidable subset of a Presburger (in the discrete case) LTL for one-cycle counter machines.

In [6], an extension of TPTL, called Presburger TPTL, is proposed and it is shown to be undecidable for discrete time. The proof in [6] does not imply (at least, not in an obvious way) the undecidability of the $\forall$-Presburger-i.o. problem and the $\forall$-Presburger-eventual problem in the paper. In that proof, the semantics of Presburger TPTL (over discrete time domain) is interpreted on timed state sequences. The transition relation of a two-counter machine can be encoded into Presburger TPTL by using $\bigcirc, U$ and the freeze quantifier. This gives the undecidability of the logic [6]. On the other hand, $\Box\Diamond P$ and $\Diamond P$ in this paper are interpreted on sequences of configurations (in contrast to timed state sequences). Formulas like $\Box\Diamond P$ and $\Diamond P$ are state formulas. That is, without using $\bigcirc$ and without introducing freeze quantifiers, we have no way to remember clock values in one configuration and use them to compare those in another configuration along p. Therefore, the transition relation of a two-counter machine cannot be encoded in our logic $\mathcal{L}_\mathcal{A}$. But we are able to show in this paper that computations of a two-counter machine can be encoded by ω-paths, restricted under $\Box\Diamond P$ or $\Diamond P$, of a *dta*, leading to the undecidability results of this paper.

We are also interested in considering the same set of liveness problems for a dense time domain. We believe that the decidability results (for the $\exists$-Presburger-i.o. problem and the $\exists$-Presburger-eventual problem) also hold for dense time when the semantics of a timed automaton is carefully defined. A possible approach is to look at Comon and Jurski's flattening construction [9]. The undecidability results in this paper can be naturally extended to the dense time domain when the ω-paths in this paper are properly redefined for dense time.

Thanks to the anonymous reviewers for a number of useful suggestions.

References

1. R. Alur, *"Timed automata"*, *CAV'99*, LNCS 1633, pp. 8-22
2. R. Alur, C. Courcoubetis, and D. Dill, *"Model-checking in dense real time,"* *Information and Computation*, **104** (1993) 2-34
3. R. Alur and D. Dill, *"Automata for modeling real-time systems,"* *Theoretical Computer Science*, **126** (1994) 183-236
4. R. Alur, T. Feder, and T. A. Henzinger, *"The benefits of relaxing punctuality,"* *J. ACM*, **43** (1996) 116-146
5. R. Alur, T. A. Henzinger, *"Real-time logics: complexity and expressiveness,"* *Information and Computation*, **104** (1993) 35-77
6. R. Alur, T. A. Henzinger, *"A really temporal logic,"* *J. ACM*, **41** (1994) 181-204
7. A. Coen-Porisini, C. Ghezzi and R. Kemmerer, *"Specification of real-time systems using ASTRAL,"* *IEEE Transactions on Software Engineering*, 23 (1997) 572-598
8. H. Comon and V. Cortier, *"Flatness is not a weakness,"* Proc. Computer Science Logic, 2000.
9. H. Comon and Y. Jurski, *"Timed automata and the theory of real numbers,"* *CONCUR'99*, LNCS 1664, pp. 242-257
10. Z. Dang, O. H. Ibarra, T. Bultan, R. A. Kemmerer, and J. Su, *"Binary reachability analysis of discrete pushdown timed automata,"* *CAV'00*, LNCS 1855, pp. 69-84
11. T. A. Henzinger, Z. Manna, and A. Pnueli, *"What good are digital clocks?,"* *ICALP'92*, LNCS 623, pp. 545-558
12. T. A. Henzinger and Pei-Hsin Ho, *"HyTech: the Cornell hybrid technology tool,"* *Hybrid Systems II*, LNCS 999, pp. 265-294
13. T. A. Henzinger, X. Nicollin, J. Sifakis, and S. Yovine, *"Symbolic model checking for real-time systems,"* *Information and Computation*, **111** (1994) 193-244
14. K. G. Larsen, P. Pattersson, and W. Yi, *"UPPAAL in a nutshell,"* *International Journal on Software Tools for Technology Transfer*, **1** (1997): 134-152
15. F. Laroussinie, K. G. Larsen, and C. Weise, *"From timed automata to logic - and back,"* *MFCS'95*, LNCS 969, pp. 529-539
16. F. Wang, *"Parametric timing analysis for real-time systems,"* *Information and Computation*, 130 (1996): 131-150
17. S. Yovine, *"A verification tool for real-time systems,"* *International Journal on Software Tools for Technology Transfer*, **1** (1997): 123-133
18. S. Yovine, *"Model checking timed automata,"* *Embedded Systems'98*, LNCS 1494, pp. 114-152

Residual Finite State Automata

François Denis, Aurélien Lemay, and Alain Terlutte

Bât. M3, GRAPPA-LIFL, Université de Lille I
59655 Villeneuve d'Ascq Cedex, France
{denis,lemay,terlutte}@lifl.fr
http://www.grappa.univ-lille3.fr

Abstract. We introduce a subclass of non deterministic finite automata (NFA) that we call Residual Finite State Automata (RFSA): a RFSA is a NFA all the states of which define residual languages of the language it recognizes. We prove that for every regular language L, there exists a unique RFSA that recognizes L and which has both a minimal number of states and a maximal number of transitions. Moreover, this canonical RFSA may be exponentially smaller than the equivalent minimal DFA but it also may have the same number of states as the equivalent minimal DFA, even if minimal equivalent NFA are exponentially smaller. We provide an algorithm that computes the canonical RFSA equivalent to a given NFA. We study the complexity of several decision and construction problems linked to the class of RFSA: most of them are PSPACE-complete.

1 Introduction

Regular languages and finite automata have been extensively studied since the beginning of formal language theory. Representation of regular languages by means of Deterministic Finite Automata (DFA) has many nice properties: there exists a unique minimal DFA that recognizes a given regular language (minimal in number of states and unique up to an isomorphism); each state q of a DFA A defines a language (composed of the words which lead to a final state from q) which is a natural component of the language L recognized by A, namely a *residual language* of L. One of the major drawbacks of DFA is that they provide representations of regular languages whose size is far to be optimal. For example, the regular language $\Sigma^*0\Sigma^n$ is represented here by a regular expression whose size is $O(\log n)$ while its minimal DFA has about 2^n states. Using Non deterministic Finite Automata (NFA) rather than DFA can drastically improve the size of the representation: the minimal NFA which recognizes $\Sigma^*0\Sigma^n$ has $n+2$ states. However, NFA have none of the two above-mentioned properties: languages associated with states have no natural interpretation and two minimal NFA can be not isomorphic.

In this paper, we study a subclass of non deterministic finite automata that we call Residual Finite State Automata (RFSA). By definition, a RFSA is a NFA all the states of which define residual languages of the language it recognizes. More precisely, a NFA $A = \langle \Sigma, Q, Q_0, F, \delta \rangle$ is a RFSA if for every state q in Q there exists a word u such that uv is recognized by A if and only if reading v, a final state can be reached from q. Clearly, all DFA are RFSA but the converse is false.

We prove that among all the RFSA which recognize a given regular language, there exists a unique element which has both a minimal number of states and a maximal

A. Ferreira and H. Reichel (Eds.): STACS 2001, LNCS 2010, pp. 144–157, 2001.
© Springer-Verlag Berlin Heidelberg 2001

number of transitions. This canonical RFSA may be exponentially smaller than the equivalent minimal DFA (for example, the canonical RFSA which recognizes $\Sigma^*0\Sigma^n$ has $n+2$ states); but it may also have the same number of states as the equivalent minimal DFA, even if minimal equivalent NFA are exponentially smaller. Another approach of canonical NFA can be found in [Car70] and [ADN92].

It is well known that for a given DFA A recognizing a language L, if we first construct the mirror automaton $\overline{A}$ and then, the deterministic automaton equivalent to $\overline{A}$ using the standard subset construction technique, we obtain the minimal DFA for $\overline{L}$. We prove a similar property for RFSA. This property provides an algorithm which computes the canonical RFSA equivalent to a given NFA. Unfortunately, we also prove that this construction problem is PSPACE-complete, as most of the constructions we define in this paper.

In section 2, we recall classical definitions and notations about regular languages and automata. We define RFSA in section 3 and we study their properties in section 4. In particular, we introduce the notion of canonical RFSA. We provide a construction of the canonical RFSA from a given NFA in section 5. In section 6, we study some particular (and pathological) RFSA. Section 7 is devoted to the study of the complexity of our constructions. Finally, we conclude by indicating where this work originates from and by describing some of its applications in the field of grammatical inference.

2   Preliminaries

In this section, we recall some definitions concerning finite automata. For more information, we invite the reader to consult [HU79,Yu97].

2.1   Automata and Languages

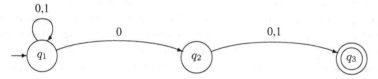

Fig. 1. A_1 Automaton Recognizes $\Sigma^*0\Sigma$ but Is neither a DFA nor a RFSA

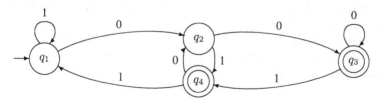

Fig. 2. A_2 Is the Minimal DFA Recognizing $\Sigma^*0\Sigma$.

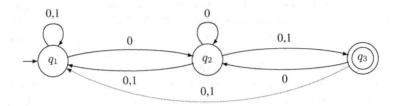

Fig. 3. A_3 Is a RFSA Recognizing $\Sigma^*0\Sigma$.

Let Σ be a finite alphabet, and let Σ^* be the set of words on Σ. We note ε the empty string and $|u|$ the length of a word u in Σ^*. A language is a subset of Σ^*.

A *non deterministic finite automaton* (NFA) is a quintuple $A = \langle \Sigma, Q, Q_0, F, \delta \rangle$ where Q is a finite set of states, $Q_0 \subseteq Q$ is the set of initial states, $F \subseteq Q$ is the set of terminal states. δ is the *transition function* of the automaton defined from a subset of $Q \times \Sigma$ to 2^Q. We also note δ the extended transition function defined from a subset of $2^Q \times \Sigma^*$ to 2^Q by:

$\delta(\{q\}, \varepsilon) = \{q\}$,
$\delta(\{q\}, x) = \delta(q, x)$,
$\delta(Q', u) = \cup \{\delta(\{q\}, u)|q \in Q'\}$ and
$\delta(\{q\}, ux) = \delta(\delta(q, u), x)$

where $Q' \subseteq Q, x \in \Sigma, q \in Q$ and $u \in \Sigma^*$.

A NFA is *deterministic* (DFA) if Q_0 contains exactly one element q_0 and if $\forall q \in Q$, $\forall x \in \Sigma, Card(\delta(q, x)) \le 1$. A NFA is *trimmed* if $\forall q \in Q, \exists w_1 \in \Sigma^*, q \in \delta(Q_0, w_1)$ and $\exists w_2 \in \Sigma^*, \delta(q, w_2) \cap F \ne \emptyset$. A state q is *reachable* by the word u if $q \in \delta(Q_0, u)$.

A word $u \in \Sigma^*$ is recognized by a NFA if $\delta(Q_0, u) \cap F \ne \emptyset$ and the language L_A recognized by A is the set of words recognized by A. We denote by $Rec(\Sigma^*)$ the class of recognizable languages. It can be proved that every recognizable language can be recognized by a DFA. There exists a unique minimal DFA that recognizes a given recognizable language (minimal with regard to the number of states and unique up to an isomorphism). Finally, the Kleene theorem [Kle56] proves that the class of regular languages $Reg(\Sigma^*)$ is identical to $Rec(\Sigma^*)$.

The *mirror* of a word $u = x_1 \ldots x_n$ ($x_i \in \Sigma$) is defined by $\bar{u} = x_n \ldots x_1$. The mirror of a language L is $\bar{L} = \{\bar{u}|u \in L\}$. The mirror of an automaton $A = \langle \Sigma, Q, Q_0, F, \delta \rangle$ is $\bar{A} = \langle \Sigma, Q, F, Q_0, \bar\delta \rangle$ where $q \in \bar\delta(q', x)$ if and only if $q' \in \delta(q, x)$. It is clear that $\overline{L_A} = L_{\bar{A}}$.

Let L be a regular language. Let $A = \langle \Sigma, Q, Q_0, F, \delta \rangle$ be a NFA that recognizes L and let $Q' \subseteq Q$. We note $L_{Q'}$ the language defined by $L_{Q'} = \{v|\delta(Q', v) \cap F \ne \emptyset\}$. When Q' contains exactly one state q, we simply denote $L_{Q'}$ by L_q.

2.2   Residual Languages

Let L be a language over Σ^* and let $u \in \Sigma^*$. The *residual* language of L with regard to u is defined by $u^{-1}L = \{v \in \Sigma^* \mid uv \in L\}$. If L is recognized by a NFA $\langle \Sigma, Q, Q_0, F, \delta \rangle$, then $q \in \delta(Q_0, u) \Rightarrow L_q \subseteq u^{-1}L$.

The Myhill-Nerode theorem [Myh57,Ner58] proves that the set of distinct residual languages of any regular language is finite. Furthermore, if $A = \langle \Sigma, Q, Q_0, F, \delta \rangle$ is the minimal DFA recognizing L, we have:

- for every non empty residual language $u^{-1}L$, there exists a unique $q \in Q$ such that $L_q = u^{-1}L$,
- $\forall q \in Q$, there exists a unique residual language $u^{-1}L$ such that $u^{-1}L = L_q$.

3   Definition of Residual Finite State Automaton

Definition 1. *A Residual Finite State Automaton (RFSA) is a NFA $A = \langle \Sigma, Q, Q_0, F, \delta \rangle$ such that, for each state $q \in Q$, L_q is a residual language of L_A. More formally, $\forall q \in Q$, $\exists u \in \Sigma^*$ such that $L_q = u^{-1}L_A$.*

Remark: Trimmed DFA have this property, and therefore are RFSA.

Definition 2. *Let $A = \langle \Sigma, Q, Q_0, F, \delta \rangle$ be a RFSA and let q be a state of A. We say that u is a* characterizing *word for q if $L_q = u^{-1}L_A$.*

Example 1. We study here the regular language $L = \Sigma^*0\Sigma$ where $\Sigma = \{0, 1\}$. One can prove that this language is recognized by the following automata A_1, A_2 and A_3 (fig. 1, 2, 3):

- A_1 is a NFA recognizing L. One can notice that A_1 is neither a DFA, nor a RFSA. Languages associated with states are $L_{q_1} = \Sigma^*0\Sigma$, $L_{q_2} = \Sigma$, $L_{q_3} = \{\varepsilon\}$. As for every u in Σ^*, we have $uL \subseteq L$ and so, $L \subseteq u^{-1}L$, we can see that neither L_2 nor L_3 are residual languages.
- A_2 is the minimal DFA that recognizes L. This automaton is also a RFSA , we have $L_{q_1} = \Sigma^*0\Sigma$, $L_{q_2} = \Sigma^*0\Sigma + \Sigma$, $L_{q_3} = \Sigma^*0\Sigma + \Sigma + \varepsilon$, $L_{q_4} = \Sigma^*0\Sigma + \varepsilon$, so, $L_{q_1} = \varepsilon^{-1}L$, $L_{q_2} = 0^{-1}L$, $L_{q_3} = 00^{-1}L$, $L_{q_4} = 01^{-1}L$.
- A_3 is a RFSA recognizing L. Indeed, we have $L_{q_1} = \varepsilon^{-1}L$, $L_{q_2} = 0^{-1}L$, $L_{q_3} = 01^{-1}L$. One can notice that this automaton is not a DFA. This automaton is the canonical RFSA of L, which is one of the smallest RFSA (regarding the number of states) recognizing L (the notion of canonical RFSA will be described later).

Example 2. To look for a characterizing word for a state q is often equivalent to look for a word u_q that only leads to q (i.e. such that $\delta(Q_0, u_q) = \{q\}$). Nevertheless, such a word does not always exist. For example, let $L = a^*b^* + b^*a^*$.

Fig. 4. A RFSA Recognizing the Language $a^*b^* + b^*a^*$.

The automaton described in figure 4 recognizes L. We have $L_{q_1} = b^*a^*$, $L_{q_2} = a^*$, $L_{q_3} = a^*b^*$, $L_{q_4} = b^*$. This automaton is a RFSA, as $L_{q_1} = b^{-1}L$, $L_{q_2} = (ba)^{-1}L$, $L_{q_3} = a^{-1}L$, $L_{q_4} = (ab)^{-1}L$. But there exists no word u such that $\delta(Q_0, u) = \{q_3\}$.

4   Properties of Residual Finite State Automata

4.1   General Properties

Definition 3. *Let L be a regular language. We say that a residual language $u^{-1}L$ is prime if it is not equal to the union of residual languages it strictly contains:*

$$u^{-1}L \text{ is prime if}$$

$$\bigcup \{v^{-1}L \mid v^{-1}L \subsetneq u^{-1}L\} \subsetneq u^{-1}L.$$

We say that a residual language is composed if it is not prime.

Notice that a prime residual language is not empty and that the set of distinct prime residual languages of a regular language is finite.

Proposition 1. *Let $A = \langle \Sigma, Q, Q_0, F, \delta \rangle$ be a RFSA. For each prime residual $u^{-1}L_A$, there exists a state $q \in Q$ such that $L_q = u^{-1}L_A$.*

Proof: Let $\delta(Q_0, u) = \{q_1, \ldots, q_s\}$ and let $v_1, \ldots, v_s$ be words such that $L_{q_i} = v_i^{-1}L_A$ for every $1 \le i \le s$. We have

$$u^{-1}L_A = \bigcup_{i=1 \text{ to } s} v_i^{-1}L_A.$$

As $u^{-1}L_A$ is prime, there exists some v_i such that $u^{-1}L_A = v_i^{-1}L_A = L_{q_i}$.   □

As a corollary, a RFSA A has at least as many states as the number of prime residuals of L_A.

4.2   Saturation Operator

We define a *saturation* operator that allows to add transitions to an automaton without modifying the language it recognizes.

Definition 4. *Let $A = \langle \Sigma, Q, Q_0, F, \delta \rangle$ be a NFA. We call* saturated *of A the automaton $S(A) = \langle \Sigma, Q, \widetilde{Q_0}, F, \widetilde{\delta} \rangle$ with $\widetilde{Q_0} = \{q \in Q \mid L_q \subseteq L_A\}$ and $\widetilde{\delta}(q, x) = \{q' \in Q \mid xL_{q'} \subseteq L_q\}$. We say that an automaton A is* saturated *if $A = S(A)$.*

Lemma 1. *Let A and A' be two NFA sharing the same set of states Q. If $L_A = L_{A'}$ and if for every state $q \in Q$, $L_q = L'_q$ (L_q and L'_q being the languages corresponding to q in both automata), then $S(A) = S(A')$.*

Proof: The state q is an initial state of $S(A)$ if and only if $L_q \subseteq L_A$, that is if and only if q is an initial state of $S(A')$.

In the same way, $q' \in \widetilde{\delta}(q, x)$ in $S(A)$ if and only if $xL_{q'} \subseteq L_q$, i.e. if and only if $q' \in \widetilde{\delta'}(q, x)$ in $S(A')$.   □

We note $\widetilde{L}_q = \{u \mid \widetilde{\delta}(q, u) \cap F \neq \emptyset\}$.

Proposition 2. *Let A be a NFA and let $S(A)$ be its saturated. For each state q of A, we have $L_q = \widetilde{L}_q$.*

Proof: Clearly, $L_q \subseteq \widetilde{L}_q$ as the saturated of an automaton is obtained by adding transitions and initial states. To prove the converse inclusion, we prove by induction that for every integer n and every state q

$$\widetilde{L}_q \cap \Sigma^{\leq n} \subseteq L_q.$$

If $n = 0$, the property is true as A and $S(A)$ have the same terminal states. Let $u = xv \in \widetilde{L}_q \cap \Sigma^{\leq n}$ with $n \geq 1$ and let $q' \in \widetilde{\delta}(q, x)$ such that $v \in \widetilde{L}_{q'}$. Because of our induction hypothesis, $v \in L_{q'}$. As $q' \in \widetilde{\delta}(q, x)$, we have $xL_{q'} \subseteq L_q$ and therefore $xv \in L_q$.   □

Corollary 1. *Let A be a NFA and $S(A)$ be its saturated. Then A and $S(A)$ recognize the same language and $S(A) = S(S(A))$.*

Proof:

- We have $L = \cup\{L_q | q \in Q_0\} = \cup\{L_q | q \in \widetilde{Q}_0\} = \cup\{\widetilde{L}_q | q \in \widetilde{Q}_0\}$ which is equal to the language recognized by $S(A)$.
- Due to the previous point and to the proposition 2, lemma 1 can be applied on A and $S(A)$ to prove that $S(S(A)) = S(A)$; the saturated of a saturated automaton is itself.

                                         □

Corollary 2. *If A is a RFSA then $S(A)$ is also a RFSA.*

Proof: The saturated of a RFSA is a RFSA as the saturation changes neither the languages associated with the states nor the language recognized by the automaton.   □

4.3   Reduction Operator ϕ

We define a *reduction* operator ϕ that deletes states in an automaton without changing the language it recognizes.

Definition 5. *Let $A = \langle \Sigma, Q, Q_0, F, \delta \rangle$ be a NFA, and let q be a state of Q. We note $R(q) = \{q' \in Q \backslash \{q\} \mid L_{q'} \subseteq L_q\}$. We say that q is erasable in A if $L_q = \bigcup\{L_{q'} \backslash q' \in R(q)\}$.*

If q is erasable, we define $\phi(A, q) = A' = \langle \Sigma, Q', Q'_0, F', \delta' \rangle$ where:

- $Q' = Q \backslash \{q\}$,
- $Q'_0 = Q_0$ *if* $q \notin Q_0$, *and* $Q'_0 = (Q_0 \backslash \{q\}) \cup R(q)$ *otherwise*,
- $F' = F \cap Q'$,
- *for every* $q' \in Q'$ *and every* $x \in \Sigma$

$$\delta'(q', x) = \begin{cases} \delta(q', x) \text{ if } q \notin \delta(q', x) \\ (\delta(q', x) \backslash \{q\}) \cup R(q) \\ \qquad \text{otherwise.} \end{cases}$$

If q is not erasable, we define $\phi(A, q) = A$.

Let $q' \in Q$ be a state different from q. We note $L_{q'}$ the language generated from q' in the automaton A and $L'_{q'}$ the language generated from q' in $A' = \phi(A, q)$.

Proposition 3. *Let A be a NFA and let q be a state of A. The automata A and $A' = \phi(A, q)$ recognize the same language and for every state $q' \neq q$, $L_{q'} = L'_{q'}$.*

Sketch of proof:
If q is not an erasable state, the proposition is straightforward. If q is an erasable state, we first prove that $L_{q'} = L'_{q'}$ using the fact that every path that allows to read a word u in A through q corresponds to a path in A' that uses an added transition and vice-versa.
Finally, we prove that $L_A = \bigcup_{q_0 \in Q_0} L_{q_0} = (\bigcup_{q_0 \in Q'_0} L'_{q_0}) = L_{A'}$.
□

Proposition 4. *The operator ϕ is an internal operator for the class of RFSA.*

Proof: Neither the language recognized by a RFSA A nor the languages associated with its states are modified by the reduction operator ϕ (c.f. previous proposition). So, languages associated with states keep being residual languages of L_A.
□
We prove now that saturation and reduction operators can be swapped.

Lemma 2. *Let $A = \langle \Sigma, Q, Q_0, F, \delta \rangle$ be a NFA and let q be a state of Q. Then the automaton $\phi(S(A), q)$ is saturated.*

Proof: We note $L'_{q'}$ (resp. $L_{q'}$) the language associated with a state q' in $\phi(S(A), q)$ (resp. in $S(A)$), δ' (resp. δ) the transition function of $\phi(S(A), q)$ (resp. in $S(A)$) and L the language recognized by the automata A, $S(A)$ and $\phi(S(A), q)$.

- If $L'_{q'} \subseteq L$ then $L_{q'} \subseteq L$ and so q' is initial in $S(A)$ and in $\phi(S(A), q)$.
- If $x L'_{q'} \subseteq L'_{q''}$ then $x L_{q'} \subseteq L_{q''}$ and so $q' \in \delta(q'', x)$ and $q' \in \delta'(q'', x)$.
□

Proposition 5. *Let $A = \langle \Sigma, Q, Q_0, F, \delta \rangle$ be a NFA and let q be a state of Q. We have*

$$S(\phi(A, q)) = \phi(S(A), q)$$

Proof: $\phi(A, q)$ and $\phi(S(A), q)$ have the same set of states. Furthermore, languages associated with every state q' in $\phi(A, q)$ and $\phi(S(A), q)$ are identical because of previous lemmas. Because of lemma 1, $S(\phi(A, q)) = S(\phi(S(A), q))$. As $\phi(S(A), q)$ is a saturated automaton (cf lemma 2), the proposition is proved.   □

Definition 6. *Let A be a NFA. If there is no erasable state in A, we say that A is reduced.*

4.4   Canonical RFSA

Definition 7. *Let L be a regular language. We define $A = \langle \Sigma, Q, Q_0, F, \delta \rangle$ the canonical RFSA of L in the following way:*

- *Σ is the alphabet of L,*
- *Q is the set of prime residuals of L, so $Q = \{u^{-1}L \mid u^{-1}L$ is prime $\}$,*
- *its initial states are prime residuals included in L, so $Q_0 = \{u^{-1}L \in Q \mid u^{-1}L \subseteq L\}$,*
- *its final states are prime residuals containing the empty word, so $F = \{u^{-1}L \in Q \mid \varepsilon \in u^{-1}L\}$,*
- *its transition function is $\delta(u^{-1}L, x) = \{v^{-1}L \in Q \mid v^{-1}L \subseteq (ux)^{-1}L\}$.*

This definition assumes that the canonical RFSA is a RFSA, we will prove this presumption below.

We have proved that the reduction operator ϕ transforms a RFSA into a RFSA, and that it could be swapped with the saturation operator. We prove now that, if A is a saturated RFSA, the reduction operator converges and that the resulting automaton is the canonical RFSA of the language recognized by A.

Proposition 6. *Let L be a regular language and let $A = \langle \Sigma, Q, Q_0, F, \delta \rangle$ be a reduced and saturated RFSA recognizing L. A is the canonical RFSA of L.*

Proof: As A is a RFSA, every prime residual $u^{-1}L$ of L can be defined as a language L_q associated with some states $q \in Q$. As there are no erasable states in A, for every state q, L_q is a prime residual and distinct states define distinct languages. As A is saturated, prime residuals contained in L correspond to initial states of Q_0. For the same reason, we can verify that the transition function is the same as in the canonical RFSA.   □

Theorem 1. *The canonical RFSA of a regular language L is a RFSA which recognizes L and which is minimal regarding the number of states.*

Proof: Let $A_0, \ldots, A_n$ be a sequence of NFA such that for every index $i \geq 1$, there exists a state q_i of A_{i-1} such that $A_i = \phi(A_{i-1}, q_i)$. Proposition 5 and 6 prove that if A_0 is a saturated RFSA and if A_n is reduced, then A_n is the canonical RFSA of the language recognized by A_0.

So the canonical RFSA can be obtained from any RFSA that recognizes L using saturation and reduction operators. Proposition 1 proves that it has a minimal number of states.   □

Remark that it is possible to find a RFSA that has as many states as the canonical RFSA of L, but fewer transitions. We have the following proposition:

Theorem 2. *The canonical RFSA of a regular language L is the unique RFSA that has a maximal number of transitions among the set of RFSA which have a minimal number of states.*

Proof: Let $A = \langle \Sigma, Q, Q_0, F, \delta \rangle$ be the canonical RFSA of a language L and let $A' = \langle \Sigma, Q', Q_0', F', \delta' \rangle$ be a RFSA which has a minimal number of states. So, A' is reduced. From proposition 6, the saturated automaton of A' is A. Therefore, A' has at most as many transitions as A.   □

5   Construction of the Canonical RFSA Using the Subset Method

In the previous section, we provided a way to build the canonical RFSA from a given NFA using saturation and reduction operators. This method requires to check whether a language is included into another one and to check whether a language is composed or not. Those checks can be very expensive, even for simple automata. We present in this section another method which stems from a classical construction of the minimal DFA of a language and which is easier to implement.

Let $A = \langle \Sigma, Q, Q_0, F, \delta \rangle$ be a NFA. The subset construction is a classical method used to build a DFA equivalent to a given NFA. It consists in building the set of reachable sets of states of A. We note $Q_{R(A)} = \{p \in 2^Q \mid \exists u \in \Sigma^* \text{ s. t. } \delta(Q_0, u) = p\}$ and we define the subset automaton $D(A) = \langle \Sigma, Q_D, Q_{D0}, F_D, \delta_D \rangle$ with

$$Q_D = Q_{R(A)},$$
$$Q_{D0} = \{Q_0\},$$
$$F_D = \{p \in Q_D \mid p \cap F \neq \emptyset\},$$
$$\delta_D(p, x) = \delta(p, x).$$

The automaton $D(A)$ is a determi-nistic automaton that recognizes the same language as A.

We remind that $\overline{L}$ (resp. $\overline{B}$) denotes the mirror of a language L (resp. of an automaton B). The following result provides a method to build the minimal DFA of L.

Theorem 3. *[Brz62] Let L be a regular language and let B be an automaton such that $\overline{B}$ is a DFA that recognizes $\overline{L}$. Then $D(B)$ is the minimal DFA recognizing L.*

We can deduce from this theorem that $D(\overline{D(\overline{A})})$ is the minimal DFA recognizing the language L_A.

We adapt the subset construction technique to deal with inclusions of sets of states. We say that a state $p \in Q_{R(A)}$ is *coverable* if there exist states $p_i \in Q_{R(A)}$, $p_i \neq p$, such that $p = \cup_i p_i$. We define the automaton $C(A) = \langle \Sigma, Q_C, Q_{C0}, F_C, \delta_C \rangle$ by

$$Q_C = \{p \in Q_{R(A)} \mid$$
$$p \text{ is not coverable }\},$$
$$Q_{C0} = \{p \in Q_C \mid p \subseteq Q_0\},$$
$$F_C = \{p \in Q_C \mid p \cap F \neq \emptyset\},$$
$$\delta_C(p, x) = \{p' \in Q_C \mid p' \subseteq \delta(p, x)\}.$$

Lemma 3. *Let A be a NFA, C(A) is a RFSA recognizing L_A such that all states are reachable.*

Sketch of proof: $C(A)$ can be obtained from $D(A)$ by using techniques which are similar to the ones used by the reduction operator. $\qquad\square$

Theorem 4. *Let L be a regular language and let B be an automaton such that $\overline{B}$ is a RFSA recognizing $\overline{L}$ such that all states are reachable. Then $C(B)$ is the canonical RFSA recognizing L.*

Sketch of proof:

Let $q_i \in Q_B$, let $\overline{L}_{q_i}$ be the language associated with q_i in $\overline{B}$ and let $v_i \in \Sigma^*$ be such that $\overline{L}_{q_i} = \overline{v_i}^{-1}\overline{L}$. Let $p, p' \in Q_{R(B)}$. We prove that:

- $v_i \in L_p$ iff $q_i \in p$.
- $L_p \subseteq L_{p'}$ iff $p \subseteq p'$.
- For every state $p, p_1, p_2 \ldots p_n \in Q_{R(B)}$, $L_p = \cup_{1\leq k\leq n}L_{p_k}$ iff $p = \cup_{1\leq k\leq n}p_k$.

From the last three statements, we can prove that $C(B)$ can be obtained from $D(B)$ by reduction and saturation. As $D(B)$ is deterministic, and using proposition 6, $C(B)$ is the canonical RFSA of L. $\qquad\square$

We can deduce from this proposition and from lemma 3 that $C(\overline{C(\overline{A})})$ is the canonical RFSA of L_A.

However, this construction also has some weaknesses. Indeed, it is possible to find examples for which $C(\overline{A})$ has an exponential number of states with regard to the number of states of A or $C(\overline{C(\overline{A})})$. We can observe this situation with the mirror of the automaton used in the proposition 8.

We can also observe that, if we are interested only in covering without saturation (if a state is covered, we delete it and we relead its transitions to covering states), we get a RFSA which has the same number of states (non-coverable states) and fewer transitions.

6   Results on Size of RFSA

We classically take the number of states of an automaton as a measure of its size. The canonical RFSA of a regular language has the size of the equivalent minimal DFA as an upper bound and the size of one of its equivalent minimal NFA as a lower bound. We show that both bounds can be reached even if there exists an exponential gap between these two bounds.

Proposition 7. *There exist languages for which the minimal DFA has a size exponentially larger than the size of the canonical RFSA, and for which the canonical RFSA has the same size as minimal NFA.*

Proof: $\Sigma^*0\Sigma^n$ languages, where n is an integer and $\Sigma = \{0,1\}$, can illustrate this proposition.

Residuals of $L = \Sigma^*0\Sigma^n$ are languages $L \cup (\bigcup_{p\in P}\Sigma^p)$ where $P \subseteq \{0,\ldots,n\}$. One can observe that there exist 2^{n+1} distinct residuals. The minimal DFA recognizing this language has 2^{n+1} states. There exist only $n+2$ prime residuals: $L, L\cup\Sigma^0, \ldots,$ $L\cup\Sigma^n$, so, the canonical RFSA of L has $n+2$ states. $\qquad\square$

Proposition 8. *There exist languages for which the size of the canonical RFSA is exponential with regard to the size of a minimal NFA.*

Proof: Let $A_n = \langle \Sigma, Q, Q_0, F, \delta \rangle$ be automata such that, for $n \geq 1$

- $\Sigma = \{a, b\}$,
- $Q = \{q_i \mid 0 \leq i \leq n - 1\}$,
- δ is defined by
  $\delta(q_i, a) = q_{i+1}$ (for $0 \leq i < n - 1$),
  $\delta(q_{n-1}, a) = q_0$,
  $\delta(q_0, b) = q_0$,
  $\delta(q_i, b) = q_{i-1}$ (for $1 < i < n$) and
  $\delta(q_1, b) = q_{n-1}$,
- $Q_0 = \{q_i \mid 0 \leq i < n/2\}$,
- $F = \{q_0\}$.

Figure 5 represents A_4.

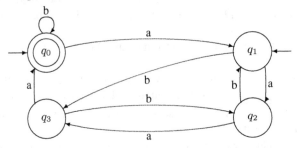

Fig. 5. An automaton A_n, $n = 4$, for which the Equivalent RFSA Is Exponentially Larger.

The mirror automata $\overline{A_n}$ are trimmed and deterministic, thus we can apply theorem 4. The automata $C(\overline{A_n})$ are canonical RFSA.

The initial state of the subset construction has $n/2$ elements. Moreover the reachable states are all the states with $n/2$ elements. So, none of them is coverable.

The canonical RFSA $C(\overline{A_n})$ are exponentially larger than the initial NFA.   □

Proposition 9. *There exist languages for which the smallest characterizing word for some state has a length exponentially bigger than the number of states of the canonical RFSA.*

Sketch of proof: Let $P = \{p_1, \ldots, p_n\}$ be a set of n distinct prime numbers. We define the NFA $A_P = \langle \Sigma, Q, Q_0, F, \delta \rangle$ by:

- $\Sigma = \{a\} \cup \{b_p \mid p \in P\}$
- $Q = \{q_i^p \mid p \in P, 0 \leq i < p\}$
- $Q_0 = \{q_0^p \mid p \in P\}$

- $F = Q_0$
- δ is defined by:

$\delta(q_i^p, a) = \{q_{(i+1)mod\ p}^p\}$
for $0 \leq i < p, p \in P$,

$\delta(q_i^p, b_{p'}) = \{q_i^p, q_{i+1}^p\}$
for $0 \leq i < p - 1, p, p' \in P$,

$\delta(q_{p-1}^p, b_{p'}) = \{q_0^{p'}\}$
for $p, p' \in P$.

The following results can be proved:

- A_P is a RFSA.
- The smallest characterizing word u_q of a state $q \in Q$ is such that $|u_q| \geq \Pi_i p_i$ which is exponential with regard to the size of A_P and therefore exponential with regard to the size of the canonical RFSA.

$\square$

Let $A = \langle \Sigma, Q, Q_0, F, \delta \rangle$ be a RFSA and let $q \in Q$ such that L_q is prime. There must exist a smallest word $u \in L_q$ such that $L_{q'} \subsetneq L_q \Rightarrow u \notin L_{q'}$. Next proposition proves that this word can be very long.

Proposition 10. *There exist languages for which the smallest word that proves that a state of the canonical RFSA is not composed has an exponential size with regard to the number of states of the minimal DFA.*

Proof: Let $p_1,...,p_n$ be distinct prime numbers. For each i, $1 \leq i \leq n$, we note $L_i = \{\varepsilon\} \cup \{a^k \mid p_i$ is not a divisor of $k\}$. Let $b_0, b_1,... b_n$ be distinct letters different from a. We consider the language $L = b_0 a^* \cup (\bigcup_{1 \leq i \leq n} b_i L_i)$.

We can easily build a minimal DFA for this language ; it contains $\sum p_i + n + 2$ states. The language $b_0^{-1}L = a^*$ is not an union of residuals $b_i^{-1}L, i \geq 1$. But the shortest word that belongs to $b_0^{-1}L \setminus \bigcup_{1 \leq i \leq n} b_i^{-1}L_i$ is $a^{p_1 \cdots p_n}$ and its length is exponential with regard to the size of the minimal DFA. $\square$

7 Complexity Results about RFSA

We have defined notions of RFSA, saturated automata, canonical RFSA ; in this section, we evaluate the complexity of our constructions and of decision problems linked to them: deciding if an automaton is saturated, building the canonical RFSA of a given language, and so on ...

Classical definitions about complexity can be found in [GJ79] and complexity results about automata can be found in [HU79]. We present here simple complexity results about RFSA, proofs of which can be found in [DLT00b].

The first notion that we defined is the notion of *saturation*. As one could guess, deciding if an automaton is saturated is easier for a DFA than for a NFA.

Proposition 11. *Deciding whether a DFA is saturated is a polynomial problem. On the other hand, deciding whether a NFA is saturated is a PSPACE-complete problem. Building the saturated of a NFA is also a PSPACE-complete problem.*

The next proposition tells us that it is not practically possible, in the worst case, to check whether a NFA is a RFSA.

Proposition 12. *Deciding if a NFA is a RFSA is a $PSPACE$-complete problem.*

Building the canonical RFSA equivalent to a given NFA is an exponential problem in general, as proved by proposition 8. The next proposition tells us that, even if the starting automaton is deterministic, this problem is $PSPACE$-complete. The problem of deciding whether the saturated of a DFA is a canonical RFSA is also $PSPACE$-complete.

Proposition 13. *Deciding if the saturated of a DFA is a canonical RFSA is a $PSPACE$-complete problem. Building the canonical RFSA equivalent to a DFA is also a $PSPACE$-complete problem.*

8   Comments and Conclusion

Ideas developed in this paper come from a work done in the domain of Grammatical Inference. A main problem in this field is to infer efficiently (a representation of) a regular language from a finite set of examples of this language. Some positive results can be proved when regular languages are represented by Deterministic Finite Automata (DFA). For example, it has been proved that Regular Languages represented by DFA can be infered from *given data* ([Gol78,Hig97]). In this framework, classical inference algorithms such as RPNI ([OG92]) need a polynomial number of examples relatively to the size of the minimal DFA that recognizes the language to be infered. So, regular languages as simple as $\Sigma^*0\Sigma^n$ cannot be infered efficiently using these algorithms since their minimal DFA have an exponential number of states. Hence, it is a natural idea to try to use other kind of representations for regular languages, such as Non deterministic Finite Automata (NFA). Unfortunately, it has been proved that Regular Languages represented by NFA cannot be efficiently infered from given data ([Hig97]). We described in [DLT00a] an inference algorithm (*DeLeTe*) that computes the canonical RFSA of a target regular language from given data. Using this algorithm, languages such as $\Sigma^*0\Sigma^n$ become efficiently learnable. So, introducing the class of RFSA in the field of grammatical inference seems to be a promising idea. However, we have to deal with the fact that most decision and construction problems linked to the class of RFSA are untractable in the worst case. What are the practical consequences of these worst-case complexity results ? Experiments we are currently leading in the field of grammatical inference let us think that they could be not too dramatic.

While achieving this work, we have felt that RFSA was a class of automata worth being studied for itself, from a language theory point of view and this is what we have done in this paper. The class of RFSA has a very simple definition. It provides a description level of regular languages which is intermediate between a representation by deterministic automata and a representation that uses the whole class of non deterministic automata. RFSA shares two main properties with the class of DFA: the existence of a canonical minimal form and the fact that states correspond to natural component of the recognized language. Moreover canonical RFSA can be exponentially smaller than the equivalent minimal DFA. All these properties show that the RFSA is an interesting class whose study must be carried on.

Acknowledgments

We would like to thank Michel Latteux for his helpful comments and advice.

References

ADN92.   A. Arnold, A. Dicky and M. Nivat. A note about minimal non-deterministic automata. *Bulletin of the EATCS*, 47:166–169, June 1992.

Brz62.   J. A. Brzozowski. Canonical regular expressions and minimal state graphs for definite events. In *Mathematical Theory of Automata*, volume 12 of *MRI Symposia Series*, pages 52–561. 1962.

Car70.   C. Carrez. On the minimalization of non-deterministic automaton. Technical report, Laboratoire de Calcul de la Faculté des Sciences de l'Université de Lille, 1970.

DLT00a.   F. Denis, A. Lemay and A. Terlutte. Apprentissage de langages réguliers à l'aide d'automates non déterministes. In *CAP'2000*, 2000.

DLT00b.   F. Denis, A. Lemay and A. Terlutte. Residual finite state automata. Technical Report LIFL 2000-08, L.I.F.L., 2000.

GJ79.   Michael R. Garey and David S. Johnson. *Computers and Intractability, a Guide to the Theory of NP-Completness*. W.H. Freeman and Co, San Francisco, 1979.

Gol78.   E.M. Gold. Complexity of automaton identification from given data. *Inform. Control*, 37:302–320, 1978.

Hig97.   Colin De La Higuera. Characteristic sets for polynomial grammatical inference. *Machine Learning*, 27:125–137, 1997.

HU79.   J.E. Hopcroft and J.D. Ullman. *Introduction to Automata Theory, Languages, and Computation*. Addison-Wesley, 1979.

Kle56.   S. C. Kleene. Representation of events in nerve nets and finite automata. In C. Shannon and J. McCarthy, editors, *Automata Studies, Annals of Math. Studies 34*. New Jersey, 1956.

Myh57.   J. Myhill. Finite automata and the representation of events. Technical Report 57-624, WADC, 1957.

Ner58.   A. Nerode. Linear automaton transformation. In *Proc. American Mathematical Society*, volume 9, pages 541–544, 1958.

OG92.   J. Oncina and P. Garcia. Inferring regular languages in polynomial update time. In *Pattern Recognition and Image Analysis*, pages 49–61, 1992.

Yu97.   Sheng Yu. *Handbook of Formal Languages, Regular Languages*, volume 1, chapter 2, pages 41–110. 1997.

Deterministic Radio Broadcasting at Low Cost

Anders Dessmark[1]* and Andrzej Pelc[2]**

[1] Department of Computer Science, Lund Institute of Technology
Box 118, S-22100 Lund, Sweden
andersd@cs.lth.se
[2] Département d'Informatique, Université du Québec à Hull
Hull, Québec J8X 3X7, Canada
pelc@uqah.uquebec.ca

Abstract. We consider the problem of distributed deterministic broadcasting in radio networks. The network is synchronous. A node receives a message in a given round if and only if exactly one of its neighbors transmits. The source message has to reach all nodes. We assume that nodes do not know network topology or even their immediate neighborhood. We are concerned with two efficiency measures of broadcasting algorithms: its execution *time* (number of rounds), and its *cost* (number of transmissions). We focus our study on execution time of algorithms which have cost close to minimum.

We consider two scenarios depending on whether nodes know or do not know global parameters of the network: the number n of nodes and the eccentricity D of the source. Our main contribution are lower bounds on time of low-cost broadcasting which show sharp differences between these scenarios.

1 Introduction

Radio networks have been extensively investigated by many researchers [1,3,5,6], [7,8,9,12,14,17,18,19]. A radio network is a collection of stations, called *nodes*, which are equipped with capabilities of transmitting and receiving messages. Every node can *reach* a given subset of other nodes, depending on the power of its transmitter and on the topography of the region. Hence a radio network can be modeled by its *reachability graph* in which the existence of a directed edge (u, v) means that node v can be reached from u. In this case u is called a *neighbor* of v.

Nodes send messages in synchronous *rounds* measured by a global clock. In every round every node either *transmits* (to all nodes within its reach) or is *silent*. A node which is silent in a given round gets a message if and only if exactly one of its neighbors transmits in this round. If at least two neighbors of u transmit simultaneously in a given round, none of the messages is received by u in this round. In this case we say that a *collision* occurred at u.

---

* This research was done during the stay of Anders Dessmark at the Université du Québec à Hull as a postdoctoral fellow.
** Andrzej Pelc was supported in part by NSERC grant OGP 0008136.

A. Ferreira and H. Reichel (Eds.): STACS 2001, LNCS 2010, pp. 158–169, 2001.

There are two models studied in the literature which differ by specifying what exactly happens during a collision. The model with *collision detection* assumes that in this case the node at which collision occurred gets a signal different from the messages transmitted but also different from the background noise, and thus the node can deduce that more than one of its neighbors transmitted. An alternative model assumes *no collision detection*, i.e., supposes that the signal obtained as a result of collision is not different from the background noise, and thus nodes cannot distinguish multiple transmissions from no transmission. A comparative discussion justifying both models can be found in [3,15]. In this paper we use the model assuming no collision detection, as e.g., in [3,7,8,9].

Broadcasting is one of the basic tasks in network communication (cf. surveys [13,16]). One node of the network, called the *source*, has to transmit a message to all other nodes. Remote nodes are informed via intermediate nodes, along directed paths in the network. We assume that there exists a directed path from the source to any other node, and we restrict attention to such graphs only.

One of the basic performance measures of a broadcasting scheme is the total time, i.e., the number of rounds it uses to inform all the nodes of the network. There is, however, another natural measure of efficiency of a broadcasting algorithm, and this is the total number of transmissions it uses. We call this number the *cost* of the broadcasting scheme. Algorithms using few transmissions to broadcast in a radio network are less expensive to run. Apart from that they may permit portions of the network which are remote from currently transmitting nodes to carry out simultaneously non broadcast related transmissions.

The aim of this paper is to study broadcasting algorithms working in radio networks of unknown topology, as in [3,7,8,9]. As opposed to [3,7,8,9], we consider not only the time but also the cost of broadcasting. The main subject of our study is execution time of *low-cost* algorithms, i.e., those whose cost is close to minimum.

We assume that nodes do not have any knowledge of network topology, and that local knowledge of every node is limited to its own label. For n-node networks we assume that labels are distinct integers from the set $\{0, ..., n-1\}$ but all our results remain valid if labels are from the set $\{0, ..., N\}$, where $N \in O(n)$. (It is well known that radio broadcasting cannot be carried out in the anonymous model, even in the 4-ring.) We consider two scenarios depending on whether nodes know or do not know global parameters of the network: the number n of nodes and the eccentricity D of the source (i.e., the maximum length of all shortest paths from the source to all other nodes). Since these parameters may be unknown to nodes, the broadcasting process may be finished but nodes may be unaware of this. In fact it was proved in [7] that broadcasting with acknowledgement is impossible in unknown networks without collision detection. Consequently, we define time of broadcasting without requiring that nodes know that the process is terminated, similarly as in [7]. A broadcasting algorithm works in t rounds on a network G, if t is the minimum integer such that after round t all nodes of G know the source message, and no messages are transmitted after round t. Likewise, an algorithm has cost c for a network G, if c is the minimum integer such that all nodes get the source message after c transmissions and no more transmissions are executed when the algorithm is run on G.

Since our algorithms run on arbitrary unknown networks with n nodes and eccentricity D of the source, we are interested in their worst-case performance on the class of all networks with these parameters. Consequently, we define the time (resp. cost) of an algorithm for networks with parameters n and D as the maximum time (resp. cost) of this algorithm over all networks in this class.

1.1 Related Work

In much of the research on broadcasting in radio networks [1,3,5,6,14,18] the network is modeled as an undirected graph, which is equivalent to the assumption that the reachability graph is symmetric. The focus of research in these papers is broadcasting time, and more precisely, finding upper and lower bounds on it, under the assumption that nodes have full knowledge of the network. In [1] the authors proved the existence of a family of n-node networks of radius 2, for which any broadcast requires time $\Omega(\log^2 n)$, while in [14] it was proved that broadcasting can be done in time $O(D + \log^5 n)$, for any n-node network of diameter D. In [17] the authors discussed broadcasting time in radio networks arising from geometric locations of nodes on the line and in the plane, under the assumption that some of the nodes may be faulty.

In the above papers, the topology of the radio network was known in advance, and broadcasting algorithms were deterministic and centralized. On the other hand, in [3] a randomized protocol was given for arbitrary radio networks of unknown topology. This randomized protocol runs in expected time $O(D \log n + \log^2 n)$. In [18] it was shown that for any randomized broadcast protocol and parameters D and n, there exists an n-node network of diameter D requiring expected time $\Omega(D \log(n/D))$ to execute this protocol.

These results suggest an interesting question concerning the efficiency of *deterministic* broadcasting algorithms working in networks of *unknown* topology. The first paper to deal with this scenario was [3]. The authors showed that any such algorithm requires time $\Omega(n)$ for some symmetric network of constant diameter. In [12] fast deterministic broadcasting algorithms were given for networks of unknown topology but of a very restricted class: the authors assumed that nodes are located in unknown points of the line, and every node can reach all nodes within a given radius from it.

Deterministic broadcasting in arbitrary radio networks of unknown topology was first investigated in [7]. The authors showed an algorithm working in time $O(n^{11/6})$ and established a lower bound $\Omega(n \log n)$ on broadcasting time (cf. [4] where this lower bound was earlier proved in a slightly different setting). Then a series of faster broadcasting algorithms have been proposed, with execution times $O(n^{5/3}(\log n)^{1/3})$ [11], $O(n^{3/2}\sqrt{\log n})$ [20], $O(n^{3/2})$ [8], and $O(n \log^2 n)$ [9]. In all these papers time was the only considered measure of efficiency of broadcasting. However, in [7] a very simple and slower broadcasting algorithm was also proposed. Its execution time is $O(n^2)$ but from our point of view it has an additional advantage: its cost is minimum, i.e., n.

To the best of our knowledge, relations between time and cost of broadcasting in radio networks have never been studied previously. However, cost of communication measured in terms of the number of transmissions has been widely studied for point-to-point networks, mostly in the context of gossiping, i.e., of all-to-all communication (cf. [16] and the literature therein). It should be stressed that, unlike in radio networks, where a node transmits to all nodes within its reach, transmissions in point-to-point networks occur between specific pairs of nodes. Tradeoffs between time and cost of communication in such networks were studied, e.g., in [10]. On the other hand, cost of broadcasting in point-to-point networks in which nodes know only their neighborhood, was the subject of [2].

1.2   Our Results

While we assume that nodes do not have any knowledge of network topology, we consider two scenarios depending on whether they know or do not know global parameters of the network: the number n of nodes and the eccentricity D of the source. We show that the minimum cost of broadcasting in an n-node network of unknown topology is n, if at least one of the above parameters is unknown, and it is $n - 1$, if both of them are known. Our main contribution are lower bounds on time of low-cost broadcasting which show sharp differences between these scenarios. We show that if nodes know neither n nor D then any broadcasting algorithm whose cost exceeds the minimum by $O(n^\beta)$, for any constant $\beta < 1$, must have execution time $\Omega(Dn \log n)$ for some network. We also show a minimum-cost algorithm that does not assume knowledge of these parameters, and works always in time $O(Dn \log n)$. On the other hand, assuming that nodes know either n or D, we show how to broadcast in time $O(Dn)$. This time cannot be improved by any low-cost algorithm even knowing both n and D. Indeed, we show that any algorithm whose cost exceeds the minimum by at most αn, for any constant $\alpha < 1$, requires time $\Omega(Dn)$. Hence we obtain asymptotically tight bounds on time of low-cost broadcasting under these two scenarios, and we show that knowing at least one of the global parameters n or D results in faster low-cost broadcasting than when none of them is known.

In addition, we show that very fast broadcasting algorithms must have high cost. We prove that every broadcasting algorithm that works in time $O(nt(n))$, where $t(n)$ is polylogarithmic in n, requires cost $\Omega(n \log n / \log \log n)$. Since the fastest known algorithm works in time $O(n \log^2 n)$ [9], its cost (as well as the cost of any faster broadcasting algorithm, if it exists) must be higher than linear.

2   Minimum-Cost Broadcasting

In this section we establish asymptotically tight upper and lower bounds on execution time of minimum-cost broadcasting algorithms in two situations: (1) when nodes know either the number n of nodes or the eccentricity D of the source, and (2) when nodes do not know any of these parameters. It turns out that optimal broadcasting time is different in each of those cases.

2.1   Lower Bounds on Cost

We first establish lower bounds on the cost of any broadcasting algorithm working without knowledge of topology. The cost of all algorithms presented in this section will match these lower bounds and thus these are minimum-cost algorithms. The minimum cost turns out to be n if at least one of the parameters n or D is unknown to nodes, and it is $n-1$, if both of them are known. The proofs of these lower bounds are omitted due to lack of space, and will appear in the full version of the paper.

Theorem 1. *If the eccentricity D of the source is unknown to nodes then every broadcasting algorithm requires cost at least n for some n-node networks, for any $n > 2$.*

Theorem 2. *If the size n of the network is unknown to nodes then, for sufficiently large integers n, every broadcasting algorithm requires cost at least n for some n-node networks.*

If nodes know both the size n of the network and the eccentricity D of the source, we will see that cost can be lower but the following result shows that the gain can only be 1.

Theorem 3. *Every broadcasting algorithm requires cost at least $n-1$ for some n-node networks, for any $n > 2$.*

2.2   Broadcasting Time with Known n or D

We first consider the case when nodes of the network know at least one of the global parameters: the number n of nodes, or the eccentricity D of the source. We begin by presenting minimum-cost broadcasting algorithms working under these assumptions. We present three different algorithms, depending on whether nodes know only n, only D, or both these parameters.

The simplest case occurs when nodes know the size n of the network but do not know the eccentricity D of the source. By Theorem 1 the lower bound on cost is n in this case, and broadcasting can be performed in time $O(Dn)$ using the following simple algorithm in which every node transmits exactly once, i.e., at cost n. Therefore this is a minimum-cost algorithm.

Algorithm Only-Size-Known
    The source transmits in the first round. A node with label l transmits the source message in the first round r after it has received the message and for which $l \equiv r \pmod{n}$, and then stops.   □

When only the eccentricity of the source is known, the situation is slightly more complicated. In this case the lower bound on cost is also n, by Theorem 2. The following algorithm performs broadcasting with the same efficiency as above (time $O(Dn)$ and cost n). The algorithm probes for n by repeatedly doubling the range of labels of nodes allowed to transmit. Again every node transmits exactly once. (Compare [7] where this technique was used with both n and D unknown, resulting in time $O(n^2)$.)

Algorithm Only-Eccentricity-Known

The source transmits in the first round. The algorithm works in stages. Stage i consists of $D2^i$ rounds. In every stage enumerate rounds from 0 to $D2^i - 1$. A node with label l transmits the source message in the first round r, after it has received the message and for which $l \equiv r \pmod{2^i}$, and then stops.   □

When both parameters D and n are known, the lower bound on cost is $n - 1$ (cf. Theorem 3), and the above algorithms which work at cost n are not minimum-cost. However Algorithm Only-Size-Known can be modified, so as to reduce cost by 1 on any n-node network with source eccentricity D.

Algorithm Both-Known

The source transmits in the first round. All other rounds are partitioned into consecutive disjoint segments of length n. Rounds in each segment are numbered $0,...,n - 1$. A node with label l that gets the source message for the first time in some round of segment $i < D - 1$, transmits it in round l of segment $i + 1$, and then stops.   □

Theorem 4. *Algorithms Only-Size-Known, Only-Eccentricity-Known, and Both-Known complete broadcasting at minimum cost, corresponding to their respective assumptions. They all work in time $O(Dn)$, in any n-node network with eccentricity D of the source.*

We now show that broadcasting at minimum cost cannot be performed asymptotically faster than done by the above algorithms, if either n or D is known to nodes. The result holds also when both parameters are known.

Theorem 5. *Suppose that nodes know either the number n of nodes or the eccentricity D of the source. Every minimum-cost broadcasting algorithm requires time $\Omega(Dn)$ for some n-node networks with eccentricity D of the source, if $1 < D \leq n - 2$.*

PROOF. Fix a minimum-cost broadcasting algorithm A working under the assumption of the theorem. (Recall that the minimum cost is either n or $n - 1$, depending on the knowledge of nodes.) In the rest of the argument we consider only n-node networks with eccentricity D of the source in which exactly one node is at distance D from the source. We will construct such a network G for which algorithm A requires time $\Omega(Dn)$. Fix a source v_1. It will have indegree 0. Assume without loss of generality that v_1 transmits in the first round.

Consider any node $u \neq v_1$ and any network with source v_1 for which v_1 is the unique neighbor of u. The behavior of u is the same for all these networks. If some such node u never transmits, the algorithm is incorrect. Indeed, there exists a network in the considered class, in which some node w has u as its unique neighbor. This node w could not get the source message. On the other hand, if some such node u transmits more than once, we get a contradiction with the fact that A is a minimum-cost algorithm.

Hence, in any network with source v_1 for which v_1 is the unique neighbor of u, node u transmits exactly once. Consider two such nodes u_1 and u_2. Their unique transmission must occur in two distinct rounds. The argument is the same as in the proof of Theorem 2.

Let v_2 be the node such that the round t in which v_2 transmits if its unique neighbor is the source, is the latest among all remaining nodes. Add this node v_2 and add the edge (v_1, v_2). v_2 will not have any other neighbors. By definition, node v_2 transmits no sooner than in round $1+(n-1)$. Next we pick node v_3 among all remaining nodes in the same way (replacing v_1 by v_2 in the construction), and we add edge (v_2, v_3). Node v_3 transmits no sooner than in round $1 + (n - 1) + (n - 2)$. We continue with nodes $v_4, ..., v_{D+1}$ in a similar manner. Finally, we attach all remaining nodes directly to the source (i.e., we add edges (v_1, v_j), for $j = D + 2, ..., n$, thus creating an n-node network with eccentricity D. The time required by algorithm A on this network is at least $1 + (n - 1) + (n - 2) + ... + (n - D) \in \Omega(Dn)$. □

2.3 Broadcasting Time with Unknown n and D

The following algorithm performs broadcasting when n and D are both unknown. This is done by repeatedly increasing Dn by a factor of 4, and for each value of Dn attempting to broadcast for different ratios of n/D. (A similar algorithm has been independently proposed in [20].)

Algorithm Unknown

The algorithm is divided into stages. Stage i consists of $i+1$ phases numbered 0 to i. Phase j of stage i has $2^{2i-j} \times 2^j = 2^{2i}$ rounds. The rounds of phase j of stage i are each assigned to one label. The kth round of the phase is assigned to label $k \bmod 2^{2i-j}$. A node v transmits the source message once, in the first round assigned to the label of v after it receives the source message. □

Theorem 6. *Algorithm Unknown performs broadcasting in $O(Dn \log n)$ rounds, at (minimum) cost n.*

The following lower bound on time shows that Algorithm Unknown is asymptotically optimal among minimum-cost broadcasting algorithms, when parameters n and D are unknown to nodes.

Theorem 7. *Every minimum-cost broadcasting algorithm requires $\Omega(Dn \log n)$ rounds for some n-node networks with eccentricity D of the source, when both n and D are unknown to nodes.*

PROOF. Fix a minimum-cost algorithm A and an even positive integer i. We will construct an n-node network G with eccentricity D of the source and with $Dn \in O(2^i)$, on which the algorithm performs broadcasting in time $\Omega(Dn \log n)$. Since D is unknown, every node must transmit at least once, as shown in the proof of Theorem 1. In fact, every node must transmit exactly once, for otherwise cost n would be exceeded.

If for a network H the algorithm schedules two distinct nodes u and v to transmit in the same round, construct a network H' by adding to H a new node w and edges (u, w) and (v, w). u and v behave identically when A is run on H and on H'. In the unique round in which u and v transmit, a collision occurs at w, and hence w cannot get any message, as u and v are its unique neighbors. Hence all nodes must transmit in separate rounds.

We now proceed with the construction of network G. The node with label 0 will be the source with indegree 0. Partition all integers between 1 and $2^i - 1$ into $i/2$ disjoint consecutive segments $S_{i/2+1}, ..., S_i$, such that $2^{j-1} \leq |S_j| \leq 2^j$. For each set S_j consider all networks whose set of labels is $S_j \cup \{0\}$. For all of these networks compute the set of rounds in which some node of the network apart from the source transmits. Let R_j be the union of the sets of rounds corresponding to all networks with set of nodes $S_j \cup \{0\}$. The sets $R_{i/2+1}, ..., R_i$ must be pairwise disjoint because two nodes cannot transmit simultaneously. Theorem 5 (with suitable relabeling of nodes and renumbering of rounds) shows that, for any $j = i/2+1, ..., i$, there exist a network G_j with set of nodes $S_j \cup \{0\}$, and eccentricity of the source $D_j \in \Theta(2^{i-j})$, on which algorithm A works in $\Omega(2^i)$ rounds from R_j. Let X_j denote the set of all rounds from R_j which are less than or equal to the last round in which A schedules a transmission when run on network G_j. Hence the sizes of all sets X_j are $\Omega(2^i)$, and all these sets are pairwise disjoint, as subsets of sets R_j. Consequently, some X_j must contain a round $\Omega(i2^i)$. This means that algorithm A requires time $\Omega(i2^i)$ on the respective network G_j.

Now we construct G by augmenting G_j as follows. Attach all nodes with labels from all sets $S_{j'}$, $j' < j$, directly to the source of G_j. The resulting network G has size $n \leq 2^{j+1}$, and nodes are numbered by consecutive integers. The eccentricity of the source remains unchanged: $D = D_j \in \Theta(2^{i-j})$. Hence $2^i \in \Omega(Dn)$ and $i \in \Omega(\log n)$, and consequently A requires time $\Omega(Dn \log n)$ on G. $\qquad\square$

3 Broadcasting at Low Cost

In this section we generalize Theorems 5 and 7 by proving the respective lower bounds on time for a larger class of algorithms: not only those with minimum cost but for all algorithms whose cost is close to minimum. This shows that in order to decrease broadcasting time, cost of broadcasting must be increased significantly. In the proofs we indicate how our previous arguments should be extended in this more general case.

Theorem 8. *Every broadcasting algorithm with cost at most αn, for a constant $\alpha < 2$, works in time $\Omega(nD)$ for some n-node networks with eccentricity D of the source, whenever $D \leq cn$, for some constant $c < 4 - 2\alpha$.*

PROOF. Fix a broadcasting algorithm A and a constant $\alpha < 2$. We will construct an n-node network G with eccentricity D of the source, for which the algorithm either works in time $\Omega(nD)$ or works at cost larger than αn. Fix a source v_1. It will have indegree 0. Assume without loss of generality that v_1 transmits in the first round. Let v_2 be the node such that the round t_1 in which v_2 first transmits the source message if its unique neighbor is v_1, is the latest among all nodes different from v_1. Add the node v_2, and the edge (v_1, v_2).

If $t_1 < 2n - \alpha n$, add the edge (v_1, u) for every other node u. We will show that for the resulting network G the algorithm has cost larger than αn. Indeed, every node except the source transmits during rounds number 2 to $2n - \alpha n - 1$. For every round r in this interval, at most one node transmits in round r and in no other round. Otherwise two nodes transmit once and during the same round which is impossible (cf. the proof of Theorem 7). Hence, at most $2n - \alpha n - 1$ nodes can transmit once and the remaining $\alpha n - n + 1$ nodes must transmit at least twice. This results in cost larger than αn.

If $t_1 \geq 2n - \alpha n$, i.e., v_2 transmits no sooner than in round $2n - \alpha n$, let v_3 be the node such that the round t_2 in which v_3 first transmits the source message if its unique neighbor is v_2, is the latest among all nodes different from v_1 and v_2. Continue the construction of G by adding node v_3, and the edge (v_2, v_3).

Similarly as above, it is either possible to construct a network yielding cost larger than αn by adding the edge (v_2, u) for every remaining node u, or v_3 transmits the source message no sooner than in round $4n - 2\alpha n - 2$. In this way, if cost does not exceed αn, we can construct a directed path of length D, for which the algorithm requires more than $(D-1)(2n - \alpha n) - D(D-1)/2 \in \Omega(nD)$ rounds. (Here we use the assumption that $D \leq cn$, for some $c < 4 - 2\alpha$). The remaining $n - D - 1$ nodes should be attached directly to the source to produce an n-node network with eccentricity D of the source, on which the algorithm requires the above number of rounds. □

Theorem 9. *Every broadcasting algorithm with cost less than $n + n^\beta$, for a positive constant $\beta < 1$, works in time $\Omega(Dn \log n)$ for some n-node networks with eccentricity D of the source, when n and D are unknown to nodes.*

PROOF. Fix a broadcasting algorithm A and an even positive integer i. We will construct an n-node network G with eccentricity D of the source and with $Dn \in O(2^i)$, on which the algorithm performs broadcasting in time $\Omega(Dn \log n)$, or a network H with $n' = 2^{i(\frac{1}{2}+\epsilon)}$ nodes (ϵ is a positive constant $< 1/2$ to be determined later), on which the algorithm requires cost $n' + n'^\beta$. Since D is unknown, every node must transmit at least once, as shown in the proof of Theorem 1.

The node with label 0 will be the source with indegree 0. Similarly as in the proof of Theorem 7, partition all integers between 1 and $2^i - 1$ into ϵi

disjoint consecutive segments $S_{i/2+1}, ..., S_{i(\frac{1}{2}+\epsilon)}$, such that $2^{j-1} \leq |S_j| \leq 2^j$. For each set S_j consider all networks whose set of labels is $S_j \cup \{0\}$. For all of these networks compute the set of rounds for which there exists a node v different from the source such that this round is the only one when v transmits. (Notice the difference from the construction in the proof of Theorem 7.) Let R_j be the union of the sets of rounds corresponding to networks with the set of nodes $S_j \cup \{0\}$. The sets $R_{i/2+1}, ..., R_{i(\frac{1}{2}+\epsilon)}$ must be pairwise disjoint because two nodes that transmit exactly once cannot transmit simultaneously, by the same argument as in the proof of Theorem 7.

For every $i/2 < j \leq i(\frac{1}{2}+\epsilon)$ we construct a network G_j with the set of nodes $S_j \cup \{0\}$. The network G_j consists of a directed path P of length $D_j \in \Theta(2^{i-j})$, in which the source is the first node, and of nodes outside of P having the source as their only neighbor and with outdegree 0. Pick the nodes $0, v_1, ..., v_p$ in path P in order of increasing indices, in the following way. At any stage of the construction, if there is a choice of v_k such that it transmits more than once, if its unique neighbor is v_{k-1}, pick any such node. Otherwise, pick the node that transmits at the latest, if its unique neighbor is v_{k-1}, among all still available nodes.

To find a lower bound on the time of algorithm A when run on G_j, we calculate the delay induced by each node of the path. We do not count delays caused by nodes picked for the first reason (transmitting more than once). A node picked for the second reason delays broadcasting by at least $|S_j| - k + 1$ rounds from the set R_j. Otherwise we would have a choice of a node transmitting more than once. Let X be the total sum over all networks $G_{i/2+1}, ..., G_{i(\frac{1}{2}+\epsilon)}$ of numbers of nodes transmitting more than once.

If $X \geq n'^\beta$ then the cost of the algorithm on network H which is the union of $G_{i/2+1}, ..., G_{i(\frac{1}{2}+\epsilon)}$, is at least $n' + n'^\beta$. This network has $n' = 2^{i(\frac{1}{2}+\epsilon)}$ nodes.

If, on the other hand, $X < n'^\beta$, then every network G_j has at least $D_j - X$ nodes contained in the directed path, that transmit exactly once. The total delay of the algorithm on network G_j is thus at least

$$(|S_j| - X) + (|S_j| - X - 1) + ... + (|S_j| - D_j + 1)$$

$$= (D_j - X)(|S_j| - X) - (D_j - X)(D_j - X - 1)/2 \in \Omega(2^i),$$

for $\epsilon < \frac{1-\beta}{2(1+\beta)}$, since $|S_j| \in \Theta(2^j)$ and $D_j \in \Theta(2^{i-j})$. Let X_j denote the set of all rounds from R_j which are less than or equal to the last round in which A schedules a transmission when run on network G_j. Hence the sizes of all sets X_j are $\Omega(2^i)$, and all these sets are pairwise disjoint, as subsets of sets R_j. Consequently, some X_j must contain a round $\Omega(i2^i)$. This means that algorithm A requires time $\Omega(i2^i)$ on the respective network G_j.

Now we construct the network G by augmenting G_j, as in the proof of Theorem 7: attach all nodes with labels from all sets $S_{j'}$, $j' < j$, directly to the source of G_j. G has size $n \leq 2^{j+1}$, and eccentricity $D = D_j \in \Theta(2^{i-j})$ of the source. Since $2^i \in \Omega(Dn)$ and $i \in \Omega(\log n)$, A requires time $\Omega(Dn \log n)$ on G.   □

4   Cost of Very Fast Broadcasting

We finally show that very fast broadcasting algorithms (in particular the fastest known, working in time $O(n \log^2 n)$) must have cost higher than linear.

Theorem 10. *Let $t(n)$ be any function polylogarithmic in n. Every broadcasting algorithm that works in time $O(nt(n))$ requires cost $\Omega(n \log n / \log \log n)$ on some n-node networks.*

PROOF. For simplicity assume that 8 divides n. Define the following n-node network G. The node with label 0 is the source of G. The remaining nodes will be divided into $n/2 - 1$ pairs. Enumerate the pairs $1, ..., n/2 - 1$. The root is the only neighbor of nodes of the first pair, and the only neighbors of nodes of pair $i + 1$ are both nodes of pair i. The last node has the source as its only neighbor. No other connections exist in G. This completes the description of the topology of G. We will later assign labels to nodes, depending on the algorithm.

Fix an algorithm A that performs broadcasting in time less than $cnt(n)$, for a fixed constant c. Labels will be chosen for nodes in the order of increasing pair numbers. Fix a pair $p \le n/4$. Both nodes of pair p receive the source message in the same round r. At this point in time all lower numbered pairs and no higher numbered pairs of nodes have received the source message. For every possible label of a node in pair p and for every round $> r$, algorithm A decides whether the node transmits or not. There are two possible cases. If there exists a pair of available labels (not used for lower numbered pairs) such that the behavior of nodes in pair p with these labels is identical up to round $r + 8ct(n)$ (by identical behavior we mean that in each of these rounds either both nodes transmit or both remain silent) then assign these labels to the nodes of pair p. Otherwise assign the two available labels that will result in the highest number of transmissions in rounds $r + 1, ..., r + 8ct(n)$.

If more than $n/8$ pairs among pairs $1, ..., n/4$, are assigned labels by the first choice (identical behavior for many rounds), the total broadcasting time exceeds $cnt(n)$, which is a contradiction. Hence, at least $n/8$ pairs are assigned labels by the second choice (maximum number of transmissions). Fix a pair $p' \le n/4$ whose nodes were assigned labels by the second choice. Associate with each available label a binary sequence of length $8ct(n)$, where 1 in position i represents the decision to transmit in round $r + i$ and 0 represents the decision to keep silent. Let S be the set of those binary sequences. The chosen labels will be the labels associated with the binary sequences from S with most occurrences of 1's. All the binary sequences in S must be different, otherwise the first choice of labels would be possible for nodes of pair p'.

Let x be the maximum number of 1's in a sequence of S. The number of binary sequences of length $8ct(n)$ with at most x 1's is $\sum_{i=0}^{x} \binom{8ct(n)}{i} < x\binom{8ct(n)}{x} < x(8ct(n))^x$. As the number of available labels is at least $n/2 - 2$, we have $x(8ct(n))^x > n/2 - 2$. Hence $x \in \Omega(\log n / \log \log n)$, because $t(n)$ is polylogarithmic in n. Since at least $n/8$ pairs will be assigned labels by the second choice, and thus contribute $\Omega(\log n / \log \log n)$ to the cost, the total cost is $\Omega(n \log n / \log \log n)$.   □

References

1. N. Alon, A. Bar-Noy, N. Linial and D. Peleg, A lower bound for radio broadcast, J. of Computer and System Sciences 43, (1991), 290–298.
2. B. Awerbuch, O. Goldreich, D. Peleg and R. Vainish, A Tradeoff Between Information and Communication in Broadcast Protocols, J. ACM 37, (1990), 238-256.
3. R. Bar-Yehuda, O. Goldreich, and A. Itai, On the time complexity of broadcast in radio networks: An exponential gap between determinism and randomization, Journal of Computer and System Sciences 45 (1992), 104-126.
4. D. Bruschi and M. Del Pinto, Lower bounds for the broadcast problem in mobile radio networks, Distr. Comp. 10 (1997), 129-135.
5. I. Chlamtac and S. Kutten, On broadcasting in radio networks - problem analysis and protocol design, IEEE Transactions on Communications 33 (1985), 1240-1246.
6. I. Chlamtac and S. Kutten, Tree based broadcasting in multihop radio networks, IEEE Trans. on Computers 36, (1987), 1209-1223.
7. B.S. Chlebus, L. Gąsieniec, A. Gibbons, A. Pelc and W. Rytter Deterministic broadcasting in unknown radio networks, Proc. 11th Ann. ACM-SIAM Symposium on Discrete Algorithms, SODA'2000, 861-870.
8. B.S. Chlebus, L. Gąsieniec, A. Östlin and J.M. Robson, Deterministic radio broadcasting, Proc. 27th Int. Coll. on Automata, Languages and Programming, ICALP'2000, July 2000, Geneva, Switzerland, LNCS 1853, 717-728.
9. M. Chrobak, L. Gąsieniec and W. Rytter, Fast broadcasting and gossiping in radio networks, Proc. FOCS 2000, to appear.
10. A. Czumaj, L. Gąsieniec and A. Pelc, Time and cost trade-offs in gossiping, SIAM J. on Discrete Math. 11 (1998), 400-413.
11. G. De Marco and A. Pelc, Faster broadcasting in unknown radio networks, Inf. Proc. Letters, to appear.
12. K. Diks, E. Kranakis, D. Krizanc and A. Pelc, The impact of knowledge on broadcasting time in radio networks, Proc. 7th Annual European Symposium on Algorithms, ESA'99, Prague, Czech Republic, July 1999, LNCS 1643, 41-52.
13. P. Fraigniaud and E. Lazard, Methods and problems of communication in usual networks, Disc. Appl. Math. 53 (1994), 79-133.
14. I. Gaber and Y. Mansour, Broadcast in Radio Networks, Proc. 6th Ann. ACM-SIAM Symp. on Discrete Algorithms, SODA'95, 577-585.
15. R. Gallager, A Perspective on Multiaccess Channels, IEEE Trans. on Information Theory 31 (1985), 124-142.
16. S.M. Hedetniemi, S.T. Hedetniemi and A.L. Liestman, A survey of Gossiping and Broadcasting in Communication Networks, Networks 18 (1988), 319-349.
17. E. Kranakis, D. Krizanc and A. Pelc, Fault-tolerant broadcasting in radio networks, Proc. 6th Annual European Symposium on Algorithms, ESA'98, Venice, Italy, August 1998, LNCS 1461, 283-294.
18. E. Kushilevitz and Y. Mansour, An $\Omega(D\log(N/D))$ Lower Bound for Broadcast in Radio Networks, Proc. 12th Ann. ACM Symp. on Principles of Distributed Computing (1993), 65-73.
19. E. Kushilevitz and Y. Mansour, Computation in noisy radio networks, Proc. 9th Ann. ACM-SIAM Symposium on Discrete Algorithms (SODA'98), 157-160.
20. D. Peleg, Deterministic radio broadcast with no topological knowledge, manuscript (2000).

The Existential Theory of Equations with Rational Constraints in Free Groups is PSPACE–Complete

Volker Diekert[1], Claudio Gutiérrez[2], and Christian Hagenah[1]

[1] Inst. für Informatik, Universität Stuttgart
Breitwiesenstr. 20-22, D-70565 Stuttgart
diekert@informatik.uni-stuttgart.de, christian@hagenah.de
[2] Centro de Mod. Matemático y
Depto. de Ciencias de la Computación, Universidad de Chile
Blanco Encalada 2120, Santiago, Chile
cgutierr@dcc.uchile.cl

Abstract. This paper extends extends known results on the complexity of word equations and equations in free groups in order to include the presence of rational constraints, i.e., such that a possible solution has to respect a specification given by a rational language. Our main result states that the existential theory of equations with rational constraints in free groups is PSPACE–complete.

Keywords: Formal languages, equations, regular language, free group.

1 Introduction

In 1977 (resp. 1983) Makanin proved that the existential theory of equations in free monoids (resp. free groups) is decidable by presenting algorithms which solve the satisfiability problem for a single word equation (resp. group equation) with constants [13,14,15]. These algorithms are very complex: For word equations the running time was first estimated by several towers of exponentials and it took more than 20 years to lower it down to the best known bound for Makanin's original algorithm, which is to date EXPSPACE [7]. For equations in free groups Kościelski and Pacholski have shown that the scheme of Makanin is not primitive recursive.

Recently Plandowski found a different approach to solve word equations and showed that the satisfiability problem for word equations is in PSPACE, [18]. Roughly speaking, his method uses data compression (first introduced for word equations in [19]) plus properties of factorization of words. Gutiérrez extended this method to the case of free groups, [9]. Thus, a non-primitive recursive scheme for solving equations in free groups was replaced by a polynomial space bounded algorithm.

In this paper we extend the results [18,9] above in order to include the presence of rational constraints. Rational constraints mean that a possible solution has to respect a specification which is given by a regular word language. Our main result states that the existential theory of equations in free groups with rational constraints is PSPACE–complete. The corresponding PSPACE–completeness for

A. Ferreira and H. Reichel (Eds.): STACS 2001, LNCS 2010, pp. 170–182, 2001.

word equations with regular constraints has been announced by first Rytter, see [18, Thm. 1] and [20].

The idea to consider regular constraints in the case of word equations is due to Schulz [21]. The importance of this concept, pointed out firstly by Schulz, can be exemplified by: the application of Schulz' result to monadic simultaneous rigid E-unification [6]; the use of regular constraints in [5] as a basic (an necessary) tool when showing that Makanin's result holds in free partially commutative monoids; the proof, in a forthcoming paper of Diekert and Muscholl, of the decidability of the existential theory of equations in graph groups (open problem stated in [5]) by using the present result; and the positive answer, by Diekert and Lohrey [4], to the question (cf [16]) about the existential theory of equations in free products of free and finite groups is decidable by relying on the general form of Theorem 2 below (we allow fixed points for the involution on Γ).

Our paper deals with the existential theory. For free groups it is also known that the positive theory without constraints is decidable, see [15]. Thus, one can allow also universal quantifiers but no negations. Note that we cannot expect that the positive theory of equations with rational constraints in free groups be decidable, since we can code the word case (with regular constraints) which is known to be undecidable. On the other hand, a negation leads to a positive constraint of a very restricted type, so it is a interesting question under which type of constraints the positive theory remains decidable.

Our proof of Theorem 1 is in the first step a reduction to the satisfiability problem of a single equation with regular constraints in a free monoid with involution. In order to avoid an exponential blow-up, we do not use a reduction as in [15], but a much simpler one. In particular, we can handle negations simply by a positive rational constraints. In the second step we show that the satisfiability problem of a single equation with regular constraints in a free monoid with involution is still in PSPACE. We extend the method of [18,9] such that it copes with the involution and with rational constraints. There seems to be no direct reduction to the word case or to the case of free groups without constraints. So we cannot use these results as black boxes. Because there is not enough space to present the whole proof in this extended abstract, we focus on those parts where there is a substantial difference to the case without constraints. In particular, we develop the notion of maximal free interval, a concept which can be used even when there are no constraints, but when one is interested in other solutions rather than the one of minimal length. The missing proofs can be found in [10] which is available on the web.[1]

2   Equations with Rational Constraints in Free Groups

Rational Languages, Equations. Let Σ be a finite alphabet and let $\overline{\Sigma} = \{\, \overline{a} \mid a \in \Sigma \,\}$. We use the convention that $\overline{\overline{a}} = a$. Define $\Gamma = \Sigma \cup \overline{\Sigma}$. Hence $^-: \Gamma \to \Gamma$ is an involution which is extended to Γ^* by $\overline{a_1 \cdots a_n} = \overline{a_n} \cdots \overline{a_1}$ for $n \geq 0$ and $a_i \in \Gamma$. We usually will write just Γ instead of $(\Gamma, ^-)$. A word $w \in \Gamma^*$ is *freely reduced*, if it contains no factor of the form $a\overline{a}$ with $a \in \Gamma$.

---

[1]   In http://inf.informatik.uni-stuttgart.de/ifi/ti/veroeffentlichungen/psfiles is the file HagenahDiss2000.ps

The elements of the free group $F(\Sigma)$ are represented by freely reduced words in Γ^*. We read $\bar{a}$ as a^{-1} in $F(\Sigma)$. There is a canonical homomorphism $\hat{}: \Gamma^* \to F(\Sigma)$, which eliminates all factors of the form $a\bar{a}$ from a word.

The class of *rational languages* in $F(\Sigma)$ is inductively defined as follows: Every finite subset of $F(\Sigma)$ is rational. If $P_1, P_2 \subseteq F(\Sigma)$ are rational, then $P_1 \cup P_2$, $P_1 \cdot P_2$, and P_1^* are rational. Hence, $P \subseteq F(\Sigma)$ is rational if and only if $P = \{\hat{w} : w \in P'\}$ for some regular language $P' \subseteq \Gamma^*$. It is well-known that the family of rational group languages is an effective Boolean algebra, in particular, it is closed under complementation [1]. (See also [2, Sect. III. 2].)

In the following Ω denotes a finite set of variables (or unknowns) and we let $^-: \Omega \to \Omega$ be an involution without fixed points. An *equation with rational constraints in free groups* is an equation $W = 1$ in free groups plus constraints on the variables of the type $X \in P$, for P a rational language. The existential fragment of these equations is the set of closed formulas of the form $\exists X_1 \ldots \exists X_n B$, where $X_i \in \Omega$ and B is a Boolean combination of atomic formulas which are either of the form $(W = 1)$ or $(X_i \in P)$, where $W \in (\Gamma \cup \Omega)^*$ and $P \subseteq F(\Sigma)$ is a rational language. The *existential theory of equations with rational constraints in free groups* is the set of such formulas which are valid in the free group $F(\Sigma)$.

Theorem 1. *The existential theory of equations with rational constraints in free groups is* PSPACE–*complete.*

Proof (Sketch). The PSPACE–hardness follows easily from [12] and is not discussed further. The proof for the inclusion in PSPACE is a reduction to the corresponding problem over free monoids with involution. It goes as follows.

First, we may assume that the input is given by some propositional formula which is in fact a conjunction of formulae of type $W = 1$, $X \in P$, $X \notin P$ with $W \in (\Gamma \cup \Omega)^*$, $X \in \Omega$, and $P \subseteq F(\Sigma)$ rational.[2] This is achieved by using DeMorgan rules to push negations to the level of atomic formulas, then replacing $W \neq 1$ by $\exists X : WX = 1 \wedge X \notin \{1\}$ (and pushing the quantifier to the out-most level), and finally eliminating the disjunctions by replacing non-deterministically every subformula of type $A \vee B$ by either A or B.

It is not difficult to see that we may also assume that $|W| = 3$ (use the equivalence of $x_1 \ldots x_n = 1$ and $\exists Y : x_1 x_2 Y = 1 \wedge \overline{Y} x_3 \cdots x_n = 1$).

Finally, we switch to the existential theory of equations with regular constraints in free monoids with involution. The key point of the translation here is the fact that rational languages P are in essence regular word languages over Γ such that $P \subseteq N$, where $N \subseteq \Gamma^*$ is the regular set of all freely reduced words. The language N is accepted by a deterministic finite automaton with $|\Gamma| + 1$ states. Then a positive constraint has just the interpretation over words and for a negative constraint we replace $X \notin P$ by $X \notin P \wedge X \in N$. Details are left to the reader.

As for the formulas $xyz = 1$, note that they have a solution if and only if they have a solution in freely reduced words. Then we can replace each subformulae $xyz = 1$ by the conjunction $\exists P \exists Q \exists R : x = PQ \wedge y = \overline{Q}R \wedge z = \overline{R}\,\overline{P}$ using simple arguments.

---

[2] The reason for keeping $X \notin P$ instead of $X \in \tilde{P}$ where $\tilde{P} = F(\Sigma) \setminus P$ is that complementation may involve an exponential blow-up of the state space.

Using a standard procedure to replace a conjunction of word equations by a single word equation we may assume that our input is given by a single equation $L = R$ with $L, R \in (\Gamma \cup \Omega)^+$ and by two lists $(X_j \in P_j, 1 \leq j \leq m)$ and $(X_j \notin P_j, m < j \leq k)$ where each $P_j \subseteq \Gamma^*$ is specified by some non-deterministic automaton $\mathcal{A}_j = (Q_j, \Gamma, \delta_j, I_j, F_j)$.

The question is whether the input is satisfiable, i.e. whether there is a solution. At this point, Boolean matrices are a better representation than finite automata. Let Q be the disjoint union of the state spaces Q_j, assume $Q = \{1, \ldots, n\}$. Let $\delta = \bigcup_j \delta_j$, then $\delta \subseteq Q \times \Gamma \times Q$ and with each $a \in \Gamma$ we can associate a Boolean matrix $g(a) \in \mathbb{B}^{n \times n}$ such that $g(a)_{i,j}$ is the truth value of "$(i, a, j) \in \delta$".

Since our monoids need an involution, we will work with $2n \times 2n$-Boolean matrices. Henceforth M denotes the following monoid with involution,

$$M = \{ \begin{pmatrix} A & 0 \\ 0 & B \end{pmatrix} \mid A, B \in \mathbb{B}^{n \times n} \}$$

where $\overline{\begin{pmatrix} A & 0 \\ 0 & B \end{pmatrix}} = \begin{pmatrix} B^T & 0 \\ 0 & A^T \end{pmatrix}$ and where the operator ^{T} means transposition.

We define a homomorphism $h : \Gamma^* \to M$ by $h(a) = \begin{pmatrix} g(a) & 0 \\ 0 & g(\bar{a})^T \end{pmatrix}$ for $a \in \Gamma$, where the mapping $g : \Gamma \to \mathbb{B}^{n \times n}$ is defined as above. The homomorphism h can be computed in polynomial time and it respects the involution. Now, for each regular language P_j we compute vectors $I_j, F_j \in \mathbb{B}^{2n}$ such that for all $w \in \Gamma^*$ we have the equivalence: $w \in P_j \Leftrightarrow I_j^T h(w) F_j = 1$. Having done these computations we make a non-deterministic guess $\rho(X) \in M$ for each variable $X \in \Omega$. We verify $\rho(\overline{X}) = \overline{\rho(X)}$ for all $X \in \Omega$ and whenever there is a constraint of type $X \in P_j$ (resp. $X \notin P_j$) then we verify $I_j^T \rho(X) F_j = 1$ (resp. $I_j^T \rho(X) F_j = 0$).

Let us make a formal definition. Let $d, n \in \mathbb{N}$. We consider an equation of the length d over some Γ and Ω with constraints in M being specified by a list E containing the following items:

- The alphabet $(\Gamma, ^-)$ with involution.
- A mapping $h : \Gamma \to M$ such that $h(\bar{a}) = \overline{h(a)}$ for all $a \in \Gamma$.
- The alphabet $(\Omega, ^-)$ with involution without fixed points.
- A mapping $\rho : \Omega \to M$ such that $\rho(\overline{X}) = \overline{\rho(X)}$ for all $X \in \Omega$.
- The equation $L = R$ where $L, R \in (\Gamma \cup \Omega)^+$ and $|LR| = d$.

If no confusion arise, we will denote this list simply by

$$E = (\Gamma, \Omega, h, \rho, L, R).$$

A *solution* is a mapping $\sigma : \Omega \to \Gamma^*$ (being extended to a homomorphism $\sigma : (\Gamma \cup \Omega)^* \to \Gamma^*$ by leaving the letters from Γ invariant) such that the following three conditions are satisfied: $\sigma(L) = \sigma(R)$, $\sigma(\overline{X}) = \overline{\sigma(X)}$, and $h\sigma(X) = \rho(X)$ for all $X \in \Omega$. We refer to the list E as an equation with constraints (in M). By the reduction above, Theorem 1 is a consequence of:

Theorem 2. *The following problem can be solved in* PSPACE.
   INPUT: An equation $E_0 = (\Gamma_0, \Omega_0, h_0, \rho_0, L_0, R_0)$.
   QUESTION: Is there a solution $\sigma : \Omega \to \Gamma^*$?

3 Equations with Regular Constraints over Free Monoids with Involution

During the procedure which solves Theorem 2 one has to consider various other equations with constraints in M. Following Plandowski we will use data compression for words in $(\Gamma \cup \Omega)^*$ in terms of exponential expressions.

Exponential Expressions. Exponential expressions (their evaluation and their size) are inductively defined as follows:

- Every word $w \in \Gamma^*$ denotes an exponential expression. The evaluation $\mathrm{eval}(w)$ is equal to w, its size $\|w\|$ is equal to the length $|w|$.
- If e, e' are exponential expressions, so is ee', the evaluation is the concatenation, $\mathrm{eval}(ee') = \mathrm{eval}(e)\mathrm{eval}(e')$, and $\|ee'\| = \|e\| + \|e'\|$.
- If e be an exponential expression and $k \in \mathbb{N}$, then $(e)^k$ is an exponential expression, and $\mathrm{eval}((e)^k) = (\mathrm{eval}(e))^k$ and $\|(e)^k\| = \log(k) + \|e\|$.

It is not difficult to show that the length of $\mathrm{eval}(e)$ is at most exponential in the size of e. Moreover, let $u \in \Gamma^*$ be a factor of a word $w \in \Gamma^*$ which can be represented by some exponential expression of size p. Then we find an exponential expression of size at most $2p^2$ that represents the factor u.

We say that an exponential expression e is *admissible*, if its size $\|e\|$ is bounded by some fixed polynomial in the input size of the equation E_0. Let $E = (\Gamma, \Omega, h, \rho, L, R)$ and e_L, e_R be exponential expressions with $\mathrm{eval}(e_L) = L$ and $\mathrm{eval}(e_R) = R$. We say that $E_e = (\Gamma, \Omega, h, \rho, e_L, e_R)$ is *admissible*, if $e_L e_R$ is admissible, $|\Gamma \setminus \Gamma_0| \leq \|e_L e_R\| + 2d$, $\Omega \subseteq \Omega_0$, and $h(a) = h_0(a)$ for $a \in \Gamma \cap \Gamma_0$. We say that E_e *represents* the equation E. For two admissible equations with constraints E and E' we write $E \equiv E'$, if E and E' represent the same object.

Because of regular constraints, we have to formalize carefully the basic operations over these equations in order to move from one equation to another.

Base Changes. Let $E' = (\Gamma', \Omega, h', \rho, L', R')$ be an equation. A mapping $\beta : \Gamma' \to \Gamma^*$ is a *base change* if both $\beta(\overline{a}) = \overline{\beta(a)}$ and $h'(a) = h\beta(a)$ for all $a \in \Gamma'$. The new equation is $\beta_*(E') = (\Gamma, \Omega, h, \rho, \beta(L), \beta(R))$. We say that β is *admissible* if $|\Gamma \cup \Gamma'|$ has polynomial size and if for each $a \in \Gamma'$, $\beta(a)$ has an admissible exponential representation.

If $\beta : \Gamma' \to \Gamma^*$ is an admissible base change and if $L' = R'$ is given by a pair of admissible exponential expressions, then we can represent $\beta_*(E')$ by some admissible equation with constraints which is computable in polynomial time.

Lemma 1. *Let E' be an equation with constraints in M and $\beta : \Gamma' \to \Gamma^*$ be a base change. If $\sigma' : \Omega \to \Gamma'^*$ is a solution of E', then $\sigma = \beta\sigma' : \Omega \to \Gamma^*$ is a solution of $\beta_*(E')$.*

Projections. Let $\Gamma \subseteq \Gamma'$ be alphabets with involution. A *projection* is a homomorphism $\pi : \Gamma'^* \to \Gamma^*$ preserving the involution and leaving Γ fixed. If $h : \Gamma \to M$ is given, then a projection π defines also $h' : \Gamma' \to M$ by $h' = h\pi$.

For an equation $E = (\Gamma, h, \Omega, \rho, L, R)$ we define $\pi^*(E) = (\Gamma', h\pi, \Omega, \rho, L, R)$. Note that every projection $\pi : \Gamma'^* \to \Gamma^*$ defines also a base change π_* such that $\pi_* \pi^*(E) = E$.

Lemma 2. *Let $\Gamma \subseteq \Gamma'$ be as above and let $E = (\Gamma, \Omega, h, \rho, L, R)$ and $E' = (\Gamma', \Omega, h', \rho, L, R)$. Then there is a projection $\pi : \Gamma'^* \to \Gamma^*$ such that $\pi^*(E) = E'$, if and only if both $h'(\Gamma') \subseteq h(\Gamma^*)$ and $a = \overline{a}$ implies $h'(a) \in h(\{w \in \Gamma^* \mid w = \overline{w}\})$ for all $a \in \Gamma'$. Moreover, if σ' is a solution of E', then we effectively find a solution σ for E with $|\sigma(L)| \leq 2|M||\sigma'(L)|$.*

Lemma 2 says that in order to test whether there exists a projection $\pi : \Gamma'^* \to \Gamma^*$ such that $\pi^*(E) = E'$, we need only space to store some Boolean matrices of $\mathbb{B}^{2n \times 2n}$, we do not need an explicit description of $\pi : \Gamma'^* \to \Gamma^*$ itself. Only if n becomes a substantial part of the input size, then we might need the full power of PSPACE (PSPACE–hardness of the satisfiability problem).

Shifts. Let $\Omega' \subseteq \Omega$ be a subset of the variables which is closed under involution, and let $\rho' : \Omega' \to M$ with $\rho'(\overline{x}) = \overline{\rho'(x)}$ (we do not require that ρ' is the restriction of ρ). A *shift* is a mapping $\delta : \Omega \to \Gamma^* \Omega' \Gamma^* \cup \Gamma^*$ such that the following conditions are satisfied:

i) $\delta(X) \in \Gamma^* X \Gamma^*$ for all $X \in \Omega'$,
ii) $\delta(X) \in \Gamma^*$ for all $X \in \Omega \setminus \Omega'$,
iii) $\delta(\overline{X}) = \overline{\delta(X)}$ for all $X \in \Omega$.

The mapping δ is extended to a homomorphism $\delta : (\Gamma \cup \Omega)^* \to (\Gamma \cup \Omega')^*$ by leaving the elements of Γ invariant. For and equation $E = (\Gamma, h, \Omega, \rho, L, R)$, we define the equation $\delta_*(E) = (\Gamma, \Omega', h, \rho', \delta(L), \delta(R))$ where ρ' is such that $\rho(X) = h(u)\rho'(X)h(v)$ for $\delta(X) = uXv$, and $\rho(X) = h(w)$ for $\delta(X) = w \in \Gamma^*$. We say that $\delta_*(E)$ is a *shift* of E.

Lemma 3. *In the notation of above, let $E' = \delta_*(E)$ for some shift $\delta : \Omega \to \Gamma^* \Omega \Gamma^* \cup \Gamma^*$. If $\sigma' : \Omega' \to \Gamma^*$ is a solution of E', then $\sigma = \sigma' \delta : \Omega \to \Gamma^*$ is a solution of E. Moreover, we have $\sigma(L) = \sigma'(L')$.*

Lemma 4. *The following problem can be solved in* PSPACE.

   INPUT: Two equations with constraints E and E'.
   QUESTION: Is there some shift $\delta : \Omega \to \Gamma^* \Omega \Gamma^* \cup \Gamma^*$ such that $\delta_*(E) \equiv E'$?

   Moreover, if $\delta_*(E) \equiv E'$, then we have $\delta(X) = \mathrm{eval}(e_u)X\mathrm{eval}(e_v)$ for all $X \in \Omega$ and for suitable admissible exponential expressions e_u, e_v. Similarly, $\delta(X) = \mathrm{eval}(e_w)$ for all $X \in \Omega \setminus \Omega'$.

Remark 1. We can think of a shift $\delta : \Omega \to \Gamma^* \Omega' \Gamma^* \cup \Gamma^*$ as a partial solution in the following sense. Assume we have an idea about $\sigma(X)$ for some $X \in \Omega$. Then we might guess $\sigma(X)$ entirely. In this case we can define $\delta(X) = \sigma(X)$ and we have $X \notin \Omega'$. For some other X we might guess only some prefix u and some suffix v of $\sigma(X)$. Then we define $\delta(X) = uXv$ and we have to guess some $\rho'(X) \in M$ such that $\rho(x) : h(u)\rho'(X)h(v)$. If our guess was correct, then such $\rho'(X)$ must exist. We have partially specified the solution and we continue this process by replacing the equation $L = R$ by the new equation $\delta(L) = \delta(R)$.

4   The Search Graph and Plandowski's Algorithm

The nodes of the search graph are admissible equations with constraints in M. Let E, E' be two nodes. We define an arc $E \to E'$, if there are a projection π, a shift δ, and an admissible base change β such that $\delta_*(\pi^*(E)) \equiv \beta_*(E')$.

Lemma 5. *The following problem can be decided in* PSPACE.
   INPUT: Admissible equations with constraints E and E'.
   QUESTION: Is there an arc $E \to E'$ in the search graph?

Proof. (Sketch) We first guess some alphabet $(\Gamma'', {}^-)$ of polynomial size together with $h'' : \Gamma'' \to M$. Then we guess some admissible base change $\beta : \Gamma' \to \Gamma''^*$ such that $h' = h'' \beta$ and we compute $\beta_*(E')$ in polynomial time. Next we check using Remark 1 and Lemma 4 that there is projection $\pi : \Gamma'' \to \Gamma$ and that there is a shift $\delta : \Omega \to \Gamma''^* \Omega' \Gamma''^* \cup \Gamma''^*$ such that $\delta_*(\pi^*(E)) \equiv \beta_*(E')$. □

Plandowski's algorithm works on $E_0 = (\Gamma_0, \Omega_0, h_0, \rho_0, L_0, R_0)$ as follows:

1. $E := E_0$
2. **while** $\Omega \neq \emptyset$ **do**
       Guess an admissible equation E' with constraints in M.
       Verify that $E \to E'$ is an arc in the search graph.
       $E := E'$
3. **return** "eval(e_L) = eval(e_R)"

By Lemmata 1, 2, and 3, if $E \to E'$ is an arc in the search graph and E' is solvable, then E is solvable, too. Thus, if the algorithm returns *true*, then E_0 is solvable. The proof of Theorem 2 is therefore reduced to the statement that if E_0 is solvable, then the search graph contains a path to some node without variables and the exponential expressions defining the equation evaluate to the same word (called a terminal node).

Remark 2. If $E \to E'$ is due to some $\pi : \Gamma''^* \to \Gamma^*$, $\delta : \Omega \to \Gamma''^* \Omega' \Gamma''^* \cup \Gamma''^*$, and $\beta : \Gamma'^* \to \Gamma''^*$, then a solution $\sigma' : \Omega' \to \Gamma'^*$ of E' yields the solution $\sigma = \pi(\beta\sigma')\delta$. Hence we may assume that the length of a solution has increased by at most an exponential factor. Since we are going to perform the search in a graph of at most exponential size, we get automatically a doubly exponential upper bound for the length of a minimal solution by backwards computation on such a path. This is still the best known upper bound (although an singly exponential bound is conjectured), see [17].

5   The Search Graph Contains a Path to a Terminal Node

This section is a proof of the existence of a path to a solvable solution in the Search Graph. The technique used is a generalization of the one used in [18] for word equations, in [9] for free group equations, and in [3] for word equations with regular constraints. Due to lack of space in this extended abstract we focus only on some few points where the technique differs substantially from those papers. For the other parts we will just refer the reader to the papers above.

The Exponent of Periodicity. Let $w \in \Gamma^*$ be a word. The exponent of periodicity $\exp(w)$ is defined as the supremum of the $\alpha \in \mathbb{N}$ such that $w = up^\alpha v$ for suitable $u, v, p \in \Gamma^*$ and $p \neq 1$. It is clear that $\exp(w) > 0$ if w is not empty. For an equation $E = (\Gamma, \Omega, h, \rho, L, R)$ the exponent of periodicity, denoted by $\exp(E)$, is defined as

$$\exp(E) = \inf\{\{\exp(\sigma(L)) \mid \sigma \text{ is a solution of } E\} \cup \{\infty\}\}.$$

The well-known result from word equations [11] transfers to the situation here: in order to prove Theorem 2 we may assume that E_0 is solvable and $\exp(E_0) \in 2^{\mathcal{O}(d+n\log n)}$. The case of word equations with regular constraints in done in [3] and for monoids with involution in [8]. A combinations of these methods give what we need here. The detailed proof has been given in [10].

Free Intervals. The following development will be fully justified at the end of the subsection and has to do with handling the constraints. Without constraints, free intervals of length more than one do not appear in a minimal solutions, making this notion unnecessary. This is not true in the presence of constraints. Free intervals handle this case and moreover, tell us that the bounds on the exponent of periodicity are the only restriction we need on solutions.

Given a word $w \in \Gamma^*$, let $\{0, \ldots, |w|\}$ be the set of its *positions*. An *interval* on these positions is a formal object denoted $[\alpha, \beta]$ with $0 \leq \alpha, \beta \leq |w|$, and $\overline{[\alpha, \beta]} = [\beta, \alpha]$. For $w = a_1 \cdots a_m$, we define $w[\alpha, \beta] = a_{\alpha+1} \cdots a_\beta$ if $\alpha < \beta$, $w[\alpha, \beta] = \overline{a_{\alpha+1} \cdots a_\beta}$ if $\alpha > \beta$, and the empty word if $\alpha = \beta$. Observe that these notations are consistent so that $\overline{w[\alpha, \beta]} = w\overline{[\alpha, \beta]}$.

Let σ_0 be a solution of $L = R$, where $L_0 = x_1 \cdots x_g$ and $R_0 = x_{g+1} \cdots x_d$ and $x_i \in (\Gamma_0 \cup \Omega_0)$. Then we have $w_0 = \sigma_0(L_0) = \sigma_0(R_0)$. Denote $m_0 = |w_0|$. For each $i \in \{1, \ldots, d\}$ we define positions $l(i)$ and $r(i)$ as follows:

$$l(i) = |\sigma_0(x_1 \cdots x_{i-1})| \bmod m_0 \in \{0, \ldots, m_0 - 1\},$$
$$r(i) = |\sigma_0(x_{i+1} \cdots x_d)| \bmod m_0 \in \{1, \ldots, m_0\}.$$

In particular, we have $l(1) = l(g+1) = 0$ and $r(g) = r(d) = m_0$. The set of l and r positions is called the set of *cuts*. There are at most d cuts which cut the word w_0 in at most $d - 1$ factors. We say that $[\alpha, \beta]$ contains a cut γ if $\min\{\alpha, \beta\} < \gamma < \max\{\alpha, \beta\}$.

For convenience we henceforth assume $2 \leq g < d < m_0$ whenever necessary and make the assumption that $\sigma_0(x_i) \neq 1$ for all $1 \leq i \leq d$ (e.g. a guess in some preprocessing).

We have $\sigma_0(x_i) = w_0[l(i), r(i)]$ and $\sigma_0(\overline{x_i}) = w_0[r(i), l(i)]$ for $1 \leq i \leq d$. By our assumption, the interval $[l(i), r(i)]$ is positive.

Let us consider $i, j \in 1, \ldots, d$ and $x_i = x_j$ or $x_i = \overline{x_j}$. For $0 \leq \mu, \nu \leq r(i) - l(i)$, we define a relation $\sim$ among intervals as follows:

$$[l(i) + \mu, l(i) + \nu] \sim [l(j) + \mu, l(j) + \nu], \text{ if } x_i = x_j,$$
$$[l(i) + \mu, l(i) + \nu] \sim [r(j) - \mu, r(j) - \nu], \text{ if } x_i = \overline{x_j}.$$

Note that $\sim$ is a symmetric relation and $[\alpha, \beta] \sim [\alpha', \beta']$ implies both $[\beta, \alpha] \sim [\beta', \alpha']$ and $w_0[\alpha, \beta] = w_0[\alpha', \beta']$. By $\approx$ we denote the equivalence relation obtained by the reflexive and transitive closure of $\sim$.

An interval $[\alpha, \beta]$ is called *free* if none of its $\approx$-equivalent intervals contains a cut. Clearly, the set of free intervals is closed under involution and whenever $|\beta - \alpha| \leq 1$ then $[\alpha, \beta]$ is free. It is also closed under taking subintervals:

Lemma 6. *Let $[\alpha, \beta]$ be a free interval and $\min\{\alpha, \beta\} \leq \mu, \nu \leq \max\{\alpha, \beta\}$. Then the interval $[\mu, \nu]$ is also free.*

If $[\alpha, \beta]$ (assume $\alpha < \beta$) is not free, then by definition there is some interval $[\alpha', \beta'] \approx [\alpha, \beta]$ which contains a cut γ'. The propagation of that cut to $[\alpha, \beta]$, that is the position γ such that $\gamma - \alpha = |\gamma' - \alpha'|$ is called an *implicit cut* of $[\alpha, \beta]$.

The following observation will be used throughout: If we have $\alpha \leq \mu < \gamma < \nu \leq \beta$ and γ is an implicit cut of $[\alpha, \beta]$, then γ is also an implicit cut of $[\mu, \nu]$. (The converse is not necessarily true.)

Lemma 7. *Let $0 \leq \alpha \leq \alpha' < \beta \leq \beta' \leq m_0$ be such that $[\alpha, \beta]$ and $[\alpha', \beta']$ are free intervals. Then the interval $[\alpha, \beta']$ is free, too.*

A free interval $[\alpha, \beta]$ is called *maximal free* if no free interval properly contains it, i.e., if $\alpha' \leq \min\{\alpha, \beta\} \leq \max\{\alpha, \beta\} \leq \beta'$ and $[\alpha', \beta']$ free, then and $\beta' - \alpha' = |\beta - \alpha|$. So Lemma 7 states a key point that maximal free intervals do not overlap.

Lemma 8. *Let $[\alpha, \beta]$ be a maximal free interval. Then there are intervals $[\gamma, \delta]$ and $[\gamma', \delta']$ such that $[\alpha, \beta] \approx [\gamma, \delta] \approx [\gamma', \delta']$ and γ and δ' are cuts.*

Proposition 1. *Let Γ be the set of words $w \in \Gamma_0^*$ such that there is a maximal free interval $[\alpha, \beta]$ with $w = w_0[\alpha, \beta]$. Then Γ is a subset of Γ_0^+ of size at most $2d - 2$. The set Γ is closed under involution.*

Proof. Let $[\alpha, \beta]$ be maximal free. Then $|\beta - \alpha| \geq 1$ and $[\beta, \alpha]$ is maximal free, too. Hence $\Gamma \subseteq \Gamma_0^+$ and Γ is closed under involution. By Lemma 8 we may assume that α is a cut. Say $\alpha < \beta$. Then $\alpha \neq m_0$ and there is no other maximal free interval $[\alpha, \beta']$ with $\alpha < \beta'$ because of Lemma 7. Hence there are at most $d - 1$ such intervals $[\alpha, \beta]$. Symmetrically, there are at most $d - 1$ maximal free intervals $[\alpha, \beta]$ where $\beta < \alpha$ and α is a cut. $\qquad \square$

Why Free Intervals Are Needed. For a moment let us put $\Delta = \Gamma_0 \cup \Gamma$ where Γ is the set defined in Proposition 1. Observe that $\Delta \subseteq \Gamma_0^+$, and so it defines a natural projection $\pi : \Gamma_0^* \to \Delta$ and a mapping $h' : \Gamma_0^* \to M$ by $h' = h_0 \pi$. (Note that here we need the fact that there is no overlapping among maximal intervals.) Consider the equation with constraints $\pi^*(E_0)$. There is an arc from E_0 to $\pi^*(E_0)$ since we may always allow the base change to be the identity and the shift to be an inclusion.

The reason to switch from Γ_0 to Δ is that, due to the constraints, the word w_0 may have long free intervals. Over Δ this can be avoided. Formally, we replace w_0 by a solution w_0' where $w_0' \in \Gamma^*$, whose definition is based on a factorization of w_0 in maximal free intervals. Recall that there is a unique sequence $0 = \alpha_0 < \alpha_1 < \cdots < \alpha_k = m_0$ such that $[\alpha_{i-1}, \alpha_i]$ are maximal free intervals and

$$w_0 = w_0[\alpha_0, \alpha_i] \cdots w_0[\alpha_k - 1, \alpha_k].$$

Moreover, all cuts occur as some α_p, so we can think of the factors $w_0[\alpha_{i-1}, \alpha_i]$ as letters in Γ. Because all constants which appear in L_0, R_0 are elements of Γ, the equation $L_0 = R_0$ appears identical in $\pi^*(E_0)$.

So, replacing w_0 by the word $w_0' \in \Gamma^*$, we can define $\sigma : \Omega \to \Gamma^*$ such that both $\sigma(L_0) = \sigma(R_0) = w_0'$ and $\rho_0 = h_0'\sigma$, that is, σ is a solution of $\pi^*(E)$. Clearly we have $w_0 = \pi(w_0')$ and $\exp(w_0') \leq \exp(w_0)$. The crucial point is that w_0' has no long free intervals anymore. (With respect to w_0' and Γ_0', all maximal free intervals have length exactly one.)

We can assume that Plandowski's algorithm follows in a first step exactly the arc from E_0 to $\pi^*(E_0)$. Phrased in a different way, we may assume that $E_0 = \pi^*(E_0)$, hence Γ is a subset Γ_0.

Moreover, the inclusion $\beta : \Gamma \to \Gamma_0^*$ defines an admissible base change. Consider $E_0' = \beta_*(\pi^*(E_0))$. Then we have $E_0' = (\Gamma, \Omega_0, h, \rho_0, L_0, R_0)$ where h is the restriction of $h_0 : \Gamma_0 \to M$. The search graph contains an arc from E_0 to E_0' and E_0' has a solution σ with $\sigma(L_0) = w_0'$ with $\exp(w_0') \leq \exp(w_0)$.

In summary, in order to save notations we may assume for simplicity that $E_0 = E_0'$ and $w_0 = w_0'$. We can make the following assumptions:

$$L_0 = x_1 \cdots x_g \text{ and } g \geq 2,$$
$$R_0 = x_{g+1} \cdots x_d \text{ and } d > g,$$
$$\Gamma_0 = \Gamma \text{ and } |\Gamma| \leq 2d - 2,$$
$$|\Omega_0| \leq 2d,$$
$$M \subseteq \mathbb{B}^{2n \times 2n}.$$

All variables X occur in $L_0 R_0 \overline{L_0} \overline{R_0}$. There is a solution $\sigma : \Omega_0 \to \Gamma$ such that $w_0 = \sigma(L_0) = \sigma(R_0)$ with $\sigma(X_i) \neq 1$ for $1 \leq i \leq d$ and $\rho_0 = h\sigma = h_0\sigma$. We have $|w_0| = m_0$ and $\exp(w_0) \in 2^{\mathcal{O}(d+n\log n)}$. All maximal free intervals have length exactly one, i.e., every positive interval $[\alpha, \beta]$ with $\beta - \alpha > 1$ contains an implicit cut.

The ℓ-Factorization. For each integer ℓ, $1 \leq \ell \leq m_0$, we define the set of *critical words* C_ℓ as the closure under involution of set of all words $w_0[\gamma - \ell, \gamma + \ell]$ where γ is a cut with $\ell \leq \gamma \leq m_0 - \ell$.

A triple $(u, w, v) \in (\{1\} \cup \Gamma^\ell) \times \Gamma^+ \times (\{1\} \cup \Gamma^\ell)$ is called a *block* if, first, first, up to a possible prefix or suffix no other factor of the word uwv is a critical word, second, $u \neq 1$ if and only if a prefix of uwv of length 2ℓ belongs to C_ℓ, and third, $v \neq 1$ if and only if a suffix of uwv of length 2ℓ belongs to C_ℓ. The set of blocks is denoted by B_ℓ and can be viewed (as a possibly infinite) alphabet with involution defined by $\overline{(u, w, v)} = (\overline{v}, \overline{w}, \overline{u})$.

We can define a homomorphism $\pi_\ell : B_\ell^* \to \Gamma^*$ by $\pi_\ell(u, w, v) = w \in \Gamma^+$ being extended to a projection $\pi_\ell : (B_\ell \cup \Gamma)^* \to \Gamma^*$ by leaving Γ invariant. We define $h_\ell : (B_\ell \cup \Gamma) \to M$ by $h_\ell = h\pi_\ell$. In the following we shall consider finite subsets $\Gamma_\ell \subseteq B_\ell \cup \Gamma$ which are closed under involution. Then by $\pi_\ell : \Gamma_\ell^* \to \Gamma^*$ and $h_\ell : \Gamma_\ell^* \to M$ we understand the restrictions of the respective homomorphisms.

For every non-empty word $w \in \Gamma^+$ we define its ℓ-*factorization* as:

$$F_\ell(w) = (u_1, w_1, v_1) \cdots (u_k, w_k, v_k) \in B_\ell^+ \tag{1}$$

where $w = w_1 \cdots w_k$ and for $1 \leq i \leq k$ the following conditions are satisfied:

- v_i is a prefix of $w_{i+1} \cdots w_k$ and $v_i = 1$ if and only if $i = k$.
- u_i is a suffix of $w_1 \cdots w_{i-1}$ and $u_i = 1$ if and only if $i = 1$.

Note that the ℓ-factorization of a word w is unique. For a factorization (1), we define $\mathrm{head}_\ell(w) = w_1$, $\mathrm{body}_\ell(w) = w_2 \cdots w_{k-1}$ and $\mathrm{tail}_\ell(w) = w_k$. Similarly for $\mathrm{Head}_\ell(w) = (u_1, w_1, v_1)$, $\mathrm{Body}_\ell(w) = (u_2, w_2, v_2) \cdots (u_{k-1}, w_{k-1}, v_{k-1})$, and $\mathrm{Tail}_\ell(w) = (u_k, w_k, v_k)$. For $k \geq 2$ (in particular, if $\mathrm{body}_\ell(w) \neq 1$) we have

$$F_\ell(w) = \mathrm{Head}_\ell(w)\mathrm{Body}_\ell(w)\mathrm{Tail}_\ell(w) \quad \text{and} \quad w = \mathrm{head}_\ell(w)\mathrm{body}_\ell(w)\mathrm{tail}_\ell(w).$$

Moreover, u_2 is a suffix of w_1 and v_{k-1} is a prefix of w_k.

Assume $\mathrm{body}_\ell(w) \neq 1$ and let $u, v \in \Gamma^*$ be any words. Then we can view w in the context uwv and $\mathrm{Body}_\ell(w)$ appears as a proper factor in the ℓ-factorization of uwv. More precisely, let $F_\ell(uwv) = (u_1, w_1, v_1) \cdots (u_k, w_k, v_k)$. Then there are unique $1 \leq p < q \leq k$ such that:

$$F_\ell(uwv) = (u_1, w_1, v_1) \cdots (u_p, w_p, v_p)\mathrm{Body}_\ell(w)(u_q, w_q, v_q) \cdots (u_k, w_k, v_k)$$
$$w_1 \cdots w_p = u\,\mathrm{head}_\ell(w) \quad \text{and} \quad w_q \cdots w_k = \mathrm{tail}_\ell(w)v$$

Finally, we note that the above definitions are compatible with the involution. We have $F_\ell(\overline{w}) = \overline{F_\ell(w)}$, $\mathrm{Head}_\ell(\overline{w}) = \overline{\mathrm{Tail}_\ell(w)}$, and $\mathrm{Body}_\ell(\overline{w}) = \overline{\mathrm{Body}_\ell(w)}$.

The ℓ-Transformation. Recall that $E_0 = (\Gamma, \Omega_0, h, \rho_0, x_1 \cdots x_g, x_{g+1} \cdots x_d)$ is our equation with constraints. We start with the ℓ-factorization of $w_0 = \sigma(x_1 \cdots x_g) = \sigma(x_{g+1} \cdots x_d)$. Let

$$F_\ell(w_0) = (u_1, w_1, v_1) \cdots (u_k, w_k, v_k).$$

A sequence $S = (u_p, w_p, v_p) \cdots (u_q, w_q, v_q)$ with $1 \leq p \leq q \leq k$ is called an ℓ-factor. We say that S is a cover of a positive interval $[\alpha, \beta]$, if both $|w_1 \cdots w_{p-1}| \leq \alpha$ and $|w_{q+1} \cdots w_k| \leq m_0 - \beta$. Thus, $w_0[\alpha, \beta]$ becomes a factor of $w_p \cdots w_q$. It is called a minimal cover if neither $(u_{p+1}, w_{p+1}, v_{p+1}) \cdots (u_q, w_q, v_q)$ nor $(u_p, w_p, v_p) \cdots (u_{q-1}, w_{q-1}, v_{q-1})$ is a cover of $[\alpha, \beta]$. The minimal cover exists and it is unique.

We let $\Omega_\ell = \{ X \in \Omega_0 \mid \mathrm{body}_\ell(\sigma(X)) \neq 1 \}$, and we are going to define a new left-hand side $L_\ell \in (B_\ell \cup \Omega_\ell)^*$ and a new right-hand side $R_\ell \in (B_\ell \cup \Omega_\ell)^*$. For L_ℓ we consider those $1 \leq i \leq g$ where $\mathrm{body}_\ell(\sigma(x_i)) \neq 1$. Note that this implies $x_i \in \Omega_\ell$ since $\ell \geq 1$ and then the body of a constant is always empty. Recall the definition of $l(i)$ and $r(i)$, and define $\alpha = l(i) + |\mathrm{head}_\ell(\sigma(x_i))|$ and $\beta = r(i) - |\mathrm{tail}_\ell(\sigma(x_i))|$. Then we have $w_0[\alpha, \beta] = \mathrm{body}_\ell(\sigma(x_i))$. Next consider the ℓ-factor $S_i = (u_p, w_p, v_p) \cdots (u_q, w_q, v_q)$ which is the minimal cover of $[\alpha, \beta]$. Then we have $1 < p \leq q < k$ and $w_p \cdots w_q = w_0[\alpha, \beta] = \mathrm{body}_\ell(\sigma(x_i))$. The definition of S_i depends only on x_i, but not on the choice of the index i.

We replace the ℓ-factor S_i in $F_\ell(w_0)$ by the variable x_i. Having done this for all $1 \leq i \leq g$ with $\mathrm{body}_\ell(\sigma(x_i)) \neq 1$ we obtain the left-hand side $L_\ell \in (B_\ell \cup \Omega_\ell)^*$ of the ℓ-transformation E_ℓ. For R_ℓ we proceed analogously by replacing those ℓ-factors S_i where $\mathrm{body}_\ell(\sigma(x_i)) \neq 1$ and $g + 1 \leq i \leq d$.

For E_ℓ we cannot use the alphabet B_ℓ, because it might be too large or even infinite. Therefore we let Γ'_ℓ be the smallest subset of B_ℓ which is closed under involution and which satisfies $L_\ell R_\ell \in (\Gamma'_\ell \cup \Omega_\ell)^*$. We let $\Gamma_\ell = \Gamma'_\ell \cup \Gamma$.

The projection $\pi_\ell : \Gamma_\ell^* \to \Gamma^*$ and the mapping $h_\ell : \Gamma_\ell \to M$ are defined by the restriction of $\pi_\ell : B_\ell \to \Gamma^*$, $\pi_\ell(u, w, v) = w$ and $h_\ell(u, w, v) = h(w) \in M$ and by $\pi_\ell(a) = a$ and $h_\ell(a) = h(a)$ for $a \in \Gamma$.

Finally, we define the mapping $\rho_\ell : \Omega_\ell \to M$ by $\rho_\ell(X) = h(\mathrm{body}_\ell(\sigma(X)))$. This yields the definition of the ℓ-transformation: $E_\ell = (\Gamma_\ell, \Omega_\ell, h_\ell, \rho_\ell, L_\ell, R_\ell)$.

The ℓ-Transformation E_ℓ Is Admissible. The proof of the following proposition uses standard techniques like those in [18] and [9] and it is therefore omitted.

Proposition 2. *There is a polynomial of degree four such that each E_ℓ is admissible for all $\ell \geq 1$.*

At this stage we know that all ℓ-transformations are admissible. Thus, the equations $E_1, \ldots, E_{m_0}$ are nodes of the search graph. What is left to prove is that the search graph contains arcs $E_0 \to E_1$ and $E_\ell \to E_{\ell+1}$ for $1 \leq \ell < \ell' \leq 2\ell$. This involves again the concept of base change, projection, and shift. But the presence of constraints does not interfere very much anymore.t Thus, the technical details are similar to those of Plandowski's paper [18] as generalized in [9].

Acknowledgment

C. Gutiérrez was supported by FONDAP, Matemáticas Aplicadas.

References

1. M. Benois. Parties rationelles du groupe libre. *C. R. Acad. Sci. Paris, Sér. A,* 269:1188–1190, 1969.
2. J. Berstel. *Transductions and context-free languages.* Teubner Studienbücher, Stuttgart, 1979.
3. V. Diekert. Makanin's Algorithm. In M. Lothaire, *Algebraic Combinatorics on Words.* Cambridge University Press, 2001. To appear. A preliminary version is on the web: `http://www-igm.univ-mlv.fr/~berstel/Lothaire/index.html`.
4. V. Diekert and M. Lohrey. A note on the existential theory of plain groups. Submitted for publication, 2000.
5. V. Diekert, Yu. Matiyasevich, and A. Muscholl. Solving word equations modulo partial commutations. *Th. Comp. Sc.,* 224:215–235, 1999. Special issue of LFCS'97.
6. Yu. Gurevich and A. Voronkov. Monadic simultaneous rigid E-unification and related problems. In P. Degano et al., editor, *Proc. 24th ICALP, Bologna (Italy) 1997*, number 1256 in Lect. Not. Comp. Sc., pages 154–165. Springer, 1997.
7. C. Gutiérrez. Satisfiability of word equations with constants is in exponential space. In *Proc. of the 39th Ann. Symp. on Foundations of Computer Science, FOCS'98*, pages 112–119, Los Alamitos, California, 1998. IEEE Computer Society Press.
8. C. Gutiérrez. Equations in free semigroups with anti-involution and their relation to equations in free groups. In G. H. Gonnet et al., editor, *Proc. Lat. Am. Theor. Inf., LATIN'2000*, number 1776 in LNCS, pages 387–396. Springer, 2000.
9. C. Gutiérrez. Satisfiability of equations in free groups is in PSPACE. In *32nd ACM Symp. on Theory of Computing (STOC'2000)*, pages 21–27. ACM Press, 2000.
10. Ch. Hagenah. *Gleichungen mit regulären Randbedingungen über freien Gruppen.* PhD-thesis, Institut für Informatik, Universität Stuttgart, 2000.

11. A. Kościelski and L. Pacholski. Complexity of Makanin's algorithm. *Journal of the Association for Computing Machinery*, 43(4):670–684, 1996. Preliminary version in *Proc. of the 31st Ann. Symp. on Foundations of Computer Science, FOCS 90*, pages 824–829, Los Alamitos, 1990. IEEE Computer Society Press.

12. D. Kozen. Lower bounds for natural proof systems. In *Proc. of the 18th Ann. Symp. on Foundations of Computer Science, FOCS 77*, pages 254–266, Providence, Rhode Island, 1977. IEEE Computer Society Press.

13. G. S. Makanin. The problem of solvability of equations in a free semigroup. *Math. Sbornik*, 103:147–236, 1977. English transl. in Math. USSR Sbornik 32 (1977).

14. G. S. Makanin. Equations in a free group. *Izv. Akad. Nauk SSR*, Ser. Math. 46:1199–1273, 1983. English transl. in Math. USSR Izv. 21 (1983).

15. G. S. Makanin. Decidability of the universal and positive theories of a free group. *Izv. Akad. Nauk SSSR*, Ser. Mat. 48:735–749, 1984. In Russian; English translation in: *Math. USSR Izvestija, 25*, 75–88, 1985.

16. P. Narendran and F. Otto. The word matching problem is undecidable for finite special string-rewriting systems that are confluent. In P. Degano et al., editor, *Proc. 24th ICALP, Bologna (Italy) 1997*, number 1256 in Lect. Not. Comp. Sc., pages 638–648. Springer, 1997.

17. W. Plandowski. Satisfiability of word equations with constants is in NEXPTIME. In *Proc. 31st Ann. Symp. on Theory of Computing, STOC'99*, pages 721–725. ACM Press, 1999.

18. W. Plandowski. Satisfiability of word equations with constants is in PSPACE. In *Proc. of the 40th Ann. Symp. on Foundations of Computer Science, FOCS 99*, pages 495–500. IEEE Computer Society Press, 1999.

19. W. Plandowski and W. Rytter. Application of Lempel-Ziv encodings to the solution of word equations. In Kim G. Larsen et al., editors, *Proc. of the 25th ICALP, 1998*, number 1443 in Lect. Not. Comp. Sc., pages 731–742. Springer, 1998.

20. W. Rytter. On the complexity of solving word equations. Lecture given at the 16th British Colloquium on Theoretical Computer Science, Liverpool (http://www.csc.liv.ac.uk/~bctcs16/abstracts.html), 2000.

21. K. U. Schulz. Makanin's algorithm for word equations — Two improvements and a generalization. In Klaus U. Schulz, editor, *Word Equations and Related Topics*, number 572 in Lect. Not. Comp. Sc., pages 85–150. Springer, 1991.

Recursive Randomized Coloring Beats Fair Dice Random Colorings

Benjamin Doerr* and Anand Srivastav

Mathematisches Seminar II, Christian–Albrechts–Universität zu Kiel
Ludewig–Meyn–Str. 4, D–24098 Kiel, Germany,
{bed,asr}@numerik.uni-kiel.de
http://www.numerik.uni-kiel.de/~{bed,asr}/

Abstract. We investigate a refined recursive coloring approach to construct balanced colorings for hypergraphs. A coloring is called balanced if each hyperedge has (roughly) the same number of vertices in each color. We provide a recursive randomized algorithm that colors an arbitrary hypergraph (n vertices, m edges) with c colors with discrepancy at most $\mathcal{O}(\sqrt{\frac{n}{c}\log m})$. The algorithm has expected running time $\mathcal{O}(nm\log c)$. This result improves the bound of $\mathcal{O}(\sqrt{n\log(cm)})$ achieved with probability at least $\frac{1}{2}$ by a random coloring that independently chooses a random color for each vertex (fair dice coloring).

Our approach also lowers the current best upper bound for the c–color discrepancy in the case $n = m$ to $\mathcal{O}(\sqrt{\frac{n}{c}\log c})$ and extends the algorithm of Matoušek, Welzl and Wernisch for hypergraphs having bounded dual shatter function to arbitrary numbers of colors.

1 Introduction and Results

One problem in the field of combinatorial optimization well-known for its hardness is the problem of balanced hypergraph colorings, also called combinatorial discrepancy problem. Our goal is to color the vertices of a given hypergraph in such a way that all hyperedges (simultaneously) are colored in a balanced manner. Balanced in this context shall mean that each edge has roughly the same number of vertices in each color. Equivalently, we may ask for a partition of the vertex set which induces a fair partition on all hyperedges.

So far, the discrepancy problem has mainly been investigated for two colors. It has found several applications. Most notably is the connection to uniformly distributed sets and sequences which play a crucial role in numerical integration in higher dimensions (quasi-Monte Carlo methods). This area is also called geometric discrepancy theory. An excellent reference on geometric discrepancies, their connection to combinatorial ones and applications is the book of Matoušek [Mat99]. The notion of linear discrepancy of matrices describes how well a solution of a linear program can be rounded to an integer solution (lattice approximation problem). Due to work of Beck and Spencer [BS84] and Lovász et al.

---

* Supported by the graduate school 'Effiziente Algorithmen und Multiskalenmethoden', Deutsche Forschungsgemeinschaft.

A. Ferreira and H. Reichel (Eds.): STACS 2001, LNCS 2010, pp. 183–194, 2001.
© Springer-Verlag Berlin Heidelberg 2001

[LSV86], the linear discrepancy can be bounded (in a constructive manner) by combinatorial discrepancies. Further applications are found in computational geometry. For this and other applications of discrepancies in theoretical computer science we refer to the new book of Chazelle [Cha00].

Recent work in communication complexity theory [BHK98] motivates the study of balanced colorings in arbitrary numbers of colors. This was begun in [DS99]. It turned out that information about the 2–coloring problem in general does not yield any information on the coloring problem in c colors, $c \in \mathbb{N}_{>2}$. On the other hand, a recursive method was given that yields for any number of colors a coloring with imbalance not larger than roughly twice the maximum 2–color discrepancy among all subhypergraphs.

In this paper we extend this approach to make use of the additional assumption that subhypergraphs on fewer vertices have smaller discrepancy. This is a natural assumption justified by many examples. Roughly speaking we show that if the 2–color discrepancy of the subhypergraphs on n_0 vertices is bounded by $\mathcal{O}(n_0^\alpha)$ for some constant $\alpha \in \,]0,1[$, then the c–color discrepancy is bounded by $\mathcal{O}((\frac{n}{c})^\alpha)$. It seems surprising that this bound is achievable by a recursive approach, as the first step in the recursion will find a 2–coloring for the whole hypergraph with discrepancy guarantee $\mathcal{O}(n^\alpha)$ only. We still get the $\mathcal{O}((\frac{n}{c})^\alpha)$–discrepancy for the final coloring due to the fact that imbalances inflicted in earlier rounds of the recursion are split up in a balanced manner in later steps. It turns out that this effect even exceeds the effect of decreasing discrepancies of smaller subhypergraphs. Crucial therefore is the last step of the recursion where colorings for hypergraphs on roughly $\frac{2n}{c}$ vertices are looked for.

There are some further difficulties, like how to handle numbers of colors that are not a power of 2, and how to guarantee that the color classes become significantly smaller, but we manage to do this in a way that the result is applicable to several problems.

For the general case of an arbitrary hypergraph having n vertices and m edges the 2–color case is well understood. A fair dice coloring, that is, one that colors each vertex independently with a random color has discrepancy $\mathcal{O}(\sqrt{n \log m})$. For m significantly larger than n, this is known to be tight apart from constant factors. Extending this approach to c–colors, we found in [DS99] that a fair dice c–coloring has discrepancy at most $\sqrt{\frac{1}{2}n \ln(4cm)}$ with probability at least $\frac{1}{2}$. In this paper we show that better random colorings can be constructed by combining the 2–color fair dice colorings with a recursive approach. This allows to compute a c–coloring with discrepancy $\mathcal{O}(\sqrt{\frac{n}{c} \log(m)})$ in expected time $\mathcal{O}(nm \log c)$.

Our recursive approach can be applied to several further multi-color discrepancy problems of which we mention two. It shows that for $n = \mathcal{O}(m)$ there is a c–coloring with discrepancy $\mathcal{O}(\sqrt{\frac{n}{c} \log c})$ (instead of $\mathcal{O}(\sqrt{n})$ as shown in [DS99]). This extends a famous result of Spencer [Spe85]. We also extend an algorithm due to Matoušek, Welzl and Wernisch [MWW84] for hypergraphs having bounded dual shatter function to arbitrary numbers of colors.

2   Preliminaries

We shortly review the key definitions of traditional 2–color discrepancy theory and the multi-colors ones from [DS99].

Let $\mathcal{H} = (X, \mathcal{E})$ denote a finite *hypergraph*, i. e. X is a finite set (of *vertices*) and $\mathcal{E}$ is a family of subsets of X (called *hyperedges*). A partition into two classes can be represented by a *coloring* $\chi : X \rightarrow \{-1, +1\}$. We call -1 and $+1$ *colors*. The color classes $\chi^{-1}(-1)$ and $\chi^{-1}(+1)$ form the partition. The imbalance of a hyperedge $E \in \mathcal{E}$ is expressed by $\chi(E) := \sum_{x \in E} \chi(x)$. The *discrepancy* of $\mathcal{H}$ with respect to χ is defined by $\mathrm{disc}(\mathcal{H}, \chi) = \max_{E \in \mathcal{E}} |\chi(E)|$.

For $X_0 \subseteq X$ we call $\mathcal{H}_{|X_0} := (X_0, \{E \cap X_0 | E \in \mathcal{E}\})$ an induced subhypergraph of $\mathcal{H}$. As the discrepancy of an induced subhypergraph cannot be bounded in terms of the discrepancy itself, it makes sense to define the *hereditary discrepancy* $\mathrm{herdisc}(\mathcal{H})$ to be the maximum discrepancy of all induced subhypergraphs.

This is all we need from the classical theory, so let us turn to c–color discrepancies. For technical reasons we need a slight extension of the c–color discrepancy notion, which refers to the problem of coloring a hypergraph in a balanced way with respect to a given ratio. A vector $p \in [0, 1]^c$ such that $\|p\|_1 = \sum_{i \in [c]} p_i = 1$ shall be called a *weight* for c colors. A c–*coloring* of $\mathcal{H}$ is simply a mapping $\chi : X \rightarrow M$, where M is any set of cardinality c. For convenience, normally one has $M = [c] := \{1, \ldots, c\}$. Sometimes a different set M will be of advantage. Note that in applications to communication complexity M can be a finite Abelian group [BHK98]. The basic idea of measuring the deviation from the aim motivates the definitions of the *discrepancy of an edge* $E \in \mathcal{E}$ *in color* $i \in M$ *with respect to* χ *and* p by

$$\mathrm{disc}_{\chi,i,p}(E) := \left| |\chi^{-1}(i) \cap E| - p_i |E| \right|.$$

We call

$$\mathrm{disc}(\mathcal{H}, \chi, i, p) := \max_{E \in \mathcal{E}} \mathrm{disc}_{\chi,i,p}(E)$$

the *discrepancy of* $\mathcal{H}$ *with respect to* χ *and* p *in color* i. The *discrepancy of* $\mathcal{H}$ *with respect to* χ *and* p then is

$$\mathrm{disc}(\mathcal{H}, \chi, p) := \max_{i \in M, E \in \mathcal{E}} \mathrm{disc}_{\chi,i,p}(E),$$

and finally the *discrepancy of* $\mathcal{H}$ *with respect to the weight* p is

$$\mathrm{disc}(\mathcal{H}, c, p) := \min_{\chi : X \rightarrow [c]} \mathrm{disc}(\mathcal{H}, \chi, p).$$

We return to our original problem of balanced coloring if we take $p = \frac{1}{c} \mathbf{1}_c$ (the c–dimensional vector with entries $\frac{1}{c}$ only) as weight. In this case we will simply omit the extra p in the definitions above, i. e. $\mathrm{disc}(\mathcal{H}, c) := \mathrm{disc}(\mathcal{H}, c, \frac{1}{c} \mathbf{1}_c)$. In this notation we have $\mathrm{disc}(\mathcal{H}, 2) = \frac{1}{2} \mathrm{disc}(\mathcal{H})$. The reason for this slightly strange relation is that the usual 2–color discrepancy notion does not compare the number of points of an hyperedge in one color with half the cardinality of the hyperedge, but twice this value due to the $-1, +1$ sums.

We further note that in the case of 2 colors the discrepancies in both colors are equal. A consequence of the relation between linear discrepancy and hereditary discrepancy discovered by [BS84] and [LSV86] is

Lemma 1. *For all hypergraphs* $\mathcal{H} = (X, \mathcal{E})$ *and all 2–color weights* $p \in [0, 1]^2$ *we have*
$$\text{disc}(\mathcal{H}, 2, p) \leq \text{herdisc}(\mathcal{H}).$$

This is constructive in the following sense: For all $h \in \mathbb{R}_{\geq 0}$ *a 2–coloring* χ *such that* $\text{disc}(\mathcal{H}, \chi, p) \leq h + \varepsilon |X|$ *holds can be computed by* $\mathcal{O}(\log \varepsilon^{-1})$ *times computing a coloring having discrepancy at most* h *for some induced subhypergraph.*

An excellent survey of classical and recent results in combinatorial discrepancy theory is the article of Beck and Sós [BS95], which also contains a proof of Lemma 1. For very recent developments we refer to Chapter 4 of Matoušek's book on geometric discrepancies [Mat99].

3   General Approach

The basic idea of recursive coloring is simple: Color the vertices of the whole hypergraph with two colors in such a way that the discrepancy is small, then iterate this on the resulting color classes. There are two points that need further attention:

Firstly, this simple approach only works if the number of colors is a power of 2. This is the reason why we use a discrepancy notion respecting weights. Thus in the case of 3 colors for example, we would look for a 2–coloring respecting the ratio $(\frac{1}{3}, \frac{2}{3})$ and then further split the second color class in the ratio $(\frac{1}{2}, \frac{1}{2})$. There is no general connection between ordinary discrepancy and discrepancy respecting a particular weight (for the same reason, as there is no general connection between the discrepancies in different numbers of colors). If the hereditary discrepancy is not too large, then Lemma 1 allows to compute low discrepancy coloring with respect to a given weight. As we even assume that the discrepancy decreases for subhypergraphs on fewer vertices, we can apply this bound without greater loss.

A second point is that to use this assumption of decreasing discrepancies we need to make sure that the vertex sets considered actually become smaller. Unfortunately, in general we do not know the size of the color classes generated by a low discrepancy coloring. If the whole vertex set is a hyperedge, we know at least that the sizes of the color classes deviate from the aimed at value by at most the discrepancy guarantee. This is not too bad if the discrepancy is relatively small, but even then keeping track of these deviations during the recursion is tedious. Better bounds seem achievable by the cleaner approach of only investigating *fair* colorings, that is, those which have discrepancy less than one on the set of all vertices.

To ease notation let us agree the following. Let $p \in [0,1]^c$ be a c–color weight and $\mathcal{H} = (X, \mathcal{E})$ a hypergraph. We say that χ *is a fair p–coloring of $\mathcal{H}$ having discrepancy at most d_i in color $i \in [c]$* to denote that

- χ is a c–coloring of $\mathcal{H}$,
- χ is fair with respect to p, that is, for all $i \in [c]$ we have $||\chi^{-1}(i)| - p_i|X|| \leq 1$,
- the discrepancy of $\mathcal{H}$ with respect to χ and p in color $i \in [c]$ is at most d_i.

One remark that eases work with the fractional parts: Let us call a weight $p \in [0,1]^c$ *integral* with respect to $\mathcal{H}$ (or $\mathcal{H}$–integral for short) if all $p_i, i \in [c]$ are multiples of $\frac{1}{|X|}$. From the definition it is clear that a fair coloring χ with respect to an integral weight p fulfills $|\chi^{-1}(i)| = p_i|X|$ for all colors $i \in [c]$. On the other hand, suppose that we know that for a given hypergraph and for all integral weights p there is a fair p–coloring that has discrepancy at most k. Then there are fair colorings having discrepancy at most $k + 1$ for any weight: For an arbitrary weight p there is an integral weight p' such that $|p_i - p'_i| < \frac{1}{|X|}$ holds for all $i \in [c]$. Therefore, a fair coloring with respect to p' is also fair with respect to p, and its discrepancy with respect to p is larger (if at all) than the one with respect to p' by less than one. For these reasons we may restrict ourselves to the more convenient case that all weights are integral.

Using the following recoloring argument we can transform arbitrary colorings into fair colorings.

Lemma 2. *Let $\mathcal{H} = (X, \mathcal{E})$ be a hypergraph such that $X \in \mathcal{E}$. Let p be a 2–color weight. Then any 2–coloring χ of $\mathcal{H}$ can be modified in $\mathcal{O}(|X|)$ time into a fair p–coloring χ' such that $\mathrm{disc}(\mathcal{H}, \chi', p) \leq 2\,\mathrm{disc}(\mathcal{H}, \chi, p)$.*

We omit the proof. To analyze our recursive algorithm we need the following constants. Let $\alpha \in]0, 1[$. For each $p \in]0, 1[$ define $v_\alpha(p)$ to be

$$\max \left\{ \sum_{i=1}^{k} \prod_{j=1}^{i} q_j^\alpha \prod_{j=i+1}^{k} q_j \;\middle|\; k \in \mathbb{N}, q_1, \ldots, q_{k-1} \in [0, \tfrac{2}{3}], q_k \in [0, 1], \prod_{j=1}^{k} q_j = p \right\}.$$

Set $c_\alpha := \frac{2}{2^{1-\alpha} - 1} \left(1 + \sum_{i=0}^{\infty} \left(\frac{2}{3} \right)^{(1-\alpha)i} \right)$. Then we have

Lemma 3. *Let $\alpha \in]0, 1[$.*

(i) Let $0 < p < q \leq \frac{2}{3}$. Then $q^\alpha v_\alpha(\frac{p}{q}) + q^\alpha \frac{p}{q} \leq v_\alpha(p)$.
(ii) For all $p \in [0, 1]$, $\frac{2}{2^{1-\alpha} - 1} v_\alpha(p) \leq c_\alpha p^\alpha$.

Proof. We skip the first claim which is not too difficult.

Let $k \in \mathbb{N}, q_1, \ldots, q_{k-1} \in [0, \frac{2}{3}], q_k \in [0, 1]$ such that $\prod_{j=1}^{k} q_j = p$ and $v_\alpha(p) = \sum_{i=1}^{k} \prod_{j=1}^{i} q_j^\alpha \prod_{j=i+1}^{k} q_j$. For $i \in [k]$ set $x_i := \prod_{j=1}^{i} q_j^\alpha \prod_{j=i+1}^{k} q_j$. Then $x_k = p^\alpha$ and $x_{k-1} \leq x_k$. For $i \in [k-2]$ we have

$$\frac{x_{k-1-i}}{x_{k-1-i+1}} = \frac{q_{k-1-i}}{q_{k-1-i}^\alpha} = q_{k-1-i}^{1-\alpha} \leq \left(\frac{2}{3} \right)^{1-\alpha},$$

and hence $x_{k-1-i} \leq \left(\frac{2}{3}\right)^{(1-\alpha)i} x_k$. Thus

$$\tfrac{2}{2^{1-\alpha}-1} v_\alpha(p) = \tfrac{2}{2^{1-\alpha}-1} \sum_{i=1}^{k} x_i \leq \tfrac{2}{2^{1-\alpha}-1} \left(1 + \sum_{i=0}^{\infty} \left(\tfrac{2}{3}\right)^{(1-\alpha)i}\right) p^\alpha = c_\alpha p^\alpha.$$

$\square$

Here is the precise setting we investigate in this section:

Assumption 1. *Let $\mathcal{H} = (X, \mathcal{E})$ be a hypergraph. Set $n := |X|$. Let $p_0, \alpha \in]0, 1[$ and $D > 0$. For all $X_0 \subseteq X$ such that $|X_0| \geq p_0|X|$ and all $q \in [0, 1]$ such that $(q, 1 - q)$ is $\mathcal{H}_{|X_0}$–integral there is a fair $(q, 1 - q)$–coloring χ of $\mathcal{H}_{|X_0}$ having discrepancy at most $D|X_0|^\alpha$.*

In addition to what we already explained there is one further detail involved in our assumption. As we do recursive partitioning, we never need a discrepancy result concerning induced subhypergraphs on fewer than $\frac{n}{c}$ vertices (in the equi-weighted case). This observation will be useful in some applications, e. g. in the case $|\mathcal{E}| = |X|$.

Concerning the complexity there are two possible measures. We can count how many 2–colorings have to be computed, or how often a 2–coloring for a vertex has to be found. The latter is useful if the complexity of computing the 2–colorings is proportional to the number of vertices of the induced subhypergraph as in Section 4.

Theorem 2. *Suppose that Assumption 1 holds. Then for each $\mathcal{H}$–integral weight $p \in [0, 1]^c$ there is a fair p–coloring χ of $\mathcal{H}$ such that the discrepancy is at most $\frac{2}{2^{1-\alpha}-1} D v_\alpha(p_i) n^\alpha \leq D c_\alpha (p_i n)^\alpha$ in all those colors $i \in [c]$ such that $p_i \geq p_0$. Such colorings can be obtained by computing at most $(c-1) \left\lceil \log_2(\frac{1}{p_0}) \right\rceil$ colorings as in Assumption 1. At most $3n \log_{1.5}(\frac{1}{p_0})$ times a color for a vertex has to be computed.*

For the proof we first show a stronger bound for the 2–color discrepancy with respect to a weight $(q, 1 - q)$, if q is small.

Lemma 4. *Suppose that Assumption 1 holds. Then for each $\mathcal{H}$–integral weight $p = (2^{-k}, 1 - 2^{-k})$, $2^{-k} \geq p_0$, a fair p–coloring χ having discrepancy at most*

$$\mathrm{disc}(\mathcal{H}, \chi, p) \leq \sum_{i=0}^{k-1} 2^{-k+1+i} 2^{-\alpha i} D n^\alpha$$

can be computed from k colorings as in Assumption 1. This requires $\sum_{i=0}^{k-1} 2^{-i} n \leq 2n$ times computing a color for a vertex.

We omit the proof here. From our assumptions on $\mathcal{H}$ it is clear that the assertion of Lemma 4 also holds for any induced subhypergraph $\mathcal{H}_{|X_0}$ of $\mathcal{H}$ as long as $2^{-k}|X_0| \geq p_0|X|$. We use this fact to extend Lemma 4 to arbitrary weights.

Lemma 5. *Suppose that Assumption 1 holds. For each $\mathcal{H}$–integral weight $(q, 1 - q)$, $p_0 \leq q \leq \frac{1}{2}$, there is a fair $(q, 1 - q)$–coloring having discrepancy at most $\frac{2}{2^{1-\alpha}-1} D(qn)^\alpha$. A coloring of this kind can be computed by $\left\lceil \log_2(\frac{1}{q}) \right\rceil$ times computing a coloring as in Assumption 1. This requires at most $3n$ times computing a color for a vertex.*

Proof. Let $k \in \mathbb{N}_0$ be maximal subject to the condition that $q' = 2^k q \leq 1$. Since $(q, 1 - q)$ is $\mathcal{H}$–integral, so is $(q', 1 - q')$. According to our assumptions there is a fair $(q', 1 - q')$–coloring $\chi_0 : X \to [2]$ having discrepancy at most Dn^α. From $|\chi_0^{-1}(1)| = q'|X|$ we have $\frac{q}{q'}|\chi_0^{-1}(1)| = q|X| \in \mathbb{N}_0$. Hence $(\frac{q}{q'}, 1 - \frac{q}{q'})$ is $(\mathcal{H}_{|\chi_0^{-1}(1)})$–integral. By Lemma 4 we may compute a fair $(\frac{q}{q'}, 1 - \frac{q}{q'})$–coloring $\chi_1 : \chi_0^{-1}(1) \to [2]$ that has discrepancy at most $\sum_{i=0}^{k-1} 2^{-k+1+i} 2^{-\alpha i} D(q'n)^\alpha$. Define a coloring $\chi : X \to [2]$ by $\chi(x) = 1$ if and only if $\chi_0(x) = 1$ and $\chi_1(x) = 1$. Then χ is a fair $(q, 1 - q)$–coloring. For an edge $E \in \mathcal{E}$ we compute its discrepancy in color 1:

$$
\begin{aligned}
&\left| |E \cap \chi^{-1}(1)| - q|E| \right| \\
&= \left| |E \cap \chi_0^{-1}(1) \cap \chi_1^{-1}(1)| - q|E| \right| \\
&\leq \left| |E \cap \chi_0^{-1}(1) \cap \chi_1^{-1}(1)| - \frac{q}{q'}|E \cap \chi_0^{-1}(1)| \right| + \left| \frac{q}{q'}|E \cap \chi_0^{-1}(1)| - q|E| \right| \\
&= \left| |E \cap \chi_0^{-1}(1) \cap \chi_1^{-1}(1)| - 2^{-k}|E \cap \chi_0^{-1}(1)| \right| + 2^{-k} \left| |E \cap \chi_0^{-1}(1)| - q'|E| \right| \\
&\leq \sum_{i=0}^{k-1} 2^{-k+1+i} 2^{-\alpha i} D(q'n)^\alpha + 2^{-k} Dn^\alpha \\
&< 2q'^\alpha \frac{2^{-\alpha k}}{2^{1-\alpha} - 1} Dn^\alpha = \frac{2}{2^{1-\alpha} - 1} D(qn)^\alpha.
\end{aligned}
$$

Note that if $q' = 1$, then we may compute χ directly using Lemma 4. Therefore the computation of χ requires $\left\lceil \log_2(\frac{1}{q}) \right\rceil$ times computing a coloring assured by Assumption 1. Computing χ_0 means computing a color for n vertices. By Lemma 4, χ_1 can be computed by at most $2q'n$ times computing a color for a vertex. To get χ we therefore computed at most $3n$ times a color for a vertex. This proves Lemma 5. □

Proof (of Theorem 2). To make the recursion work properly we need to fix a set C of colors at the beginning. A weight then is a vector $p = (p_i)_{i \in C}$ indexed by the colors, or, more formally, a function $p : C \to [0, 1]$, such that $\|p\|_1 = \sum_{i \in C} p_i = 1$. To avoid trivial cases we shall always assume that no color $i \in C$ has the weight $p_i = 0$.

We analyze the following recursive algorithm:

Input: A hypergraph $\mathcal{H} = (X, \mathcal{E})$ fulfilling Assumption 1, a set C of at least 2 colors and an $\mathcal{H}$–integral weight function $p : C \to [0, 1]$.
Output: A coloring $\chi : X \to C$ as in Theorem 2.

1. Choose a partition $\{C_1, C_2\}$ of the set of colors C such that $\|p_{|C_1}\|_1, \|p_{|C_2}\|_1 \le \frac{2}{3}$ or C_1 contains a single color with weight at least $\frac{1}{3}$. Set $(q_1, q_2) := (\|p_{|C_1}\|_1, \|p_{|C_2}\|_1)$.

2. Following Lemma 5, compute a fair (q_1, q_2)–coloring $\chi_0 : X \to [2]$ that has discrepancy at most $\frac{2}{2^{1-\alpha}-1} D(q_i n)^\alpha$ in color $i = 1, 2$ if $q_i \ge p_0$. Set $X_i := \chi^{-1}(i)$ for $i = 1, 2$.

3. For $i = 1, 2$ do
   if $|C_i| > 1$,
   then by recursion compute a fair coloring $\chi_i : X_i \to C_i$ with respect to the weight $\frac{1}{q_i} p_{|C_i}$ having discrepancy at most $\frac{2}{2^{1-\alpha}-1} Dv_\alpha(\frac{p_j}{q_i})(q_i n)^\alpha$ in each color $j \in C_i$, $p_j \ge p_0$
   else if $C_i = \{j\}$ for some $j \in C$, set $\chi_i : X_i \mapsto \{j\}$.

4. Return $\chi : X \to C$ defined by $\chi(x) := \chi_1(x)$, if $x \in X_1$, and $\chi(x) := \chi_2(x)$, if $x \in X_2$, for all $x \in X$.

We prove that our algorithm produces a coloring as claimed in Theorem 2 and also fulfills the complexity statements. Suppose by induction that this holds for sets of less than c colors. We analyze the algorithm being started on an input as above with $|C| = c$.

We first show correctness. For Step 1 note that both C_1 and C_2 are non-empty and that $q_2 \le \frac{2}{3}$ holds. Therefore by Lemma 5 and induction the colorings χ_i, $i = 0, 1, 2$ can be computed as desired in Step 2 and 3. Let $E \in \mathcal{E}$, $i \in [2]$ and $j \in C_i$ such that $p_j \ge p_0$. If $|C_i| > 1$, then

$$
\begin{aligned}
& \left| |E \cap \chi^{-1}(j)| - p_j|E| \right| \\
&= \left| |E \cap \chi_0^{-1}(i) \cap \chi_i^{-1}(j)| - p_j|E| \right| \\
&\le \left| |E \cap \chi_0^{-1}(i) \cap \chi_i^{-1}(j)| - \frac{p_j}{q_i}|E \cap \chi_0^{-1}(i)| \right| + \left| \frac{p_j}{q_i}|E \cap \chi_0^{-1}(i)| - p_j|E| \right| \\
&\le \left| |(E \cap X_i) \cap \chi_i^{-1}(j)| - \frac{p_j}{q_i}|E \cap X_i| \right| + \frac{p_j}{q_i} \left| |E \cap \chi_0^{-1}(i)| - q_i|E| \right| \\
&\le \frac{2}{2^{1-\alpha}-1} Dv_\alpha(\frac{p_j}{q_i})(q_i n)^\alpha + \frac{p_j}{q_i} \frac{2}{2^{1-\alpha}-1} D(q_i n)^\alpha \\
&\le \frac{2}{2^{1-\alpha}-1} Dv_\alpha(p_j)n^\alpha
\end{aligned}
$$

by Lemma 3 (i). On the other hand, if C_i contains a single color j, then $p_j = q_i$ and

$$
\begin{aligned}
\left| |E \cap \chi^{-1}(j)| - p_j|E| \right| &= \left| |E \cap \chi_0^{-1}(i)| - q_i|E| \right| \\
&\le \frac{2}{2^{1-\alpha}-1} D(q_i n)^\alpha \\
&\le \frac{2}{2^{1-\alpha}-1} Dv_\alpha(p_j)n^\alpha.
\end{aligned}
$$

This is the correctness statement.

Concerning the complexity note that the computation of χ_0 takes at most $\lceil \log_2(\frac{1}{p_0}) \rceil$ and the one of the χ_i takes at most $(|C_i| - 1) \lceil \log_2(\frac{q_i}{p_0}) \rceil$ colorings as in Assumption 1. These are not more than $(c-1) \lceil \log_2(\frac{1}{p_0}) \rceil$ colorings altogether.

By Lemma 5 we compute at most $3n$ times a color for a vertex in Step 2. If $|C_i| > 1$ for both $i = 1, 2$, then $q_i \leq \frac{2}{3}$ and computing χ_i involves at most $3q_i n \log_{1.5}(\frac{q_i}{p_0}) \leq 3q_i n \log_{1.5}(\frac{2}{3p_0})$ times computing a color for a vertex. Altogether this makes at most $3n + 3q_1 n \log_{1.5}(\frac{q_1}{p_0}) + 3q_2 n \log_{1.5}(\frac{q_2}{p_0}) \leq 3n(1 + \log_{1.5}(\frac{2}{3p_0})) = 3n \log_{1.5}(\frac{1}{p_0})$ times computing a color for a vertex. If $|C_i| = 1$ then there is nothing to do to get χ_i and the respective term just vanishes in the calculation above.   □

4   A Randomized Algorithm for Arbitrary Hypergraphs

Let $\mathcal{H} = (X, \mathcal{E})$ denote an arbitrary hypergraph. Set $n := |X|$ and $m := |\mathcal{E}|$ for convenience. In [DS99] it was shown that a random coloring generated by coloring each vertex independently with each color with probability $\frac{1}{c}$ has discrepancy at most $\sqrt{\frac{1}{2}n \ln(4mc)}$ with probability at least $\frac{1}{2}$. This can be used to design a randomized algorithm computing such a coloring by repeatedly generating and testing such a random coloring until its discrepancy is at most $\sqrt{\frac{1}{2}n \ln(4mc)}$.

In this section we show that via the recursive approach of Theorem 2 a better bound can be achieved. This also proves that the discrepancy decreases for larger numbers of colors.

Theorem 3. *For each $\mathcal{H}$–integral c–color weight p a c–coloring χ having discrepancy at most $\mathrm{disc}(\mathcal{H}, \chi, p, i) \leq 45\sqrt{p_i n \ln(4m)}$ in color $i \in [c]$ can be computed in expected time $\mathcal{O}(nm \log(\min\{p_i | i \in [c]\}))$. In particular, a c–coloring χ such that*

$$\mathrm{disc}(\mathcal{H}, \chi, c) \leq 45\sqrt{\tfrac{n}{c} \ln(4m)} + 1$$

can be computed in expected time $\mathcal{O}(nm \log c)$.

Proof. There is little to do for $m = 1$, so let us assume that $m \geq 2$. We show that the colorings required by Assumption 1 can be computed in expected time $\mathcal{O}(|X_0|m)$. Denote by $\overline{\mathcal{H}}$ the hypergraph obtained from $\mathcal{H}$ by adding the whole vertex set as an additional hyperedge. Let $X_0 \subseteq X$ and $(q, 1-q)$ be a 2–color weight. Let $\chi : X_0 \rightarrow [2]$ be a random coloring independently coloring the vertices with probabilities $P(\chi(x) = 1) = q$ and $P(\chi(x) = 2) = 1 - q$ for all $x \in X_0$. A standard application of the Chernoff inequality (cf. [AS00]) shows that

$$(*)\qquad \mathrm{disc}(\overline{\mathcal{H}}_{|X_0}, \chi, (q, 1-q)) \leq \sqrt{\tfrac{1}{2}|X_0| \ln(4m)}$$

holds with probability at least $\frac{m-1}{2m}$. Hence by repeatedly generating and testing these random colorings until $(*)$ holds we obtain a randomized algorithm computing such a coloring with expected running time $\mathcal{O}(nm)$. By Lemma 2 we get a fair $(q, 1-q)$–coloring for $\mathcal{H}_{|X_0}$ having discrepancy at most $\sqrt{2|X_0| \ln(4m)}$. Hence for $\alpha = \frac{1}{2}$, $D = \sqrt{2 \ln(4m)}$ and arbitrary p_0 the colorings required in Assumption 1 can be computed in expected time $\mathcal{O}(|X_0|m)$.

Therefore we may apply Theorem 2 with $p_0 = \min\{p_i | i \in [c]\}$. The discrepancy bounds follow from $c_\alpha \leq 31.4$. Computing such a coloring involves $\mathcal{O}(\log(\frac{1}{p_0})n)$ times computing a color for a vertex. As this can be done in expected time $\mathcal{O}(m)$, we have the claimed bound of $\mathcal{O}(nm \log(\frac{1}{p_0}))$. $\square$

Some remarks concerning the theorem and its proof above. For the complexity guarantee we assumed that the complexity contribution of computing the 2–colorings dominates the remaining operations of the recursive algorithm of Section 3. This is justified by the fact that we may assume $c \leq n$ since integrality ensures $p_i \geq \frac{1}{|X|}$ for all colors $i \in C$.

A second point is that the constant of 45 could be improved by a more careful way of generating the random 2–colorings. In particular by taking a random fair coloring we could avoid the extra factor of 2 inflicted by Lemma 2. This though requires an analysis of the hypergeometric distribution, which is considerably more difficult that ours.

Finally let us remark that the construction of the 2–colorings can be derandomized through an algorithmic version of the Chernoff-Hoeffding inequality (cf. [SS96]). Thus the colorings in Theorem 3 can be computed by a deterministic algorithm as well.

5   Further Results

In this section we give two more applications of Theorem 2 that extend 2–color bounds or algorithms to c colors.

5.1   Six Standard Deviations

The famous "Six Standard Deviations" result due to Spencer [Spe85] states that there is a constant K such that for all hypergraphs $\mathcal{H} = (X, \mathcal{E})$ having n vertices and $m \geq n$ edges

$$\text{disc}(\mathcal{H}) \leq K\sqrt{n \ln(\frac{2m}{n})}$$

holds.

The interesting case is of course the one where $m = \mathcal{O}(n)$ and thus $\text{disc}(\mathcal{H}) = \mathcal{O}(\sqrt{n})$. For m significantly larger than n this result is outnumbered by the simple fair coin flip random coloring. The title "Six Standard Deviations Suffice" of this paper comes from the fact that for $n = m$ large enough, $\text{disc}(\mathcal{H}) \leq 6\sqrt{n}$ holds. Using the relation between discrepancies respecting a particular weight and hereditary discrepancy (Lemma 1) and the recoloring argument (Lemma 2), we derive from Spencer's result (without proof)

Lemma 6. *For any* $X_0 \subseteq X$ *and* $\mathcal{H}_{|X_0}$*–integral weight* $(q, 1 - q)$ *there is a fair* $(q, 1 - q)$*–coloring of* $\mathcal{H}_{|X_0}$ *that has discrepancy at most*

$$2K\sqrt{|X_0| \ln(\frac{2m+2}{|X_0|})}.$$

Lemma 6 and Theorem 2 yield

Theorem 4. *Let $\mathcal{H} = (X, \mathcal{E})$ denote a hypergraph having n vertices and $m \geq n$ edges and $p \in [0,1]^c$ an integral weight. Set $p_0 := \min_{i \in [c]} p_i$. Then there is a fair p–coloring having discrepancy at most $63K \sqrt{p_i n \ln(\frac{2m+2}{p_0 n})}$ in color i. In consequence, for $|X| = |\mathcal{E}| = n$ we have*

$$\mathrm{disc}(\mathcal{H}, c) \leq \mathcal{O}\left(\sqrt{\frac{n}{c} \ln c}\right).$$

Proof. By Lemma 6 we may apply Theorem 2 with $\alpha = \frac{1}{2}$, $D = 2K\sqrt{\ln(\frac{2m+2}{p_0 n})}$ and p_0. This yields a fair p–coloring having discrepancy at most $Dc_\alpha\sqrt{p_i n}$ in color $i \in [c]$. The claim follows from $c_\alpha \leq 31.4$. □

This is quite close to the optimum. An extension of Spencer's [Spe87] proof shows that hypergraphs arising from Hadamard matrices have c–color discrepancy $\Omega(\sqrt{\frac{n}{c}})$. It is a famous open problem already for 2 colors whether colorings having discrepancy $\mathcal{O}(\sqrt{n \log(\frac{m}{n})})$ can be computed efficiently. Therefore a constructive version of Theorem 4 is not to be expected at the moment.

5.2 Dual Shatter Function Bound

If we have some more structural information about the hypergraph, in many cases there are constructive solutions to the discrepancy problem. Matoušek, Welzl and Wernisch showed that if the dual shatter function $\pi_{\mathcal{H}}^*$ of $\mathcal{H}$ is bounded by $\pi_{\mathcal{H}}^* = \mathcal{O}(m^d)$ for some constant $d \geq 2$, then a 2–coloring χ such that $\mathrm{disc}(\mathcal{H}, \chi) = \mathcal{O}(n^{1/2-1/2d}\sqrt{\log n})$ can be computed by a randomized polynomial time algorithm. The dual shatter function is monotone with respect to induced subhypergraph, that is, for all $X_0 \subseteq X$ we have $\pi_{\mathcal{H}|X_0}^* \leq \pi_{\mathcal{H}}^*$. Hence we conclude from Lemma 1 that Assumption 1 is fulfilled with $\alpha = \frac{1}{2} - \frac{1}{2d}$. We derive

Theorem 5. *If the dual shatter function $\pi_{\mathcal{H}}^*$ of $\mathcal{H}$ is bounded by $\pi_{\mathcal{H}}^* = \mathcal{O}(m^d)$ for some constant $d \geq 2$, then a c–coloring χ such that*

$$\mathrm{disc}(\mathcal{H}, \chi, c) = \mathcal{O}((\tfrac{n}{c})^{1/2-1/2d}\sqrt{\log n})$$

can be computed by a randomized polynomial time algorithm.

6 Conclusion and Discussion

In this paper we presented a recursive method to construct c–colorings from 2–colorings with respect to a given weight. Our approach uses the fact that induced subhypergraphs on fewer vertices often have smaller discrepancies. We extend several 2–color results to arbitrary numbers of colors. In particular, we show that a clever extension of the 2–color approach of independently choosing a color for each vertex is not doing the same with c colors, but combining the 2–color result with a recursive algorithm. This seems strange at first, but the gain of a $\sqrt{c}$ factor is convincing.

We should remark that this gain is 'real', that is, it evolves from the difference in the random experiments rather than a weak analysis of the fair dice colorings.

We believe that there are three reasons explaining the behavior. First, it is a general result of [DS99] that 2–coloring has the significant advantage that the discrepancy in both colors is the same. Therefore one actually has to take care of just one color. Second, instead of having just one random experiment which has to yield a 'good' coloring with sufficiently large probability, here we have a series of random experiments that are executed one after another. Third, all colorings generated by our algorithm are fair.

Acknowledgments

We would like to thank the referees of STACS 2001 for their thorough work which improved the presentation of our result significantly.

References

AS00.   N. Alon and J. Spencer. *The Probabilistic Method.* John Wiley & Sons, Inc., 2nd edition, 2000.

BHK98.   L. Babai, T. P. Hayes, and P. G. Kimmel. The cost of the missing bit: Communication complexity with help. In *Proceedings of the 30th STOC*, pages 673–682, 1998.

BS84.   J. Beck and J. Spencer. Integral approximation sequences. *Math. Programming*, 30:88–98, 1984.

BS95.   J. Beck and V. T. Sós. Discrepancy theory. In R. Graham, M. Grötschel, and L. Lovász, editors, *Handbook of Combinatorics.* 1995.

Cha00.   B. Chazelle. *The Discrepancy Method.* Princeton University, 2000.

DS99.   B. Doerr and A. Srivastav. Approximation of multi-color discrepancy. In D. Hochbaum, K. Jansen, J. D. P. Rolim, and A. Sinclair, editors, *Randomization, Approximation and Combinatorial Optimization*, volume 1671 of *Lecture Notes in Computer Science*, pages 39–50, Berlin–Heidelberg, 1999. Springer Verlag.

LSV86.   L. Lovász, J. Spencer, and K. Vesztergombi. Discrepancies of set-systems and matrices. *Europ. J. Combin.*, 7:151–160, 1986.

Mat99.   J. Matoušek. *Geometric Discrepancy.* Springer-Verlag, Berlin, 1999.

MWW84.   J. Matoušek, E. Welzl, and L. Wernisch. Discrepancy and approximations for bounded VC–dimension. *Combinatorica*, 13:455–466, 1984.

Spe85.   J. Spencer. Six standard deviations suffice. *Trans. Amer. Math. Soc.*, 289:679–706, 1985.

Spe87.   J. Spencer. *Ten Lectures on the Probabilistic Method.* SIAM, 1987.

SS96.   Anand Srivastav and Peter Stangier. Algorithmic Chernoff-Hoeffding inequalities in integer programming. *Random Structures & Algorithms*, 8:27–58, 1996.

Randomness, Computability, and Density[*]

Rod G. Downey[1], Denis R. Hirschfeldt[1], and André Nies[2]

[1] School of Mathematical and Computing Sciences, Victoria University of Wellington
[2] Department of Mathematics, The University of Chicago

Abstract. We study effectively given positive reals (more specifically, computably enumerable reals) under a measure of relative randomness introduced by Solovay [32] and studied by Calude, Hertling, Khoussainov, and Wang [7], Calude [3], Slaman [28], and Coles, Downey, and LaForte [14], among others. This measure is called *domination* or *Solovay reducibility*, and is defined by saying that α dominates β if there are a constant c and a partial computable function φ such that for all positive rationals $q < \alpha$ we have $\varphi(q) \downarrow < \beta$ and $\beta - \varphi(q) \leqslant c(\alpha - q)$. The intuition is that an approximating sequence for α generates one for β whose rate of convergence is not much slower than that of the original sequence. It is not hard to show that if α dominates β then the initial segment complexity of α is at least that of β.

In this paper we are concerned with structural properties of the degree structure generated by Solovay reducibility. We answer a long-standing question in this area of investigation by establishing the density of the Solovay degrees. We also provide a new characterization of the random c.e. reals in terms of splittings in the Solovay degrees. Specifically, we show that the Solovay degrees of computably enumerable reals are dense, that any incomplete Solovay degree splits over any lesser degree, and that the join of any two incomplete Solovay degrees is incomplete, so that the complete Solovay degree does not split at all. The methodology is of some technical interest, since it includes a priority argument in which the injuries are themselves controlled by randomness considerations.

1 Introduction

In this paper we are concerned with effectively generated reals in the interval $(0, 1]$ and their relative randomness. In what follows, *real* and *rational* will mean positive real and positive rational, respectively. It will be convenient to work modulo 1, that is, identifying $n + \alpha$ and α for any $n \in \omega$ and $\alpha \in (0, 1]$, and we do this below without further comment.

Our basic objects are reals that are limits of computable increasing sequences of rationals. We call such reals *computably enumerable* (c.e.), though

---

[*] Downey and Hirschfeldt's research supported by the Marsden Fund for Basic Science. Nies and Downey's research supported by a US/NZ cooperative science grant. Nies's research supported by NSF grant DMS-9803482.

A. Ferreira and H. Reichel (Eds.): STACS 2001, LNCS 2010, pp. 195–205, 2001.
© Springer-Verlag Berlin Heidelberg 2001

they have also been called *recursively enumerable, left computable* (by Ambos-Spies, Weihrauch, and Zheng [1]), and *left semicomputable*.[1] If, in addition to the existence of a computable increasing sequence $q_0, q_1, \ldots$ of rationals with limit α, there is a total computable function f such that $\alpha - q_{f(n)} < 2^{-n}$ for all $n \in \omega$, then α is called *computable*. These and related concepts have been widely studied. In addition to the papers and books mentioned elsewhere in this introduction, we may cite, among others, early work of Rice [26], Lachlan [21], Soare [29], and Ceĭtin [9], and more recent papers by Ko [18,19], Calude, Coles, Hertling, and Khoussainov [6], Ho [17], Boldi and Vigna [2], and Downey and LaForte [16].

A computer M is *self-delimiting* if, for each binary string σ, $M(\sigma)\downarrow$ implies that $M(\sigma')\uparrow$ for all σ' properly extending σ. It is *universal* if for each self-delimiting computer N there is a constant c such that, for each binary string σ, if $N(\sigma)\downarrow$ then $M(\tau)\downarrow = N(\sigma)$ for some τ with $|\tau| \leqslant |\sigma| + c$.

Fix a self-delimiting universal computer M. We can define Chaitin's number $\Omega = \Omega_M$ via

$$\Omega = \sum_{M(\sigma)\downarrow} 2^{-|\sigma|} \ .$$

The properties of Ω relevant to this paper are independent of the choice of M. A c.e. real is an *Ω-number* if it is Ω_M for some self-delimiting universal computer M.

The c.e. real Ω is random in the canonical Martin-Löf sense. Recall that a *Martin-Löf test* is a uniformly c.e. sequence $\{V_e : e > 0\}$ of c.e. subsets of $\{0,1\}^*$ such that for all $e > 0$,

$$\mu(V_e\{0,1\}^\omega) \leqslant 2^{-e} \ ,$$

where μ denotes the usual product measure on $\{0,1\}^\omega$. The string $\sigma \in \{0,1\}^\omega$ and the real $0.\sigma$ are *random*, or more precisely, 1-*random*, if $\sigma \notin \bigcap_{e>0} V_e\{0,1\}^\omega$ for every Martin-Löf test $\{V_e : e > 0\}$.

An alternate characterization of the random reals can be given via the notion of a *Solovay test*. We give a somewhat nonstandard definition of this notion, which will be useful below. A Solovay test is a c.e. multiset $\{I_i : i \in \omega\}$ of intervals with rational endpoints such that $\sum_{i\in\omega} |I_i| < \infty$, where $|I|$ is the length of the interval I. As Solovay [32] showed, a real α is random if and only if $\{i \in \omega : \alpha \in I_i\}$ is finite for every Solovay test $\{I_i : i \in \omega\}$.

Many authors have studied Ω and its properties, notably Chaitin [11,12,13] and Martin-Löf [25]. In the very long and widely circulated manuscript [32] (a fragment of which appeared in [33]), Solovay carefully investigated relationships between Martin-Löf-Chaitin prefix-free complexity, Kolmogorov complexity, and properties of random languages and reals. See Chaitin [11] for an account of some of the results in this manuscript.

---

[1] We recognize that the term *computably enumerable real* is not ideal, but it is the one used by Solovay, Chaitin, Soare, and others in this tradition (modulo the recent terminological move from *recursive* to *computable*), and the alternatives are also problematic; for instance, *semicomputable* has an unrelated meaning in computability theory.

Solovay discovered that several important properties of Ω (whose definition is model-dependent) are shared by another class of reals he called Ω-like, whose definition is model-independent. To define this class, he introduced the following reducibility relation among c.e. reals, called *domination* or *Solovay reducibility*.

Definition 1.1. Let α and β be c.e. reals. We say that α *dominates* β, and write $\beta \leqslant_s \alpha$, if there are a constant c and a partial computable function $\varphi : \mathbb{Q} \to \mathbb{Q}$ such that for each rational $q < \alpha$ we have $\varphi(q)\downarrow < \beta$ and

$$\beta - \varphi(q) \leqslant c(\alpha - q) .$$

We write $\beta <_s \alpha$ if $\beta \leqslant_s \alpha$ and $\alpha \not\leqslant_s \beta$, and we write $\alpha \equiv_s \beta$ if $\alpha \leqslant_s \beta$ and $\beta \leqslant_s \alpha$.

The notation $\leqslant_{\mathrm{dom}}$ has sometimes been used instead of $\leqslant_s$.

Recall that the prefix-free complexity $H(\tau)$ of a binary string τ is the length of the shortest binary string σ such that $M(\sigma)\downarrow = \tau$, where M is a fixed self-delimiting universal computer. (The choice of M does not affect the prefix-free complexity, up to a constant additive factor.) Most of the statements about $H(\tau)$ made below also hold for the standard Kolmogorov complexity $K(\tau)$. For more on the the definitions and basic properties of $H(\tau)$ and $K(\tau)$, see Chaitin [13], Calude [4], and Li and Vitanyi [24]. Among the many works dealing with these and related topics, and in addition to those mentioned elsewhere in this paper, we may cite Solomonoff [30,31], Kolmogorov [20], Levin [22,23], Schnorr [27], Chaitin [10], and the expository article Calude and Chaitin [5].

Solovay reducibility is naturally associated with randomness because of the following fact. (We identify a real $\alpha \in (0, 1]$ with the infinite binary string σ such that $\alpha = 0.\sigma$. The fact that certain reals have two different dyadic expansions need not concern us here, since all such reals are rational.)

Theorem 1.2 (Solovay [32]). *Let $\beta \leqslant_s \alpha$ be c.e. reals. There is a constant $O(1)$ such that $H(\beta \restriction n) \leqslant H(\alpha \restriction n) + O(1)$ for all $n \in \omega$.*

Solovay observed that Ω dominates all c.e. reals, and Theorem 1.2 implies that if a c.e. real dominates all c.e. reals then it must be random. This led Solovay to define a c.e. real to be Ω-*like* if it dominates all c.e. reals. The point is that the definition of Ω-like seems quite model-independent (in the sense that it does not require a choice of self-delimiting universal computer), as opposed to the model-dependent definition of Ω. However, Calude, Hertling, Khoussainov, and Wang [7] showed that the two notions coincide.

Theorem 1.3 (Calude, Hertling, Khoussainov, and Wang). *A c.e. real is Ω-like if and only if it is an Ω-number.*

This circle of ideas was completed recently by Slaman [28], who proved the converse to the fact that Ω-like reals are random.

Theorem 1.4 (Slaman). *A c.e. real is random if and only if it is Ω-like.*

It is natural to seek to understand the c.e. reals under Solovay reducibility. A useful characterization of this reducibility is given by the following lemma.

Lemma 1.5. *Let α and β be c.e. reals. Then $\beta \leqslant_s \alpha$ if and only if for every computable sequence of rationals $a_0, a_1, \ldots$ such that*

$$\alpha = \sum_{n \in \omega} a_n$$

there are a constant c and a computable sequence of rationals $\varepsilon_0, \varepsilon_1, \ldots < c$ such that

$$\beta = \sum_{n \in \omega} \varepsilon_n a_n .$$

Phrased another way, Lemma 1.5 says that the c.e. reals dominated by a given c.e. real α essentially correspond to splittings of α under arithmetic addition.

Corollary 1.6. *Let α and β be c.e. reals. Then $\beta \leqslant_s \alpha$ if and only if there is a c.e. real γ and a rational c such that $c\alpha = \beta + \gamma$.*

Solovay reducibility has a number of other beautiful interactions with arithmetic, as we now discuss.

The relation $\leqslant_s$ is symmetric and transitive, and hence $\equiv_s$ is an equivalence relation on the c.e. reals. Thus we can define the *Solovay degree* $[\alpha]$ of a c.e. real α as its $\equiv_s$ equivalence class. (When we mention Solovay degrees below, we always mean Solovay degrees of c.e. reals.) The Solovay degrees form an uppersemilattice (a partial ordering in which every pair of elements has a least upper bound, called the *join* of these elements), with the join of $[\alpha]$ and $[\beta]$ being $[\alpha + \beta]=[\alpha\beta]$, a fact observed by Solovay and others ($\oplus$ is definitely not a join operation here). We note the following slight improvement of this result. Recall that an uppersemilattice U is *distributive* if for all $a_0, a_1, b \in U$ with $b \leqslant a_0 \vee a_1$ there exist $b_0, b_1 \in U$ such that $b_0 \vee b_1 = b$ and $b_i \leqslant a_i$ for $i = 0, 1$.

Lemma 1.7. *The Solovay degrees of c.e. reals form a distributive uppersemilattice with $[\alpha] \vee [\beta] = [\alpha + \beta] = [\alpha\beta]$.*

There is a least Solovay degree, the degree of the computable reals, as well as a greatest one, the degree of Ω. For proofs of these facts and more on c.e. reals and Solovay reducibility, see for instance Chaitin [11,12,13], Calude, Hertling, Khoussainov, and Wang [7], Calude and Nies [8], Calude [3], Slaman [28], and Coles, Downey, and LaForte [14].

Despite the many attractive features of the Solovay degrees, their structure is largely unknown. Coles, Downey, and LaForte [14] have shown that this structure is very complicated by proving that it has an undecidable first order theory.

One question addressed in the present paper, open since Solovay's original 1974 notes, is whether the structure of the Solovay degrees is dense. Indeed, up to now, it was not known even whether there is a *minimal* Solovay degree. That is, intuitively, if a c.e. real α is not computable, must there be a c.e. real that is also not computable, yet is strictly less random that α?

In this extended abstract, we sketch part of a proof that the Solovay degrees of c.e. reals are dense. In the full version of this paper, we give the complete proof of this result. This proof is divided into two parts: we show that if $\alpha <_s \Omega$ then there is a c.e. real γ with $\alpha <_s \gamma <_s \Omega$, and we also show that every *incomplete* Solovay degree splits over each lesser degree.

The nonuniform nature of the argument is essential given the techniques we use, since, in the splitting case, we have a priority construction in which the control of the injuries is directly tied to the enumeration of Ω. The fact that if a c.e. real α is Solovay-incomplete then Ω must grow more slowly than α is what allows us to succeed. (We will discuss this more fully in Sect. 2.) This unusual technique is of some technical interest, and clearly cannot be applied to proving upwards density, since in that case the top degree is the degree of Ω itself. To prove upwards density, we use a different technique, taking advantage of the fact that, however we construct a c.e. real, it is automatically dominated by Ω.

In light of these results, and further motivated by the general question of how randomness can be produced, it is natural to ask whether the complete Solovay degree can be split, or in other words, whether there exist nonrandom c.e. reals α and β such that $\alpha + \beta$ is random. We give a negative answer to this question, thus characterizing the random c.e. reals as those c.e. reals that cannot be written as the sum of two c.e. reals of lesser Solovay degrees.

We remark that there are (non-c.e.) nonrandom reals whose sum is random; the following is an example of this phenomenon. Define the real α by letting $\alpha(n) = 0$ if n is even and $\alpha(n) = \Omega(n)$ otherwise. (Here we identify a real with its dyadic expansion as above.) Define the real β by letting $\beta(n) = 0$ if n is odd and $\beta(n) = \Omega(n)$ otherwise. Now α and β are clearly nonrandom, but $\alpha + \beta = \Omega$ is random.

Before turning to the precise statements of our main results and sketches of some of their proofs, we point out that there are other reducibilities one can study in this context. Coles, Downey, and LaForte [14,15] introduced one such reducibility, called *sw-reducibility*; it is defined as follows. For sets of natural numbers A and B, we say that $A \leqslant_{sw} B$ if there are a computable procedure Γ and a constant c such that $\Gamma^B = A$ and the use of Γ on argument x is bounded by $x + c$. For reals $\alpha, \beta \in (0, 1]$, we say that $\alpha \leqslant_{sw} \beta$ if there are sets A and B such that $\alpha = 0.\chi_A$, $\beta = 0.\chi_B$, and $A \leqslant_{sw} B$.

As in the case of Solovay reducibility, it is not difficult to argue that if $\alpha \leqslant_{sw} \beta$ then $H(\alpha \upharpoonright n) \leqslant H(\beta \upharpoonright n) + O(1)$ for all $n \in \omega$. However, Coles, Downey, and LaForte [14] showed that Solovay reducibility and sw-reducibility are different, since there are c.e. reals α, β, γ, and δ such that $\alpha \leqslant_s \beta$ but $\alpha \not\leqslant_{sw} \beta$ and $\gamma \leqslant_{sw} \delta$ but $\gamma \not\leqslant_s \delta$, and that there are no minimal sw-degrees of c.e. reals.

Question 1.8. Is every random c.e. real sw-complete?

Question 1.9. Are the sw-degrees of c.e. reals dense?

Ultimately, the basic reducibility we seek to understand is H-reducibility, where $\sigma \leqslant_H \tau$ if there is a constant $O(1)$ such that $H(\sigma \upharpoonright n) \leqslant H(\tau \upharpoonright n) + O(1)$ for all $n \in \omega$. Little is known about this directly.

2   Main Results

The following lemma, implicit in [32] and proved in [14], provides an alternate characterization of Solovay reducibility, which is the one that we will use below.

Lemma 2.1. *Let α and β be c.e. reals, and let $\alpha_0, \alpha_1, \ldots$ and $\beta_0, \beta_1, \ldots$ be computable increasing sequences of rationals converging to α and β, respectively. Then $\beta \leqslant_s \alpha$ if and only if there are a constant d and a total computable function f such that for all $n \in \omega$,*

$$\beta - \beta_{f(n)} < d(\alpha - \alpha_n) \ .$$

Whenever we mention a c.e. real α, we assume that we have chosen a computable increasing sequence $\alpha_0, \alpha_1, \ldots$ converging to α. The previous lemma guarantees that, in determining whether one c.e. real dominates another, the particular choice of such sequences is irrelevant. For convenience of notation, we adopt the convention that, for any c.e. real α mentioned below, the expression $\alpha_s - \alpha_{s-1}$ is equal to α_0 when $s = 0$.

We begin by sketching the proof that every incomplete Solovay degree can be split over any lesser Solovay degree.

Theorem 2.2. *Let $\gamma <_s \alpha <_s \Omega$ be c.e. reals. There are c.e. reals β^0 and β^1 such that $\gamma <_s \beta^0, \beta^1 <_s \alpha$ and $\beta^0 + \beta^1 = \alpha$.*

Proof Sketch. We want to build β^0 and β^1 so that $\gamma \leqslant_s \beta^0, \beta^1 \leqslant_s \alpha$ and $\beta^0 + \beta^1 = \alpha$, while satisfying the following requirement for each $e, k \in \omega$ and $i < 2$:

$$R_{i,e,k} : \Phi_e \text{ total} \ \Rightarrow \exists n(\alpha - \alpha_{\Phi_e(n)} \geqslant k(\beta^i - \beta_n^i)) \ .$$

It is not hard to check that, since $\gamma <_s \alpha$, there are a constant c and a computable increasing sequence $\gamma_0, \gamma_1, \ldots$ of rationals converging to γ such that $\gamma_s - \gamma_{s-1} < c(\alpha_s - \alpha_{s-1})$ for all $s \in \omega$. Since multiplying a c.e. real by a positive integer does not change its Solovay degree, we may assume without loss of generality that $2(\gamma_s - \gamma_{s-1}) \leqslant \alpha_s - \alpha_{s-1}$ for all $s \in \omega$.

Most of the essential features of our construction are already present in the case of two requirements $R_{i,e,k}$ and $R_{1-i,e',k'}$, so we limit our discussion to this case. We assume that $R_{i,e,k}$ has priority over $R_{1-i,e',k'}$ and that both Φ_e and $\Phi_{e'}$ are total. We will think of the β^j as being built by adding amounts to them in stages. Thus β_s^j will be the total amount added to β^j by the end of stage s. At each stage s we begin by adding $\gamma_s - \gamma_{s-1}$ to the current value of each β^j; in the limit, this ensures that $\beta^j \geqslant_s \gamma$.

We will say that $R_{i,e,k}$ is satisfied through n at stage s if $\Phi_e(n)[s] \downarrow$ and $\alpha_s - \alpha_{\Phi_e(n)} > k(\beta_s^i - \beta_n^i)$. The strategy for $R_{i,e,k}$ is to act whenever either it is not currently satisfied or the least number through which it is satisfied changes. Whenever this happens, $R_{i,e,k}$ initializes $R_{1-i,e',k'}$, which means that the amount of $\alpha - 2\gamma$ that $R_{1-i,e',k'}$ is allowed to funnel into β^i is reduced. More specifically, once $R_{1-i,e',k'}$ has been initialized for the mth time, the total amount that it is thenceforth allowed to put into β^i is reduced to 2^{-m}.

The above strategy guarantees that if $R_{1-i,e',k'}$ is initialized infinitely often then the amount put into β^i by $R_{1-i,e',k'}$ (which in this case is all that is put into β^i except for the coding of γ) adds up to a computable real. In other words, $\beta^i \equiv_s \gamma <_s \alpha$. But it is not hard to argue that this means that there is a stage s after which $R_{i,e,k}$ is always satisfied and the least number through which it is satisfied does not change. So we conclude that $R_{1-i,e',k'}$ is initialized only finitely often, and that $R_{i,e,k}$ is eventually permanently satisfied.

This leaves us with the problem of designing a strategy for $R_{1-i,e',k'}$ that respects the strategy for $R_{i,e,k}$. The problem is one of timing. To simplify notation, let $\hat{\alpha} = \alpha - 2\gamma$ and $\hat{\alpha}_s = \alpha_s - 2\gamma_s$. Since $R_{1-i,e',k'}$ is initialized only finitely often, there is a certain amount 2^{-m} that it is allowed to put into β^i after the last time it is initialized. Thus if $R_{1-i,e',k'}$ waits until a stage s such that $\hat{\alpha} - \hat{\alpha}_s < 2^{-m}$, adding nothing to β^i until such a stage is reached, then from that point on it can put all of $\hat{\alpha} - \hat{\alpha}_s$ into β^i, which of course guarantees its success. The problem is that, in the general construction, a strategy working with a quota 2^{-m} cannot effectively find an s such that $\hat{\alpha} - \hat{\alpha}_s < 2^{-m}$. If it uses up its quota too soon, it may find itself unsatisfied and unable to do anything about it.

The key to solving this problem (and the reason for the hypothesis that $\alpha <_s \Omega$) is the observation that, since the sequence $\Omega_0, \Omega_1, \ldots$ converges much more slowly than the sequence $\hat{\alpha}_0, \hat{\alpha}_1, \ldots$, we can use Ω to modulate the amount that $R_{1-i,e',k'}$ puts into β^i. More specifically, at a stage s, if $R_{1-i,e',k'}$'s current quota is 2^{-m} then it puts into β^i as much of $\hat{\alpha}_s - \hat{\alpha}_{s-1}$ as possible, subject to the constraint that the total amount put into β^i by $R_{1-i,e',k'}$ since the last stage before stage s at which $R_{1-i,e',k'}$ was initialized must not exceed $2^{-m}\Omega_s$. It can be shown that the fact that $\Omega >_s \alpha$ implies that there is a stage v after which $R_{1-i,e',k'}$ is allowed to put all of $\hat{\alpha} - \hat{\alpha}_v$ into β^i.

In general, at a given stage s there will be several requirements, each with a certain amount that it wants (and is allowed) to direct into one of the β^j. We work backwards, starting with the weakest priority requirement that we are currently considering. This requirement is allowed to direct as much of $\hat{\alpha}_s - \hat{\alpha}_{s-1}$ as it wants (subject to its current quota, of course). If any of $\hat{\alpha}_s - \hat{\alpha}_{s-1}$ is left then the next weakest priority strategy is allowed to act, and so on up the line. □

It is also possible to show that the Solovay degrees are upwards dense, that is, that if $\gamma <_s \Omega$ is a c.e. real then there is a c.e. real β such that $\gamma <_s \beta <_s \Omega$. We omit the proof in the interest of space. Together with the previous theorem, this result implies that the Solovay degrees are dense.

Theorem 2.3. *The Solovay degrees of c.e. reals are dense.*

We finish by sketching a proof that the hypothesis that $\alpha <_s \Omega$ in the statement of Theorem 2.2 is necessary. This fact will follow easily from a stronger result which shows that, despite the upwards density of the Solovay degrees, there is a sense in which the complete Solovay degree is very much above all other Solovay degrees. We begin by noting the following lemma, which gives a useful sufficient condition for domination.

Lemma 2.4. *Let f be an increasing total computable function and let $k > 0$ be a natural number. Let α and β be c.e. reals for which there are infinitely many $s \in \omega$ such that $k(\alpha - \alpha_s) > \beta - \beta_{f(s)}$, but only finitely many $s \in \omega$ such that $k(\alpha_t - \alpha_s) > \beta_{f(t)} - \beta_{f(s)}$ for all $t > s$. Then $\beta \leqslant_s \alpha$.*

Theorem 2.5. *Let α and β be c.e. reals, let f be an increasing total computable function, and let $k > 0$ be a natural number. If β is random and there are infinitely many $s \in \omega$ such that $k(\alpha - \alpha_s) > \beta - \beta_{f(s)}$ then α is random.*

Proof Sketch. By taking $\beta_{f(0)}, \beta_{f(1)}, \ldots$ instead of $\beta_0, \beta_1, \ldots$ as an approximating sequence for β, we may assume that f is the identity. If α is rational then we can replace it with a nonrational computable real α' such that $\alpha' - \alpha'_s \geqslant \alpha - \alpha_s$ for all $s \in \omega$, so we may assume that α is not rational.

We assume that α is nonrandom and there are infinitely many $s \in \omega$ such that $k(\alpha - \alpha_s) > \beta - \beta_s$, and show that β is nonrandom. The idea is to take a Solovay test $A = \{I_i : i \in \omega\}$ such that $\alpha \in I_i$ for infinitely many $i \in \omega$ and use it to build a Solovay test $B = \{J_i : i \in \omega\}$ such that $\beta \in J_i$ for infinitely many $i \in \omega$.

Let

$$U = \{s \in \omega : k(\alpha - \alpha_s) > \beta - \beta_s\} \ .$$

It is not hard to show that U is Δ_2^0, except in the trivial case in which $\beta \equiv_s \alpha$. Thus a first attempt at building B could be to run the following procedure for all $i \in \omega$ in parallel. Look for the least t such that there is an $s < t$ with $s \in U[t]$ and $\alpha_s \in I_i$. If there is more than one number s with this property then choose the least among such numbers. Begin to add the intervals

$$[\beta_s, \beta_s + k(\alpha_{s+1} - \alpha_s)], [\beta_s + k(\alpha_{s+1} - \alpha_s), \beta_s + k(\alpha_{s+2} - \alpha_s)], \ldots \quad (*)$$

to B, continuing to do so as long as s remains in U and the approximation of α remains in I_i. If the approximation of α leaves I_i then end the procedure. If s leaves U, say at stage u, then repeat the procedure (only considering $t \geqslant u$, of course).

If $\alpha \in I_i$ then the variable s in the above procedure eventually assumes a value in U. For this value, $k(\alpha - \alpha_s) > \beta - \beta_s$, from which it follows that $k(\alpha_u - \alpha_s) > \beta - \beta_s$ for some $u > s$, and hence that $\beta \in [\beta_s, \beta_s + k(\alpha_u - \alpha_s)]$. So β must be in one of the intervals $(*)$ added to B by the above procedure.

Since α is in infinitely many of the I_i, running the above procedure for all $i \in \omega$ guarantees that β is in infinitely many of the intervals in B. The problem is that we also need the sum of the lengths of the intervals in B to be finite, and the above procedure gives no control over this sum, since it could easily be the case that we start working with some s, see it leave U at some stage t (at which point we have already added to B intervals whose lengths add up to $\alpha_{t-1} - \alpha_s$), and then find that the next s with which we have to work is much smaller than

t. Since this could happen many times for each $i \in \omega$, we would have no bound on the sum of the lengths of the intervals in B.

This problem would be solved if we had an infinite computable subset T of U. For each I_i, we could look for an $s \in T$ such that $\alpha_s \in I_i$, and then begin to add the intervals $(*)$ to B, continuing to do so as long as the approximation of α remained in I_i. (Of course, in this easy setting, we could also simply add the single interval $[\beta_s, \beta_s + k|I|]$ to B.) It is not hard to check that this would guarantee that if $\alpha \in I_i$ then β is in one of the intervals added to B, while also ensuring that the sum of the lengths of these intervals is less than or equal to $k|I_i|$. Following this procedure for all $i \in \omega$ would give us the desired Solovay test B. Unless $\beta \leqslant_s \alpha$, however, there is no infinite computable $T \subseteq U$, so we use Lemma 2.4 to obtain the next best thing.

Let

$$S = \{s \in \omega : \forall t > s(k(\alpha_t - \alpha_s) > \beta_t - \beta_s)\} \ .$$

If $\beta \leqslant_s \alpha$ then β is nonrandom, so, by Lemma 2.4, we may assume that S is infinite. Furthermore, S is co-c.e. by definition, but it has the additional useful property that if a number s leaves S at stage t then so do all numbers in the interval (s, t).

To construct B, we run the following procedure P_i for all $i \in \omega$ in parallel. Note that B is a multiset, so we are allowed to add more than one copy of a given interval to B.

1. Look for an $s \in \omega$ such that $\alpha_s \in I_i$.
2. Let $t = s + 1$. If $\alpha_t \notin I_i$ then terminate the procedure.
3. If $s \notin S[t]$ then let $s = t$ and go to step 2. Otherwise, add the interval

$$[\beta_s + k(\alpha_{t-1} - \alpha_s), \beta_s + k(\alpha_t - \alpha_s)]$$

to B, increase t by one, and repeat step 3.

This concludes the construction of B. It is not hard to show that the sum of the lengths of the intervals in B is finite and that β is in infinitely many of the intervals in B.   □

Corollary 2.6. *If α and β are c.e. reals such that $\alpha + \beta$ is random then at least one of α and β is random.*

Combining Theorem 2.2 and Corollary 2.6, we have the following results, the second of which also depends on Theorem 1.4.

Theorem 2.7. *A c.e. real γ is random if and only if it cannot be written as $\alpha + \beta$ for c.e. reals $\alpha, \beta <_s \gamma$.*

Theorem 2.8. *Let $\mathbf{d}$ be a Solovay degree. The following are equivalent:*

1. *$\mathbf{d}$ is incomplete.*
2. *$\mathbf{d}$ splits.*
3. *$\mathbf{d}$ splits over any lesser Solovay degree.*

References

1. K. Ambos-Spies, K. Weihrauch, and X. Zheng, Weakly computable real numbers, to appear.
2. P. Boldi and S. Vigna, Equality is a jump, Theoret. Comput. Sci. 219 (1999) 49–64.
3. C. S. Calude, A characterization of c.e. random reals, to appear in Theoret. Comput. Sci.
4. C. S. Calude, Information and Randomness, an Algorithmic Perspective, Monographs in Theoretical Computer Science (Springer–Verlag, Berlin, 1994).
5. C. S. Calude and G. Chaitin, Randomness everywhere, Nature 400 (1999) 319–320.
6. C. S. Calude, R. Coles, P. H. Hertling, and B. Khoussainov, Degree-theoretic aspects of computably enumerable reals, in S. B. Cooper and J. K. Truss (eds.), Models and Computability (Leeds, 1997), vol. 259 of London Math. Soc. Lecture Note Ser. (Cambridge Univ. Press, Cambridge, 1999) 23–39.
7. C. S. Calude, P. H. Hertling, B. Khoussainov, and Y. Wang, Recursively enumerable reals and Chaitin Ω numbers, Centre for Discrete Mathematics and Theoretical Computer Science Research Report Series 59, University of Auckland (October 1997), extended abstract in STACS 98, Lecture Notes in Comput. Sci. 1373, Springer, Berlin, 1998, 596–606.
8. C. S. Calude and A. Nies, Chaitin Ω numbers and strong reducibilities, J. Univ. Comp. Sci. 3 (1998) 1162–1166.
9. G. S. Ceïtin, A pseudofundamental sequence that is not equivalent to a monotone one, Zap. Naučn. Sem. Leningrad. Otdel. Mat. Inst. Steklov. (LOMI) 20 (1971) 263–271, 290, in Russian with English summary.
10. G. J. Chaitin, A theory of program size formally identical to information theory, J. Assoc. Comput. Mach. 22 (1975) 329–340, reprinted in [13].
11. G. J. Chaitin, Algorithmic information theory, IBM J. Res. Develop. 21 (1977) 350–359, 496, reprinted in [13].
12. G. J. Chaitin, Incompleteness theorems for random reals, Adv. in Appl. Math. 8 (1987) 119–146, reprinted in [13].
13. G. J. Chaitin, Information, Randomness & Incompleteness, vol. 8 of Series in Computer Science, 2nd ed. (World Scientific, River Edge, NJ, 1990).
14. R. J. Coles, R. Downey, and G. LaForte, Randomness and reducibility I, to appear.
15. R. Downey, Computability, definability, and algebraic structures, to appear in Proceedings of the 7th Asian Logic Conference, Taiwan.
16. R. Downey and G. LaForte, Presentations of computably enumerable reals, to appear.
17. C. Ho, Relatively recursive reals and real functions, Theoret. Comput. Sci. 210 (1999) 99–120.
18. K.-I. Ko, On the definition of some complexity classes of real numbers, Math. Systems Theory 16 (1983) 95–100.
19. K.-I. Ko, On the continued fraction representation of computable real numbers, Theoret. Comput. Sci. 47 (1986) 299–313.
20. A. N. Kolmogorov, Three approaches to the quantitative definition of information, Internat. J. Comput. Math. 2 (1968) 157–168.
21. A. H. Lachlan, Recursive real numbers, J. Symbolic Logic 28 (1963) 1–16.
22. L. Levin, On the notion of a random sequence, Soviet Math. Dokl. 14 (1973) 1413–1416.
23. L. Levin, The various measures of the complexity of finite objects (an axiomatic description), Soviet Math. Dokl. 17 (1976) 522–526.

24. M. Li and P. Vitanyi, An Introduction to Kolmogorov Complexity and its Applications, 2nd ed. (Springer–Verlag, New York, 1997).

25. P. Martin-Löf, The definition of random sequences, Inform. and Control 9 (1966) 602–619.

26. H. Rice, Recursive real numbers, Proc. Amer. Math. Soc. 5 (1954) 784–791.

27. C.-P. Schnorr, Process complexity and effective random tests, J. Comput. System Sci. 7 (1973) 376–388.

28. T. Slaman, Randomness and recursive enumerability, to appear.

29. R. Soare, Cohesive sets and recursively enumerable Dedekind cuts, Pacific J. Math. 31 (1969) 215–231.

30. R. J. Solomonoff, A formal theory of inductive inference I, Inform. and Control 7 (1964) 1–22.

31. R. J. Solomonoff, A formal theory of inductive inference II, Inform. and Control 7 (1964) 224–254.

32. R. M. Solovay, Draft of a paper (or series of papers) on Chaitin's work ... done for the most part during the period of Sept.–Dec. 1974 (May 1975), unpublished manuscript, IBM Thomas J. Watson Research Center, Yorktown Heights, NY, 215 pages.

33. R. M. Solovay, On random r.e. sets, in A. Arruda, N. Da Costa, and R. Chuaqui (eds.), Non-Classical Logics, Model Theory, and Computability. Proceedings of the Third Latin-American Symposium on Mathematical Logic, Campinas, July 11–17, 1976, vol. 89 of Studies in Logic and the Foundations of Mathematics (North-Holland, Amsterdam, 1977) 283–307.

On Multipartition Communication Complexity[*]
(Extended Abstract)

Pavol Ďuriš[1,2], Juraj Hromkovič[1], Stasys Jukna[3],
Martin Sauerhoff[4], and Georg Schnitger[3]

[1] Lehrstuhl für Informatik I, RWTH Aachen
Ahornstraße 55, 52074 Aachen, Germany
jh@i1.informatik.rwth-aachen.de
[2] Department of Informatics, Comenius University
Mlynska dolina, 84215 Bratislava, Slovakia
[3] Fachbereich Informatik, Johann Wolfgang Goethe-Universität Frankfurt
Robert-Mayer-Straße 11–15, 60054 Frankfurt am Main, Germany
{georg,jukna}@thi.informatik.uni-frankfurt.de
[4] FB Informatik, LS 2, Universität Dortmund
44221 Dortmund, Germany
sauerhoff@ls2.cs.uni-dortmund.de

Abstract. We study *k-partition communication protocols*, an extension of the standard two-party best-partition model to k input partitions. The main results are as follows.

1. A strong explicit *hierarchy* on the degree of non-obliviousness is established by proving that, using $k+1$ partitions instead of k may decrease the communication complexity from $\Theta(n)$ to $\Theta(\log k)$.
2. Certain linear codes are hard for k-partition protocols even when k may be exponentially large (in the input size). On the other hand, one can show that all characteristic functions of linear codes are *easy* for randomized OBDDs.
3. It is proven that there are subfunctions of the *triangle-freeness function* and the function $\oplus \text{CLIQUE}_{n,3}$ which are hard for multipartition protocols. As an application, truly exponential lower bounds on the size of nondeterministic read-once branching programs for these functions are obtained, solving an open problem of Razborov [17].

1 Introduction

The communication complexity of two-party protocols was introduced by Yao [18]. The initial goal was to develop a method for proving lower bounds on the complexity of distributed and parallel computations. In the meantime, communication complexity has been successfully applied as a tool for proving lower bounds in various other models of computation (see, e. g., [7, 12] for a survey).

Let $f\colon \{0,1\}^n \to \{0,1\}$ be a Boolean function defined on a set X of n Boolean variables, and let $\Pi = (X_1, X_2)$ be a balanced partition of X, i. e., a partition with $-1 \leqslant |X_1| - |X_2| \leqslant 1$.

[*] The work of the first and second author has been supported by DFG grant Hr 14/3-2, and of the fourth author by DFG grant We 1066/9-1.

A. Ferreira and H. Reichel (Eds.): STACS 2001, LNCS 2010, pp. 206–217, 2001.

A *deterministic two-party communication protocol P for f according to Π* is an algorithm by which two players, called Alice and Bob, can evaluate f as follows. At the beginning of the computation, Alice obtains an input $x\colon X_1 \to \{0,1\}$ and Bob an input $y\colon X_2 \to \{0,1\}$. Then the players communicate according to P by exchanging messages. The players may use unbounded resources to compute their messages. At the end, one of them has to output $f(x,y)$. A *nondeterministic protocol* allows each player to access a (private) string of *nondeterministic bits* as an additional input. It is required that there is an assignment to the nondeterministic bits such that the protocol outputs 1 if and only if $f(x,y) = 1$.

The *complexity of a nondeterministic protocol P* is the maximum of the number of exchanged bits taken over all inputs, including the nondeterministic bits. The *nondeterministic communication complexity of f according to Π*, $ncc\,(f,\Pi)$, is the minimum complexity of a nondeterministic protocol according to Π which computes f. Finally, the *(best-partition) nondeterministic communication complexity of f*, $ncc\,(f)$, is defined as the minimum of $ncc\,(f,\Pi)$ over all balanced partitions Π of the set of input variables of f.

A protocol is *oblivious* because it uses only one partition of the set of input variables for all inputs. Most applications of communication complexity are therefore restricted to oblivious models of computation. However, Borodin, Razborov, and Smolensky [5] succeeded in deriving exponential lower bounds for the non-oblivious model of computation of (syntactic) read-k-times branching programs. Their approach leads, from the perspective of communication protocols, to the following notion of *multipartition communication protocols* [8]:

Definition 1. Let f be a Boolean function defined on a set X of Boolean variables, and let k be a positive integer. A *k-partition protocol P for f* is a collection of k nondeterministic (sub-)protocols $P_1, \ldots, P_k$, each P_i with its own balanced partition of X, such that $f = P_1 \vee P_2 \vee \cdots \vee P_k$, where we use P_i also to denote the function computed by protocol P_i. If m_i is the number of submatrices of P_i (i.e., m_i is the number of leaves in the protocol tree of P_i), then the *complexity of P* is $\lceil \log(\sum_{i=1}^{k} m_i) \rceil$. The *k-partition communication complexity of f*, $k\text{-}pcc\,(f)$, is the minimum complexity of a k-partition protocol computing f. The *multipartition communication complexity of f* is $mpcc\,(f) := \min\{k\text{-}pcc\,(f) \mid k \in \mathbb{N}\}$.

To better understand the model of multipartition communication, we compare $mpcc\,(f)$ with the best-partition nondeterministic communication complexity $ncc\,(f)$. Let $f\colon \{0,1\}^n \to \{0,1\}$ be a Boolean function, $A \subseteq f^{-1}(1)$, and let Π be a partition of the variables of f. Define the distribution μ_A on $\{0,1\}^n$ by $\mu_A(x) := |A|^{-1}$ if $x \in A$, and $\mu_A(x) := 0$ otherwise. Define $B_{A,\Pi}^1(f) := \log(1/\max_M \mu_A(M))$, where the maximum extends over all all-1 submatrices M of the communication matrix of f according to Π.

We have $ncc\,(f,\Pi) = \max_{A \subseteq f^{-1}(1)} B_{A,\Pi}^1(f) + O(\log n)$ by the proof of Theorem 2.16 in [12], and consequently $ncc\,(f) = \min_\Pi \max_{A \subseteq f^{-1}(1)} B_{A,\Pi}^1(f) + O(\log n)$, where the minimum extends over all balanced partitions Π of the variables of f. A similar argument yields:

Lemma 1. *For every Boolean function* $f\colon \{0,1\}^n \to \{0,1\}$,

$$mpcc(f) = \max_{A \subseteq f^{-1}(1)} \min_{\Pi} B^1_{A,\Pi}(f) + O(\log n).$$

When dealing with multipartition communication complexity, the notion of rectangles as introduced by Borodin, Razborov, and Smolensky [5] is useful. Let X be a set of n variables and let $\Pi = (X_1, X_2)$ be a balanced partition of X. A function $r\colon \{0,1\}^n \to \{0,1\}$ defined on X is called a *rectangle (with respect to Π)* if it can be written as $r = r^1 \wedge r^2$, where the functions r^i depend only on variables from X_i, $i = 1, 2$. Given a Boolean function f defined on X, its *rectangle complexity* $R(f)$ is the minimal number t for which there exist t rectangles $r_1, r_2, \ldots, r_t$ (each with its own partition of the variables in X) such that $f = r_1 \vee r_2 \vee \cdots \vee r_t$. The *$k$-partition rectangle complexity* $R_k(f)$ of f is the minimal number of rectangles needed to cover f under the restriction that these rectangles may use at most k different partitions. Note that

$$R_k(f) = \min_{f_1, f_2, \ldots, f_k} R_1(f_1) + R_1(f_2) + \cdots + R_1(f_k),$$

where the minimum is taken over all k-tuples of Boolean functions $f_1, f_2, \ldots, f_k$ with $f_1 \vee f_2 \vee \cdots \vee f_k = f$. Furthermore, $R(f) = \min_k R_k(f)$. We obtain:

Proposition 1. *For all Boolean functions f,*

$$\lceil \log R_k(f) \rceil = k\text{-}pcc(f), \quad \text{and} \quad \lceil \log R(f) \rceil = mpcc(f).$$

The measure $R(f)$ can also be used to prove lower bounds on the size of nondeterministic read-once branching programs (1-n.b.p. for short): Borodin, Razborov, and Smolensky [5] have shown that every Boolean function f requires a 1-n.b.p. of size at least $R(f)^{1/4}$. In fact this lower bound is $R(f)/(2n)$ for n-input functions f due to an observation of Okolnishnikova [15].

The goal of this paper is to develop lower bounds for the fundamental measures $mpcc(f)$ and $R(f)$, resp., and apply these results to branching programs. In the following, we give an overview on the paper.

1. In [8], an exponential gap between $ncc(f) = 1\text{-}pcc(f)$ and $2\text{-}pcc(f)$ has been shown. In Section 2 (Theorem 1), we prove that for infinitely many n and for all $k = k(n)$, there is an explicitly defined function $f_{k,n}\colon \{0,1\}^n \to \{0,1\}$ such that,

$$k\text{-}pcc(f_{k,n}) = \Omega(n), \quad \text{and} \quad (k+1)\text{-}pcc(f_{k,n}) = O(\log k).$$

Thus, a small increase of the degree of non-obliviousness can result in an unbounded decrease of communication complexity.

2. In Section 3, we observe that an argument from [9, 15] yields a *linear* lower bound on the multipartition communication complexity of the characteristic function of a *random* linear code. Moreover, $mpcc(\mathrm{BCH}_n) \geqslant \log R(\mathrm{BCH}_n) = \Omega(n^{1/2})$ for the characteristic function of BCH-code of length n and designed distance $d = 2t + 1$ with $t \approx n^{1/2}$ (Theorem 2).

On the other hand, we prove that the complement of a linear code can be computed by small *randomized OBDDs* with arbitrarily small one-sided error (Theorem 3). Thus we obtain the apparently best known tradeoff between randomized and nondeterministic branching program complexity.

3. In Section 4, we consider the problem of determining whether a given graph has no triangles. The corresponding *triangle-freeness function* Δ_n has $n = \binom{m}{2}$ Boolean variables (one for each potential edge) and accepts a given graph G on m vertices if and only if G has no triangles. We prove that there is a subfunction Δ'_n of Δ_n with $R(\Delta'_n) = 2^{\Omega(n)}$ (Theorem 4).

Although this result does not imply a lower bound on the rectangle complexity (and thus the multipartition complexity) of the triangle-freeness function Δ_n itself, the result has an interesting consequence for nondeterministic read-once branching programs. Razborov ([17], Problem 11) asks whether a truly exponential lower bound holds for the function $\oplus\,\mathrm{CLIQUE}_{n,3}$ on $n = \binom{m}{2}$ variables which outputs the parity of the number of triangles in a graph on m vertices. In the case of *deterministic* read-once branching programs, such a lower bound for $\oplus\,\mathrm{CLIQUE}_{n,3}$ has been proven by Ajtai *et al.* in [2]. We solve this problem by proving that nondeterministic read-once branching programs for $\oplus\,\mathrm{CLIQUE}_{n,3}$ and for the triangle-freeness function Δ_n require size at least $2^{\Omega(n)}$. The only other truly exponential lower bounds for nondeterministic read-once programs have been proven for a class of functions based on quadratic forms in [3–5]. In the deterministic case, the recent celebrated result of Ajtai [1] gives a truly exponential lower bound for a function similar to $\oplus\,\mathrm{CLIQUE}_{n,3}$ even for linear time branching programs.

2 A Strong Hierarchy on the Degree of Non-obliviousness

The goal of this section is to prove that allowing one more partition of the input variables can lead to an unbounded decrease of the communication complexity for explicitly defined functions.

Theorem 1. *For infinitely many n and all $k = k(n)$, there is an explicitly defined function $f_{k,n} : \{0,1\}^n \to \{0,1\}$ such that,*

$$k\text{-pcc}\,(f_{k,n}) = \Omega(n), \quad \text{and} \quad (k+1)\text{-pcc}\,(f_{k,n}) = O(\log k).$$

Furthermore, the upper bound can even be achieved by using $(k+1)$-partition protocols where each protocol is deterministic.

We describe how the functions used in the proof of Theorem 1 are constructed. The idea is to take some function h which is known to be "hard" even if *arbitrarily* many partitions are allowed. From h, a new function f_k is constructed which will be "easy" for $(k + 1)$-partition protocols, but "hard" for k-partition protocols.

For $h : \{0,1\}^m \to \{0,1\}$, the respective function f_k is defined on vectors of variables $x = (x_1, \ldots, x_{2m})$, $y = (y_0, \ldots, y_{\ell-1})$, and $z = (z_0, \ldots, z_{\ell-1})$, where $\ell := \lceil \log(k+1) \rceil$. We use a fixed set $\mathcal{P} = \{\Pi_1^*, \ldots, \Pi_{k+1}^*\}$ of balanced partitions of the x-variables (described later on). For a given value i from $\{1, \ldots, k+1\}$ represented by the y-variables, the vector x is divided into two halves $x^1(i)$, $x^2(i)$ of length m according to the partition Π_i^*. The function f_k is defined by $f_k(x, y, z) := h(x^1(i))$. (Observe that the z-variables are only used for "padding" the input.)

It is obvious that f_k has $(k+1)$-partition protocols of small complexity:

Proof of Theorem 1 – Upper Bound. The protocol for f_k uses $k+1$ partitions which divide the x-vector according to the partitions in $\mathcal{P}$, and which give all y-variables to the first player and all z-variables to the second player. In the ith subprotocol, the first player outputs $h\big(x^1(i)\big)$ if i is the value represented by the y-variables, and 0 otherwise. The second player does nothing. The complexity of the whole protocol is obviously $\lceil \log(2(k+1)) \rceil = \lceil \log(k+1) \rceil + 1$. □

In the following, we can only give an outline of the proof of the lower bound. We first describe the main combinatorial idea. If we can ensure that all the sets occurring as halves of partitions in $\mathcal{P}$ (where $|\mathcal{P}| = k+1$) are "very different," then the partitions in $\mathcal{P}$ cannot be "approximated" by only k partitions, as the following lemma shows.

Lemma 2. *Define the (Hamming) distance between two sets $A, B \subseteq \{1,\dots,n\}$ by $d(A,B) := |A \cap \overline{B}| + |\overline{A} \cap B|$. Let $\mathcal{A}$ and $\mathcal{B}$ be families of subsets of $\{1,\dots,n\}$ with $D \leqslant d(A,A') \leqslant n - D$ for all different $A, A' \in \mathcal{A}$ and $D/2 \leqslant |B| \leqslant n - D/2$ for all $B \in \mathcal{B}$. If $|\mathcal{A}| \geqslant |\mathcal{B}| + 1$ then there exists an $A_0 \in \mathcal{A}$ and $S \in \{A_0, \overline{A_0}\}$ such that $D/4 \leqslant |S \cap B| \leqslant n - D/4$ for all $B \in \mathcal{B}$.*

Proof. We first show that there is an $A_0 \in \mathcal{A}$ such that $D/2 \leqslant d(A_0, B) \leqslant n - D/2$ for all $B \in \mathcal{B}$. Assume to the contrary that for each $A \in \mathcal{A}$ there is a $B \in \mathcal{B}$ such that $d(A,B) < D/2$ or $d(\overline{A}, B) = n - d(A,B) < D/2$. Since $|\mathcal{A}| \geqslant |\mathcal{B}| + 1$, the pigeonhole principle implies that there exists $B \in \mathcal{B}$ such that $d(S_1, B) < D/2$ and $d(S_2, B) < D/2$ for some $S_1 \in \{A_1, \overline{A_1}\}$, $S_2 \in \{A_2, \overline{A_2}\}$ and $A_1 \neq A_2 \in \mathcal{A}$. But then $d(S_1, S_2) \leqslant d(S_1, B) + d(B, S_2) < D$, a contradiction.

Now fix some $B \in \mathcal{B}$. Define the real-valued 2×2 matrix $M = (m_{rs})$ by setting $m_{11} := |A_0 \cap B|$, $m_{12} := |\overline{A_0} \cap B|$, $m_{21} := |A_0 \cap \overline{B}|$, and $m_{22} := |\overline{A_0} \cap \overline{B}|$. We have $m_{11} + m_{12} = |B| \geqslant D/2$ and $m_{21} + m_{22} = |\overline{B}| \geqslant D/2$. Furthermore, $m_{11} + m_{22} = d(\overline{A_0}, B) \geqslant D/2$ and $m_{12} + m_{21} = d(A_0, B) \geqslant D/2$. It follows that there is at least one column of M for which both elements are at least $D/4$. □

In order to meet the requirements of Lemma 2, we choose $\mathcal{P}$ such that the characteristic vectors of the Π_i^* form a code $C \subseteq \{0,1\}^{2m}$ with the following properties: (i) All $x \in C$ have exactly m ones and m zeros, i.e., C is a so-called *balanced code*. (ii) Any two different codewords have Hamming distance at least $D = 2\delta m$ and at most $2m - D = 2(1-\delta)m$, $\delta > 0$ a constant. To construct a code with these properties and exponentially many codewords, we start with a Justesen code (see, e. g., [13]), which is a linear code with appropriate lower *and* upper bounds on the weight of its codewords, and then "balance" the codewords by "padding."

Let $\Pi_i^* = (\Pi_{i,1}^*, \Pi_{i,2}^*)$, for $i = 1, \dots, k+1$. Let $\Pi_i = (\Pi_{i,1}, \Pi_{i,1})$, for $i = 1, \dots, k$, be arbitrary balanced partitions. We apply Lemma 2 to $\mathcal{A} = \{\Pi_{i,1}^* \mid i = 1, \dots, k+1\}$ and $\mathcal{B} = \{X \cap \Pi_{i,1} \mid i = 1, \dots, k\}$, where $X = \{x_1, \dots, x_{2m}\}$. This yields an index i_0 such that at least one half of the partition $\Pi_{i_0}^*$ has at least $D/4$ variables on both sides of all partitions Π_i, $i = 1, \dots, k$. It is now easy to prove the following.

Lemma 3. *Let $\beta := D/(4m) = \delta/2$. There are partitions $\Pi_1', \dots, \Pi_k'$ of the variables of h which are β-balanced, i. e. $|\Pi_{i,1}'|, |\Pi_{i,2}'| \geqslant \lfloor \beta m \rfloor$ for $i = 1, \dots, k$, and a k-partition protocol for h with these partitions which has complexity at most k-pcc (f_k).*

To obtain the desired lower bound for f_k, we require an explicitly defined function h which has large multipartition complexity even if the given partitions are only β-balanced for some small constant $\beta > 0$. A linear lower bound of this type is contained, e. g., in the results of Beame, Saks, and Thathachar ([4], Lemma 4) or in [11].

3   The Multipartition Communication Complexity of Linear Codes

A (binary) code of length n and distance d is a subset of vectors $C \subseteq \{0,1\}^n$ for which the Hamming distance between any two vectors in C is at least d. The following lemma is implicit in [9, 15], where a stronger version has been used to show that linear codes are hard for read-k-times branching programs:

Lemma 4 ([9,15]). *Let $C \subseteq \{0,1\}^n$ be a code of distance $2t + 1$. Let P be a multipartition protocol computing the characteristic function of C. Then P uses at least* $\log\left(|C| \cdot \binom{\lfloor n/2 \rfloor}{t}^2 \cdot 2^{-n}\right)$ *bits of communication.*

The number of codewords and the distance of *random* linear codes are known to meet the Gilbert-Varshamov bound [13]. As a consequence, the above lemma gives *linear* lower bounds for the characteristic functions of such codes. To give a constructive example, we consider binary BCH-codes with length $n = 2^m - 1$ and designed distance $d = 2t+1$; such a code has at least $2^n/(n+1)^t$ vectors and distance at least d. Let BCH_n be the characteristic function of such a BCH code with $t \approx n^{1/2}$. Using Lemma 4, we obtain:

Theorem 2. *Each multipartition protocol for BCH_n has complexity at least $\Omega(n^{1/2})$.*

On the other hand, all linear codes have small randomized communication complexity even in the fixed-partition model (we omit the easy proof):

Proposition 2. *Let f_C be a characteristic function of a linear binary code of length n. Then the two-party fixed-partition one-round bounded error communication complexity of f_C is $O(1)$ with public coins and $O(\log n)$ with private coins.*

The characteristic functions f_C of linear codes are known to be hard for different models of branching programs, including k-n.b.p.'s – nondeterministic read-k-times branching programs where along any path no variable appears more than k times [9], and $(1, +k)$-b.p.'s – deterministic branching programs where along each *consistent* path at most k variables are allowed to be tested more than once [10]. On the other hand, the negation $\neg f_C$ is just an OR of at most n scalar products of an input vector with the rows of the corresponding parity-check matrix. Hence, for every linear code, the characteristic function $\neg f_C$ of its complement has a small *nondeterministic OBDD* (an OBDD is a read-once branching program where the variables along every path appear according to a fixed order). We can strengthen this observation even to *randomized OBDDs with one-sided error.*

Theorem 3. *Let $C \subseteq \{0,1\}^n$ be a linear code and let f_C be its characteristic function. Then, for every integer $r \geqslant 2$, $\neg f_C$ can be computed by a randomized OBDD of size $O(n^{4r})$ with one-sided error at most 2^{-r}.*

Sketch of Proof. Let H be the $m \times n$ parity-check matrix of C. Let $\mathbf{w}$ be chosen uniformly at random from $\{0, 1\}^n$. The essence of the construction is the simple fact that $\mathbf{w}^\top H x \equiv 0 \bmod 2$ for $x \in C$, whereas Prob $[\mathbf{w}^\top H x \not\equiv 0 \bmod 2] = 1/2$ for $x \notin C$. We cannot use this representation of f_C directly to construct a randomized OBDD, since this OBDD would require exponentially many probabilistic nodes to randomly choose the vector $\mathbf{w}$.

To reduce the number of random bits, we apply an idea which has appeared in different disguises in several papers (see, e. g., Newman [14]): By a probabilistic argument it follows that, for all δ with $0 < \delta < 1/2$, there is a set $W \subseteq \{0, 1\}^n$ with $|W| = O(n/\delta^2)$ such that for $\mathbf{w}$ chosen uniformly at random from W and all $x \notin C$, Prob $[\mathbf{w}^\top H x \not\equiv 0 \bmod 2] \geqslant 1/2 - \delta$. Choose $\delta = 1/5$ and let W be the obtained set of vectors.

Let G be the randomized OBDD which starts with a tree on $\lceil \log |W| \rceil$ probabilistic variables at the top by which an element $w \in W$ is chosen uniformly at random. At the leaf of the tree belonging to the vector w, append a deterministic sub-OBDD which checks whether $w^\top H x \equiv 0 \bmod 2$. By the above facts, this randomized OBDD computes $\neg f_C$ with one-sided error at most $7/10$. The size of G is bounded by $O(n^2)$.

To decrease the error probability, we regard G as a deterministic OBDD on all variables (deterministic and probabilistic ones). Applying the known OBDD-algorithms, we obtain an OBDD G' for the OR of $2r$ copies of G with different sets of probabilistic variables. This OBDD G' has one-sided error at most $(7/10)^{2r} < 2^{-r}$ and size $O(n^{4r})$. □

Apparently, this result gives the strongest known tradeoff between nondeterministic and randomized branching program complexity.

4   A Lower Bound for Triangle-Freeness

The *triangle-freeness function* Δ_n is a function on $n = \binom{m}{2}$ Boolean variables (encoding the edges on an m-vertex graph) which, given a graph G on m vertices, accepts it if and only if G has no triangles. The function $\oplus \mathrm{CLIQUE}_{n,3}$ has the same set of variables and outputs the parity of the number of triangles in G.

Theorem 4. *There is a subfunction* Δ_n' *of* Δ_n *such that* $R(\Delta_n') = 2^{\Omega(n)}$. *The same holds also for* $\oplus \mathrm{CLIQUE}_{n,3}$.

This result is sufficient to prove that each nondeterministic read-once branching program detecting the triangle-freeness of a graph requires truly exponential size. Since by assigning constants to some variables, we can only decrease the branching program size, the desired lower bound on the size of any 1-n.b.p. computing Δ_n follows directly from Theorem 4 and the fact that each Boolean function f on n variables requires a 1-n.b.p. of size at least $R(f)/(2n)$ (as mentioned in the introduction). We obtain the following main result which also answers Problem 11 of Razborov from [17].

Theorem 5. *Nondeterministic read-once branching programs for the triangle-freeness function* Δ_n *as well as for* $\oplus \mathrm{CLIQUE}_{n,3}$ *require size* $2^{\Omega(n)}$.

Remark. Using a similar probabilistic argument, the following has recently been proven in [11]: (i) $R(\Delta_n) = 2^{\Omega(n^{3/4})}$; (ii) $R_k(\Delta_n) = 2^{\Omega(n)}$ provided $k \leqslant 2^{c\sqrt{n}}$ for a sufficiently small constant $c > 0$; and (iii) there is a constant $C > 0$ such that syntactic nondeterministic read-k-times branching programs, detecting the absence of 4-cliques in a graph on m vertices, require size at least $2^{\Omega(m^2/C^k)}$. Moreover, it is shown that Theorem 4 remains true also for β-balanced partitions, for all constants β with $0 < \beta \leqslant 1/2$.

4.1   Outline of the Proof of Theorem 4

We give the details only for Δ_n and discuss the changes required for $\oplus \text{CLIQUE}_{n,3}$ at the end of this section. To define the desired subfunction of Δ_n, we consider graphs on m vertices partitioned into sets $U = \{1, \ldots, m/2\}$ and $V = \{m/2 + 1, \ldots, m\}$. The subfunction Δ'_n will depend only on variables corresponding to the edges in the bipartite graph $U \times V$; the variables corresponding to the edges within the parts U and V will be fixed. Hence, Δ'_n will still have $m^2/4$ variables.

The proof consists essentially of two parts: First, we probabilistically construct an assignment which fixes the subgraphs G_U and G_V on the vertex sets U and V. After fixing these graphs, we obtain a subfunction Δ'_n of Δ_n which depends only on variables belonging to edges in the bipartite graph $G_B = U \times V$. We then consider only those partitions Π which are balanced with respect to the bipartite (non-fixed) part. Our goal is to choose the graphs G_U and G_V such that none of them contains a triangle and the resulting graph $G = G_U \cup G_V \cup G_B$ contains many triangles whose bipartite edges belong to different halves of a partition.

A pair of edges in $U \times V$ is called a *test*, if they form a triangle together with an edge from G_U or G_V. Two tests are said to *collide*, if a triangle can be formed by picking one edge from the first test, one edge from the second test and an edge from $G_U \cup G_V$. In particular, tests collide if they share an edge.

Given a balanced partition $\Pi = (E_1, E_2)$ of the edges in $U \times V$, say that a test is *hard for* Π, if each part E_i of the partition contains one edge of the test. The following lemma about graph partitions is the core of our argument.

Lemma 5. Let $\Pi_1, \ldots, \Pi_k$ be $k \leqslant 2^{\alpha m^2}$ balanced partitions of $U \times V$, where $\alpha > 0$ is a sufficiently small constant. Then there exist triangle-free graphs G_U and G_V such that the resulting graph $G = G_U \cup G_V \cup G_B$ has a set T of tests such that T does not contain any colliding pairs, and T contains a subset T_i of $\Omega(m^2)$ hard tests for each Π_i, $i = 1, \ldots, k$.

Let us first show how this lemma implies the theorem; we will then sketch the proof of the lemma itself.

Choose G_U and G_V according to the lemma and let Δ'_n be the resulting subfunction on $U \times V$. We construct a set A of hard inputs for Δ'_n which will already require many rectangles to be covered. Edge variables outside of T are fixed to 0 for all inputs in A. For each test in T, we then choose exactly one edge and set the respective variable to 1, the second one is set to 0. Thus, the graph corresponding to an input in A has precisely one of the two edges of each test in T, and two graphs differ only on edges in T. Since no two tests in T collide, the graphs are triangle-free and we obtain a total of $2^{|T|}$ graphs. Hence, $|A| = 2^{|T|}$.

Now let functions $f_1, \ldots, f_k$ be given with $\Delta'_n = f_1 \vee \cdots \vee f_k$, $k \leqslant 2^{\alpha m^2}$, and $\sum_{i=1}^{k} R_1(f_i) = R_k(\Delta'_n)$, and let $\Pi_1, \ldots, \Pi_k$ be the partitions corresponding to optimal covers of $f_1, \ldots, f_k$ by rectangles. Then there is at least one function f_i with $|f_i^{-1}(1) \cap A| \geqslant |A|/k = 2^{|T|}/k$. By Lemma 5, there is a set $T_i \subseteq T$ of $h = \Omega(m^2)$ tests which are hard for the partition Π_i. Let $B \subseteq f_i^{-1}(1) \cap A$ be a set of maximum size such that two different inputs from B differ in at least one bit corresponding to a test in T_i. Then $|B| \geqslant |f_i^{-1}(1) \cap A|/2^{|T|-h} \geqslant 2^h/k$.

Since all the inputs from B are accepted by f_i, it remains to show that no rectangle $r \leqslant f_i$ with the underlying partition Π_i can accept more than one input from B. Assume that (a, b) and (a', b') are two different inputs in B accepted by r. By the choice of B, they differ in a test $t = \{e_1, e_2\}$ which is hard for Π_i, i.e., whose edges belong to different halves of the partition Π_i. By the definition of A, exactly one of the two edges e_1 and e_2 is present in each of the graphs belonging to (a, b) and (a', b'), resp., and these edges are different. Now, if $r(a, b) = 1$, then $r(a, b') = 0$ or $r(a', b) = 0$ because either the graph corresponding to (a, b') or to (a', b) will contain *both* edges e_1, e_2, which, together with the corresponding edge of G_U or G_V, forms a triangle. This is a contradiction to the fact that r is a rectangle. Altogether, we have completed the proof of the lower bound for Δ'_n.

Changes for $\oplus \text{CLIQUE}_{n,3}$. We consider the subfunction $\oplus \text{CLIQUE}'_{n,3}$ which is obtained from $\oplus \text{CLIQUE}_{n,3}$ in same way as Δ'_n from Δ_n. Let $t := |T|$. For $x, y \in \{0,1\}^t$, define $\text{IP}_t(x, y) := \sum_{i=1}^{t} x_i y_i \bmod 2$. Define the set A of hard inputs for $\oplus \text{CLIQUE}'_{n,3}$ as follows: For all $(x, y) \in \text{IP}_t^{-1}(1)$, include the input obtained by setting variables outside of T to 0 and setting the two edge variables of the ith test in T to x_i and y_i, resp. Then $|A| = |\text{IP}_t^{-1}(1)| \geqslant 2^{2t-1}$ and $A \subseteq \oplus \text{CLIQUE}_{n,3}^{-1}(1)$.

Following the proof for Δ'_n, we obtain a set B of at least $2^{t+h-1}/k$ inputs from A which are hard for one of the partitions Π_i in a cover of $\oplus \text{CLIQUE}'_{n,3}$. Using the well-known fact that $|r^{-1}(1)| \leqslant 2^t$ for each rectangle $r \leqslant \text{IP}_t$, one easily proves that no rectangle $r' \leqslant \oplus \text{CLIQUE}'_{n,3}$ can contain more than 2^t inputs from B. Thus, at least $2^{h-1}/k$ rectangles are needed to cover B. $\qquad\square$

4.2   Sketch of Proof for Lemma 5

Recall that a test is a pair of edges in $U \times V$ which form a triangle together with an edge in G_U or G_V, and that a test is hard with respect to a partition Π if its two edges lie in different halves of Π.

Lemma 6. *There exist graphs G_U and G_V such that:*

(i)   *each of the graphs G_U and G_V has $\Theta(m)$ edges, at most $O(1)$ triangles, and at most $O(m)$ paths of length 2 or 3; and*

(ii) *for every balanced partition Π of $U \times V$, there are $h = \Omega(m^2)$ tests which are hard for Π.*

Sketch of Proof. We prove the existence of the desired graphs by a probabilistic argument. In what follows, let $\mathbf{G_U}$ ($\mathbf{G_V}$) stand for the random graph on U (resp., on V) obtained by inserting the edges independently at random with probability $p = \Theta(1/m)$ each[1]. Using Markov's inequality, it is easy to show that the graphs $\mathbf{G_U}$ and $\mathbf{G_V}$ have the properties described in Part (i) of the lemma with probability at least $1/2$. It remains to prove that, with probability larger than $1/2$, for every balanced partition of $U \times V$, there are at least $\Omega(m^2)$ hard tests.

Let Π be such a balanced partition. The partition Π distributes the edges in $U \times V$ to two sets of size $m^2/8$ each which are given to the players Alice and Bob. Call a vertex *mixed* if each of the two players have at least $\frac{1}{8} \cdot \frac{m}{2}$ bipartite edges incident to it.

Claim 1. There are $\Omega(m)$ mixed vertices in each of the sets U and V.

Proof of the Claim. We use essentially the same argument as Papadimitriou and Sipser in [16]. W. l. o. g., assume that we have at most εm mixed vertices in V, where $\varepsilon > 0$ is a sufficiently small constant ($\varepsilon < 1/112$ works fine). Call a vertex v an *A-vertex* (resp. *B-vertex*) if Alice (resp. Bob) has at least $\frac{7}{8} \cdot \frac{m}{2}$ edges incident to v. Thus, vertices which are neither A- nor B-vertices are mixed. Observe first that the number of A-vertices as well as the number of B-vertices in each of the sets U and V is at most $b_{max} := \frac{4}{7} \cdot \frac{m}{2}$, since otherwise Alice or Bob would have more than $m^2/8$ edges. On the other hand, the number of A-vertices as well as the number of B-vertices *in V* is bounded *from below* by $b_{min} := \frac{3}{7} \cdot \frac{m}{2} - \varepsilon m$, since otherwise there would be more than εm mixed vertices in V, contrary to the assumption.

Now *more* than half of the edges from A-vertices in U to B-vertices in V belong to Alice, because otherwise there will be an A-vertex $u \in U$ such that Alice has at most half of the edges from u to B-vertices in V, and thus altogether at most $\frac{1}{2} \cdot b_{max} + |V| - b_{min} = \frac{1}{2} \cdot \frac{4}{7} \cdot \frac{m}{2} + \frac{m}{2} - (\frac{3}{7} \cdot \frac{m}{2} - \varepsilon m) \leqslant \frac{6}{7} \cdot \frac{m}{2} + \varepsilon m < \frac{7}{8} \cdot \frac{m}{2}$ edges incident to u. With the same reasoning, however, *more* than half of all edges from A-vertices in U to B-vertices in V belong to Bob. Contradiction. $\qquad\square$

For each mixed vertex $u \in U$, let $V_A(u)$ ($V_B(u)$) be the set of vertices $v \in V$ for which Alice (resp. Bob) has the edge $\{u, v\}$. Since u is mixed, $|V_A(u)|, |V_B(u)| \geqslant \frac{1}{8} \cdot \frac{m}{2}$. Observe that each edge between $V_A(u)$ and $V_B(u)$ leads to a hard test with respect to the given partition Π.

Claim 2. The following event has probability larger than $1/2$ with respect to the random choices of $\mathbf{G_V}$: For all pairs of disjoint sets $S_1, S_2 \subseteq V$ of size at least $m/16$ each, the number of edges in $\mathbf{G_V}$ between S_1 and S_2 is at least $p|S_1||S_2|/2$.

Proof of the Claim. The expected number of edges between fixed sets of vertices S_1 and S_2 is $p|S_1||S_2|$. By Chernoff bounds, the true number of edges is at least $p|S_1||S_2|/2$ with probability at least $1 - e^{-cm}$, where the constant $c > 0$ can be adjusted by the choice of the constant in the definition of p. Since there are at most $\left(2^{m/2}\right)^2 = 2^m$ choices for the sets $S_1, S_2 \subseteq V$, the probability of the described event is at least $1 - 2^m \cdot e^{-cm}$, which is larger than $1/2$ for appropriate c. $\qquad\square$

---

[1] For the sake of simplicity, we omit the exact constant in the definition of p here.

We apply the claim to the sets $V_A(u)$ and $V_B(u)$, where u is a mixed vertex, generated by *all* balanced partitions Π. Due to the claim, the event that, for each partition Π and all sets $V_A(u)$ and $V_B(u)$ generated by Π, these sets are connected by at least $p|V_A(u)||V_B(u)|/2 = \Omega(m)$ edges, has probability larger than $1/2$. Thus, with probability larger than $1/2$, for each partition Π there are $\Omega(m^2)$ hard tests generated by the $\Omega(m)$ mixed vertices. This completes the proof of the lemma. (Observe that it does not matter whether we carry out the above argument for mixed vertices in U or in V.)   □

We apply Lemma 6 and fix graphs G_U and G_V with the described properties. Since there are only $O(1)$ triangles, we can remove these triangles without destroying the other properties. Especially, we still have linearly many edges. By Property (ii), this pair of graphs produces a set of $h = \Omega(m^2)$ hard tests T_i for each of the partitions Π_i ($i = 1, \ldots, k$) from a given multipartition protocol for Δ_n.

Let T_0 be the set of all tests induced by G_U and G_V, and let $t = |T_0|$ be its size. Since both graphs G_U and G_V have $\Theta(m)$ edges, $t = \Omega(m^2)$. Using the properties of these graphs stated in Lemma 6 (i), it is easy to show (by case analysis) that at most $O(t)$ of all $\binom{t}{2}$ pairs of tests in T_0 will collide:

Lemma 7. *There are at most $O(t)$ pairs of colliding tests in T_0.*

To finish the proof of Lemma 5, it remains to find a subset $T \subseteq T_0$ such that: (i) there is no pair of tests from T which collide; and (ii) $|T \cap T_i| = \Omega(m^2)$ for all $i = 1, \ldots, k$. We again use a probabilistic construction. Let $\mathbf{T}$ be a set of s tests picked uniformly at random from the set T_0, where $s = \gamma t$ and γ is a constant with $0 < \gamma < 1$ chosen later on.

Lemma 8.
(i) *With probability at least $1/2$, the set $\mathbf{T}$ contains at most $O(s^2/t)$ pairs of colliding tests (where $t = |T_0|$ is the total number of tests).*
(ii) *With probability larger than $1/2$, $|\mathbf{T} \cap T_i| \geq \frac{s \cdot h}{2t}$ for all $i = 1, \ldots, k$.*

Proof. Part (i): We define the *collision graph* to have tests as vertices and edges for each collision. Let c be the number of edges in the collision graph. By Lemma 7, we know that $c = O(t)$.

Let $\mathbf{c_T}$ be the number of edges in the subgraph of the collision graph induced by the randomly chosen set $\mathbf{T}$. Since we pick tests uniformly at random, the expected number of edges is $\mathrm{E}[\mathbf{c_T}] = \frac{s(s-1)}{t(t-1)} \cdot c$. By Markov's inequality, it follows that the actual number of edges is at most $2 \cdot \mathrm{E}[\mathbf{c_T}]$ with probability at least $1/2$. Hence, the number of pairs of colliding tests in $\mathbf{T}$ is at most $2 \cdot \mathrm{E}[\mathbf{c_T}] = O((s/t)^2 \cdot c) = O(s^2/t)$ with probability at least $1/2$.

Part (ii): Consider a fixed partition Π_i. The probability to choose a hard test from T_i is h/t, $t = \Omega(m^2)$ the total number of tests. Thus the expected number of elements in $\mathbf{T} \cap T_i$ for a randomly chosen set $\mathbf{T}$ of s tests is $s \cdot h/t$. Let $\lambda := h/(2t)$. By Chernoff bounds, it follows that $\mathrm{Prob}\left[|\mathbf{T} \cap T_i| < \lambda \cdot s\right] \leq 2e^{-\lambda^2 s} = e^{-\Omega(s)}$. Hence, the probability that $\mathbf{T}$ contains at least $\lambda \cdot s = sh/(2t)$ hard tests for each of the partitions at least $1 - k \cdot 2^{-\Omega(s)}$. Since $s = \gamma t = \Theta(m^2)$, this probability is larger than $1/2$ for $k \leq 2^{\alpha m^2}$ with $\alpha > 0$ sufficiently small.   □

Lemma 8 yields the existence of a set $T \subseteq T_0$ with the following properties: (i) $|T| = s = \gamma t$; (ii) there are at most $\delta s^2/t$ pairs of tests in T which collide, $\delta > 0$ some constant; and (iii) for all $i = 1, \ldots, k$, $|T \cap T_i| \geqslant sh/(2t)$.

By deleting at most $\delta s^2/t$ tests from T, we remove all collisions, obtaining a smaller set T'. The number of hard tests for each Π_i in T' is still $sh/(2t) - \delta s^2/t = (s/t) \cdot (h/2 - \delta s) = \gamma \cdot (h/2 - \delta \gamma t)$. Since this number is of the order $\Omega(m^2)$ for $\gamma = h/(4\delta t) = O(1)$, we have completed the proof of Lemma 5.   □

Acknowledgment

Thanks to Ingo Wegener for critical comments on an earlier version of this paper.

References

1. M. Ajtai, A non-linear time lower bound for Boolean branching programs, *Proc. of 40th FOCS*, 1999, pp. 60–70.
2. M. Ajtai, L. Babai, P. Hajnal, J. Komlos, P. Pudlák, V. Rödl, E. Szemeredi, and Gy. Turán, Two lower bounds for branching programs, in: *Proc. 18th ACM STOC*, 1986, pp. 30–38.
3. P. Beame, M. Saks, X. Sun, and E. Vee, Super-linear time-space tradeoff lower bounds for randomized computation, Technical Report **25**, *Electr. Coll. on Comp. Compl.*, 2000.
4. P. Beame, M. Saks, and J. S. Thathachar, Time-space tradeoffs for branching programs, in: *Proc. of 39th FOCS*, 1998, pp. 254–263.
5. A. Borodin, A. Razborov, and R. Smolensky, On lower bounds for read-k-times branching programs, *Computational Complexity* **3** (1993), pp. 1–18.
6. A. Hajnal, W. Maass, and G. Turán, On the communication complexity of graph properties, in: *Proc. of 20th ACM STOC*, 1988, pp. 186–191.
7. J. Hromkovič, *Communication Complexity and Parallel Computing*, EATCS Texts in Theoretical Computer Science, Springer-Verlag, 1997.
8. J. Hromkovič and M. Sauerhoff, Tradeoffs between nondeterminism and complexity for communication protocols and branching programs, in: *Proc. of STACS 2000*, LNCS 1770, pp. 145–156.
9. S. Jukna, A note on read-k-times branching programs, *RAIRO Theor. Inf. and Applications* **29**:1 (1995), pp. 75–83.
10. S. Jukna and A. Razborov, Neither reading few bits twice nor reading illegally helps much, *Discrete Appl. Math.* **85**:3 (1998), pp. 223–238.
11. S. Jukna and G. Schnitger, On the complexity of graphs which lack small cliques, manuscript.
12. E. Kushilevitz and N. Nisan, *Communication Complexity*, Cambridge University Press, 1997.
13. F. J. MacWilliams and N. J. A. Sloane, *The Theory of Error-Correcting Codes*, North-Holland, 1998.
14. I. Newman, Private vs. common random bits in communication complexity, *Information Processing Letters* **39** (1991), pp. 67–71.
15. E. A. Okol'nishnikova, On Lower Bounds for Branching Programs, *Siberian Advances in Mathematics* **3**:1 (1998), pp. 152–166.
16. Ch. H. Papadimitriou and M. Sipser, Communication complexity, *J. Comput. Syst. Sci.* **28** (1984), pp. 260–269.
17. A. Razborov, Lower bounds for deterministic and nondeterministic branching programs, in: *Proc. of FCT '91*, Lecture Notes in Computer Science **529**, Springer-Verlag 1991, pp. 47–60.
18. A. Yao, The entropic limitations of VLSI computations, in: *Proc. 13th ACM STOC* (1981), pp. 308–311.

Scalable Sparse Topologies with Small Spectrum[*]

Robert Elsässer[1], Rastislav Královič[2][**], and Burkhard Monien[1]

[1] University of Paderborn, Germany
{elsa,bm}@uni-paderborn.de
[2] Comenius University, MFF-UK Bratislava, Slovakia
kralovic@dcs.fmph.uniba.sk

Abstract. One of the fundamental properties of a graph is the number of distinct eigenvalues of its adjacency or Laplacian matrix. Determining this number is of theoretical interest and also of practical impact. Graphs with small spectra exhibit many symmetry properties and are well suited as interconnection topologies. Especially load balancing can be done on such interconnection topologies in a small number of steps. In this paper we are interested in graphs with maximal degree $O(\log n)$, where n is the number of vertices, and with a small number of distinct eigenvalues. Our goal is to find scalable families of such graphs with polylogarithmic spectrum in the number of vertices. We present also the eigenvalues of the Butterfly graph.

1 Introduction

Spectral methods in graph theory have received great attention since their introduction and have proved to be a valuable tool for the theoretical and applied graph theory [7,3]. A (Laplacian or adjacency) matrix is associated to each graph. The set of the eigenvalues of this matrix is called the (Laplacian or adjacency) spectrum of the graph; it is one of the most important algebraic invariants of a graph. Although in general a graph is not characterized uniquely by its spectrum, there is a strong connection between the eigenvalues and many structural properties of the graph (diameter, bisection width, expansion etc). See [5] for a selection of results in this area.

An important parameter connected to a spectrum of a graph is its size, i.e. the number of distinct eigenvalues of the adjacency (or Laplacian) matrix of the graph. This value is correlated to the symmetry properties of it: the only graph having two distinct eigenvalues is the complete graph and its automorphism group is as rich as possible – the symmetric group. Graphs having three distinct eigenvalues are called strongly regular; their diameter is 2 and they posses many interesting properties [16,3]. Another well studied class of highly symmetric graphs are distance-regular graphs [4]. The size of their spectrum is $1 + diam(G)$ which matches a lower bound for all graphs.

In the past there have been written several papers about well structured graphs. Consider for example the hypercube $Q(d)$ as a vertex and edge symmetric graph. It has

[*] This work was partially supported by the German Research Association (DFG) within the SFB 376 "Massive Parallelität: Algorithmen, Entwurfsmethoden, Anwendungen".

[**] The research was done while the second author was visiting the Department of Mathematics and Computer Science at the University of Paderborn.

A. Ferreira and H. Reichel (Eds.): STACS 2001, LNCS 2010, pp. 218–229, 2001.
© Springer-Verlag Berlin Heidelberg 2001

2^d vertices and a diameter resp. vertex degree of d. The hypercube has $d + 1$ distinct eigenvalues and a large application as an interconnection topology. Other graphs as cliques, complete bipartite graphs or the star have only 2 or 3 distinct eigenvalues, but because of their high density are ill-suited as interconnection topologies.

There exists some graphs with an even better relation between number of vertices, vertex degree, diameter and number of eigenvalues than the hypercube. One of them is the Petersen graph, which has 10 vertices, a vertex degree of 3, diameter 2 and 3 different eigenvalues. Another one is the $Cage(6, 6)$, which has 62 vertices, vertex degree 6, a diameter of 2 and only 3 distinct eigenvalues. A family of graphs with a very good behavior is the family of the star graphs [2,1]. The star graph of order d has $d!$ vertices, a vertex degree of $d-1$, diameter $\frac{3}{2}(d-1)$ and using a previous work of Flatto, Odlyzko and Wales it turns out that it has only $2d - 1$ distinct eigenvalues [12].

In this paper we focus our attention on constructing scalable families of sparse graphs (maximal vertex degree $O(\log n)$ where n is the number of vertices) with small spectra. We use the term scalable to denote a family of graphs, which contains for each natural n an n-vertex graph. Our motivation for studying this question comes from the area of load balancing in distributed systems. Let there be given an arbitrary, undirected, connected graph $G = (V, E)$ in which node $v \in V$ contains a load of $w(v)$. The goal is to determine a schedule to move load across edges so that finally the load on each node will be the same. In each step load can be moved from any node to its neighbors. Communication between non-adjacent vertices is not allowed. This problem describes load balancing in synchronous distributed processor networks and parallel machines when we associate a node with a processor, an edge with a communication link and the load with identical, independent tasks [6,18].

Load balancing algorithms are typically based on a fixed topology which defines the load balancing partners in the system. Consider for example a bus system where each processor can communicate with any other processor in the network. To avoid high communication costs, we allow any processor to communicate only with a small number of other nodes in the system. Then we can define a topology, which has a small vertex degree and supports fast load balancing on the network. See also [8] for a practical point of view of this problem.

Now the load balancing process can be split into two phases, the flow computation phase, which computes the network flow, and the migration phase, which migrates the load items according to the computed flow. Algorithms for the flow computation phase have been extensively studied. Many of them are local iterative schemes based on diffusion or dimension exchange [10,11,13,14,15,18]. The diffusion algorithms studied in the above mentioned papers calculate an l_2-optimal flow. We are interested in topologies, for which the optimal scheme OPT [11] (see also the next section) has a small number of iteration steps (polylogarithmic in the number of vertices). Applying the optimal scheme we need only $m - 1$ iterations where m is the number of distinct eigenvalues of the Laplacian of the graph. In any iteration step a node has to communicate with all of its neighbors, so the total cost of the load balancing algorithm depends on the number of the distinct eigenvalues of the graph and on its vertex degree. The number of steps is in fact the product of both. Therefore we are interested in topologies with a small product of the vertex degree and of the number of distinct eigenvalues.

The paper is organized as follows. In section 2 we present the definitions and lemmas used in this paper to compute the eigenvalues of the graphs constructed below. In section 3 we compute the spectrum of the Butterfly graph and propose scalable families of trees of constant degree whose spectrum consists of $O(\log^2 n)$ different eigenvalues. Since the tree topology is not well suited for our application we present a scalable family of well connected graphs with at most $O(\log^3 n)$ distinct eigenvalues and a vertex degree of $O(\log n)$. In the last section we improve the previous results for the case of the adjacency spectrum using another technique to obtain $O(\log^2 n)$ distinct eigenvalues, where the vertex-degree still remains $O(\log(n))$.

Concerning the product of the vertex degree and the number of distinct eigenvalues, the star graphs are the best graphs we know. In their case this product is $O((\frac{\log n}{\log \log n})^2)$ where n is the number of vertices. We found scalable families of graphs with a good behavior, but we did not reach this bound. We also do not know, if this bound is optimal. The only lower bound is $\Omega(\log n)$. So there are a lot of open problems which are left to be solved in this important field.

2   Definitions and Lemmas

In this paper we are interested in scalable families of graphs. A family of graphs $\mathcal{G}$ is called scalable, if for each $n \in \mathbb{N}$ there is an n-vertex graph $G \in \mathcal{G}$. The identity matrix will be denoted $\mathbf{I}_n \in \mathbb{R}^{n \times n}$. Symbols $\mathbf{J}_{m,n}, \mathbf{0}_{m,n}$ denote $m \times n$ matrices containing all ones and all zeros, respectively. The spectrum of a matrix $\mathbf{A}$ is the set of its eigenvalues: $Sp(\mathbf{A}) = \{\lambda \mid \exists \mathbf{x} : \mathbf{A}\mathbf{x} = \lambda\mathbf{x}\}$

The operation "$\otimes$" denotes the *Kronecker product*: for the matrices $\mathbf{A} \in \mathbb{R}^{m \times n}$, $\mathbf{B} \in \mathbb{R}^{p \times q}$ the matrix $\mathbf{A} \otimes \mathbf{B} \in \mathbb{R}^{mp \times nq}$ is the matrix obtained from $\mathbf{A}$ by replacing every element a_{ij} by the block $a_{ij}\mathbf{B}$.

Consider a (weighted) digraph $G = (V, E)$ with $w(e)$ being the weight of an edge e. The *adjacency matrix* of G is the matrix $\mathbf{A}_G = (a_{ij})_{1 \leq i,j \leq |V|}$ where $a_{ij} = w(e_{ij})$ if an edge e_{ij} leads from a vertex v_i to a vertex v_j and $a_{ij} = 0$ otherwise (a_{ii} is the weight of a self-loop in v_i). The *Laplacian matrix* of G is the matrix $\mathbf{\Delta}_G = \mathbf{D} - \mathbf{A}_G$ where $\mathbf{D} = (d_{ij})$ is a diagonal matrix with entries $d_{ii} = \sum_j a_{ij}$. The spectrum of adjacency and Laplacian matrix of a graph G will be denoted $Sp_A(G)$ and $Sp_\Delta(G)$ and called adjacency and Laplacian spectrum of G, respectively. Note that for d-regular graphs the adjacency and Laplacian spectrum are equivalent. The Laplacian spectrum of a d-regular graph consists of all values $\lambda_\Delta = d - \lambda_A$, where λ_A is an eigenvalue of the adjacency matrix and d is the vertex degree of the graph.

We are looking in this paper for graphs where the optimal scheme OPT has a small number of iteration steps. The optimal scheme is defined as follows. Let $\lambda_1, \lambda_2, \ldots, \lambda_m$ be the m nonzero distinct eigenvalues of the Laplacian of the graph. Now, in the t-th iteration step each vertex v_i sends a load of $\frac{1}{\lambda_t}w_i^t$ to its neighbors, where w_i^t is the load of vertex v_i after the iteration step $t - 1$. So in the t-th iteration step, the load of the vertex v has the form

$$w^{t+1}(v) = w^t(v) - \sum_{\{v,u\} \in E} \frac{1}{\lambda_t}(w^t(v) - w^t(u))$$

After m steps, the load of the network will be totally balanced (see [11]). Note that this implies that the diameter of the graph is less than or equal to $m - 1$ (see also [7]).

Following there is a well known lower bound on the size of the spectrum:

Lemma 1. *The number of distinct eigenvalues of the adjacency and Laplacian matrix of an undirected connected graph G with n vertices and maximal degree d is $\Omega\left(\frac{\log n}{\log d}\right)$.*

Proof. We have seen above that for the diameter of G it holds $diam(G) \leq |Sp_A(G)| - 1$ (e.g. [7]). By an argument known as Moore's bound: a graph with maximal degree d and diameter $diam(G)$ can have at most $1 + d + d(d-1) + \cdots + d(d-1)^{diam(G)-1}$ vertices, the lemma follows. For the case of the adjacency spectrum a similar argument holds ([7]).   □

In our approach to the construction of graphs with small spectra we shall use the following lemmas. The intuition behind is that having a graph G with the corresponding (adjacency or Laplacian) matrix A, we transform A using a suitable non-singular matrix X into a block-diagonal form. The spectrum of A is given by the union of the spectra of the particular block matrices. It is often convenient to view these block components as matrices of some simpler graphs.

Lemma 2. *Let $n = p \cdot m + r$, $A \in \mathbb{R}^{n \times n}$ be a matrix of the form*

$$A = \begin{pmatrix} C & J_{1,p} \otimes S \\ J_{p,1} \otimes R & I_p \otimes B + (J_{p,p} - I_p) \otimes X \end{pmatrix}$$

where $C \in \mathbb{R}^{r \times r}$, $S \in \mathbb{R}^{r \times m}$, $R \in \mathbb{R}^{m \times r}$ and $B, X \in \mathbb{R}^{m,m}$. Then the spectrum of A can be written as the union

$$Sp(A) = Sp(B - X) \cup Sp\begin{pmatrix} C & \sqrt{p} \cdot S \\ \sqrt{p} \cdot R & B + (p-1)X \end{pmatrix} \tag{1}$$

Proof. Consider the matrices

$$W = \begin{pmatrix} I_r & 0 \\ 0 & U \otimes I_m \end{pmatrix}, U = \frac{1}{\sqrt{p}}\begin{pmatrix} I_{p-1} & -J_{p-1,1} \\ J_{1,p-1} & 1 \end{pmatrix} \tag{2}$$

and

$$U^{-1} = \frac{1}{\sqrt{p}}\begin{pmatrix} (p-1) \cdot I_{p-1} - J_{p-1,p-1} & J_{p-1,1} \\ -J_{1,p-1} & 1 \end{pmatrix} \tag{3}$$

Using the transformation $Sp(A) = Sp(WAW^{-1})$ we get

$$Sp(A) = Sp\begin{pmatrix} C & 0 & \sqrt{p} \cdot S \\ 0 & I_{p-1} \otimes (B - X) & 0 \\ \sqrt{p} \cdot R & 0 & B + (p-1)X \end{pmatrix}$$

The lemma follows by interchanging the second and the third row and column block.   □

As an example we present the case $p = 2$. Taking into consideration, that in this work we consider only symmetric matrices, $\boldsymbol{S} = \boldsymbol{R}^T$ holds and the matrix A has the form

$$\begin{pmatrix} \boldsymbol{C} & \boldsymbol{R}^T & \boldsymbol{R}^T \\ \boldsymbol{R} & \boldsymbol{B} & \boldsymbol{X} \\ \boldsymbol{R} & \boldsymbol{X} & \boldsymbol{B} \end{pmatrix} \text{ and is transformed by Lemma 2 to } \begin{pmatrix} \boldsymbol{C} & \sqrt{2} \cdot \boldsymbol{R}^T & \boldsymbol{0} \\ \sqrt{2} \cdot \boldsymbol{R} & \boldsymbol{B} + \boldsymbol{X} & \boldsymbol{0} \\ \boldsymbol{0} & \boldsymbol{0} & (\boldsymbol{B} - \boldsymbol{X}) \end{pmatrix}$$

The matrix A above can be viewed as the adjacency matrix of a graph, which is constructed from a core (with the adjacency matrix C) and 2 copies of a graph with the adjacency matrix B. The subgraph B appears 2 times in the graph and therefore some eigenvalues appear also more than once.

From now on we will not distinguish between the notation of a matrix and an edge weighted graph. The vertices of A are connected to some vertices of both copies of B and these edges are represented by the matrix R. Now the spectrum of this is the union of the spectrum of two other graphs. The first is constructed from C and one copy of B, where the matrix $\sqrt{2} \cdot R$ represents the edges between C and $B + X$. The second graph will be $B - X$.

In the following sections we consider $\mathbf{I}$ or $\mathbf{0}$ as the matrix $\boldsymbol{X}$. We define Q as the smallest matrix with the property: $\boldsymbol{R}^T = \begin{pmatrix} \boldsymbol{Q}^T & \boldsymbol{0} \end{pmatrix}$. Note that Q^T has the same number of rows as the matrix C. Since not any vertex of B will be connected to C, the number of columns of Q^T equals to the number of vertices of B, which have an adjacent vertex in C.

We denote with $T_{1,A}(C, B, Q, p)$ resp. with $T_{2,A}(C, B, Q, p)$ the graph described by $T_{1,A}$ resp. by $T_{2,A}$, where

$$T_{1,A} = \begin{pmatrix} \boldsymbol{C} & \mathbf{J}_{1,p} \otimes \boldsymbol{R}^T \\ \mathbf{J}_{p,1} \otimes \boldsymbol{R} & \mathbf{I}_p \otimes \boldsymbol{B} \end{pmatrix}, T_{2,A} = \begin{pmatrix} \boldsymbol{C} & \mathbf{J}_{1,p} \otimes \boldsymbol{R}^T \\ \mathbf{J}_{p,1} \otimes \boldsymbol{R} & \mathbf{I}_p \otimes \boldsymbol{B} + (\mathbf{J}_{p,p} - \mathbf{I}_p) \otimes \mathbf{I} \end{pmatrix}.$$

In the following we define a sequence of graphs, where each graph is of the form $T_{1,A}(C, B, Q, p)$ or $T_{2,A}(C, B, Q, p)$. Note, that R describes Q in a unique way, so $T_{1,A}$ is well-defined.

Definition 1. *Let* $(G_k)_{1 \leq k < \infty}$ *be a sequence of graphs defined as follows. There exists a sequence of matrices* $(C_k)_{1 \leq k < \infty}, (Q_k)_{1 \leq k < \infty}$ *resp. of integers* $(p_k)_{1 \leq k < \infty}$ *for any* $1 \leq k < \infty$, *where* $G_k = T_{1,A}(C_k, Q_k, G_{k-1}, p_k)$. Q_k *represents the edges between the core* C_k *of* G_k *and the core* C_{k-1} *of* G_{k-1}; Q_k *has the same number of rows as* C_k *and the same number of columns as* C_{k-1}. *There are no edges between* C_k *and* $G_{k-1} \setminus C_{k-1}$.

The sequence $(G'_k)_{1 \leq k < \infty}$ *is defined in a similar way as* $(G_k)_{1 \leq k < \infty}$. *For any* $1 \leq k < \infty$ *there exists a sequence of matrices* $(C'_k)_{1 \leq k < \infty}, (Q'_k)_{1 \leq k < \infty}$ *and a sequence of integers* $(p_k)_{1 \leq k < \infty}$ *so that* $G'_k = T_{2,A}(C'_k, Q'_k, G'_{k-1}, p_k)$ *where* Q'_k *represents the edges between the core* C'_k *of* G'_k *and the core* C'_{k-1} *of* G'_{k-1}. *There is no edge between* C'_k *and* $G'_{k-1} \setminus C'_n$.

To show how to calculate the eigenvalues of the graphs G_n resp. G'_n defined above we have to define the graph class $\mathcal{M}_A(C_1, Q_{1,2}, C_2, Q_{2,3} \dots, C_n)$.

Definition 2. *Let $C_1, C_2, \ldots C_k$ be a sequence of matrices with $1 \leq k < \infty$. Let $Q_{1,2}, Q_{2,3}, \ldots, Q_{k-1,k}$ be also a sequence of matrices.*
$\mathcal{M}_A(C_1, Q_{1,2}, C_2, \ldots, Q_{k-1,n}, C_k)$ denotes the graph with a block tri-diagonal adjacency matrix of the form

$$\begin{pmatrix} C_k & Q_{k-1,k}^T & 0 & \cdots & \cdots & 0 \\ Q_{k-1,k} & C_{k-1} & Q_{k-2,k-1}^T & \cdots & \cdots & 0 \\ 0 & Q_{k-2,k-1} & C_{k-2} & \cdots & \cdots & 0 \\ \vdots & \vdots & \vdots & \ddots & \vdots & \vdots \\ 0 & 0 & 0 & \cdots & C_2 & Q_{1,2}^T \\ 0 & 0 & 0 & \cdots & Q_{1,2} & C_1 \end{pmatrix}.$$

In the following lemma we show that the eigenvalues of G_k defined in definition 1 can be reduced to the eigenvalues of some graphs $\mathcal{M}_A$ defined in definition 2.

Lemma 3. *The spectrum of $G_k = T_{1,A}(C_k, Q_k, G_{k-1}, p_k)$ is the union of the spectra of n graphs $M_1, \ldots, M_k$, where $M_i = \mathcal{M}_A(C_i, \sqrt{p_i} \cdot Q_i, C_{i-1}, \sqrt{p_{i-1}} \cdot Q_{i-1}, \ldots, C_1)$ for any $1 \leq i \leq k$.*
Furthermore the spectrum of $G'_k = T_{2,A}(C'_k, Q'_k, G'_{k-1}, p)$ is the union of the spectra of some graphs $M_{i,j}$, $1 \leq i \leq k$ and $1 \leq j \leq k - i$ for $i \neq k$ and $j = 1$ for $i = k$. The graph $M_{i,j} = \mathcal{M}_A(C_{i,j,0}, \sqrt{p} \cdot Q'_i, C_{i,j,1}, \ldots, C_{i,j,i-1})$, where $C_{i,j,l} = C'_{i-l} + (i - k + p(j - 1) + l(p - 1)) \cdot I$.

Proof. The first statement of the lemma can be proved by induction. We can apply lemma 2 on G_k and obtain, that its eigenvalues are the union of the spectra of G_{k-1} and of the graph $T_{1,A}(C'_{k,k-1}, Q_{k-1}, G_{k-2}, p_{k-1})$, where $C'_{k,k-1} = \mathcal{M}_A(C_k, \sqrt{p_k} \cdot Q_k, C_{k-1})$. As an example we present the matrix after applying 2 times lemma 2 on G_k for $p_{k-2} = 2$ and obtain

$$\begin{pmatrix} C_k & \sqrt{p_k}Q_k^T & 0 & 0 & 0 \\ \sqrt{p_k}Q_k & C_{k-1} & \sqrt{p_{k-1}}Q_{k-1}^T & 0 & 0 \\ 0 & \sqrt{p_{k-1}}Q_{k-1} & C_{k-2} & R_{k-2}^T & R_{k-2}^T \\ 0 & 0 & R_{k-2} & G_{k-3} & 0 \\ 0 & 0 & R_{k-2} & 0 & G_{k-3} \end{pmatrix}.$$

Now we assume that if we apply i times lemma 2, the eigenvalues are the union of the spectra of $G_{k-1} \ldots G_{k-i}$ and of the graph

$$T_{1,A}(C^*_{k,k-1,\ldots,k-i}, Q^*_{k-i}, G_{k-i-1}, p_{k-i}),$$

with

$$C^*_{k,k-1,\ldots,k-i} = \mathcal{M}_A(C_k, \sqrt{p_k} \cdot Q_k, C_{k-1}, \ldots, \sqrt{p_{k-1}} \cdot Q_{k-1}C_{k-i})$$

and $Q^{*T}_{k-i} = \begin{pmatrix} 0 \\ Q^T_{k-i} \end{pmatrix}$. After applying lemma 2 once again we obtain, that the spectrum of G_k can be obtained from the spectra of $G_{k-1} \ldots G_{k-i-1}$ and of the spectrum of the graph

$$T_{1,A}(C^*_{k,k-1,\ldots,k-i-1}, Q^*_{k-i-1}, G_{k-i-2}, p_{k-i-1}),$$

where

$$C^*_{k,k-1,\ldots,k-i-1} = \mathcal{M}_A(C_k, \sqrt{p_k} \cdot Q_k, C_{k-1}, \ldots, C_{k-i-1})$$

and $Q^{*T}_{k-i-1} = \begin{pmatrix} 0 \\ Q^T_{k-i-1} \end{pmatrix}$. After k such steps we obtain the first statement of the
lemma. The second statement can be proved in a similar way. $\qquad\square$

A similar lemma can be also formulated for the Laplacian eigenvalues of such graphs.

3   Spectra of Scalable and Non-scalable Topologies

In this section we deal with some known families of graphs used as interconnection networks. We begin with the Butterfly network, which is designed to have many favorable properties for distributed computing, such as small maximal degree and diameter, large connectivity etc. Next, we turn our attention to scalable families of graphs. First we construct the graphs $T_{(d,n)}$ with $O\left(\left(\frac{\log n}{\log d}\right)^2\right)$ distinct eigenvalues of both adjacency and Laplacian matrix. Since trees are not well suited as interconnection topologies we define a new graph class with much better network properties. To compute the eigenvalues, we use the technique presented in the previous section.

3.1   The Spectrum of the Butterfly Graph

The spectral properties of some interconnection networks like rings, tori, hypercubes [7] or DeBruijn graphs [9] have been investigated. Now we present the adjacency and Laplacian spectrum of the Butterfly without wrap-around edges.

The Butterfly graph $BF_{(k)}$ consists of $k + 1$ columns, each column containing 2^k vertices labeled with unique binary strings of length k. An edge connects two vertices in $BF_{(k)}$ if and only if they are in consecutive i-th and $(i + 1)$-st columns and their labels are either equal or differ only in the i-th bit.

Theorem 1. *The adjacency spectrum of the Butterfly graph* $BF_{(k)}$ *is*

$$Sp_A(BF_{(k)}) = \left\{ 4\cos\left(\frac{\pi i}{j+1}\right) \mid 1 \le i \le j \le k+1 \right\}$$

Proof. The Butterfly graph $BF_{(k)}$ can be recursively constructed by taking two copies of $BF_{(k-1)}$, adding one column of 2^k vertices and connecting these vertices to both $BF_{(k-1)}$'s. Thus the adjacency matrix of the Butterfly graph $BF_{(k)}$ has the structure from Lemma 3 of $\mathcal{T}_{1,A}(C_k, BF_{k-1}, Q_k, 2)$ with $C_k = 0$ and $Q^T_k = \begin{pmatrix} I_{2^{k-1}} \\ I_{2^{k-1}} \end{pmatrix}$. Using Lemma 3, we get $Sp_A(BF_{(k)}) = \cup_{i=0}^k \{Sp(M_i)\}$. The matrix M_i can be viewed as an adjacency matrix of a weighted binary tree with all edges of weight $\sqrt{2}$. After permuting the vertices, Lemma 3 can be used again with $C_k = (0)$, $p_k = 2$ and $Q_k = (\sqrt{2})$. Thus $Sp_A(BF_{(k)}) = \cup_{i=0}^k \{Sp(M_i)\} = \cup_{i=0}^k \cup_{j=0}^i \{Sp(M'_j)\}$. The matrix M'_j can be viewed as an adjacency matrix of a weighted path with $j + 1$ vertices where all edges have weight 2, thus using [7] $Sp(M'_j) = 2Sp(P_{j+1}) = 4\left\{\cos\left(\frac{\pi l}{j+2}\right) \mid l = 1, ..., j+1\right\}$. $\square$

Using similar techniques the Laplacian spectrum of the Butterfly can be also computed.

Theorem 2. *The Laplacian spectrum of the Butterfly graph* $BF_{(k)}$ *is*

$$Sp_\Delta(BF_{(k)}) = \left\{4 - 4\cos\left(\frac{\pi i}{j+1}\right) \mid 0 \leq i \leq j \leq k\right\} \cup$$
$$\cup \left\{4 - 4\cos\left(\frac{\pi(2i-1)}{2j+1}\right) \mid 1 \leq i \leq j \leq k\right\}$$

3.2 Adjacency and Laplacian Spectrum of Scalable Families of Graphs

Now we turn our attention to scalable families of graphs. In the sequel, n will always denote the number of vertices. Let us define the graph class $T_{(d,n)}$ as follows.

Definition 3. *The tree* $T_{(d,n)}$ *is a rooted tree defined recursively.* $T_{(d,1)}$ *contains only the root vertex. For* $n < d+1$, $T_{(d,n)}$ *is a star with one root and* $n - 1$ *leaves connected to it. For* $n > d$, $T_{(d,n)} = (V, E)$ *is constructed as follows. Let* $s = \lfloor \frac{n-1}{d} \rfloor$ *and* $q = n - 1 - ds$. *Let* $(V_1, E_1), ..., (V_d, E_d)$ *be* d *disjoint copies of* $T_{(d,s)}$ *with respective roots* $r_1,...,r_d$. *Then* $V = \{r\} \cup \{v_1, ..., v_q\} \cup \bigcup_{i=1}^{d} V_i$ *and* $E = \{(r, v_i) \mid i = 1...q\} \cup \{(r, r_i) \mid i = 1..d\} \cup \bigcup_{i=1}^{d} E_i$.

Informally, constructing a tree $T_{(d,n)}$ involves setting one vertex r as a root, then dividing the remaining $n - 1$ vertices evenly and constructing a number of copies of $T_{(d,s)}$. The roots of these copies are connected to r. Remaining vertices are added as vertices with degree 1 connected to r.

Remark 1. The graph $T_{(d,n)}$ has n vertices and the maximal degree at most $2d + 1$.

Theorem 3. *The graph* $T_{(d,n)}$ *has at most* $O\left(\left(\frac{\log n}{\log d}\right)^2\right)$ *different eigenvalues.*

The proof of this theorem follows from Lemma 3 and it will be omitted because of space limitations.

The choice of the parameter d results in graphs with different properties. Depending on the application, one may ask for either minimal number of eigenvalues or for the minimal value of $deg \cdot |Sp|$, where deg is the maximal degree of the graph. For getting a small number of eigenvalues, some reasonable large value of d should be chosen (e.g., $d = O(\log n)$), and for setting a small value of the product, some small value should be chosen (e.g., $d = 2$).

However, trees are extremely ill-suited for the application of load balancing because of their poor connectivity properties. To overcome this weak point, we give another construction of a graph $H_{(n)}$ with $O\left((\log n)^3\right)$ distinct eigenvalues which is much better suited as a topology for load balancing.

Definition 4. *Every graph $H_{(n)}$ has a set of distinguished vertices (core) denoted by $C(H_{(n)})$. The graph $H_{(n)}$ is defined recursively as follows. $H_{(1)} = C(H_{(1)}) = K_1$, $H_{(2)} = C(H_{(2)}) = K_2$. For $H_{(n)} = (V, E)$, $n > 2$, let $H_{(n)}$ be of the form $T_{2,A}(C_n, H_{(\lfloor \frac{n-1}{2} \rfloor)}, Q_n, 2)$, where*

$$C_n = \begin{cases} K_2 \ if \ n \ is \ even \\ K_1 \ if \ n \ is \ odd \end{cases},$$

and $Q_n = J$.

Informally, to construct $H_{(n)}$, first construct two copies of $H_{(\lfloor \frac{n-1}{2} \rfloor)}$ and connect the corresponding vertices. The remaining 1 or 2 vertices of $H_{(n)}$ form its core and are connected mutually and to all vertices from the cores of both $H_{(\lfloor \frac{n-1}{2} \rfloor)}$s (see Figure 3.2). In consequence, $H_{(n)}$ has a vertex degree of at most $\log n + 5$.

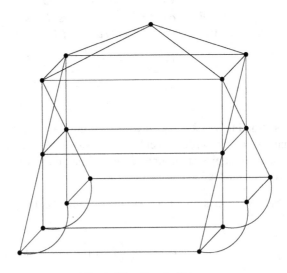

Fig. 1. The Graph $H_{(17)}$.

Theorem 4. *The graph $H_{(n)}$ has at most $O(\log^3 n)$ distinct eigenvalues.*

This theorem can be proved using Lemma 3 and it will be omitted because of space limitations.

4 Adjacency Spectrum

In this section we improve the previous results for the case of the adjacency spectrum. Using another technique we construct a scalable family of graphs $G_{(n)}$ with $O\left((\log n)^2\right)$ distinct eigenvalues.

The graph $G_{(n)}$ is constructed by taking a hypercube, subdividing each edge and then replicating the vertices from the original hypercube a number of times (see Figure 4).

Definition 5. *Let d ($d > 0$) satisfy the inequality $2^{d-1}(d+2) \leq n < 2^d(d+3)$. Let $\{\Delta_i\}_{i=0}^d$ be a sequence defined as follows:*

$$\Delta_i = \begin{cases} 0 & \text{for } i > \lfloor \frac{d}{2} \rfloor \\ \left(n - 2^{d-1}(d+2)\right) \bmod 2^d & \text{for } i = \lfloor \frac{d}{2} \rfloor \\ \Delta_{i+1} \bmod \binom{d}{i+1} & \text{for } 0 \leq i < \lfloor \frac{d}{2} \rfloor \end{cases}$$

where $a \bmod b = a - b \cdot \lfloor \frac{a}{b} \rfloor$.

Let $Q(d)$ be a hypercube. We shall refer to the vertices as binary strings from $\{0,1\}^d$. The k-th level is defined as $\mathcal{L}_k = \{x \in \{0,1\}^d \mid \#_1(x) = k\}$.

The graph $G_{(n)}$ is defined as follows. Consider the graph $S(Q(d))$ with $2^{d-1}(d+2)$ vertices obtained from the hypercube $Q(d)$ by subdivision of each edge. For each node x from the original graph $Q(d)$ add a set V_x of isolated vertices of cardinality $|V_x| = \left\lfloor \frac{\Delta_k}{\binom{d}{k}} \right\rfloor + \left\lfloor \frac{n - 2^{d-1}(d+2)}{2^d} \right\rfloor$ where $x \in \mathcal{L}_k$. For each $y \in V_x$ add edges (y, v) for all edges (x, v) from $S(Q(d))$.

Now, the diameter of a hypercube $Q(n)$ is $\log(n)$ and from the construction of the graph follows:

Remark 2. The graph $G_{(n)}$ ($n > 2$) has n vertices and $diam(G) \leq 2 \log n$.

The maximal degree of $G_{(n)}$ follows also from its definition.

Remark 3. The maximal degree of $G_{(n)}$ ($n > 2$) is at most $3 \log n + o(\log n)$.

To compute the number of distinct eigenvalues of $G_{(n)}$ we need the further two lemmas.

Lemma 4. *[7] Let M be a non-singular square matrix, then*

$$\begin{vmatrix} M & N \\ P & Q \end{vmatrix} = |M| \cdot |Q - PM^{-1}N|$$

Lemma 5. *Let G be a graph obtained from the hypercube $Q(d)$ as follows. For every vertex x add a self-loop with weight w_k^2, where $x \in \mathcal{L}_k$. Each edge connecting vertices on levels k, $k+1$ has weight $w_k \cdot w_{k+1}$. Then the graph G has at most $O(d^2)$ distinct eigenvalues.*

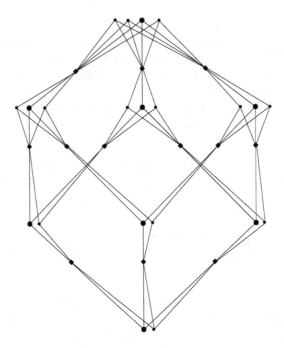

Fig. 2. The Graph $G_{(33)}$.

The proof of this lemma is omitted because of space limitations.

Theorem 5. *The adjacency matrix of* $G_{(n)}$ *$(n > 2)$ has $O(\log^2 n)$ different eigenvalues*

.

Proof. Let A be the adjacency matrix of $G_{(n)}$. Consider an arbitrary vertex x from the original hypercube $Q(d)$ together with all vertices from $\mathcal{V}_x$. Let C be the adjacency matrix of a graph obtained from $G_{(n)}$ by removing vertices $\{x\} \cup \mathcal{V}_x$. Clearly A is of the form of Lemma 2. After applying Lemma 2 iteratively to the sets $\{x\} \cup \mathcal{V}_x$ for each $x \in Q(d)$, we get a matrix A' with the same spectrum as A (except possibly the eigenvalue 0). The matrix A' can be viewed as the adjacency matrix of a weighted graph G' defined as follows. Consider a subdivided hypercube $S(Q(d))$. Each edge incident with a vertex x from $Q(d)$ has a weight $\sqrt{1 + |\mathcal{V}_x|}$.

Clearly, A' is of the form $A' = \begin{pmatrix} 0 & R^\mathsf{T} D \\ DR & 0 \end{pmatrix}$, where $R \in \mathbb{R}^{2^d \times d2^{d-1}}$ is

the vertex-edge incidence matrix of $Q(d)$ and $D \in \mathbb{R}^{2^d \times 2^d}$ is a diagonal matrix $D = (d_i \delta_{ij})_{i,j} = \left(\sqrt{1 + |\mathcal{V}_i|}\delta_{ij}\right)_{i,j}$. Now using Lemma 4 we get $|\lambda \mathbf{I} - A'| =$

$\begin{vmatrix} \lambda \mathbf{I} & -R^\mathsf{T} D \\ -DR & \lambda \mathbf{I} \end{vmatrix} = |\lambda \mathbf{I}| \cdot |\lambda \mathbf{I} - DR\frac{1}{\lambda}R^\mathsf{T} D| = \lambda^{d2^{d-1}} \cdot |\frac{1}{\lambda}(\lambda^2 \mathbf{I} - DRR^\mathsf{T} D)| =$

$\lambda^{d2^{d-1}-2^d} \cdot |\lambda^2 \mathbf{I} - DRR^\mathsf{T} D|$. Using the fact that $RR^\mathsf{T} = Q + d\mathbf{I}$, where Q is the adjacency matrix of hypercube $Q(d)$, we can conclude that $G_{(n)}$ has at most $2m+1$ distinct eigenvalues, where m is the number of distinct eigenvalues of a graph G'' defined as follows. Consider a hypercube $Q(d)$. For each $k = 0, ..., d$, let $w_k = \sqrt{1 + |\mathcal{V}_k|}$. Add to each vertex x from $Q(d)$ a self-loop with weight w_k^2, where $x \in \mathcal{L}_k$. Each edge connecting vertices in levels k and $k + 1$ has weight $w_k \cdot w_{k+1}$. Now using Lemma 5 we can show, that G'' has only $O(\log^2 n)$ distinct eigenvalues.

Acknowledgment

The authors would like to thank Jean-Pierre Tillich and David Wales for the fruitful discussions and their assistance w.r.t. the eigenvalues of the Star graph.

References

1. Akers, S.B., Harel, D., Krishnamurthy, B.: *The Star Graph: An Attractive Alternative to the n-Cube*, Proc. of the International Conference on Parallel Processing, 1987, pp.393–400
2. Akers, S.B., Krishnamurthy, B.: *A Group-Theoretic Model for Symmetric Interconnection Networks*, IEEE Transactions on Computers **38**, 1989, pp. 555–565
3. Biggs, N. L.:*Algebraic Graph Theory*, (2nd ed.), Cambridge University Press, Cambridge, 1993
4. Brouwe, A. E., Cohen, A. M., Neumaier A.: *Distance-Regular Graphs*, Springer Verlag 1989
5. Chung, F.R.K.: *Spectral Graph Theory*, American Mathematical Society, 1994
6. Cybenko, G: *Load balancing for distributed memory multiprocessors* J. of Parallel and Distributed Computing **7**, 1989, pp. 279–301
7. Cvetković, D. M., Doob, M., Sachs, H.: *Spectra of graphs, Theory and Application*, Academic Press, 1980
8. Decker, T., Monien, B., Preis, R.: *Towards Optimal Load Balancing Topologies*, Proceedings of the 6th EuroPar Conference, LNCS, 2000, to appear
9. Delorme, C., Tillich, J. P.: *The spectrum of DeBruijn and Kautz Graphs*, European Journal of Combinatorics **19**, 1998, pp. 307–319
10. Diekmann R., Frommer, A., Monien, B.: *Efficient schemes for nearest neighbor load balancing*, Parallel Computing **25**, 1999, pp. 789–812
11. Elsässer, R., Frommer, A., Monien, B., Preis, R.: *Optimal and Alternating-Direction Load Balancing*, EuroPar'99, LNCS **1685**, 1999, pp. 280–290
12. Flatto, L., Odlyzko, A.M., Wales, D.B.: *Random Shuffles and Group Representations*, The Annals of Probability **13**, 1985 pp. 154–178
13. Gosh, B., Muthukrishnan, S., Schultz, M.H.: *First and Second order diffusive methods for rapid, coarse, distributed load balancing*, SPAA, 1996, pp. 72–81
14. Hu, Y.F., Blake, R.J.: *An improved diffusion algorithm for dynamic load balancing*, Parallel Computing **25**, 1999, pp. 417–444
15. Hu, Y.F., Blake, R.J., Emerson, D.R.: *An optimal migration algorithm for dynamic load balancing*, Concurrency: Prac. and Exp. **10**, 1998, pp. 467–483
16. Hubaut, X. L.:*Strongly Regular Graphs*, Discrete Math. **13**, 1975, pp. 357–381
17. Tillich, J.-P.: *The spectrum of the double-rooted tree*, personal communication
18. Xu, C., Lau, F.C.M.: *Load Balancing in Parallel Computers*, Kluwer, 1997

Optimal Preemptive Scheduling on Uniform Processors with Non-decreasing Speed Ratios

Leah Epstein

School of Computer and Media Sciences
The Interdisciplinary Center, P.O.B. 167, 46150 Herzliya, Israel
epstein.leah@idc.ac.il

Abstract. We study preemptive scheduling on uniformly related processors, where jobs are arriving one by one in an on-line fashion.
We consider the class of machine sets where the speed ratios are non-decreasing as speed increases. For each set of machines in this class, we design an algorithm of optimal competitive ratio. This generalizes the known result for identical machines, and solves other interesting cases.
Keywords: Algorithms, scheduling.

1 Introduction

We consider on-line scheduling on m uniformly related machines. Jobs arrive on-line, and each job has to be assigned before the next job arrives. This scheduling model is called "scheduling jobs one by one" (see [9]). Preemption is allowed, hence each job may be cut into a few pieces. These pieces are to be assigned to possibly different machines, in non-overlapping time slots. (Non-preemptive algorithms are not allowed to cut the job and have to assign it continuously to one machine.)

Each job j is associated with a weight $w(j)$ and each machine i has a speed s_i. The processing time of a job (or a part of a job) of weight w, on machine i is w/s_i. The machines are sorted so that $s_i \leq s_{i+1}$ for $1 \leq i \leq m-1$ and $s_m = 1$. The last condition is general since it is possible to scale any set of speeds and job weights into this form.

For a given set of speeds, let $x = \sum_{i=1}^{m} s_i / (\sum_{i=1}^{m} s_i - 1)$. Note that $\sum_{i=1}^{m} s_i = \frac{x}{x-1}$ and $\sum_{i=1}^{m} s_i - 1 = \frac{1}{x-1}$.

The load of machine i, L_i, is equal to the processing times of all parts of jobs assigned to machine i, on this machine.

The goal of an algorithm is to minimize the makespan, which is the maximum load on any machine.

The quality of an on-line algorithm is measured by the competitive ratio that is the worst case ratio between C_{on} which is the cost (the makespan, in our case) of the on-line algorithm and C_{opt}, which is the cost of an optimal off-line algorithm which knows all the sequence in advance.

A. Ferreira and H. Reichel (Eds.): STACS 2001, LNCS 2010, pp. 230–237, 2001.

In this paper we solve the case of non-decreasing speed ratios, i.e. $\frac{s_{i-1}}{s_i} \leq \frac{s_i}{s_{i+1}}$ for $2 \leq i \leq m - 1$.

We give an algorithm of optimal competitive ratio for every set of speeds. Specifically, we design a deterministic preemptive algorithm of competitive ratio $\frac{x^m}{(x-1)\sum_{i=1}^{m} s_i x^{i-1}}$ and a matching lower bound. The lower bounds are valid for deterministic or randomized preemptive algorithms.

Note that non-decreasing speed ratios for related machines were already considered by Vestjens [10]. He studied a different preemptive on-line scheduling model where jobs arrive over time, instead of one by one. He showed that for this model, an algorithm with competitive ratio 1, which used a finite number of preemptions can be given if and only if speed ratios are non-decreasing.

Our results generalize a few previous results. Chen, Van Vliet and Woeginger gave a preemptive optimal algorithm for identical machines [3]. Some ideas of our results are based on that paper. They show that the best competitive ratio for identical machines is $m^m/(m^m - (m-1)^m)$.

A lower bound of the same value on the competitive ratio of non-preemptive randomized algorithms is also known. The proofs use similar sequences as the ones in [3] and were given independently by [2] and [8]. However, no optimal non-preemptive randomized algorithm is know for $m > 3$. (For m=2, such an optimal algorithm is given in [1].)

Preemptive scheduling on two related machines was studied independently by [4] and by [11]. Both papers show that the optimal competitive ratio is $1 + s_1/(s_1^2 + s_1 + 1)$.

Preemptive scheduling on related machines was also considered by Epstein and Sgall [5]. The paper gives a constant competitive algorithm for any m and set of speeds. That paper also gives a lower bound of 2 on the competitive ratio of any algorithm with an unbounded number of machines and specific lower bounds for constant values of m. Those lower bounds are valid for randomized preemptive or non-preemptive algorithms. Our lower bounds are the general case of the lower bound in [5] for unbounded m. Even though our result does not hold for non-preemptive algorithms, in Section 3 we mention some cases where it holds, and one of these cases is an exponential set of speeds ($s_i = y^{m-i}$ for some $0 < y < 1$) which is also used in [5]. The tight competitive ratio in this case is $\frac{x^m(x-y)}{(x-1)(x^m-y^m)}$.

We start the paper with definitions and proofs of the optimal algorithms and prove the lower bounds in Section 3.

2   Algorithms

We describe the preemptive algorithm. Note that it is easy to compute the optimal preemptive off-line load at every step. The formula was given by [7] and by [6]. The optimal load is the maximum of the following m values; the total weight of all jobs divided by the sum of speeds, and for $1 \leq j \leq m - 1$, the sum of weights of the largest j jobs divided by the sum of largest j speeds of machines.

Our algorithm, similarly to [3] tries to maintain a ratio of x between loads of subsequent machines.

We use the following notations; the load of machine l after the arrival of t jobs is denoted by L_l^t. The optimal load at that time is denoted by C_{opt}^t and the sum of weights of the first t jobs is denoted by W^t.

Let

$$r = \frac{x^m}{(x-1)\sum\limits_{i=1}^{m} s_i x^{i-1}} \quad .$$

The algorithm maintains the following three invariants.

- At any time t, $L_1^t \leq L_2^t \leq \cdots \leq L_m^t$.

- At any time t, $L_m^t \leq r \cdot C_{opt}^t$.

- At any time t, for every $1 \leq k \leq m$ $\sum\limits_{i=1}^{k} s_i L_i^t \leq \dfrac{\sum\limits_{i=1}^{k} s_i x^{i-1}}{\sum\limits_{i=1}^{m} s_i x^{i-1}} \cdot W^t$

A new job J_{t+1} (which arrives at time $t+1$) is assigned as follows. The new optimal off-line is computed using its weight $w(J_{t+1})$. Then the following intervals are reserved. On machine m, the interval:

$$I_m = [L_m^t \ , \ r \ C_{opt}^{t+1}] \quad ;$$

and on machine j $(1 \leq j \leq m-1)$, the interval $I_j = [L_j^t \ , L_{j+1}^t]$. Those intervals are disjoint. The intervals relate to load and not to weight, the weight that can be assigned on I_j $(1 \leq j \leq m-1)$ is $s_j(L_{j+1}^t - L_j^t)$.

To assign J_{t+1}, go from I_m to I_1, putting a part of the job, as large as possible in each interval, until all the job is assigned. After the assignment there will be some fully occupied intervals $I_{z+1}, \cdots, I_m$, some empty intervals $I_1, \cdots, I_{z-1}$ and a partially or fully occupied interval I_z.

Next, we show that it is always possible to partition a job among those intervals.

For convenience define $s_0 = 0$ and $s_{m+1} = 1$. Then $s_{i-1}/s_i \leq s_i/s_{i+1}$ holds for all $1 \leq i \leq m$.

Lemma 1. *If the invariants are fulfilled at step t, then the reserved intervals are sufficient to assign J_{t+1}.*

Proof. The total weight that can be assigned to all intervals is

$$A = (r \ C_{opt}^{t+1} - L_m^t)s_m + \sum_{j=1}^{m-1}(L_{j+1}^t - L_j^t)s_j$$

$$= r \ C_{opt}^{t+1} + \sum_{j=1}^{m}(s_{j-1} - s_j)L_j^t$$

$$= r\ C_{opt}^{t+1} + \sum_{j=1}^{m} \left(\frac{s_{j-1}}{s_j} - \frac{s_j}{s_{j+1}} \right) \sum_{i=1}^{j} s_i L_i^t$$

Since $s_{j-1}/s_j \le s_j/s_{j+1}$ we can use the third invariant for each value of j and get that the above is at least

$$A \ge \frac{1}{\sum_{i=1}^{m} s_i x^{i-1}} \left[\frac{x^m}{x-1} C_{opt}^{t+1} + W^t \sum_{j=1}^{m} \left(\frac{s_{j-1}}{s_j} - \frac{s_j}{s_{j+1}} \right) \sum_{i=1}^{j} s_i x^{i-1} \right]$$

$$= \frac{1}{\sum_{i=1}^{m} s_i x^{i-1}} \left[\frac{x^m}{x-1} C_{opt}^{t+1} + \left((x-1) \sum_{j=1}^{m} s_j x^{j-1} - x^m \right) W^t \right]$$

We consider two cases:

1. $w(J_{t+1}) \ge \frac{W^{t+1}}{\sum_{i=1}^{m} s_i}$ and then $C_{opt}^{t+1} \ge w(J_{t+1})$.

2. $w(J_{t+1}) \le \frac{W^{t+1}}{\sum_{i=1}^{m} s_i}$ and then $C_{opt}^{t+1} \ge W^{t+1} / \sum_{i=1}^{m} s_i$.

We show that the assignment is successful in both cases.

Case 1 :
Since the term multiplied by W^t is non-positive we can substitute $W^t \le w(J_{t+1})$ $\left(\sum_{i=1}^{m} s_i - 1 \right)$ and get that

$$A \ge \frac{1}{\sum_{i=1}^{m} s_i x^{i-1}} \left[\frac{x^m}{x-1} w(J_{t+1}) \right.$$

$$\left. + w(J_{t+1})(\sum_{i=1}^{m} s_i - 1)((x-1) \sum_{i=1}^{m} s_j x^{j-1} - x^m) \right]$$

simplifying this gives

$$A \ge w(J_{t+1}).$$

Case 2:
In this case, $C_{opt}^{t+1} \ge \frac{w(J_{t+1}) + W^t}{\sum_{i=1}^{m} s_i}$. Substituting this we get that the term multi-

plied by W^t is $\dfrac{x^m}{(x-1) \sum_{i=1}^{m} s_i} - x^m + (x-1) \sum_{i=1}^{m} s_i x^{i-1}$, which is non-negative. By using $W^t \geq w(J_{t+1})\left(\sum_{i=1}^{m} s_i - 1\right)$ we get

$$
A \geq \frac{w(J_{t+1})}{\sum_{i=1}^{m} s_i x^{i-1}} \cdot \left[\frac{1}{\sum_{i=1}^{m} s_i}\left(\frac{x^m}{x-1} + \frac{x^m}{x-1}\cdot\left(\sum_{i=1}^{m} s_i - 1\right)\right) + \right.
$$
$$
\left.\left(\sum_{i=1}^{m} s_i - 1\right)\left((x-1)\sum_{i=1}^{m} s_j x^{j-1} - x^m\right)\right]
$$

simplifying this also gives $A \geq w(J_{t+1})$.

To complete the proof of the algorithm, we need to show that all invariants are kept after an assignment of a job. This is clear for the first two invariants, from the definition of the algorithm.

Lemma 2. *If invariants are fulfilled after step t, then they are also kept after step $t+1$*

This would be sufficient since at the start, all loads are zero.

Proof. We only need to show that the third invariant holds for every $1 \leq k \leq m$.

According to the definition of the algorithm, there exists a machine z such that for $i < z$, $L_i^{t+1} = L_i^t$, for $z < i \leq m$, $L_i^{t+1} = L_{i+1}^t$, and $L_z^t < L_z^{t+1} \leq L_{z+1}^t$ (for convenience $L_{m+1}^t = rC_{opt}^{t+1}$).

If $k < z$, then

$$
\sum_{i=1}^{k} s_i L_i^{t+1} = \sum_{i=1}^{k} s_i L_i^t \leq \frac{\sum_{i=1}^{k} s_i x^{i-1}}{\sum_{i=1}^{m} s_i x^{i-1}} W^t \leq \frac{\sum_{i=1}^{k} s_i x^{i-1}}{\sum_{i=1}^{m} s_i x^{i-1}} W^{t+1} .
$$

If $k \geq z$ then we need to show

$$
\sum_{i=k+1}^{m} s_i L_i^{t+1} \geq \frac{\sum_{i=k+1}^{m} s_i x^{i-1}}{\sum_{i=1}^{m} s_i x^{i-1}} W^{t+1} \tag{1}
$$

Since $k \geq z$, $L_i^{t+1} = L_{i+1}^t$ and the left hand size is equal to

$$
\sum_{i=k+2}^{m} s_{i-1} L_i^t + rC_{opt}^{t+1}
$$

$$
= rC_{opt}^{t+1} + \sum_{i=k+2}^{m} s_i L_i^t \left(\frac{s_{k+1}}{s_{k+2}}\right) + \sum_{j=k+3}^{m}\left(\sum_{i=j}^{m} s_i L_i^t\right)\left(\frac{s_{j-1}}{s_j} - \frac{s_{j-2}}{s_{j-1}}\right)
$$

since $\frac{s_{j-1}}{s_j} - \frac{s_{j-2}}{s_{j-1}} \geq 0$ for all j, we can use the invariants of step t, and get that this value is at least

$$\frac{1}{\sum_{i=1}^{m} s_i x^{i-1}} \left[\frac{x^m}{x-1} C_{opt}^{t+1} + \left(\sum_{i=k+1}^{m-1} s_i x^i \right) W^t \right]$$

Let $w(J_{t+1}) = \mu W^{t+1}$, then $C_{opt}^{t+1} = \max\left\{ \mu, \frac{1}{\sum_{i=1}^{m} s_i} \right\} W^{t+1}$ and $W^t = (1 - \mu)W^{t+1}$. Simple calculations show that inequality (1) holds.

3   Matching Lower Bounds

To prove a matching lower bound, we use the following lemma, given in [5].

Lemma 3. *Consider a sequence of at least m jobs, where $J_0, J_1, \cdots, J_{m-1}$ are the last m jobs, Let $C_{opt}(J_i)$ be the preemptive optimal off-line cost after the arrival of J_i.*

The competitive ratio of any preemptive randomized on-line algorithm is at least

$$W/\left(\sum_{i=1}^{m} s_i C_{opt}(J_{i-1}) \right)$$

where W is the total weight of all jobs in the sequence.

Note that if we consider non-preemptive optimal off-line costs, the same expression lower bounds the competitive ratio of non-preemptive randomized algorithms.

Next, we construct the lower bound sequence. The construction is somewhat similar to the proofs in [2,3,8].

The sequence starts with an amount of $\sum_{i=1}^{m-1} s_i - 1$ of very small jobs (sand). These jobs are followed by $m - 1$ jobs $J_1, \cdots, J_{m-1}$ where $W(J_i) = x^{i-1}$.

Theorem 1. *The competitive ratio of any preemptive randomized on-line algorithm is at least*

$$x^m / \left((x-1) \sum_{i=1}^{m} (s_i x^{i-1}) \right)$$

The proof of the theorem follows from Lemma 3 and the following lemma

Lemma 4. *The preemptive optimal off-line cost after the arrival of i big jobs is x^{i-1}, this is true for $i = 0, \cdots, m - 1$.*

Proof. We start by proving the following claim

Claim. For each $1 \le k \le m - 1$

$$\sum_{i=1}^{k} x^{i-1} / \left(\sum_{i=1}^{k} s_{m-k+i} \right) \le x^{k-1} .$$

Let j be a maximum index such that $s_j < \frac{1}{x^{m-j}}$. If no such index exists then $j = 0$. Note that $j < m$ since $s_m = 1$. Hence for $k > j$, $s_k \ge \frac{1}{x^{m-k}}$. We consider two cases.

Case 1: $j < m - k + 1$

Then $x^{k-1} \left(\sum_{i=1}^{k} s_{m-k+i} \right) \ge x^{k-1} \sum_{i=1}^{k} \frac{1}{x^{k-i}} = \sum_{i=1}^{k} x^{i-1}.$

Case 2: $j \ge m - k + 1$

Since $\frac{s_j}{s_{j+1}} < \frac{1}{x^{m-j}} \cdot \frac{x^{m-j-1}}{1} = \frac{1}{x}$ then by induction for all $p \le j$, $s_p \le \frac{1}{x^{m-p}}$. Hence

$$\sum_{i=1}^{k} s_{m-k+i} = \sum_{i=1}^{m} s_i - \sum_{i=1}^{m-k} s_i$$

$$\ge \frac{x}{x-1} - \sum_{i=1}^{m-k} \frac{1}{x^{m-i}} \ge \frac{x}{x-1} - \frac{1}{x^{k-1}} \cdot \frac{1}{x-1}$$

$$\ge \frac{1}{x^{k-1}} \sum_{i=1}^{k} x^{i-1} .$$

The lemma is satisfied for $i = 0$, since splitting the sand evenly gives

$$C_{opt} = \frac{\sum_{i=1}^{m} s_i - 1}{\sum_{i=1}^{m} s_i} = \frac{1}{x} .$$

For $i > 0$ let W^i be the total weight of jobs arriving no later than J_i. C_{opt} after the arrival of J_i is the maximum between

$W^i / \sum_{i=1}^{m} s_i$ and $\max_{1 \le k \le i} \left(\sum_{p=k}^{i} W(J_p) \right) / \left(\sum_{p=m+i-k}^{m} s_p \right)$. The first value is x^{i-1}.

According to the claim, each term in the second value is at most x^{i-1}. This proves the lemma.

Note that if only case 1 of the claim occurs, i.e. for all $i \ge 2$, $s_i \ge \frac{1}{x^{m-i}}$, then the optimal off-line does not use preemptions, and then the lower bound is valid for randomized non-preemptive on-line algorithms.

This is the case for two basic sets of machines.

1. All machines but the slowest have speed 1. The lower bound value in this case is $x^m/(x^m - 1 + (x-1)(s_1 - 1))$. This gives the result for m identical machines for $s_1 = 1$ and for $m - 1$ identical machines for $s_1 = 0$. (see [3]).

2. Machines speeds are powers of some number $0 < y < 1$, i.e. $s_i = y^{m-i}$. In this case $x = \frac{1-y^m}{y-y^m}$ hence $y \geq \frac{1}{x}$. The lower bound is $x^m(x-y)/((x-1)(x^m-y^m))$ which tends to $(y+1)$ as m goes to infinity. This lower bound is given in [5].

4   Conclusions and Open Problems

We have given optimal algorithms for a class of uniformly related machines. It would be interesting to give optimal algorithms for other classes. It is also unknown what the best competitive ratio for general related machines is. There is a large gap between the lower bound of 2 given in [5], and the algorithm given there which has competitive ratio above 20.

References

1. Y. Bartal, A. Fiat, H. Karloff, and R. Vohra. New algorithms for an ancient scheduling problem. *J. Comput. Syst. Sci.*, 51(3):359–366, 1995.
2. B. Chen, A. van Vliet, and G. J. Woeginger. Lower bounds for randomized online scheduling. *Information Processing Letters*, 51:219–222, 1994.
3. B. Chen, A. van Vliet, and G. J. Woeginger. An optimal algorithm for preemptive on-line scheduling. *Operations Research Letters*, 18:127–131, 1995.
4. L. Epstein, J. Noga, S.S. Seiden, J.Sgall, and G.J. Woeginger. Randomized online scheduling on two uniform machines. In *Annual ACM-SIAM Symposium on Discrete Algorithms*, pages 317–326, 1999. To appear in Journal of Scheduling.
5. L. Epstein and J. Sgall. A lower bound for on-line scheduling on uniformly related machines. *Oper. Res. Lett.*, 26(1):17–22, 2000.
6. T. F. Gonzales and S. Sahni. Preemptive scheduling of uniform processor systems. *J. Assoc. Comput. Mach.*, 25:92–101, 1978.
7. E. Horwath, E. C. Lam, and R. Sethi. A level algorithm for preemptive scheduling. *J. Assoc. Comput. Mach.*, 24:32–43, 1977.
8. J. Sgall. A lower bound for randomized on-line multiprocessor scheduling. *Inf. Process. Lett.*, 63(1):51–55, 1997.
9. J. Sgall. On-line scheduling. In *A. Fiat and G. J. Woeginger, editors,* Online Algorithms: The State of the Art, volume 1442 of *LNCS*, pages 196–231. Springer-Verlag, 1998.
10. A. P. A. Vestjens. Scheduling uniform machines on-line requires nondecreasing speed ratios. Technical Report Memorandum COSOR 94-35, Eindhoven University of Technology, 1994. To appear in *Math. Programming*.
11. J. Wen and D. Du. Preemptive on-line scheduling for two uniform processors. *Oper. Res. Lett.*, 23:113–116, 1998.

The UPS Problem

Cristina G. Fernandes[1],* and Till Nierhoff[2],**

[1] Departamento de Ciência da Computação
Instituto de Matemática e Estatística
Universidade de São Paulo - Brazil
cris@ime.usp.br
[2] Institut für Informatik
Humboldt-Universität zu Berlin
nierhoff@informatik.hu-berlin.de

Abstract. The UPS Problem consists of the following: given a vertex set V, vertex probabilities $(p_v)_{v \in V}$, and distances $l : V^2 \to R^+$ that satisfy the triangle inequality, find a Hamilton cycle such that the expected length of the shortcut that skips each vertex v with probability $1 - p_v$ (independently of the others) is minimum. This problem appears in the following context. Drivers of delivery companies visit customers daily to deliver packages. For the company, the shorter the distance traversed, the better. For a driver, routes that change dramatically from one day to the other are inconvenient; it is better if one only has to shortcut a fixed route. The UPS problem, whose objective captures these two points of view, is at least as hard to approximate as the Metric TSP. Given that one of the vertices has probability one, we show that the performance ratio of a TSP tour for the UPS problem is $1/p_{\min}$, where $p_{\min} := \min_{v \in V} p_v$. We also show that this is tight. Consequently, Christofides' algorithm for the TSP has a performance ratio of $3/(2p_{\min})$ for the UPS problem and the approximation threshold for the UPS problem is at most $1/p_{\min}$ times the one for the TSP.

1 Introduction

1.1 Motivation

Package delivery companies, like the *United Parcel Service (UPS)*, have to deliver packages daily to several of their customers. The order of delivery is chosen so that to minimize the distance traversed by the drivers. Each delivery concerns only a subset of the customers. Therefore each delivery could be optimized individually. It is, however, easier for a driver to shortcut a fixed route than to travel each time a completely different route. In this paper we study a variation of the *Traveling Salesman Problem (TSP)* which captures the issue described above.

---

* Research partially done while at Humboldt-Universität zu Berlin, supported in part by CAPES/DAAD Proc. 089/99, CNPq Proc. 301174/97-0, FAPESP Proc. 96/04505-2 and ProNEx 107/97 – MCT/FINEP (Brazil).

** Research supported in part by Deutsche Forschungsgemeinschaft, Pr296/6-1.

A. Ferreira and H. Reichel (Eds.): STACS 2001, LNCS 2010, pp. 238–246, 2001.
© Springer-Verlag Berlin Heidelberg 2001

The delivery company has the information on how often each customer receives a package. From this information one can estimate the probability that a customer receives a package per day. Roughly speaking, the here called *UPS Problem* consists of the following: find an ordering of all customers that minimizes the expected length of the route that starts and ends at a company location and visits in this ordering a randomly chosen (according to the customer's estimated probabilities) subset of the customers. The problem, as well as the described application, was proposed by [8]. Also according to [8], even the special case where customers are divided into two clusters—the customers who receive packages often and the ones who receive packages not so often—is of interest. The setup is conceivable in other delivery systems as well. We therefore expect the study of this problem to have several applications.

1.2   Notation and Problem Statement

Let $G = (V, E)$ be the complete graph on n vertices. A *path* is a sequence $\langle v_0, v_1, \ldots, v_k \rangle$ of distinct vertices of G. A *cycle* is a sequence $\langle v_0, v_1, \ldots, v_k \rangle$ of vertices of G, where $v_0, v_1, \ldots, v_{k-1}$ are distinct and $v_k = v_0$. For a path (cycle) $P = \langle v_0, v_1, \ldots, v_k \rangle$, we denote by $V(P)$ the set $\{v_0, v_1, \ldots, v_k\}$ and we say that P is a path (cycle) on $V(P)$. A *tour* (or a *Hamilton cycle*) is a cycle on V.

We denote the set $\{(v_0, v_1), (v_1, v_2), \ldots, (v_{k-1}, v_k)\}$ by $E(P)$. The *length of P with respect to a function* $l : V^2 \to R^+$ is denoted by $l(P)$, and is given by

$$l(P) := \sum_{e \in E(P)} l(e).$$

The *Traveling Salesman Problem (TSP)* is the following: given a complete graph $G = (V, E)$ and a function $l : V^2 \to R^+$, find a tour of minimum length. We refer to such a tour as a *TSP tour*.

Unless specified otherwise, we consider in the following only functions $l : V^2 \to R^+$ that satisfy the *triangle inequality*: for any x, y, z in V, $l(x, y) \le l(x, z) + l(z, y)$. Under this condition, TSP is called *Metric TSP*.

Note that l may be given partially. Then the length of an edge is considered to be the infimum of the lengths of all paths between its end vertices. This *closure* satisfies the triangle inequality.

Given a path P and a subset S of $V(P)$, the *shortcut of P induced by S*, denoted by $sc_S(P)$, is the path on S given by the subsequence of P containing exactly the vertices of S. Similarly we can define the *shortcut of a cycle C induced by a subset S of V(C)* and denote it by $sc_S(C)$.

Assume each vertex v in V has an associated probability p_v and let $\mathbf{p} := (p_v)_{v \in V}$. These probabilities induce a probability distribution on the subsets of V: for each $S \subseteq V$,

$$\Pr[S] := \prod_{v \in S} p_v \prod_{v \notin S} (1 - p_v).$$

Let $\mathbf{p}$ and $\mathbf{q}$ be two sets of vertex probabilities. We say that $\mathbf{p}$ *dominates* $\mathbf{q}$ if $p_v \ge q_v$ for all $v \in V$.

Given a cycle C, denote by $lsc_{\mathbf{p}}(C)$ the expected value of the length of the shortcut of C induced by a vertex subset which is randomly chosen according to $\mathbf{p}$:

$$lsc_{\mathbf{p}}(C) := \sum_{S \subseteq V} \Pr[S]\, l(scs(C)).$$

Now we are ready to state the UPS problem:

Definition 1. *Given the complete graph $G = (V, E)$, a function $l : V^2 \to R^+$ satisfying the triangle inequality, and probabilities $\mathbf{p} = (p_v)_{v \in V}$, the UPS problem asks for a tour C that minimizes $lsc_{\mathbf{p}}(C)$.*

The performance ratio of a tour C for the UPS problem is the ratio $lsc_{\mathbf{p}}(C)/$ opt, where opt denotes the optimal value of the UPS problem, that is, opt $=$ $lsc_{\mathbf{p}}(C^{\mathrm{UPS}})$ for some optimal tour C^{UPS} of the UPS problem.

Throughout the paper, we consider the UPS problem under the additional assumption that $p_u = 1$ for at least one vertex u (representing, say, a UPS location).

1.3   Results

We start studying a restricted class of vertex probabilities for the UPS problem. Let G, l, and $\mathbf{p}^*$ be the input of the UPS problem, where we assume that, for some $0 \le p \le 1$, $p_v^* \in \{p, 1\}$ for all v. Our first result is a lower bound on the objective function for this particular case in terms of the TSP optimum.

Theorem 1. *Let C be a tour and C^{TSP} be a TSP tour. Then $lsc_{\mathbf{p}^*}(C) \ge p \cdot l(C^{TSP})$.*

The assumption that the vertex probabilities only attain the values p and 1 can be removed with the help of the following proposition.

Proposition 1. *Let $\mathbf{p}$ and $\mathbf{q}$ be two sets of vertex probabilities, where $\mathbf{p}$ dominates $\mathbf{q}$. Then $lsc_{\mathbf{q}}(C) \le lsc_{\mathbf{p}}(C)$ for any tour C.*

The performance ratio of C^{TSP} as a solution for the general UPS problem follows from Theorem 1, using Proposition 1:

Corollary 1. *Let C^{TSP} be a TSP tour, let C^{UPS} be an optimal solution of the UPS problem, and let $p_{\min} := \min_{v \in V} p_v$. Denote by opt the optimal value $lsc_{\mathbf{p}}(C^{UPS})$. Then*

$$\frac{lsc_{\mathbf{p}}(C^{TSP})}{\mathrm{opt}} \le \frac{1}{p_{\min}}.$$

Our second result is that this bound is tight.

Theorem 2. *For every $\epsilon > 0$ there is an instance I_ϵ to the UPS problem such that there are two TSP tours C_1 and C_2 with $lsc_{\mathbf{p}}(C_2) \le (p_{\min} + \epsilon) lsc_{\mathbf{p}}(C_1)$.*

The tightness then follows from opt $\leq lsc_{\mathbf{p}}(C_2)$.

When studying approximations for a computational problem, it is certainly necessary to explore the complexity theoretical limitations of that approach. We prove the following hardness of approximation result.

Theorem 3. *The approximation threshold of the UPS problem with any constantly bounded probability set is not less than the approximation threshold of the Metric TSP.*

The proof of Theorem 1 will be given in Section 2. In Section 3 we will sketch the proof of Proposition 1 and give more details on Corollary 1. Theorems 2 and 3 will be proved in Sections 4 and 5, respectively.

1.4   Conclusions and Open Problems

The UPS problem extends the TSP in that not only the length of the tour, but also the lengths of the subtours determine its objective value. As one might expect, the tradeoff depends on the vertex probabilities. We give matching upper and lower bounds on the rate in Theorems 1 and 2.

This result, being interesting in its own right, has several consequences for the approximation properties of the UPS problem. The currently best known approximation algorithm for the Metric TSP, Christofides' algorithm [3], has a performance ratio of $\frac{3}{2}$. As a consequence of Corollary 1, the same algorithm has a performance ratio of $\frac{3}{2p_{\min}}$ for the UPS problem.

Similarly, every other approximation algorithm for the Metric TSP can be applied to the UPS problem, while the performance ratio is multiplied by a factor of $\frac{1}{p_{\min}}$. Thus, the approximation threshold of the UPS problem is at most $\frac{\theta}{p_{\min}}$, where θ is the approximation threshold for the Metric TSP (for the definition of the approximation threshold and related notions see, e.g., Chapter 13 in [5]). This fact is complemented by Theorem 3, which states that it is at least θ.

One of the first questions that one might ask in this context concerns the influence of different probabilities. The factor of $\frac{1}{p_{\min}}$ might seem too pessimistic, if there were only few vertices with probability $p_{\min}$ and lots of vertices with much larger probabilities. However, the tight examples given in the proof of Theorem 2 can be modified so as to show that the bound given in Theorem 1 is very accurate.

The situation is less clear in the case of approximation algorithms. Here it is conceivable that an algorithm takes into account the distances given by l and combines them with the individual vertex probabilities in a clever way. The non-approximability result in Theorem 3 does not set any limit for that, however it is not less conceivable that the hardness result could be improved to show that the approximation threshold is actually $\frac{\theta}{p_{\min}}$.

The same consideration applies to the important special case of Euclidean instances. An instance of the TSP is called Euclidean if there is a point in the plane for every vertex such that the distance given by l is the Euclidean distance between the points. For this special case, there exist polynomial-time approximation schemes (PTAS) [2,4] for TSP. (A PTAS consists of a polynomial-time algorithm for the problem with a performance ratio of at most $1 + \epsilon$, for each $\epsilon > 0$.) Thus, for each $\epsilon > 0$, by Theorem 1 there exists a polynomial time algorithm for the UPS problem with Euclidean instances with a performance ratio of at most $\frac{1+\epsilon}{p_{\min}}$. On the other hand, it is conceivable both that there is a PTAS and that the approximation threshold is up to $\frac{1}{p_{\min}}$.

2   The UPS Problem with Probabilities 1 or p

The aim of this section is to prove Theorem 1. Recall that there is a $u \in V$ with $p_u^* = 1$ and that, for that theorem, the vertex probabilities are restricted to values of p and 1.

Let C^{TSP} be a TSP tour and C be a tour. To bound $lsc_{\mathbf{p}^*}(C)$ in terms of $l(C^{\text{TSP}})$ we first need another formulation for the corresponding expectation. To this end we introduce some more notation.

Let U be the set of vertices of probability 1, $\bar{U} := V \setminus U$, and let t be the number of vertices in $\bar{U}$. If $t = 0$ then $U = V$ and $lsc_{\mathbf{p}^*}(C) = l(C)$. Since $l(C) \geq l(C^{\text{TSP}})$, the theorem clearly holds in this case. So we may assume $t \geq 1$.

For every $u, v \in \bar{U}$, let P_{uv} be the subsequence of C beginning at u and ending at v (circularly). Let P_{uv}^1 be the shortcut of P_{uv} induced by $\{u, v\} \cup U$. Note that P_{vv}^1 denotes a cycle—the shortcut of C induced by $\{v\} \cup U$.

Denote by $v_0, v_1, \ldots, v_{t-1}$ the vertices of $\bar{U}$ in the order given by C. For $i = 0, \ldots, t-1$, set $C_i := \{P_{v_j v_{j+i+1}}^1 : 0 \leq j < t\}$, where indices are taken modulo t, and $l(C_i) := \sum_{P \in C_i} l(P)$. Each C_i is a collection of paths (cycles if $i = t - 1$) in G whose concatenation results in an Eulerian subgraph of G. Because $U \neq \emptyset$ and each vertex in $\bar{U}$ appears in some path (cycle if $i = t-1$) in C_i, each of these Eulerian subgraphs is connected and spanning. Therefore, for each i,

$$l(C_i) \geq l(C^{\text{TSP}}). \tag{1}$$

Using the notation above we can give the following characterization of $lsc_{\mathbf{p}^*}(C)$:

Lemma 1. $lsc_{\mathbf{p}^*}(C) = (1 - p)^t l(sc_U(C)) + p(1 - p)^{t-1} l(C_{t-1}) + \sum_{i=0}^{t-2} p^2 (1 - p)^i l(C_i)$.

Proof. By definition, $lsc_{\mathbf{p}^*}(C) = \sum_{U \subseteq S \subseteq V} \Pr[S] \, l(sc_S(C))$. Note that the summands where $|S \setminus U| \leq 1$ contribute with $(1 - p)^t l(sc_U(C)) + p(1 - p)^{t-1} l(C_{t-1})$ to $lsc_{\mathbf{p}^*}(C)$.

Let $\mathcal{S} := \{S : U \subseteq S \subseteq V, |S \setminus U| \geq 2\}$ be the collection of the other vertex subsets. For any $S \in \mathcal{S}$, let $E_S := E(sc_{S \setminus U}(C))$. Then, adding indices modulo t,

$$\sum_{S \in \mathcal{S}} \Pr[S]\, l(sc_S(C)) = \sum_{S \in \mathcal{S}} \Pr[S] \sum_{(v_i, v_j) \in E_S} l(P^1_{v_i v_j})$$

$$= \sum_{i=0}^{t-1} \sum_{j=0, j \neq i}^{t-1} l(P^1_{v_i v_j}) \sum_{\substack{S \in \mathcal{S} \\ (v_i, v_j) \in E_S}} \Pr[S]$$

$$= \sum_{i=0}^{t-2} \sum_{j=0}^{t-1} l(P^1_{v_j v_{j+i+1}}) \Pr\left[(v_j, v_{j+i+1}) \in E_S\right]$$

$$= \sum_{i=0}^{t-2} l(C_i) p^2 (1-p)^i$$

and the lemma holds. □

Putting (1) and Lemma 1 together, we have that

$$lsc_{\mathbf{p}^*}(C) \geq p(1-p)^{t-1} l(C^{\mathrm{TSP}}) + \sum_{i=0}^{t-2} p^2 (1-p)^i l(C^{\mathrm{TSP}})$$

$$= l(C^{\mathrm{TSP}}) \left(p(1-p)^{t-1} + \sum_{i=0}^{t-2} p^2 (1-p)^i \right).$$

Straightforward calculation shows that the right hand side is equal to $l(C^{\mathrm{TSP}})p$, concluding the proof of Theorem 1. □

3   Arbitrary Probabilities

Our result on the UPS problem with arbitrary probabilities, i.e. Corollary 1, is a consequence of Proposition 1. The proof of Proposition 1 is based on the FKG-Inequality [1, p. 75] and we only sketch it here.

Let $\mathbf{p}$ and $\mathbf{q}$ be two sets of vertex probabilities and assume that $\mathbf{p}$ dominates $\mathbf{q}$. Let C be any tour. For $S \subseteq V$, let $f(S) := l(sc_S(C))$ and $g(S) := \prod_{v \in S}(p_v/q_v) \cdot \prod_{v \notin S}(1 - p_v)/(1 - q_v)$. Observe that $S \mapsto \Pr_{\mathbf{q}}[S]$ is log-super-modular, that f is increasing because of the triangle inequality, and that g is increasing because $\mathbf{p}$ dominates $\mathbf{q}$. Thus, the requirements of the FKG inequality are met and we have

$$\sum_{S \subseteq V} f(S) \Pr_{\mathbf{q}}[S] \cdot \sum_{S \subseteq V} g(S) \Pr_{\mathbf{q}}[S] \leq \sum_{S \subseteq V} f(S) g(S) \Pr_{\mathbf{q}}[S] \cdot \sum_{S \subseteq V} \Pr_{\mathbf{q}}[S].$$

Since $\Pr_{\mathbf{p}}[S] = \Pr_{\mathbf{q}}[S] \cdot g(S)$, this implies Proposition 1. □

Corollary 1 follows by sandwiching $\mathbf{p}$ between two appropriate sets of vertex probabilities. More precisely, let

$$p_v^* := \begin{cases} p_{\min} & \text{if } p_v < 1, \\ 1 & \text{otherwise.} \end{cases}$$

Then $\mathbf{p}^*$ is dominated by $\mathbf{p}$, which in turn is dominated by the all-ones probability set. The corollary follows from

$$\text{opt} = lsc_{\mathbf{p}}(C^{\text{UPS}}) \geq lsc_{\mathbf{p}^*}(C^{\text{UPS}}) \geq p_{\min}\, l(C^{\text{TSP}}) \geq p_{\min}\, lsc_{\mathbf{p}}(C^{\text{TSP}}),$$

where the first and the last inequality are implied by Proposition 1 and the second inequality by Theorem 1, applied to $C = C^{\text{UPS}}$.

4   Tight Examples

Assume that $p < 1$ and let $\epsilon > 0$. In this section we give the construction of an instance I_ϵ of the UPS problem with probabilities p and 1. It has two TSP tours C_1 and C_2 and (3) states that their UPS values differ by a factor of at least $\frac{1}{p+\epsilon}$. This implies Theorem 2.

We assume w.l.o.g. that $\epsilon < 1 - p$. Let k be a positive integer, large enough so that $k + \log_{1-p}(8k^2) \geq \log_{1-p} \epsilon$. Let n be a prime such that $2k^2 < n < 4k^2$. Then

$$2n(1-p)^k \leq 8k^2(1-p)^k \leq \epsilon. \tag{2}$$

Let $V := \{0, \ldots, n-1\}$ and let $H := (V, E)$, where

$$E := \{ij : j - i \pmod{n} \leq k\}.$$

That is, $H = C_n^k$ is a cycle on n vertices plus all chords of length at most k. Let $l(e) = 1$ for all $e \in E$. Then two TSP tours for H and l are (indices are taken modulo n, as usual)

$$C_1 := \langle 0, k, \ldots, ik, \ldots, nk \rangle \text{ and}$$
$$C_2 := \langle 0, 1, \ldots, i, \ldots, n \rangle.$$

Note that C_1 is a tour because of the primality of n.

Let $p_0 := 1$ and $p_i := p$ for $i \geq 1$. Then I_ϵ consists of (the closure of) H, l, and $\mathbf{p}$. In the rest of this section we shall prove that

$$lsc_{\mathbf{p}}(C_2) \leq (p + \epsilon)lsc_{\mathbf{p}}(C_1). \tag{3}$$

Let S be a randomly chosen subset of V. Call S *dense for* C_1 if S intersects any set of k consecutive vertices of C_1. The probability of that event is at most $n(1-p)^k$, where $n(1-p)^k \leq \epsilon/2$ by (2). The event that S is *dense for* C_2 is defined analogously, and its probability is the same.

Assume that S is dense for C_1 and let ik and $jk > ik$ be two subsequent vertices of $scs(C_1)$. Then $j - i \leq k$ because S is dense for C_1. But then $jk - ik \leq$

$k^2 < n/2$, and by the choice of H the distance between ik and jk is $j - i$. Assume that $scs(C_1) = \langle i_1 k, \ldots, i_{|S|} k \rangle$ and let $i_0 := i_{|S|}$. Then

$$n = \sum_{j=1}^{|S|} (i_j - i_{j-1} \pmod{n}) = \sum_{j=1}^{|S|} l(i_j k, i_{j-1} k) = l(scs(C_1)),$$

and therefore

$$lsc_{\mathbf{p}}(C_1) \geq n \Pr[S \text{ is dense for } C_1] \geq (1 - \epsilon/2)n. \tag{4}$$

If S is dense for C_2, then there is a chord between any two subsequent vertices of $scs(C_2)$ and thus $l(scs(C_2)) = |S|$. This implies that

$$lsc_{\mathbf{p}}(C_2) \leq n \Pr[S \text{ is not dense for } C_2] + \sum_{S \subseteq V} |S| \Pr[S] \leq (p + \epsilon/2)n. \tag{5}$$

As a consequence of (4) and (5),

$$\frac{lsc_{\mathbf{p}}(C_2)}{lsc_{\mathbf{p}}(C_1)} \leq \frac{p + \epsilon/2}{1 - \epsilon/2},$$

which implies (3), using $\epsilon < 1 - p$. $\qquad\square$

5   Hardness of Approximation

The Metric TSP is APX-complete [7] and the currently best lower bound on the approximation threshold is $\frac{41}{40}$ in the asymmetric case and $\frac{129}{128}$ in the symmetric case [6]. It is trivial that the UPS problem has the same lower bounds, because the objective functions coincide for the all-ones probability set $p_v \equiv 1$. It might, however, be interesting to verify that the same holds for the probability set $p_v \equiv p$, where $0 < p < 1$. This, together with Proposition 1, proves Theorem 3.

Next we present a reduction from Metric TSP to the UPS problem in instances with probability set $p_v \equiv p$ for each $\epsilon > 0$. The reduction preserves the approximation ratio up to a factor of $1 - \epsilon$.

Let V be a vertex set and let $l : V^2 \to R^+$ be an instance of the Metric TSP on V. The corresponding instance of the UPS problem consists of the following. Add $c_{p,\epsilon}(n)$ copies of every vertex to get V' and let l' be the extension of l to V' such that all copies of the same vertex have distance 0 to each other and copies of different vertices have the same distance as the original vertices. Here $c_{p,\epsilon}(n) = O(\log_{\frac{1}{1-p}} n)$ is chosen large enough that $(1 - (1 - p)^{c_{p,\epsilon}(n)})^n \geq 1 - \epsilon$. Note that this can be done in polynomial time.

Any tour on V' whose performance ratio (for the UPS instance) is at most η can be converted into a tour on V whose performance ratio (for the original TSP instance) is at most $\eta/(1 - \epsilon)$. Indeed, let C' be a tour on V' whose performance ratio is at most η for the UPS problem. We may assume w.l.o.g. that all copies of each original vertex occur subsequently in C'. Then $l(scs(C')) = l(C')$ as long as

S contains at least one copy of each original vertex. Since the probability for that event is at least $1-\epsilon$, we have that $lsc_{\mathbf{p}}(C') \geq (1-\epsilon)l'(C') = (1-\epsilon)l(C)$, where C is the ordering of V induced by C'. Now let C^{TSP} be a TSP tour on V. If we extend it to a tour on V' by visiting all copies of every vertex subsequently, we know that $lsc_{\mathbf{p}}(C^{\mathrm{TSP}}) \leq l'(C^{\mathrm{TSP}}) = l(C^{\mathrm{TSP}})$. Therefore the optimal UPS solution has length at most $l(C^{\mathrm{TSP}})$. Thus $(1 - \epsilon)l(C) \leq lsc_{\mathbf{p}}(C) \leq \eta lsc_{\mathbf{p}}(C^{\mathrm{UPS}}) \leq \eta l(C^{\mathrm{TSP}})$. This completes the analysis of the reduction, and, together with Proposition 1, the proof of Theorem 3. □

Acknowledgment

We thank Martin Savelsbergh for suggesting the problem and Deryk Osthus and Hanno Lefmann for fruitful discussions. Carlos Eduardo Rodrigues Alves gave a simple proof of a weaker version of Theorem 1 (with p^2 instead of p), which inspired our proof. An anonymous referee gave several useful hints on the presentation of the paper.

References

1. N. Alon, J. Spencer, P. Erdős, *The Probabilistic Method*, Wiley, New York, 1992.
2. S. Arora, *Polynomial Time Approximation Schemes for Euclidean TSP and other Geometric Problems*, Proceedings of the 37th Symposium on the Foundations of Computer Science, 2–11 (1996).
3. N. Christofides, *Worst-case Analysis of a new Heuristic for the Travelling Salesman Problem*, Technical Report (Graduate School of Industrial Administration, Carnegie-Mellon University, Pittsburgh, PA), 1976.
4. J. Mitchell, *Guillotine Subdivisions Approximate Polygonal Subdivisions: A Simple Polynomial-Time Approximation Scheme for Geometric TSP, k-MST, and Related Problems*, SIAM Journal on Computing 28, 1298–1309 (1999).
5. C.H. Papadimitriou, *Computational Complexity*, Addison-Wesley, 1994.
6. C.H. Papadimitriou and S. Vempala, *On the Approximability of the Traveling Salesman Problem*, Proceedings of the 32nd Annual ACM Symposium on Theory of Computing, 126–133 (2000).
7. C.H. Papadimitriou and M. Yannakakis, *The Traveling Salesman Problem with Distances One and Two*, Mathematics of Operations Research 18, 1–11 (1993).
8. M. Savelsbergh, personal communication (1999).

Gathering of Asynchronous Oblivious Robots with Limited Visibility

Paola Flocchini[1], Giuseppe Prencipe[2], Nicola Santoro[3], and Peter Widmayer[4]

[1] University of Ottawa, `flocchin@site.uottawa.ca`
[2] Università di Pisa, `prencipe@di.unipi.it`
[3] Carleton University, `santoro@scs.carleton.ca`
[4] ETH Zürich, `pw@inf.ethz.ch`

Abstract. We consider a collection of robots which are identical (anonymous), have limited visibility of the environment, and no memory of the past (oblivious); furthermore, they are totally asynchronous in their actions, computations, and movements. We show that, even in such a totally asynchronous setting, it is possible for the robots to gather in the same location in finite time, provided they have a compass.
Keywords: Distributed algorithms, coordination, control, mobile robots.

1 Introduction

In current robotics research, both from engineering and behavioral viewpoints, the trend has been to move away from the design and deployment of few, rather complex, usually expensive, application-specific robots. Instead, the interest has shifted towards the design and use of a large number of "generic" robots which are very simple, with very limited capabilities and, thus, relatively inexpensive.

In particular, each robot is only capable of sensing its immediate surrounding, performing computations on the sensed data, and moving towards the computed destination; its behavior is an (endless) cycle of sensing, computing, moving and being inactive (e.g., see [2,7,8,9]). On the other hand, the robots should be able, together, of performing rather complex tasks. Examples of typical basic tasks are *gathering*, *leader election*, *pattern formation*, *scattering*, etc.

A very important set of questions refer to determining the robots capabilities; that is how "simple" the robots can be to perform the required task [3]. In computational terms, this question is to identify the factors which influence solvability of a given problem (the task).

These questions have been extensively studied both experimentally and theoretically in the *unlimited visibility* setting, that is assuming that the robots are capable to sense ("see") the entire space (e.g., see [4,6,10,12]). In general and more realistically, robots can sense only a surrounding with a radius of bounded size. This setting, called the *limited visibility* case, is understandably more difficult, and only few algorithmic results are known [1,11].

In this paper we are interested in *gathering*: the basic task of having the robots meet in a same location (the choice of the location is arbitrary). Since

A. Ferreira and H. Reichel (Eds.): STACS 2001, LNCS 2010, pp. 247–258, 2001.
© Springer-Verlag Berlin Heidelberg 2001

the robots are modeled as points in the plane, the task of robots gathering is also called the *point formation problem*. Gathering (or point formation) has been investigated both experimentally and theoretically. In particular, in the limited visibility setting, Ando *et al.* [1] presented a gathering algorithm for indistinguishable robots which are placed on a plane without any common coordinate system; their algorithm does not require the robots to remember observations nor computations performed in the previous steps. Their result implies that gathering can be performed with limited visibility by very simple robots: *anonymous*, *oblivious* and *disoriented*.

Their solution, however, is based on a very strong "atemporal" assumption on the duration of the robots' actions: their robots must be capable in every cycle to perform all the sensing, computing and moving *instantaneously*.

This assumption has many consequences crucial for its correctness. For example, since movement is instantaneous, a robot can *not* be seen by the others while moving (and its temporary position mistaken for a destination location); since sensing and computing is instantaneous, a robot always has available the correct current situation of its neighborhood. Note that, since instantaneous movement is not physically realizable, their solution is only of theoretical interest.

In this paper, we study the gathering problem in the most general case of an *asynchronous* system of robots with limited visibility, where both their computations and their movement requires a *finite* but otherwise unpredictable amount of time. The question motivating our investigation is whether point formation is possible in such a system. Since in these systems gathering is *unsolvable* if the robots are disoriented (i.e., have no common system of coordinates), we shall restrict ourselves to systems with sense of direction (i.e., the robots share the same coordinate system).

In this paper we show that indeed anonymous oblivious robots with limited visibility can gather within a finite number of moves even if they are fully asynchronous. In fact, we describe a new algorithm for solving the point formation problem in the asynchronous setting by anonymous oblivious robots with limited visibility. We then prove its correctness showing that the robots will gather in a point within a finite amount of time. This result holds not only allowing each activity and inactivity of the robots to be totally unpredictable (but finite) in duration, but also making their movement towards a destination unpredictable in length (but not infinitesimally small). In other words, we show that gathering can be performed by simpler robots with fewer restrictions than known before, provided they have a common coordinate system.

From a theoretical point of view, this result proves that, with respect to the gathering problem, "sense of direction" has the same computational power as "instantaneous actions". From a practical point of view, this result has fundamental consequences. In fact, it allows to substitute a theoretically interesting but physically unrealizable motorial and computing capability requirement (instantaneous actions) with a property (sense of direction) which is both simple and inexpensive to provide (e.g., by a compass).

The paper is organized as follows. In Section 2 the model under study is formally presented. In Section 3 the notations used in the paper and some useful geometric lemmas are introduced. The gathering algorithm is described in Section 4, and in Section 5 its correctness is proven. Due to space limitations, some of the proofs are omitted and can be found in [5].

2   The Model

We consider a system of autonomous mobile robots. Each robot is capable of sensing its immediate surrounding, performing computations on the sensed data, and moving towards the computed destination; its behavior is an (endless) cycle of sensing, computing, moving and being inactive.

The robots are modeled as units with computational capabilities, which are able to freely move in the plane. They are viewed as points, and are equipped with sensors that let each robot observe the positions of the others with respect to its local coordinate system. Each robot can see only a portion of the plane; more precisely, it can observe whatever is at most at a fixed distance V from it (*limited visibility*).

Each robot has its own *local view* of the world. This view includes a local Cartesian coordinate system with origin, unit of length, and the *directions* of two coordinate axes, together with their *orientations*, identified as the positive and negative sides of the axes. In this paper we assume that the robots share the same coordinate system (*sense of direction*); however, they do not necessarily agree on the location of the origin (that we can assume, without loss of generality, to be placed in the view of a robot in its own current position), nor on the unit distance.

The robots are *oblivious*, meaning that they do not remember any previous observation nor computations performed in the previous steps. The robots are *anonymous*, meaning that they are a priori indistinguishable by their appearances, and they do not have any kind of identifiers that can be used during the computation. Moreover, there are no explicit direct means of communication: the communication occurs in a totally implicit manner. Specifically, it happens by means of observing the change of its fellows' positions in the plane while they execute the algorithm.

Summarizing, the robots are *oblivious*, *anonymous*, and with limited visibility; they do however have a common coordinate system.

They execute the same deterministic algorithm, which takes as input the observed positions of the robots within the visibility radius, and returns a destination point towards which the executing robot moves. A robot is initially in a waiting state (*Wait*); at any point in time, asynchronously and independently from the other robots, it observes the environment in its area of visibility (*Look*), it calculates its destination point based only on the current locations of the observed robots (*Compute*), it then moves towards that point (*Move*) and goes back to a waiting state. The sequence: *Wait (W) - Look (L) - Compute (C) - Move (M)* will be called a *computation cycle* of a robot.

The robots are fully *asynchronous*. In particular, the amount of time spent in a computation, in a movement, and in inactivity is finite but otherwise unpredictable. Moreover, a robot moving towards the computed destination can stop after an unpredictable amount of space, provided is neither infinite, nor infinitesimally small (unless it reaches its destination). More precisely, the only assumptions made are the following:

Assumption A1. Any robot will complete its cycle in an amount of time which is finite and bounded from below.

Assumption A2. The distance traveled by a robot in a move is finite and bounded from below (unless the destination is closer than the bound).

As a consequence, the (global) time that passes between two successive movements of the same robot is finite; furthermore, while a robot is moving, it can be seen an unpredictable but finite number of times by another robot.

3   Notations and Geometric Lemmas

We first define sets related to which state a robot is at a given time during the computation.

$W(t)$ and $L(t)$ are the set of all the robots that are respectively in state W and L at time t.

$C(t) = C_{\emptyset}(t) \cup C_+(t)$ is the set of all the robots that at time t are computing. The set $C_{\emptyset}$ contains those robots whose computation's result is to stay still (we say that they execute a *null movement*), while C_+ contains those robots whose computation's result is some destination point (we say that they will execute a *real movement*).

$M(t) = M_{\emptyset}(t) \cup M_+(t)$ is the set of all the robots that at time t are executing a movement. The set $M_{\emptyset}(t)$ contains the robots executing a *null movement* (they stay still); $M_+(t)$ contains those executing a *real movement* (they are effectively moving towards a destination).

We define *circle of visibility* $\mathcal{C}_i(t)$ of a robot r_i at time t the circle of radius V centered in r_i, if $r_i \in L(t)$. Otherwise $\mathcal{C}_i(t) = \mathcal{C}_i(t')$, where $t' = \max\{\bar{t} | r_i \in L(\bar{t})\}$.

In other words, if a robot is *Observing*, its circle of visibility is the circle of radius V centered in itself; otherwise, it is the circle of radius V centered in the location of its most recent *Look phase*. Where no ambiguity arises, the parameter t in $\mathcal{C}_i(t)$ will be omitted.

We now introduce some notations and geometrical lemmas which will be needed later. Let A and B be two points; with $\overline{AB}$ we will indicate the segment starting in A and terminating in B. When no ambiguity arises we will also use the notation $\overline{AB}$ to denote the length of such a segment. Let A and B be two points on a circle; with arc(AB) we indicate the smallest arc on the circle passing through A and B. r indicates a generic robot in the system (when no ambiguity arises, r is used also to represent the point in the plane occupied by robot r); capital italic letters indicate regions (e.g. $\mathcal{L}$, $\mathcal{R}$); given a region, we denote by $|\cdot|$ the number of robots in that region.

Lemma 1. *Every internal chord of a general triangle has length less or equal to the longest side of the triangle.*

Lemma 2. *Let Q be a convex quadrilateral. If all the sides and the two internal diagonals have length less or equal to V then every internal chord of Q is less or equal to V.*

Lemma 3. *Let $\overline{OB}$ be the radius of a circle centered in O and D be a point on the circle such that $B\widehat{O}D = \beta$, with $0 \leq \beta \leq 90°$. Then $\overline{pC} \leq \overline{BC}$, $\forall p \in arc(BD)$ and $\forall C \in \overline{OD}$. (see figure 1.b)*

4   The Algorithm

Let us call *Universe* (U) the smallest isothetic rectangle containing the initial configuration of the robots and let us call *Right* and *Bottom* respectively, the rightmost and the bottom most side of U.

   The idea of the algorithm is to make the robots move either towards the bottom or towards the right of the Universe (a robot will never move up or to its left), in such a way that, after a finite number of steps, they will gather at the bottom most lower most corner of the Universe.

   A robot r can move only if it does not see any robot neither to its left nor above on its vertical axis. Several situations could arise depending on the positions of the robots in its area of visibility:

- If r does not see any robot, it does not move;
- If r sees robots only below on its vertical axis, it moves down towards the nearest robot;
- If r sees robots only to its right, it moves horizontally towards the vertical axis of the nearest robot
- If r sees robots both below on its axis and on its right, it computes a destination point and performs a diagonal move towards the right.

   Recall that C_i is the circle of visibility of robot r_i. Let $\overline{AA'}$ be the vertical diameter of such region; let $\mathcal{R}_i$ and $\mathcal{L}_i$ denote the regions to the right and to the left of r_i, respectively (see Figure 1). Let $S_p = \overline{r_i A'}$ and $S_o = \overline{r_i A}$.

Algorithm 1 (Gathering).
$\quad Extrem := (|\mathcal{L}_i| = 0 \wedge |S_p| = 0);$
$\quad$ **If** I am $\neg Extrem$ **Then**
$\quad\quad$ Do_nothing();
$\quad$ **Else**
$\quad\quad$ **If** $(|\mathcal{R}_i| = 0 \wedge |S_o| = 0)$ **Then**
$\quad\quad\quad$ Do_nothing();
$\quad\quad$ **If** $|\mathcal{R}_i| = 0$ **Then**
$\quad\quad\quad$ $r_j :=$ nearest visible robot on S_o;

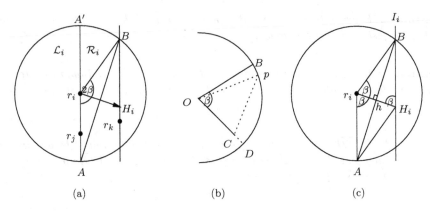

Fig. 1. (a) The Notation Used in Algorithm 1; (b) Lemma 3; (c) Lemma 6.

> Move(r_j).
> **If** ($|\mathcal{R}_i| \neq 0 \wedge |S_o| = 0$) **Then**
> $I_i :=$ Nearest();
> $H_i :=$ H_Destination(I_i);
> Move(H_i).
> **If** $|\mathcal{R}_i| \neq 0$ **Then**
> $I_i :=$ Nearest();
> Diagonal_Movement(I_i).

Nearest() returns the vertical axis on which the robot in $\mathcal{R}_i$ with the nearest axis to r_i lies.

H_Destination(I_i) returns the intersection between I_i and a line parallel to the x direction and passing through r_i.

Move(p) terminates the local computation of the calling robot and moves it towards p.

In the last case of the Algorithm 1, r_i sees somebody below it and somebody to its right, therefore, to avoid losing some robots, it has to move diagonally, as indicated by the following routine.

Algorithm 2 (Diagonal_Movement(I_i)).

1: $B :=$ upper intersection between $\mathcal{C}_i$ and I_i;
2: $A :=$ point on S_o at distance V from me;
3: $2\beta = A\widehat{r_i}B$;
4: **If** $\beta < 60°$ **Then**
5:     $B :=$ Rotate(r_i, B).
6: $H_i :=$ D_Destination(V, I_i, A, B);
7: Move(H_i).

Rotate(r_i, B) rotates the segment $\overline{r_i B}$ in such a way that $\beta = 60°$ and returns the new position of B.

With D_Destination(V, I_i, A, B), r_i computes its destination in the following way: the direction of its movement is given by the perpendicular to the segment $\overline{AB}$; $H_i = \min\{ V, \text{the distance of } I_i \text{ according the direction of movement}\}$.

5   Correctness

In this section we will prove the correctness of the algorithm by first showing that the robots which are mutually visible at any point of the computation, will stay mutually visible until the end of the computation, and concluding that at the end of the computation all robots will gather in one point. We first introduce some lemmas. From Assumptions A1 and A2 it directly follows that:

Lemma 4. *Let r_i and r_j be two generic robots and let t and $t' > t$ two moment of the computation. If $r_i \in L(t)$, $r_i \in L(t')$, $r_j \in M(t)$, $r_j \in M(t')$, $r_j \in C_i(t)$ and $r_j \in C_i(t')$, then r_j can not be in the same point in t and t'.*

Moreover, from the Gathering algorithm it follows that:

Lemma 5. *Let r_j and r_i two arbitrary robots, with r_i to the right of r_j at time t. If $r_j \in L(t)$ and $\overline{r_j r_i} \leq V$, then r_j can not pass r_i in one step.*

Let us consider a generic robot r_i executing the algorithm. Let β be the angle between the vertical axis of r_i and the direction of its movement ($A\widehat{r_i}H_i$ in Figure 1.c).

Lemma 6. *The segment $\overline{r_i H_i}$ is always smaller or equal to V. Moreover, $\overline{BH_i} = \overline{AH_i} = V$ and $\overline{pH_i} \leq V, \forall p \in \overline{r_i A}$.*

Thus, $\Diamond(A, r_i, B, H_i)$ is a parallelogram. We now introduce the definition of visibility graph. The *visibility graph* $G = (N, E)$ of the robots is a graph whose node set N is the set of the input robots and, $\forall r_i, r_j \in N, (r_i, r_j) \in E$ iff r_j and r_j are initially at distance smaller than the visibility radius V. We first show that the visibility graph must be connected in order for the algorithm to be correct.

Lemma 7. *If the visibility graph G is disconnected, the problem is unsolvable.*

Thus, in the following we will always assume that G is connected.

5.1   Preserved Visibility

In this section we prove that the visibility graph is preserved during the entire execution of the algorithm. We prove so by introducing the notion of mutual visibility and by showing that the robots which are connected in the visibility graph (i.e., those which are initially within distance V) will eventually become mutually visible, and that two robots that are mutually visible at some point in the algorithm will stay mutually visible until the end of the computation.

Informally speaking, we say that two robots are mutually visible if each robot includes the other one in its computation, namely each of them had seen the other one during its observation phase. Formally, two robots r_1 and r_2 are *mutually visible* at time t iff

- $r_1 \in (L(t) \cup C_{\emptyset}(t) \cup M_{\emptyset}(t)) \wedge r_2 \in C_1(t) \wedge r_2 \in (W(t) \cup L(t))$, or
- $r_2 \in (L(t) \cup C_{\emptyset}(t) \cup M_{\emptyset}(t)) \wedge r_1 \in C_2(t) \wedge r_1 \in (W(t) \cup L(t))$.

Since all the robots at the beginning are in W, from the above definition we have that the robots that at the beginning are within distance V will become mutually visible in finite time. That is, the following lemma holds:

Lemma 8. *Let r_i and r_j be two robots that at the beginning are within distance V. Robots r_i and r_j will become mutually visible in a finite number of steps.*

We now introduce a couple of lemmas which will be useful to prove that mutually visible robots will stay so until the end of the algorithm. Let r_i be a generic robot on an axis S. Let S' and S'' be two vertical axes to the right of S. We will denote by $\overline{SS'}$ and $\overline{SS''}$ the distances between the corresponding axis. Then we have:

Lemma 9. $\overline{SS'} < \overline{SS''} \Leftrightarrow \beta_{S'} > \beta_{S''}$, *where $\beta_{S'}$ and $\beta_{S''}$ are respectively the angles computed by the routines* Diagonal_Movement(S') *and* Diagonal_Movement(S'') *(Figure 2.a).*

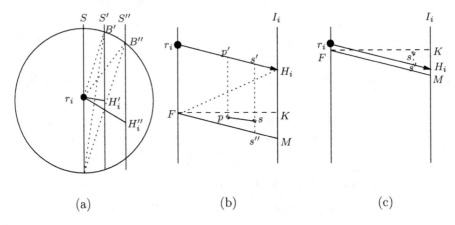

Fig. 2. (a) Lemma 9; (b) and (c) Lemma 10.

Lemma 10. *Let us consider the situation depicted in Figure 2.b, where F is a point at distance $\leq V$ from r_i on its axis (with $F \neq r_i$), H_i is the destination point of r_i. Let $\overline{ps}$ be a segment in $\triangle(F, M, K)$, with s to the right of p, and s' the projection of s over $\overline{r_i H_i}$. Then we have $\overline{ll'} \leq V$, $\forall l \in \overline{ps}$, $\forall l' \in \overline{s'H_i}$.*

We are now ready to show that, as soon as two robots becomes mutually visible, they will stay mutually visible. We first prove that this property holds when two mutually visible robots lie on the same vertical axis; and then we prove that it holds for two robots lying on different vertical axes. In the next lemma we will refer to the notation introduced in Figure 1.a and Lemma 10.

Lemma 11. *Let r_i and r_j be robots which are mutually visible at time t; moreover, let them lie, at time t, on the same vertical axis with r_j being below r_i. There is a time $t' > t$ when r_i and r_j are mutually visible. Moreover, between t and t' $\overline{r_i r_j} \leq V$.*

Proof. Let us first consider the case when $\mathcal{R}_i$ is empty. In such a case, r_i would clearly move towards r_j (shortening their distance), while r_j would not move. Since by Algorithm 1 r_i can not pass r_j, the first time r_i stops while it is moving towards r_j the mutual visibility definition holds, and the lemma follows.

Let us now consider the more interesting case when $\mathcal{R}_i$ is not empty. In the following we shall consider several situations:

Case i: r_j does not look until r_i reaches its destination H_i. We have that $r_i \in W$ while r_i is moving towards H_i. Since $\overline{AH_i} = V$ (Lemma 6) and $\overline{r_i H_i} \leq V$ (Lemma 6), we have that, by Lemma 1 on $\triangle(r_i A H_i)$, the distance between r_i and r_j is always $\leq V$ while r_i is moving. Therefore, the first time r_i stops along its path (at most on H_i), the mutual visibility definition applies and the lemma follows.

Case ii: r_j looks while r_i is moving towards its destination H_i. Since r_i is on r_j's right, r_j can not perform a Vertical Move. Hence, r_j can either decide not to move (because it sees some robots above) or to move. In the first case the proof reduces to the one of Case i. On the other hand, r_j can decide to move after having looked. From Case i we know that r_j can see r_i on its right. Moreover, it might also see some other robots below it, that can be either on the same axis (r_j perform a Diagonal Move) or not (r_j performs an Horizontal Move). The following applies to both situations (Figure 3).

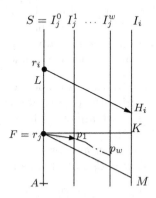

Fig. 3. Case ii of Lemma 11.

Let us call I_j^w the wth axis, counting from S, from where r_j looks while r_i is still on its way towards H_i, and p_w the points on this axis from where r_j performs the *look* phases. Clearly $I_j^0 = S$ and $F = p_0$ coincides with the position of r_j on S. In the following we will prove by induction that

a. I_j^w is to the left of I_j^{w+1},

b. The destination point d_{w+1} that r_j computes when it is on I_j^w is inside $\triangle(F, K, M)$,

c. $\overline{p_{w+1}r_i} \leq V$, and I_j^{w+1} is to the left of r_i.

Basis. Let d_1 be the first destination point r_j computes. Since r_i is on its right, r_j can only decide to perform a Diagonal Movement, therefore d_1 must be to the right of I_j^0, and as a consequence I_j^0 is to the left of I_j^1. Moreover, by Lemma 9 we know that $\overline{r_j d_1}$ must lie above $\overline{r_j M}$, hence p_1 (that is on $\overline{r_j d_1}$) must be within $\triangle(F, K, M)$. Finally, r_j can see r_i by hypothesis and at the beginning I_j^0 is to the left of r_i, and the basis of the induction follows.

Inductive Step. Let us assume that all the statements are true for $1, \ldots, w$. Since by inductive hypothesis I_j^w is to the left of r_i and r_j can see r_i from I_j^w, r_j can only decide to perform a Diagonal Movement, therefore d_{w+1} must be to the right of I_j^w and can not be after r_i (because of how Diagonal_Movement($\cdot$) works), and, as a consequence, I_j^w is to the left of I_j^{w+1}, and a. follows.

Moreover, since $\overline{I_j^w I_i} < \overline{SI_i}$ and , by Lemma 9, we have that $\overline{d_w p_{d+1}}$ must be above $\overline{FM}$ but cannot be above $\overline{FK}$ (because the algorithm does not allow "up" movements). Therefore the point b. follows.

Furthermore, since b. holds and I_j^{w+1} can not be after d_{w+1}, by Lemma 10 c. follows, and the induction is proved.

Now we know that all the stop r_j does while r_i is moving towards H_i are inside $\triangle(F, K, M)$, hence, by Lemma 10, within distance V from r_i. Thus we have that, when r_i reaches H_i, it can see r_j on its left, therefore, it can not move further. It follows that, until r_j is before it, r_i can be only in $L(\cdot)$, $C_\emptyset(\cdot)$, or $M_\emptyset(\cdot)$. Therefore, the first time that r_j stops after r_i reached H_i, say at time $t' > t$, r_i and r_j will be mutual visible. Moreover, between t and t', by Lemma 10 $\overline{r_i r_j} \leq V$, and the lemma follows. $\square$

In the following lemma we show that if a robot sees some robots on its right, then it will never lose them during the computations. Let r_i be a robot in the system, R be the set of robots which are mutually visible with r_i at time t and that are located to the right of I_i, and r_k a robot in R (Figure 4). Moreover, let B and C be respectively the upper and lower intersection between I_i and C_i, and H_i' be the intersection between C_i and the line passing through $\overline{r_i H_i}$.

Lemma 12. *There exists a time $t' > t$ after which r_i will be always mutually visible with the robots in R. Moreover, $\overline{r_i r^*} \leq V, \forall r^* \in R$.*

Proof. From Algorithm 1, we know that robots in R cannot perform any movement while r_i is on their left. Let t^* the time when r_i enters its Look phase and p be the destination point it computes. Clearly, p can not be to the right of any robot in R. In the following, we first prove that $\overline{lr^*} \leq V, \forall r^* \in R$ and $\forall l \in \overline{r_i p}$.

From Lemma 3, it follows that: $\forall p \in arc(BH_i'), pH_i \leq \overline{BH_i} = V$ (1). Moreover, $\overline{H_i C} = \overline{BC} - \overline{BH_i} \leq 2V - V = V$ and from Lemma 2 we have:

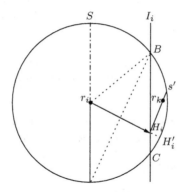

Fig. 4. Lemma 12.

$\forall p \in arc(H_i'C), pH_i \leq H_iC \leq V$ (2). Plugging (1) and (2) we obtain: $\forall p \in arc(BC), pH_i \leq V$ (3).

Let us now consider a robot $r_k \in sector(BCB)$ (that is in the area to the right of I_i and in C_i) and let s' be the intersection between $arc(BC)$ and the line passing through H_i and r_k. We have that $\overline{H_ir_k} \leq \overline{H_is'} \leq V$ (from (3)), $\overline{r_ir_k} \leq V$, and $\overline{r_iH_i} \leq V$. Therefore, applying Lemma 1 to $\triangle(r_i, r_k, H_i)$ we have that $\overline{qr_k} \leq V, \forall q \in \overline{r_iH_i}$. In conclusion, when r_i stops in p, say at time $t' > t$, it will see all the robots in R, that can only be in $L(t')$, $C_\emptyset(t')$, or $M_\emptyset(t')$, and the lemma follows. □

By Lemma 8, 11 and 12 we can conclude that:

Theorem 1. *The visibility graph G is preserved during the execution of the algorithm.*

5.2 Finiteness

In this section we will prove that, after a finite number of steps, the robots will gather in a point.

Lemma 13. *Let us suppose to have several robots on a vertical axis A and no robots to the left of A. If r is the topmost robot on A that can see a robot to the right of A, then, in a finite number of steps, either all the robots above r on A will reach r, or one of them will leave A.*

The next two lemmas show that all the robots in the system converge to the *Right* axis of the Universe, and actually reach it.

Lemma 14. *For any given vertical axis I before Right which is at any distance $d > 0$ from it, all the robots that are on the left of I at the beginning of the algorithm, will pass I in a finite number of steps.*

Lemma 15. *After a finite number of steps, all the robots in the system reach Right.*

The following lemma states what happens when all the robots lie on the same vertical axis: they will reach the bottom most robot on that axis in a finite number of steps.

Lemma 16. *If all the robots of the system lie on the same vertical axis A, then in a finite number of steps all the robots will reach the bottom most robot on A.*

We can finally conclude that:

Theorem 2. *In a finite number of steps, all the robots in the system gather in a point; the rightmost and bottom most corner of the universe.*

References

1. H. Ando, Y. Oasa, I. Suzuki, and M. Yamashita. A Distributed Memoryless Point Convergence Algorithm for Mobile Robots with Limited Visibility. *IEEE Trans. on Robotics and Autom.*, 15(5):818–828, 1999.
2. G. Beni and S. Hackwood. Coherent Swarm Motion Under Distributed Control. In *Proc. DARS'92*, pages 39–52, 1992.
3. E. H. Durfee. Blissful Ignorance: Knowing Just Enough to Coordinate Well. In *ICMAS*, pages 406–413, 1995.
4. P. Flocchini, G. Prencipe, N. Santoro, and P. Widmayer. Hard Tasks for Weak Robots: The Role of Common Knowledge in Pattern Formation by Autonomous Mobile Robots. In *ISAAC '99*, pages 93–102.
5. P. Flocchini, G. Prencipe, N. Santoro, and P. Widmayer. Limited Visibility Gathering by a Set of Autonomous Mobile Robots. Technical Report TR-00-09, Carleton University, 2000.
6. D. Jung, G. Cheng, and A. Zelinsky. Experiments in Realising Cooperation between Autonomous Mobile Robots. In *5th International Symposium on Experimental Robotics (ISER)*, June 1997.
7. Y. Kawauchi and M. Inaba and. T. Fukuda. A Principle of Decision Making of Cellular Robotic System (CEBOT). In *Proc. IEEE Conf. on Robotics and Automation*, pages 833–838, 1993.
8. M. J Matarić. *Interaction and Intelligent Behavior*. PhD thesis, MIT, May 1994.
9. S. Murata, H. Kurokawa, and S. Kokaji. Self-Assembling Machine. In *Proc. IEEE Conf. on Robotics and Autom.*, pages 441–448, 1994.
10. K. Sugihara and I. Suzuki. Distributed Algorithms for Formation of Geometric Patterns with Many Mobile Robots. *J. of Robotics Systems*, 13:127–139, 1996.
11. I. Suzuki and M. Yamashita. Distributed anonymous mobile robots. In *Proc. of 3rd International Colloquium on Structural Information and Communication Complexity*, pages 313–330, Siena, 1996.
12. I. Suzuki and M. Yamashita. Distributed Anonymous Mobile Robots: Formation of Geometric Patterns. *SIAM J. Comput.*, 28(4):1347–1363, 1999.

Generalized Langton's Ant: Dynamical Behavior and Complexity

Anahí Gajardo[1,2], Eric Goles[1], and Andrés Moreira[1]

[1] Center for Mathematical Modeling and Departamento de Ingeniería Matemática
FCFM, U. de Chile, Casilla 170/3-Correo 3, Santiago, Chile
{agajardo,egoles,anmoreir}@dim.uchile.cl
[2] Laboratoire de l'Informatique du Parallélisme, ENS-Lyon
46, allée d'Italie, 69364 Lyon, France

Abstract. Langton's ant is a simple discrete dynamical system, with a surprisingly complex behavior. We study its extension to general planar graphs. First we give some relations between characteristics of finite graphs and the dynamics of the ant on them. Then we consider the infinite bi-regular graphs of degrees 3 and 4, where we prove the universality of the system, and in the particular cases of the square and the hexagonal grids, we associate a P-hard problem to the dynamics. Finally, we show strong spatial restrictions on the trajectory of the ant in infinite bi-regular graphs with degrees strictly greater than 4, which contrasts with the high unpredictability on the graphs of lower degrees.

1 Introduction

The virtual ant defined by Chris Langton ([1], [2]) is a simple system where an agent, the "ant", moves on the square grid. Each cell is in one of two states, *to-left* or *to-right*, and the ant is represented as an arrow between two adjacent cells. It moves one cell forward at each time step, turning according to the state of the cells, and switching these states thereafter. Interesting behavior follows: a single ant, starting with all cells in the *to-left* state, has a more or less symmetric trajectory in the first 500 steps; then it goes seemingly randomly for about 10,000 steps, until it suddenly starts building an infinite diagonal "highway" (a periodic motion with drift).

As [3] points out, the ant is so "natural" that it has been independently invented at least three times. Langton proposed it as a simple model of artificial life [1], and it appeared again as one of the "turmites", the two-dimensional Turing machines studied by G. Turk [4]. It has also been studied as a paradigm for signal propagation in random media, in particular as a model of a particle in two-dimensional Lorentz Lattice Gases [5]. Another source of interest is the relation with the agent-based systems (also called "ant systems") that have been intensively studied and applied for several optimization problems in the last years, with good performance but few exact mathematical results. Langton's ant shares with them the so called "stigmergy": the movement of the agent is determined by some properties of the environment next to it, and these properties are in turn modified by that movement.

A. Ferreira and H. Reichel (Eds.): STACS 2001, LNCS 2010, pp. 259–270, 2001.

The ant has motivated several studies, both experimental and analytical. It has been analyzed in the other regular grids, like the triangular ([6], [5], [7]) and the hexagonal grids [7]. The case of bi-regular graphs of degree 3 was studied in [8], and some possible definitions for the ant on the line where examined in [9] and [10]. There have been some generalizations of the ant, to allow more than two states of the cell and to consider several ants.

The dynamics of the ant is strongly related to the topology of the underlying graph. On the triangular grid, the trajectory is always restricted to two rows, and is easily predicted [6]. The hexagonal grid is again different: when starting with all cells in the same state, the ant follows paths that are bilaterally symmetric with respect to the starting position, and no highway appears [8].

The most important result concerning the dynamics of the ant on the square grid, due to [3], states that the set of the cells that are visited infinitely often by the ant (for a given initial configuration) has no *corners*. A *corner* of a set is a cell where at least two neighbors are not in the set, and these are not opposite to each other. The main consequence of this is the following fact (already demonstrated in [11]): *For any initial configuration, the trajectory of the ant is unbounded.* These unboundness is also true on the triangular grid [5]. On the other hand, bounded trajectories are known to exist on the hexagonal grid [7].

Unfortunately, this result does not tell us anything else about the behavior of the ant in the long term. The experiments, however, suggest that the long-term behavior of the ant, although unbounded, is unbounded in a highly repetitive way. Specifically, the following conjecture has been open for at least ten years: "For any initial configuration with finite support, the ant eventually starts building the periodic highway, in some unobstructed direction". If this conjecture is true, then any problem associated with the ant, whose input is an initial configuration with finite support, turns out to be decidable, since in that case it suffices to iterate on the configuration until the highway appears; the question may be answered at that point, since the future dynamics is easily predicted.

The Present Work

We consider the natural extension of the ant to general planar graphs, where the nodes in the graph take the place of the cells in the square grid, and the neighbors of a node are the nodes to which it is connected. We generalize the rule in the most obvious way: the states at the nodes are still *to-left* and *to-right*, and the ant changes these states each time it goes through a node. Furthermore, the ant turns to the indicated direction at each time step; for this purpose, "turning to the left" is defined as leaving the node through the edge which is found moving clockwise, starting from the edge which was used by the ant to arrive. The square grid becomes a particular case of regular planar graph.

In Section 2 we study the ant on finite graphs. In a restricted family (graphs where no edge belongs to more than one simple cycle) the period of the system is linearly bounded in the number of nodes, but in the general case, we show a family where the periods grow exponentially with the number of nodes.

We consider next the case of the infinite bi-regular graphs $\Gamma(k, d)$: these are graphs where all the nodes have d neighbors, and all the faces (the smallest cycles) have k neighboring faces. They generalize the original system (the square grid corresponds to $\Gamma(4, 4)$), and were chosen as an intermediate point between it and general infinite graphs. They allow us to study the dependence of the dynamics of the ant on the degree of the graph and the length of the faces.

In Section 3, we show how to calculate boolean circuits with the trajectory of the ant. [1] The construction is embedded in any infinite bi-regular graph of degree 3 or 4, and since it is finite, it can be also embedded in finite graphs. In the particular cases of the square grid, the hexagonal grid and finite graphs, the construction uses an appropriately bounded amount of space in the configuration, and the following questions are thus found to be P-hard problems: Given a finite initial configuration[2] of $\Gamma(4, 4)$ ($\Gamma(6, 3)$) and two nodes α, β, does the ant visit α before β? Given a finite graph, an initial configuration, and two nodes α and β, does the ant visit α before β?

The construction of circuits has further and important consequences, which are presented in Section 3.3. First, the ant can draw the space-time diagram of any one-dimensional cellular automata (for finite configurations). It follows that the system is universal, since it may simulate a universal Turing machine. Finally, there are undecidable problems related to the dynamics of the ant.

In Section 4 we consider the case of infinite bi-regular graphs which have degree strictly greater than 4. In spite of the higher connectivity of these graphs, the system seems to be less complex on them. The trajectory of the ant is restricted to a low connected sub graph, a fractal tree of faces, and the construction of circuits of Section 3 cannot be carried over. The restrictions do not depend on the exact degree, but only on the lower bound (5), provided that the lengths of the faces are constant. The conjecture stated in the previous section is proved to be true on these graphs: for any finite initial configuration, the ant falls in a periodic motion with drift.

1.1   Definitions

A *non-directed simple graph* (the only kind we will use) is a pair $G=(U,E)$, where U is the set of nodes, and E is a set of edges of the form $\{u, v\}$, $u \neq v \in U$. A *path* $\mathcal{D}$ is a list of nodes of the form $(u_0, u_1, \ldots, u_k)$, such that $\forall i \; \{u_i, u_{i+1}\} \in E$. The *length* of $\mathcal{D}$ is the integer k. A *cycle* is a path whose extreme nodes coincide. A path (a cycle) is *simple* if it does not repeat nodes (other than the extreme nodes, in the case of cycles). Two cycles are *tangent* if they have a unique common node. The *distance* between two nodes is the length of the shortest path connecting them (if there is no such path, it is infinite). The *diameter* of the graph is the

---

[1]   As far as we know, the method used to calculate boolean circuits presented here is original, and differs completely from the classical methods introduced by [12] and [13] for two-dimensional systems.

[2]   In infinite graphs we say that a configuration of the system is finite if all but a finite number of nodes are in the same state.

maximum distance between its nodes. A graph is *connected* if there are paths connecting any two nodes. A *tree* is a connected graph which has no cycles. An *isthmus* is an edge whose removal disconnects the graph. The *neighbors* of a node u are the nodes $N(u) = \{v \in U : \{u, v\} \in E\}$. The *degree* of u is $|N(u)|$. A graph is k-*regular* if all the nodes have degree k. A *leaf* is a node with degree 1.

A graph is *planar* if it may be injected in $\mathbb{R}^2$, the nodes being represented by points and the edges by simple curves, so that the curves *do not intersect*. A graph is *locally finite* if any sphere in $\mathbb{R}^2$ contains a finite number of nodes. In a planar graph, a *face* is one of the regions of the partition induced by the graph. The *dual graph* of G, G', is defined as the graph $G' = (U', E')$, where U' is the set of faces of G, and $\{i', j'\} \in E'$ iff i' and j' have a common edge.

We will be interested both in general finite graphs, and in some regular infinite graphs. The *bi-regular graph* $\Gamma(k, d)$ is the locally finite planar d-regular graph whose dual is k-regular. $\Gamma(k, d)$ is finite for $k = 3$ and $d < 6$, for $k \in 4, 5$ and $d < 4$, and for $k = 6$ and $d < 3$. $\Gamma(6, 3)$, $\Gamma(4, 4)$ and $\Gamma(3, 6)$ can be embedded in $\mathbb{R}^2$ with edges of constant lengths, and correspond to the hexagonal, square and triangular grids, respectively. The rest of the cases corresponds to the so called "hyperbolic graphs", that can be embedded in the hyperbolic plane.

A *decision problem* is one where the solution, for a given instance, is yes or no. It is said to be *decidable* if there is an algorithm which answers the question in a finite time. Decidable problems are classified in complexity classes, which describe the amount of work needed to solve them. An important class is P: problems where the answer can be found in polynomial time. A problem to which any problem in P may be reduced (with the reduction satisfying logarithmic conditions: see [14], p.160), is called P-hard; if it also belongs to P, is called P-complete. Thus, to show that a problem is P-hard, it is enough to reduce a P-complete problem to it.

We say that a system is *universal* if it may simulate a universal Turing machine. This notion of universality implies, in particular, the existence of undecidable problems. The complexity and undecidability of problems associated to a dynamical system, as well as the existence of some kind of universality in it, are ways to measure its complexity. For Complexity Theory, see [14].

1.2   Some Basic Facts About the Ant

We consider a connected, simple, planar, non-directed graph $G = (U, E)$. Planarity provides an order of the edges inciding a node u, and the rule of Langton's ant is naturally extended in the way already explained in the introduction. A *configuration* of the system is defined as the assignation of states to the nodes at a given time, together with the position of the ant.

The first thing to notice is that the rule is invertible (in the finite case, this implies that any configuration belongs to a periodic trajectory of the system). Moreover, the ant is its own inverse: if the ant turns back at some moment (for instance, when it comes to a leaf), the path to be followed afterwards will be exactly the reverse of the path it had followed before.

We note also that at nodes with degree 1, the ant reflects (the only edge it may use to leave is the same it used to arrive). At nodes with degree 2, it will go on, since the next-to-the-right and the next-to-the-left edges are the same.

2   The Ant on Finite Graphs

First we consider the case of trees. The idea is the following: the ant goes on, until it finds a leaf. At that moment, it will turn and undo its path, until it finds another leaf. It will oscillate between these two leaves, forever.

Theorem 1. *On a tree with diameter D, periods are bounded by $4D$, and the set of edges visited by the ant forms a simple path.*

We consider next a graph with no string in its cycles, i.e., such that each edge belongs to at most one simple cycle. Such a graph consists of a collection of simple cycles, which may be tangent to each other, or may be connected by paths. Two cases are to be considered: if there are no isthmuses, then the ants goes exactly twice through all the edges, in each period. If there are isthmuses, the graph is analyzed as a tree, where each node represents a component without isthmuses, and the result is a combination of the first case and Theorem 1.

Theorem 2. *On a graph without strings, periods are bounded by $20|U|$.*

Theorem 3. *There is a family of planar graphs $G_n = (U_n, E_n)$, with $|U_n| = 2n$, such that for each G_n there is a configuration with period greater than 2^n.*

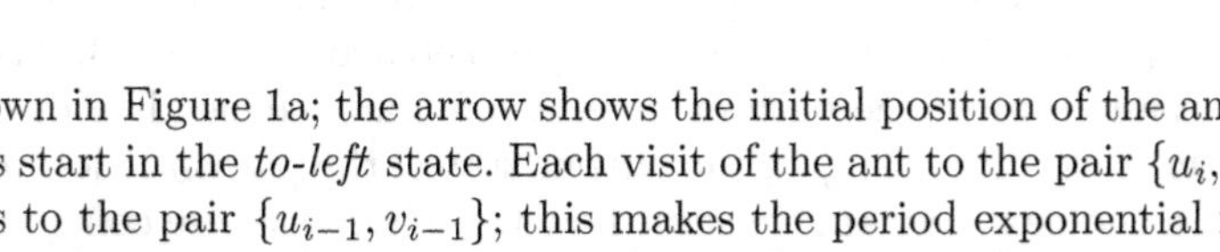

Fig. 1. The Period Grows Exponentially in the Size of these Graphs

G_n is shown in Figure 1a; the arrow shows the initial position of the ant, and all the nodes start in the *to-left* state. Each visit of the ant to the pair $\{u_i, v_i\}$ takes two visits to the pair $\{u_{i-1}, v_{i-1}\}$; this makes the period exponential in n, and thus we see that we can have exponential periods once we drop the condition of Theorem 2 (absence of strings). That condition may seem too restrictive; nevertheless, we find the following: if we keep that condition, but drop the condition of planarity -not even of the graph, but of the representation determined by the local left-right orientation- we may get exponential periods: the behavior of the ant over the graph of Figure 1b is analogous to the case in 1a.

3   Circuit Construction

We can impose a path to the ant, by putting the appropriate states in the nodes. If, in addition, we define the states of certain nodes as our logical variables, the ant will "read" them and choose its path accordingly. Now we will show how to use this to build a logical gate, where the output is "calculated" by the ant. The general form of the gate is described in Figure 2a: at the top, we have some nodes whose states represent the input. At the bottom, some nodes represent the output; at the beginning, all output nodes are in the *to-left* state, which will represent the logical value *false*. The ant enters the gate at the left, and exits at the right. While being in the gate, it visits the input nodes, and visits (and switches) the correct output nodes, according to the function which the gate represents. The changes are done *from inside*, thus allowing the output nodes to be used as the input for other gates.

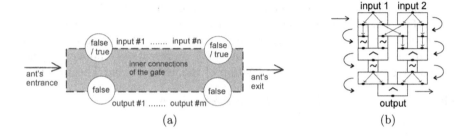

Fig. 2.   (a) Sketch of a Gate (b) XOR Function, Built as $(\sim (i_1 \wedge i_2) \wedge \sim (\sim i_1 \wedge \sim i_2))$

To compute a boolean circuit we put the input variables in some nodes at the top of the configuration (see Figure 2b), and for the consecutive stages of evaluation we put consecutive rows of logical gates. The ant goes through every row, starting with the upper one. After going through the last row, the state of the last output node contains the evaluation of the circuit for the given input.

To write a boolean circuit it is enough to have the NOT and the AND functions. To construct the circuit we also use gates that allow us to duplicate, cross and copy variables. All these gates are sketched in Figure 3. The general scheme is the following: the path of the ant bifurcates, depending on the input states. After (possibly) changing the output states, the paths are joined and the ant exits. In the next section, we show how to construct these configurations in bi-regular graphs with degree 3 or 4.

3.1   Embedding in Infinite Regular Graphs

First of all, we need to have paths for the ant to follow. With *ant-path* we will refer to a path which may be walked by the ant, provided that it encounters

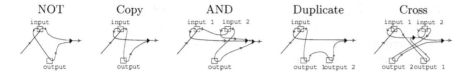

Fig. 3. Simplified Schemes of the Gates

the appropriate states on it. In a 3-regular graph, any path is an ant-path, but in the general case this is not true. The next lemma shows that in $\Gamma(k, 4)$, it is always possible to bring the ant from any location to any other location. We will see in Section 4 that this simple fact is not true in the $\Gamma(k, d)$ graphs if $d \geq 5$.

Lemma 1. *Let $P = v_0, v_1, .., v_n$ be a simple path in $\Gamma(k, 4)$. Then there is a simple ant-path $a_0, ...a_m$ that begins at v_0 and ends at v_n. It is composed by edges that share a face with those of P, and it arrives to v_n through an edge that is to the right or to the left from (v_{n-1}, v_n), or is (v_{n-1}, v_n) itself.*

For the schemes of Figure 3, we need to cross and join paths. To do it, we built *Crossings* and *Junctions*, which may be inserted at the places where they are needed. They are shown in Figure 4. In the *Junction*, if the ant enters at 1 or at 2, it exits at 3. In the *Crossing*, if the ant first enters at 1, it exits at 2. If afterwards it enters at 3, it exits at 4. But if it enters first at 3, it exits at 5.

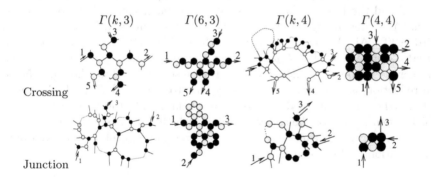

Fig. 4. White Stands for *to-left*, Black for *to-right*

Following Figure 3 and using the configurations of Figure 4, and simple paths, we define configurations that simulate the AND, NOT, Cross, Copy and Duplicate gates. We can choose the dimensions of these gates and the positions of their inputs and outputs arbitrarily, and this can be done in an automatic way. A procedure that takes a boolean circuit and writes the corresponding configuration in a $\Gamma(k, d)$ graph can thus be defined.

Figure 5 shows a Duplicate gate for $\Gamma(4,4)$ and $\Gamma(6,3)$ (the square and the hexagonal grids, respectively). The construction of the other gates for $\Gamma(4,4)$ may be found in [15]. The fast growth of the configurations in hyperbolic graphs does not allow us to show them on Euclidean paper.

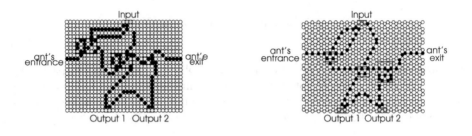

Fig. 5. *Duplicate* Gate in the Square and Hexagonal Grids

3.2   Computational Complexity

The problem (CIRCUIT-VALUE) of determining, given a boolean circuit C and a truth assignment t, whether C outputs *true* with input t, is known to be *P*-complete ([14], p.168). Now, fix (k,d), with $d = 3, 4$. From 3.1, for any pair (C,t) we can build a configuration in $\Gamma(k,d)$ representing them, so that the ant will end the last row having visited or not having visited the output node of that row, depending on the outcome of C with input t. Thus the problem (CIRCUIT-VALUE) is being reduced to the problem (P) of knowing, for a finite initial configuration of $\Gamma(k,d)$, whether the ant visits a given node α before another given node β, or not. For $\Gamma(4,4)$ (the square grid) we show in [15] that the reduction satisfies the conditions needed to make (P) *P*-hard; the case of $\Gamma(6,3)$ (the hexagonal grid) is analogous. Taking only the part of the graphs which is being used for the construction of each circuit, we see that the problem (P') of answering the same question for a given finite graph and a given initial configuration is also *P*-hard.

3.3   Universality

In a cellular automata (CA), a *quiescent state* is defined by the following property: if a cell and all its neighbors are in the quiescent state, the cell remains in it at the next iteration. Hence, all the dynamics of the system takes place at the cells in non-quiescent states and their neighbors. An initial configuration with a finite support (i.e., a finite number of non-quiescent states) will keep this property through the iterations of the CA.

The transition rule of a CA can be calculated with a multi-output finite boolean circuit. So, for a given one-dimensional CA with quiescent state, we can

define an initial configuration on the grid consisting of infinitely many copies of
this circuit, arranged in an infinite trapezoidal array with top row of length L, as
shown in Figure 6. Any initial configuration of the CA whose support has width
less than L can be written as the input of the first row, and the ant simulates the
CA. For widths bigger than L, we just put the initial configuration in a lower
row, and let the ant start running from the appropriate node.

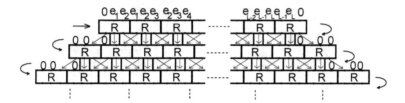

Fig. 6. The ant simulates each iteration of the CA in a row of gates, crosses the
repetitions of the outputs (preparing the next input) and goes to the next row.
R stands for the circuit that calculates the rule.

The undecidability of some CA problems is inherited by the ant system. For
instance, the problem of knowing whether a given (finite) word will ever appear in
the evolution of a given one-dimensional CA, for a given initial configuration with
finite support, is reduced to the problem of deciding whether a given finite block
ever appears in the evolution of the ant, for a given infinite initial configuration
of the grid. Since any Turing machine, in particular a universal one, can be
simulated by a one-dimensional CA with quiescent state, the ant is also universal.

4 Limitations in Highly Connected Graphs

When the underlying graph has degree strictly greater than 4, the ant cannot
reach all the nodes of the graph, given a fixed starting position. In the triangular
grid, for instance, the unique simple path is a zigzagging line.

For most of the following results, we will consider a generalization of the
bi-regular graphs: *A graph is said to verify (H), if all its nodes have degree $\geq d$,
and its dual graph is k-regular, with $d = 5$ and $k \geq 4$, or $d \geq 6$ and $k \geq 3$.*

The proof of the following lemma is based on [16], and uses relations between
the number of nodes, edges and faces enclosed by the cycle. Here we call *ant-cycle*
an ant-path that is a cycle.

Lemma 2. *For a graph verifying (H), the unique simple ant-cycles are the faces.*

If one tries to design an ant-path so as to form a cycle different from a
face, soon it is noticed that the origin, as well as an infinity of other nodes, are
impossible to reach. This is exactly what Lemma 3 says.

Lemma 3. *Let us consider a $\Gamma(k,5)$ graph and Figure 7a. Then, no simple ant-path starting with the nodes (a_0, b) may exit the shaded zone.*

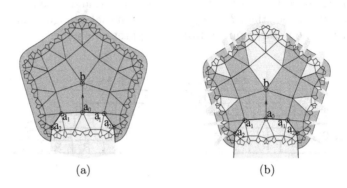

(a)   (b)

Fig. 7. The boundary in (a) is composed by the two edges incident to a_0 to the right and to the left of (a_0, b), and the edges found by adding recursively the edges adjacent to the last ones, so as to leave two edges inside of the zone.

Applying recursively Lemma 3 to each edge of each simple ant-path, we obtain that in fact the simple ant-paths are restricted to the sub-graph shown in Figure 7b. This graph is a fractal tree of tiles, with degree $k-1$. The nodes have degree 4 or 5 in this sub-graph, except a_0, which has degree 3. Moreover, this fact is not only true for simple paths, and with the help of following lemma we can apply it to arbitrary ant-paths.

Lemma 4. *Let G be a graph verifying (H). If the ant begins between two nodes in the same state, then, a node that can be reached by the ant, can be also reached through simple ant-paths.*

But the ant is frequently between two nodes in the same state. Indeed, it cannot avoid this situation for more than k steps, for k consecutive equal states would bring it back to the first node. The ant will therefore be always restricted to a subgraph like the one described above. Since this sub-graph is defined independently of the degree of the graph (only the lower bound is required), the result requires only (H), and we obtain the next theorem.

Theorem 4. *Let G be a graph verifying (H). Then, each time the ant is between two nodes in the same state, its future trajectory is restricted to the sub-graph depicted in Figure 7b.*

The following theorem shows that any problem related to the behavior of the ant over a finite configuration is decidable. In general, there are many cases where the trajectory of the ant turns out to be easily predicted, due to the restricted behavior, and in particular, the trajectory of the ant is found to be unbounded for any initial configuration.

Theorem 5. *Let G be a graph verifying (H). Starting over a finite initial configuration, the ant always falls in a periodic motion with drift. The period of this eventual behavior is $(k-1)(k+1)$.*

5 Conclusions

We studied the generalization of Langton's ant to different planar graphs, with special emphasis in the point of view of complexity. Several constructions and formal results were obtained, and can be useful in future studies.

In the general cases, a high degree of unpredictability was seen. A hint for this is the existence of families of finite graphs were the period of the system may grow exponentially with the size of the graph. A further hint is the existence of P-hard problems included in the prediction of the dynamics of the system, for the family of the finite graphs, and for the square and hexagonal grids.

Infinite bi-regular graphs were studied, dividing them in two cases: first, the graphs with degree 3 or 4, and second, the graphs with degree equal or greater than 5. A natural reason for this division is found in Lemma 1 and Theorem 4: in the graphs of the second case, the ant cannot go from any location to any other location, whereas in the first case this is always possible. This difference seems to have deep implications, for the results obtained in the different cases, even if not directly contradictory, point towards different levels of complexity.

In the first case (low degrees), which includes the classical square grid, a method for the evaluation of boolean circuits was found. This was used to show the universality of the system, and to show the existence of undecidable problems related to the trajectory of the ant.

In the second case, there are strong restrictions for the trajectory of the ant, who can only walk on a tree of tiles. This forbids the construction of circuits like the ones in the first case. Moreover, its behavior is decidable for initial configurations with finite support.

Acknowledgments

A. Gajardo and A. Moreira are partially supported by CONICYT Ph.D. fellowships. A. Gajardo and E. Goles are partially supported by ECOS. E. Goles is also partially supported by the FONDAP program in Applied Mathematics, and by the FONDECYT project nr. 1970398. We wish to thank J. Mazoyer and C. Papazian for their useful comments.

References

1. Langton, C.: Studying Artificial Life with Cellular Automata. Physica D **22** (1986) 120–149
2. Gale, D. Tracking the Automatic Ant and Other Mathematical Explorations. Springer Verlag, New York (1998)

3. Troubetzkoy, S.: Lewis–Parker Lecture 1997 The Ant. Alabama J. Math. **21(2)** (1997) 3–13
4. Dewdney, A.K.: Computer Recreations: Two-dimensional Turing Machines and Tur-mites Make Tracks on a Plane. Sci. Am. (September 1989) 124–127
5. Kong, X., Cohen, E.: Diffusion and propagation in a triangular Lorentz lattice gas cellular automata. J. of Stat. Physics **62** (1991) 737–757
6. Grosfils, P., Boon, J.P., Cohen, E., Bunimovich, L.: Propagation and organization in a lattice random media. J. of Stat. Physics **97 (3-4)** (1999) 575-608
7. Wang, F.: Ph.D. Dissertation. The Rockefeller University (1995)
8. Gajardo, A., Mazoyer, J.: Langton's ant on graphs of degree 3. Preprint (1999)
9. Gajardo, A., Goles, E.: Ant's evolution in a one-dimensional lattice. Preprint (1998)
10. Bunimovich, L.: Walks in Rigid Environments. Preprint (2000)
11. Bunimovich, L., Troubetzkoy, S.: Recurrence properties of Lorentz Lattice Gas Cellular Automata. J. Stat. Phys. **67** (1992) 289–302
12. Burks, A.W.: Essays on Cellular Automata. Univ. Illinois Press (1970)
13. Margolus, N.: Physics-like models of computations. Physica D **10** (1984) 81-95
14. Papadimitriou, C. Computational Complexity. Addison-Wesley (1994)
15. Gajardo, A., Moreira, A., Goles, E.: Complexity of Langton's Ant. To appear in Discrete Appl. Math.
16. Papazian, C., Rémila, E.: Some Properties of Hyperbolic Discrete Planes. Accepted for DGCI 2000

Optimal and Approximate Station Placement in Networks[*]

(With Applications to Multicasting and Space Efficient Traversals)

Clemente Galdi[1], Christos Kaklamanis[2],
Manuela Montangero[1], and Pino Persiano[1]

[1] Dipartimento di Informatica ed Applicazioni
Università di Salerno, 84081, Baronissi (SA), Italy
{clegal,montange,giuper}@dia.unisa.it
[2] Computer Technology Institute
Dept. of Computer Engineering and Informatics
University of Patras, 26500 Rio, Greece
kakl@cti.gr

Abstract. In this paper we study the *k-station placement* problem (*k*-SP problem, in short) on graphs. This problem has application to efficient multicasting in circuit-switched networks and to space efficient traversals. We show that the problem is NP-complete even for 3-stage graphs and give an approximation algorithm with logarithmic approximation ratio. Moreover we show that the problem can be solved in polynomial time for trees.

Keywords: Multicasting, approximation algorithms, distributed systems, networks.

1 Introduction

In this paper we introduce and study the *k*-station placement problem on graphs. Consider a communication network modeled by a weighted directed graph $G = (V, E)$ where the length $l(e)$ of an edge e is the cost of sending a message along that edge. Any vertex u of G that needs to communicate with another vertex v does so by first establishing a virtual circuit p from u to v along a path connecting u and v and then by sending the message. The cost of establishing a virtual circuit p is the sum of the lengths of the edges of p. Suppose now, that we are given a distinguished *source vertex s* and a set D of *destination vertices* and that s needs to multicast a message to the vertices of D. One possible way of performing this task would be for s to establish a virtual circuit with each vertex v of D along the shortest path between s and v. This approach has the advantage that transmission is achieved in one step but its cost might be very high: one

---

[*] Partially supported by a grant from the Università di Salerno (Italy) and by Progetto cofinanziato MURST 40% Resource Allocation in Communication Networks.

A. Ferreira and H. Reichel (Eds.): STACS 2001, LNCS 2010, pp. 271–282, 2001.

edge might be used by multiple circuits with the effect that its cost would be added with same multiplicity to the total cost. A different approach would be to identify a set S_1 of intermediate stations and assign each vertex of D to a vertex of S_1. The communication takes places in two steps. First, s establishes a virtual circuit with each of the vertices of S_1 and transfers the message. In the second phase, each vertex of S_1 establishes circuits with its assigned destinations so that the message is finally delivered to the destination vertices. In general, one can have k sets $S_1, \cdots, S_k$ of intermediate stations. The communication takes place in k steps: the source vertex establishes a circuit with each of the vertices of S_1 (level 1 stations); vertices of S_i (level i stations), $1 \leq i < k$, establish circuits with the vertices of S_{i+1} (level $i+1$ stations) and finally vertices of S_k establish circuits with the vertices of D.

There is a clear tradeoff between the number of steps needed to complete the multicasting (that is the number of intermediate stations along a path from s to a destination vertex) and the cost of the transmission: 1-step communication incurs in high cost; having each vertex as an intermediate destination yields minimum cost multicasting but it completely wastes the performance offered by circuit-switched network.

In this paper we study the problem of allocating intermediate stations in a graph so that at most k station are encountered on a path from the source to a destination and the cost of multicasting is minimum.

The k-SP problem has also applications to the problem of of traversing an ordered binary tree T. Each vertex of the tree has pointers to the left and right child only and we are provided with a pointer to the root of the tree. The inorder traversal of the tree reaches all the leaves of the tree in time $O(n)$ (here n is the number of vertices of T). However, in the worst case, it needs $\Omega(n)$ registers to store the addresses of the vertices of T for which the traversal has not been completed yet. This is hidden by the recursive approach often used to present the inorder traversal. Alternatively, one might consider the following approach. Each leaf v of a binary tree of depth h is uniquely identified by the binary string PATH(v) of length at most h that describes the path from the root to v (0 stands for a link to a left child and 1 for a link to a right child). Thus, we have the following very simple algorithm: for each leaf v, start from the root of T and follow the path specified by PATH(v). As it is easily seen this algorithm does not need any additional register to store addresses of vertices of the tree but, on the other hand, its running time is proportional to the path length of the tree that can be $\Omega(n^2)$. In general, one might ask to perform the fastest traversal of the tree given that only k registers are available. As it will be clear in the sequel, this problem is closely related to the k-SP problem.

The k-Station Placement Problem. We now formally define the k-Station Placement (k-SP) problem.

Definition 1 (The k-SP Problem). *An instance of the k-SP problem consists of a directed graph $G = (V, E)$, a length function ℓ defined over the edges of G, an integer k, a source vertex s and a set of destination vertices D.*

A feasible solution to the k-SP problem (or k-placement) consists of $k+1$ sets of stations $S_1, \cdots, S_k, S_{k+1}$, with $S_{k+1} = D$, and k assignments $\phi_1, \cdots, \phi_k$, with ϕ_i mapping vertices of S_{i+1} into vertices of $S_i \cup \{s\}$. For every i, function ϕ_i must satisfy the following property (that we will call strictness): for any $v \in S_{i+1}$, the lightest path from v to $\phi_i(v)$ does not contain any vertex in S_{i+1} other than v.

The cost of a feasible solution $P = (S_1, \cdots, S_k, \phi_1, \cdots, \phi_k)$ is

$$Cost(P) = \sum_{i=0}^{k} \sum_{v \in S_{i+1}} w(\phi_i(v), v),$$

where $\phi_0(v) = s$ for all $v \in S_1$ and $w(u, v)$ is the cost of the shortest path from u to v according to ℓ.

The task is to compute a feasible solution of minimum cost.

The *strictness* property guarantees that the shortest path from the source to any destination node contains no more than k stations. If the property does not hold, we have a new problem we call the k-Unrestricted Station Placement problem, or k-USP problem.

Going back to the example of multicasting in a network G, we observe that the minimum-cost k-hop multicasting is obtained by solving the k-SP problem on G. The k-placement gives the k sets $S_1, \cdots, S_k$ of intermediate destinations and specifies, by means of the ϕ_i's, the virtual connections each vertex of S_i has to establish (*i.e.*, $v \in S_i$ has to establish a virtual circuit with each vertex $u \in S_{i+1}$ such that $\phi_i(u) = v$). The cost of the k-placement is the sum of the lengths of the circuits that are established to accomplish multicasting.

Let us now briefly discuss how the k-SP problem can be used to design the fastest traversal of a tree T using at most k registers to store pointers to vertices. Suppose we have the solution to the k-SP problem on T with s equal to the root of T and the set D equal to the set of leaves of T. The traversal proceeds in the following way. From the root, we reach each of the vertices of S_1. While at $s_1 \in S_1$, we recursively traversal the tree rooted at s_1 using $k - 1$ registers (one register is used to keep a pointer to s_1). It is easy to see that the cost of the k-placement is equal to the time spent to perform the visit.

Missing proofs and other generalizations of the problems can be found into the final version of the paper.

Related Problem. The Steiner tree problem defined as follows shows some similarity to the k-SP problem.

Definition 2. Steiner Tree Problem
INSTANCE: a simple graph $G = (V, E)$, a weight function $w(e) \in N$ for each edge $e \in E$, a target subset $D \subseteq V$ of vertices.
TASK: find a minimum weight subtree of G that covers all vertices in D.

It is well known that this problem is NP-Complete ([ND12] in [4]). The k-SP problem differs from the Steiner tree problem because of different cost functions.

However, in Section 4, we show how an approximate solution to the k-SP can be derived from an approximate solution to a special variation on the Steiner tree problem.

Roadmap. In Section 2, we show that the k-SP problem is NP-complete for any value of k, even if we consider multi-stage directed graphs with all the edges having the same length. In Section 3.1, we present a polynomial-time algorithm k-SP-TREE for the special case of trees and in Section 3.3 we present an algorithm based on dynamic programming for the 1-USP problem on constant degree trees. In Section 4, we give approximation algorithms for the k-USP problem.

2   Hardness Result for Multi-stage Graphs

In this section we first prove that the 1-SP problem is NP-Complete even on 3-stage directed graphs by reducing Set Cover (SC for short) to this problem. Based on this, we also prove that for any k, the k-SP problem is NP-Complete on $(k + 2)$-stage directed graphs. We further show similar results for undirected multi-stage graphs. The decisional version of k-SP problem is the following:

Definition 3. (Decisional k-Station Placement (k-DSP))
INSTANCE: (G, s, D, ℓ, B) where $G = (V, E)$ is a simple connected graph, $s \in V$ is the source, $D \subseteq V$ is a set of destinations, $\ell : E \to \mathcal{N}$ is a positive function, representing length of edges, and B is a positive integer.
QUESTION: is there a feasible solution $(S_1, \ldots, S_k, \phi_1, \ldots, \phi_k)$ to the k-SP problem on G such that $Cost(S_1, \ldots, S_k, \phi_1, \ldots, \phi_k) \leq B$?

We now briefly recall the definition of decisional-SC and p-stage graphs and then we show the reduction of decisional-SC to 1-SP problem.

Definition 4. Decisional Set Cover
INSTANCE: (T, C, B) where C is a collection of subsets of a finite set T and B is an integer.
QUESTION: is there a set cover for S (i.e., a subset $C' \subseteq C$ such that every element in S belongs to at least one member of C') of cardinality less or equal to B?

The problem has been shown to be approximable within $1 + \ln |T|$ in [5] and not approximable within $(1 - \epsilon) \ln |T|$ for any $\epsilon > 0$ unless $NP \subset DTIME(n^{\log \log n})$ in [3].

Definition 5 (p-Staged Graphs). *A p-stage graph $G = (V, E)$ is a directed graph whose vertices can be partitioned into p sets $V_1, \ldots, V_p$ such that for every edge $(u, v) \in E$, $u \in V_i$ and $v \in V_{i+1}$ or vice-versa for some $i = 1, \ldots, p - 1$. A weighted p-stage graph is a p-stage graph with weights on edges. A strong p-stage graph is a p-stage graph with edges directed from V_i to V_{i+1} for $i = 1, \ldots, p - 1$.*

Theorem 1. 1-*DSP on weighted strong 3-stage graphs is NP-Complete even if all the edges have the same weight.*

Proof. Obviously, the 1-DSP problem is in NP. We reduced Decisional-SC to 1-DSP. Suppose $I = (T, C, B)$ is an instance of Decisional-SC in which $C = \{C_1, \ldots, C_n\}$ and, for $i = 1, \ldots, n$, $C_i \subseteq T = \{t_1, \ldots, t_m\}$. We construct a quintuple $I' = (G, s, D, \ell, B')$ such that if I' belongs to 1-DSP then I belongs to Decisional-SC, where $G = (V, E)$ is a strong the following 3-stage graph:

$$V = \{s\} \cup \{C_1, \ldots, C_n\} \cup \{t_1, \ldots, t_m\}$$
$$E = \{(s, C_i) \mid i = 1, \ldots n\} \cup \{(C_i, t_j) \mid t_j \in C_i\}$$

$D = T$ and, w.l.o.g., we assume that for any $e \in E$, $\ell(e) = 1$.

Let (S^*, ϕ^*) be a 1-placement for G such that $Cost(S^*, \phi^*) \leq B + |D|$ for source s and set of destinations $D = \{t_1, \ldots, t_m\}$.

Suppose, at first, that $S^* \subseteq C$, then, by definition, S^* is a set cover for C. Moreover, the cardinality of the set cover is less or equal to B:

$$B + |D| \geq Cost(S^*, \phi^*) = \sum_{v \in S^*} w(s, v) + \sum_{d \in D} w(\phi^*(d), d) = \sum_{v \in S^*} 1 + \sum_{d \in D} 1 = |S^*| + |D|.$$

Suppose, now, that $S^* \not\subseteq \{C_1, \ldots, C_n\}$. Then, by the *strictness* property, either $S^* = \{s\}$ or S^* contains some vertex in $\{t_1, \ldots, t_m\}$. In both cases, we show how to construct, in polynomial time, a new feasible 1-placement (S, ϕ) for G such that $S \subseteq C$ and $Cost(S, \phi) \leq Cost(S^*, \phi^*)$:

1. $S^* = \{s\}$: Notice that $Cost(S^*, \phi^*) = 2|D|$ because, for every $d \in D$, $w(s, d) = 2$. Define a new function $\phi : D \to \{C_1, \ldots, C_n\}$ that associates, to every $d \in D$, a vertex C_i on one path from s to d. Let $S = \{\phi(d) | d \in D\}$ and consider the new 1-placement (S, ϕ); by construction

$$Cost(S, \phi) = |S| + |D| \leq 2|D| = Cost(S^*, \phi^*).$$

2. There exists a node t such that $t \in S^* \cap \{t_1, \ldots, t_m\}$: Also in this case, we construct a new feasible solution by substituting t in S^* with one of its parent in C. As t has no outgoing edges, there is no destination vertex $d \in D$, different from t, such that $\phi^*(d) = t$ and thus the cost of this new placement is less or equal to the cost of (S^*, ϕ^*).   □

We can prove that the problem is NP-complete also for undirected graphs.

Theorem 2. 1-*DSP on undirected weighted 3-stage graphs is NP-Complete.*

Moreover, by the known non-approximability results [3] of Set Cover we have the following corollary.

Corollary 1. 1-*DSP on weighted 3-stage graphs is not approximable within* $(1 - \epsilon) \ln |D|$ *for any* $\epsilon > 0$, *unless* $NP \subset DTIME(n^{\log \log n})$, *even if all edges have the same weight.*

Lemma 1. *For all $i \geq 1$, i-DSP reduces to $(i+1)$-DSP.*

Proof. Let $\mathcal{I} = (G = (V, E), s \in V, D \subseteq V, \ell : E \to \mathcal{N}, B \in \mathcal{N})$ be an instance of the i-DSP problem. Construct graph $G' = (V \cup \{z_1, z_2\}, E \cup \{(z_1, s), (s, z_2)\})$ and let $\ell' : E' \to \mathcal{N}$ be the natural extension of ℓ to G' such that $\ell'(z_1, s) = |V| \sum_{v,u \in V} w(v, u)$ and $\ell'(s, z_2) > 0$. Consider the following instance of $(i+1)$-DSP problem: $\mathcal{I}' = (G', z_1, D \cup \{z_2\}, \ell', B + \ell'(z_1, s) + \ell'(s, z_2))$.

We now show how to derive a feasible solution $P = (S_1, S_2, \ldots, S_i, \phi_1, \ldots, \phi_i)$ to instance $\mathcal{I}$ given a feasible solution $P' = (S'_1, S'_2, \ldots, S'_{i+1}, \phi'_1, \ldots, \phi'_{i+1})$ to instance $\mathcal{I}'$. By the *strictness* property we deduce that if $s \in S'_1$ then $S'_1 = \{s\}$ and, thus, given P' only the following cases can arise:

1. $S'_1 = \{s\}$: if there is only one station at the first level and this station is exactly s, restricting P' to graph G, we obtain a feasible solution P for $\mathcal{I}$.

2. $S'_1 \neq \{s\}$: we can construct a new feasible solution $P'' = (S''_1, S'_2, \ldots, S'_{i+1}, \phi''_1, \phi'_2, \ldots, \phi'_{i+1})$ to $\mathcal{I}'$, such that $S''_1 = \{s\}$ and, for every $v \in S'_2$, $\phi''_1(v) = s$. P'' is a feasible solution to $\mathcal{I}'$, and $Cost(P') \geq Cost(P'')$:

$$Cost(P') = w(z_1, z_2) + \sum_{v \in S'_1} w(z_1, v) + \sum_{i=1}^{k} \sum_{v \in S'_{i+1}} w(\phi'_i(v), v)$$

$$= \ell'(z_1, s) + \ell'(s, z_2) + |S'_1|\ell'(z_1, s) + \sum_{v \in S'_1} (s, v) + \sum_{i=1}^{k} \sum_{v \in S'_{i+1}} w(\phi'_i(v), v)$$

$$\geq \ell'(z_1, s) + \ell'(s, z_2) + |S'_1|\ell'(z_1, s) + \sum_{v \in S'_2} w(\phi'_1(v), v) + \sum_{i=2}^{k} \sum_{v \in S'_{i+1}} w(\phi'_i(v), v) \quad (1)$$

$$\geq \ell'(z_1, s) + \ell'(s, z_2) + \sum_{v \in S'_2} w(s, v) + \sum_{i=2}^{k} \sum_{v \in S'_{i+1}} w(\phi'_i(v), v) = Cost(P'') \quad (2)$$

where, to go from (1) to (2), we used the fact that $\ell'(z_1, s) \geq \sum_{v \in S'_2} w(s, \phi'_1(v))$. Restricting P'' to graph G, we have a feasible solution P for $\mathcal{I}$ and

$$B + \ell'(z_1, s) + \ell'(s, z_2) \geq Cost(P'') = \ell'(z_1, s) + \ell'(s, z_2) + Cost(P).$$

$\square$

From Theorem 1 and Lemma 1 we obtain:

Corollary 2. *k-DSP on weighted strong $(k+2)$-stage graphs is NP-Complete even if all edges have the same weight.*

From Theorem 2 and Lemma 1 we obtain:

Corollary 3. *The k-DSP problem is NP-Complete on $(k+2)$-stage graphs.*

Finally, by Lemma 1, Corollary 1, we have the following non-approximability result:

Theorem 3. *The k-DSP problem is not approximable within* $(1 - \epsilon) \ln |D|$ *for any* $\epsilon > 0$ *unless* $NP \subset DTIME(n^{\log \log n})$.

3   Optimal Placement on Trees

In this section we present polynomial-time algorithms for the k-SP problem and the k-USP on directed trees. We first present an algorithm for the 1-SP problem and, then, extend it to the general case $k > 1$. We then show a simple dynamic-programming algorithm for the 1-USP problem.

W.l.o.g., we can assume that the set D of destinations is the set of leaves in the tree T. Indeed, starting from T we can remove the leaves that are not in D and for any internal vertex $v \in D$, we can add a new leaf d_v, that becomes a new destination, and a new edge (v, d_v) with cost zero, obtaining a new tree T'. It is easy to see that solving the problem on T is equivalent to solve the problem on T'.

In the following, for any vertex v, we denote by $T(v)$ the subtree rooted at v, by $L(v)$ the set of leaves in $T(v)$ and by $p(v)$ the parent of v in T.

3.1   Optimal 1-Station Placement on Directed Trees

We present algorithm 1-SP-TREE for the 1-SP problem on a n-vertex tree T with source s and set of destinations D consisting of all the leaves of T.

The algorithm associates, in $O(n)$ time, a cost $c(u, v)$ to every edge (u, v) of the tree in the following way:

$$c(u, v) \stackrel{\text{def}}{=} w(s, v) + \sum_{d \in L(v)} w(v, d). \qquad (3)$$

Referring to the k-hop multicasting example, we can think that $c(u, v)$ corresponds to the cost of multicasting to the vertices of $D \cap T(v)$ by placing one station at node v: the first term is the cost of sending the message from the source to v and the second is the cost of sending messages from v to the vertices of $D \cap T(v)$ without using any other intermediate station.

Next, algorithm 1-SP-TREE constructs a graph G by adding a new vertex t to T and by connecting the vertices of D to t using infinite-cost edges. Then the algorithm computes a minimum cut of G with respect to source s and sink t. Let C be the computed cut; the algorithm outputs placement (S_1, ϕ_1) such that: S_1 consists of all the vertices v such that the edge $(p(v), v)$ belongs to C; function ϕ_1 assigns to each vertex $v \in D$ its closest ancestor that belongs to S.

Theorem 4. *Algorithm 1-SP-TREE, on input an n-vertex tree T, outputs an optimal solution to the 1-SP problem on T in time $O(M(n))$, where $M(n)$ is the running time of the fastest MIN-CUT algorithm on n vertex graphs.*

Proof. The analysis of the running time is obvious. Correctness follows from two observations: first, by construction, placement (S_1, ϕ_1) output by 1-SP-TREE is a feasible solution. Second, for every 1-placement $P = (S, \phi)$ on T, we can derive a legal cut C' for G such that $Cost(P) = Cost(C')$: cut C' is simply composed by the edges $(p(v), v)$ for every $v \in S$. The proof now simply follows by contradiction.   □

3.2   Optimal k-Station Placement on Trees

In the following, we say that, given a k-placement $P = (S_1, \cdots, S_k, \phi_1, \cdots, \phi_k)$, the subplacement of P with respect to a vertex $t \in S_i$, denoted by $P|_t$, is the restriction of P to $T(t)$. We start with the following lemma.

Lemma 2. *Let $P = (S_1, \cdots, S_k, \phi_1, \cdots, \phi_k)$ be an optimal k-placement for a tree T, then for every station $t \in S_1$ the subplacement $P|_t$ is an optimal $(k-1)$-placement for the subtree rooted in t.*

Proof. By the way of contradiction, assume that P is an optimal k-placement for a tree T and that there exists one vertex $t \in S_1$ for which $P|_t$ is not an optimal $(k-1)$-placement and, thus, there exists a new placement $P'|_t$ with lower cost. We can, thus, construct a new placement P' for T, substituting $P'|_t$ to $P|_t$, such that $Cost(P') < Cost(P)$ contradicting the hypothesis.   □

Algorithm k-SP-TREE works in k phases: the first phase computes optimal solution for the 1-SP problem for $T(v)$, for each vertex v; phase $j > 1$, computes, for every vertex v, an optimal j-placement for $T(v)$ using the optimal $(j-1)$-placements computed at the previous phase. In details:

Phase 1: For every v in T compute an optimal 1-placement $P_1(v)$ for $T(v)$ and its corresponding cost. This is done by running algorithm 1-SP-TREE.

Phase $1 < j < k$: For every node v in T compute an optimal j-placement $P_j(v)$ for $T(v)$ by defining costs for every edge (x, y) in $T(v)$ in the following way: $c(x, y) = w(v, y) + Cost(P_{j-1}(y))$. Compute, then, a MIN-CUT on this subtree (notice that $Cost(P_{j-1}(y))$ has been computed for every y during the previous phase).

Phase k: Compute an optimal k-placement for T defining new costs for every edge (u, v) in T in the following way: $c(u, v) = w(s, v) + Cost(P_{k-1}(v))$. Compute, then, a MIN-CUT on T.

The correctness of algorithm k-SP-TREE follows directly from Lemma 2. Moreover observe that k-SP-TREE has to solve $O(k \cdot n)$ min-cut problem on n vertex graphs and, thus, its running time is $O(k \cdot n \cdot M(n))$.

Theorem 5. *Algorithm k-SP-TREE, on input a weighted tree with n vertices, a distinguished vertex s, a set of destinations D and an integer k, outputs a k-placement of minimum cost in time $O(k \cdot n \cdot M(n))$.*

3.3   Optimal 1-Unrestricted Station Placement on Trees

We will now present algorithm 1-USP-TREE based on dynamic programming to solve the 1-USP problem on a binary directed tree T where the source s is the root of T. For the sake of presentation, we assume the tree to be binary, but the results can be easily extended to constant degree trees. In the following we will say that a destination d is served by a station v if $\phi(d) = v$.

Notice, first, that an optimal 1-placement has an optimal substructure; in fact, with a proof analogous to the one of Lemma 2, we can prove that:

Lemma 3. *Given a tree T rooted in s and an optimal 1-placement OPT for T, let u_1 and u_2 be the children of s. Then, $OPT|_{u_1}$ and $OPT|_{u_2}$ are optimal 1-placements for $T(u_1) \cup \{s\}$ and $T(u_2) \cup \{s\}$, respectively, where the source is s for both placements and $L(u_1)$ and $L(u_2)$ are the destination sets, respectively.*

The proof is analogous to the one of Lemma 2.

We now define the value of an optimal solution recursively in terms of the optimal solutions to subproblems. A subproblem is defined as determining the cost of a placement in a subtree rooted in v placing no more than k stations and serving all the leaves of $T(v)$, except a given number λ.

Given tree T, let $\mathcal{C}(v, k, \lambda)$ be the minimum cost of serving destinations in $L(v)$, using at most k stations and knowing that there are exactly λ leaves of $T(v)$ that are served by some station placed in the up going path from v to s; i.e., these leaves will not be served by the k stations we will place in $T(v)$. We define $\mathcal{C}(v, k, \lambda)$ in the recursive following way:

If v is a leaf, then $\mathcal{C}(v, k, 1) = 0$, while for $\lambda \neq 1$, we have $\mathcal{C}(v, k, \lambda) = +\infty$.

If v is not a leaf, let u_1 and u_2 be its children. $\mathcal{C}(v, k, \lambda)$ is calculated choosing the cheapest solution between placing or not placing a station in v and looking for the optimal solutions for $T(u_1)$ and $T(u_2)$. This is done according to the following constraints:

1. if we place (respectively not place) a station in v, then we can not place more than $k - 1$ (resp. k) stations in the subtrees;

2. if we place a station in v, then this station will serve $f > 0$ leaves of $T(v)$, $f_1 \leq |L(u_1)|$ of them in $L(u_1)$ and $f_2 \leq |L(u_2)|$ in $L(u_2)$.

3. if $\lambda > 0$, then $\lambda = \lambda_1 + \lambda_2$ leaves are served by a station ancestor of v such that $0 \leq \lambda_1 \leq |L(u_1)|$ are in $T(u_1)$ and $0 \leq \lambda_2 \leq |L(u_2)|$ are in $T(u_2)$.

Now, let $\mathcal{C}_Y(v, k, \lambda)$ (respectively $\mathcal{C}_N(v, k, \lambda)$) be the cost of placing (resp. non placing) a station in v and placing $k - 1$ (resp. k) stations in the subtrees, knowing that $\lambda \geq 0$ leaves are served by an ancestor station. Then,

$$\mathcal{C}(v, k, \lambda) = \begin{cases} \min \{\mathcal{C}_Y(v, k, \lambda), \mathcal{C}_N(v, k, \lambda)\} & \text{if } \lambda \leq |L(v)| \\ +\infty & \text{otherwise} \end{cases}$$

Both $\mathcal{C}_Y(v, k, \lambda)$ and $\mathcal{C}_N(v, k, \lambda)$ are the sum of two terms: the first counts how many times edges (v, u_1) and (v, u_2) are traversed; i.e., edge (v, u_i), $i = 1, 2$, is traversed once for every station placed in $T(u_i)$, once for every leaf in $L(u_i)$ served by a station ancestor of v and once for every leaf in $L(u_i)$ served by v.

The second is the recursive call on $T(u_1)$ and $T(u_2)$ with the proper value of the parameters. In details:

1. If $\lambda = 0$ then $\mathcal{C}_N(v, k, 0)$ is equal to:

$$\min_{0 \le x+y \le k} \{\ell(v, u_1) \cdot x + \ell(v, u_2) \cdot y + \mathcal{C}(u_1, x, 0) + \mathcal{C}(u_2, y, 0)\}$$

2. If $\lambda \ne 0$ then $\mathcal{C}_N(v, k, \lambda)$ is equal to:

$$\min_{\substack{0 \le x+y \le k, \\ \lambda_1 + \lambda_2 = \lambda, \\ \lambda_i \le |L(u_i)|, i = 1, 2}} \{\ell(v, u_1)(x + \lambda_1) + \ell(v, u_2)(y + \lambda_2) + \mathcal{C}(u_1, x, \lambda_1) + \mathcal{C}(u_2, y, \lambda_2)\}$$

In fact, if we do not place a station in v and we do not serve any leaf of an ancestor station, we simply look for the best way to distribute up to k stations in the subtrees. If we serve some leaves of an ancestor station, we have to find the cheapest way to distribute these too.

Before giving the definition of $\mathcal{C}_Y(v, k, \lambda)$, we need the following lemma:

Lemma 4. *Given any optimal solution $OPT = (S, \phi)$ to the $1 - USP$ problem on binary tree T and a leaf l, there does not exist a vertex $v \in S$, different from $\phi(l)$, that belongs to the unique path going from $\phi(l)$ to l.*

It is important to notice that the previous Lemma does not state that two stations cannot be on the same root-leaf path, but only that, in the optimal solution, leaves are served by the closest station.

As a consequence of the Lemma 4, we define $\mathcal{C}_Y(v, k, \lambda) \ne +\infty$ only when $\lambda = 0$. Thus, if $\lambda = 0$, $\mathcal{C}_Y(v, k, 0)$ is equal to

$$\min_{\substack{0 \le x+y \le k-1, \\ f_1 + f_2 > 0, \\ 0 \le f_i \le |L(u_i)|, i = 1, 2}} \{\ell(v, u_1)(x + f_1) + \ell(v, u_2)(y + f_2) + \mathcal{C}(u_1, x, f_1) + \mathcal{C}(u_2, y, f_2)\}$$

In fact, if we place a station in v we only have up to $k - 1$ station to place in the subtrees and we have to find most convenient set of leaves of $T(v)$ to be served by v.

Finally, the cost of an optimal solution OPT for tree T rooted in s is calculated in the following way:

$$Cost(OPT) = \mathcal{C}(s, n, 0).$$

To recover OPT, it is sufficient to remember the vertices in which we placed the stations and this gives us set S. Function ϕ is easily determinate using Lemma 4: given leaf l, $\phi(l)$ is the first vertex of S we find on the upgoing path from l to the root of the tree.

Theorem 6. *Algorithm 1-USP-TREE on binary trees runs in $O(n^5)$.*

Proof. For every vertex v in T and each of the $O(n^2)$ pairs (x, y) such that $x + y \le n$ we have to consider the $O(n^2)$ pairs (λ_1, λ_2) such that $\lambda_1 + \lambda_2 \le n$. and the $O(n^2)$ pairs (f_1, f_2) such that $f_1 + f_2 > 0$ and $f_1, f_2 \le n$. □

4   Approximation Algorithm on General Graphs

In Section 2 we have shown that the k-SP problems is NP-Complete. This implies the following:

Corollary 4. *The k-DUSP problem is NP-Complete for every k.*

In this section we show an approximation algorithm for this problem on general graphs. The key idea of the algorithm is to reduce the k-USP to the problem of computing a Steiner tree with bounded depth on a graph. Let $K_{|V|}$ be the complete graph over $|V|$ vertices in which the weight of edge (x, y) is the cost of the shortest path from x to y in the graph G. The cost of the $(k+1)$-depth Steiner minimum tree of $K_{|V|}$ is equal to the cost of the optimal solution of the k-USP problem on G. For the sake of presentation, we will present only the case $k = 1$, but similar arguments can be used for the case $k > 1$.

Lemma 5. *Let T be a Steiner tree rooted at s for graph $K_{|V|}$ with depth at most 2 and target D. There exists a 1-placement P for the graph G with source s on destination set D such that $Cost(P) = Cost(T)$.*

Proof. The placement P is constructed as follows: the set of stations consists of the vertices at level 1 in the tree T. For each vertex in $v \in D$, $\phi_1(v)$ is the parent of v in the tree T.   □

Using a dual argument it is possible to prove the following:

Lemma 6. *Let P be a 1-placement for the graph G on destination set D and source s. There exists a Steiner tree T rooted at s with maximum height 2 on the clique $K_{|V|}$ with target D such that $Cost(P) = Cost(T)$.*

Lemma 7. *Let P^* be an optimal 1-placement for a graph G on destination set D and let T^* be a minimum Steiner tree on the complete graph of shortest paths in G. It holds that:*

$$Cost(P^*) = Cost(T^*)$$

Proof. Assume, by contradiction, that $Cost(P^*) < Cost(T^*)$. By Lemma 6, it is possible to construct a new Steiner tree T' on $K_{|V|}$ with destination D such that $Cost(T') = Cost(P^*) < Cost(T^*)$. But this contradicts the hypothesis that T^* was a minimum Steiner tree. A dual argument can be used to prove that if $Cost(P^*) > Cost(T^*)$, then P^* is not an optimal placement.

Since the problem of computing a Minimum Steiner tree is NP-Complete, we use an approximation algorithm for this problem in order to obtain an approximation algorithm for the k-USP. We recall the following results by Kortsarz and Peleg:

Theorem 7. *[6] Let $G = (V, E)$ be a graph and let $D \subseteq V$ be a set of destinations. There is an approximation algorithm for the minimum Steiner tree problem on G with destination D and maximum depth d with approximation ratio $O(\log |D|)$, for any constant d, and $O(|D|^\epsilon)$, for any $\epsilon > 0$, for general d.*

Given the discussion above, the following theorems can be easily proven:

Theorem 8. *For any constant k, there exists approximation algorithm for the k-USP with optimal approximation ratio $O(\log |D|)$.*

Corollary 5. *For any k, and for any $\epsilon > 0$ there exists an $O(|D|^\epsilon)$ approximation algorithm for the k-USP.*

5   Open Problems

The immediate open problem left in our work is the design of better approximation algorithms for general graphs and d. Also, we do not know of any natural class of graphs (other than trees) for which the problem can be solved efficiently.

From a more combinatorial point of view it would be interesting to ask if there exists a function $f(\cdot)$ such that for all trees with n nodes the cost of the best $f(n)$-SP is $O(n)$. It is obvious that any function $f(n) = \Omega(n)$ will do (just put a station in any vertex). We can show that $f(n) = \log^* n$ works for complete binary trees but could not extend this result to general trees.

Acknowledgments

We would like to thank the anonymous referee for his useful comments used to improve the non-approximabily result.

Bibliography

1. G. Ausiello, P. Crescenzi, G. Gambosi, V. Kann, A. Marchetti Spaccamela and M. Protasi, *Complexity and Approximations - Combinatorial Optimalization Problems and their Approximability Properties*, Springer-Verlag, 1999.
2. J. Bar-Ilan and G. Kortsarz and D. Peleg, *Generalized Submodular Cover Problem and Applications*, Proc. 4th Israel Symposium on Theory of Computing and Systems, pp. 110-118, 1996.
3. U. Feige, *A threshold of ln n for approximating set cover*, Proc. 28th ACM Symposium on Theory of computation, pp. 314-318, 1996.
4. M.R. Garey and D.S. Johnson, *Computers and Intractability - A Guide to the Theory of NP-Completeness*, W.H. Freeman and Company, 1979, New York.
5. D. S. Johnson, *Approximation Algorithms for Combinatorial Problems*, J. Computers and System Sciences, Vol 9, pp. 256-278, 1974.
6. G. Kortsarz and D. Peleg, *Approximating the Weight of Shallow Steiner Tree*, Proc. SODA 97, 1997.
7. J. Plesní k, *The Complexity of Designing a Network with Minimum Diameter*, Networks, Vol 11, pp. 77-85, 1981.

Learning Expressions over Monoids
(Extended Abstract)

Ricard Gavaldà[1][*] and Denis Thérien[2][**]

[1] Department of Software (LSI), Universitat Politècnica de Catalunya
gavalda@lsi.upc.es
[2] School of Computer Science, McGill University
denis@cs.mcgill.ca

Abstract. We study the problem of learning an unknown function represented as an expression over a known finite monoid. As in other areas of computational complexity where programs over algebras have been used, the goal is to relate the computational complexity of the learning problem with the algebraic complexity of the finite monoid. Indeed, our results indicate a close connection between both kinds of complexity. We focus on monoids which are either groups or aperiodic, and on the learning model of exact learning from queries. For a group G, we prove that expressions over G are easily learnable if G is nilpotent and impossible to learn efficiently (under cryptographic assumptions) if G is nonsolvable. We present some partial results for solvable groups, and point out a connection between their efficient learnability and the existence of lower bounds on their computational power in the program model. For aperiodic monoids, our results seem to indicate that the monoid class known as DA captures exactly learnability of expressions by polynomially many Evaluation queries.

1 Introduction

Formal models of the process of learning have been proposed since the 60s to give mathematical foundation to machine learning tasks, mostly to concept learning. The first models, such as Gold's identification in the limit, were of recursion-theoretic flavor and did not emphasize efficient use of computational resources. In the mid 80s, several models that take time and memory into consideration were proposed, thus allowing the use of concepts and tools from computational complexity theory in the study of learnability.

The resulting area, known as Computational Learning Theory, has produced an important number of results stating that certain classes of functions are or are

[*] Partially supported by the EC ESPRIT Working group NeuroCOLT2, the EC ESPRIT AlcomFT project, and by FRESCO PB98-0937-C04-04. Part of this research was done while this author visited McGill University, supported by NSERC and FCAR.

[**] Partially supported by NSERC and FCAR grants and by the von Humboldt Foundation.

A. Ferreira and H. Reichel (Eds.): STACS 2001, LNCS 2010, pp. 283–293, 2001.

not efficiently learnable in rigorously defined learning models. But, in general, the properties that determine whether a class is easy or hard to learn are not yet well identified. We propose an algebraic approach that, in the long run, might help in clarifying such properties.

Programs over finite monoids and other algebraic structures are models of computation that have been successfully used to expose the deep reasons why some computational problems can or cannot be solved within certain resources. In this paper we initiate the study of formal models of function learning from an algebraic point of view, i.e., we would like to determine the complexity of learning a class of functions from the classes of algebras that are powerful enough to compute the class of functions (in the program model or related ones). We concentrate on the case where the algebra M consists of an associative operation with an identity, i.e., when M is a monoid.

So far, programs over monoids have been studied mostly as devices to compute boolean functions, and many circuit complexity classes have been shown to admit characterizations as those problems solved by programs over particular algebraic structures [4,6,5]. To avoid replicating this study, we look instead at programs over M as computing functions from M^n into M. *Expressions* over M are a particular type of programs that appear very naturally in this context.

We study the problem of learning an unknown *target* function from M^n to M, for a fixed and known finite monoid M. It is assumed only that the function is computed by some expression on n variables over M. We work mostly in Angluin's query-based model of exact learning [1,2], where algorithms can ask Evaluation queries, or Equivalence queries, or both.

Note that this problem is not obviously comparable to the problem of learning the class of boolean circuits corresponding to programs over monoid M. This is because, on the one hand, the problem might be harder because the function class is richer. On the other hand, the answers to the queries provide finer information than in the boolean setting, namely, elements of the monoid, and this could help in learning.

We present several results on the complexity of learning expressions over specific classes of monoids. We concentrate on monoids that are either *groups* or *aperiodic*, as these two classes are known to be the building blocks of all monoids via the so-called *wreath product* operation.

Along the paper, we often say "a monoid M is learnable" for the sake of brevity, meaning "expressions over monoid M are learnable", and similarly for a class of monoids.

For the case of groups, we prove:

- Expressions over Abelian groups are polynomial-time learnable both from a linear number of Evaluation queries and from a linear number of Equivalence queries.
- Nilpotent groups (a generalization of Abelian groups) are learnable in polynomial time from Evaluation queries.
- Solvable groups formed as extensions of a group Z_p by a group Z_q (for any p and q) are polynomial-time learnable from Evaluation and Equivalence

queries in the form of Multiplicity Automata, and learnable in probabilistic polynomial time from Evaluation queries alone.

- A slightly larger subclass of solvable groups can be identified by a probabilistic strategy using polynomially many Evaluation queries, and by a deterministic strategy using quasipolynomially many Evaluation queries (though no claims are made regarding computation time). We show that, in fact, these or similar results will hold for any group for which we can prove a certain type of lower bound on its computation power; in other words, such lower bounds on computation power provide upper bounds on learning complexity.
- Expressions over nonsolvable groups are not learnable unless NC^1 circuits also are, even in the very strong learning model of PAC-prediction with Membership queries [3]. Recall that under plausible cryptographic assumptions, NC^1 circuits are not learnable in this or any other standard model of learning [3].

In the aperiodic case, our results involve the class of aperiodic monoids known as DA [15]. For some algorithmic problems on monoids, it is known that feasibility depends essentially on membership to DA. For example, the membership problem is known to be PSPACE-complete for any aperiodic monoid outside of DA [7]. Also, the word problem for an aperiodic monoid can be resolved in sublinear communication complexity (in the 2-player setting) iff the monoid belongs to DA [14]. Our results are:

- Expressions over DA monoids are identifiable from polynomially many Evaluation queries (though possibly not in polynomial time).
- For a subclass of DA, idempotent R-trivial monoids, we can in fact give a polynomial-time learning algorithm using Evaluation queries.
- It is known that there are exactly two minimal aperiodic monoids not belonging to DA. We show that expressions over any of these two minimal monoids are not learnable with subexponentially many Evaluation queries, even with arbitrary computation time. We conjecture the same is true for any monoid outside of DA, because it is also known that every monoid outside of DA is divided by at least one of these two minimal monoids.

Certainly the picture is still partial, with many upper and lower bounds missing. But our results seem to indicate a very close connection between the complexity of the monoid (in the algebraic sense) and its learning complexity.

Finally, let us comment on the relevance of these results for the mainstream of computational learning theory. A good deal of the effort in this theory has been on learning classes of boolean functions, and especially those inside NC^1 since it seems hopeless to try to learn any larger one. As mentioned, many of the central complexity classes definable by small-depth circuits can be also defined by programs over central classes of monoids. For example, polynomial-length programs over aperiodic monoids compute exactly the functions in AC^0, programs over solvable groups compute the functions in CC^0 (polynomial-size circuits made of mod_q gates for a fixed q), and nonsolvable groups (or monoids) compute all functions in NC^1 [4,6]. Formally, we do not know how to translate neither positive nor negative results in our model to the boolean case. On the other hand, in the learning context, various small fragments of AC^0 and CC^0 are known to be learnable, and NC^1 is known not be learnable in a strong sense, in fact a situation quite in analogy with the results in this paper. We believe they are interesting as they provide a new perspective on learning problems near the borderline of current knowledge.

Due to lack of space, this extended abstract does not contain the proofs of our results. The full version of the paper can be found on the respective home pages of the authors, http://www.lsi.upc.es/gavalda and http://www.cs.mcgill.ca/denis.

2 Preliminaries

2.1 Expressions and Programs over Monoids

A *monoid* is a pair $(M, \cdot)$ where M is a set and $\cdot$, the product over M, is a binary associative operation on M with an identity element. A *group* is a monoid where each element has a (two-sided) inverse with respect to $\cdot$. A monoid is *aperiodic* if it has no submonoid which is a non-trivial group. All monoids considered in this paper will be finite. We look at two computation models based on products over a monoid: expressions and programs.

An *expression* over monoid M and variables $x_1, \ldots, x_n$ is a string over the alphabet $M \cup \{x_1, \ldots, x_n\}$. Such an expression defines quite naturally a function from M^n to M: to evaluate the (function represented by the) expression over a vector or assignment $(w_1, w_2, \ldots, w_n)$ in M^n, replace with w_i each occurrence of each variable x_i in the expression, then multiply out the resulting string of monoid elements to obtain a single monoid element. For example, assume that a, b, c, d are four elements in M. Then the value of expression $ax_2x_1x_2bx_3cx_1x_3d$ on the assignment (c, b, a) is the element $a \cdot b \cdot c \cdot b \cdot b \cdot a \cdot c \cdot c \cdot a \cdot d$, where $\cdot$ is the product in M.

Expressions are a particular case of programs. A *program* over M with domain D is a list of instructions of the form (i, f), where $i \in \{1, \ldots, n\}$ and f is a function $D \mapsto M$. Instruction (i, f) is interpreted as follows: read the value of variable x_i and append $f(x_i)$ to the string of monoid elements to be multiplied. Hence, expressions are programs whose instructions use only constant functions and the identity function.

In the literature, programs have been used mostly to compute boolean functions [4,6]. In these *boolean programs*, domain D is {true,false} and M is partitioned into True and False sets to interpret a boolean result. Expressions could also be used to compute boolean functions, say by encoding true and false inputs by distinct elements of the monoid. We will be here mainly interested in programs and expressions computing functions from M^n into M, rather than boolean functions.

2.2   Learning Expressions over Monoids: Problem Statement

In this section we define the learning problem we study. We use mostly Angluin's query-based model of learning [1,2], although Valiant's PAC model [17] or related ones are mentioned occasionally. All definitions are standard for function learning in these models, although we give them only in the terms of our specific problem, learning expressions over monoids. For background in computational learning theory, the reader is referred to the books, surveys, and bibliography in the recently created server [13].

The task of a learning algorithm (or *learner*) is to find out a *target function* M^n to M fixed by a *teacher* in an adversary way. The function is assumed to be representable as an expression over M and variables $x_1, \ldots, x_n$, though not all variables are necessarily used. The learning algorithm is initially given n and some upper bound m on the length of a shortest expression for the target.[1] Monoid M is fixed and known both to the teacher and the learning algorithm.

The learning algorithm must produce an expression (or some other representation of the target function, as discussed later) equivalent to the target one on all of M^n.

To achieve learning, teacher and learner exchange information on the target function following some protocol; some specific protocols will be defined later.

Resources used by the algorithm are measured as a function of n and m. The class of expressions over M is learnable in time $t(n, m)$ in a given learning protocol if there is an algorithm that learns every expression over M in time $t(n, m)$. In particular, we study mostly whether expressions over a class of monoids is *polynomial-time learnable*, meaning whether for each monoid M in the class there is a polynomial $p(n, m)$ such that expressions over M are learnable in time $p(n, m)$.

Similarly, we say that expressions over a monoid M are *identifiable* with interaction $s(n, m)$ if there is an algorithm that learns every expression over M using an amount of interaction $s(n, m)$ with the teacher (and arbitrary computation time). The meaning of "amount of interaction" may be different in each learning protocol, but in general it has to be bounded by the number of bits of information exchanged by the teacher and the learning algorithm. Identifiability thus represents the information-theoretic cost of solving a learning task, ignoring the computational complexity of the problems that the learning algorithm has to solve internally at each stage of the process.

Let us stress that we assume that monoid M is fixed and known to the learning algorithm, and thus we regard $|M|$ as a constant. We often present algorithmic schemes to learn whole classes of monoids whose running time is exponential (or more) in the size of the monoid. We still call these algorithms "polynomial-time" as long as their dependence on n and m is polynomial. A stronger notion of "polynomial-time learnability" of a monoid class would ask for an algorithmic scheme depending only polynomially on $|M|$. In an even stricter

---

[1] Symbols n and m will always have this meaning in the paper, i.e., number of variables and an upper bound on the length of shortest expression for the target.

sense, we could ask for an algorithm that receives the multiplication table of M as part of the input, with the promise that M belongs to the monoid class, and is polynomial in $|M|$, n, and m; this would be truly "uniformly" learning of class, in the sense that the algorithm does not rely on hardwired information for each monoid.

As for interaction between teacher and learner, we consider two standard query types in Angluin's model: Evaluation and Equivalence queries. Let the target f be a function from set A to set B. In an Evaluation query, the learning algorithm produces an element $a \in A$ and the teacher must return $f(a)$.[2] In an Equivalence query, the learning algorithm produces a hypothesis h, representing a function $A \mapsto B$ in some way; the teacher must return Yes if $h \equiv f$ on A, or else a counterexample: an element $a \in A$ such that $f(a) \neq h(a)$, together with the value of $f(a)$. If hypotheses issued by the algorithm always belong to the same syntactic class of functions that is being learned, the algorithm is called *proper*. Otherwise, hypotheses may belong to a different and possibly richer class, and the algorithm is called *nonproper*. An important requirement on any such hypothesis class is that it must be polynomial-time evaluatable, i.e., that a given hypothesis can be evaluated on a given input in polynomial time.

Before investigating the learnability of specific classes of monoids, let us consider a quite general question. There are several constructions for building monoids from other monoids. The most natural ones are direct product, submonoid, and homomorphic image. It is a natural question whether learnability is preserved by these operations. We only have very partial answers so far.

Proposition 1.

1. *For the three models of query learning (Evaluation queries only, Equivalence queries only, or both) the following is true. If expressions over monoids S and T are polynomial-time learnable, then expressions over $S \times T$ are learnable (possibly nonproperly).*

2. *If expressions over T are polynomial-time learnable from Equivalence queries and S is a submonoid of T, then expressions over S are polynomial-time learnable from Equivalence queries.*

We do not know whether learnability under Evaluation queries is preserved by taking submonoids. The difficulty is that the algorithm for the larger monoid may ask queries on $T^n \setminus S^n$; the teacher, knowing only a target function $S^n \mapsto S$, is not able to answer these. In fact, the query is ill-posed as the answer may be different for different T-expressions defining the same target function over S. Under homomorphic image and either type of query, the problem lies in inverting the homomorphism on answers to queries in a way that is guaranteed to be consistent with some expression over the larger monoid.

---

[2] Evaluation queries generalize Membership queries [1,2] for functions with non-binary range.

3   Abelian Groups

Abelian or commutative groups are the simplest from an algebraic point of view. Quite naturally, they are easiest from the learning point of view, in the sense that they are learnable with a linear number of either Evaluation or Equivalence queries.

Theorem 1.

a) *Expressions over an Abelian group G are learnable with $n + 1$ Evaluation queries.*

b) *Expressions over an Abelian group G are learnable with $O(n)$ Equivalence queries.*

4   Nilpotent Groups

Let a and b be elements of a group G and let $[a, b] = aba^{-1}b^{-1}$ denote the commutator of a and b. These are the commutators of weight 2. Commutators of weight 3 are $[a, [b, c]]$ and $[[a, b], c]$ and commutators of weight k are defined inductively in the obvious way. We say that G is nilpotent of class-k iff all commutators of weight $k + 1$ are the identity, and observe that any commutator of any weight involving the identity it itself the identity.

It is clear that nilpotent groups of class 1 are exactly the Abelian groups. And indeed several properties of nilpotent groups are natural generalizations of those for Abelian groups. For example, it can be shown that n-variable functions that are realizable by programs over nilpotent groups of class k can always be represented (in the sense of [5]) by polynomials of degree k (with coefficients in an appropriate ring). Expressions over nilpotent groups are learnable from Evaluation queries alone. As in the Abelian case, the learning algorithm is based on the fact that programs can be rewritten to a normal form, although the transformation is more involved in this case.

Theorem 2. *Expressions over a nilpotent class-k group G are learnable with $|G|^k n^k$ Evaluation queries and $n^k |G|^{O(|G|^k)}$ time.*

For Equivalence queries, an approach like that in Theorem 1, part (b), seems difficult since it would involve solving polynomial equations over cyclic groups.

5   Solvable Groups

In this section we will present some partial results on learnability in solvable groups. For any two subsets A and B of a group G, denote by $[A, B]$ the subgroup generated by all commutators $[a, b]$, with $a \in A, b \in B$. We can then form the descending chain of subgroups $G_0, G_1, \ldots$ by setting $G_0 = G$ and $G_i = [G_{i-1}, G_{i-1}]$. The group G is solvable iff this chain goes to the trivial subgroup.

In order to present our results concerning solvable groups, we have to recall the following notion. Let G be a group, and suppose H is a normal subgroup of G. Then G is said to be an extension of H by the quotient group G/H and it admits the following representation.

We view elements of G as pairs in $H \times G/H$. For any pairs $(h_1, g_1), (h_2, g_2) \in G$, product in G can be expressed as:

$$(h_1, g_1) \cdot_G (h_2, g_2) = (h_1 \cdot_H f_{g_1, g_2}(h_2), g_1 \cdot_{G/H} g_2).$$

Functions $f_{g_1, g_2} : H \mapsto H$ for each $g_1, g_2 \in G/H$ are called the "twist functions" for G and describe the interaction between the two components of the group.

Our results in this section concern solvable groups which are extensions of Z_p by Z_q, where Z_p and Z_q are any cyclic groups. Note that, for example, the group S_3 of permutations on three points is an extension of Z_3 by Z_2.

From now on, we view elements of a group G as above as pairs in $Z_p \times Z_q$. For an element $g \in G$, we use notations $Z_p(g)$ and $Z_q(g)$ to denote the first and second elements of the pair associated to g, that is, we identify g with the pair $(Z_p(g), Z_q(g))$.

The learning algorithm for these groups uses Multiplicity Automata as hypothesis class. Multiplicity Automata over rings (MA for short) are an important generalization of classical automata. They were first used in the context of learning by Bergadano and Varricchio [10], who gave a polynomial-time algorithm for learning MA over fields by Equivalence and Evaluation (there called Multiplicity) queries. Later, Bshouty, Tamon, and Wilson [12] extended the algorithm to work over a large class of rings instead of fields, including all finite integer rings.

The algorithm for MA has been used to learn several other classes of functions [9,8]. In particular, [9] uses the MA learning algorithm to learn some classes of boolean circuits with modular gates and boolean permutation branching programs of width at most 4. These results are probably related to the connection between some solvable groups and MA that we find here.

We give here a working definition of Multiplicity Automata; for more systematic presentations see [10,12]. An MA over an alphabet Σ and a ring K is a nondeterministic finite automata where each transition triple (q, a, q') $(a \in \Sigma)$ is additionally labeled by an element of K. To each path in the automata we associate the value in K given by the product of all the labels along the path. The MA computes a function $M : \Sigma^* \mapsto K$ in the following way: for each input $w \in \Sigma^*$, $M(w)$ is the sum of the values of all nondeterministic paths defined by input w on the MA. A particular case of MA that we use here is when $\Sigma = K$ and the MA is evaluated on inputs of a fixed length n, so that $M : K^n \mapsto K$.

Multiplicity Automata are able to simulate expressions over the groups above. The proof is based on a somewhat careful study of the structure of their twist functions.

Theorem 3. *Let G be a group which is an extension of Z_p by Z_q. Then there is a function $f : Z_q \mapsto Z_p$ such that for every expression $E(x_1, \ldots, x_n)$ over G there is a multiplicity automata M over Z_p of size $O(n|E|)$ such that for all $a_1, \ldots, a_n \in G$,*

$$Z_p(E(a_1, \ldots, a_n)) = M(Z_p(a_1), \ldots, Z_p(a_n), f(Z_q(a_1)), \ldots, f(Z_q(a_n))).$$

Furthermore, for all vectors $u_1, \ldots, u_n \in Z_p$ and $v_1, \ldots, v_n \in Z_p$ such that some v_i is not in the range of f,

$$M(u_1, \ldots, u_n, v_1, \ldots, v_n) = 0.$$

Then, using the learning algorithm for MA over rings [12], we can prove:

Theorem 4. *For every group G as above, expressions over G are learnable in polynomial time using Evaluation and Equivalence queries. The Equivalence queries and the output of the algorithm are pairs formed by a Multiplicity Automata over the ring Z_p and an expression over Z_q.*

When p is prime, the groups we have been working with in this section can in particular be viewed as special cases of wreath products of Abelian groups by p-groups. This larger family has been studied in [5] where it was shown that any group of that form could not possibly compute the AND function via a program of subexponential length. Moreover, these groups were also shown to have the following property.

Say that a solvable group G is non-narrowing if there is a polynomial $p(m)$ such that for all program P and all $a \in G$, if there is an assignment $w \in G^n$ such that $P(w) = a$ then there are at least $|G|^n / p(length(P))$ such assignments. This property gives an identification strategy by polynomially many Evaluation queries (not necessarily a polynomial-time algorithm as we don't know how to efficiently obtain a hypothesis from the answers).

The remaining results in this section were obtained through discussions with Cris Moore.

Theorem 5. *If G is non-narrowing, then programs (hence, expressions) over G can be identified probabilistically from Evaluation queries in polynomial time.*

With the same argument, it is easy to show that an Equivalence query to the groups above can be simulated with high probability by polynomially many random Evaluation queries. We can combine this observation with the Equivalence and Evaluation query algorithm in Theorem 4.

Corollary 1. *Any solvable group which is an extension of Z_p (p prime) by Z_q is learnable from Evaluation queries in probabilistic polynomial time.*

We finally observe that an exponential lower bound on the length of programs over G that compute the AND function translates into a quasipolynomial upper bound on the number of Evaluation queries needed to identify a program over that group.

Theorem 6. *If programs over G cannot compute the AND function in subexponential length, then programs over G can be identified from $n^{O(\log m)}$ Evaluation queries.*

Note that it is conjectured in [5] that the exponential lower bound on the AND function holds for all solvable groups.

6   Hardness of Nonsolvable Groups

In this section we now show that expressions over nonsolvable groups are not polynomial-time learnable unless NC^{1}circuits are polynomial-time learnable too. Hence, under the cryptographic assumptions in [3], they are not learnable at all.

Theorem 7. *Expressions over nonsolvable groups are not learnable from Evaluation and Equivalence queries with any polynomial-time evaluatable class, unless NC^{1}circuits also are*

The proof works by reducing learnability of nonsolvable groups to that of boolean programs over simple non-Abelian groups, which are polynomially equivalent to NC^{1}circuits [4].

Conceptually, there are three parts in the reduction: 1) learning expressions over some nonsolvable group implies learning expressions over some simple non-Abelian group; this does not follow trivially from the fact that a nonsolvable group contains a simple non-Abelian one, because the learning algorithm for nonsolvable might conceivably use queries outside the simple non-Abelian one to learn it; 2) over a simple non-Abelian group, programs and expressions compute the same functions; 3) over a simple non-Abelian group, programs simulate boolean programs in a prediction-preserving sense.

7   Aperiodic Monoids

Theorem 8. *Expressions over a monoid in DA are learnable from a polynomial number of Evaluation queries and unbounded computation time.*

Although the answers to these many Evaluation queries identify uniquely the target, they give no obvious way to predict the value of the target on a different input. For a small subclass of DA we know how to reconstruct efficiently an expression for the target from these answers.

Theorem 9. *Expressions over idempotent R-trivial monoids are learnable in polynomial time from Evaluation queries. The same is true for aperiodic commutative monoids.*

It is known that there are exactly two minimal monoids outside of DA, named U and BA_2. Monoid U is the syntactic monoid of the language $(aa^\star b)^\star$, has 6 elements, and is known to be universal because programs over it can simulate DNF formulas [16]. Monoid BA_2 is the syntactic monoid of $(ab)^\star$, has 6 elements also, and is provably not universal.

We reduce the problem of learning expressions over U from Evaluation queries reduces to the problems of learning monotone DNF formulas from Membership queries, which requires exponentially many queries [11]. Similarly, learning expressions over BA_2 reduces to learning singleton sets by Membership queries.

Theorem 10. *Expressions over monoid U are not learnable from subexponentially many Evaluation queries, even using unbounded computation time. The same holds for monoid BA_2.*

Acknowledgments

We are indebted to Cris Moore for ideas and discussions that lead to Theorems 5, 1, and 6. We thank Pascal Tesson and José L. Balcázar for helpful discussions.

References

1. D. Angluin. Learning regular sets from queries and counterexamples. *Information and Computation*, 75:87–106, 1987.
2. D. Angluin. Queries and concept learning. *Machine Learning*, 2:319–342, 1988.
3. D. Angluin and M. Kharitonov. When won't membership queries help? *Journal of Computer and System Sciences*, 50:336–355, 1995.
4. D. Barrington. Bounded-width polynomial-size branching programs recognize exactly those languages in NC^1. *Journal of Computer and System Sciences*, 38:150–164, 1989.
5. D. M. Barrington, H. Straubing, and D. Thérien. Non-uniform automata over groups. *Information and Computation*, 89:109–132, 1990.
6. D. M. Barrington and D. Thérien. Finite monoids and the fine structure of NC^1. *Journal of the ACM*, 35:941–952, 1988.
7. M. Beaudry, P. McKenzie, and D. Thérien. The membership problem in aperiodic transformation monoids. *Journal of the ACM*, 39(3):599–616, 1992.
8. A. Beimel, F. Bergadano, N. Bshouty, E. Kushilevitz, and S. Varricchio. Learning functions represented as multiplicity automata. *Journal of the ACM*, 47:506–530, 2000.
9. F. Bergadano, N. Bshouty, C. Tamon, and S. Varricchio. On learning branching programs and small depth circuits. *Proc. 3rd European Conference on Computational Learning Theory (EuroCOLT'97), Springer-Verlag LNCS*, 1208:150–161, 1997.
10. F. Bergadano and S. Varricchio. Learning behaviors of automata from multiplicity and equivalence queries. *SIAM Journal on Computing*, 25:1268–1280, 1996.
11. N. Bshouty, R. Cleve, R. Gavaldà, S. Kannan, and C. Tamon. Oracles and queries that are sufficient for exact learning. *Journal of Computer and System Sciences*, 52:421–433, 1996.
12. N. Bshouty, C. Tamon, and D. Wilson. Learning matrix functions over rings. *Algorithmica*, 22:91–111, 1998.
13. The learning theory server. http://www.learningtheory.org.
14. J.-F. Raymond, P. Tesson, and D. Thérien. An algebraic approach to communication complexity. *Proc. ICALP'98, Springer-Verlag LNCS*, 1443:29–40, 1998.
15. M. Schützenberger. Sur le produit de concaténation non ambigu. *Semigroup Forum*, 13:47–75, 1976.
16. D. Thérien. Programs over aperiodic monoids. *Theoretical Computer Science*, 64(3):271–280, 29 1989.
17. L. Valiant. A theory of the learnable. *Communications of the ACM*, 27:1134–1142, 1984.

Efficient Recognition of Random Unsatisfiable
k-SAT Instances by Spectral Methods

Andreas Goerdt[1] and Michael Krivelevich[2*]

[1] Fakultät für Informatik, TU Chemnitz
09107 Chemnitz, Germany
`goerdt@informatik.tu-chemnitz.de`
[2] Department of Mathematics, Faculty of Exact Sciences, Tel Aviv University
Tel Aviv 69978, Israel
`krivelev@math.tau.ac.il`

Abstract. It is known that random k-SAT instances with at least cn clauses where $c = c_k$ is a suitable constant are unsatisfiable (with high probability). We consider the problem to certify efficiently the unsatisfiability of such formulas. A result of Beame et al. shows that k-SAT instances with at least $n^{k-1}/\log n$ clauses can be certified unsatisfiable in polynomial time. We employ spectral methods to improve on this: We present a polynomial time algorithm which certifies random k-SAT instances for k even with at least $2^k \cdot (k/2)^7 \cdot (\ln n)^7 \cdot n^{k/2} = n^{(k/2)+o(1)}$ clauses as unsatisfiable (with high probability).

Introduction

We study the complexity of certifying unsatisfiability of random k-SAT instances (or k-CNF formulas) over n propositional variables. (All our discussion refers to k fixed and then letting n be sufficiently large.) The probability space of random k-SAT instances has been widely studied in recent years for several good reasons. The most recent literature is [A2000],[Fr99], [Be et al98].

One of the reasons for studying random k-SAT instances is that they have the following sharp threshold behaviour [Fr99]: There exists a constant $c = c_k$ such that for any $\varepsilon > 0$ formulas with at most $(1 - \varepsilon) \cdot c \cdot n$ clauses are satisfiable whereas formulas with at least $(1 + \varepsilon) \cdot c \cdot n$ are unsatisfiable with high probability (that means with probability tending to 1 when n goes to infinity). In fact, it is by now not proven that c_k is a constant. It might be that $c_k = c_k(n)$ depends on n. However, it is known that c_k is at most $2^k \cdot \ln 2$ and the general conjecture is that c_k converges to a constant. For formulas with at least $2^k \cdot (\ln 2) \cdot n$ clauses the expected number of satisfying assignments of a random formula tends to 0 and the formulas are unsatisfiable with high probability. For 3-SAT instances much effort is spent to approximate the value of c_3. The currently best results are that c_3 is at least 3.125 [A2000] and at most 4.601 [KiKrKrSt98]. In [Du et al2000] it is claimed that $c_3 \leq 4.501$. (For $k = 2$ we have $c_2 = 1$ [ChRe92], [Go96].)

* Partially supported by a USA-Israeli BSF grant.

The algorithmic interest in this threshold is due to the empirical observation that random k-SAT instances at the threshold, i.e. with around $c_k n$ random clauses are hard instances. The following behaviour has been reported consistently in experimental studies with suitably optimised backtracking algorithms searching for a satisfying assignment,see for example [SeMiLe96] [CrAu96]: The average running time is quite low for instances below the threshold. For 3-SAT instances we observe: Formulas with at most $4n$ clauses are satisfiable and it is quite easy to find a satisfying assignment. A precipitous increase in the average running time is observed at the threshold. For 3-SAT: About half of the formulas with $4.2n$ clauses are satisfiable and it is difficult to decide if a formula is satisfiable or not. Finally a speedy decline to lower complexity is observed beyond the threshold. For 3-SAT: All formulas with $4.5n$ clauses are unsatisfiable and the running time decreases again (in spite of the fact that now always the whole backtracking tree must be searched.)

There are no general complexity theoretical results relating the threshold to hardness. The following observation is trivial: If we can efficiently certify almost all instances with dn clauses where d is above the threshold as unsatisfiable, then we can efficiently certify almost all instances with $d'n$ clauses where $d' > d$ as unsatisfiable by simply chopping off the superfluous clauses. The analogous fact holds below the threshold, where we extend a given formula by some random clauses. Analogous observations apply to the number of literals in clauses.

The relationship of hardness and thresholds is rather general and not restricted to satisfiability. It is known for k-colourability of random graphs with a linear number of edges. In [PeWe89] a peak in running time seemingly related to the threshold is reported. The existence of a threshold is proved in [AcFr99] but again the value and convergence to a constant are only known experimentally. For the subset sum problem which is of a quite different nature we have also this relationship between threshold and hardness: The threshold is known and some discussion related to hardness is found in [ImNa96].

Abandoning the general complexity theoretic point of view and looking at concrete algorithms the following results are known for random k-SAT instances: All progress approximating the threshold from below is based on the analysis of rather simple polynomial time heuristics. In fact the most advanced heuristic being analysed [A2000] only finds a satisfying assignment with probability of at least ε where $\varepsilon > 0$ is a small constant for 3-SAT formulas with at most $3.145n$ clauses. The heuristic in [FrSu96] finds a satisfying assignment for 3-SAT almost always for 3-SAT instances with at most $3.003n$ clauses. On the other hand the progress made in approximating the threshold from above does not provide us at all with efficient algorithms certifying the unsatisfiability of the formula at hand. Only the expectation of the number of satisfying assignments is calculated and is shown to tend to 0.

In fact beyond the threshold we have negative results: For arbitrary but fixed $d \geq 2^k \cdot \ln 2$ random k-SAT instances with dn clauses (are unsatisfiable and) have only resolution proofs with an exponential number, that is with at least $(1+\varepsilon)^n = 2^{\Omega(n)}$ clauses [ChSz88]. This has been improved upon by [Fu98],

[BePi96], and [Be et al98] all proving (exponential) lower bounds for somewhat larger clause/variable ratios. Note that a lower bound on the size of resolution proofs provides a lower bound on the number of nodes in *any* classical backtracking tree as generated by any variant of the well known Davis-Putnam procedure.

Provably polynomial time results beyond the threshold are rather limited by now: In [Fu98] it is shown that k-SAT formulas with at least n^{k-1} clauses allow for polynomial size resolution proofs and thus can be certified unsatisfiable efficiently. This is strengthened in [Be et al98] to the best result known by now: For at least $n^{k-1}/\log n$ random clauses a backtracking based algorithm proves unsatisfiability in polynomial time with high probability. (In fact the result of Beame et al. is slightly stronger as it applies to formulas with $\Omega(n^{k-1}/\log n)$ random clauses.)

We extend the number of clauses for which a provably polynomial time algorithm exists. We give an algorithm which works when the number of clauses is only n to a constant fraction on k (with high probability). Our algorithm certifies k-SAT instances for k even with at least $2^k \cdot (k/2)^7 \cdot (\ln n)^7 \cdot n^{k/2} = n^{(k/2)+o(1)}$ clauses as unsatisfiable. We thus get the first improvement of existing bounds for $k = 4$. To obtain our result we leave the area of strictly combinatorial algorithms considered by now. Instead we associate a graph with a given formula and show how to certify unsatisfiability of the formula with the help of the eigenvalue spectrum of a certain matrix associated to this graph. Note that the eigenvalue spectrum can be calculated in polynomial time by standard linear algebra methods.

Eigenvalues are used in two ways in the algorithmic theory of random structures: They can be used to find a solution of an NP-hard problem in a random instance generated in such a way that it has a solution (not known to the algorithm). An example for 3-colourability is [AlKa94]. They can also be used to prove the absence of a solution of an NP-problem. However these applications are somewhat rare at the moment. The most prominent example here is the expansion property of random regular graphs [AlSp92]. Note that the expansion property is coNP-complete [Bl et al81] and the eigenvalues certify the absence of a non-expanding subset of vertices (which is the solution in this case). Our result is an example of the second kind.

1   From Random Formulas to Random Graphs

We use the following notation throughout: $\mathrm{Form}_{n,k,m}$ is our probabilistic model of k-SAT instances with m clauses over n propositional variables. Most of the time we assume that k is even. $\mathrm{Form}_{n,k,m}$ is defined as follows: The probability space of clauses of size k, $\mathrm{Clause}_{n,k}$, is the set of ordered k-tuples of literals over n propositional variables $v_1, \ldots, v_n$. We write $l_1 \vee \ldots \vee l_k$ with $l_i = x$ or $l_i = \neg x$ where x is one of our variables. Our definition of $\mathrm{Clause}_{n,k}$ allows for clauses containing the same literal twice and clauses which contain a variable and its negation in order to simplify the subsequent presentation. We consider $\mathrm{Clause}_{n,k}$ as endowed with the uniform probability distribution: The probability of a clause is given by $P(l_1 \vee \ldots \vee l_k) = (1/(2n))^k$. $\text{Form}_{n,k,m}$ is the m-fold Cartesian product space of $\text{Clause}_{n,k}$. We write $F = C_1 \wedge \ldots \wedge C_m$ and $P(F) = (1/(2n))^{k \cdot m}$. There are several ways of defining k-SAT probability spaces. Our results refer to these spaces, too. We discuss this matter after the presentation of the algorithm.

Our algorithm uses the following graphs:

Definition 1. *Let $F \in \text{Form}_{n,k,m}$ be given. The graph $G = G_F$ depends only on the sequence of all-positive clauses of F:*

- *The set of vertices of G is $V = V_F = \{x_1 \vee \ldots \vee x_{k/2} \mid x_i \text{ a variable}\}$. We have $|V| = n^{k/2}$ and V is independent of F.*
- *The set of edges of G, $E = E_F$ is given by: For two different vertices $x_1 \vee \ldots \vee x_{k/2}$ and $y_1 \vee \ldots \vee y_{k/2}$ we have that $\{x_1 \vee \ldots \vee x_{k/2}, \; y_1 \vee \ldots \vee y_{k/2}\} \in E$ iff $x_1 \vee \ldots \vee x_{k/2} \vee y_1 \vee \ldots \vee y_{k/2}$ (or $y_1 \vee \ldots \vee y_{k/2} \vee x_1 \vee \ldots \vee x_{k/2}$) is a clause of F. Note that it is possible that $|E| < m$ as a clause might induce no edge or two clauses induce the same edge. Our definition does not allow for loops or multiple edges.*

The graph H_F is defined in a totally analogous way for the all-negative clauses of F. □

Recall that an *independent set* of a graph G is a subset of vertices W of G such that we have no edge $\{v, w\}$ in G where both $v, w \in W$. The independence number of the graph G denoted by $\alpha(G)$ is the number of vertices in a largest independent set. It is NP-hard to determine $\alpha(G)$.

Lemma 2. *If $F \in \text{Form}_{n,k,m}$ is satisfiable then*

$$\alpha(G_F) \geq (n/2)^{k/2} = (1/2)^{k/2} \cdot |V| \text{ or } \alpha(H_F) \geq (1/2)^{k/2} \cdot |V|.$$

As k remains constant when n gets large this means that we have independent sets consisting of a constant fraction of all vertices of G_F of H_F.

Proof. Let $\mathcal{A}$ be an assignment of the n underlying variables with the truth values $0, 1$ (where $0 = $ false and $1 = $ true) which makes F true. We assume that $\mathcal{A}$ assigns 1 to at least $n/2$ variables. Let S be this set of variables then F has no all-negative clause consisting only of literals over S. Therefore H_F has an independent set with at least $|S|^{k/2} \geq (1/2)^{k/2} \cdot n^{k/2}$ vertices. If the assignment assigns more than half of the variables a 0 the analogous statement applies to G_F. □

In the subsequent discussion we refer mainly to G_F. Of course everything applies also to H_F. We need to show that the distribution of G_F is just the distribution of a usual random graph. To this end let be $G_{n,m}$ be the probability space of random graphs with n labelled vertices and m different edges. Each graph is equally likely, that is the probability of G is $P(G) = 1/\binom{\binom{n}{2}}{m}$.

Lemma 3. *(1) Conditional on the event in $Form_{n,k,m}$ that $|E_F| = r$ the graph G_F is a random member of the space $G_{\nu,r}$ where $\nu = n^{k/2}$ is the number of vertices of G_F.*
(2) Let $\varepsilon > 0$. For $F \in Form_{n,k,m}$ the number of edges of G_F is between $m \cdot (1/2)^k \cdot (1 - \varepsilon)$ and $m \cdot (1/2)^k \cdot (1 + \varepsilon)$ with high probability.

Proof. (1) Let $V = \{x_1 \vee \ldots \vee x_{k/2} \mid x_i \text{ a variable}\}$ be the set of vertices. Let $G = (V, E)$ be a graph with $|E| = r$. We show further below that the probability of the event that $F \in Form_{n,k,m}$ induces the edges set E, denoted by $P(F; E_F = E)$, depends only on r, but is independent of the actual edge set E. This implies the claim because

$$P(F; E_F = E \mid |E_F| = r) = \frac{P(F; E_F = E)}{P(|E_F| = r)} = 1 / \binom{\binom{\nu}{2}}{r},$$

where the last equation holds because $P(F; E_F = E)$ is independent of E and therefore must be the same for all E with r edges.

It remains to show that $P(F; E_F = E)$ is independent of E. To this end we show that

$$P(F; E_F = E \mid F \text{ has exactly } s \text{ all-positive clauses})$$

is independent of E. This implies the claim because by conditioning

$$P(F; E_F = E) = \sum_{s \geq 0} P(F; F \text{ has } s \text{ positive clauses})$$

$$\cdot P(F; E_F = E \mid F \text{ has } s \text{ positive clauses}).$$

The distribution of $Form_{n,k,m}$ conditional on the set of formulas with exactly s all-positive clauses is the uniform one. We therefore just need to count the number of formulas F with exactly s positive clauses such that $E_F = E$. Each such formula F with $E_F = E$ is obtained exactly once by the following choosing process: 1. Pick a sequence of s positive clauses with k literals $(C_1, \ldots, C_s) \in (Clause_{n,k})^s$ which induce the edge set E. 2. Pick s positions from the m positions available and put the clauses $(C_1, \ldots, C_s)$ from left to right into the corresponding slots. 3. Fill the remaining $m - s$ positions of F with clauses containing at least one negative literal.

For 2 edge sets E, E' with r edges there is a natural (but technically not easy to describe) bijective correspondence between the $(C_1, \ldots, C_s)$ for E and the $(C_1', \ldots, C_s')$ for E' picked in step 1. Therefore the number of choosing possibilities is independent of the actual set E and we are done.

(2) The claim follows from the following statements which we prove further below:

- Let $\varepsilon > 0$ be fixed. The number of all-positive clauses of $F \in Form$ is between $(1 - \varepsilon) \cdot (1/2)^k \cdot m$ and $(1 + \varepsilon) \cdot (1/2)^k \cdot m$ with high probability.

- The number of all-positive clauses like $x_1 \vee \ldots \vee x_{k/2} \vee x_1 \vee \ldots \vee x_{k/2}$, that is with the same first and second half, is $o(m)$.
- The number of unordered pairs of positions of F on which we have positive clauses which induce only one edge, that is pairs of clauses $\{x_1 \vee \ldots \vee x_k, \; y_1 \vee \ldots \vee y_k\}$ where $\{x_1 \vee \ldots \vee x_{k/2}, \; x_{k/2+1} \vee \ldots \vee x_k\} = \{y_1 \vee \ldots \vee y_{k/2}, \; y_{k/2+1} \vee \ldots \vee y_k\}$ is also $o(m)$ with high probability.

This implies the claim of the lemma with ε slightly lower than the ε from the first statement above because we have only $o(m)$ clauses inducing no additional edge.

The first statement: This statement follows with Chernoff bounds because the probability that a clause at a fixed position is all-positive is $(1/2)^k$ and clauses at different positions are independent. The second statement: The probability that the clause at position i has the same first and second half is $(1/n)^{k/2}$. The expected number of such clauses in a random F is therefore $m \cdot (1/n)^{k/2} = o(m)$. The third statement: We fix 2 positions $i \neq j$ of F. The probability that the clauses at these positions have the same set of first and second halves is $2 \cdot (1/n)^k$ and the expected number of such unordered pairs is at most $m^2 \cdot 2 \cdot (1/n)^k = O(m/n)$ provided $m = O(n^{k-1})$ which we can assume. Let X be the random variable counting the number of unordered pairs of positions with clauses with the same first and second half and let $\varepsilon > 0$. Markov's inequality gives us

$$P(X > n^\varepsilon \cdot EX) \leq EX/(n^\varepsilon \cdot EX) = 1/n^\varepsilon.$$

Therefore we get that with high probability $X \leq n^\varepsilon \cdot (m/n) = o(m)$. $\qquad\square$

2 Spectral Considerations

Eigenvalues of matrices associated with general graphs are somewhat less common at least in Computer Science applications than those of regular graphs. The monograph [Ch97] is a standard reference for the general case. The easier regular case is dealt with in [AlSp92]. The necessary Linear Algebra details cannot all be given here. They are very well presented in the textbook [St88].

Let $G = (V, E)$ be an undirected graph (loopless and without multiple edges) with $V = \{1, \ldots, n\}$ being a standard set of n vertices. For $0 < p < 1$ we consider the matrix $A = A_{G,p}$ as in [KrVu2000] and [Ju82] which is defined as follows:

The $(n \times n)$-matrix $A = A_{G,p} = (a_{i,j})_{1 \leq i,j \leq n}$ has $a_{i,j} = 1$ iff $\{i, j\} \notin E$ and $a_{i,j} = -(1-p)/p = 1 - 1/p$ iff $\{i, j\} \in E$. In particular $a_{i,i} = 1$. As A is real and symmetric A has n real eigenvalues when counting them with their multiplicities. We denote these eigenvalues by $\lambda_1(A) \geq \lambda_2(A) \geq \cdots \geq \lambda_n(A)$.

Now we have an efficiently computable upper bound for $\alpha(G)$:

Lemma 4. *(Lemma 4 of [KrVu2000])* For any possible p $\lambda_1(A_{G,p}) \geq \alpha(G)$.

Proof. Proof: Let $l = \alpha(G)$. Then the matrix $A_{G,p}$ has an $l \times l$-block which contains only 1's. This block of course is indexed with the vertices from a largest independent set. It follows from interlacing with a suitable $l \times n$-matrix N (cf. Lemma 31.5, page 396 of [vLWi]) that $\lambda_1(A_{G,p})$ is at least as large as l. This is the claim.   □

In order to bound the size of the eigenvalues of $A_{G,p}$ when G is a random graph we rely on a suitably modified version of the following theorem:

Theorem 5. *(Theorem 2 of [FuKo81]) Let for $1 \leq i, j \leq n$ and $i \leq j$ $a_{i,j}$ be independent , real valued random variables (not necessarily identically distributed) satisfying the following conditions:*

- *$|a_{i,j}| \leq K$ for all $i \leq j$,*
- *the expectation $E a_{i,i} = \nu$ for all i,*
- *the expectation $E a_{i,j} = 0$ for all $i < j$,*
- *the variance $V a_{i,j} = E[a_{i,j}^2] - (E a_{i,j})^2 = \sigma^2$ for all $i < j$,*

where the values K, ν, σ are constants independent of n.

For $j \geq i$ let $a_{j,i} = a_{i,j}$ and let $A = (a_{i,j})_{1 \leq i,j \leq n}$ be the random $(n \times n)$-matrix defined by the $a_{i,j}$. Let the eigenvalues of A be $\lambda_1(A) \geq \lambda_2(A) \geq \cdots \geq \lambda_n(A)$. With probability at least $1 - (1/n)^{10}$ the matrix A is such that

$$\mathrm{Max}\{|\lambda_i(A)| \,|\, 1 \leq i \leq n\} = 2 \cdot \sigma \cdot \sqrt{n} + O(n^{1/3} \cdot \log n) = 2 \cdot \sigma \cdot \sqrt{n} \cdot (1 + o(1)).$$

□

We intend to apply this theorem to a random matrix $A = A_{G,p}$ where G is a random graph from the probability space $G_{n,m}$. However, in this case the entries of A are not strictly independent and Theorem 5 cannot be directly applied. We first consider random graphs from the space $G_{n,p}$ and proceed to $G_{n,m}$ later on. Recall that a random graph G from $G_{n,p}$ is obtained by inserting each possible edge with probability p independently of other edges.

For p constant and G a random member from $G_{n,p}$ the assumptions of Theorem 5 can easily be checked to apply to $A_{G,p}$. However, for sparser random graphs that is $p = p(n) = o(1)$ the situation changes. We have that $a_{i,j}$ can assume the value $-1/o(1) + 1$ and thus is not any more bounded above by a constant. The same applies to the variance: $\sigma^2 = (1-p)/p = 1/o(1) - 1$.

It can however be checked that the proof of Theorem 5 as given in [FuKo81] goes through as long as we consider matrices $A_{G,p}$ where $p = (\ln n)^7/n$. In this case we have that $K = n/(\ln n)^7 - 1$ and $\sigma = n/(\ln n)^7 - 1$. With this modification and the other assumptions just as before the proof of [FuKo81] leads to:

Corollary 6. *With probability at least $1 - (1/n)^{10}$ the random matrix A satisfies*

$$\begin{aligned} \mathrm{Max}\,\{|\lambda_i(A)| \,|\, 1 \leq i \leq n\} &= 2 \cdot \sigma \cdot \sqrt{n} + O(n/(\ln n)^{22/6}) \\ &= 2 \cdot (1/(\ln n)^{7/2}) \cdot n \cdot (1 + o(1)). \end{aligned}$$

Proof. We sketch the changes which need to be applied to the proof of Theorem 2 in [FuKo81]. These changes refer to the final estimations of the proof on page 237. We set

$$k := (\sigma/K)^{1/3} \cdot n^{1/6} = (\ln n)^{7/6}(1 + o(1)),$$

in fact k should be the closest even number. We set the error term

$$v := 50 \cdot n/(\ln n)^{22/6}.$$

We have

$$2 \cdot \sigma \cdot \sqrt{n} = 2 \cdot n/(\ln n)^{7/2} = 2 \cdot n/(\ln n)^{21/6}$$

which implies that $v = o(2 \cdot \sigma \cdot \sqrt{n})$. Concerning the error estimate we get

$$\frac{v \cdot k}{2 \cdot \sigma \cdot \sqrt{n} + v} = \frac{50 \cdot (\ln n)^{7/6}}{(\ln n)^{1/6}} \cdot (1 + o(1)) = 50 \cdot \ln n \cdot (1 + o(1)).$$

This implies the claim.   □

Together with Lemma 4 we now get an efficiently computable certificate bounding the size of independent sets in random graphs from $G_{n,m}$.

Corollary 7. *Let G be a random member from $G_{n,m}$ where $m = ((\ln n)^7/2) \cdot n$. and let $p = m/\binom{n}{2} = (\ln n)^7/(n-1)$. We have with high probability that*

$$\lambda_1(A_{G,p}) \le 2 \cdot (1/(\ln n)^{7/2}) \cdot n \cdot (1 + o(1)).$$

Proof. The proof is a standard transfer from the random graph model $G_{n,p}$ to $G_{n,m}$. For G random from $G_{n,p}$ the induced random matrix $A_{G,p}$ satisfies the assumptions of the last corollary. We have that with probability at least $1 - (1/n)^{10}$ the eigenvalues of $A_{G,p}$ are bounded by $2 \cdot (1/(\ln n)^{7/2}) \cdot n \cdot (1 + o(1))$.

By the Local Limit Theorem for the binomial distribution the probability that a random graph from $G_{n,p}$ has *exactly* m edges is of $\Omega(1/(n \cdot p)^{1/2}) = \Omega(1/(\ln n)^{7/2})$. This implies the claim as the probability in $G_{n,p}$ that the eigenvalue is not bounded as claimed is $O((1/n)^{10}) = o(1/(\ln n)^{7/2})$. (We omit the formal conditioning argument.)   □

3   The Algorithm

We consider the probability space of formulas Form $=$ Form$_{n,k,m}$ where k is even and the number of clauses is

$$m = 2^k \cdot (\ln n^{k/2})^7 \cdot n^{k/2} = 2^k \cdot (k/2)^7 \cdot (\ln n)^7 \cdot n^{k/2}.$$

Given a random formula F from Form the algorithm first considers the all-positive clauses from F and constructs the graph G_F. From Lemma 3 we know that $G = G_F$ is a random member of $G_{\nu,\mu}$ where $\nu = n^{k/2}$ and $\mu \ge m \cdot (1/2)^k \cdot (1 - \varepsilon) = (\ln \nu)^7 \cdot \nu \cdot (1 - \varepsilon)$, where we fix $\varepsilon > 0$ sufficiently small, in fact $\varepsilon = 1/2$

will do. In case the number of edges is smaller than this bound the algorithm fails.

The algorithm determines the matrix $A = A_{G,p}$ where $p = \mu/\binom{\nu}{2} \geq (\ln \nu)^7/(\nu - 1)$. ¿From Corollary 7 we get that with high probability

$$\lambda_1(A) \leq 2 \cdot (1/(\ln \nu)^{7/2}) \cdot \nu \cdot (1 + o(1)) < (1/2)^{k/2} \cdot \nu$$

for n sufficiently large. In case the second inequality does not hold the algorithm fails. By Lemma 4 G_F has no independent set with $(1/2)^{k/2} \cdot \nu$ vertices.

The algorithm proceeds in the same way for the all negative clauses and the graph H_F. In case it succeeds (which happens with high probability) we have that F is unsatisfiable by Lemma 2.

In case we want to apply this algorithm when the number of literals per clause k is odd we first extend each clause by a random literal. The algorithm succeeds when the number of clauses is $2^{k+1} \cdot ((k+1)/2)^7 \cdot (\ln n)^7 \cdot n^{(k+1)/2}$.

Some technical matters come up when this algorithm is applied to other k−SAT probability spaces used in the literature. The first problem arises when the formulas are defined such that clauses are not allowed to contain the same literal several times. This implies that certain edges are excluded from the graph G_F and we cannot any more speak of a random graph.

The probability that a random clause from our space $\text{Clause}_{n,k}$ has the same literal several times is bounded from above by $O(k^2 \cdot (1/n)) = O(1/n)$ and bounded from below by $1/n$. Thus the expected number of clauses with the same literal several times in a formula from the space $\text{Form}_{n,k,m}$ is $O(m/n)$. Recall that $m > n^{k/2}$ for our algorithm to work so there are quite a few clauses with double occurrences. By the Local Limit Theorem for the binomial distribution with parameters m and $\Theta(1/n)$ the probability that a formula from $\text{Form}_{n,k,m}$ has exactly the expected number of clauses with double occurrences is $\Omega(1/(m/n)^{1/2})$. Let $\nu = n^{k/2}$ and $m = \Theta((\ln \nu)^7 \cdot \nu)$ then still we have that $O(1/\nu)^{10}) = o(1/(\ln \nu)^{7/2} \cdot 1/(m/n)^{1/2})$ cf. the proof of Corollary 7.

Now, given a random sequence of clauses F' without double occurrences of literals we add randomly exactly the expected number of clauses with double occurrences of literals to get the formula F. Then we apply our algorithm to the resulting formula. With high probability (by the above local limit consideration) the algorithm certifies that G_F has only independent sets with $o(\nu)$ vertices given the number of clauses of F is $m = 2^k \cdot (\ln \nu)^7 \cdot \nu$. After deleting the edges of G_F which are induced by the $O(m/n)$ double occurrence clauses any independent set can only increase by $O(2m/n)$ vertices. This implies that we still have no linear size independent set of vertices in G'_F. This and the same consideration for the graph $H_{F'}$ certifies the unsatisfiability of F'.

The remaining variants of probability spaces (clauses as sets, formulas as sets, picking each clause with a probability p) can more easily be dealt with.

Conclusion

By now a large part of the algorithmic theory of random structures is concerned with efficient algorithms *finding* solutions to an NP-problem. Often the probability spaces used are designed in such a way that we know a solution is present and the algorithm then must find it (or any other solution).

The present paper is concerned with the complementary aspect. We certify efficiently the *absence* of a solution to an NP-problem which we know not to be present by non-efficient means. It seems that this aspect is by now somewhat neglected in the algorithmic theory of random structures. We think it deserves more attention as it may lead to natural questions about natural probability spaces. Spectral methods are one way to deal with these problems. A paper in the same spirit is the recent [KrVu2000] where the *non-existence* of a colouring with a given number of colours is certified by spectral methods.

One problem which can be directly treated based on the ideas developed here is the 3-colouring problem of sparse random graphs: For random graphs with $c \cdot n$ edges the following facts are known: For $c \leq 1.932$ graphs are 3-colourable with high probability [AcMo97]. For $c \geq 2.522$ graphs are not 3-colourable [DuZi98] [AcMo]. There is a sharp threshold [AcFr99] with experimental hardness.

The results of the present paper imply that for $c = c(n) \geq 1/2 \cdot (\ln n)^7$ we can efficiently certify that we do not have any more an independent set with $n/3$ vertices (with high probability). Therefore we have no 3-colouring.

The following two problems however seem to require new ideas: First, the efficient certification of unsatisfiability of formulas with less than $n^{k/2}$ clauses. The problem here is that the average degree in the graph G_F now is $o(1)$ and the bounds on the eigenvalues make no sense. Second, to improve the bound of $n^2/\log n$ known for 3-SAT.

References

[A2000]   Dimitris Achlioptas. Setting 2 variables at a time yields a new lower bound for random 3-SAT. In Proceedings SToC 2000, ACM.

[AcFr99]   Dimitris Achlioptas, Ehud Friedgut. A threshold for random k-colourability. Random Structures and Algorithms 1999.

[AcMo97]   Dimitris Achlioptas, Mike Molloy. Analysis of a list colouring algorithm on a random graph. In Proceedings FoCS 1997, IEEE.

[AcMo]   Dimitris Achlioptas, Mike Molloy. Almost all graphs with 2.522n edges are not 3-colourable. Undated manuscript.

[AlKa94]   Noga Alon, Nabil Kahale. A spectral technique for colouring random 3-colourable graphs (preliminary version). In Proceedings 26th SToC, 1994, ACM, 346-355.

[AlSp92]   Noga Alon, Joel H. Spencer. The Probabilistic Method. Wiley & Sons Inc., 1992.

[Be et al98]   Paul Beame, Richard Karp, Toniann Pitassi, Michael Saks. On the complexity of unsatisfiability proofs for random k-CNF formulas. In Proceedings 30th SToC, 1998, ACM, 561-571.

[BePi96]   Paul Beame, Toniann Pitassi. Simplified and improved resolution lower bounds. In Proceedings 37th FoCS, 1996, IEEE, 274-282.

[Bo85]   Bela Bollobas. Random Graphs. Academic Press, 1985.

[Bl et al81]   Manuel Blum, Richard Karp, Oliver Vornberger, Christos H. Papadim-itriou, Mihalis Yannakakis. The complexity of testing whether a graph is a superconcentrator. Information Processing Letters 13, 1981, 164-167.

[Ch97]   Fan R. K. Chung. Spectral Graph Theory. American Mathematical Society, 1997.

[ChRe92]   Vasek Chvatal, Bruce Reed. Mick gets some (the odds are on his side). In Proceedings 33nd FoCS, 1992, IEEE, 620-627.

[ChSz88]   Vasek Chvatal, Endre Szemeredi. Many hard examples for resolution. Journal of the ACM 35(4), 1988, 759-768.

[CrAu96]   J. M. Crawford, L. D. Auton. Experimental results on the crossover point in random 3-SAT. Artificial Intelligence 81, 1996.

[Du et al2000]   Olivier Dubois, Yacine Boufkhad, Jacques Mandler. Typical random 3-SAT formulae and the satisfiability threshold. In Proceedings SoDA 2000, SIAM.

[DuZi98]   Paul E. Dunne, Michele Zito. An improved upper bound for the non-3-colourability threshold. Information Processing Letters 1998.

[Fu98]   Xudong Fu. The complexity of the resolution proofs for the random set of clauses. Computational Complexity 1998.

[Fr99]   Ehud Friedgut. Necessary and sufficient conditions for sharp thresholds of graph properties and the k-SAT problem. Journal of the American Mathematical Society 12, 1999, 1017-1054.

[FrSu96]   Alan M. Frieze, Stephen Suen. Analysis of two simple heuristics on a random instance of k-SAT. Journal of Algorithms 20(2), 1996, 312-355.

[FuKo81]   Z. Furedi, J. Komlos. The eigenvalues of random symmetric matrices. Combinatorica 1(3), 1981, 233-241.

[Go96]   Andreas Goerdt. A threshold for unsatisfiability. Journal of Computer and System Sciences 53, 1996, 469-486.

[ImNa96]   Russel Impagliazzo, Moni Naor. Efficient cryptographic schemes provably as secure as subset sum. Journal of Cryptology 9, 1996, 199-216.

[Ju82]   Ferenc Juhasz. The asymptotic behaviour of Lovasz theta function for random graphs. Combinatorica 2(2), 1982, 153-155.

[KrVu2000]   Michael Krivelevich, Van H. Vu. Approximating the independence number and the chromatic number in expected polynomial time. In Proceedings ICALP 2000, LNCS 1853, 13-24.

[KiKrKrSt98]   Lefteris M. Kirousis, Evangelos Kranakis, Danny Krizanc, Yiannis Sta-matiou. Approximating the unsatisfiability threshold of random formulas. Random Structures and Algorithms 12(3), 1998, 253-269.

[PeWe89]   A. D. Petford, Dominic Welsh. A Randomised 3-colouring algorithm. Discrete Mathematics 74, 1989, 253-261.

[SeMiLe96]   Bart Selman, David G. Mitchell, Hector J. Levesque. Generating hard satisfiability problems. Artificial Intelligence 81(1-2), 1996, 17-29.

[St88]   Gilbert Strang. Linear Algebra and its Applications. Harcourt Brace Jovanovich, Publishers, San Diego, 1988.

[vLWi]   J. H. van Lint, R. M. Wilson. A Course in Combinatorics. Cambridge University Press, 1992.

On the Circuit Complexity of Random Generation Problems for Regular and Context-Free Languages[*]

Massimiliano Goldwurm[1], Beatrice Palano[2], and Massimo Santini[1]

[1] Dipartimento di Scienze dell'Informazione, Università degli Studi di Milano
Via Comelico, 39/41, 20135 Milano – Italia
{goldwurm, santini}@dsi.unimi.it

[2] Dipartimento di Informatica, Università degli Studi di Torino
C.so Svizzera, 185, 10149 Torino – Italia
beatrice@di.unito.it

Abstract We study the circuit complexity of generating at random a word of length n from a given language under uniform distribution. We prove that, for every language accepted in polynomial time by 1-NAuxPDA of polynomially bounded ambiguity, the problem is solvable by a logspace-uniform family of probabilistic boolean circuits of polynomial size and $O(\log^2 n)$ depth. Using a suitable notion of reducibility (similar to the NC^1-reducibility), we also show the relationship between random generation problems for regular and context-free languages and classical computational complexity classes such as DIV, L and DET.

Keywords: Uniform random generation, ambiguous context-free languages, auxiliary pushdown automata, circuit complexity.

1 Introduction

Given a formal language L, the uniform random generation problem for L consists of computing, for an instance $n > 0$, a word of length n in L uniformly at random. We study the circuit complexity of this problem for several classes of languages including regular, context-free (c.f. for short) and more generally languages accepted by one-way nondeterministic auxiliary push-down automata (1-NAuxPDA).

Several sequential algorithms have been proposed for the random generation of strings in regular and context-free languages [12, 10, 9, 11]. The problem is particularly interesting in the c.f. case because these languages can codify a wide variety of combinatorial structures; moreover, sampling words from c.f. languages is naturally motivated by other applications such as testing parsers of programming languages [12] or evaluating the performance of algorithms which process DNA sequences [20, 19].

In the case of unambiguous c.f. languages the best known algorithm for random generation works in $O(n \log n)$ arithmetic time [10]; this is a special case of more general procedures for the random generation of so called "labelled combinatorial structures". In the case of general (possibly ambiguous) c.f. languages a subexponential time algorithm is described in [11] for the (almost uniform) random generation of strings of given

---

[*] This work has been partially supported by MURST Research Program "Unconventional computational models: syntactic and combinatorial methods".

A. Ferreira and H. Reichel (Eds.): STACS 2001, LNCS 2010, pp. 305–316, 2001.

length. The problem is solvable in polynomial time if the language is generated by a c.f. grammar of polynomially bounded ambiguity [4]. This result also holds for languages accepted by polynomial time 1-NAuxPDA of polynomially bounded ambiguity and, under suitable hypotheses, a similar approach can be applied to the combinatorial structures that admit an ambiguous specification (in the sense that the same object may have several distinct descriptions).

In this work we give a classication of the circuit complexity of these problems which includes languages described by possibly ambiguous specifications.

Our most general result states that for every language accepted by a polynomial time 1-NAuxPDA of polynomially bounded ambiguity the uniform random generation problem can be solved by a log-space uniform family of probabilistic boolean circuits of polynomial size and $O(\log^2 n)$ depth. This, in particular, emphasizes the difference between counting and random generation: indeed, for some finitely ambiguous c.f. languages the counting problem is $\#P_1$ complete [3].

Stronger results can be obtained for less general and well-known classes of languages such as regular and context-free languages. To compare the complexity of our problem for such classes, we give a natural extension of the usual NC^1 reducibility [7]. We say that the uniform random generation problem for a language L is RNC_g^1-reducible to a class $\mathscr{C}$ of boolean functions if it can be solved by a logspace-uniform family of probabilistic boolean circuits of polynomial size and $O(\log n)$ depth using oracle nodes in $\mathscr{C}$. Using this notion we show the relationship between our problem and classical computational complexity classes such as DIV, DET and $\#SAC^1$ [7, 21] (here defined in Section 2).

We show that, for every regular language the problem of uniform random generation is RNC_g^1-reducible to the class DIV; moreover, in case of unambiguous c.f. languages the problem is RNC_g^1-reducible to DIV $\cup$ L and, for polynomially ambiguous c.f. languages it is RNC_g^1-reducible to $\#SAC^1$. Finally, we consider a general version of the uniform random generation problem for regular languages, where the deterministic finite automaton describing the language is part of the input; in this case, the problem is RNC_g^1-reducible to DET. These results are obtained by combining the complexity of counting and recognition problem with the study of some reachability problems on certain random graphs arising from the design of the circuits.

2   Probabilistic Circuits for Random Generation

We assume some familiarity with (bounded fan-in) *boolean circuits* as defined in [7, 22]. We say that a family $\{c_n\}_{n>0}$ of boolean circuits is *uniform* if there exists a log-space bounded Turing machine which on input 1^n computes a description of c_n. The class NC^k is the set of boolean functions computable by uniform families of boolean circuits of polynomial size and $O(\log^k n)$ depth, where n is the input size. A boolean function f is NC^1-*reducible* to a boolean function g, if f can be computed by a uniform family of boolean circuits of polynomial size and $O(\log n)$ depth equipped with *oracle* nodes for computing g; here, the depth of any oracle node with fan-in i and fan-out o counts for $\lceil \log(i + o) \rceil$. Given a class $\mathscr{C}$ of boolean functions, we denote by $NC^1(\mathscr{C})$ the closure of $\mathscr{C}$ under NC^1 reducibility.

Let *intdet* and *intdiv* be the problems of computing respectively the determinant of $n \times n$ matrix of n-bit integers and the division of two n-bit integers. As usual, we denote by L (NL) the class of languages recognized in $O(\log n)$ space by a deterministic (nondeterministic) Turing machine. Hence, the classes L^*, NL^*, DET and DIV are defined respectively by $L^* = NC^1(L)$, $NL^* = NC^1(NL)$, $DET = NC^1(\{intdet\})$ and $DIV = NC^1(\{intdiv\})$. The following relations are known [7]:

$$NC^1 \subseteq \begin{matrix} L^* \subseteq NL^* \\ DIV \end{matrix} \subseteq DET \subseteq NC^2.$$

Finally, by $\#SAC^1$ we denote the set of functions computing the number of accepting subtrees in a uniform family of semi-unbounded circuits of polynomial size and $O(\log n)$ depth [21]; we also recall that $\#SAC^1 \subseteq NC^2$.

In this work we use boolean circuits to solve uniform random generation problems. To this end we use the notion of probabilistic boolean circuit as introduced in [7]. This is a boolean circuit equipped in addition with independent and identically distributed random input bits: each of them assumes a value in $\{0, 1\}$ with probability $1/2$.

Example 1. Consider the problem of generating at random an integer according to some specified distribution. Let $a_1, a_2, \ldots, a_n$ be n-bit positive integers, we design a probabilistic boolean circuit c_n which, on input $a_1, a_2, \ldots, a_n$, outputs a $k \in \{1, 2, \ldots, n\} \cup \{\bot\}$ such that:

1. $\Pr\{k = \bot\} \leq 1/4$,
2. for every $1 \leq i \leq n$, $\Pr\{k = i \mid k \neq \bot\} = a_i/a$, where $a = \sum_{i=1}^{n} a_i$.

First of all, the circuit computes in parallel all $s_i = \sum_{j \leq i} a_j$, for $1 \leq i \leq n$; then it computes $\ell = \min\{i : s_n < 2^i\}$. Let now $r_1, r_2 \in \{1, 2, \ldots, 2^\ell\}$ be two random integers defined by two distinct sets of ℓ random input bits each. The circuit computes in parallel $k_j = \min\{i : r_j \leq s_i\}$ for $j = 1, 2$ (where we assume $\min \varnothing = \bot$). Finally it outputs k_1 if this is different from $\bot$, else it outputs k_2.

Clearly, the probability of giving $\bot$ as output is less than or equal to $1/4$ while, if this is not the case, the output has the required distribution. Recalling the circuit complexity of elementary arithmetic operations [22], one can conclude that the size of the circuit is polynomial and its depth is $O(\log n)$.

Notice that, by taking $m = n^{O(1)}$ parallel copies of the same circuit, one can solve the problem, still in polynomial size and $O(\log n)$ depth, reducing the probability of answering $\bot$ to $1/4^m$ at most. □

We now introduce a parallel hierarchy to classify the uniform random generation problem for formal languages.

Definition 1. *A uniform family of probabilistic boolean circuits* $\{c_n\}_{n>0}$ *is a uniform random generator (u.r.g.) for a formal language* $L \subseteq \Sigma^*$, *if each* c_n, *on input* 1^n, *computes a value* ω_n *in* $\Sigma^n \cup \{\bot\}$ *such that, if* $L \cap \Sigma^n \neq \varnothing$, *then:*

1. $\Pr\{\omega_n = \bot\} \leq 1/4$,
2. $\Pr\{\omega_n = x \mid \omega_n \neq \bot\} = 1/\#(L \cap \Sigma^n)$, *for every* $x \in L \cap \Sigma^n$.

Moreover, we say that the uniform random generation problem for L belongs to the class RNC_g^k *if there exists a u.r.g. for L of polynomial size and $O(\log^k n)$ depth.*

Observe that this class is not the usual class RNC^k [7], since here we are not interested in computing a boolean function bounding the probability of a wrong answer, but we rather want to produce a random output with a given distribution explicitly notifying the possible failure of the computation (due to the restriction to unbiased random bits).

We say that the uniform random generation problem for a language L is RNC_g^1-*reducible* to a class $\mathscr{C}$ of boolean functions if there exists a u.r.g. for L of polynomial size and $O(\log n)$ depth which uses oracle nodes in $\mathscr{C}$ (again, the depth of any oracle node with fan-in i and fan-out o counts for $\lceil \log(i + o) \rceil$); we denote by $\text{RNC}_g^1(\mathscr{C})$ the class of uniform random generation problems RNC_g^1-reducible to $\mathscr{C}$.

3   Regular Languages

In this section we study the circuit complexity of the uniform random generation problem for regular languages. We show the problem to be RNC_g^1-reducible to *intdiv*.

Let $\mathscr{A} = \langle \Sigma, Q, q_0, F, \delta \rangle$ be a deterministic finite automaton and define, for $q \in Q$ and $0 \leq \ell \leq n$, the language $L_q^\ell = \{x \in \Sigma^\ell : \delta(q, x) \in F\}$ and set $\eta(q, \ell) = \#L_q^\ell$ (where, as usual, $\Sigma^0 = \{\epsilon\}$).

We start by defining a family of (random) graphs which allows to design the circuits for solving our problem. For every integer $n > 0$, define the (direct acyclic) labelled graph $G_n(\mathscr{A}) = \langle V_n, E_n \rangle$ such that $V_n = \{(q, \ell) : q \in Q, 0 \leq \ell \leq n\}$ and E_n is built according to the following procedure: for every $v = (q, \ell) \in V_n$ with $\ell > 0$ pick $\sigma_v \in \Sigma$ at random such that, for every $\sigma \in \Sigma$,

$$\Pr\{\sigma_v = \sigma\} = \frac{\eta(\delta(q, \sigma), \ell - 1)}{\eta(q, \ell)}$$

and add to E_n the edge $((q, \ell), (\delta(q, \sigma_v), \ell - 1))$ with label σ_v.

Since $G_n(\mathscr{A})$ is acyclic and all nodes (q, ℓ) with $\ell > 0$ have out-degree 1, for every $(q, \ell) \in V_n$ and $0 < m \leq \ell$ there exists just one node reachable from (q, ℓ) through a path of length m. Let $\omega(q, \ell)$ be the word consisting of the labels along the path leaving (q, ℓ) of length ℓ: i.e. $\omega(q, \ell) = \sigma_1 \cdots \sigma_\ell$, where $q_1 = q$, $q_{i+1} = \delta(q_i, \sigma_i)$ and $((q_i, \ell - i + 1), (q_{i+1}, \ell - i)) \in E_n$, for $1 \leq i < \ell$. Reasoning by induction on $1 \leq \ell \leq n$, one can prove that $\Pr\{\omega(q, \ell) = x\} = 1/\eta(q, \ell)$, for every $L_q^\ell \neq \varnothing$ and every $x \in L_q^\ell$. Hence, we obtain the following

Lemma 1. *For every $n > 0$ such that $L(\mathscr{A}) \cap \Sigma^n \neq \varnothing$,*

$$\Pr\{\omega(q_0, n) = x\} = \frac{1}{\#(L(\mathscr{A}) \cap \Sigma^n)},$$

for every $x \in L(\mathscr{A}) \cap \Sigma^n$.

We now show that, if the automaton $\mathscr{A}$ is fixed, given 1^n and $G_n(\mathscr{A})$ as input, computing the word $\omega(q_0, n)$ belongs to NC^1. To this aim, we need some preliminary tools.

We say that a $nd \times nd$ boolean matrix A is (d, t)-upper–diagonal if A is a block matrix of the form $A = (A_{i,j})$, where all $A_{i,j}$ are $d \times d$ matrices such that $A_{i,j} \neq 0$ (the zero matrix) iff $j = i + t$ $(d, t > 0, i, j = 1, \ldots, n)$.

Observe that, for every pair of $nd \times nd$ boolean matrices A, B, if A is (d, s)-upper–diagonal and B is (d, t)-upper–diagonal, then the product AB is $(d, s + t)$-upper–diagonal:

$$(AB)_{i,j} = \begin{cases} A_{i,i+s} B_{i+s,i+(s+t)} & \text{if } j = i + (s + t), \\ 0 & \text{otherwise;} \end{cases}$$

moreover, AB can be obtained by computing in parallel $n - (s + t)$ many products of $d \times d$ matrices. For this reason, we can prove the following

Lemma 2. *Let $d > 0$ be a fixed integer. If A is a (d, s)-upper–diagonal boolean matrix of size $nd \times nd$, then computing the boolean power A^n on input A belongs to NC^1.*

Proof. Observe that A^2 is $(d, 2s)$-upper–diagonal and can be computed by a boolean circuit of polynomial size and constant depth. So, for every $i > 0$, A^{2^i} is a $(d, 2^i s)$-upper–diagonal matrix and can be computed in polynomial size and $O(i)$ depth. Then

$$A^n = \prod_{i:b_i=1} A^{2^i},$$

where $b_i \in \{0, 1\}$, for $0 \leq i \leq \lfloor \log n \rfloor$, are the digits of the binary expansion of n, i.e. $n = \sum_i b_i 2^i$. Hence A^n can be obtained by a product of a logarithmic number of upper-diagonal matrices. Such a product can be computed in polynomial size and $O(\log \log n)$ depth.   □

Since all the edges of $G_n(\mathscr{A})$ are of the form $((q, \ell), (q', \ell - 1))$ for some $q, q' \in Q$ and $0 < \ell \leq n$, its adjacency matrix of $G_n(\mathscr{A})$ is $(\#Q, 1)$-upper–diagonal (where each block corresponds to a set of nodes with the same second component).

Lemma 3. *For a fixed automaton $\mathscr{A}$, given $G_n(\mathscr{A})$ as input, the computation of $\omega(q_0, n)$ belongs to NC^1.*

Proof. Let M be the adjacency matrix of $G_n(\mathscr{A})$. Recall that for every $v = (q, \ell) \in V_n$ and $0 < m \leq \ell$ there exists exactly one node that can be reached from v by a path of lenght m, hence the row corresponding to v in M^m contains exactly one 1. Hence, for $0 \leq i < n - 1$, all the nodes $(q_i, n - i)$ reachable from (q_0, n) by a path of length i can be computed in parallel as in Lemma 2.   □

Now let us describe the probabilistic boolean circuit c_n which on input 1^n computes a word in $L(\mathscr{A}) \cap \varSigma^n$ under uniform distribution. First the circuit computes in parallel all the coefficients $\eta(q, \ell)$ for $q \in Q$ and $0 \leq \ell \leq n$. This computation belongs to DIV as proven in [2]. Then the circuit computes the graph $G_n(\mathscr{A})$ by generating in parallel all random symbols σ_v for $v \in V_n$. As shown in Example 1, this step can be executed in $O(\log n)$ depth so that, for each $v \in V_n$, $\Pr\{\sigma_v = \bot\} \leq 2^{-(2+\lceil \log(n\#Q) \rceil)}$ and hence, the probability that $\sigma_v = \bot$ for some $v \in V_n$ is at most $1/4$. Thus, if all labels of $G_n(\mathscr{A})$ are in Σ the circuit outputs the string $\omega(q_0, n)$ computed in $O(\log n)$ depth as shown in Lemma 3; in this case, by Lemma 1, the distribution of the output is uniform. Otherwise, if $\sigma_v = \bot$ for some $v \in V_n$, the circuit outputs $\bot$. This proves the following

Theorem 1. *For every regular language, the uniform random generation problem belongs to* $\mathrm{RNC}_g^1(\mathrm{DIV})$.

4   Context Free Languages

In this section we study the uniform random generation problem for context-free languages. We first show that for unambiguous c.f. languages the problem is RNC_g^1-reducible to $L^* \cup \mathrm{DIV}$. Then we prove that, for all inherently ambiguous c.f. languages having polynomial ambiguity degree, the problem is RNC_g^1-reducible to $\#\mathrm{SAC}^1$ and hence belongs to RNC_g^2.

4.1   Unambiguous Context-Free Languages

Let $\mathscr{G} = \langle N, \Sigma, S, P \rangle$ be an unambiguous c.f. grammar in Chomsky normal form without useless variables, where N is the set of variables, Σ the set of terminals, S the initial variable and P the set of productions. For every $A \in N$ and every $1 \leq \ell \leq n$, define $\eta(A, \ell)$ as the number of derivation trees of $\mathscr{G}$ rooted at A and deriving a word in Σ^ℓ. Moreover, let $L_A^\ell = \{x \in \Sigma^\ell : A \overset{*}{\Rightarrow} x\}$; since $\mathscr{G}$ is unambiguous, $\eta(A, \ell) = \#L_A^\ell$.

As in the regular language case, we start by defining a family of (random) graphs which allows to design the circuits for solving our problem. For every integer $n > 0$, define the (direct acyclic) graph $G_n(\mathscr{G}) = \langle V_n, E_n \rangle$ such that $V_n = \{(A, r, s) : A \in N, 1 \leq r \leq s \leq n\} \cup \{(\sigma, r) : \sigma \in \Sigma, 1 \leq r \leq n\}$ and E_n is built according to the following procedure:

- for each $v = (A, r, r) \in V_n$, pick $p_v \in P$ at random such that, for every $(A \rightarrow \sigma) \in P$

$$\Pr\{p_v = (A \rightarrow \sigma)\} = \frac{1}{\eta(A, 1)}$$

  and add to E_n the edge $((A, r, r), (\sigma, r))$;
- for each $v = (A, r, s) \in V_n$ with $s > r$, pick $p_v \in P \times \{1, \ldots, s - r\}$ at random such that, for every $(A \rightarrow BC) \in P$ and $1 \leq k \leq s - r$,

$$\Pr\{p_v = (A \rightarrow BC, k)\} = \frac{\eta(B, k)\eta(C, s - r + 1 - k)}{\eta(A, s - r + 1)}$$

  and add to E_n the edges $((A, r, s), (B, r, r + k - 1))$ and $((A, r, s), (C, r + k, s))$.

Clearly $G_n(\mathcal{G})$ is acyclic, all its nodes $(A, r, s) \in V_n$ with $s > r$ have out-degree 2, and the subgraph of $G_n(\mathcal{G})$ induced by the set of nodes reachable from any (A, r, s) is a binary tree with $s - r + 1$ leaves of the form $(\sigma, r) \in V_n$. Let $\omega(A, r, s) = \sigma_r \cdots \sigma_s$, where the nodes (σ_i, i), for $r \le i \le s$, are the leaves of the subtree of $G_n(\mathcal{G})$ rooted at (A, r, s). Reasoning by induction on $1 \le \ell \le n$, one can prove that for every $L_A^\ell \ne \varnothing$ and every $x \in L_A^\ell$, if $1 \le r \le s \le n$ and $s - r + 1 = \ell$, then $\Pr\{\omega(A, r, s) = x\} = 1/\eta(A, \ell)$. As a consequence, we obtain the following

Lemma 4. *For every $n > 0$ such that $L(\mathcal{G}) \cap \Sigma^n \ne \varnothing$,*

$$\Pr\{\omega(S, 1, n) = x\} = \frac{1}{\#(L(\mathcal{G}) \cap \Sigma^n)},$$

for every $x \in L(\mathcal{G}) \cap \Sigma^n$.

We now consider the problem of computing $\omega(S, 1, n)$.

Lemma 5. *Let $\mathcal{G} = \langle N, \Sigma, S, P \rangle$ be a fixed unambiguous c.f. grammar in Chomsky normal form without useless variables. Given $G_n(\mathcal{G})$ as input, the computation of $\omega(S, 1, n)$ belongs to L^*.*

Proof. First observe that every $(A, r, s) \in V_n$ with $r < s$ has only two out-neighbours $(B, r, r + k - 1)$ and $(C, r + k, s)$, for some $1 \le k \le s - r$ and some $B, C \in N$; hence, for every $r \le i \le s$, a node (σ, i) is reachable from (A, r, s) iff it is reachable either from $(B, r, r + k - 1)$ in the case $i < r + k$, or from $(C, r + k, s)$ otherwise. Thus a log-space bounded deterministic Turing machine can be designed which tests whether a node $(\sigma, i) \in V_n$ is reachable from $(S, 1, n)$. Then the word $\omega(S, 1, n)$ can be computed by testing in parallel the reachability of (σ, i) from $(S, 1, n)$ for all $1 \le i \le n$ and all $\sigma \in \Sigma$. $\square$

Now, reasoning as in Section 3, a probabilistic boolean circuit can be designed which, on input 1^n, first computes in parallel all the coefficients $\eta(A, r, s)$, then determines the graph $G_n(\mathcal{G})$ and finally it generates the string $\omega(S, 1, n)$. The first step can be done in DIV [2] while the last one is in L^* as shown in Lemma 5. This, together with Lemma 4, yields the following

Theorem 2. *For every unambiguous context-free language, the uniform random generation problem belongs to $\mathrm{RNC}_g^1(\mathrm{DIV} \cup \mathrm{L})$.*

4.2   Polynomially Ambiguous Context-Free Languages

In this section we study the uniform random generation problem for inherently ambiguous context-free languages. Let $\mathcal{G} = \langle N, \Sigma, S, P \rangle$ be a c.f. grammar in Chomsky normal form without useless variables; for every $x \in \Sigma^*$, we denote by $\mathrm{amb}_\mathcal{G}(x)$ the *ambiguity* of x, i.e., the number of derivation trees of x in $\mathcal{G}$. We call *ambiguity degree* of $\mathcal{G}$ the function $d_\mathcal{G} : \mathbf{N} \to \mathbf{N}$ defined by $d_\mathcal{G}(n) = \max\{\mathrm{amb}_\mathcal{G}(x) : x \in \Sigma^n\}$, for every $n \in \mathbf{N}$. Then, $\mathcal{G}$ is said *polynomially ambiguous* if, for some polynomial $p(n)$, we have $d_\mathcal{G}(n) \le p(n)$ for every $n > 0$.

One can easily prove that if $\mathscr{G}$ is an ambiguous c.f. grammar the circuit designed for Theorem 2, on input 1^n, gives output w_n such that $\Pr\{w_n = \bot\} \leq 1/4$ and, for every $x \in \Sigma^n$

$$\Pr\{w_n = x \mid w_n \neq \bot\} = \frac{\mathrm{amb}_{\mathscr{G}}(x)}{\sum_{y \in \Sigma^n} \mathrm{amb}_{\mathscr{G}}(y)}; \tag{1}$$

the main change, in this case, is that $\eta(A, \ell)$ and $\#L_A^{\ell}$ may be different.

In order to obtain the uniform distribution we use a "rejection method" [15], giving a parallel version of a procedure described in [4]. Assume now that $\mathscr{G}$ is polynomially ambiguous and let $p(n)$ be a polynomial such that $d_{\mathscr{G}}(n) \leq p(n)$ for every $n > 0$. A probabilistic boolean circuit can be designed which on input 1^n first computes $m = p(n)!$ and then executes $4 \cdot p(n)$ times in parallel (and independently of one another) the following computation:

- $y = \bot$;
- generate w_n at random in $L(\mathscr{G}) \cap \Sigma^n$ according to the distribution given by (1);
- if $w_n \neq \bot$, then
    compute $a = \mathrm{amb}_{\mathscr{G}}(w_n)$;
    generate r uniformly at random in $\{1, \ldots, 2^{\lceil \log m \rceil}\}$;
    if $a \cdot r \leq m$ then $y = w_n$;
- return y.

Then the circuit outputs $\bot$ if all the $4 \cdot p(n)$ computations return $\bot$, otherwise it outputs the first $y \neq \bot$. Reasoning as in [4], it can be proven that the probability of getting $\bot$ is at most $1/4$, otherwise, the output is distributed uniformly at random in $L(\mathscr{G}) \cap \Sigma^n$. Evaluating the complexity of the circuit, we observe that the computation of $\mathrm{amb}_{\mathscr{G}}(x)$ for all $x \in \Sigma^*$ belongs to $\#SAC^1$ [21]. Hence, since both L and DIV are included in $\#SAC^1$, we obtain the following

Theorem 3. *For every language generated by a polynomially ambiguous context-free grammar, the uniform random generation problem belongs to* $\mathrm{RNC}_g^1(\#SAC^1)$.

5   One-Way Nondeterministic Auxiliary Pushdown Automata

In this section we describe a family of probabilistic boolean circuits to solve our problem in the case of languages accepted by *one-way nondeterministic auxiliary pushdown automata* (1-NAuxPDA, for short). These circuits are based on the computation of the ambiguity of terminal strings with respect to different c.f. grammars. For this reason we first study the problem of computing the value $\mathrm{amb}_{\mathscr{G}}(x)$ having in input a c.f. grammar $\mathscr{G}$ in Chomsky normal form and a word $x \in \Sigma^*$.

5.1   The General Ambiguity Problem

We start by recalling a result given in [18] to evaluate arithmetic circuits of size n and degree d in $O(\log n \log(nd))$ parallel time (see also [16]). Here, by *arithmetic circuit* over a semiring R we mean a labelled directed acyclic graph with three kinds of vertices: input nodes of fan-in 0 with labels in R, addition nodes of fan-in greater than 1 labelled by $+$, and multiplication nodes of fan-in 2 labelled by $\times$; we also assume that there is no edge between two multiplication nodes. The *degree* of the circuit is the maximum degree of its nodes, defined by induction as follows: every input node has degree 1, the degree of every multiplication node is the sum of the degrees of its two inputs and the degree of every addition node is the maximum of the degrees of its inputs. The *value* of a node can be defined in the standard way: all input nodes take as value their labels, the value of an addition (multiplication) node is the sum (product) of the values of its inputs.

Proposition 1 ([18]). *The values of all nodes in any arithmetic circuit over R of size n and degree d can be computed in $O(\log n \log(nd))$ parallel time using $M(n)$ processors, where $M(n)$ is the number of processors required to multiply two $n \times n$ matrices over R in $O(\log n)$ time.*

Now, in order to compute $\mathrm{amb}_\mathcal{G}(x)$ on input $\mathcal{G} = \langle N, \Sigma, S, P \rangle$ and $x = \sigma_1 \sigma_2 \cdots \sigma_n \in \Sigma^n$ we define an arithmetic circuit $C(\mathcal{G}, x)$ on $\mathbf{N}$ implementing a counting version of the traditional CYK algorithm. The input nodes of $C(\mathcal{G}, x)$ are (A, i, i), where $A \in N$, $1 \leq i \leq n$ and they are labelled by 1 if $(A \to \sigma_i) \in P$ and 0 otherwise. Addition nodes are (A, i, j) with $A \in N$, $1 \leq i < j \leq n$, and multiplication nodes are (B, C, i, k, j) with $(D \to BC) \in P$ for some $D \in N$, $1 \leq i \leq k < j \leq n$. The inputs of every addition node (A, i, j) are the nodes (B, C, i, k, j) such that $(A \to BC) \in P$; the inputs of every multiplication node (B, C, i, k, j) are the nodes (B, i, k) and $(C, k+1, j)$. It is easy to show that the value of node $(S, 1, n)$ is $\mathrm{amb}_\mathcal{G}(x)$.

Lemma 6. *The problem of computing $\mathrm{amb}_\mathcal{G}(x)$ given as input a terminal string x and a context-free grammar $\mathcal{G}$ in Chomsky normal form, can be solved by a uniform family of boolean circuits of $(nm)^{O(1)}$ size and $O((\log n + \log m)^2)$ depth, where $n = |x|$ and m is the size of $\mathcal{G}$.*

Proof (sketch). We observe that Proposition 1 is based on a parallel algorithm which, for an input arithmetic circuit of size $\hat{n}$ and degree d, executes $O(\log \hat{n} d)$ times a cycle of operations, the most expensive one being the product of two $\hat{n} \times \hat{n}$ matrices over R.

In our case, $\hat{n} = O(n^3 \cdot m)$, $d = n$ and the value of the nodes is bounded by m^{3n}. Hence, the above matrix product can be computed by a boolean circuit of polynomial size and $O(\log n + \log m)$ depth. □

Using the same approach, one can compute on input $\mathcal{G} = \langle N, \Sigma, S, P \rangle$, $A \in N$ and $\ell > 0$, the number $\eta_\mathcal{G}(A, \ell)$ of derivation trees of $\mathcal{G}$ rooted at A and deriving a word in Σ^ℓ. It is sufficient to map all terminal symbols $\sigma \in \Sigma$ into the unique symbol z, so defining a new c.f. grammar $\mathcal{G}'_A = \langle N, \{z\}, A, P' \rangle$, where P' is obtained from P by replacing all productions $(B \to \sigma) \in P$ with $B \to z$ and labelling every input node (B, i, i) of the circuit $C(\mathcal{G}'_A, z^\ell)$ with the cardinality of $\{(B \to \sigma) \in P : \sigma \in \Sigma\}$. Hence, $\eta_\mathcal{G}(A, \ell) = \mathrm{amb}_{\mathcal{G}'_A}(z^\ell)$ and the computation can be carried out as in Lemma 6.

This allows to apply the approach presented in Section 4.1 to generate uniformly at random a word of length n, according to the distribution given in (1), assuming the grammar as a part of the input.

5.2   Polynomially Ambiguous 1-NAuxPDA

We recall that a 1-NAuxPDA is a nondeterministic Turing machine having a one-way read-only input tape, a pushdown tape and a *log-space bounded* two-way read-write work tape [6, 5]. It is known that the class of languages accepted by 1-NAuxPDA working in polynomial time coincides with the class of decision problems reducible to context-free recognition via one-way log-space reduction [17].

Given a 1-NAuxPDA $\mathcal{M}$, we denote by $\text{amb}_{\mathcal{M}}(x)$ the number of accepting computations of $\mathcal{M}$ on input $x \in \Sigma^*$, and call *ambiguity degree* of $\mathcal{M}$ the function $d_{\mathcal{M}} : \mathbf{N} \rightarrow \mathbf{N}$ defined by $d_{\mathcal{M}}(n) = \max\{\text{amb}_{\mathcal{M}}(x) : x \in \Sigma^n\}$, for every $n \in \mathbf{N}$. Then, $\mathcal{M}$ is said *polynomially ambiguous* if, for some polynomial $p(n)$, we have $d_{\mathcal{M}}(n) \leq p(n)$ for every $n > 0$.

It is known that, if $\mathcal{M}$ works in polynomial time, given an integer input $n > 0$, a c.f. grammar $\mathcal{G}_n$ in Chomsky normal form, of size polynomial in n, can be built such that $L(\mathcal{G}_n) \cap \Sigma^n = L(\mathcal{M}) \cap \Sigma^n$ [6]. This construction can be refined in such a way that the ambiguity degree of $\mathcal{G}_n$ does not increase with respect to the ambiguity degree of $\mathcal{M}$, i.e., for every $n \in \mathbf{N}$, the number of derivation trees of any word $x \in \Sigma^n$ in $\mathcal{G}_n$ is less or equal to the number of accepting computations of $\mathcal{M}$ on input x [4]. Moreover, the problem of computing such a refined $\mathcal{G}_n$ on input 1^n belongs to NC^2 as shown in [1].

Therefore, the random generation problem for the language accepted by a polynomial time $\mathcal{M}$ is reduced to generating words of length n from the grammar $\mathcal{G}_n$ uniformly at random. This can be done by a general version of the algorithm described in Subsection 4.2 where the c.f. grammar $\mathcal{G}_n$ is part of the input. Thus, if the ambiguity of $\mathcal{M}$ is polynomial, by Lemma 6, the overall computation can be carried out in $O(\log^2 n)$ depth and polynomial size.

Theorem 4. *For every language accepted by a polynomially ambiguous 1-NAuxPDA working in polynomial time, the uniform random generation problem belongs to* RNC_g^2.

6   The General Case for Regular Languages

In this section we consider the random generation problem for regular languages assuming as input both the length of the word to be generated and the deterministic finite automaton recognising the language. Using the same notation of Section 3, we say that a family of probabilistic boolean circuits $\{c_{n,m}\}_{n,m>0}$ solves the *general* problem of uniform random generation for regular languages, if each $c_{n,m}$, having in input 1^n and a deterministic finite automaton $\mathcal{A}$ of size m, computes a value $\omega_{n,m}$ in $\Sigma^n \cup \{\bot\}$ such that, if $L(\mathcal{A}) \cap \Sigma^n \neq \varnothing$, then:

1. $\Pr\{\omega_{n,m} = \bot\} \leq 1/4$,
2. $\Pr\{\omega_{n,m} = x \mid \omega_{n,m} \neq \bot\} = 1/\#(L(\mathcal{A}) \cap \Sigma^n)$, for every $x \in L(\mathcal{A}) \cap \Sigma^n$.

The problem can be solved by a family of circuits designed as in Section 3 to generate a word uniformly at random from a fixed regular language. Here, there are two main differences. First of all, since $\mathcal{A} = \langle \Sigma, Q, q_0, F, \delta \rangle$ is part of the input, the coefficients $\eta(q, \ell)$ for $q \in Q$ and $0 \leq \ell \leq n$ can be computed in DET (rather than in DIV), because such task is reducible to computing the ℓ-th power of a $m \times m$ integer matrix. Second, once the graph $G_n(\mathscr{A})$ is obtained, the computation of $\omega(q_0, n)$ belongs to L* (rather than NC^1) since it is reducible to a reachability problem in a direct acyclic graph whose nodes have out-degree at most 1 [8]. Hence, we obtain the following

Theorem 5. *The general problem of uniform random generation for regular languages is solved by a uniform family of probabilistic boolean circuits of polynomial size and $O(\log(n + m))$ depth with oracle nodes in* DET.

7   Concluding Remarks

In this paper we have studied the circuit complexity of the uniform random generation problem for several classical formal languages. An interesting application of the results presented here is related to counting problems, i.e. computing $\#(L \cap \Sigma^n)$ on input $n > 0$.

It is well-known that random generation is related to counting and that there are cases in which exact counting is hard, while the random uniform generation is easy and allows to obtain approximation schemes for the counting problem [14, 13]. This is for instance the case for some finitely ambiguous context-free languages, as discussed in [3, 4]. In a forthcoming paper we will show that a RNC^2 approximation scheme can be designed for the counting problem of every language accepted by a polynomial time 1-NAuxPDA of polynomially bounded ambiguity.

References

[1] E. Allender, D. Bruschi, and G. Pighizzini. The complexity of computing maximal word functions. *Computational Complexity*, 3:368–391, 1993.

[2] A. Bertoni, M. Goldwurm, and P. Massazza. Counting problems and algebraic formal power series in noncommuting variables. *Information Processing Letters*, 34(3):117–121, April 1990.

[3] A. Bertoni, M. Goldwurm, and N. Sabadini. The complexity of computing the number of strings of given length in context-free languages. *Theoretical Computer Science*, 86(2):325–342, 1991.

[4] A. Bertoni, M. Goldwurm, and M. Santini. Random generation and approximate counting of ambiguously described combinatorial structures. In Horst Reichel and Sophie Tison, editors, *Proceedings of 17th Annual Symposium on Theoretical Aspects of Computer Science (STACS)*, number 1770 in Lecture Notes in Computer Science, pages 567–580. Springer, 2000.

[5] F.-J. Brandenburg. On one-way auxiliary pushdown automata. In H. Waldschmidt H. Tzschach and H. K.-G. Walter, editors, *Proceedings of the 3rd GI Conference on Theoretical Computer Science*, volume 48 of *Lecture Notes in Computer Science*, pages 132–144, Darmstadt, FRG, March 1977. Springer.

[6] S. A. Cook. Characterizations of pushdown machines in terms of time-bounded computers. *Journal of the ACM*, 18(1):4–18, January 1971.

[7] S. A. Cook. A taxonomy of problems with fast parallel algorithms. *Information and Control*, 64:2–22, 1985.

[8] S. A. Cook and P. McKenzie. Problems complete for deterministic logarithmic space. *Journal of Algorithms*, 8(3):385–394, September 1987.

[9]   A. Denise. Génération aléatoire et uniforme de mots de langages rationnels. *Theoretical Computer Science*, 159(1):43–63, 1996.

[10] P. Flajolet, P. Zimmerman, and B. Van Cutsem. A calculus for the random generation of labelled combinatorial structures. *Theoretical Computer Science*, 132(1-2):1–35, 1994.

[11] V. Gore, M. Jerrum, S. Kannan, Z. Sweedyk, and S. Mahaney. A quasi-polynomial-time algorithm for sampling words from a context-free language. *Information and Computation*, 134(1):59–74, 10 April 1997.

[12] T. Hickey and J. Cohen. Uniform random generation of strings in a context-free language. *SIAM Journal on Computing*, 12(4):645–655, November 1983.

[13] M. Jerrum and A. Sinclair. Approximate counting, uniform generation and rapidly mixing markov chains. *Information and Computation*, 82:93–133, 1989.

[14] M. R. Jerrum, L. G. Valiant, and V. V. Vazirani. Random generation of combinatorial structures from a uniform distribution. *Theoretical Computer Science*, 43(2-3):169–188, 1986.

[15] R. M. Karp, M. Luby, and N. Madras. Monte-carlo approximation algorithms for enumeration problems. *Journal of Algorithms*, 10:429–448, 1989.

[16] R. M. Karp and V. Ramachandran. Parallel algorithms for shared-memory machines. In J. van Leeuwen, editor, *Handbook of Computer Science*. MIT Press/Elsevier, 1992.

[17] C. Lautemann. On pushdown and small tape. In K. Wagener, editor, *Dirk-Siefkes, zum 50. Geburststag (proceedings of a meeting honoring Dirk Siefkes on his fiftieth birthday)*, pages 42–47. Technische Universität Berlin and Universität Ausgburg, 1988.

[18] G. L. Miller, V. Ramachandran, and E. Kaltofen. Efficient parallel evaluation of straight-line code and arithmetic circuits. *SIAM Journal on Computing*, 17(4):687–695, August 1988.

[19] D. B. Searls. The computational linguistics of biological sequences. In Larry Hunter, editor, *Artificial Intelligence and Molecular Biology*, chapter 2, pages 47–120. AAAI Press, 1992.

[20] R. Smith. A finite state machine algorithm for finding restriction sites and other pattern matching applications. *Comput. Appl. Biosci.*, 4:459–465, 1988.

[21] V. Vinay. Counting auxiliary pushdown automata and semi-unbounded arithmetic circuits. In Christopher Balcázar, José; Borodin, Alan; Gasarch, Bill; Immerman, Neil; Papadimitriou, Christos; Ruzzo, Walter; Vitányi, Paul; Wilson, editor, *Proceedings of the 6th Annual Conference on Structure in Complexity Theory (SCTC '91)*, pages 270–284, Chicago, IL, USA, June 1991. IEEE Computer Society Press.

[22] I. Wegener. *The Complexity of Boolean Functions*. B. G. Teubner, Stuttgart, 1987.

Efficient Minimal Perfect Hashing
in Nearly Minimal Space

Torben Hagerup and Torsten Tholey

Institut für Informatik
Johann Wolfgang Goethe-Universität Frankfurt
D-60054 Frankfurt am Main, Germany

Abstract. We consider the following problem: Given a subset S of size n of a universe $\{0, \ldots, u-1\}$, construct a minimal perfect hash function for S, i.e., a bijection h from S to $\{0, \ldots, n-1\}$. The parameters of interest are the space needed to store h, its evaluation time, and the time required to compute h from S. The number of bits needed for the representation of h, ignoring the other parameters, has been thoroughly studied and is known to be $n \log e + \log \log u \pm O(\log n)$, where "log" denotes the binary logarithm. A construction by Schmidt and Siegel uses $O(n + \log \log u)$ bits and offers constant evaluation time, but the time to find h is not discussed. We present a simple randomized scheme that uses $n \log e + \log \log u + o(n + \log \log u)$ bits and has constant evaluation time and $O(n + \log \log u)$ expected construction time.
Keywords: Computational and structural complexity, algorithms and data structures, perfect hashing, sparse tables, space complexity.

1   Introduction

Suppose that S is a subset of size n of the universe $\{0, \ldots, u-1\}$ for some $n, u \in \mathbb{N} = \{1, 2, \ldots\}$. A function h defined on $\{0, \ldots, u-1\}$ is said to be *perfect* for S if it is injective on S. If, moreover, the range of h is the set $\{0, \ldots, n-1\}$, then h is called a *minimal perfect hash function* for S. We consider the problem of constructing minimal perfect hash functions for given sets of nonnegative integers.

Let $\mathcal{A}$ be an algorithm that inputs an arbitrary set S of nonnegative integers and outputs a minimal perfect hash function h for S. Several performance parameters of $\mathcal{A}$ are of interest:

- *Encoding size:* The number of bits of storage occupied by the representation of h output by $\mathcal{A}$.
- *Evaluation time:* The time needed to compute $h(x)$ for an arbitrary x in the domain of h.
- *Construction time:* The time needed to compute h from S.
- *Working space:* The amount of space needed to compute h from S.

We view these parameters as functions of $n = |S|$ and $u = 1 + \max S$. Fredman, Komlós and Szemerédi described a randomized construction that achieves

A. Ferreira and H. Reichel (Eds.): STACS 2001, LNCS 2010, pp. 317–326, 2001.

$O(n \log u)$ encoding size, $O(1)$ evaluation size and $O(n)$ expected construction time [4]. Strictly speaking, their scheme yields a function h whose range is of size $O(n)$ rather than n, but is it easy to obtain a minimal perfect hash function within the same resource bounds. Using a counting argument, Fredman and Komlós proved a worst-case lower bound of $n \log e + \log \log u - O(\log n)$ bits for the encoding size of a minimal perfect hash function for a subset of size n of a universe of size u, provided that $u \geq n^{2+\epsilon}$ for some fixed $\epsilon > 0$ [3] (an easy alternative proof was given by Radhakrishnan [10]). That this bound is almost tight follows by comparing it with an upper bound of $n \log e + \log \log u + O(\log n)$ bits given by Mehlhorn [8, Sect. III.2.3, Thm. 8]. His construction, however, has $n^{\Theta(ne^n u \log u)}$ construction and evaluation time. Schmidt and Siegel showed the existence of minimal perfect hash functions combining an encoding size of $O(n + \log \log u)$ bits with $O(1)$ evaluation time, but the time needed to find such functions was not discussed [11]. We present a new construction that not only works in almost linear expected time while still offering constant-time evaluation, but also reduces the encoding size to within lower-order terms of the lower bound.

Our model of computation is a unit-cost word RAM [5] with an instruction set including multiplication and integer division. We denote the word length of the machine by w and assume that every input set S consists of numbers representable in single words, i.e., $\max S < 2^w$. We will measure the encoding size of a hash function in bits, but the working space needed for its construction in w-bit words. Our main result is expressed in the following theorem.

Theorem 1. *For all integers $n, u, w \geq 4$ with $u \leq 2^w$ and for every given subset S of size n of $\{0, \ldots, u-1\}$, a minimal perfect hash function for S that can be evaluated in $O(1)$ time and stored in $n \log e + \log \log u + O(n(\log \log n)^2 / \log n + \log \log \log u)$ bits can be constructed in $O(n + \log \log u)$ expected time using $O(n)$ words of working space on a unit-cost word RAM with a word length of w bits and an instruction set including multiplication and integer division.*

Our approach is very simple. Suppose that we are given an input set S of size n. Repeatedly replacing the elements of S by their remainders modulo suitable primes, we begin by mapping S bijectively to a set S' whose elements are either bounded by a polynomial in n or far smaller than $\max S$. In the former and more interesting case, we proceed to partition S' into groups of elements small enough to be handled by the doubly exponential algorithm of Mehlhorn mentioned above. The division into groups is done in two stages, each of which defines a group as the set of elements mapped to a common value by a suitable hash function, a so-called *bucket* of the hash function. The hash functions employed for this purpose have to be chosen rather carefully, as the maximum bucket size must be within a constant factor of the average bucket size. Essential in achieving a construction time that is linear and not merely almost linear in n is the observation that the superlinear component in the running time of Mehlhorn's algorithm can be amortized over all groups.

2   Reducing the Size of the Universe

We denote by the term *range reduction* the process of reducing an instance of the problem at hand, namely computing a minimal perfect hash function for a given set S, to an instance that involves smaller input numbers, i.e., the process of reducing the size of the universe. We employ a range reduction based on the following lemma, proved essentially as [4, Lemma 2].

Lemma 2. *There is a constant $\beta \in \mathbb{N}$ such that for every nonempty set S of nonnegative integers and for every $m \geq \beta|S|^2 \log(1 + \max S)$, the function $x \mapsto x \bmod p$ is injective on S for at least half of the primes p bounded by m.*

Let S be an input set and take $n = |S|$, $u = 1 + \max S$, $\lambda = \beta n^2 \lceil \log u \rceil$ and $D = \{p \in \mathbb{N} \mid p \leq \lambda$ and the function $x \mapsto x \bmod p$ is injective on $S\}$. We assume that $n \geq 4$. In order to put Lemma 2 to use, we need a way to compute an element of D.

If $\log u$ and therefore λ are polynomial in n, we can pick an integer p uniformly at random from $M = \{1, \ldots, \lambda\}$, apply Rabin's randomized primality test [9] to p $\lceil \log n \rceil$ times and, if p passes this test—which happens with probability at most $1/n$ if p is composite—proceed to test directly whether $p \in D$ by means of radix sorting. If p fails any test, we immediately discard it and pick a new random integer, continuing until an element of D is encountered. By Lemma 2, the expected number of trials in which p is prime is bounded by a constant, and the expected time spent in such trials is $O(n)$. By the prime number theorem, the density of primes in M is $\Omega(1/\log \lambda) = \Omega(1/\log n)$, so that the expected total number of trials is $O(\log n)$. Since Rabin's test works in $(\log n)^{O(1)}$ time, the total expected time is $O(n)$.

For $\log u \geq n^3$, we sketch a different method and allow an expected time of $O(n + \log \lambda) = O(n + \log \log u)$. Note first that λ can be computed within this time bound. We pick a set R of $\lceil \log \lambda \rceil$ random elements of M and store these, each replicated $\lfloor \sqrt{\lambda} \rfloor$ times, together in a single computer word A. The condition $\log u \geq n^3$ ensures that the word length is sufficient for this to be possible (unless u is smaller than some constant). We also create a word B containing the sequence $1, \ldots, \lfloor \sqrt{\lambda} \rfloor$, replicated $\lceil \log \lambda \rceil$ times, and proceed to divide each number in A by the corresponding number in B. Simulating the school method for long division, this can be carried out simultaneously for all pairs of numbers in $O(\log \lambda)$ time; a more detailed discussion of similar computations can be found in [5]. As a result, we learn for each element of R whether it has a divisor bounded by $\lfloor \sqrt{\lambda} \rfloor$, i.e., whether it is composite. The set R was chosen sufficiently large to ensure that with probability $\Omega(1)$ it contains at least one prime. If this is the case, we pick such a prime p and test whether $p \in D$. Because p is much smaller than u, this can be done in $O(n)$ time by sorting [5]; alternatively, it can be done in $O(n)$ expected time using universal hashing [1]. If no element of D is found, we repeat the entire procedure. Since each trial takes $O(n + \log \log u)$ time and succeeds with probability $\Omega(1)$, the overall expected time is $O(n + \log \log u)$.

Faced with an input set S with $|S| = n$, we repeatedly apply the reduction based on Lemma 2 and discussed above until we reach a set S' with $\max S' \leq n^3$, but at most four times. The expected time to do this is $O(n + \log\log u)$, and $O(n)$ words of working space suffice. The first reduction requires a prime of at most $\log \lambda = \log\log u + O(\log n)$ bits to be stored as part of the representation of the final minimal perfect hash function. The number of bits required for all following reductions is $O(\log n + \log\log\log u)$.

After four reduction steps, we have a set S' with $\max S' = O(n^2(\log^{(3)} n + \log^{(4)} u))$. If the condition $\max S' \leq n^3$ is still not satisfied, $n = O(\log^{(4)} u)$, and we can simply store S' using the method of Fredman, Komlós and Szemerédi [4], which requires $O(n \log \max S') = O((\log^{(4)} u)^4)$ bits of storage, for a total of $\log\log u + o(\log\log\log u)$ bits. In the following, we can therefore assume without loss of generality that the input set S satisfies $\max S \leq n^3$.

3   Splitting into Groups

Our goal in this section is to partition the set S into $O(n/\hat{n})$ groups of at most $\hat{n}$ elements each, where $\hat{n} = \gamma \log n / \log\log n$ for a constant $\gamma > 0$ to be chosen later. Our main tool is a class $\mathcal{R}$ of hash functions introduced by Dietzfelbinger and Meyer auf der Heide [2] (another possibility would be to use a class defined by Siegel [12]). For our purposes, the distinguishing feature of $\mathcal{R}$ is that a function drawn at random from $\mathcal{R}$ is likely to spread a key set about evenly over its range.

We begin by defining the class $\mathcal{R}$. Fix a prime $p \geq u$, let $U = \{0, \ldots, p-1\}$ and, for $d, s \in \mathbb{N}$, take

$$\mathcal{H}_s^d = \{h_a \mid a = (a_0, \ldots, a_d) \in U^{d+1}\},$$

where, for $a = (a_0, \ldots, a_d) \in U^{d+1}$, $h_a : U \to \{0, \ldots, s-1\}$ is the function given by

$$h_a(x) = \left(\sum_{i=0}^{d} a_i x^i \bmod p\right) \bmod s$$

for all $x \in U$. Informally, $\mathcal{H}_s^d$ is known as the class of polynomials of degree d. The class $\mathcal{R}$ depends on four parameters $r, s, d_1, d_2 \in \mathbb{N}$, a dependence made explicit by writing $\mathcal{R}$ as $\mathcal{R}(r, s, d_1, d_2)$. For $r, s, d_1, d_2 \in \mathbb{N}$,

$$\mathcal{R}(r, s, d_1, d_2) = \{h_{(f,g,a_0,\ldots,a_{r-1})} \mid f \in \mathcal{H}_r^{d_1}, g \in \mathcal{H}_s^{d_2} \text{ and } 0 \leq a_0, \ldots, a_{r-1} < s\},$$

where, for $f \in \mathcal{H}_r^{d_1}$, $g \in \mathcal{H}_s^{d_2}$ and $a_0, \ldots, a_{r-1} \in \{0, \ldots, s-1\}$, $h_{(f,g,a_0,\ldots,a_{r-1})} : U \to \{0, \ldots, s-1\}$ is the function given by

$$h_{(f,g,a_0,\ldots,a_{r-1})}(x) = (g(x) + a_{f(x)}) \bmod s,$$

for all $x \in U$. One way to visualize $\mathcal{R}$ is as follows: A key $x \in U$ is first mapped to row $f(x)$ and column $g(x)$ of an $r \times s$ table. Then row i is rotated cyclically a distance of a_i, for $i = 0, \ldots, r-1$, and the resulting column number is taken as the final function value.

The nontrivial fact about $\mathcal{R}$ of interest to us is expressed in the following lemma, related to Lemma 4.4 and Theorem 4.6 of [2].

Lemma 3. *For every $c > 0$, there is a $C > 0$ such that for all $r, s, d_1, d_2 \in \mathbb{N}$ with $r \leq n$, $s \leq cn/\log n$, $rs \geq n^{1+1/c}$, $d_1 \geq C$ and $d_2 \geq C$, the relation*

$$\forall i \in \{0, \ldots, s-1\}: \quad |\{x \in S \mid h(x) = i\}| \leq Cn/s$$

holds with probability at least $1 - n^{-1}$ if h is chosen uniformly at random from $\mathcal{R}(r, s, d_1, d_2)$.

Informally, the lemma says that if r and s are chosen so that $rs = \Omega(n^{1+\epsilon})$ for some fixed $\epsilon > 0$ and $s = O(n/\log n)$, then for sufficiently large d_1 and d_2, the maximum bucket size of a random function from $\mathcal{R}(r, s, d_1, d_2)$ will be within a constant factor of the average bucket size, except with negligible probability.

We prove Lemma 3 using several auxiliary lemmas. Note that we can assume n to be larger an arbitrary constant, since the maximum bucket size is trivially bounded by Cn/s if $C \geq kn$. In particular, we assume that $s \leq n$.

Let $\xi = n^{1+1/(2c)}/r$. We begin by showing, using the following lemma, that if $f \in \mathcal{H}_r^{d_1}$ is chosen uniformly at random and d_1 is sufficiently large, then the maximum bucket size $\max_{0 \leq j < r} |\{x \in S \mid f(x) = j\}|$ of f is bounded by 2ξ, except with negligible probability.

Lemma 4. *Let $n, d \in \mathbb{N}$, let $X_1, \ldots, X_n$ be d-independent, equidistributed 0-1-variables, let $\mu \geq E(X_1)$ and assume that $n\mu \geq d$. Then, for some α that depends only on d and for every $\xi > 0$,*

$$\Pr\left(\sum_{i=1}^n (X_i - \mu) > \xi\right) \leq \frac{\alpha(n\mu)^{d/2}}{\xi^d}.$$

The lemma is essentially [7, Corollary 4.20]. We generalize the original formulation in a trivial way by allowing $\mu \geq E(X_1)$ instead of taking $\mu = E(X_1)$ and replace the original condition $n \geq d/(2\mu)$ by the stronger condition $n\mu \geq d$, which seems called for by the proof.

In our context, with $S = \{x_1, \ldots, x_n\}$, we fix $j \in \{0, \ldots, r-1\}$ and take

$$X_i = \begin{cases} 1, & \text{if } f(x_i) = j \\ 0, & \text{otherwise,} \end{cases}$$

for $i = 1, \ldots, n$. Then $X_1, \ldots, X_n$ satisfy the conditions of Lemma 4 with $d = d_1$ and $\mu = 2/r + d_1/n$. For every d_1 and for sufficiently large n, we have $\xi \geq n\mu$, and therefore the quantity $|\{x \in S \mid f(x) = j\}| = \sum_{i=1}^n X_i$ is bounded by 2ξ, except with probability at most $\alpha\xi^{-d_1/2}$, where α depends only on d_1. For d_1 and subsequently n chosen sufficiently large, the latter probability is at most n^{-3}, so $\max_{0 \leq j < r} |\{x \in S \mid f(x) = j\}| > 2\xi$ with probability at most $rn^{-3} \leq n^{-2}$.

Assuming that f has been chosen so that its maximum bucket size is indeed bounded by 2ξ, we next show that if $g \in \mathcal{H}_s^{d_2}$ is chosen uniformly at random and d_2 is sufficiently large, then for each application of g to a bucket of f, the maximum bucket size is bounded by d_2, except with negligible probability.

Lemma 5 ([2, Fact 2.2(b)]). *For all $m, s, d \in \mathbb{N}$ and for every subset B of U of size m, if g is chosen uniformly at random from $\mathcal{H}_s^d$, then $\max_{0 \le i < s} |\{x \in B \mid g(x) = i\}| \le d$ with probability at least $1 - m \cdot (2m/s)^d$.*

We use the lemma for $j = 0, \ldots, r - 1$ with $B = B_j = \{x \in S \mid f(x) = j\}$. In our case, $d = d_2$, and $m \le 2\xi$, so that $2m/s \le 4n^{1+1/(2c)}/(rs) \le 4n^{-1/(2c)}$. Thus, if d_2 and subsequently n are chosen sufficiently large, then $\Pr(|\{x \in B \mid g(x) = i\}| \ge d_2) \le n^{-3}$ for $i = 0, \ldots, s - 1$ and $\Pr(\max_{0 \le i < s} |\{x \in B \mid g(x) = i\}| \ge d_2) \le sn^{-3} \le n^{-2}$.

We now come to the proof of Lemma 3 itself. Concerning the random choice of h, we assume that f and g have already been selected, so that only $a_0, \ldots, a_{r-1}$ remain to be chosen. Then, for each fixed $i \in \{0, \ldots, s - 1\}$, the quantity $Z_i = |\{x \in S \mid h(x) = i\}|$ is the sum of independent random variables $X_0, \ldots, X_{r-1}$ where, for $j = 0, \ldots, r - 1$, $X_j = |\{x \in B_j \mid h(x) = i\}|$. It is easy to see that $E(Z_i) = n/s$. We will assume that $X_j \le d_2$ for $j = 0, \ldots, r - 1$; by what was shown above, this ignores an event of negligible probability.

Lemma 6 (Hoeffding; see [6, p. 104]). *Let Z be a sum of independent non-negative random variables, each bounded by $z > 0$, and take $\mu = E(Z)$. Then, for all $t > 0$,*

$$\Pr(Z \ge \mu + t) \le \left[\left(\frac{\mu}{\mu + t} \right)^{\mu + t} e^t \right]^{1/z}.$$

Using the lemma with $z = d_2$, $\mu = n/s$ and $t = (C - 1)\mu$, we obtain

$$\Pr(Z_i \ge Cn/s) \le (e/C)^{Cn/(d_2 s)} \le (e/C)^{(C \log n)/(d_2 c)}.$$

For sufficiently large C, we have $\Pr(Z_i \ge Cn/s) \le n^{-3}$ and $\Pr(\max_{0 \le i < s} Z_i \ge Cn/s) \le sn^{-3} \le n^{-2}$. Adding the three "error" probabilities identified above, we see that the assertion of Lemma 3 holds with probability at least $1 - 3n^{-2} \ge 1 - n^{-1}$. This ends the proof of Lemma 3.

The condition $s = O(n/\log n)$ of Lemma 3 prevents us from splitting S into groups of size at most $\hat{n}$ in one go. We therefore begin by splitting S into groups of size $O((\log n)^3)$. We take $r = \Theta(\sqrt{n})$, $s = \Theta(n/(\log n)^3)$ and C, d_1 and d_2 according to Lemma 3 (for $c = 3$, say) and repeatedly choose $h \in \mathcal{R}(r, s, d_1, d_2)$ uniformly at random until $\max_{0 \le i < s} |\{x \in S \mid h(x) = i\}| \le Cn/s$. By Lemma 3, the expected number of trials is $O(1)$, and the computation can be carried out in $O(n)$ expected time using $O(n)$ words of working space. By the assumption $\max S \le n^3$, the chosen function h can be represented in $O(r \log n) = O(\sqrt{n} \log n)$ space, and its evaluation takes $O(1)$ time.

For each of the resulting groups of size at most $l = \Theta((\log n)^3)$, we use a single range reduction based on Lemma 2 to force all integers in the group below an integer $\hat{v}$ with $\hat{v} = (\log n)^{O(1)}$. This requires the storage of one prime of $O(\log \log n)$ bits per group, for a total of $o(n/\log n)$ bits.

Our remaining task is to reduce the group size further from at most l to at most $\hat{n} = \gamma \log n / \log \log n$. We do this by another application of Lemma 3 with $r = \Theta((\log n)^2)$ and with $s \geq Cl/\hat{n}$ (ensuring that each group is split into subgroups of at most $\hat{n}$ elements each), but $s = O(l/\hat{n})$ (ensuring that the total number of subgroups is $O(n/\hat{n})$). The space needed for storing the required functions in $\mathcal{R}$ is $O(r \log \log n)$ bits per group, for a total of $O(n \log \log n / \log n)$ bits.

4 Perfect Hashing by Brute Force

In the previous section we reduced the original problem of size n to a collection of subproblems of size $O(\log n / \log \log n)$ and involving numbers of size polylogarithmic in n. In this section we discuss how to solve a single such subproblem using nearly minimal space.

Fix integers $m, v \geq 2$ and let $\mathcal{F}$ be the set of all functions from $V = \{0, \ldots, v-1\}$ to $\{0, \ldots, m-1\}$. We call a (multi)subset F of $\mathcal{F}$ *perfect* if for every subset B of V of size m, F contains a (minimal) perfect hash function for B. For $t \in \mathbb{N}$, denote by $q(t)$ the probability that a multiset of t random functions drawn independently from the uniform distribution over $\mathcal{F}$ is not perfect. In the proof of [8, Sect. III.2.3, Thm. 7], Mehlhorn argues that

$$q(t) \leq \binom{v}{m}\left(1 - \frac{m!}{m^m}\right)^t$$

for all $t \in \mathbb{N}$ and proves that the right-hand side is smaller than 1 for $t = t^* = \lceil m e^m \ln v \rceil$.

It follows that there exists a perfect set of size t^*, and Mehlhorn proceeds to define a canonical such set F^* as the first perfect set encountered in some fixed enumeration of the subsets of $\mathcal{F}$ of size t^*. Because F^* can be recalculated for every query, it need not be stored, which is crucial in the original setting. In our application, however, m and v are tiny, and storing a perfect set is feasible. This allows us to replace the deterministic procedure of [8], which runs in doubly-exponential time, by a randomized procedure whose running time is merely singly exponential.

We first observe that since $m! \geq (m/e)^m$,

$$q(t^* + 1) \leq q(t^*)\left(1 - \frac{m!}{m^m}\right) \leq 1 - e^{-m}.$$

It follows that if we repeatedly draw a multiset of $t^* + 1$ random functions from $\mathcal{F}$ until a perfect multiset is encountered, then the expected number of trials is $O(e^m)$. Moreover, each multiset can be tested for perfectness in time $O\left(\binom{v}{m} m(t^* + 1)\right) = O(v^m e^m \ln v)$, so a perfect multiset F^* of $t^* + 1$ functions from $\mathcal{F}$ can be found in $v^{O(m)}$ expected time. As a by-product of the computation, we discover for each subset B of V of size m a function in F^* that is perfect for B.

Lemma 7. *Given integers $m, v \geq 2$ and a subset B of size m of $\{0, \ldots, v-1\}$, $v^{O(m)}$ expected time and $v^{O(m)}$ words of working space suffice to compute a* *minimal perfect hash function for B that can be evaluated in constant time and whose representation consists of $v^{O(m)}$ bits that depend only on m and v and $m \log e + \log \log v + O(\log m)$ bits that depend also on B.*

Proof. Carry out the construction described above and store the set F^*, which depends only on m and v, as well as a pointer of $\log |F^*| = m \log e + \log \log v + O(\log m)$ bits to a function in F^* that is perfect for B.

5   The Complete Construction

In Section 3 the problem of hashing the original set S was reduced to that of hashing groups $G_1, \ldots, G_k$, where $|G_i| \leq \hat{n}$ and $\max G_i < \hat{v}$ for $i = 1, \ldots, k$ and $k = \Theta(n/\hat{n})$. More precisely, we showed that, within the resource bounds of Theorem 1, we can map S injectively to a set $\hat{S}$ of pairs $(i, j) \in \{1, \ldots, k\} \times \{0, \ldots, \hat{v} - 1\}$ such that for $i = 1, \ldots, k$, $|G_i| \leq \hat{n}$, where $G_i = \hat{S} \cap (\{i\} \times \{0, \ldots, \hat{v} - 1\})$. Moreover, for a certain constant $\rho \geq 1$, we showed in Section 4 how to compute a minimal perfect hash function h_i for G_i in at most $\hat{v}^{\rho\hat{n}}$ steps, for $i = 1, \ldots, k$, such that the representation space of h_i consists of at most $\hat{v}^{\rho\hat{n}}$ bits that depend only on $|G_i|$ (the *shared part*) and $|G_i| \log e + O(\log \log n)$ bits that depend also on G_i (the *individual part*). We still need to describe how to combine the solution for single groups to a solution for the full set S.

Fix a constant ν so that $\hat{v} \leq (\log n)^{\nu}$ and recall that $\hat{n} = \gamma \log n/\log \log n$ for a constant $\gamma > 0$ that can still be chosen freely. Now

$$\hat{v}^{\rho\hat{n}} \leq 2^{\nu \log \log n \cdot \rho \cdot \gamma \log n/\log \log n} = n^{\nu\rho\gamma}.$$

We choose $\gamma = 1/(3\nu\rho)$, which makes $\hat{v}^{\rho\hat{n}} \leq n^{1/3}$. Then, since the number of possible distinct groups is bounded by $1 + \hat{v}^{\hat{n}} = O(n^{1/3})$, we can compute minimal perfect hash functions for all possible groups in $O(n^{2/3})$ time. We compute a table mapping each possible group size to the corresponding public part and another table mapping each group, represented as an integer of size $O(n^{1/3})$, to the corresponding individual part. The space needed for these tables is negligible. We create another table L mapping each $i \in \{1, \ldots, k\}$ to $|G_i|$, which allows the public part of h_i to be accessed, and to the individual part of h_i. The entries in L can be computed in $O(n)$ time, and their total size is

$$\sum_{i=1}^{k} (|G_i| \log e + O(\log \log n)) = n \log e + O(n(\log \log n)^2/\log n)$$

bits, in accordance with Theorem 1. We still have to solve two problems, however:

(1) We cannot use the functions $h_1, \ldots, h_k$ directly, because their ranges overlap. Rather, we would like to replace h_i by $h_i + \sum_{j=1}^{i-1} |G_j|$, for $i = 1, \ldots, k$.
(2) Because the entries in L are not all of the same length, it is not clear how to access the ith entry in constant time.

The following lemma provides a solution to these problems.

Lemma 8. *For all integers $m, N \geq 4$, given m integers $a_1, \ldots, a_m$ with $\sum_{i=1}^{m} a_i$ $\leq N$, a data structure that occupies $O(m(\log \log m + \log(1 + N/m)))$ bits and allows the computation of $b_i = \sum_{j=1}^{i} a_j$ from i in constant time, for $i = 1, \ldots, m$, can be constructed in $O(m)$ time and space.*

Proof. If $N > m^2$, we can simply store $b_1, \ldots, b_m$ in a table with m entries of $\lceil \log(N+1) \rceil$ bits each. Assume therefore that $N \leq m^2$. Our data structure is a tree T of depth 2 with at least m leaves in which every node of depth 1 has $d = O(\log m)$ children and the root has $O(m/\log m)$ children. Conceptually, we label the ith leaf of T, counted from the left, with a_i, for $i = 1, \ldots, m$, and the remaining leaves with zero. For every node v of T, denote by $s(v)$ the sum of the labels at leaves that are descendants of left siblings of v or equal to v. For $i = 1, \ldots, m$, the prefix sum b_i is the sum of $s(v)$ over all ancestors v of the ith leaf of T.

We call a leaf v *good* if $s(v) \leq (N/m)(\log m)^2$, and *bad* otherwise, If a leaf v is good, we store $s(v)$ in a field of $O(\log \log m + \log(1 + N/m))$ bits associated with v. Similarly, for each internal node v, we store $s(v)$ in a field of $O(\log m)$ bits associated with v. Together, these fields occupy $O(m(\log \log m + \log(1 + N/m)))$ bits.

Call a node v of depth 1 good if all of its children are good, and bad otherwise. For each bad node v of depth 1, we store all the values $s(y)$, where y is a child of v, in a table with d fields of $O(\log m)$ bits each in an overflow area and store a pointer to this table with v. Since the number of bad nodes of depth 1 is bounded by $m/(\log m)^2$, an overflow area of size $O(m)$ suffices. Altogether, the space needed is $O(m(\log \log m + \log(1 + N/m)))$ bits, and it is easy to see that b_i can be computed in constant time from i for $i = 1, \ldots, m$.

In order to solve Problem (1), we store the groups sizes $|G_1|, \ldots, |G_k|$ using the method of Lemma 8. Since $\sum_{i=1}^{k} |G_i| = n$, the space needed comes to $O(k(\log \log n + \log(1 + n/k))) = O(n(\log \log n)^2 / \log n)$. In order to solve Problem (2), we store the individual parts of $h_1, \ldots, h_k$ as one contiguous bit string W and store the sizes of these individual parts using the method of Lemma 8, which allows us to pick out any individual part from W in constant time. Since the total size of all individual parts is $O(n)$, the space needed again is $O(n(\log \log n)^2 / \log n)$. This ends the proof of Theorem 1.

References

1. J. L. Carter and M. N. Wegman, Universal Classes of Hash Functions, *J. Comput. System Sci.* **18** (1979), pp. 143–154.
2. M. Dietzfelbinger and F. Meyer auf der Heide, A new universal class of hash functions and dynamic hashing in real time, Proc. 17th International Colloquium on Automata, Languages and Programming (ICALP 1990), Lecture Notes in Computer Science, Vol. 443, Springer-Verlag, Berlin, pp. 6–19.

3. M. L. Fredman and J. Komlós, On the size of separating systems and families of perfect hash functions, *SIAM J. Alg. Disc. Meth.* **5** (1984), pp. 61–68.

4. M. L. Fredman, J. Komlós and E. Szemerédi, Storing a sparse table with $O(1)$ worst case access time, *J. ACM* **31** (1984), pp. 538–544.

5. T. Hagerup, Sorting and searching on the word RAM, Proc. 15th Annual Symposium on Theoretical Aspects of Computer Science (STACS 1998), Lecture Notes in Computer Science, Vol. 1373, Springer-Verlag, Berlin, pp. 366–398.

6. M. Hofri, *Probabilistic Analysis of Algorithms*, Springer-Verlag, New York, 1987.

7. C. P. Kruskal, L. Rudolph and M. Snir, A complexity theory of efficient parallel algorithms, *Theoret. Comput. Sci.* **71**, (1990), pp. 95–132.

8. K. Mehlhorn, *Data Structures and Algorithms, Vol. 1: Sorting and Searching*, Springer-Verlag, Berlin, 1984.

9. M. O. Rabin, Probabilistic algorithm for testing primality. *J. Number Theory* **12**, (1980), pp. 128–138.

10. J. Radhakrishnan, Improved bounds for covering complete uniform hypergraphs, *Inform. Process. Lett.* **41** (1992), pp. 203–207.

11. J. P. Schmidt and A. Siegel, The spatial complexity of oblivious k-probe hash functions, *SIAM J. Comput.* **19** (1990), pp. 775–786.

12. A. Siegel, On universal classes of fast high performance hash functions, their time-space tradeoff, and their applications, Proc. 30th Annual IEEE Symposium on Foundations of Computer Science (FOCS 1989), pp. 20–25.

Small PCPs with Low Query Complexity

Prahladh Harsha and Madhu Sudan*

Laboratory for Computer Science, Massachusetts Institute of Technology
545 Technology Square, Cambridge, MA 02139, USA
{prahladh,madhu}@mit.edu

Abstract. Most known constructions of probabilistically checkable proofs (PCPs) either blow up the proof size by a large polynomial, or have a high (though constant) query complexity. In this paper we give a transformation with slightly-super-cubic blowup in proof size, with a low query complexity. Specifically, the verifier probes the proof in 16 bits and rejects every proof of a false assertion with probability arbitrarily close to $\frac{1}{2}$, while accepting corrects proofs of theorems with probability one. The proof is obtained by revisiting known constructions and improving numerous components therein. In the process we abstract a number of new modules that may be of use in other PCP constructions.

1 Introduction

Probabilistically checkable proofs (PCP) have played a major role in proving the hardness of approximation of various combinatorial optimization problems. Constructions of PCPs have been the subject of active research in the last ten years. In the last decade, there have been several "efficient" construction of PCPs which in turn have resulted in tighter inapproximability results. Arora et al. [1] showed that it is possible to transform any proof into a probabilistically checkable one of polynomial size, such that it is verifiable with a constant number of queries. Valid proofs are accepted with probability one (this parameter is termed the completeness of the proof), while any purported proof of an invalid assertion is rejected with probability $1/2$ (this parameter is the soundness of the proof). Neither the proof size, nor the query complexity is explicitly described there; however the latter is estimated to be around 10^6.

Subsequently much success has been achieved in improving the parameters of PCPs, constructing highly efficient proof systems either in terms of their size or their query complexity. The best result in terms of the former is a result of Polishchuk and Spielman [12]. They show how any proof can be transformed into a probabilistically checkable proof with only a mild blowup in the proof size, of $n^{1+\epsilon}$ for arbitrarily small $\epsilon > 0$ and that is checkable with only a constant number of queries. This number of queries however is of the order of $O(1/\epsilon^2)$, with the constant hidden by the big-Oh being some multiple of the query complexity of [1]. On the other hand, Håstad [10] has constructed PCPs for arbitrary NP statements where the query complexity is a mere three bits (for completeness almost 1 and soundness $1/2$). However the blowup in the proof size of Håstad's PCPs has an exponent proportional to the query complexity of the PCP of [1]. Thus neither of these "nearly-optimal" results provides simultaneous optimality of the

---

* Supported in part by a Sloan Foundation Fellowship and NSF Career Award CCR-9875511.

A. Ferreira and H. Reichel (Eds.): STACS 2001, LNCS 2010, pp. 327–338, 2001.
© Springer-Verlag Berlin Heidelberg 2001

two parameters. It is reasonable to wonder if this inefficiency in the combination of the two parameters is inherent; and our paper is motivated by this question.

We examine the size and query complexity of PCPs jointly and obtain a construction with reasonable performance in both parameters. The only previous work that mentions the joint size vs. query complexity of PCPs is a work of Friedl and Sudan [8], who indicate that NP has PCPs with nearly quadratic size complexity and in which the verifier queries the proof for 165 bits. The main technical ingredient in their proof was an improved analysis of the "low-degree test". Subsequent to this work, the analysis of low-degree tests has been substantially improved. Raz and Safra [13] and Arora and Sudan [3] have given highly efficient analysis of different low-degree tests. Furthermore, techniques available for "proof composition" have improved, as also have the construction for terminal "inner verifiers". In particular, the work of Håstad [9,10], has significantly strengthened the ability to analyze inner verifiers used at the final composition step of PCP constructions.

In view of these improvements, it is natural to expect the performance of PCP constructions to improve. Our work confirms this expectation. However, our work exposes an enormous number of complications in the natural path of improvement. We resolve most of these, with little loss in performance and thereby obtain the following result: Satisfiability has a PCP verifier that makes at most 16 oracle queries to a proof of size at most $n^{3+o(1)}$, where n is the size of the instance of satisfiability. Satisfiable instances have proofs that are accepted with probability one, while unsatisfiable instances are accepted with probability arbitrarily close to $1/2$. (See Main Theorem 1.)

We also raise several technical questions whose positive resolution may lead to a PCP of nearly quadratic size and query complexity of 6. Surprisingly, no non-trivial limitations are known on the joint size + query complexity of PCPs. In particular, it is open as to whether nearly linear sized PCPs with query complexity of 3 exist for NP statements.

2   Overview

We first recall the standard definition of the class $\mathrm{PCP}_{c,s}[r, q]$.

Definition 1. *For functions $r, q : \mathbb{Z}^+ \to \mathbb{Z}^+$, a probabilistic oracle machine (or verifier) V is (r, q)-restricted if on input x of length n, the verifier tosses at most $r(n)$ random coins and queries an oracle π for at most $q(n)$ bits. A language $L \in \mathrm{PCP}_{c,s}[r, q]$ if there exists an (r, q)-restricted verifier V that satisfies the following properties on input x.*

Completeness. *If $x \in L$ then there exists π such that V on oracle access to π accepts with probability at least c.*

Soundness. *If $x \notin L$ then for every oracle π, the verifier V accepts with probability strictly less than s.*

While our principal interest is in the size of a PCP and not in the randomness, it is well-known that the size of a probabilistically checkable proof (or more precisely, the number of distinct queries to the oracle π) is at most $2^{r(n)+q(n)}$. Thus the size is implicitly governed by the randomness and query complexity of a PCP. The main result of this paper is the following.

Main Theorem 1. *For every* $\varepsilon, \mu > 0$, SAT $\in$ PCP$_{1,\frac{1}{2}+\mu}[(3+\varepsilon)\log n, 16]$.

Remark: Actually the constants ε and μ above can be replaced by some $o(1)$ functions; but we don't derive them explicitly.

It follows from the parameters that the associated proof is of size at most $O(n^{3+\varepsilon})$. Cook [6] showed that any language in NTIME$(t(n))$ could be reduced to SAT in $O(t(n)\log t(n))$ time such that instances of size n are mapped to Boolean formulae of size at most $O(t(n)\log t(n))$. Combining this with the Main Theorem 1, we have that every language in NP has a PCP with at most a slightly super-cubic blowup in proof size and a query complexity as low as 16 bits.

2.1 MIP and Recursive Proof Composition

As pointed out earlier, the parameters we seek are such that no existing proof system achieves them. Hence we work our way through the PCP construction of Arora et al. [1] and make every step as efficient as possible. The key ingredient in their construction (as well as most subsequent constructions) is the notion of recursive composition of proofs, a paradigm introduced by Arora and Safra [2]. The paradigm of recursive composition is best described in terms of multi-prover interactive proof systems (MIPs).

Definition 2. *For integer p, and functions $r, a : \mathbb{Z}^+ \to \mathbb{Z}^+$, an MIP verifier V is (p, r, a)-restricted if it interacts with p mutually-non-interacting provers $\pi_1, \ldots, \pi_p$ in the following restricted manner. On input x of length n, V picks a random $r(n)$-bit string R and generates p queries $q_1, \ldots, q_p$ and a circuit C of size at most $a(n)$. The verifier then issues query q_i to prover π_i. The provers respond with answers $a_1, \ldots, a_p$ each of length at most $a(n)$ and the verifier accepts x iff $C(a_1, \ldots, a_p) = $ true. A language L belongs to MIP$_{c,s}[p, r, a]$ if there exists a (p, r, a)-restricted MIP verifier V such that on input x:*

Completeness. *If $x \in L$ then there exist $\pi_1, \ldots, \pi_p$ such that V accepts with probability at least c.*

Soundness. *If $x \notin L$ then for every $\pi_1, \ldots, \pi_p$, V accepts with probability less than s.*

It is easy to see that MIP$_{c,s}[p, r, a]$ is a subclass of PCP$_{c,s}[r, pa]$ and thus it is beneficial to show that SAT is contained in MIP with nice parameters. However, much stronger benefits are obtained if the containment has a small number of provers, even if the answer size complexity (a) is not very small. This is because the verifier's actions can usually be simulated by a much more efficient verification procedure, one with much smaller answer size complexity, at the cost of a few more provers. Results of this nature are termed proof composition lemmas; and the efficient simulators of the MIP verification procedure are usually called "inner verification procedures".

The next three lemmas divide the task of proving Main Theorem 1 into smaller subtasks. The first gives a starting MIP for satisfiability, with 3 provers, but polylogarithmic answer size. We next give the composition lemma that is used in the intermediate stages. The final lemma gives our terminal composition lemma – the one that reduces answer sizes from some slowly growing function to a constant.

Lemma 2. *For every $\varepsilon, \mu > 0$, SAT $\in$ MIP$_{1,\mu}[3, (3+\varepsilon)\log n, \text{poly}\log n]$.*

Lemma 2 is proven in Sect. 3. This lemma is critical to bounding the proof size. This lemma follows the proof of a similar one (the "parallelization" step) in [1]; however various aspects are improved. We show how to incorporate advances made by Polishchuk and Spielman [12], and how to take advantage of the low-degree test of Raz and Safra [13]. Most importantly, we show how to save a quadratic blowup in this phase that would be incurred by a direct use of the parallelization step in [1].

The first composition lemma we use is an off-the-shelf product due to [3]. Similar lemmas are implicit in the works of Bellare et al. [5] and Raz and Safra [13].

Lemma 3 ([3]). *For every $\epsilon > 0$ and $p < \infty$, there exist constants c_1, c_2, c_3 such that for every $r, a : \mathbb{Z}^+ \to \mathbb{Z}^+$,*

$$\mathrm{MIP}_{1,\epsilon}[p, r, a] \subseteq \mathrm{MIP}_{1,\epsilon^{1/(2p+2)}}[p + 3, r + c_1 \log a, c_2(\log a)^{c_3}] \ .$$

The next lemma shows how to truncate the recursion. This lemma is proved in Sect. 4 using a "Fourier-analysis" based proof, as in [9,10]. This is the first time that this style of analysis has been applied to MIPs with more than 2 provers. All previous analyses seem to have focused on composition with canonical 2-prover proof systems at the outer level. Our analysis reveals surprising complications and forces us to use a large number (seven) of extra bits to effect the truncation.

Lemma 4. *For every $\epsilon > 0$ and $p < \infty$, there exists a $\gamma > 0$ such that for every $r, a : \mathbb{Z}^+ \to \mathbb{Z}^+$,*

$$\mathrm{MIP}_{1,\gamma}[p, r, a] \subseteq \mathrm{PCP}_{1,\frac{1}{2}+\epsilon}[r + O\left(2^{pa}\right), p + 7] \ .$$

Proof (of Main Theorem 1). The proof is straightforward given the above lemmas. We first apply Lemma 2 to get a 3-prover MIP for SAT, then apply Lemma 3 twice to get a 6- and then a 9-prover MIP for SAT. The answer size in the final stage is poly log log log n. Applying Lemma 4 at this stage we obtain a 16-query PCP for SAT; and the total randomness in all stages remains $(3 + \varepsilon) \log n$. □

Organization of the Paper: In Section 3, we prove Lemma 2. For this purpose, we present the Polynomial Constraint Satisfaction problem in Section 3.2 and discuss its hardness. We then discuss the Low degree Test in Section 3.3. Most aspects of the proofs in Section 3 are drawn from previous works of [1,3,12,13]. Hence, we abstract the main results in this section and leave the detailed proofs to the full version of the paper[1] In Section 4, we present the proof of Lemma 4. In section 5 we suggest possible approaches for improvements in the joint size-query complexity of PCPs.

3   A Randomness Efficient MIP for SAT

In this section, we use the term "length-preserving reductions", to refer to reductions in which the length of the target instance of the reduction is nearly-linear ($O(n^{1+\epsilon})$ for arbitrarily small ϵ) in the length of the source instance.

---

[1] A full version of this paper can be found at ftp://ftp.eccc.uni-trier.de/pub/ eccc/reports/2000/TR00-061/index.html.

To prove membership in SAT, we first transform SAT into an algebraic problem. This transformation comes in two phases. First we transform it to an algebraic problem (that we call AP for lack of a better name) in which the constraints can be enumerated compactly. Then we transform it to a promise problem on polynomials, called Polynomial Constraint Satisfaction (PCS), with a large associated gap. We then show how to provide an MIP verifier for the PCS problem.

Though most of these results are implicit in the literature, we find that abstracting them cleanly significantly improves the exposition of PCPs. The first problem, AP, could be proved to be NP-hard almost immediately, if one did not require length-preserving reductions. We show how the results of Polishchuk and Spielman [12] imply a length preserving reduction from SAT to this problem. We then reduce this problem to PCS. This step mimics the sum-check protocol of Lund et al. [11]. The technical importance of this intermediate step is the fact that it does *not* refer to "low-degree" tests in its analysis. Low-degree tests are primitives used to test if the function described by a given oracle is close to some (unknown) multivariate polynomial of low-degree. Low-degree tests have played a central role in the constructions of PCPs. Here we separate (to a large extent) their role from other algebraic manipulations used to obtain PCPs/MIPs for SAT .

In the final step, we show how to translate the use of state-of-the-art low-degree tests, in particular the test of Raz and Safra [13], in conjunction with the hardness of PCS to obtain a 3-prover MIP for SAT. This part follows a proof of Arora et al. [1] (their parallelization step); however a direct implementation would involve $6 \log n$ randomness, or an n^6 blow up in the size of the proof. Part of this is a cubic blow up due to the use of the low-degree test and we are unable to get around this part. Direct use of the parallelization also results in a quadratic blowup of the resulting proof. We save on this by creating a variant of the parallelization step of [1] that uses higher dimensional varieties instead of 1-dimensional ones.

3.1 A Compactly Described Algebraic NP-Hard Problem

Definition 3. *For functions* $m, h : \mathbb{Z}^+ \to \mathbb{Z}^+$, *the problem* $\text{AP}_{m,h}$ *has as its instances* $(1^n, H, T, \psi, \rho_1, \ldots, \rho_6)$ *where:* H *is a field of size* $h(n)$, $\psi : H^7 \to H$ *is a constant degree polynomial,* T *is an arbitrary function from* H^m *to* H *and the* ρ_i's *are linear maps from* H^m *to* H^m, *for* $m = m(n)$. *(T is specified by a table of values, and* ρ_i's *by* $m \times m$ *matrices.)* $(1^n, H, T, \psi, \rho_1, \ldots, \rho_6) \in \text{AP}_{m,h}$ *if there exists an assignment* $A : H^m \to H$ *such that for every* $x \in H^m$, $\psi(T(x), A(\rho_1(x)), \ldots, A(\rho_6(x))) = 0$.

The above problem is just a simple variant of standard constraint satisfaction problems, the only difference being that its variables and constraints are now indexed by elements of H^m. The only algebra in the above problem is in the fact that the functions ρ_i, which dictate which variables participate in which constraint, are linear functions. The following statement, abstracted from [12], gives the desired hardness of AP.

Lemma 5. *There exists a constant c such that for any pair of functions* $m, h : \mathbb{Z}^+ \to \mathbb{Z}^+$ *satisfying* $h(n)^{m(n)-c} \geq n$ *and* $h(n)^{m(n)} = O(n^{1+o(1)})$, *SAT reduces to* $\text{AP}_{m,h}$ *under length preserving reductions.*

We note that Szegedy [16] has given an alternate abstraction of the result of [12] which focuses on some different aspects and does not suffice for our purposes.

3.2 Polynomial Constraint Satisfaction

We next present an instance of an algebraic constraint satisfaction problem. This differs from the previous one in that its constraints are "wider", the relationship between constraints and variables that appear in it is arbitrary (and not linear), and the hardness is not established for arbitrary assignment functions, but only for low-degree functions. All the above changes only make the problem harder, so we ought to gain something – and we gain in the gap of the hardness. The problem is shown to be hard even if the goal is only to separate satisfiable instances from instances in which only ϵ fraction of the constraints are satisfiable. We define this gap version of the problem first.

Definition 4. *For* $\epsilon : \mathbb{Z}^+ \to \mathbb{R}^+$, *and* $m, b, q : \mathbb{Z}^+ \to \mathbb{Z}^+$ *the promise problem* $\mathrm{GapPCS}_{\epsilon,m,b,q}$ *has as instances* $(1^n, d, k, s, \mathbb{F}; C_1, \ldots, C_t)$, *where* $d, k, s \leq b(n)$ *are integers and* $\mathbb{F}$ *is a field of size* $q(n)$ *and* $C_j = (A_j; x_1^{(j)}, \ldots, x_k^{(j)})$ *is an algebraic constraint, given by an algebraic circuit* A_j *of size* s *on* k *inputs and* $x_1^{(j)}, \ldots, x_k^{(j)} \in \mathbb{F}^m$, *for* $m = m(n)$. $(1^n, d, k, s, \mathbb{F}; C_1, \ldots, C_t)$ *is a YES instance if there exists a polynomial* $p : \mathbb{F}^m \to \mathbb{F}$ *of degree at most* d *such that for every* $j \in \{1, \ldots, t\}$, *the constraint* C_j *is satisfied by* p, *i.e.,* $A_j(p(x_1^{(j)}), \ldots, p(x_k^{(j)})) = 0$. $(1^n, d, k, s, \mathbb{F}; C_1, \ldots, C_t)$ *is a NO instance if for every polynomial* $p : \mathbb{F}^m \to \mathbb{F}$ *of degree at most* d *it is the case that at most* $\epsilon(n) \cdot t$ *of the constraints* C_j *are satisfied.*

Lemma 6. *There exist constants* c_1, c_2 *such that for every choice of functions* ϵ, m, b, q *satisfying* $(b(n)/m(n))^{m(n)-c_1} \geq n$, $q(n)^{m(n)} = O\left(n^{1+o(n)}\right)$, $q(n) \geq c_2 b(n)/\epsilon(n)$, *SAT reduces to* $\mathrm{GapPCS}_{\epsilon,m,b,q}$ *under length preserving reductions.*

(The problem $\mathrm{AP}_{m,h}$ is used as an intermediate problem in the reduction. However we don't mention this in the lemma, since the choice of parameters m, h may confuse the statement further.) The proof of this lemma is inspired by the sum-check protocol of Lund et al. [11] while the specific steps in our proof follow the proof in Sudan [15].

3.3 Low-Degree Tests

Using GapPCS it is easy to produce a simple probabilistically checkable proof for SAT. Given an instance of SAT, reduce it to an instance $\mathcal{I}$ of GapPCS ; and provide as proof the polynomial $p : \mathbb{F}^m \to \mathbb{F}$ as a table of values. To verify correctness a verifier first "checks" that p is close to some polynomial and then verifies that a random constraint C_j is satisfied by p. Low-degree tests are procedures designed to address the first part of this verification step – i.e., to verify that an arbitrary function $f : \mathbb{F}^m \to \mathbb{F}$ is close to some (unknown) polynomial p of degree d.

Low-degree tests have been a subject of much research in the context of program checking and PCPs. For our purposes, we need tests that have very low probability of error. Two such tests with analyses are known, one due to Raz and Safra [13] and another due to Rubinfeld and Sudan [14] (with low-error analysis by Arora and Sudan [3])

For our purposes the test of Raz and Safra is more efficient. We describe their results first and then compare its utility with the result in [3].

A plane in $\mathbb{F}^m$ is a collection of points parametrized by two variables. Specifically, given $a, b, c \in \mathbb{F}^m$ the plane $\wp_{a,b,c} = \{\wp_{a,b,c}(t_1, t_2) = a + t_1 b + t_2 c | t_1, t_2 \in \mathbb{F}\}$. Several parameterizations are possible for a given plane. We assume some canonical one is fixed for every plane, and thus the plane is equivalent to the set of points it contains. The low-degree test uses the fact that for any polynomial $p : \mathbb{F}^m \to \mathbb{F}$ of degree d, the function $p_\wp : \mathbb{F}^2 \to \mathbb{F}$ given by $p_\wp(t_1, t_2) = p(\wp(t_1, t_2))$ is a bivariate polynomial of degree d. The verifier tests this property for a function f by picking a random plane through $\mathbb{F}^m$ and verifying that there *exists* a bivariate polynomial that has good agreement with f restricted to this plane. The verifier expects an auxiliary oracle f_{planes} that gives such a bivariate polynomial for every plane. This motivates the test below.

Low-Degree Test (Plane-Point Test)
Input: A function $f : \mathbb{F}^m \to \mathbb{F}$ and an oracle f_{planes}, which for each plane in $\mathbb{F}^m$ gives a bivariate degree d polynomial.
  1. Choose a random point in the space $x \in_R \mathbb{F}^m$.
  2. Choose a random plane $\wp$ passing through x in $\mathbb{F}^m$.
  3. Query f_{planes} on $\wp$ to obtain the polynomial $h_\wp$. Query f on x.
  4. Accept iff the value of the polynomial $h_\wp$ at x agrees with $f(x)$.

It is clear that if f is a degree d polynomial, then there exists an oracle f_{planes} such that the above test accepts with probability 1. It is non-trivial to prove any converse and Raz and Safra give a strikingly strong converse. (see Theorem 7)

First some more notation. Let $\text{LDT}^{f, f_{\text{planes}}}(x, \wp)$ denote the outcome of the above test on oracle access to f and f_{planes}. Let $f, g : \mathbb{F}^m \to \mathbb{F}$ have agreement δ if $\Pr_{x \in \mathbb{F}^m}[f(x) = g(x)] = \delta$.

Theorem 7. *There exist constants c_0, c_1 such that for every positive real δ, integers m, d and field $\mathbb{F}$ satisfying $|\mathbb{F}| \geq c_0 d (m/\delta)^{c_1}$, the following holds: Fix $f : \mathbb{F}^m \to \mathbb{F}$ and f_{planes}. Let $\{P_1, \ldots, P_l\}$ be the set of all m-variate polynomials of degree d that have agreement at least $\delta/2$ with the function $f : \mathbb{F}^m \to \mathbb{F}$. Then*

$$\Pr_{x, \wp}[f(x) \notin \{P_1(x), \ldots, P_l(x)\} \text{ and } \text{LDT}^{f, f_{\text{planes}}}(x, \wp) = \text{accept}] \leq \delta.$$

Remarks:
1. The actual theorem statement of Raz and Safra differs in a few aspects. The main difference being that the exact bound on the agreement probability described is different; and the fact that the claim may only say that if the low-degree test passes with probability greater than δ, then there exists some polynomial that agrees with f in some fraction of the points. The full version of this paper will include a proof of the above theorem from the statement of Raz and Safra.
2. The cubic blowup in our proof size occurs from the oracle f_{planes} which has size cubic in the size of the oracle f. A possible way to make the proof shorter would be to use an oracle for f restricted only to lines. (i.e., an analogous line-point test to the above test) The analysis of [3] does apply to such a test. However they require the field size to be (at least) a fourth power of the degree; and this results in a blowup in the proof to (at least) an eighth power. Note that the above theorem only needs a linear relationship between the degree and the field size.

3.4   Putting them Together

As pointed out earlier a simple PCP for GapPCS can be constructed based on the low-degree test above. A proof would be an oracle f representing the polynomial and the auxiliary oracle f_{planes}. The verifier performs a low-degree test on f and then picks a random constraint C_j and verifies that C_j is satisfied by the assignment f. But the naive implementation would make k queries to the oracle f and this is too many queries. The same problem was faced by Arora et al. [1] who solved it by running a curve through the k points and then asking a new oracle f_{curves} to return the value of f restricted to this curve. This solution cuts down the number of queries to 3, but the analysis of correctness works only if $|\mathbb{F}| \geq kd$. In our case, this would impose an additional quadratic blowup in the proof size and we would like to avoid this. We do so by picking r-dimensional varieties (algebraic surfaces) that pass through the given k points. This cuts down the degree to $rk^{1/r}$. However some additional complications arise: The variety needs to pass through many random points, but not at the expense of too much randomness. We deal with these issues below.

A variety $\mathcal{V} : \mathbb{F}^r \to \mathbb{F}^m$ is a collection of m functions, $\mathcal{V} = \langle \mathcal{V}_1, \ldots, \mathcal{V}_m \rangle$, $\mathcal{V}_i : \mathbb{F}^r \to \mathbb{F}$. A variety is of degree D if all the functions $\mathcal{V}_1, \ldots, \mathcal{V}_m$ are polynomials of degree D. For a variety $\mathcal{V}$ and function $f : \mathbb{F}^m \to \mathbb{F}$, the restriction of f to $\mathcal{V}$ is the function $f|_{\mathcal{V}} : \mathbb{F}^r \to \mathbb{F}$ given by $f|_{\mathcal{V}}(a_1, \ldots, a_r) = f(\mathcal{V}(a_1, \ldots, a_r))$. Note that the restriction of a degree d polynomial $p : \mathbb{F}^m \to \mathbb{F}$ to an r-dimensional variety $\mathcal{V}$ of degree D is an r-variate polynomial of degree Dd.

Let $S \subseteq \mathbb{F}$ be of cardinality $k^{1/r}$. Let $z_1, \ldots, z_k$ be some canonical ordering of the points in S^r. Let $\mathcal{V}^{(0)}_{S, x_1, \ldots, x_k} : \mathbb{F}^r \to \mathbb{F}^m$ denote a canonical variety of degree $r|S|$ that satisfies $\mathcal{V}^{(0)}_{S, x_1, \ldots, x_k}(z_i) = x_i$ for every $i \in \{1, \ldots, k\}$. Let $Z_S : \mathbb{F}^r \to \mathbb{F}$ be the function given by $Z_S(y_1, \ldots, y_r) = \prod_{i=1}^{r} \prod_{a \in S}(y_i - a)$; i.e. $Z_S(z_i) = 0$. Let $\alpha = \langle \alpha_1, \ldots, \alpha_m \rangle \in \mathbb{F}^m$. Let $\mathcal{V}^{(1)}_{S, \alpha}$ be the variety $\langle \alpha_1 Z_S, \ldots, \alpha_m Z_S \rangle$. We will let $\mathcal{V}_{S, \alpha, x_1, \ldots, x_k}$ be the variety $\mathcal{V}^{(0)}_{S, x_1, \ldots, x_k} + \mathcal{V}^{(1)}_{S, \alpha}$. Note that if α is chosen at random, $\mathcal{V}_{S, \alpha, x_1, \ldots, x_k}(z_i) = x_i$ for $z_i \in S^r$ and $\mathcal{V}_{S, \alpha, x_1, \ldots, x_k}(z)$ is distributed uniformly over $\mathbb{F}^m$ if $z \in (\mathbb{F} - S)^r$. These varieties will replace the role of the curves of [1]. We note that Dinur et al. also use higher dimensional varieties in the proof of PCP-related theorems [7]. Their use of varieties is for purposes quite different from ours.

We are now ready to describe the MIP verifier for $\text{GapPCS}_{\epsilon, m, b, q}$. (Henceforth, we shall assume that t, the number of constraints in $\text{GapPCS}_{\epsilon, m, b, q}$ instance is at most q^{2m}. In fact, for our reduction from SAT (Lemma 6), t is exactly equal to q^m.)

MIP Verifier$^{f, f_{planes}, f_{varieties}}(1^n, d, k, s, \mathbb{F}; C_1, \ldots, C_t)$.
Notation: r is a parameter to be specified. Let $S \subseteq \mathbb{F}$ be such that $|S| = k^{1/r}$.
1. Pick $a, b, c \in \mathbb{F}^m$ and $z \in (\mathbb{F} - S)^r$ at random.
2. Let $\wp = \wp_{a,b,c}$. Use b, c to compute $j \in \{1, \ldots, t\}$ at random (i.e., j is fixed given b, c, but is distributed uniformly when b and c are random.) Compute α such that $\mathcal{V}(z) = a$ for $\mathcal{V} = \mathcal{V}_{S, \alpha, x_1^{(j)}, \ldots, x_k^{(j)}}$.
3. Query $f(a)$, $f_{planes}(\wp)$ and $f_{varieties}(\mathcal{V})$.
   Let $g = f_{planes}(\wp)$ and $h = f_{varieties}(\mathcal{V})$.
4. Accept if all the conditions below are true:
   (a) g and f agree at a.

(b) h and f agree at a.

(c) A_j accepts the inputs $h(z_1), \ldots, h(z_k)$.

Complexity: Clearly the verifier V makes exactly 3 queries. Also, exactly $3m \log q + r \log q$ random bits are used by the verifier. The answer sizes are at most $O((drk^{1/r} + r)^r \log q)$ bits.

Now to prove the correctness of the verifier. Clearly, if the input instance is a YES instance then there exists a polynomial P of degree d that satisfies all the constraints of the input instance. Choosing $f = P$ and constructing f_{planes} and $f_{\text{varieties}}$ to be restrictions of P to the respective planes and varieties, we notice that the MIP verifier accepts with probability one. We now bound the soundness of the verifier.

Claim 8. *Let δ be any constant that satisfies the conditions of Theorem 7 and $\delta \geq 4\sqrt{\frac{d}{q}}$ where $q = |\mathbb{F}|$. Then the soundness of the MIP Verifier is at most $\delta + 4\epsilon/\delta + 4rk^{\frac{1}{r}}d/\delta(q - k^{\frac{1}{r}})$.*

Proof. Let $P_1, \ldots, P_l$ be all the polynomials of degree d that have agreement at least $\delta/2$ with f. (Note $l \leq 4/\delta$ since $\delta/2 \geq 2\sqrt{d/q}$) Now suppose, the MIP Verifier had accepted a NO instance. Then one of the following events must have taken place.

Event 1: $f(a) \notin \{P_1(a), \ldots, P_l(a)\}$ and $\mathrm{LDT}^{f, f_{\text{planes}}}(a, \wp) = \text{accept}$.

We have from Theorem 7, that Event 1 could have happened with probability at most δ.

Event 2: There exists an $i \in \{1, \ldots, l\}$, such that constraint C_j is satisfiable with respect to polynomial P_i. (i.e., $A_j(P_i(x_1^{(j)}), \ldots, P_i(x_k^{(j)})) = 0$).

As the input instance is a NO instance of $\mathrm{GapPCS}_{\epsilon, m, b, q}$, this events happens with probability at most $l\epsilon \leq 4\epsilon/\delta$.

Event 3: For all $i \in \{1, \ldots, p\}$, $P_i|_{\mathcal{V}} \neq h$, but the value of h at a is contained in $\{P_1(a), \ldots, P_l(a)\}$.

To bound the probability of this event happening, we reinterpret the randomness of the MIP verifier. First pick $b, c, \alpha \in \mathbb{F}^m$. From this we generate the constraint C_j and this defines the variety $\mathcal{V} = \mathcal{V}_{S, \alpha, x_1^{(j)}, \ldots, x_k^{(j)}}$. Now we pick $z \in (\mathbb{F} - S)^r$ at random and this defines $a = \mathcal{V}(z)$. We can bound the probability of the event in consideration after we have chosen $\mathcal{V}$, as purely a function of the random variable z as follows. Fix any i and $\mathcal{V}$ such that $P_i|_{\mathcal{V}} \neq h$. Note that the value of h at a equals $h(z)$ (by definition. of a, z and $\mathcal{V}$). Further $P_i(a) = P_i|_{\mathcal{V}}(z)$. But z is chosen at random from $(\mathbb{F} - S)^r$. By the Schwartz-Zippel lemma, the probability of agreement on this domain is at most $rk^{1/r}d/(|\mathbb{F}| - |S|)$. Using the union bound over the i's we get that this event happens with probability at most $lrk^{1/r}d/(|\mathbb{F}| - |S|) \leq 4rk^{\frac{1}{r}}d/\delta(q - k^{\frac{1}{r}})$.

We thus have that the probability of the verifier accepting a NO instance is at most $\delta + 4\epsilon/\delta + 4rk^{\frac{1}{r}}d/\delta(q - k^{\frac{1}{r}})$. $\qquad\square$

We can now complete the construction of a 3-prover MIP for SAT and give the proof of Lemma 2.

Proof (of Lemma 2). Choose $\delta = \frac{\mu}{3}$. Let c_0, c_1 be the constants that appear in Theorem 7. Choose $\varepsilon' = \varepsilon/2$ where ε is the soundness of the MIP, we wish to prove. Choose $\epsilon = \min\{\delta\mu/12, \varepsilon'/3(9 + c_1)\}$. Let n be the size of the SAT instance. Let $m = \epsilon \log n / \log \log n$, $b = (\log n)^{3 + \frac{1}{\epsilon}}$ and $q = (\log n)^{9 + c_1 + \frac{1}{\epsilon}}$. Note that this choice of parameters satisfies the requirements of Lemma 6. Hence, SAT reduces to $\mathrm{GapPCS}_{\epsilon, m, b, q}$

under length preserving reductions. Combining this reduction with the MIP verifier for GapPCS, we have a MIP verifier for SAT. Also δ satisfies the requirements of Claim 8. Thus, this MIP verifier has soundness as given by Claim 8. Setting $r = \frac{1}{\epsilon}$, we have that for sufficiently large n, $4rk^{\frac{1}{r}}d/\delta(q - k^{\frac{1}{r}}) \leq 8rk^{\frac{1}{r}}d/q\delta \leq \mu/3$. Hence, the soundness of the MIP verifier is at most $\delta + 4\epsilon/\delta + \mu/3 \leq \mu$. The randomness used is exactly $3m \log q + r \log q$ which with the present choice of parameters is $(3 + \varepsilon') \log n + \text{poly} \log n \leq (3 + \varepsilon) \log n$. The answer sizes are clearly poly $\log n$. Thus, SAT $\in \text{MIP}_{1,\frac{1}{2}+\mu}[(3 + \varepsilon) \log n, \text{poly} \log n]$.   □

4   Constant Query Inner Verifier for MIPs

In this section, we truncate the recursion by constructing a constant query "inner verifier" for a p-prover interactive proof system. An inner verifier is a subroutine designed to simplify the task of an MIP verifier. Say an MIP verifier V_{out}, on input x and random string R, generated queries $q_1, \ldots, q_p$ and a linear sized circuit C. In the standard protocol the verifier would send query q_i to prover Π_i and receive some answer a_i. The verifier accepts if $C(a_1, \ldots, a_p) = \text{true}$. An inner verifier reduces the answer size complexity of this protocol by accessing oracles $A_1, \ldots, A_p$, which are supposedly encodings of the responses $a_1, \ldots, a_p$, and an auxiliary oracle B, and probabilistically verifying that the A_i's really correspond to some commitment to strings $a_1, \ldots, a_p$ that satisfy the circuit C. The hope is to get the inner verifier to do all this with very few queries to the oracles $A_1, \ldots, A_p$ and B and we do so with one (bit) query each to the A_i's and seven queries to B. For encoding the responses $a_1, \ldots, a_p$, we use the *long code* of Bellare et al. [4]. We then adapt the techniques of Håstad [9,10] to develop and analyze a protocol for the inner verifier.

Let $\mathcal{A} = \{+1, -1\}^a$ and $\mathcal{B} = \{(a_1, \ldots, a_p)|C(a_1, \ldots, a_p) = -1\}$. Let π_i be the projection function $\pi_i : \mathcal{B} \to \mathcal{A}$ which maps $(a_1, \ldots, a_p)$ to a_i. By abuse of notation, for $\beta \subseteq \mathcal{B}$, let $\pi_i(\beta)$ denote $\{\pi_i(x)|x \in \beta\}$. Queries to the oracle A_i will be functions $f : \mathcal{A} \to \{+1, -1\}$. Queries to the oracle B will be functions $g : \mathcal{B} \to \{+1, -1\}$. The inner verifier expects the oracles to provide the long codes of the strings $a_1, \ldots, a_p$, i.e., $A_i(f) = f(a_i)$ and $B(g) = g(a_1, \ldots, a_p)$. Of course, we can not assume these properties; they need to be verified explicitly by the inner verifier. We will assume however that the tables are "folded", i.e., $A_i(f) = -A_i(-f)$ and $B(g) = -B(-g)$ for every i, f, g. (This is implemented by issuing only one of the queries f or $-f$ for every f and inferring the other value, if needed by complementing it.) We are now ready to specify the inner verifier.

$V_{\text{inner}}{}^{A_1,\ldots,A_p,B}(\mathcal{A}, \mathcal{B}, \pi_1, \ldots, \pi_p)$.
1. For each each $i \in \{1, \ldots, p\}$, choose $f_i : \mathcal{A} \to \{+1, -1\}$ at random.
2. Choose $f, g_1, g_2, h_1, h_2 : \mathcal{B} \to \{+1, -1\}$ at random and independently.
3. Let $g = f(g_1 \wedge g_2)(\Pi f_i \circ \pi_i)$ and $h = f(h_1 \wedge h_2)(\Pi f_i \circ \pi_i)$.
4. Read the following bits from the oracles $A_1, \ldots, A_p, B$
    $y_i = A_i(f_i)$, for each $i \in \{1, \ldots, p\}$.
    $w = B(f)$.
    $u_1 = B(g_1); u_2 = B(g_2); v_1 = B(h_1); v_2 = B(h_2)$
    $z_1 = B(g); z_2 = B(h)$
5. Accept iff $w \prod_{i=1}^{p} y_i = (u_1 \wedge u_2)z_1 = (v_1 \wedge v_2)z_2$

It is clear that if $a_1, \ldots, a_p$ are such that $C(a_1, \ldots, a_p) = -1$ and for every i and f, $A_i(f) = f(a_i)$ and for every g, $B(g) = g(a_1, \ldots, a_p)$, then the inner verifier accepts with probability one. The following lemma gives a soundness condition for the inner verifier, by showing that if the acceptance probability of the inner verifier is sufficiently high then the oracles $A_1, \ldots, A_p$ are non-trivially close to the encoding of strings $a_1, \ldots, a_p$ that satisfy $C(a_1, \ldots, a_p) = -1$. The proof uses, by now standard, Fourier analysis.

Note that the oracle A_i can be viewed as a function mapping the set of functions $\{\mathcal{A} \to \{+1, -1\}\}$ to the reals. Let the inner product of two oracles A and A' be defined as $\langle A, A' \rangle = 2^{-|\mathcal{A}|} \sum_f A(f) A'(f)$. For $\alpha \subseteq \mathcal{A}$, let $\chi_\alpha(f) = \prod_{a \in \alpha} f(a)$. Then the χ_α's give an orthonormal basis for the space of oracles A. This allows us to express $A(\cdot) = \sum_\alpha \hat{A}_\alpha \chi_\alpha(\cdot)$, where $\hat{A}_\alpha = \langle A, \chi_\alpha \rangle$ are the Fourier coefficients of A. In what follows, we let $\hat{A}_{i,\alpha}$ denote the αth Fourier coefficient of the table A_i. Similarly one can define a basis for the space of oracles B and the Fourier coefficients of any one oracle.

Our next claim lays out the precise soundness condition in terms of the Fourier coefficients of the oracles $A_1, \ldots, A_p$.

Claim 9. *For every $\epsilon > 0$, there exists a $\delta > 0$ such that if $V_{\mathrm{inner}}^{A_1, \ldots, A_p, B}(\mathcal{A}, \mathcal{B}, \pi_1, \ldots, \pi_p)$ accepts with probability at least $\frac{1}{2} + \epsilon$, then there exist $a_1, \ldots, a_p \in \mathcal{A}$ such that $C(a_1, \ldots, a_p) = -1$ and $|\hat{A}_{i,\{a_i\}}| \geq \delta$ for every $i \in \{1, \ldots, p\}$.*

There is a natural way to compose a p-prover MIP verifier V_{out} with an inner verifier such as V_{inner} above so as to preserve perfect completeness. The number of queries issued by the composed verifier is exactly that of the inner verifier. The randomness is the sum of the randomness. The analysis of the soundness of such a verifier is also standard and in particular shows that if the composed verifier accepts with probability $\frac{1}{2} + 2\epsilon$, then there exist provers $\Pi_1, \ldots, \Pi_p$ such that V_{out} accepts them with probability at least $\epsilon \cdot \delta^{2p}$, where δ is from Claim 9 above. Thus we get a proof of Lemma 4.

5 Scope for Further Improvements

The following are a few approaches which would further reduce the size-query complexity in the construction of PCPs described in this paper.

1. An improved low-error analysis of the low-degree test of Rubinfeld and Sudan [14] in the case when the field size is linear in the degree of the polynomial. (It is to be noted that the current best analysis [3] requires the field size to be at least a fourth power of the degree.) Such an analysis would reduce the proof blowup to nearly quadratic.
2. It is known that for every $\epsilon, \delta > 0$, $\mathrm{MIP}_{1,\epsilon}[1, 0, n] \subseteq \mathrm{PCP}_{1-\delta, \frac{1}{2}}[c \log n, 3]$ from the results of Håstad [10]. Traditionally, results of this nature have led to the construction of inner verifiers for p-prover MIPs and thus showing that for every $\delta > 0$ and p there exists $\epsilon > 0$ and c such that

$$\mathrm{MIP}_{1,\epsilon}[p, r, a] \subseteq \mathrm{PCP}_{1-\delta, \frac{1}{2}}[r + c \log a, p + 3] .$$

Proving a result of this nature would reduce the query complexity of the small PCPs constructed in this paper to 6.

338   Prahladh Harsha and Madhu Sudan

References

1. ARORA, S., LUND, C., MOTWANI, R., SUDAN, M., AND SZEGEDY, M. Proof verification and the hardness of approximation problems. *Journal of the ACM 45*, 3 (May 1998), 501–555.

2. ARORA, S., AND SAFRA, S. Probabilistic checking of proofs: A new characterization of NP. *Journal of the ACM 45*, 1 (Jan. 1998), 70–122.

3. ARORA, S., AND SUDAN, M. Improved low degree testing and its applications. In *Proc. 29th ACM Symp. on Theory of Computing* (El Paso, Texas, 4–6 May 1997), pp. 485–495.

4. BELLARE, M., GOLDREICH, O., AND SUDAN, M. Free bits, PCPs, and nonapproximability—towards tight results. *SIAM Journal of Computing 27*, 3 (June 1998), 804–915.

5. BELLARE, M., GOLDWASSER, S., LUND, C., AND RUSSELL, A. Efficient probabilistically checkable proofs and applications to approximation. In *Proc. 25th ACM Symp. on Theory of Computing* (San Diego, California, 16–18 May 1993), pp. 294–304.

6. COOK, S. A. Short propositional formulas represent nondeterministic computations. *Information Processing Letters 26*, 5 (11 Jan. 1988), 269–270.

7. DINUR, I., FISCHER, E., KINDLER, G., RAZ, R., AND SAFRA, S. PCP characterizations of NP: Towards a polynomially-small error-probability. In *Proc. 31th ACM Symp. on Theory of Computing* (Atlanta, Georgia, 1–4 May 1999), pp. 29–40.

8. FRIEDL, K., AND SUDAN, M. Some improvements to total degree tests. In *Proc. 3rd Israel Symposium on Theoretical and Computing Systems* (1995).

9. HÅSTAD, J. Clique is hard to approximate within $n^{1-\epsilon}$. In *Proc. 37nd IEEE Symp. on Foundations of Comp. Science* (Burlington, Vermont, 14–16 Oct. 1996), pp. 627–636.

10. HÅSTAD, J. Some optimal inapproximability results. In *Proc. 29th ACM Symp. on Theory of Computing* (El Paso, Texas, 4–6 May 1997), pp. 1–10.

11. LUND, C., FORTNOW, L., KARLOFF, H., AND NISAN, N. Algebraic methods for interactive proof systems. In *Proc. 31st IEEE Symp. on Foundations of Comp. Science* (St. Louis, Missouri, 22–24 Oct. 1990), pp. 2–10.

12. POLISHCHUK, A., AND SPIELMAN, D. A. Nearly-linear size holographic proofs. In *Proc. 26th ACM Symp. on Theory of Computing* (Montréal, Québec, Canada, 23–25 May 1994), pp. 194–203.

13. RAZ, R., AND SAFRA, S. A sub-constant error-probability low-degree test, and a sub-constant error-probability PCP characterization of NP. In *Proc. 29th ACM Symp. on Theory of Computing* (El Paso, Texas, 4–6 May 1997), pp. 475–484.

14. RUBINFELD, R., AND SUDAN, M. Robust characterizations of polynomials with applications to program testing. *SIAM Journal of Computing 25*, 2 (Apr. 1996), 252–271.

15. SUDAN, M. *Efficient Checking of Polynomials and Proofs and the Hardness of Approximation Problems.* PhD thesis, University of California, Berkeley, Oct. 1992.

16. SZEGEDY, M. Many-valued logics and holographic proofs. In *Automata, Languages and Programming, 26st International Colloquium* (Prague, Czech Republic, 11–15 July 1999), J. Wiedermann, P. van Emde Boas, and M. Nielsen, Eds., vol. 1644 of *Lecture Notes in Computer Science*, Springer-Verlag, pp. 676–686.

Space Efficient Algorithms for Series-Parallel Graphs

Andreas Jakoby*, Maciej Liśkiewicz**, and Rüdiger Reischuk

Universität zu Lübeck
Inst. für Theoretische Informatik
Wallstr. 40, D-23560 Lübeck, Germany
{jakoby,liskiewi,reischuk}@tcs.mu-luebeck.de

Abstract. The subclass of directed *series-parallel graphs* plays an important role in computer science. To determine whether a graph is series-parallel is a well studied problem in algorithmic graph theory. Fast sequential and parallel algorithms for this problem have been developed in a sequence of papers. For series-parallel graphs methods are also known to solve the reachability and the decomposition problem time efficiently. However, no dedicated results have been obtained for the space complexity of these problems – the topic of this paper.

For this special class of graphs, we develop deterministic algorithms for the *recognition, reachability, decomposition* and the *path counting problem* that use only logarithmic space. Since for arbitrary directed graphs reachability and path counting are believed not to be solvable in log-space the main contribution of this work are novel deterministic path finding routines that work correctly in series-parallel graphs, and a characterisation of series-parallel graphs by forbidden subgraphs that can be tested space-efficiently. The space bounds are best possible, i.e. the decision problems is shown to be $\mathcal{L}$-complete with respect to $\mathcal{AC}^0$-reductions, and they have also implications for the parallel time complexity of series-parallel graphs. Finally, we sketch how these results can be generalised to extension of the series-parallel graph family: to graphs with multiple sources or multiple sinks and to the class of *minimal vertex series-parallel graphs*.

1 Introduction

All graphs $G = (V, E)$ that will be considered in this paper are directed. n denotes the number of vertices V of G and m the number of edges E. A well studied subclass of graphs are the *series-parallel graphs,* for which different definitions and characterisations have been given [6]. We will consider the basic class, sometimes also called *two terminal series-parallel graphs,* that are most important for applications in program analysis.

Definition 1. $G = (V, E)$ *is a series-parallel graph, SP-graph for short, if either G is a line graph of length 1, that is a pair of nodes connected by a single edge, or there exist two disjoint series-parallel graphs $G_i = (V_i, E_i)$, $i = 1, 2$, with sources $v_{in,i}$, and*

---

* Part of this research was done while visiting the Depart. of Computer Science, Univ. of Toronto, Canada.
** On leave of Instytut Informatyki, Uniwersytet Wrocławski, Poland.

A. Ferreira and H. Reichel (Eds.): STACS 2001, LNCS 2010, pp. 339–352, 2001.
© Springer-Verlag Berlin Heidelberg 2001

sinks $v_{out,i}$ such that $V = V_1 \cup V_2$, $E = E_1 \cup E_2$, and either

(A) **parallel composition**: $v_{in} = v_{in,1} = v_{in,2}$ and $v_{out} = v_{out,1} = v_{out,2}$, or

(B) **series composition**: $v_{in} = v_{in,1}$ and $v_{out} = v_{out,2}$ and $v_{out,1} = v_{in,2}$.

Since the sources and sinks of the G_i are merged every series-parallel graph G has a unique source and a unique sink. G it is specified by a list of edges, but we put no restrictions on the ordering of the edges. In particular, it is not required that this ordering reflects the structure of the series-parallel composition operations. Otherwise, recognising and handling series-parallel graphs becomes quite easy. The correctness and efficiency of the algorithms presented below will not depend on the representation of the input graphs. For example, one could use adjacency-matrices as well.

Series-parallel graphs are suitable to describe the information flow within a program that is based on sequential and parallel composition. The graphical description of a program helps to decide whether it can be parallelised and to generate schedules for a parallel execution.

To determine if a given graph G belongs to the class of series-parallel graphs is a basic problem in algorithmic graph theory. An optimal linear time sequential algorithm for this problem has been developed by Valdes, Tarjan, and Lawler in [15] long time ago. Also, fast parallel algorithms have been published. He and Yesha have presented an EREW PRAM algorithm working in time $O(\log^2 n)$ while using $n+m$ processors [12]. Eppstein has reduced the time bound constructing an algorithm that takes only $O(\log n)$ steps on the stronger PRAM model with concurrent instead of exclusive read and write, that requires $C(m, n)$ processors [11]. Here $C(m, n)$ denotes the number of processors necessary to compute the connected components of a graph in logarithmic time. Finally, Bodlaender and de Fluiter have presented an EREW PRAM algorithm using $O(\log n \cdot \log^* n)$ time and $O(n + m)$ operations [5].

The space complexity of this problem, however, has not been investigated successfully so far. In this paper we give an answer to this question.

The *decompositon of a series-parallel graph* is quite useful to decide other graph properties. Hence, another important task is to compute such a decomposition efficiently. In [15] a linear-time sequential algorithm for decomposing series-parallel graphs has been given. We will show that this task can be done in small space as well.

For general graphs, the *reachability problem,* that is the question whether there exists a path between a given pair of nodes, is the classical $\mathcal{NL}$-complete problem. By well known simulations, for the parallel time complexity one can infer a logarithmic upper bound on CRCW PRAMs. The reachability problem restricted to series-parallel graphs, however, can be solved in logarithmic time already by an EREW PRAM using the minimal number $(n + m)/\log n$ of processors [15]. Certain graph properties like acyclicity are also complete for $\mathcal{NL}$, while for other problems their computational complexity is still unsolved. Recently, Allender and Mahajan have made a major step in classifying the computational complexity of planarity testing showing that this problem is hard for $\mathcal{L}$ and belongs to $\mathcal{SL}$ (symmetric Logspace) [3]. They leave as an open problem to close the gap between the lower bound and the upper bound. In this paper we determine the computational complexity of a nontrivial subproblem of planarity testing precisely. For series-parallel graphs this question is $\mathcal{L}$-complete.

For $\mathcal{L}$ several simple graph problems are known to be complete with respect to $\mathcal{AC}^0$-reductions: for example, whether a graph is a forest or even a tree, or whether in a given forest G two nodes belong to the same tree (for a complete list see [9,13]). In this paper we will prove three problems for series-parallel graphs to be $\mathcal{L}$-complete as well: the *recognition problem,* the *reachability problem,* and *counting the number of paths mod 2.* While the hardness of these problems can be obtained in a straightforward way, it requires a lot of algorithmic effort to prove that the lower bound can actually be achieved. Thus, the main technical contribution of this paper are new graph-theoretical notions and algorithmic methods that allow us to solve these problems using only logarithmic space.

Furthermore, not only decision problems for series-parallel graphs turn out to be tractable. A decomposition of such graphs can be computed within the same space bound as well. For general graphs counting the number of paths is one of the generic complete problems for the class $\#\mathcal{L}$ [2]. Thus, this problem is not computable in $\mathcal{FL}$, the functional deterministic log-space complexity class, unless certain hierarchies collapse. We will prove that restricting to series-parallel graphs the counting problem can be solved in $\mathcal{FL}$. This will be achieved by combining our space efficient reachability decision procedure with a modular representation of numbers requiring only little space, and the recent result that a Chinese Remainder Representation can be converted to the standard binary one in logarithmic space [8].

Because of the relation between $\mathcal{L}$ and parallel time complexity classes defined by the EREW PRAM model (see [14]) these new algorithms can be modified to solve these problems in logarithmic time on EREW PRAMs as well. Finally, these results can also be extended to generalizations of series-parallel graphs: multiple source or multple sink, and minimal vertex-series-parallel graphs.

This paper is organized as follows. In Section 2 we will prove the $\mathcal{L}$-hardness of the reachability and the recognition problem. Procedures solving these problems within logarithmic space will be described in detail in Section 3. Section 4 outlines an algorithm that generates an edges-ordering that reflects the structure of a given series-parallel graph. Based on this ordering we sketch a decomposition algorithm in Section 5. In Section 6, we combine the methods presented so far to solve the path counting problem. Finally, in Section 7 it will be indicated how this results can be extended to generalizations of series-parallel graphs. The paper ends with some conclusions and open problems.

2   Hardness Results

To establish meaningful lower bounds for the deterministic space complexity class $\mathcal{L}$ one has to restrict the concept of polynomial time many-one reductions to simpler functions. We consider the usual requirement that the reducing function f can be computed in $\mathcal{AC}^0$. The $\mathcal{L}$-hardness for series-parallel graphs can be shown in a direct way.

Theorem 1. *The following problems are hard for $\mathcal{L}$ under $\mathcal{AC}^0$ reducibility: (i) recognition of series-parallel graphs, (ii) reachability in series-parallel graphs, and (iii) counting the number of paths mod 2.*

Proof: Let L be a language in $\mathcal{L}$ and M a logarithmic space-bounded deterministic Turing machine accepting L by taking at most n^k steps on inputs X of length n, where k is a fixed exponent. We may assume that M has unique final configurations C_{acc}, the accepting one, and C_{rej}, the rejecting one. In addition, all configurations C of M on X are time-stamped, that means are actually tuples (C, t) with $0 \leq t \leq n^k$. Then the successor configuration of (C, t) is $(C', t + 1)$ if $t < n^k$ and M can move in one step from C to C'. If C is a final configuration and $t < n^k$ then $(C, t + 1)$ is the successor of (C, t). For input X we construct a directed graph G_X, where the time-stamped configurations (C, t) are the vertices of G_X and edges represent (the inverse of) the successor relation: G_X contains the edge $((C', t+1), (C, t))$ iff $(C', t + 1)$ is a successor of (C, t). Obviously, G_X is a forest consisting of trees with roots of the form (C, n^k). To prove the hardness of the rechability problem we augment G_X by two new nodes u and v. For every configuration (C, n^k) the edge $(u, (C, n^k))$ is added, and for every leaf (C, t) the edge $((C, t), v)$. It is easy to see that the resulting graph is series-parallel with source u and sink v. Furthermore, it contains a path from (C_{acc}, n^k) to $(C_{init}, 0)$, where $(C_{init}, 0)$ represents the starting configuration of M iff M accepts X. The reduction itself can be computed in $\mathcal{AC}^0$.

The hardness of the recognition problem and the counting problem can be shown in a similar way. ∎

3   Recognition and Reachability in Logspace

Establishing corresponding upper bounds is not obvious at all. We will give a space efficient characterization of series-parallel graphs by forbidden subgraphs and exploit the structure of internal paths very thoroughly. Assume that the nodes of the input graph G are represented by the set of numbers $\{1, 2, \ldots, n\}$. G is given by a list of edges $(i_1, j_1), (i_2, j_2), \ldots (i_m, j_m)$, where i_k, j_k are binary representations of the names of the nodes. Let **pred**(v) denote the set of direct predecessors of v, and **pred**(v, i) the i-th direct predecessor of v according to the ordering implicitly given by the specification of G. Similarly, let **succ**(v) and **succ**(v, i) be the set of direct successors of v, resp. its i-th direct successor. $\text{succ}(v, i)$ and $\text{pred}(v, i)$ can be computed in deterministic logarithmic space: the Turing machine searches through the list of edges looking for the i-th entry that starts (resp. ends) with v.

Define **pred**$^+(v)$, resp. **succ**$^+(v)$, as the transitive closure of $\text{pred}(v)$, resp. $\text{succ}(v)$, not containing v, and **pred**$^*(v) := \text{pred}^+(v) \cup \{v\}$ and **succ**$^*(v) := \text{succ}^+(v) \cup \{v\}$. To shorten the notation, let us introduce the predicate **PATH**(u, v) being true iff the given graph G possesses a path from node u to v. Thus,

$$\text{PATH}(u, v) \quad \Longleftrightarrow \quad u \in \text{pred}^*(v) \quad \Longleftrightarrow \quad v \in \text{succ}^*(u).$$

Remember that deciding PATH for arbitrary graphs is $\mathcal{NL}$-complete. To construct a deterministic space efficient algorithm solving this problem for series-parallel graphs we introduce the following concepts:

lm-down(v) := the max. acyclic path $v = u_1, u_2, \ldots, u_l$ with $u_{i+1} = \text{succ}(u_i, 1)$,
lm-up(v) := the max. acyclic path $v = u_1, u_2, \ldots, u_l$ with $u_{i+1} = \text{pred}(u_i, 1)$,
lm-pred$^*(v)$:= $\{u \mid \text{lm-down}(u) \cap \text{lm-up}(v) \neq \emptyset\}$, and
lm-succ$^*(v)$:= $\{u \mid \text{lm-down}(v) \cap \text{lm-up}(u) \neq \emptyset\}$.

Here, "lm" stands for *left-most*, that means in each node u_i the path follows the first edge as specified by the representation of G. A path being acyclic requires that all its nodes u_i are different. Thus, a maximal acyclic down-path either ends in a sink or stops immediately before hitting a node as the left-most successor a second time. These sets can be decided by the procedure membership-test1. The algorithm for testing whether $u \in$ lm-up(v) is just the symmetric dual.

```
procedure  membership-test1[u ∈ lm-down(v)]
1 let n be the number of nodes in G; x := v; i := 1;
2 while x ≠ u and |succ(x)| > 0 and i ≤ n do
3    let x := succ(x, 1); i := i + 1 od
4 if x = u then return TRUE else return FALSE
```

Fig. 1. $u \notin$ lm-down(v) and $v \notin$ lm-up(u).

Fig. 2. $v \in$ lm-pred$^*(u)$ and $u \in$ lm-succ$^*(v)$.

To check if $u \in$ lm-pred$^*(v)$ one can use the procedure membership-test2, which uses membership-test1 to decide lm-up and lm-down.

```
procedure  membership-test2[u ∈ lm-pred*(v)]
1 result := FALSE
2 forall nodes x in G do
3    if x ∈ lm-down(u) and x ∈ lm-up(v) then let  result := TRUE od
4 return result
```

In the dual way we can test whether $u \in$ lm-succ$^*(v)$. Hence it follows:

Lemma 1. *For an arbitrary graph G and node v the membership problem for the sets* lm-down(v), lm-up(v), lm-pred$^*(v)$, *and* lm-succ$^*(v)$ *can be solved deterministically in logarithmic space.*

A graph G is called ***st*-connected** if G is has a unique source named s and a unique sink named t, and for every node v it holds: PATH(s, v) and PATH(v, t).

We start with the procedure preliminary-test. For an acyclic graph G it returns TRUE iff G is st-connected. If G contains a cycle $C = (v_1, v_2, \ldots, v_l)$ the procedure will detect the cycle if it is on a left-most path. In such a case the procedure outputs FALSE. Cycles that are not of this form will not be detected, and the procedure erroneously may output TRUE.

<u>procedure</u>   preliminary-test(G)
1 <code>if not</code> [G has a unique source s and a unique sink t]
2   <code>then return</code> FALSE <code>and exit</code>
3 <code>forall</code> nodes v in G <code>do</code>
4   <code>if</code> $t \notin$ lm-down(v) <code>or</code> $s \notin$ lm-up(v) <code>then return</code> FALSE <code>and exit</code>
5 <code>return</code> TRUE

Lemma 2. *The procedure* preliminary-test *can be implemented deterministically in log-space. Moreover, if* preliminary-test(G) *outputs* TRUE *then G is st-connected. If it outputs* FALSE *then at least one of the following conditions holds: (i) G has more then one source or more than one sink, or (ii) G is st-connected, but it has a cycle.*

The proof of this lemma is straightforward and we omit it. Note that a graph G with output TRUE can still have a cycle. To detect this property is difficult for deterministic machines since this question can easily be shown to be $\mathcal{NL}$-complete. Therefore, we look for a simpler task.

Let W denote the graph shown in Fig. 3. A graph W' is *homeomorphic* to W if it contains four distinct vertices a, b, c, d and pairwise internally vertex disjoint paths $P_{ab}, P_{ac}, P_{bd}, P_{cd}$ and P_{bc}. If G contains a homeomorphic image of W as a subgraph then W is called a *minor* of G. The following characterization of series-parallel graphs by forbidden minors has been known for long [10,15]. Let G be an st-connected acyclic graph. Then G is series-parallel iff W is not a minor of G.

Fig. 3. The forbidden Minor W.   **Fig. 4.** The Forbidden Induced Subgraph H.

To make series-parallel graph recogniton space efficient, instead of searching for the forbidden minor W we will use the following characterization. Let H be a graph with four distinct nodes z_1, z_2, z_3, z_4 such that

1. $(z_1, z_2), (z_3, z_4)$ are edges of H and PATH(z_1, z_4),
2. $\neg$ PATH(z_1, z_3) and $\neg$ PATH(z_2, z_4).

These conditions are illustrated in Fig. 4. In the following we will show how H can be used to determine whether a graph is series-parallel. We say that H is an *induced subgraph* of G if G contains four nodes z_1, z_2, z_3, z_4 which fulfil these connectivity conditions.

Theorem 2. *Let G be an st-connected acyclic graph. Then G is series-parallel iff it does not contain H as an induced subgraph.*

This follows by showing that a st-connected acyclic graph G contains H as an induced subgraph iff W is a minor of G.

Now, we will deduce the key property that makes reachability in series-parallel graphs easier compared to arbitrary graphs. Although the parallel composition operator introduces a lot of nondeterminism into the structure of these graphs when trying to find a path from a node u to a node v this question can be solved by considering the unique lm-down-path starting at u and the unique lm-up-path starting in v and deciding whether these two intersect. In other words, it holds:

Theorem 3. *If G is series-parallel then* $\text{pred}^*(v) = \text{lm-pred}^*(v)$ *for every node v.*

Proof: Assume that $\text{pred}^*(v) \neq \text{lm-pred}^*(v)$ for some node v of G. We will show that then H has to be an induced subgraph of G – a contradiction to Theorem 2.

Obviously, v cannot be the source s of G. Since G is st-connected and acyclic every lm-down-path from an arbitrary node u has to terminate in the sink t. Thus, for t holds $\text{pred}^*(t) = \text{lm-pred}^*(t) = V$, and hence $v \neq t$.

Let $u_1 \in \text{pred}^*(v) \setminus \text{lm-pred}^*(v)$. Since every lm-up-path terminates in the source s we can conclude $u_1 \neq s$. Let $u_1, \ldots, u_k = t$ be the leftmost down-path lm-down(u_1). $u_1 \notin \text{lm-pred}^*(v)$ implies that $u_i \neq v$ for all $i \in [1..k]$. Furthermore, let $v_1 = v, v_2, \ldots, v_\ell = s$ be the leftmost up-path lm-up(v) from v.

Since $u_1 \in \text{pred}^*(v_1)$ there exists a non-trivial path from u_1 to v_1. On the other hand, because of $v_\ell = s$ and $v_1 = v \neq s$ it holds $\neg \text{PATH}(v_1, v_\ell)$, and similarly because of $u_k = t$ and $v_1 \neq t$, $\neg \text{PATH}(u_k, v_1)$. Hence, there exist $i \in [1..k - 1]$ and $j \in [1..\ell - 1]$ such that $\text{PATH}(u_i, v_j)$, $\neg \text{PATH}(u_i, v_{j+1})$, and $\neg \text{PATH}(u_{i+1}, v_j)$.

The nodes $z_1 := u_i$, $z_2 := u_{i+1}$, $z_3 := v_{j+1}$, and $z_4 := v_j$ prove that H is an induced subgraph of G. ∎

Thus, if for some node v of G the relation $\text{pred}^*(v) = \text{lm-pred}^*(v)$ is violated one can conclude that G is not series-parallel. This equality, however, can be tested space efficiently.

Lemma 3. *There exists a deterministic logarithmic space-bounded Turing machine that for arbitrary $v \in G$ decides whether* $\text{pred}^*(v) = \text{lm-pred}^*(v)$.

Proof: Assume that $\text{pred}^*(v) \neq \text{lm-pred}^*(v)$. First, we claim that there has to be an edge $(u, w) \in E$ such that $u \in \text{pred}^*(v) \setminus \text{lm-pred}^*(v)$ and $w \in \text{lm-pred}^*(v)$. To see this, let $x \in \text{pred}^*(v) \setminus \text{lm-pred}^*(v)$ and $u_1 = x, u_2, u_3, \ldots, u_k = v$ be a down-path from x to v. Obviously, $u_1, u_2, \ldots, u_k \in \text{pred}^*(v)$, $u_1 \notin \text{lm-pred}^*(v)$, and $u_k \in \text{lm-pred}^*(v)$. Therefore, there exists an index $i \in [1..k - 1]$ such that $u_i \in \text{pred}^*(v) \setminus \text{lm-pred}^*(v)$ and $u_{i+1} \in \text{lm-pred}^*(v)$. This proves our claim. Now it is easy to see that the following algorithm answers the question whether $\text{pred}^*(v) = \text{lm-pred}^*(v)$:

```
procedure  equality-test[pred*(v) = lm-pred*(v)]
1 result := TRUE
2 forall edges (u, w) in G do
3     if (u ∉ lm-pred*(v)) ∧ (w ∈ lm-pred*(v)) then result := FALSE od
4 return result
```

∎

Corollary 1. *Let G be an st-connected graph with* $\text{pred}^*(v) = \text{lm-pred}^*(v)$ *for every node v. Then reachability within G can be decided in $\mathcal{L}$.*

<u>procedure</u> SER-PAR(G)

```
 1 if preliminary-test(G) returns FALSE then return FALSE and exit
 2 forall nodes v in G do
 3    if pred*(v) ≠ lm-pred*(v) then return FALSE and exit od
 4 forall pairs of nodes x, y in G do
 5    if x ∈ lm-pred*(y) ∧ y ∈ lm-pred*(x)
 6       then return FALSE and exit od
 7 forall pairs of edges (z₁, z₂), (z₃, z₄), with z₁ ≠ z₄ do
 8    if z₁ ∈ lm-pred*(z₄) ∧ z₁ ∉ lm-pred*(z₃) ∧ z₂ ∉ lm-pred*(z₄)
 9       then return FALSE and exit od
10 return TRUE
```

The procedure SER-PAR specified above decides for an arbitrary graph G whether it is series-parallel. To prove its correctness we argue as follows. From Lemma 2 one can conclude that the algorithm stops at line 1 and outputs FALSE if G has more then one source or more than one sink, or G is st-connected, but it has a cycle. Hence, G is not series-parallel and the answer FALSE is correct. On the other hand, if the procedure does not stop at line 1 then G is st-connected.

Furthermore, if SER-PAR(G) outputs FALSE in line 3 then $\text{pred}^*(v) \neq \text{lm-pred}^*(v)$ for some node v. By Theorem 3 it follows that this answer is correct, too. If the algorithm continues, we can presuppose at the beginning of line 4 that G is st-connected and for any v it holds $\text{pred}^*(v) = \text{lm-pred}^*(v)$. In lines 4-6 we check whether G is acyclic, and stop if not. The answer will be correct since $\text{lm-pred}^*(y)$ contains all predecessors of a node y.

Let us recapitulate the conditions a graph G has to fulfil such that SER-PAR(G) does not stop before line 7: G has to be st-connected, acyclic and for every pair of nodes x, y in G it holds: $\text{PATH}(y, x) \iff y \in \text{lm-pred}^*(x)$. This guarantees that in lines 7-9 the existence of H as an induced subgraph is tested correctly. Finally, since all tests applied can be performed in deterministic logarithmic space we can conclude:

Theorem 4. *The question whether a graph is series-parallel can be decided in* $\mathcal{L}$.

4   An Edge Ordering Algorithm

For a graph specified by a list of edges we have made no assumptions about their ordering. In particular, this ordering is not required to reflect the construction process of the series-parallel graph in any way. In this section we present a log-space algorithm that given a series-parallel graph G outputs a special ordering called **SP-ordering**. The crucial property of this ordering is that for any series-parallel component C with source v all direct successors of v in C are enumerated with consecutive integers. Speaking formally, for a node v the sequence **SP-succ(v)** is a permutation of $\text{succ}(v)$ such that for all $u \in \text{succ}^+(v)$ the set $\{\, i \mid \texttt{SP-succ}(v, i) \in \text{pred}^*(u)\,\}$ consists of consecutive integers. Here, for $1 \le i \le |\text{succ}(v)|$ the value $\texttt{SP-succ}(v, i)$ denotes the i'th vertex in the SP-ordering of $\text{succ}(v)$. Recall that $\text{succ}(v, i)$ is the i'th direct successor of v according to the ordering implicitly given by the input specification. Hence, in general $\texttt{SP-succ}(v, i)$ will be different from $\text{succ}(v, i)$. To compute the SP-ordering we will use $\text{succ}(v, i)$ and the following function for a node v different from s and t:

START(u) := nearest $v \in \text{pred}^+(u)$, such that any path from s to u contains v.

It it is easy to see that START(u) gives the source v of a smallest series-parallel component containing u and its direct predecessors in pred(u). If pred(u) contains only a single node v then START(u) = v. Otherwise, START(u) can be computed by finding the nearest common predecessor of the left-most up-paths from any direct predecessor of u to the source of G. Let START$^{-1}(v) := \{u \mid \text{START}(u) = v\}$. Both START($u$) and START$^{-1}(v)$ can be computed in logarithmic space.

Let us now introduce an important notion, which arises from our analysis of series-parallel graphs in the previous section. Let $v_1 \neq v_2$ be two arbitrary nodes of G. Then we define the set of **bridge nodes** between v_1 and v_2 as follows:

$$\text{BRIDGES}(v_1, v_2) := \{u \in \text{START}^{-1}(v_1) \cap \text{pred}^+(v_2) \mid$$
$$\forall w \in \text{succ}^+(u) \cap \text{pred}^+(v_2) : w \notin \text{START}^{-1}(v_1)\}.$$

Obviously, BRIDGES(v_1, v_2) $\neq \emptyset$ if START$^{-1}(v_1) \cap \text{pred}^+(v_2) \neq \emptyset$.

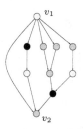

Fig. 5. Marked: The Nodes in START$^{-1}(v_1)$; Black: The subset BRIDGES(v_1, v_2).

Using the functions succ(v, i) and START(v), the set BRIDGES(v_1, v_2) can be computed deterministically in logarithmic space. Let BRIDGES(v_1, v_2, i) be the i-th element of such an enumeration of the nodes in this set. Furthermore, given a node $v_3 \in$ BRIDGES(v_1, v_2) and v_1, the lower endpoint v_2 can be determined in logarithmic space as well.

We now describe a recursive procedure SP-sequence(v, u) that outputs the sequence of direct successors of a node v in SP-ordering. This procedure will initially be called with the successor u of v in START$^{-1}(v)$ that is furthest from v. By definition of START(v) this successor is unique.

<u>procedure</u> SP-sequence(v, u)
1 if $u \in$ succ(v) then output u
2 for $i = 1$ to $|$BRIDGES(v, u)$|$ do SP-sequence(v, BRIDGES(v, u, i)) od

Using a log-space algorithm to compute BRIDGES(v, u, i) one can implement this procedure with logarithmic space as well. Furthermore, SP-succ(v, i) can be computed by counting the nodes in the output of the procedure SP-sequence(v, FINAL(v)).

5   The Decomposition in Log-Space

The *decomposition tree* of a series-parallel graph provides information how this graph has been built using the parallel and serial constructors.

Definition 2. *A binary tree $T = (V_T, E_T)$ with a labeling function $\sigma : V_T \to \{p, s\} \cup E$ is called a **decomposition tree** of an SP-graph $G = (V, E)$ iff leaves of T are labeled with elements of E, internal nodes with p or s and G can be generated recursively using T as follows: If T is a single node v then G consists of the single edge $\sigma(v)$. Otherwise, let T_1 (resp. T_2) be the right (resp. left) subtree of T and G_i be SP-graphs with decomposition tree T_i: if $\sigma(v) = p$ (resp. s) then G is the parallel (resp. serial) composition of G_1 and G_2.*

The algorithm to generate a decomposition tree is based on the functions START(v), and BRIDGES(u, v, i) described in the previous section. Given a series-parallel graph G with source s and sink t, the procedure SP-DECOMP(s, t) outputs the root DTR of a decomposition tree of G. As an example for such a decomposition see Fig. 6.

```
procedure SP-DECOMP(u, v)
 1 if |BRIDGES(u,v)| ≥ 1 then do
 2     getnode(r); σ(r) :=s;
 3     left(r) := SP-DECOMP(u, BRIDGES(u,v,1))
 4     right(r) := SP-DECOMP(BRIDGES(u,v,1), v)
 5     DTR:=r;
 6     for i:= 2 to |BRIDGES(u,v)| do
 7        getnode(c); σ(c) :=s;
 8        left(c) := SP-DECOMP(u, BRIDGES(u,v,i))
 9        right(c) := SP-DECOMP(BRIDGES(u,v,i), v)
10        getnode(b); σ(b) :=p; left(b) := DTR; right(b) := c;
11        DTR := b endfor
12 endif
13 if (u,v) ∈ E then do
14     getnode(c); σ(c) := (u,v);
15     if |BRIDGES(u,v)| > 0 then do
16             getnode(b); σ(b) :=p; left(b) := DTR; right(b) := c;
17             DTR := b od
18         else DTR := c endif
19 endif
20 return DTR.
```

To achieve space efficiency we do not want to store the values of the variables for each recursive activation of SP-DECOMP as it is done in standard implementation of recursion. In our special situation these values of the calling activation of SP-DECOMP can be recomputed when returning from a recursive call, thus we don't have to store them explicitly.

Theorem 5. *The decomposition tree of a SP-graph can be computed in $\mathcal{FL}$.*

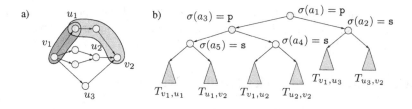

Fig. 6. An example of a ecomposition tree generated by SP-DECOMP(v_1, v_2): $T_{v,u}$ are subtrees generated by SP-DECOMP that consist of a single edge or a larger component.

6 Path Counting Problems

In this section we show that for series-parallel graphs the classical problem to count the number of paths can be solved in $\mathcal{FL}$. For general graphs counting the number of paths is not solvable in $\mathcal{FL}$ – unless certain hierarchies collapse – since this problem is one of the generic complete problems for the class $\#\mathcal{L}$ [2]. Speaking more precisely, let us define the functional problem #PATH as follows: given a graph G and nodes a, b estimate the number of different paths from a to b in G.

Theorem 6. *Restricted to series-parallel graphs* #PATH *can be computed in* $\mathcal{FL}$.

Proof: Consider the subgraph G_{ab} of G induced by $V' := \text{succ}^*(a) \cap \text{pred}^*(b)$. It is either empty (and then #PATH$(a, b) = 0$), or it is a series-parallel graph with source a and sink b. This follows from the fact that the predicate PATH restricted to nodes in V' is identical on G and G_{ab}. Furthermore, all paths from a to b in G occur in G_{ab} as well, thus the number of paths is identical. A simple induction shows that #PATH(a, b) can be bounded by 2^{n+m}. Using the reachability algorithm presented in Section 3 we can also decide in log-space whether an arbitrary edge of G belongs to G_{ab}.

Let $T = (V_T, E_T)$ be the decomposition tree of G_{ab} and $z \leq n + m$ be its size. We interpret the tree as an arithmetic expression as follows. Every leaf represents the integer 1. An internal node v of T labeled by $\sigma(v) = \mathsf{s}$ (resp. p) corresponds to a multiplication (resp. addition) of the expressions given by the sons of v. It is easy to see that the value ρ of the root of T equals #PATH(a, b).

Below we sketch how ρ can be computed in logarithmic space. Let $p_1 < p_2 < \ldots$ be the standard enumeration of primes. The prime number theorem implies

$$\prod_{p_i \leq n+m} p_i = e^{(n+m)(1+o(1))} > \#\text{PATH}(a, b) . \tag{1}$$

Using the log-space algorithm of [7] one can transform T into a binary tree T' of depth $O(\log z)$ representing an arithmetic expression with the same value ρ as T.

We evaluate $T' \bmod p_i$ using the algorithm in [4]. For $p_i \leq n + m$ this algorithm works in space $O(\log z + \log(n + m)) \leq O(\log n)$. By inequality (1), taking all $p_i \leq n + m$ the values $\rho \bmod p_i$ give a *Chinese Remainder Representation* of ρ. Using the recent result of Chiu, Davida, and Litow [8] that such a representation can be converted to the ordinary binary representation in log-space, finishes the proof. ∎

Using the hardness result shown in Section 2 it follows, that the problem to compute #PATH mod 2 is $\mathcal{L}$-complete. Using the techniques presented so far one can also solve some other counting problems, like determining the size of G_{ab} in $\mathcal{FL}$.

7   Generalisations of Series-Parallel Graphs

First we will consider graphs with several sources, but still with a unique sink.

Definition 3. *The family of multiple source series-parallel graphs, MSSP-graphs for short, are an extension of series-parallel graphs, adding the following constructor:*
   (C) **In-Tree composition***: a graph $G = (V, E)$ is generated from MSSP-graphs $G_i = (V_i, E_i)$ for $i = 1, 2$ by selecting a node $\hat{v}$ in G_1, identifying it with the sink $v_{out,2}$ of G_2 and forming the union of both graphs: $V := V_1 \cup V_2$ and $E := E_1 \cup E_2$.*

An in-tree composition may be applied several times, but only at the end. As soon as a graph has several sources the series and the parallel constructor can no longer be used. *Multiple sink series-parallel graphs* with a unique source can be defined in the dual way. In the following, we will restrict ourselves to the first extension – the main results hold for both classes. Unlike ordinary series-parallel graphs, the reachability problem for MSSP-graphs cannot be solved by the following leftmost path. PATH(x, y) not longer implies lm-down$(x) \cap$ lm-up$(y) \neq \emptyset$. To solve this problem we have to use a more sophisticated strategy. Define
   Elude$(v) := \{ u \mid \exists\, w_1, w_2 \in$ succ$(u)\ :\ v \in$ lm-down$(w_1) \setminus$ lm-down$(w_2) \}$
and minElude(v) as the closest predecessor u of v contained in Elude(v). It can be shown that such a unique node always exists and that it lies on lm-up(v). If Elude$(v) = \emptyset$ we set minElude$(v) := v$.

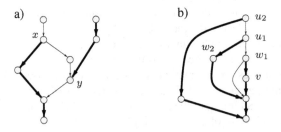

Fig. 7. a) Left-Most Paths that Do not Meet, b) Computing Elude(v)

Let minElude$(v, 0) := v$ and minElude$(v, i) :=$ minElude$($minElude$(v, i-1))$ for $i > 0$. Finally, define

$$\text{minElude}^*(v) := \bigcup_{i \in \mathbb{N}} \text{minElude}(v, i)$$
$$\text{Elude--pred}^*(v) := \{\, u \mid \exists w \in \text{minElude}^*(v)\ :\ w \in \text{lm-down}(u) \,\}\,.$$

Lemma 4. *If* $\text{Elude}(v) \neq \emptyset$ *then there exists a unique* $u \in \text{Elude}(v)$ *fulfilling the conditions in the definition of* minElude. *It can be computed in deterministic log-space. Furthermore, for every MSSP-graph G holds:* $\text{pred}^*(v) = \text{Elude} - \text{pred}^*(v)$ *for all* v. *For an arbitrary graph and node* v, *the equality* $\text{pred}^*(v) = \text{Elude} - \text{pred}^*(v)$ *can be checked in* $\mathcal{L}$.

To verify whether a given graph is a MSSP we can make use of the forbidden subgraph H again. However, a set of nodes fulfilling its PATH conditions can occur in MSSP-graph G, but only if z_3 belongs to another component G_2 of G than z_1 and z_2, which is connected to the rest graph of G via an in-tree composition step at z_4. Since this can also be verified in log-space we obtain:

Theorem 7. *Recognition and reachability for MSSP-graphs is in* $\mathcal{L}$.

The notion of decomposition trees can be extended to this graph family: first generate nodes that represent the in-tree composition steps, then the subtrees that describes the decomposition of the basic series-parallel graphs.

Theorem 8. *A decomposition tree of a MSSP-graph can be computed in* $\mathcal{FL}$.

The counting algorithm for SP-graphs can be extended to this class as well.

Theorem 9. *For MSSP-graphs the function* #PATH *can be computed in* $\mathcal{FL}$. *The same holds for the size of subgraphs of the form* G_{ab}.

Finally, let us remark another generalisation of series-parallel graphs: the minimal vertex-series-parallel graphs, MVSP for short. It has been shown that the *line graphs* of MVSP-graphs are closely related to series-parallel graphs (see [15] for the definition of MVSPs, line graphs and further details). Using a slight modification of our algorithms and the line graph of a MVSP graph we can extend all results shown in this paper to MVSP-graph

8   Conclusions and Open Problems

A deterministic Turing machine working in space $S \geq \log$ can be simulated by an EREW PRAM in time $O(S)$ (see e.g. [14]). The machine may use an exponential number of processors with respect to S. Therefore, we get immediately that the graph problems investigated in this paper can be solved in logarithmic parallel time. The simulation of space-bounded Turing machines by PRAMs can even be performed by the EROW model (exclusive-read owner-write). Hence we can deduce

Corollary 2. *For series-parallel graphs and their extensions considered above, recognition, reachability, decomposition, and path counting can be done in logarithmic time on EROW PRAMs with a polynomial number of processors.*

The exact number of processors depends on the time complexity of the Turing machine. Since our basic log-space algorithms require time $O(n^c)$ for some constant c significantly larger than 1, we probably will not achieve a linear number of processors this way. For the reachability problem in series-parallel graphs it is known that an EREW PRAM can solve it in logarithmic time using $n/\log n$ processors [15]. But it is still open whether also recognition and decomposition can be done in logarithmic time using at most a linear number of processors.

If we switch to *undirected* graphs the problems considered here seem to be inherently more difficult. In the undirected case series-parallel graphs can be characterized as the set of graphs containing no clique of size 4 as a minor [10].

In contrast to the series-parallel graph family, the reachability problem for arbitrary graphs seems to be easier in the undirected case than in the directed case. From [1] we know that the undirected version can be solved by a randomized log-space bounded machine, whereas no randomized algorithm is known for the directed case. Are there other distinctions of this kind?

Acknowledgement

Thanks are due to Eric Allender and Markus Bläser for helpful comments and pointers to the literature. Furthermore we would thank Faith Fich, Pierre McKenzie, and Christian Schindelhauer for fruitful discussions.

References

1. R. Aleliunas, R. Karp, R. Lipton, L. Lovasz, C. Rackoff, *Random Walks, Universal Sequences and the Complexity of Maze Problems,* Proc. 20. FOCS, 1979, 218-223.
2. C. Álvarez, B. Jenner, *A Very Hard Log-space Counting Classes,* TCS 107, 1993, 3-30.
3. E. Allender, M. Mahajan, *The Complexity of Planarity Testing,* Proc. 17. STACS, LNCS 1770, 2000, 87-98.
4. M. Ben-Or, R. Cleve *Computing Algebraic Formulas Using a Constant Number of Registers,* SIAM J. Comput. 21, 1992, 54-58.
5. H. Bodlaender, B. de Fluiter, *Parallel Algorithms for Series Parallel Graphs,* Proc. 4. ESA, LNCS 1136, 1996, 277-289.
6. A. Brandstädt, V. Bang Le, J. Spinrad, *Graph Classes: A Survey,* SIAM 1999.
7. S. Buss, S. Cook, A. Gupta, V. Ramachandran, *An Optimal Parallel Algorithm for Formula Evaluation,* SIAM J. Comput. 21, 1992, 755-780.
8. A. Chiu, G. Davida, B. Litow, NC^1 *Division,* unpublished manuscript, November 1999.
9. S. Cook, P. McKenzie, *Problems Complete for Deterministic Logarithmic Space,* J. Algo. 8, 1987, 385-394.
10. R. Duffin, *Topology of Series-Parallel Networks,* J. Math. Analysis Appl. 10, 1965, 303-318.
11. D. Eppstein, *Parallel Recognition of Series-Parallel Graphs,* Inf. & Comp. 98, 1992, 41-55.
12. X. He, Y. Yesha, *Parallel Recognition and Decomposition of Two Terminal Series Parallel Graphs,* Inf. & Comp. 75, 1987, 15-38.
13. B. Jenner, K.-J. Lange, P. McKenzie, *Tree Isomorphism and Some Other Complete Problems for Deterministic Logspace,* publication #1059, DIRO, Université de Montréal, 1997.
14. R. Karp, V. Ramachandran, *Parallel Algorithms for Shared-Memory Machines,* in: J. van Leeuwen (Ed.): Handbook of Theoretical Computer Science, Volume A, 1990, 869-941.
15. J. Valdes, R. Tarjan, E. Lawlers *The Recognition of Series Parallel Digraphs,* SIAM J. Comput. 11, 1982, 298-313.

A Toolkit for First Order Extensions of Monadic Games

David Janin[1] and Jerzy Marcinkowski[2*]

[1] Laboratoire Bordelais de Recherche en Informatique, Université de Bordeaux I
351, cours de la Libération, 33 405 Talence cedex
janin@labri.u-bordeaux.fr
[2] Institute of Computer Science, University of Wrocław
Przesmyckiego 20, 51151 Wrocław, Poland
jma@tcs.uni.wroc.pl

Abstract. In 1974 R. Fagin proved that properties of structures which are in NP are exactly the same as those expressible by existential second order sentences, that is sentences of the form: *there exist $\vec{P}$ such that φ*, where $\vec{P}$ is a tuple of relation symbols and φ is a first order formula. Fagin was also the first to study monadic NP: the class of properties expressible by existential second order sentences where all the quantified relations are unary.

In [AFS00] Ajtai, Fagin and Stockmeyer introduce closed monadic NP: the class of properties which can be expressed by a kind of monadic second order existential formula, where the second order quantifiers can interleave with first order quantifiers. In order to prove that such alternation of quantifiers gives substantial additional expressive power they construct graph properties $\mathcal{P}_1$ and $\mathcal{P}_2$: $\mathcal{P}_1$ is expressible by a sentence with the quantifier prefix in the class $(\exists\forall)^* \mathbf{\exists}^* (\exists\forall)^*$ [1] but not by a boolean combination of sentences from monadic NP (i.e with the prefix of the form $\mathbf{\exists}^* (\exists\forall)^*$) and $\mathcal{P}_2$ is expressible by a sentence $\mathbf{\exists}^* (\exists\forall)^* \mathbf{\exists}^* (\exists\forall)^*$ but not by a Boolean combination of sentences of the form $(\exists\forall)^* \mathbf{\exists}^* (\exists\forall)^*$. A natural question arises here whether the hierarchy inside closed monadic NP, defined by the number of blocks of second order existential quantifiers, is strict.

In this paper we present a technology for proving some non expressibility results for monadic second order logic. As a corollary we get a new, easy, proof of the two results from [AFS00] mentioned above. With our technology we can also make a first small step towards an answer to the hierarchy question by showing that the hierarchy inside closed monadic NP does not collapse on a first order level. The monadic complexity of properties definable in Kozen's mu-calculus is also considered as our technology also applies to the mu-calculus itself.

1 Introduction

1.1 Previous Works

In 1974 R. Fagin proved that the properties of structures which are in $\mathcal{NP}$ are exactly the same as those expressible by existential second order sentences, known also as Σ_1^1

---

* This paper has been written while the author was visiting Laboratoire Bordelais de Recherche en Informatique, in Bordeaux, France. I was also supported by Polish KBN grant 2 PO3A 018 18.

[1] In this paper we use the symbols $\exists$, $\forall$ for the first order quantifiers and $\mathbf{\exists}$, $\mathbf{\forall}$ for the monadic second order quantifiers

---

A. Ferreira and H. Reichel (Eds.): STACS 2001, LNCS 2010, pp. 353–364, 2001.
© Springer-Verlag Berlin Heidelberg 2001

sentences, i.e. sentences of the form: *there exist relations $\vec{P}$ such that φ*, where $\vec{P}$ is a tuple of relation symbols (possibly of high arity) and φ is a first order formula.

Fagin was also the first to study *monadic* NP: the class of properties expressible by existential second order sentences where all quantified relations are unary. The reason for studying this class was the belief that it could serve as a training ground for attacking the "real problems" like whether NP equals co-NP. It is not hard to show ([F75]) that monadic NP is different from monadic co-NP. A much stronger result has even been proved by Matz and Thomas ([MT97]). They show that the monadic hierarchy, the natural monadic counterpart of the polynomial hierarchy, is strict (a property is in the k-th level of the monadic hierarchy if it is expressible by a sentence of monadic second order logic where all the second order quantifiers are at the beginning and there are at most $k-1$ alternations between second order existential and second order universal quantifiers).

An important part of research in the area of monadic NP is devoted to the possibility of expressing different variations of graph connectivity. Already Fagin's proof that monadic NP is different from monadic co-NP is based on the fact that connectivity of undirected graphs is not expressible by a sentence in monadic Σ_1^1, while nonconnectivity obviously is. Then de Rougemont [dR87] and Schwentick [S95] proved that connectivity is not in monadic NP even in the presence of various built-in relations.

However, as observed by Kanellakis, the property of reachability (for undirected graphs) is in monadic NP (reachability is the problem if, for a given graph and two distinguished nodes s and t, there is a path from s to t in this graph). It follows that connectivity, although not in monadic NP, is expressible by a formula of the form $\forall x \forall y \, \exists \vec{P} \varphi$. This observation leads to the study of *closed monadic NP*, the class of properties expressible by a sentence with quantifier prefix of the form $(\, \exists^*(\exists\forall)^*)^*$, and of the *closed monadic hierarchy*, the class of properties expressible by a sentence with quantifier prefix of the form $((\, \exists^*(\exists\forall)^*)^*(\, \forall^*(\exists\forall)^*)^*)^*$.

In [AFS00] and [AFS98] Ajtai, Fagin and Stockmeyer argue that closed monadic NP is even a more interesting object of study than monadic NP: it is still a subclass of NP (and also the k-th level of closed monadic hierarchy is still a subclass of the k-th level of polynomial hierarchy), it is defined by a simple syntax and it is closed under first order quantifications. In order to prove that such alternation of quantifiers gives substantial additional expressive power they construct graph properties $\mathcal{P}_1$ and $\mathcal{P}_2$ such that $\mathcal{P}_1$ is expressible by a sentence with the quantifier prefix in the class $(\exists\forall)^* \, \exists^*(\exists\forall)^*$, but not by a Boolean combination of sentences from monadic NP (i.e with the prefix of the form $\exists^*(\exists\forall)^*$) and $\mathcal{P}_2$ is expressible by a sentence $\exists^*(\exists\forall)^* \, \exists(\exists\forall)^*$ but not by a Boolean combination of sentences of the form $(\exists\forall)^* \, \exists^*(\exists\forall)^*$. The non expressibility results for $\mathcal{P}_1$ and $\mathcal{P}_2$ in [AFS00] are by no means easy and constitute the main technical contribution of this long paper. As the authors write: *Our most difficult result is the fact that there is an undirected graph property that is in closed monadic NP but not in the first order/Boolean closure of monadic NP. In the game corresponding to the first order/Boolean closure of monadic NP, played over graphs G_0 and G_1, the spoiler not only gets to choose which of G_0 and G_1 he wishes to color , but he does not have to make his selection until after a number of pebbling moves had been played. Thus, not only are we faced with the situation where the spoiler gets to choose which structure* *to color, but apparently also for the first time, we are being forced to consider a game where there are pebbling rounds both before and after the coloring round.*

There are many natural open questions in the area, most of them stated in [AFS00]: is the hierarchy inside closed monadic NP strict ? We mean here the hierarchy defined by the number of blocks of second order existential quantifiers, alternating with first order quantifiers. Is there any property in the monadic hierarchy (or, equivalently, in the closed monadic hierarchy) which is not in closed monadic NP ? Is the closed monadic hierarchy strict ? These questions seem to be quite hard: so far we do not know any property in the (closed) monadic hierarchy which would not be expressible by a sentence with quantifier prefix $\exists^*(\forall\exists)^* \exists^*(\forall\exists)^*$.

1.2 Our Contribution

In this paper we present an inductive and compositional technology for proving some non expressibility results for monadic second order logic. In particular, our technology gives an alternative simple solution to all the technical problems described in the citation from [AFS00] above. But unlike the construction in [AFS00], which is specific for first order/Boolean closure of monadic NP, our technology is universal: it deals with first order/Boolean closure of most monadic classes.

To be more precise, we show how to construct, for any given property S not expressible by a sentence with quantifier prefix in some non trivial[2] class W, two properties $bool(S)$ and $reach(S)$ which are not much harder than S and such that (1) property $bool(S)$ cannot be expressed by boolean combination of sentences with quantifier prefix in W and (2) property $reach(S)$ cannot be expressed by a sentence with quantifier prefix vw where $v \in (\exists + \forall)^*$ is a block of first order quantifiers and $w \in W$. Saying that $bool(S)$ and $reach(S)$ are *not much harder than* S we mean that if S is expressible by a sentence with quantifier prefix in some class V then $bool(S)$ is expressible by a sentence with the prefix of the form $\exists\exists v$ where $v \in V$ and $reach(S)$ is expressible by a sentence with the prefix of the form $\exists\forall v$ where $v \in V$. The non expressibility proof for $reach$ generalizes the second author's proof of the fact that directed reachability is not expressible by a sentence with the prefix of the form $(\forall\exists)^* \exists^*(\forall\exists)^*$ [M99].

Our lower bounds are proved in the language of Ehrenfeucht-Fraïssé games. To show that, for example, $reach(S)$ cannot be expressed by a sentence with a prefix of the form $\forall\exists w$ where $w \in W$ we assume as (inductive) hypothesis that there are two structures $P \in S$ and $R \notin S$ such that Duplicator has a winning strategy in the game (corresponding to the prefix w) on (P, R). Then we show how to apply some graph composition methods to get, from P and R, new structures $P_1 \in reach(S)$ and $R_1 \notin reach(S)$ such that Duplicator has a winning strategy in the game (corresponding to the new prefix $\forall\exists w$) on (P_1, R_1). But since we know nothing about P and R our knowledge about P_1 and R_1 is quite limited, so the strategy for Duplicator uses as a black box the unknown Duplicator's strategy in a game on (P, R).

With our technology we can make the first small step answering the hierarchy questions. To be more precise, we show that the hierarchy inside closed monadic NP does

---

[2] See definition below.

not collapse on any first order level. Since we do not need to care if the w (the prefix which does not express S) contains, or not, universal second order quantifiers a variety of results of this kind can also be proved with our technology about the structure of closed monadic hierarchy.

A new, very easy, proof of the results from [AFS00] is just a corollary of our method.

It also appears that - with minor modifications - the above inductive constructions can also be applied inside Kozen's mu-calculus [Ko83]. This constitutes a first small step towards trying to understand, over finite models, the (descriptive) complexity (in terms of patterns of FO and/or monadic quantifiers' prefix) of properties definable in the mu-calculus.

2 Technical Part

2.1 Structures

All the structures we consider in this paper are finite graphs (directed or not). The signature of the structures may also contain some additional unary relations ("colors") and constants (s and t).

2.2 Games

Definition 1. *1. A pattern of a monadic game (or just pattern) is any word over the alphabet $\{\forall, \exists, \mathbf{\forall}, \mathbf{\exists}, \oplus\}$.*

2. If w is a pattern then the pattern $\bar{w}$ (dual to w) is inductively defined as $\forall\bar{v}, \exists\bar{v}, \mathbf{\forall}\bar{v}, \mathbf{\exists}\bar{v}$ or $\oplus\bar{v}$ if w equals $\exists v, \forall v, \mathbf{\exists}v, \mathbf{\forall}v$ or $\oplus v$ respectively. The dual of the empty word is the empty word.

$\forall$ and $\exists$ still keep the meaning of universal and existential first order quantifiers, while $\mathbf{\forall}$ and $\mathbf{\exists}$ are universal and existential monadic second order (set) quantifiers. As you will soon see $\oplus$ should be understood as a sort of boolean closure of a game. We will use the abbreviation FO for the regular expression $(\forall + \exists)$.

Definition 2. *Let P and R be two relational structures over the same signature. Let w be some pattern. An Ehrenfeucht-Fraïssé game with pattern w over (P, R) is then the following game between 2 players, called Spoiler and Duplicator:*

1. If w is the empty word then the game is over and Duplicator wins if the substructures induced in P and in R by all the constants in the signature are isomorphic. Spoiler wins if they are not isomorphic.

2. If w is nonempty then:

(a) If $w = \exists v$ ($w = \forall v$) for some v then a new constant symbol c is added to the signature, Spoiler chooses the interpretation of c in P (R resp.) and then Duplicator chooses the interpretation of c in R (P resp.). Then they play the game with pattern v on the enriched structures.

(b) If $w = \exists v$ $(w = \forall v)$ for some v then a new unary relation symbol C is added to the signature, Spoiler chooses the interpretation of C in P (R resp.) and then Duplicator chooses the interpretation of C in R (P resp.) Then they play the game with pattern v on the enriched structures.

(c) If $w = \oplus v$ for some v then Spoiler can decide if he prefers to continue with the game with pattern v or rather with $\bar{v}$. Then they play the game with the pattern chosen by Spoiler.

The part of the game described by item (a) is called a first order round, or pebbling round. The part described by item (b) is a second order round, or coloring round.

Definition 3. *We say that a property (i.e a class of structures) S is expressible by a pattern w if for each two structures $P \in S$ and $R \notin S$ Spoiler has a winning strategy in the game with pattern w on (P, R). If W is a set of patterns then we say that S is expressible in W if there exists a $w \in W$ such that S is expressible by w.*

The following theorem illustrates the links between games and logics. We skip its proof as well known (see for example [EF] and [AFS00]):

Theorem 1. *1. Monadic NP is exactly the class of properties expressible by $\exists^* FO^*$;*

2. The boolean closure of monadic NP is exactly the class of properties expressible by $\oplus \exists^* FO^*$;

3. The first order closure of monadic NP is exactly the class of properties expressible by $FO^* \oplus \exists^* FO^*$;

4. 2k-th level of the monadic hierarchy is exactly the class of properties expressible by $(\exists^* \forall^*)^k FO^*$;

5. 2k-th level of the closed monadic hierarchy is exactly the class of properties expressible by $(FO^* \exists^* \forall^*)^k FO^*$;

6. Closed monadic NP is exactly the class of properties expressible by $(FO^* \exists^*)^*$;

The last theorem motivates:

Definition 4. A non trivial class of game patterns *(or just* class*) is a set of game patterns denoted by a regular expression without union over the alphabet* $\{\oplus, \exists, \forall, FO\}$*, which ends with FO^* and contains at least one $\forall^*$ or $\exists^*$*

In the sequel, all classes of game patterns we consider are non trivial.

2.3 Graph Operations

The techniques we are going to present are inductive and compositional. *Inductive* means here that we will assume as a hypothesis that there is a property expressible by some class of patterns W_1 but not by W and then, under this hypothesis, we will prove that there is a property expressible in the class $V_1 W_1$ but not in the class VW where V_1 and V will be some (short) prefixes. The word *compositional* means here that the pair of structures (P_{VW}, R_{VW}) (on which Duplicator has a winning strategy in a VW game) will be directly constructed from the pair of structures (P_W, R_W) (on which Duplicator has a winning strategy in a W game). For this construction we do not need to know anything about the original structures.

In the sequel, we will assume that all our structures are connected and that the signature contains a constant s (for *source*). This is possible thanks to the following natural definition and obvious lemma:

Definition 5. *Let S be a property of structures (with the signature without constant s). Then $cone(S)$ is the property of structures (with the same signature, enriched with constant s): For every x distinct from s there is an edge from s to x and the substructure induced by all the vertices distinct from s has the property S.*

Lemma 1. *If S is expressible by w then $cone(S)$ also is. If S is not expressible by w then there is a pair of connected structures (P, R) (see Definition 6 below) such that P has the property $cone(S)$, R does not, and Duplicator has a winning strategy in the w game on (P, R).* ■

Now we introduce some notations for graph operations. As we just mentioned we assume that all the graphs we are dealing with are connected and have some distinguished node s. Some of them will also have another distinguished node t (for *target*).

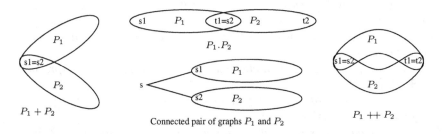

Fig. 1. Some Graph Operations.

Definition 6. *1.* Let U denote the graph containing just two vertices, s and t, and one edge $E(s,t)$.
   2. If A is a set of graphs, then $\Sigma^s_{P \in A} P$ ($\Sigma^{st}_{P \in A} P$) is the union of all graphs in A with all the s vertices identified (resp. and all the t vertices identified). We will use also the notation $\Sigma^s_c P$ ($\Sigma^{st}_c P$) if A contains just c copies of the same structure P. If there are only two elements, say P and R in A, then we write $P+R$ (or $P++R$) instead of of $\Sigma^s_{P \in A} P$ (or $\Sigma^{st}_{P \in A} P$).
   3. If P is a graph with constants s and t then $P.R$ (or PR for short) is the graph being a union of P and R with t of P identified with s of R (so that s of the new graph is the s of P and the t of the new graph is the t of R if it exists.
   4. If A is a set of graphs then the graph $\Sigma^s_{P \in A}(UP)$ will be called a connected set of graphs. If there are just two elements in A then we will call it a connected pair of graphs.

2.4   Some Simple Lemmas about Games

Let us start with an obvious lemma, which would remain true even without the assumption that the relations introduced during the second order rounds are unary:

Lemma 2. *If the graphs P and R are isomorphic then Duplicator has a winning strategy in the w game on (P, R) whatever w is.* ∎

The following Lemmas 3-5 are not much harder to prove that Lemma 2 but the assumption that games are monadic is crucial here:

Lemma 3. *If Duplicator has winning strategies in w games on (P_1, R_1) and on (P_2, R_2) then he also has winning strategies in w games on (P_1+P_2, R_1+R_2), on $(P_1 P_2, R_1 R_2)$ and on $(P_1 \mathbin{+\!\!+} P_2, R_1 \mathbin{+\!\!+} R_2)$.* ∎

Lemma 4. *For every structure P and pattern w there exists a number n such that provided $m \geq n$ then Duplicator has winning strategies in the w games on $(\Sigma_m^s P, \Sigma_{m+1}^s P)$ and $(\Sigma_m^{st} P, \Sigma_{m+1}^{st} P)$*

Proof. Induction on the structure of w. Use the fact that for a structure P of some fixed size there are only finitely many colorings of it, so if we have enough copies some colorings must repeat many times. ∎

Lemma 5. *Let P be a connected pair of structures P_1 and P_2 and let R be a connected pair of structures R_1 and R_2. Suppose for some (non trivial[3]) class V there exists $v \in V$ such that Spoiler has a winning strategy on the v games on (P_1, R_1) and on (P_1, R_2). Then there exists $w \in \exists V$ such that Spoiler has a winning strategy in the w games on (P, R).*

Proof. The strategy of Spoiler is to take as his first constant the source of P_1 in P. Duplicator must answer either with the source of R_1 or of R_2, and so he must make a commitment on which of the two structures is going to play the role of P_1 in R now. The cases are symmetric, so let us assume he decides on R_1. Then Spoiler uses his strategy for the v game on (P_1, R_1) to win the game. Actually, Spoiler must force Duplicator to move only inside the structures P_1 and R_1. This can achieved with one more coloring round (at any time in the v game) subsequently playing a w-game for some $w \in V$ since V is non trivial. The next remark makes this observation more precise. ∎

Remark 6. *After the first round, when Spoiler picks the source of P_1 and Duplicator answers by the source of R_1, Spoiler must force Duplicator to restrict the moves of the remaining game only to the structures P_1 and R_1. In other words, Spoiler needs to be sure that each time he picks a constant inside P_1 (R_1) Duplicator actually answers with a constant inside R_1 (P_1). This can be secured with the use of an additional coloring round: Spoiler paints P_1 (or R_1, he is as happy with a $\exists$ round as with a $\forall$ one) with some color leaving the rest of P unpainted. Duplicator must answer by painting R_1 (P_1) with this color, leaving the rest of R unpainted. Otherwise, this will be detected by Spoiler with the use of the final first order rounds. Notice that the additional coloring round can take place at any moment of the game, and so that the strategy is available for Spoiler for some $\exists V$ game since V is a nontrivial class of patterns.*

---

[3] See Definition 4.

2.5   A Tool for the Boolean Closure

Let S be any property. Then, a connected pair of structures $UP+UR$ will be called SS if both the structures P and R belong to S, $\bar{S}S$ if exactly one of them belongs to S and $\bar{S}\bar{S}$ otherwise.

Definition 7. *For a property S define bool(S) as the property:* the structure is a connected set of connected pairs of structures, and at least one of those pairs is $\bar{S}\bar{S}$.

Lemma 7. *Suppose a property S is not expressible in class W, but both S and its complement $\bar{S}$ are expressible in some other class V. Then bool(S) is not expressible in $\oplus W$ but is expressible in $\exists\exists V$.*

Proof. Let us first show that there exists $w \in V$ such that, provided $P \in bool(S)$ and $R \notin bool(S)$, Spoiler has a winning strategy in the $\exists\exists w$ game on (P, R). This will prove that property $bool(S)$ is expressible by $\exists\exists V$.

First observe that if R is not a connected set of pairs then either the vertices of R at distance less than 2 from s do not form a tree, or there is a vertex at distance 2 from s whose degree is not 3, or R is not connected, or there is a vertex x at distance 2 from s such that the structure resulting from removing x (and all the three adjacent edges) from R has less than 3 connected components. In each of those cases Spoiler can win some game in $\exists V$ for every nontrivial V.

If R is a connected set of pairs then in his first move Spoiler takes as his constant the source of some $\bar{S}\bar{S}$ pair in P. Duplicator must answer by showing a source of some pair in R. There are two cases: either Duplicator shows a source of some SS pair in R or a source of some $\bar{S}\bar{S}$ pair in R. In each of the two cases we may think that one pair of structures has been selected in P and one in R. Spoiler can restrict the game to the two selected pairs (see Remark 6). Then we use Lemma 5 to finish the proof.

Now we will show that whatever a pattern $\oplus w$ is, where $w \in W$, there exist two structures $P \in bool(S)$ and $R \notin bool(S)$ such that Duplicator has a winning strategy in the $\oplus w$ game on (P, R). Let (P_1, R_1) be such a pair of structures that $P_1 \in S$, $R_1 \notin S$ and Duplicator has a winning strategy in the w game on (P_1, R_1). Let c be some huge constant. Let $R = \Sigma_c^s (U(UP_1+UP_1)+U(UR_1+UR_1))$. So R is a connected set of $2c$ connected pairs, c of them are $\bar{S}\bar{S}$ and c are SS. Obviously, $R \notin bool(S)$. Let $P = R+U(UP_1+UR_1)$ be R with one more pair, a $S\bar{S}$ one, so that $P \in bool(S)$.

Now, if Spoiler in his first move decides to play the game w on P and R then remark that P is $Q_1+Q_2+Q_3$ where $Q_1 = \Sigma_c^s (U(UP_1+UP_1))$, $Q_2 = \Sigma_c^s (U(UR_1+UR_1))$ and $Q_3 = U(UR_1+UP_1)$ while R is $Q_4+Q_5+Q_6$ where $Q_4 = \Sigma_c^s (U(UP_1+UP_1))$, $Q_5 = \Sigma_{c-1}^s (U(UR_1+UR_1))$ and $Q_6 = U(UR_1+UR_1)$. We know that Duplicator has a winning strategies in w games on (Q_1, Q_4) (by Lemma 2), on (Q_2, Q_5) (by Lemma 4) and on (Q_3, Q_6) (by Lemma 3, since he has a winning strategy in a w game on (P_1, R_1)). So, again by Lemma 3 he has a winning strategy in w game on (P, R).

If Spoiler decides in his first round to continue with $\bar{w}$ rather than w then take Q_1, Q_2, Q_3 as before but $Q_4 = \Sigma_{c-1}^s (U(UP_1+UP_1))$, $Q_5 = \Sigma_c^s (U(UR_1+UR_1))$ $Q_6 = U(UP_1+UP_1)$ and use the same reasoning, using the fact that Duplicator has a winning strategy in the $\bar{w}$ game on (R_1, P_1). ∎

2.6   A Tool for First Order Quantifiers

Now the signature of our structures will contain additional unary relation symbol G (for *gate*). For a given structure P, and for two its vertices x, y, such that $G(y)$ holds let $P_{x,y}$ be the structure consisting of the connected component of $P - \{x\}$, containing y as its source. $P - \{x\}$ is here understood to be the structure resulting from P after removing x and all its adjacent edges. So $P_{x,y}$ could be read as "the structure you enter from x crossing the gate y" (see Figure 2).

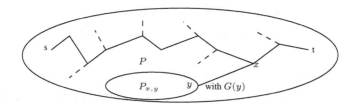

Fig. 2. $P_{x,y}$ Is the Structure You Enter from x Crossing the Gate y.

Definition 8. *Let S be some property of structures. Then reach(S) will be the following property (of a structure P): there is a path from s to t such that for every x on this path it holds that (i) $x \notin G$ and (ii) for every y such that $E(x, y)$ and $G(y)$ the structure $P_{x,y}$ has the property S.*

By a *path from s to t* we mean a subset H of the set of vertices of the structure such that $s, t \in H$, each of s and t has exactly one adjacent vertex in H and each element of H which is neither s nor t has exactly 2 adjacent vertices in H. The fact that H is a path is expressible by FO^*.

Lemma 8.   *1. Suppose a property S is not expressible in some class W. Then reach(S) is not expressible in FO^*W;*
2. Suppose a property S is expressible in some class W. Then reach(S) is expressible in the class $\exists \forall \forall W$.

Proof. 1. First of all we will show that if S is not expressible in W, then also reach(S) is not expressible in W. For a given $w \in W$ there are structures P and R such that $P \in S$, $R \notin S$ and Duplicator has a winning strategy in the w game on (P, R). Consider a structure T whose only elements are s, t, x, y, whose edges are $E(s, x)$, $E(x, t)$, $E(x, y)$ and for which $G(y)$ holds. Let P_0 be the union of T and P, with y of T identified with s of P. The s and t of P_0 are s and t of T. Let R_0 be the structure constructed in the same way from T and R. Then obviously $P_0 \in reach(S)$, $R_0 \notin reach(S)$ and Duplicator has a winning strategy in the w game on (P_0, R_0). Notice that both P_0 and R_0 have the following property :

(*) (property of structure Q) if x is reachable from s or from t by a path disjoint from G and if y is such that $G(y)$ and $E(x, y)$ then Q_{xy} contains neither s of Q nor t of Q.

Now let P and R be structures, both satisfying (*) and such that $P \in reach(S)$, $R \notin reach(S)$ and Duplicator has a winning strategy in a w game on (P, R). In order to prove our claim it is enough (by induction) to construct structures (P_1, R_1) both satisfying (*) and such that $P_1 \in reach(S)$, $R_1 \notin reach(S)$ and Duplicator has a winning strategy in a $\forall\exists w$ game on (P_1, R_1). Let n be a huge enough constant. Define: $R_1 = (\Sigma_n^{st}(PR)) + (\Sigma_n^{st}(RP))$ and $P_1 = R_1 + PP$. Obviously $P_1 \in reach(S)$ and $R_1 \notin reach(S)$ hold. Now will show a winning strategy for Duplicator in a $\forall\exists w$ game on (P_1, R_1). In his first round Spoiler selects some constant in R_1. Duplicator answers with the same constant in P_1 (this is possible since R_1 can be viewed as a subset of P_1). Now notice that after this first round R_1 can be seen as

$$RP + PR + (\Sigma_{n-1}^{st}(PR)) + (\Sigma_{n-1}^{st}(RP))$$

and P_1 as

$$RP + PR + (\Sigma_{n-1}^{st}(PR)) + (\Sigma_{n-1}^{st}(RP)) + PP$$

where the constant selected in the first round is in the first $RP + PR$, both in R_1 and in P_1. By Lemma 2 and Lemma 3 it is now enough to show that Duplicator has a winning strategy in the remaining $\exists w$ game on (P_2, R_2) where

$$P_2 = \Sigma_{n-1}^{st}(PR)) + (\Sigma_{n-1}^{st}(RP)) + PP$$

and

$$R_2 = \Sigma_{n-1}^{st}(PR)) + (\Sigma_{n-1}^{st}(RP))$$

Let Spoiler select some constant in P_2.

If Spoiler selects a constant in $\Sigma_{n-1}^{st}(PR)) + (\Sigma_{n-1}^{st}(RP))$ then Duplicator answers with the same constant in R_2 and then wins easily. The only interesting case is when Spoiler selects his constant in PP. Suppose it is selected in the first P (the other case is symmetric). Then Duplicator answers by selecting the same constant in the P of some PR in R_2. Notice that $P_2 = Q_1 + Q_2 + (\Sigma_{n-1}^{st}(RP))$ and $R_2 = Q_3 + Q_4 + (\Sigma_{n-1}^{st}(RP))$, where $Q_1 = PP$, $Q_2 = \Sigma_{n-1}^{st}(PR))$, $Q_3 = PR$ and $Q_4 = \Sigma_{n-2}^{st}(PR))$, and where some constant is already fixed in the first P of Q_1 and in the P of Q_3. Now the w game remains to be played. But since Duplicator has a winning strategy in the w game on (P, R) he also has (by Lemmas 2 and 3) a winning strategy in a w game on (Q_1, Q_3). By Lemma 4 he has a winning strategy in a w game on (Q_2, Q_4) and so, again by Lemma 3 we get a winning strategy for Duplicator in the $\exists w$ game on (P_2, R_2).

2. Suppose $P \in reach(S)$ and $R \notin reach(S)$. Spoiler, in his first move fixes a path in P, as in the definition of $reach(S)$. Duplicator answers selecting a set in R. If the set selected by Duplicator is not a path from s to t then Spoiler only needs some fixed number of first order rounds to win. If it is such a path then there must be some x on the path, and some y such that $E(x, y)$, $G(y)$ hold in R and $R_{x,y} \notin S$. Now Spoiler uses his two first order universal rounds to fix those x and y. Duplicator answers with some two points z, t in P such that $E(z, t)$ and $G(t)$ hold in P. But, since $P \in reach(S)$ it turns out that $P_{z,t} \in S$, so Spoiler can use rounds of the remaining w game to secure a win (a trick from Remark 6 will be needed here to restrict the w game to $P_{x,y}, R_{z,t}$).

∎

Remark 9. *The role of predicate G is not crucial for the construction above. It could be replaced by a graph gadget if the reader wishes to see $\mathcal{P}_2$ being a property of undirected uncolored graphs.*

Another way to avoid the unary relation G (as suggested by Larry Stockmeyer) is to define reach(S) as: there is a path from s to t such that for every x on this path and every y such that $E(x, y)$ and y is not on this path, the structure $P_{x,y}$ has the property S.

2.7   Corollaries

As the first application of our toolkit we reprove the results from [AFS00]:

Theorem 2. *There exists property $\mathcal{P}_1$ expressible in $FO^* \, \exists^* FO^*$ but not in $\oplus \, \exists^* FO^*$. There exists property $\mathcal{P}_2$ expressible in $\exists FO^* \, \exists^* FO^*$ but not in $FO^* \oplus \, \exists^* FO^*$.*

Proof. Let *Cted* be the property of connectivity. It is well known that *Cted* is not expressible in $\exists^* FO^*$ but both *Cted* and its complement are expressible in $\forall\forall \, \exists^* FO^*$. now take $\mathcal{P}_1 = bool(cone(Cted))$ and $\mathcal{P}_2 = reach(bool(cone(Cted)))$. Use Lemmas 7 and 8 to finish the proof. ∎

A new result we can prove is that even if the hierarchy inside closed monadic NP collapses, it does not collapse on a first order level:

Theorem 3. *If there is a property expressible in FO^*W but not in W, where $W = (\, \exists^* FO^*)^k$ then there is a property expressible in $\exists FO^*W$ but not in FO^*W.*

Proof. This follows immediately from Lemma 8 ∎

Several similar results can be proved for the closed monadic hierarchy or reproved for the monadic hierarchy (see [MT97] and [Ma99] sections 4.4 and 4.5).

It is interesting to remark that the inductive constructions presented here are also definable (with minor and insignificant variations) inside Kozen's propositional μ-calculus [Ko83].

More precisely, given some unary predicates S, one may define in the μ-calculus the new predicates that depend on S: $Bool(S) = \Diamond(\Diamond S \wedge \Diamond \neg S)$ and $Reach(S) = \mu X.(\Box(G \Rightarrow S) \wedge (\Diamond X \vee T))$ which almost denote the same constructions (here the "target" constant t is replaced by the set of "possible targets" T and the "source" constant s is the implicit free FO variable in any mu-calculus formula).

From Lemmas 7 and 8 (which extend to these definitions inside the mu-calculus) and the fact that (the mu-calculus version of) directed reachability: $dreach = \mu X.(\Diamond X \vee T)$ is not expressible in $\exists^* FO^*$ while both $dreach$ and its complement are expressible in $\exists_\forall \, \exists^* FO^*$, one has:

Corollary 1. *There are properties $\mathcal{R}_1$ and $\mathcal{R}_2$ definable in monadic μ-calculus such that $\mathcal{R}_1$ is expressible in $FO^* \, \exists FO^* \, \exists^* FO^*$ but not in $\oplus \, \exists^* FO^*$ and $\mathcal{R}_2$ is expressible in $\exists FO^* \, \exists FO^* \, \exists^* FO^*$ but not in $FO^* \oplus \, \exists^* FO^*$.*

Proof. Take $\mathcal{R}_1 = Bool(dreach)$ and $\mathcal{R}_2 = Reach(dreach)$ and apply Lemmas 7 and 8 to finish the proof. ∎

Acknowledgment

The authors thank Oskar Miś who wrote for us the TeX macros for ∃,∀, ∃ and ∀, an anonymous referee who even found a bug (corrected in the present version) in our proofs, Larry Stockmeyer for numerous and detailed comments on the first draft of this paper, and Mike Robson for his help in debugging our international English writing.

References

AFS98.   M.Ajtai, R.Fagin, L.Stockmeyer *The Closure of Monadic $\mathcal{NP}$*, (extended abstract of [AFS00]) Proc. of 13th STOC, pp 309-318, 1998;

AFS00.   M.Ajtai, R.Fagin, L.Stockmeyer *The Closure of Monadic $\mathcal{NP}$*, Journal of Computer and System Sciences, vol. 60 (2000), pp. 660-716;

F75.   R. Fagin *Monadic Generalized spectra*, Zeitschrift fuer Mathematische Logik und Grundlagen der Mathematik, 21;89-96, 1975;

dR87.   M. de Rougemont *Second-order and inductive definability on finite structures*, Zeitschrift fuer Mathematische Logik und Grundlagen der Mathematik, 33:47-63, 1987;

E61.   A. Ehrenfeucht *an application of games to the completeness problem for formalized theories*, Fund. Math. 49:129-141,1961;

EF.   H-D. Ebinghaus, J. Flum *Finite Model Theory*, Springer 1995;

Fr54.   R. Fraïssé *Sur quelques classifications des systemes de relations*, Publ. Sci. Univ. Alger. Ser. A, 1:35-182, 1954;

Ko83.   D. Kozen, *Results on The Propositional μ-calculus* Theor. Comp. Science, 27:333-354, 1983;

M99.   J. Marcinkowski *Directed Reachability: From Ajtai-Fagin to Ehrenfeucht-Fraisse games*, Proceedings of the Annual Conference of the European Association of Computer Science Logic (CSL 99) Springer LNCS 1683, pp 338-349;

MT97.   O. Matz, W. Thomas *The monadic quantifier alternation hierarchy over graphs is infinite*, Proc. 12th IEEE LICS 1997, pp 236-244;

Ma99.   Oliver Matz, *Dot-Depth and Monadic Quantifier Alternation over Pictures*, PhD thesis, report 99-8, Aachener Informatik-Berichte, RWTH Aachen, 1999;

S95.   T. Schwentick *Graph connectivity, monadic $\mathcal{NP}$ and built-in relations of moderate degree*, Proceedings of 22nd ICALP: 405-416,1995;

Polynomial Time Approximation Schemes for MAX-BISECTION on Planar and Geometric Graphs

Klaus Jansen[1], Marek Karpinski[2], Andrzej Lingas[3], and Eike Seidel[1]

[1] Institute of Computer Science and Applied Mathematics
Christian-Albrechts-University of Kiel, 24 098 Kiel, Germany
{kj,ese}@informatik.uni-kiel.de
[2] Department of Computer Science, University of Bonn, 53117 Bonn, Germany
marek@cs.uni-bonn.de
[3] Department of Computer Science, Lund University, 22100 Lund, Sweden
Andrzej.Lingas@cs.lth.se

Abstract. The Max-Bisection and Min-Bisection problems are to find a partition of the vertices of a graph into two equal size subsets that respectively maximizes or minimizes the number of edges with endpoints in both subsets.

We design the first polynomial time approximation scheme for the Max-Bisection problem on arbitrary planar graphs solving a long time standing open problem. The method of solution involves designing exact polynomial time algorithms for computing optimal partitions of bounded treewidth graphs, in particular Max- and Min-Bisection, which could be of independent interest.

Using similar method we design also the first polynomial time approximation scheme for Max-Bisection on unit disk graphs (which could be easily extended to other geometrically defined graphs).

1 Introduction

The max-bisection and min-bisection problems, i.e., the problems of constructing a halving of the vertex set of a graph that respectively maximizes or minimizes the number of edges across the partition, belong to the basic combinatorial optimization problems.

The best known approximation algorithm for max-bisection yields a solution whose size is at least 0.701 times the optimum [16] whereas the best known approximation algorithm for min-bisection achieves "solely" a log-square approximation factor [11]. The former factor for max-bisection is considerably improved for regular graphs to 0.795 in [10] whereas the latter factor for min-bisection is improved for graphs excluding any fixed minor (e.g., planar graphs) to a logarithmic one in [11]. For dense graphs, Arora, Karger and Karpinski give polynomial time approximation schemes for max- and min-bisection in [2].

In this paper, we study the max-bisection and min-bisection problems on bounded treewidth graphs and on planar graphs. Both graph families are known

A. Ferreira and H. Reichel (Eds.): STACS 2001, LNCS 2010, pp. 365–375, 2001.

to admit exact polynomial time algorithms for max-cut, i.e., for finding a bi-partition that maximizes the number of edges with endpoints in both sets in the partition [9,14].

Our first main result are exact polynomial time algorithms for finding a partition of a bounded treewidth graph into two sets of a priori given cardinalities, respectively maximizing or minimizing the number of edges with endpoints in both sets. Thus, in particular, we obtain polynomial time algorithms for max-bisection and min-bisection on bounded treewidth graphs.

The complexity and approximability status of max-bisection on planar graphs have been long-standing open problems. Contrary to the status of planar max-cut, planar max-bisection has been proven recently to be NP-hard in exact setting by Jerrum [17]. Karpinski et al. observed in [18] that the max-bisection problem for planar graphs does not fall directly into the Khanna-Motwani's syntactic framework for planar optimization problems [19]. On the other hand, they provided a polynomial time approximation scheme (PTAS) for max-bisection in planar graphs of sublinear maximum degree. (In fact, their method implies that the size of max-bisection is very close to that of max-cut in planar graphs of sublinear maximum degree.)

Our second main result is the first polynomial time approximation scheme for the max-bisection problem for arbitrary planar graphs. It is obtained by combining (via tree-typed dynamic programming) the original Baker's method of dividing the input planar graph into families of k-outerplanar graphs [4] with our method of finding maximum partitions of bounded treewidth graphs.

Note that the NP-hardness of exact planar max-bisection makes our PTAS result best possible under usual assumptions.

Interestingly, our PTAS for planar max-bisection can be easily modified to a PTAS for the problem of min-bisection on planar graphs in the very special case where the min-bisection is relatively large, i.e., cuts $\Omega(n \log \log n / \log n)$ edges.

Unit disk graphs are another important class of graphs defined by the geometric conditions on a plane. An undirected graph is a unit disk graph if its vertices can be put in one to one correspondence with disks of equal radius in the plane in such a way that two vertices are joined by an edge if and only if the corresponding disks intersect. Tangent disks are considered to intersect.

Our third main result is the first polynomial time approximation scheme for the max-bisection problem on unit disk graphs. The scheme can be easily generalized to include other geometric intersection graphs. It is obtained by combining (again via tree-typed dynamic programming) the idea of Hunt et al. of dividing the input graph defined by plane conditions into families of subgraphs [15] with the aforementioned known methods of finding maximum partitions of dense graphs [2].

The structure of our paper is as follows. The next section complements the introduction with basic definitions and facts. In Section 3, the algorithms for optimal partitions of bounded treewidth graphs are given. Section 4 presents the PTAS for planar max-bisections. In Section 5, we make several observations on the approximability of planar min-bisection. Finally, Section 6 describes the

PTAS for max-bisection on unit disk graphs. In conclusion we notice that same technique can be applied also for other geometric intersection graphs.

2 Preliminaries

We start with formulating the underlying optimal graph partition problems.

Definition 1. *A partition of a set of vertices of an undirected graph G into two sets X, Y is called an $(|X|, |Y|)$-partition of G. The edges of G with one endpoint in X and the other in Y are said to be cut by the partition. The size of an (l, k)-partition is the number of edges which are cut by it. An (l, k)-partition of G is said to be a* maximum (l, k)-partition *of G if it has the largest size among all (l, k)-partitions of G. An (l, k)-partition of G is a* bisection *if $l = k$. A bisection of G is a* max bisection *or a* min bisection *of G if it respectively maximizes or minimizes the number of cut edges. An (l, k)-partition of G is a* max cut *of G if it has the largest size among all (l', k')-partitions of G. The* max-cut problem *is to find a max cut of a graph. Analogously, the* max-bisection problem *is to find a max bisection of a graph. The* min-cut problem *and the* min-bisection problem *are defined analogously.*

The notion of treewidth of a graph was originally introduced by Robertson and Seymour [21]. It has turned out to be equivalent to several other interesting graph theoretic notions, e.g., the notion of partial k-trees [1,5].

Definition 2. *A* tree-decomposition *of a graph $G = (V, E)$ is a pair $(\{X_i \mid i \in I\}, T = (I, F))$, where $\{X_i \mid i \in I\}$ is a collection of subsets of V, and $T = (I, F)$ is a tree, such that the following conditions hold:*

1. *$\bigcup_{i \in I} X_i = V$.*
2. *For all edges $(v, w) \in E$, there exists a node $i \in I$, with $v, w \in X_i$.*
3. *For every vertex $v \in V$, the subgraph of T, induced by the nodes $\{i \in I \mid v \in X_i\}$ is connected.*

The treewidth *of a tree-decomposition $(\{X_i \mid i \in I\}, T = (I, F))$ is $\max_{i \in I} |X_i| - 1$. The* treewidth *of a graph is the minimum treewidth over all possible tree-decompositions of the graph. A graph which has a tree-decomposition of treewidth $O(1)$ is called a* bounded treewidth graph.

Fact 1[6]: *For a bounded treewidth graph, a tree decomposition of minimum treewidth can be found in linear time.*

To state our approximation results on max-bisection we need the following definition.

Definition 3. *A real number α is said to be an* approximation ratio *for a maximization problem, or equivalently the problem is said to be* approximable within *a ratio α, if there is a polynomial time algorithm for the problem which always produces a solution of size at least α times the optimum. If a problem is approximable for arbitrary $\alpha < 1$ then it is said to admit a* polynomial time approximation scheme *(a PTAS for short).*

An approximation ratio and a PTAS for a minimization problem are defined analogously.

2.1   Optimal Partitions for Graphs of Bounded Treewidth

Let G be a graph admitting a tree-decomposition $T = (I, F)$ of treewidth at most k, for some constant k. By [9], one can easily modify T, without increasing its treewidth, such that one can see T as a rooted tree, with root $r \in I$, fullfiling the following conditions:

1. T is a binary tree.
2. If a node $i \in I$ has two children j_1 and j_2, then $X_i = X_{j_1} = X_{j_2}$.
3. If a node $i \in I$ has one child j, then either $X_j \subset X_i$ and $|X_i \setminus X_j| = 1$, or $X_i \subset X_j$ and $|X_j \setminus X_i| = 1$.

Provided a tree-decomposition of width k is given such a modified tree-decomposition of the same width can be constructed in linear time whereby the new decomposition-tree has at most $O(|V(G)|)$ nodes. We will assume in the remainder that such a modified tree-decomposition T of G is given.

For each node $i \in I$, let Y_i denote the set of all vertices in a set X_j with $j = i$ or j is a descendant of i in the rooted tree T. Our algorithm computes for each $i \in I$, an array $maxp_i$ with $O(2^k |Y_i|)$ entries. For each $l \in \{0, 1, ..., |Y_i|\}$ and each subset S of X_i, the entry $maxp_i(l, S)$ is set to $\max_{S' \subseteq Y_i, |S'| = l, S' \cap X_i = S} |\{(v, w) \in E | v \in S' \& w \in Y_i \setminus S'\}|$. In other words, $maxp_i(l, S)$ is set to the maximum number of cut edges in an $(l, |Y_i| - l)$-partition of Y_i where S and $X_i \setminus S$ are in the different sets of the partition and the set including S is of cardinality l. For convention, if such a partition is impossible, $maxp_i(l, S)$ will be set to $-\infty$.

The entries of the array are computed following the levels of the tree-decomposition T in a bottom-up manner. The following lemma shows how the array can be determined efficiently.

Lemma 1.

- *Let i be a leaf in T. Then for all $l \in \{0, 1, ..., |X_i|\}$ and $S \subseteq X_i$ where $|S| = l$, $maxp_i(l, S) = |\{(v, w) \in E | v \in S, w \in X_i \setminus S\}|$. The remaining entries of $maxp_i$ are set to $-\infty$.*
- *Let i be a node with one child j in T. If $X_i \subseteq X_j$ then for all $l \in \{0, 1, ..., |Y_i|\}$ and $S \subseteq X_i$, $maxp_i(l, S) = \max_{S' \subseteq X_j, S' \cap X_i = S} maxp_j(l, S')$.*
- *Let i be a node with one child j in T. If $X_j \cup \{v\} = X_i$ where $v \notin X_j$ then for all $l \in \{0, 1, ..., |Y_i|\}$ and $S \subseteq X_i$, if $v \in S$ then $maxp_i(l, S) = maxp_j(l-1, S \setminus \{v\}) + |\{(v, s) | s \in X_i \setminus S\}|$ else $maxp_i(l, S) = maxp_j(l, S) + |\{(v, s) | s \in S\}|$.*
- *Let i be a node with two children j_1, j_2 in T, with $X_i = X_{j_1} = X_{j_2}$. For all $l \in \{0, 1, ..., |Y_i|\}$ and $S \subseteq X_i$, $maxp_i(l, S) = \max_{l_1 + l_2 - |S| = l \& l_1 \geq |S| \& l_2 \geq |S|} (maxp_{j_1}(l_1, S) + maxp_{j_2}(l_2, S) - |\{(v, w) \in E | v \in S, w \in X_i \setminus S\}|)$.*

It follows that computing an array $maxp_i$ on the basis of the arrays computed for the preceding level of T can be done in time $O(2^k |Y_i|^2)$. Consequently, one can compute the array $maxp_r$ for the root r of T in cubic time.

Theorem 1. *All maximum $(l, n-l)$-partitions of a graph on n nodes given with a tree-decomposition of treewidth k can be computed in time $O(2^k n^3)$.*

By substituting *min* for *max*, we can analogously compute all minimum $(l, n - l)$-partitions of a graph with constant treewidth.

Theorem 2. *All minimum $(l, n-l)$-partitions of a graph on n nodes given with a tree-decomposition of treewidth k can be computed in time $O(2^k n^3)$.*

By Fact 1 we obtain the following corollary.

Corollary 1. *All maximum and minimum $(l, n - l)$-partitions of a bounded treewidth graph on n vertices can be computed in time $O(n^3)$.*

Since a tree-decomposition of a planar graph on n vertices with treewidth $O(\sqrt{n})$ can be found in polynomial time by the planar separator theorem [7], we obtain also the following corollary.

Corollary 2. *All maximum and minimum $(l, n-l)$-partitions of a planar graph on n vertices can be computed in time $2^{O(\sqrt{n})}$.*

3   A PTAS for Max-Bisection of an Arbitrary Planar Graph

The authors of [18] observed that the requirements of the equal size of the vertex subsets in a two partition yielding a max bisection makes the max-bisection problem hardly expressible as a maximum planar satisfiability formula. For this reason we cannot directly apply Khanna-Motwani's [19] syntactic framework yielding PTASs for several basic graph problems on planar graphs (e.g., max cut). Instead, we combine the original Baker's method [4] with our algorithm for optimal maximum partitions on graphs of bounded treewidth via tree-type dynamic programming in order to derive the first PTAS for max-bisection of an arbitrary planar graph.

Algorithm 1
input: a planar graph $G = (V, E)$ on n vertices and a positive integer k;
output: $(1 - \frac{1}{k+1})$-approximations of all maximum $(l, n - l)$-partitions of G

1. Construct a plane embedding of G;
2. Set the level of a vertex in the embedding as follows: the vertices on the outer boundary have level 1, the vertices on the outer boundary of the subgraph obtained by deleting the vertices of level $i - 1$ have level i, for convention extend the levels by k empty ones numbered $-k + 1, -k + 2, ..., 0$;
3. For each level j in the embedding construct the subgraph H_j of G induced by the vertices on levels $j, j + 1, ..., j + k$;
4. For each level j in the embedding set n'_j to the number of vertices in H_j and compute all maximum $(l, n'_j - l)$-partitions of H_j;

5. For each i, $0 \leq i \leq k$, set G_i to the union of the subgraphs H_j where j (mod $k+1$) $= i$;
6. For each i, $0 \leq i \leq k$, set n_i to the number of vertices in G_i and compute all maximum $(l, n_i - l)$-partitions of G_i by dynamic programming in a tree fashion, i.e., first compute all maximum partitions for pairs of "consecutive" H_j where j (mod $k+1$) $= i$, then for quadruples of such H_j etc.;
7. For each l, $1 \leq l < n$, output the largest among the maximum $(l, n - l)$-partitions of G_i, $0 \leq i \leq k$.

Lemma 2. *For each l, $1 \leq l < n$, Algorithm 1 outputs an $(l, n - l)$-partition of G within $k/(k+1)$ of the maximum.*

Proof. Let P be a maximum $(l, n - l)$-partition of G. For each edge e in P, there is at most one i, $0 \leq i \leq k$, such that e is not an edge of G_i. Consequently, there is i', $0 \leq i' \leq k$, such that $G_{i'}$ does not include at most $|P|/(k+1)$ edges of P. It follows that a maximum $(l, n - l)$-partition of such a $G_{i'}$ cuts at least $k|P|/(k+1)$ edges. Algorithm 1 outputs an $(l, n - l)$-partition of G cutting at least so many edges as a maximum $(l, n - l)$-partition of $G_{i'}$. □

Lemma 3. *Algorithm 1 runs in $O(k2^{3k-1}n^3)$ time.*

Proof. The time complexity of the algorithm is dominated by that of step 4 and 6.

The subgraphs H_j of G are so called k-outerplanar graphs and have bounded treewidth $3k - 1$ [7]. Hence, for a given i, $0 \leq i \leq k$, all maximum $(l, n'_j - l)$-partitions of H_j where j (mod $k+1$) $= i$ can be computed in time $O(2^{3k-1}n^3)$ by Lemma 1, the pairwise disjointness of the subgraphs and $j \leq n$. It follows that the whole step 4 can be implemented in time $O(k2^{3k-1}n^3)$.

In step 6, a maximum $(l, n_i - l)$-partition of the union of 2^{q+1} "consecutive" H_j's satisfying j (mod $k+1$) $= i$ can be determined on the basis of appropriate maximum partitions of its two halves, each being the union of 2^q of the H_j's, in time $O(n)$. Hence, since $l \leq n_i$ and the number of nodes in the dynamic programming tree is $O(n)$, the whole step 6 takes $O(kn^3)$ time. □

Theorem 3. *Algorithm 1 yields a PTAS for all maximum $(l, n - l)$-partitions of a planar graph.*

Corollary 3. *The problem of max-bisection on planar graphs admits a PTAS.*

4   Observations on Min-Bisection for Planar Graphs

We can easily obtain an analogous PTAS for min-bisection of planar graphs in the very special case when the size of min-bisection is $\Omega(n)$. Simply, at least one of the subgraphs G_i of G misses at most $|E|/(k+1)$ edges of G. Therefore, the number of edges cut by a min-bisection of such a G_i can increase at most by $|E|/(k+1)$ in G. By picking k sufficiently large we can guarantee an arbitrarily close approximation of min-bisection in G.

In fact, we can obtain even a slightly stronger result on min-bisection for planar graphs by observing that our method runs in polynomial time even for non-constant k (up to $O(\log n)$) provided that a tree-decomposition of graphs with treewidth equal to such a k can be determined in polynomial time. At present, the best tree-decomposition algorithms have the leading term k^k [8] so we can set k to $O(\log n/\log\log n)$ keeping the polynomial time performance of our method. In this way, we obtain the following theorem.

Theorem 4. *The min-bisection problem on planar graphs in which the size of min-bisection is $\Omega(n\log\log n/\log n)$ admits a PTAS.*

Observe that the presence of large degree vertices in a planar graph can cause the large size of min-bisection, e.g., in a star graph. For bounded-degree planar graphs the size of min-bisection is $O(\sqrt{n})$ by the following argument.
For a planar graph of maximum degree d construct a separator tree by applying the planar separator theorem [20] recursively. Next, find a path in the tree from the root down to the median leaf. By deleting the edges incident to the vertex separators along the path and additionally $O(1)$ edges, we can easily halve the set of vertices of the graph such that none of the remaining edges connects a pair of vertices from the opposite halves. The number of deleted edges is clearly $O(d\sqrt{n})$. In fact, we do not have to construct the whole separator tree, but just the path, and this can be easily done in time $O(n\log n)$ [20].

Theorem 5. *For a planar graph on n vertices and maximum degree d, a bisection of size $O(d\sqrt{n})$ can be found in time $O(n\log n)$.*

Clearly, if a graph has an $O(1)$-size bisection, it can be found by exhaustive search in polynomial time. We conclude that at present we have efficient methods for at least $O(1)$-approximation of min-bisection in planar graphs if its size is either $\Omega(n\log\log n/\log n)$ or $O(1)$, or $O(\sqrt{n})$ and the maximum degree is constantly bounded. These observations suggest that a substantial improvement of the logarithmic approximation factor for min-bisection on planar graphs given in [11] might be possible.

5   PTAS for Max-Bisection of a Unit Disk Graph

In this section we design a PTAS for max-bisection of unit disk graphs, another important class of graphs defined by the geometric conditions on a plane.

Recall that an undirected graph G is a unit disk graph if its vertices can be put in one to one correspondence with disks of equal radius in the plane in such a way that two vertices are joined by an edge if and only if the corresponding disks intersect. Tangent disks are considered to intersect. We may assume w.l.o.g that the radius of each disk is one. Since the recognition problem for unit disk graph is NP-hard, we shall also assume that a geometric representation of the graph is given as input.

Our technique works in a similar way as in the case for planar graphs. The input graph G is divided into families of subgraphs $H_{i,j}$ using the ideas of Hunt et al. given in [15]. Next, approximative solution to all $(l, n_{i,j} - l)$-partitions of every subgraph $H_{i,j}$, where $n_{i,j}$ denotes the number of vertices in $H_{i,j}$, are computed by the methods given in [2]. Via a tree-type dynamic programming these solutions are used to obtain an overall solution for G.

In order to divide the graph G, we impose a grid of horizontal and vertical lines on the plane, that are 2 apart of each other. The v-th vertical line, $-\infty < v < \infty$, is at $x = 2v$. The h-th horizontal line, $-\infty < h < \infty$, is at $y = 2h$. We say, that the v-th vertical line has index v and that the h horizontal line has index h. Further we denote the vertical strip between the v-th and the $(v+1)$-th vertical line as the strip with index v and analogue for the horizontal strip between the h-th and the $(h+1)$-th horizontal line.

Each vertical strip is left closed and right open, each horizontal strip is closed at the top and open at the bottom. A disk is said to lie in a given strip if its center lies in that strip. Note that every disk lies in exactly one horizontal and vertical strip.

For a fixed k consider the subgraph $H_{i,j}$ of G, $-\infty < i, j < \infty$, induced by the disks that lie in the intersection of the horizontal strips $i, i+1, \ldots, i+k$ and the vertical strips $j, j+1, \ldots, j+k$. Let $n_{i,j}$ be the number of vertices of $H_{i,j}$. By a packing argument it can be shown that for fixed $k > 0$, the size of a maximum independent set of such a subgraph is at most $2(k+3)^2\pi$.

Lemma 4. *There is a positive constant c such that if $n_{i,j} > c\log n$ then the subgraph $H_{i,j}$ of G is dense.*

Proof. Partition the vertex-set of $H_{i,j}$ successively into maximal independent sets by determining a maximal independent set I_1, remove its vertices and again determine a maximal independent set I_2 and so on. As described above the number of independent sets is at least $n_{i,j}/2(k+3)^2\pi$. Since each I_j is maximal there is at least one edge from a vertex of I_j to every $I_{j'}$, $j < j'$. If we understand the set of independent sets as a complete graph on $n_{i,j}/2(k+3)^2\pi$ vertices it follows that $H_{i,j}$ has $\Omega(n_{i,j}^2)$ edges and hence $H_{i,j}$ is dense. □

Corollary 4. *If $n_{i,j} > c\log n$ then the size of a maximum bisection of $H_{i,j}$ is $\Omega(n_{i,j}^2)$.*

Proof. Partition the vertex-set of $H_{i,j}$ as before and use the maximum independent sets to build up the sets of the bisection. Since all independent sets are maximal there are $\Omega(n_{i,j}^2)$ edges between the sets of bisection. □

Consequently the techniques given in [2] are applicable to the subgraph $H_{i,j}$.

Algorithm 2

input: a unit disk graph $G = (V, E)$ specified by a set V of disks in the plane and the coordinates of their centers and a positive integer k;

output: $(1 - \frac{1}{k+1}^2)(1 - \delta)$-approximations of maximum bisection of G

1. Divide the plane by imposing a grid of width two;
2. Construct the subgraphs $H_{i,j}$ of G as described above;
3. For each i and each j set $n'_{i,j}$ to the number of vertices in $H_{i,j}$ and compute all $(l, n'_{i,j} - l)$-partitions of $H_{i,j}$ either approximatively or optimal if $n'_{i,j} = O(\log n)$;
4. For each r and s, $0 \le r, s \le k$, set $G_{r,s}$ to the union of the subgraphs $H_{i,j}$ where $i \pmod{k+1} = r$ and $j \pmod{k+1} = s$;
5. For each r and s, $0 \le r, s \le k$, set $n_{r,s}$ to the number of vertices in $G_{r,s}$ and compute a bisection of $G_{r,s}$ within $(1 - \delta)$ of its maximum by dynamic programming in a tree fashion. Therefore enumerate the subgraphs in increasing order of the sum $i + j$ and compute all partitions of pairs of "consecutive" $H_{i,j}$ respectively to this ordering on the basis of the computed partitions, then for quadruples of such $H_{i,j}$ etc.;
6. Output the largest bisection of $G_{r,s}$, $0 \le r, s \le k$.

If $n_{i,j} \le c \log n$ we can find all the maximum $(l, n_{i,j} - l)$-partitions of the subgraph $H_{i,j}$ in polynomial time by enumerating all possibilities. Otherwise the problem is solvable approximatively in polynomial time by solving the following polynomial integer program:

$$\text{maximize} \sum_{\{i,j\} \in E(H_{i,j})} x_i(1 - x_j) + x_j(1 - x_i) \tag{1}$$
$$\text{subject to} \qquad \sum x_i = l, \tag{2}$$
$$x_i \in \{0, 1\} \qquad\qquad i = (1, \dots, n_{i,j}) \tag{3}$$

This program can be solved by the use of Theorem 1.10 given in [2] within an error of at most $\epsilon n_{i,j}^2$, which also satisfies the linear constraint (2) of the program within an additive error of $O(\epsilon \sqrt{n_{i,j} \log n_{i,j}})$. In order to get a subset of size l we move at most $\epsilon \sqrt{n_{i,j} \log n_{i,j}}$ in or out. This affects the number of edges included in the partition by at most $\epsilon n_{i,j} \sqrt{n_{i,j} \log n_{i,j}} \le \epsilon n_{i,j}^2$. Hence we can compute a maximum $(l, n_{i,j} - l)$-partition of a subgraph $H_{i,j}$ that has more than $c \log n$ vertices within an additive error of $2\epsilon n_{i,j}^2$ of the maximum.

Lemma 5. *Algorithm 2 outputs a bisection of G within $(1 - \frac{1}{k+1})^2(1 - \delta)$ of the maximum.*

Proof. Let P be a maximum bisection of G. For each edge $e \in P$ and a fixed r, $0 \le r \le k$, there is at most one s, $0 \le s \le k$, such that e crosses a vertical line whose index modulo $k+1$ is s. Analogously, there is for each $e \in P$ and a fixed s, $0 \le s \le k$, at most one r, $0 \le r \le k$, such that e crosses a horizontal line whose index modulo $k + 1$ is r. Consequently there is a pair (r, s), $0 \le r, s \le k$, such that a maximum $(l, n - l)$-partition of $G_{r,s}$ cuts at least $(1 - \frac{1}{k+1})^2|P|$ edges.

By Corollary 4, the size of maximum bisection of the subgraph $G'_{r,s}$ of $G_{r,s}$ that consists of all $H_{i,j}$ with more than $c \log n$ vertices is $\sum_{n_{i,j} > c \log n} \Omega(n_{i,j}^2)$. Consequently, the error caused by the solutions of the polynomial integer programs for the subgraphs $H_{i,j}$ of $G'_{r,s}$ are at most a $\delta = 2\epsilon$ fraction of an optimum solution of maximum bisection for $G'_{r,s}$. Since the partitions for each $H_{i,j}$ with at most $c \log n$ vertices are computed optimally, we obtain a bisection of $G_{r,s}$ within $(1 - \delta)$ of the maximum.

Thus algorithm 2 outputs a bisection of G within $(1 - \frac{1}{k+1})^2 (1 - \delta)$ of the maximum. □

Theorem 6. *The problem of max-bisection on unit disk graphs admits a PTAS.*

The same approach can be used to obtain a PTAS for the maximum bisection problem in geometric intersection graphs both of other regular polygons and also of regular geometric objects in higher dimensions.

Acknowledgments

We thank Andreas Björklund, Hans Bodlaender, Uri Feige, Mark Jerrum, Miroslaw Kowaluk, Mike Langberg, and Monique Laurent for many stimulating remarks and discussions.

References

1. S. Arnborg, Efficient algorithms for combinatorial problems on graphs with bounded decomposability — A survey, *BIT*, 25 (1985), pp. 2 – 23.
2. S. Arora, D. Karger and M. Karpinski. Polynomial Time Approximation Schemes for Dense Instances of NP-hard Problems, *Proceedings 27th ACM Symposium on the Theory of Computing*, pp. 284-293, 1995.
3. A.A. Ageev and M.I. Sviridenko. Approximation algorithms for Maximum Coverage and Max Cut with cardinality constraints. *Proceedings of the Conference of Integer Programming and Combinatorial Optimization 99*, LNCS 1610, pp. 17-30, 1999.
4. B.S. Baker. Approximation algorithms for NP-complete problems on planar graphs. *Proceedings of the 24th IEEE Foundation of Computer Science*, 1983, pp. 265-273.
5. H.L. Bodlaender, A tourist guide through treewidth. *Acta Cybernetica*, 11 (1993), pp. 1 – 23.
6. H.L. Bodlaender, A linear time algorithm for finding tree-decompositions of small treewidth. *SIAM Journal on Computing*, 25 (1996), pp. 1305 – 1317.
7. H.L. Bodlaender, A partial k-arboretum of graphs with bounded treewidth. Available at `http://www.cs.ruu.nl/~hansb/index.html`.
8. H.L. Bodlaender, Personal communication, August, 2000.
9. H.L. Bodlaender and K. Jansen. On the complexity of the Maximum Cut problem. *Nordic Journal of Computing*, 7(2000), pp. 14-31, 2000.
10. U. Feige, M. Karpinski and M. Langberg. A Note on Approximating MAX-BISECTION on Regular Graphs. ECCC (`http://www.eccc.uni-trier.de/eccc/`), TR00-043 (2000).
11. U. Feige and R. Krauthgamer. A polylogarithmic approximation of the minimum bisection. To appear in *Proceedings of the Foundation of Computer Science 2000*.
12. A. Frieze and M. Jerrum. Improved approximation algorithms for MAX k-CUT and MAX BISECTION. *Algorithmica 18*, pp. 67-81, 1997.

13. M.X. Goemans and D.P. Williamson. Improved approximation algorithms for maximum cut and satisfiability problems using semidefinite programming. *Journal of ACM*, 42, pp. 1115-1145, 1995.

14. F. Hadlock. Finding a maximum cut of a planar graph in polynomial time. *SIAM Journal on Computing* 4(1975), pp. 221-225.

15. H.B. Hunt, M.V. Marathe, V. Radhakrishnan, S.S. Ravi, D.S. Rosenkrantz, R.E. Stearns. NC-approximation schemes for NP- and PSPACE-hard problems for geometric graphs. *Proceedings 2nd Annual European Symposium on Algorithms, (ESA)*, LNCS 855, pp. 468-477, Springer Verlag, June, 1994

16. E. Halperin and U. Zwick, Improved approximation algorithms for maximum graph bisection problems, Manuscript, 2000.

17. M. Jerrum, Personal communication, August, 2000.

18. M. Karpinski, M. Kowaluk and A. Lingas. Approximation Algorithms for Max-Bisection on Low Degree Regular Graphs and Planar Graphs. ECCC (http://www.eccc.uni-trier.de/eccc/), TR00-051 (2000).

19. S. Khanna and R. Motwani. Towards a Syntactic Characterization of PTAS. *Proceedings of the 28th ACM Symposium on the Theory of Computing*, 1996, pp. 329-337.

20. R.J. Lipton and R.E. Tarjan. A separator theorem for planar graphs. *SIAM Journal of Applied Mathematics*, 36 (1979), pp. 177-189.

21. N. Robertson and P.D. Seymour, Graph minors. II. Algorithmic aspects of treewidth, *Journal of Algorithms*, 7 (1986), pp. 309-322.

22. Y. Ye, A O.699 - approximation algorithm for Max-Bisection, Submitted to Mathematical Programming, available at URL http://dollar.biz.uiowa.edu/col/ye, 1999

Refining the Hierarchy of Blind Multicounter Languages

Matthias Jantzen and Alexy Kurganskyy

Universität Hamburg, FB Informatik
Vogt-Kölln-Straße 30, 22527 Hamburg
jantzen@informatik.uni-hamburg.de

Abstract. We show that the families (k, r)–RBC of languages accepted (in quasi-realtime) by one-way counter automata having k blind counters of which r are reversal-bounded form a strict and linear hierarchy of semi-AFLs. This hierarchy comprises the families $BLIND = \mathcal{M}_\cap(C_1)$ of blind multicounter languages with generator $C_1 := \{w \in \{a_1, b_1\}^* \mid |w|_{a_1} = |w|_{b_1}\}$ and $RBC = \mathcal{M}_\cap(B_1)$ of reversal-bounded multicounter languages with generator $B_1 := \{a_1^n b_1^n \mid n \in I\!\!N\}$. This generalizes and sharpens the known results from [Grei 78] and [Jant 98].

1 Introduction

Hierarchies of counter automata are often proved by arguments concerning the dimension of the memory space, i.e., the number of counters, see for example [FiMR 68,Grei 76,Grei 78] or counting cycles within the computations, as in [Hrom 86]. If one does not alter the dimension, and changes only the strategy of accessing the counters, other methods have to be found. The method applied here for the first time uses techniques from linear algebra and shows that the formerly known two hierarchies of blind and of reversal-bounded multicounter languages are in fact part of one linear hierarchy of semi-AFLs.

The family of languages accepted by one-way reversal-bounded multicounter automata (in quasi-realtime) is a well known semi-AFL which is principal as an *intersection*-closed semi-AFL $\mathcal{M}_\cap(B_1)$ with generator $B_1 := \{a_1^n b_1^n \mid n \in I\!\!N\}$, which is not a principal semi-AFL, see [FiMR 68,Grei 78].

The known situation for these hierarchies, shown in [Grei 78], is as follows: $\bigcup_{i \geq 1} \mathcal{M}(C_i) = \mathcal{M}_\cap(C_1) = BLIND = \bigcup_{i \geq 1} \mathcal{M}(B_i) = \mathcal{M}_\cap(B_1) = RBC$, where $\mathcal{M}(\mathcal{L})$ denotes the least trio generated by the family $\mathcal{L}$, which is a semi-AFL if $\mathcal{L} = \{L\}$ and then we write $\mathcal{M}(L)$ instead of $\mathcal{M}(\mathcal{L})$. For all $i \geq 1$ we have $\mathcal{M}(B_i) \subsetneq \mathcal{M}(B_{i+1})$, see [Gins 75], and $\mathcal{M}(C_i) \subsetneq \mathcal{M}(C_{i+1})$, shown in [Grei 76,Grei 78]. (For the definition of the languages B_i and C_i see Definition 2.1 below).

We study the families (k, r)–RBC of languages accepted (in quasi-realtime) by one-way (or on-line) counter automata having k blind counters of which $r \leq k$ are reversal-bounded and prove (k_1, r_1)–$RBC \subsetneq (k_2, r_2)$–$RBC$ if and only if $k_1 < k_2$ or $k_1 = k_2$ and $r_1 > r_2$. Then $(k, 0)$–$RBC = \mathcal{M}(C_k)$, and $\bigcup_{k \geq 1} (k, 0)$–$RBC = \bigcup_{i \geq 1} \mathcal{M}(C_i) = \mathcal{M}_\cap(C_1)$ forms a hierarchy of *twist*-closed semi-AFLs (see [Jant 98]). The strict inclusions are proved here for the first time.

A. Ferreira and H. Reichel (Eds.): STACS 2001, LNCS 2010, pp. 376–387, 2001.

2   Basic Definitions

Definition 1. *For any alphabet Σ and $x \in \Sigma$ let $|w|_x$ denote the number of occurrences of the symbol x within the string $w \in \Sigma^*$, $|w| := \sum_{x \in \Sigma} |w|_x$, and $\psi : \Sigma^* \longrightarrow I\!N^n$ is the Parikh mapping, defined by $\psi(w) := (|w|_{x_1}, \ldots, |w|_{x_n})$, where $n := |\Sigma|$. The empty word is denoted by λ and $\psi(\lambda) = \mathbf{0} \in I\!N^n$ is the vector, all of whose coordinates are 0.*

The languages we use here are constructed using the specific alphabet Γ_n specified for each $n \in I\!N, n \geq 1$ by: $\Gamma_n := \{a_i, b_i \mid 1 \leq i \leq n\}$, and the homomorphisms h_i defined for $i \geq 1$ by: $h_i(x) := \begin{cases} x, & \text{if } x \in \{a_i, b_i\} \\ \lambda, & \text{else} \end{cases}$.

$$C_n := \left\{ w \in \Gamma_n^* \mid \forall 1 \leq i \leq n : |w|_{a_i} = |w|_{b_i} \right\}$$

$$B_n := \left\{ w \in C_n^* \mid \forall 1 \leq i \leq n : h_i(w) = a_i^m b_i^m, \text{ for some } m \in I\!N \right\}$$

$$D_n := \left\{ w \in \Gamma_n^* \mid \forall 1 \leq i \leq n : \left(|w|_{a_i} = |w|_{b_i} \wedge \forall w = uv : |u|_{a_i} \geq |u|_{b_i} \right) \right\}$$

The language D_1 defined above is the so-called semi-Dyck language on one pair of brackets which is often abbreviated by D'^*_1, see e.g. [Bers 80]. D_n here denotes the n-fold shuffle of disjoint copies of the semi-Dyck language D_1 and it is known, [Grei 78,Jant 79], that $\bigcup_{i \geq 1} \mathcal{M}(D_i) = \mathcal{M}_\cap(D_1) = PBLIND(n)$. The latter family consists of languages accepted in quasi-realtime by nondeterministic one-way multicounter acceptors which operate in such a way that in every computation no counter can store a negative value, and the information on whether or not the value stored in a counter is *zero* is not used for deciding the next move.

The languages C_n are the (symmetric) Dyck languages on n pairs of brackets a_i, b_i, often abbreviated by D_n^*, see again [Bers 80]. Greibach, [Grei 78], has shown that $\bigcup_{i \geq 1} \mathcal{M}(C_i) = \mathcal{M}_\cap(C_1) = BLIND = BLIND(lin) = BLIND(n) = \bigcup_{i \geq 1} \mathcal{M}(B_i) = \mathcal{M}_\cap(B_1) = RBC(n) = RBC \subsetneqq PBLIND$.

Here $BLIND$ ($BLIND(n)$, $BLIND(lin)$) denotes the family of languages accepted (in quasi-realtime, linear time, resp.) by nondeterministic one-way multicounter acceptors which operate in such a way that in every computation all counters may store arbitrary integers, and the information on the contents of the counters is not used for deciding the next move. The family RBC is the family of languages accepted by nondeterministic one-way multicounter acceptors performing at most one reversal in each computation. The formal definition is to be found in Section 3.

3   Blind k Counter Automata with $r \leq k$ Reversal-Bounded Counters.

We shall deal only with counter-automata that have a one-way read-only input tape (also known as on-line automata) and have k-blind counters of which precisely r counters are reversal-bounded.

Definition 2. *A blind k-counter automaton $M := (Q, \Sigma, \delta, q_0, Q_{fin})$ consists of a finite set of states Q, a designated initial state $q_0 \in Q$, a designated set of final states $Q_{fin} \subseteq Q$, a finite input alphabet Σ, and a transition function $\delta : Q \times (\Sigma \cup \{\lambda\}) \to 2^{Q \times \{+1, 0, -1\}^k}$.*

An instantaneous description (ID) of M is an element of $Q \times \Sigma^* \times \mathbb{Z}^k$. We write $(q_1, aw, z_1, \ldots, z_k) \vdash_M (q_2, w, z_1 + \Delta(1), \ldots, z_k + \Delta(k))$ if $(q_2, \Delta) \in \delta(q_1, a)$ where $(\Delta(1), \ldots, \Delta(k)) = \Delta'$ is the transpose of vector Δ and we omit the subscript M if no confusion will arise. $\vdash_M^*$ denotes the reflexive transitive closure of the computation relation $\vdash_M$ and is defined as usual from the n-step computation relations $\vdash_M^n := \vdash_M^{n-1} \circ \vdash_M$ by $\vdash_M^* := \bigcup_{i \geq 0} \vdash_M^i$, where $\vdash_M^0$ is the identity relation on the ID's of the nondeterministic automaton M.

$ID_i \vdash_M^* ID_j$ is an accepting computation for w iff $ID_i := (q_0, w, 0, \ldots, 0))$ and $\exists q_e \in Q_{fin}$ such that $ID_j := (q_e, \lambda, 0, \ldots, 0))$.

$L(M) := \{w \in \Sigma^* \mid M$ has an accepting computation for $w\}$ is the language accepted by M.

A specific k-counter automaton M can most easily be described by a finite state transition diagram in which a directed arc from state q_1 to q_2 is inscribed by the input symbol x to be processed and a vector $\Delta \in \{+1, 0, -1\}^k$ used for updating the counters by adding the component $\Delta(i)$ of Δ to the current contents z_i of the i-th counter. This will be written as $q_1 \xrightarrow[\Delta]{x} q_2$.

Definition 3. *A blind k-counter automaton $M := (Q, \Sigma, \delta_M, q_0, Q_{fin})$ accepts $L(M)$ in in linear time with factor $d \in \mathbb{N}$, if for any $w \in L(M)$ there exists an accepting n-step computation $ID_0 \vdash_M^n ID_1$ for w such that $n \leq d \cdot max(|w|, 1)$.*

If there exists $d \in \mathbb{N}$ such that $(q_1, \lambda, z_1, \ldots, z_k) \vdash_M^n (q_2, \lambda, z_1', \ldots, z_k')$ implies $n \leq d$, then the automaton M is said to work in quasi-realtime of delay d. If in this case $d = 0$ then M works in realtime.

The i-th counter $(1 \leq i \leq k)$ of some blind k-counter automaton M is reversal-bounded iff for any subcomputation $(q_0, w, 0, \ldots, 0) \vdash_M^* (q_1, w_1, x_1, \ldots, x_k)$ $\vdash_M^* (q_2, w_2, y_1, \ldots, y_k) \vdash_M^* (q_3, w_3, z_1, \ldots, z_k)$ $x_i > y_i$ implies $y_i \geq z_i$.

By this definition, a reversal-bounded counter has to be increased first and decreased after its reversal. Counters that are first decreased and solely increased after one reversal can be replaced by those required by Definition 3 above. In addition, reversal bounded counters are forced by the finite control to perform at most one reversal on each computation, even in the non-accepting ones!

Definition 4. *For all $k, r \in \mathbb{N}$ let (k, r)–RBC denote the family of languages accepted by (k, r)-counter automata, i.e., are accepted by on-line counter automata having k blind counters of which r are reversal-bounded.*

Obviously we have $\mathcal{M}(C_k) = (k, 0)$–RBC and $\mathcal{M}(B_k) = (k, k)$–RBC.

Definition 5. *$L_{k,r} := \{w \in \Gamma_k^* \mid \forall_{1 \leq i \leq r} : h_i(w) = a_i^m b_i^m$ for some $m \in \mathbb{N} \wedge \forall_{r+1 \leq i \leq k} : |h_i(w)|_{a_i} = |h_i(w)|_{b_i}\}$.*

By results from Ginsburg and Greibach ([GiGr 70], Corr. 3, and [Gins 75], Prop. 3.6.1,) one can deduce that the language $L_{k,r}$ is a generator of the family (k,r)–RBC. We do not give a detailed explanation using these standard techniques and state Lemma 1 without proof:

Lemma 1. (k,r)–$RBC = \mathcal{M}(L_{k,r})$.

Greibach showed $C_1 \in \mathcal{M}(B_3)$, (Lemma 1 in [Grei 78]). It was shown in [Jant 98] that it is sufficient to accept C_k using only k+1 reversal-bounded counters, which is stated in Lemma 2.

Lemma 2. $\forall k \in \mathbb{N}, k \geq 1 : \mathcal{M}(C_k) \subsetneqq \mathcal{M}(B_{k+1})$.

Ginsburg ([Gins 75] Example 4.5.2) has shown $\mathcal{M}(B_k) \subsetneqq \mathcal{M}(B_{k+1})$. And $\mathcal{M}(C_i) \subsetneqq \mathcal{M}(C_{i+1})$ has been shown in [Grei 76], [Grei 78].

We will obtain the sharpening of the above results by proving Lemma 6 and the main result Theorem 1.

For the formulation and usage of techniques from linear algebra to prove these results we need some more notation that in most cases applies only to those (k,r)-counter automata which accept languages from Γ_k^*.

Definition 6. *For any (k,r)-counter automaton $A := (Q, \Sigma, \delta_A, q_0, Q_{\mathit{fin}})$ let $G_A \subseteq Q \times \Sigma \times \{+1, 0, -1\}^k \times Q$ be the finite set defined by $G_A := \{(p, x, \Delta, q) \mid (q, \Delta) \in \delta_A(p, x)\}$, which is in bijection with the arcs of A's state diagram. For later use let $n_A := |G_A|$ be the number of elements in the arbitrarily but fixed ordered set $G_A = \{g_1, g_2, \ldots, g_{n_A}\}$. (The ordering that is actually used depends on $L(A)$ and will be described later.)*

The four projections $\pi_i, 1 \leq i \leq 4$, $\pi_1, \pi_4 : G_A \to Q$, $\pi_2 : G_A \to \Sigma \cup \{\lambda\}$, and $\pi_3 : G_A \to \{+1, 0, -1\}^k$, are defined by: $\pi_1((p, x, \Delta, q)) := p$, $\pi_2((p, x, \Delta, q)) := x$, $\pi_3((p, x, \Delta, q)) := \Delta$, $\pi_4((p, x, \Delta, q)) := q$.

The mappings π_1 and π_4 are mere coding, whereas π_2 and π_3 are canonically extended to homomorphisms, by mild abuse of notation: For all strings $u, v \in G_A^*$ let $\pi_2 : G_A^* \to \Sigma^*$ with $\pi_2(uv) = \pi_2(u)\pi_2(v)$ and $\pi_3 : G_A^* \to \mathbb{Z}^k$ with $\pi_3(uv) = \pi_3(u) + \pi_3(v)$, where $+$ is the componentwise addition of the vectors $\pi_3(u)$ and $\pi_3(v)$. For an easier readability let $\Delta_g := \pi_3(g)$ denote the counter update induced by the transition $g \in G_A$ of A.

Let $R_A := \{g_{i_0} g_{i_1} \cdots g_{i_t} \mid t \in \mathbb{N} \wedge \forall \mu \in \{0, \ldots t\} : (g_{i_\mu} \in G_A) \wedge (\pi_1(g_{i_0}) = q_0) \wedge (\pi_4(g_{i_t}) \in Q_{\mathit{fin}}) \wedge (\pi_4(g_{i_\mu}) = \pi_1(g_{i_{\mu+1}})$ for $\mu \neq t)\} \subseteq G_A^*$ be the regular set describing all the accepting paths in A's state diagram, interpreted as finite automaton with input alphabet G.

Of course, $w \in R_A$ does not imply that $\pi_2(w)$ will be accepted by the counter automaton A, since the final counter values may not be equal to *zero*. Note, that the number of reversals of the reversal-bounded counters are handled by the finite control and can never be wrong.

On the basis of a (k,r)-counter automaton $A := (Q, \Gamma_k, \delta_A, q_0, Q_{\mathit{fin}})$ two matrices A_Δ and A_Γ are defined.

Definition 7. *$A_\Delta \in \mathbb{Z}^{k \times n_A}$ is defined for each component, $1 \leq i \leq k$, $1 \leq j \leq n_A$ by:*

$$A_\Delta(i,j) := \Delta_{g_j}(i).$$

Hence A_Δ can be written as composite matrix as follows:

$$A_\Delta = \left(\Delta_{g_1} \Delta_{g_2} \cdots \Delta_{g_{n_A}} \right)$$

With the notation from Definition 6 we see that $A_\Delta \cdot \psi(v) = \pi_3(v)$ for each $v \in G_A^*$ and the following is a consequence of the definition of acceptance for (k,r)-counter automata:

Lemma 3. *Let $A := (Q, \Sigma, \delta_A, q_0, Q_{fin})$ be some (k,r)-counter automaton then*

$$\forall v \in R_A : \quad A_\Delta \cdot \psi(v) = \mathbf{0} \quad \textit{iff} \quad \pi_2(v) \in L(A).$$

Proof: $v \in R_A$ ensures that there exists a path in the state diagram of A beginning in q_0 and ending in some final state of Q_{fin}. If in addition $\pi_3(v) = A_\Delta \cdot \psi(v) = 0$, then $\pi_2(v) \in L(A)$. Conversely, for any $w \in L(A)$ there exists an accepting path in A having a corresponding string $v' \in R_A$ with $w = \pi_2(v')$. Since a (k,r)-counter automaton accepts if the k-counters are empty at the beginning and at the end, it follows that $\pi_3(v') = A_\Delta \cdot \psi(v') = \mathbf{0}$. □

Definition 8. *For each (k,r)-counter automaton $A := (Q, \Gamma_k, \delta_A, q_0, Q_{fin})$ the following matrix $A_\Gamma \in \{+1, 0, -1\}^{k \times n_A}$ is defined for each component $A_\Gamma(i,j)$, $1 \le i \le k$, $1 \le j \le n_A$, by:*

$$A_\Gamma(i,j) := \begin{cases} 1 & , \text{ if } \pi_2(g_j) = a_i \\ -1 & , \text{ if } \pi_2(g_j) = b_i \\ 0 & , \text{ if } \pi_2(g_j) \notin \{a_i, b_i\}. \end{cases}$$

Without loss of generality the ordering of the elements in G_A is such, that

$$A_\Gamma = \begin{pmatrix} 1 & \cdots & 1 & -1 & \cdots & -1 & \cdots & 0 & \cdots & 0 & 0 & \cdots & 0 & 0 & \cdots & 0 \\ 0 & \cdots & 0 & 0 & \cdots & 0 & \cdots & \vdots & \ddots & \vdots & \vdots & \ddots & \vdots & \vdots & \ddots & \vdots \\ \vdots & \ddots & \vdots & \vdots & \ddots & \vdots & \cdots & 0 & \cdots & 0 & 0 & \cdots & 0 & \vdots & \ddots & \vdots \\ 0 & \cdots & 0 & 0 & \cdots & 0 & \cdots & 1 & \cdots & 1 & -1 & \cdots & -1 & 0 & \cdots & 0 \end{pmatrix}.$$

$$= \left(\gamma_1 \gamma_2 \cdots \gamma_{n_A} \right),$$

where γ_j denotes the j-th column $A_\Gamma(:, j)$ of A_Γ.
The next fact is obvious from the definitions and formulated without proof:

Lemma 4. *Let $A := (Q, \Gamma_k, \delta_A, q_0, Q_{fin})$ be some (k,r)-counter automaton then*

$$\forall v \in G_A^* : \quad A_\Gamma \cdot \psi(v) = \mathbf{0} \quad \textit{iff} \quad \pi_2(v) \in C_k.$$

We combine the preceding Lemmas (3 and 4) to get an equality which is independent from the number of reversal bounded counters but, through R_A, not independent of the language accepted:

Lemma 5. *Let $A := (Q, \Gamma_k, \delta_A, q_0, Q_{fin})$ be some (k, r)-counter automaton accepting $L(A) \subseteq C_k$, and let $\begin{pmatrix} A_\Delta \\ A_\Gamma \end{pmatrix}$ denote the compound matrix of dimension $2k \times n_A$ then*

$$\{v \in R_A \mid A_\Delta \cdot \psi(v) = \mathbf{0}\} = \{v \in R_A \mid \begin{pmatrix} A_\Delta \\ A_\Gamma \end{pmatrix} \cdot \psi(v) = \mathbf{0}\}.$$

Definition 9.

$$B_{k,r} := \{a_1^{i_1} b_1^{i_1} \cdots a_r^{i_r} b_r^{i_r} a_{r+1}^{i_{r+1}} b_{r+1}^{i_{r+1}+j_{r+1}} a_{r+1}^{j_{r+1}} \cdots a_k^{i_k} b_k^{i_k+j_k} a_k^{j_k} \mid \forall \mu : i_\mu, j_\mu \in I\!N\}.$$

Lemma 6. $(k, r+1)\text{-}RBC \neq (k, r)\text{-}RBC$ *for all $k \in I\!N$ and $0 \leq r < k$.*

We will in fact prove $B_{k,r} \notin (k, r+1)\text{-}RBC$, where the subset $B_{k,r} \subsetneq L_{k,r}$ is defined above (Def. 9). We will see, that the equation of Lemma 5 cannot be satisfied if $B_{k,r}$ is accepted by using r+1 reversal bounded counters. That this suffices is obvious, since $B_{k,r}$ is obtained from $L_{k,r}$ by intersection with an appropriate bounded regular set, hence $B_{k,r} \in (k, r)\text{-}RBC$ is easily seen.

The proof of Lemma 6 is quite involved and needs a lot of definitions first. For the sake of contradiction, let us assume $B_{k,r} \in (k, r+1)\text{-}RBC$ and let $A := (S_A, \Gamma_k, \delta_A, q_0, Q_{fin})$ be a blind k-counter automaton having $r+1$ reversal bounded counters that accepts $B_{k,r} = L(A)$. Without loss of generality, we assume that the first $r+1$ counters are reversal-bounded.

Definition 10. *For each $k \in I\!N, k \neq 0$ and each $l, 1 \leq l < k$ let $\xi_k^l \subseteq G_A \times G_A$ be defined by:*

$$(g, g') \in \xi_k^l \text{ iff } \begin{cases} \exists x \in \Gamma_k : \pi_2(g), \pi_2(g') \in \{x, \lambda\}, & and \\ \forall 1 \leq j \leq l : \Delta_g(j) > 0 \Longrightarrow \Delta_{g'}(j) \geq 0, \\ \qquad\qquad \Delta_{g'}(j) > 0 \Longrightarrow \Delta_g(j) \geq 0. \end{cases}$$

$(g, g') \in \xi_k^l$ means that the counter automaton A does not read two different symbols from the input by using g and g', if any at all, and these arcs do not force a reversal on any of the counters with index less or equal to l. The remaining counters with index strictly larger than l do not have any restriction on their updating. The relation ξ_k^l is obviously symmetric and reflexive but not necessarily transitive. So we can only find subsets of $C \subseteq G_A \times G_A$ which are transitively closed. Any such set will be called a ξ_k^l-clique.

Within the set R_A we identify a certain non-regular subset K_1 to be used for the proof of Lemma 7 below.

Definition 11. *The set*

$$K_0 := \{w \in R_A \mid \exists i \in I\!N : \pi_2(w) = a_1^i b_1^i a_2^i b_2^i \cdots a_r^i b_r^i a_{r+1}^i b_{r+1}^{2i} a_{r+1}^i \cdots a_k^i b_k^{2i} a_k^i\}$$

is a non-regular subset of R_A of which we select the set $K_1 \subseteq K_0 \subsetneq R_A$ where no two different strings have an identical π_2-projection:

$$K_1 := \{w \in K_0 \mid \forall w' \in K_0 : \pi_2(w) = \pi_2(w') \text{ implies } w = w'\}.$$

By w_i we denote the unique string in K_1, for which

$$\pi_2(w_i) = a_1^i b_1^i a_2^i b_2^i \cdots a_r^i b_r^i a_{r+1}^i b_{r+1}^{2i} a_{r+1}^i \cdots a_k^i b_k^{2i} a_k^i.$$

For a step-by-step definition of a specific non-regular subset of R_A which contains infinitely many strings from the set K_1 we use the property $p_{\xi_k^{r+1}}$ to specify certain strings within the set K_1.

Definition 12. *$p_{\xi_k^{r+1}} : G_A^* \times I\!N \to \{true, false\}$ is defined by:*

$p_{\xi_k^{r+1}}(w, p) = true \quad iff \quad \exists u_1, \ldots, u_p \in G_A^*:$

1. $w = u_1 u_2 \cdots u_p$ *and*
2. $\forall j, 1 \le j \le p : \forall g, g' \in G_A : g, g' \sqsubseteq u_j \Longrightarrow (g, g') \in \xi_k^{r+1}$
3. $\forall j, 1 \le j < p : \exists g, g' \in G_A : g \sqsubseteq u_j \wedge g' \sqsubseteq u_{j+1} \wedge (g, g') \notin \xi_k^{r+1}$

For each $u \in G_A^*$ let $G(u) := \{g \in G_A \mid g \sqsubseteq u\}$, where $\sqsubseteq$ denotes the substring relation.

Here, $G(u_j)$ forms a (maximal) ξ_k^{r+1}-clique for each u_j of the decomposition $w = u_1 u_2 \cdots u_p$. If two arcs $g, g' \in G_A$ are in the same ξ_k^{r+1}-clique, then there exists $x \in \Gamma_k$ such that $\pi_2(g), \pi_2(g') \in \{\lambda, x\}$ and their π_3-projections do not lead to a reversal on one of the first $r + 1$ counters. The change between two ξ_k^{r+1}-cliques can thus be forced either by changing the symbols ($\neq \lambda$) of the π_2-projections or by performing a reversal on one of the first $r + 1$ counters.

For each $w_i \in K_1$ we have $p_{\xi_k^{r+1}}(w_i, p) = true$ implies $p \le 3k + 1$. This is seen as follows: There exist at most $2r + 3(k - r)$ different ξ_k^{r+1}-cliques with a component from Γ_k, since there are at most that many different blocks of consecutive identical symbols. Because $(g, g') \in \xi_k^{r+1}$ also allows $\pi_2(g) = \pi_2(g') = \lambda$, some of these arcs may fall into the neighboring ξ_k^{r+1}-clique, as long as these arcs do not force a reversal on one of the first $r + 1$ counters. At most $r + 1$ reversals may fall into the $2r + 3(k - r)$ different blocks, which allows for $r + 1$ additional substrings in the decomposition of $w_i = u_1 u_2 \cdots u_p$ and $p \le 2r + 3(k - r) + r + 1 = 3k + 1$.

Since k is a constant and K_1 is infinite, there exists some $p \le 3k + 1$ such that infinitely many strings $w \in K_1$ satisfy $p_{\xi_k^{r+1}}(w, p) = true$. This gives rise to the subset $K_2 \subseteq K_1$ defined next.

Definition 13. *Let $p \le 3k + 1$ be fixed and such that $K_2 := \{w \in K_1 \mid p_{\xi_k^{r+1}}(w, p) = true\}$ is infinite. Let $\#(K_2) := \{i \in I\!N \mid w_i \in K_2\}$ denote the index set for the strings in K_2.*

Since G_A is finite there exists a fixed string $w_g := g_{l,1} g_{l,2} \cdots g_{l,p} \in G_A^*$ where $g_{l,j}$ is the leftmost symbol of u_j for each $1 \le j \le p$ in the decomposition of $w = u_1 u_2 \cdots u_p$ for infinitely many strings $w \in K_2$. These strings are collected in the set $K_3 \subseteq K_2$:

Definition 14. *Let $w_g = g_{l,1} g_{l,2} \cdots g_{l,p} \in G_A^*$ be fixed and such, that $K_3 := K_2 \cap \{g_{l,1}\} G(u_1)^* \{g_{l,2}\} G(u_2)^* \cdots \{g_{l,p}\} G(u_p)^*$, is infinite. Moreover, $\#(K_3) := \{i \in I\!N \mid w_i \in K_3\}$ denotes the index set for the strings in K_3.*

The set $K_3 \subseteq K_2 \subseteq K_1$ is not regular but we shall find an infinite regular set $L \subseteq \{g_{l,1}\}G(u_1)^*\{g_{l,2}\}G(u_2)^* \cdots \{g_{l,p}\}G(u_p)^*$ such that $K_3 \subsetneqq L \subsetneqq R_A$.

For each j, $1 \leq j \leq p$, let L_j be the regular set accepted by the finite Automaton $A_j := (Q_j, G(u_j), \delta_j, \pi_1(g_{l,j}), Q_{j,fin})$, where

1. $Q_j := \{\pi_1(g), \pi_4(g) \mid g \in G(u_j)\}$,
2. $\delta_j : Q_j \times G(u_j) \to Q_j$ is given by $\delta_j(\pi_1(g), g) := \pi_4(g)$,
3. $Q_{j,fin} := \pi_4(g')$, where g' is the rightmost symbol of u_j.

Since each accepting path in the automaton A_j is a part of an accepting path in $A's$ state diagram, we see that $K_3 \subseteq L \subseteq R_A$ for $L := L_1 L_2 \cdots L_p$. Moreover, at least $3k - r$ languages among the $L_1, L_2, \ldots . L_p$ must be infinite, since the projection of the elements of K_3 onto the elements of Γ_k are infinite for each of the $2r + 3(k - r)$ blocks of identical symbols. Since L is regular, the Parikh-image $\psi(L) = \sum_{1 \leq j \leq p} \psi(L_j)$ is a semilinear set and infinite, too. The sum is understood elementwise for the p semilinear sets $\psi(L_j)$. Each linear subset of $\psi(L_j)$ has a representation of the form:

$$\{C_j + P_j Y \mid Y \in I\!N^{h_j}\} \text{ for some } h_j \geq 1, C_j \in I\!N^{n_A}, \text{ and } P_j \in I\!N^{n_a \times h_j}.$$

With these preliminaries we can formulate and prove the following important result:

Lemma 7. *There exists an infinite set $K \subseteq R_A$ such that a) to c) hold:*

a) $\psi(K) = \{C + PY \mid Y \in I\!N^h\}$ for some $h \in I\!N$, $C \in I\!N^{n_A}$, and $P \in I\!N^{n_A \times h}$,

b) If $P(s, j) \cdot P(t, j) \neq 0$ for $1 \leq j \leq h$, $1 \leq s, t \leq n_A$ then $(g_s, g_t) \in \xi_k^{r+1}$,

c) $\forall n_0 \in I\!N : \exists Y_0 \in I\!N^h : (\forall j : 1 \leq j \leq h \wedge Y_0(j) > n_0) \wedge C + PY_0 \in \psi(K_1)$.

Proof: By definition of the finite automaton A_j the matrix P_j satisfies b) of Lemma 7. Given $L := L_1 L_2 \cdots L_p$ we choose for each L_j a linear subset $\{C_j + P_j Y \mid Y \in I\!N_j^h\} \subseteq \psi(L_j)$ which should be infinite whenever L_j is infinite. The set $S := \{C_S + P_S Y \mid Y \in I\!N_S^h\}$ defined by $C_S := \sum_{j=1}^p C_j$, $h_S := \sum_{j=1}^p h_j$, and the compound matrix $P_S := \left(P_1 P_2 \cdots P_p\right) \in I\!N^{n_A \times h_S}$ is linear and infinite, too. The matrix P_S satisfies property b) of Lemma 7, since each submatrix P_j fulfilled this property. Now, $K := L \cap \psi^{-1}(S)$ is an infinite subset of L containing infinitely many elements from $K_3 \subseteq K_1$, thus satisfying properties a) (by $\psi(L') = S$) and b) of Lemma 7. Of course we had to choose the appropriate linear subsets of each L_j to see that $\psi^{-1}(S)$ contains infinitely many elements of $K_3 \subsetneqq L = L_1 L_2 \cdots L_p$. Now we modify the matrix P_S by omitting certain columns to obtain a matrix P that also satisfies c) of the Lemma. First, $K \cap K_3 \subseteq R_A$ is infinite, so that there exists an infinite set $M \subseteq I\!N^{h_S}$ such that $\psi(K \cap K_3) = \{C_S + P_S Y \mid Y \in M\}$. Let $m_0, m_1, \ldots, m_i, \ldots$ be any enumeration of the elements of $M = \{m_i \mid i \in I\!N\}$. Then there exists a subset $M' = \{m_{i_j} \mid \forall j \in I\!N : i_j \in I\!N \wedge m_{i_j} \in M \wedge i_j < i_{j+1}\} \subseteq M$ such that for each j, $1 \leq j \leq h_S$:

either $m_{i_1}(j) = m_{i_2}(j)$ for all $i_1, i_2 \in \mathbb{N}$,

or $m_{i_1}(j) < m_{i_2}(j)$ for all $i_1 < i_2$.

This result is a variant of Dickson's Lemma and can be proved easily.

From M' we deduce the following index sets and constants:

$$I_{le} := \{j \mid 1 \leq j \leq h_S, \forall l \geq 1 : m_{i_l}(j) < m_{i_{l+1}}(j)\},$$

$$I_{eq} := \{j \mid 1 \leq j \leq h_S, \forall l \geq 1 : m_{i_l}(j) = m_{i_{l+1}}(j)\}, \text{ and}$$

$$c_j := m_{i_1}(j) \text{ for each } j \in I_{eq}.$$

Now,

$$\psi(K \cap K_3) = \{C_S + P_S Y \mid Y \in M\} \supseteq \{C_S + \sum_{j \in I_{eq}} P_S(:, j)c_j + PY \mid Y \in M''\},$$

where $P \in \mathbb{N}^{n_A \times h}$ is obtained from P_S by omitting the columns $P_S(:, j)$ having index $j \in I_{eq}$, $h := h_S - |I_{eq}|$, and M'' is obtained from M' by omitting all components j, where $j \in I_{eq}$. Thereby, $M'' \subsetneq \mathbb{N}^h$ is a set which can be linearly ordered by $<$ and this relation applies to all components of its elements. Thus, also property c) of Lemma 7 is satisfied, and the proof is finished. $\square$

Lemma 8.

$$rank\left(\begin{pmatrix} A_\Delta \\ A_\Gamma \end{pmatrix} \cdot P\right) \quad > \quad rank\,(A_\Delta \cdot P).$$

Proof: Let $\begin{pmatrix} Y_1 & Y_2 & \cdots & Y_h \\ Z_1 & Z_2 & \cdots & Z_h \end{pmatrix} = \begin{pmatrix} A_\Delta \\ A_\Gamma \end{pmatrix} \cdot P$, where for each $1 \leq j \leq h$, the

columns $Y_j := (A_\Delta \cdot P)(:, j)$ of $A_\Delta \cdot P$ are given by $Y_j(l) := \sum_{i=1}^{n_A} A_\Delta(l, i)P(i, j) :=$

$\sum_{i=1}^{n_A} A_{g_i}(l) \cdot P(i, j)$ for $1 \leq l \leq k$ and likewise $Z_j := (A_\Gamma \cdot P)(:, j)$ denotes the

j-th column of $A_\Gamma \cdot P$ with $Z_j(l) := \sum_{i=1}^{n_A} A_\Gamma(l, i)P(i, j)$. From b) in Lemma 7 one

concludes that each column Z_j, $1 \leq j \leq h$, has at most one non-zero component: if $Z_j(l) \neq 0$ then $Z_j(i) = 0$ for each $i \neq l$.

We still have to verify: $rank\begin{pmatrix} Y_1 & Y_2 & \cdots & Y_h \\ Z_1 & Z_2 & \cdots & Z_h \end{pmatrix} > rank\,(Y_1 \quad Y_2 \quad \cdots \quad Y_h)$.

By the definition of matrix A_Γ (Def. 8) and the construction of P (Lemma 7) one readily verifies that the rows of the compound matrix $(Z_1 \quad Z_2 \quad \cdots \quad Z_h)$ are linearly independent. For later use let $\alpha_1, \alpha_2, \ldots \alpha_k$ and $\beta_1, \beta_2, \ldots \beta_k$ denote the rows of $(Y_1 \quad Y_2 \quad \cdots \quad Y_h)$, respectively those of $(Z_1 \quad Z_2 \quad \cdots \quad Z_h)$.

Each word $w_i \in K \cap K_3$, $i \in \#(K_3)$ can be written as

$$w_i = u_{1,1}^{(i)} u_{1,2}^{(i)} u_{2,1}^{(i)} u_{2,2}^{(i)} \cdots u_{r,1}^{(i)} u_{r,2}^{(i)} w_{r+1,1}^{(i)} w_{r+1,2}^{(i)} w_{r+1,3}^{(i)} \cdots w_{k,1}^{(i)} w_{k,2}^{(i)} w_{k,3}^{(i)}, \text{ where}$$

1. $\pi_2(u_{s,1}^{(i)}) = a_s^i$ and $\pi_2(u_{s,2}^{(i)}) = b_s^i$ for each s, $1 \leq s \leq r$,

2. $\pi_2(w_{s,1}^{(i)}) = \pi_2(w_{s,3}^{(i)}) = a_s^i$ and $\pi_2(w_{s,2}^{(i)}) = b_s^{2i}$ for each s, $r + 1 \leq s \leq k$.

If $P(i, j) \neq 0$ then for each $n_0 \in \mathbb{N}$ there exists $w \in K$ such that $|w|_{g_i} > n_0$. This follows from c) in Lemma 7. Let $G_1 := \{g \in G_A \mid \forall n_0 \in \mathbb{N} : \exists w \in K : |w|_{g_i} > n_0\}$ be the set of all these arcs. Now we want to show that the row-space $\{\alpha_1, \alpha_2, \ldots \alpha_k, \beta_1, \beta_2, \ldots \beta_k\}$ contains strictly more linearly independent elements than the row-space $\{\beta_1, \beta_2, \ldots \beta_k\}$ if $B_{k,r} = L(A)$ for the counter automaton A having r+1 reversal bounded counters. We distinguish two cases:

1. Assume that one of the reversal bounded counters will be changed by an arc $g \in G(w_{l,1}^{(i)} w_{l,2}^{(i)} w_{l,3}^{(i)}) \cap G_1$ for some l, $r + 1 \leq l \leq k$. W.l.o.g. we assume that this is the first counter, hence $\Delta_g(1) = \pi_3(g)(1) \neq 0$. By choosing two more arcs from $G(w_{l,1}^{(i)} w_{l,2}^{(i)} w_{l,3}^{(i)}) \cap G_1$ we can always find three elements $g_{\mu_1}, g_{\mu_2}, g_{\mu_3} \in G_A$, $1 \leq \mu_1, \mu_2, \mu_3 \leq n_A$, such that:

1. $g \in \{g_{\mu_1}, g_{\mu_2}, g_{\mu_3}\}$,

2. $g_{\mu_1} \sqsubseteq w_{l,1}^{(i)}$, $g_{\mu_2} \sqsubseteq w_{l,2}^{(i)}$, and $g_{\mu_3} \sqsubseteq w_{l,3}^{(i)}$,

3. if $g \neq g_{\mu_j}$ for some $1 \leq j \leq 3$ then $\pi_2(g_{\mu_j}) \neq 0$.

Now consider two triples $\boldsymbol{y} := (y_1, y_2, y_3)$ and $\boldsymbol{z} := (z_1, z_2, z_3)$, where y_1, y_2, and y_3 are entries of the matrix $(\, Y_1 \quad Y_2 \quad \cdots \quad Y_h \,)$ and z_1, z_2, and z_3 are entries of the matrix $(\, Z_1 \quad Z_2 \quad \cdots \quad Z_h \,)$. $\boldsymbol{y}$ and $\boldsymbol{z}$ are specified as follows: for $1 \leq j \leq 3$ the elements y_j are located in the first row α_1 of $(\, Y_1 \quad Y_2 \quad \cdots \quad Y_h \,)$ and the elements z_j are located in the l-th row β_l of $(\, Z_1 \quad Z_2 \quad \cdots \quad Z_h \,)$ and their crossing with some column $\begin{pmatrix} Y_{m_j} \\ Z_{m_j} \end{pmatrix}$, where $1 \leq m_j \leq h$ for $1 \leq j \leq 3$, and m_j is such, that $P(\mu_j, m_j) \neq 0$. As mentioned before, each column of $(\, Z_1 \quad Z_2 \quad \cdots \quad Z_h \,)$ has at most a single entry not equal to *zero*. Since $\pi_2(g_{\mu_1}), \pi_2(g_{\mu_2}), \pi_2(g_{\mu_3}) \in \{a_l, b_l, \lambda\}$ these entries must occur in the l-th row β_l of $(\, Z_{m_1} \quad Z_{m_2} \quad Z_{m_3} \,)$, hence applies to the elements z_1, z_2, and z_3. Consequently, if $\boldsymbol{y}$ was linearly independent of $\boldsymbol{z}$ then also the first row α_1 would be linearly independent of the rows $\beta_1, \beta_2, \ldots, \beta_k$. This would imply the statement of Lemma 8. Thus it suffices to prove that indeed $\boldsymbol{y}$ and $\boldsymbol{z}$ are linearly independent.

Among the cases $g = g_{\mu_1}$, $g = g_{\mu_2}$, or $g = g_{\mu_3}$ we select $g := g_{\mu_1}$ as subcase **1.1** (the remaining cases are similar):

By the choice of g we have $\Delta_g(1) \neq 0$ which implies $y_1 \neq 0$. Now, either $\boldsymbol{z} = (0, -1, 1)$ or $\boldsymbol{z} = (1, -1, 1)$ by definition of $\{g_{\mu_1}, g_{\mu_2}, g_{\mu_3}\}$. Since $\boldsymbol{z} = (0, -1, 1)$ means independence of $\{\boldsymbol{y}, \boldsymbol{z}\}$ we proceed by assuming $\boldsymbol{z} = (1, -1, 1)$. Since the first counter is reversal bounded, only the following choices are possible for $\boldsymbol{y}$: $y_1 > 0, y_2 > 0, y_3 \neq 0$, $y_1 > 0, y_2 < 0, y_3 < 0$, or $y_1 < 0, y_2 < 0, y_3 < 0$. It is immediately verified that in all these cases $\boldsymbol{y}$ is linearly independent of $\boldsymbol{z}$.

2. We next have to consider the case, that for each $l, r + 1 \leq l \leq k$ no arc $g \in G(w_{l,1}^{(i)} w_{l,2}^{(i)} w_{l,3}^{(i)}) \cap G_1$ updates one of the r+1 reversal bounded counters. Let $G_2 := G(w_{r+1,1}^{(i)} w_{r+1,2}^{(i)} w_{r+1,3}^{(i)} \cdots w_{k,1}^{(i)} w_{k,2}^{(i)} w_{k,3}^{(i)}) \cap G_1$ be the relevant set of these arcs. Again we consider matrices defined from columns of $\begin{pmatrix} Y_1 & Y_2 & \cdots & Y_h \\ Z_1 & Z_2 & \cdots & Z_h \end{pmatrix}$ as follows:

Let $(\, Y_{m_1} \quad Y_{m_2} \quad \cdots \quad Y_{m_q} \,)$ and $(\, Z_{m_1} \quad Z_{m_2} \quad \cdots \quad Z_{m_q} \,)$ be the matrices consisting of those columns of $(\, Y_1 \quad \cdots \quad Y_h \,)$, respectively of $(\, Z_1 \quad \cdots \quad Z_h \,)$, for which $P(j, m_i) \neq 0$ and $g_j \in G_2$ where $1 \leq j \leq n_A$, and $1 \leq m_i \leq h$ for all $1 \leq i \leq q$. Now, $g \in G_2$ implies $\Delta_g(j) = 0$ for $1 \leq j \leq r+1$, since none of the reversal bounded counters is modified by an arc from the set G_2. Consequently $Y_{m_i}(j) = 0$

for $1 \leq j \leq r+1$ and each $1 \leq i \leq q$, so that $rank\,(\,Y_{m_1}\quad Y_{m_2}\quad \cdots \quad Y_{m_q}\,) \leq k - (r+1)$. On the other hand, $rank\,(\,Z_{m_1}\quad Z_{m_2}\quad \cdots \quad Z_{m_q}\,) = k - r$, since $rank\,(\,Z_1\quad Z_2\quad \cdots \quad Z_h\,) = k$ and r rows of $(\,Z_{m_1}\quad Z_{m_2}\quad \cdots \quad Z_{m_q}\,)$ have an entry equal to *zero* (recall $\pi_2(g) \notin \Gamma_r$ for $g \in G_2$). Also in case **2.** the statement of the lemma has been proved. $\qquad\square$

Let $A \in \mathbb{Z}^{n \times h}$ be of rank r, $B \in \mathbb{Z}^h$, and $L := \{x \in \mathbb{N}^h \mid Ax = B\}$ be the set of all non-negative solutions of the linear equation $Ax = B$. It is known from Linear Algebra that each subset $M \subseteq L$ of linearly independent elements has cardinality of at most $(h - r)$.

Lemma 9. *If* $L := \{x \in \mathbb{N}^h \mid Ax = B\}$ *for some* $A \in \mathbb{Z}^{n \times h}$ *of rank r and* $B \in \mathbb{Z}^h$. *If for each* $n \in \mathbb{N}$ *there exists* $x \in L$ *such that* $x(i) > n$ *for each i, $1 \leq i \leq h$, then L contains a subset* $M = \{x_1, x_2, \ldots, x_{h-r}\}$ *of linearly independent elements.*

Proof: Let $y_1, y_2, \ldots, y_{h-r} \in \mathbb{Z}^h$ be linearly independent solutions of the homogenous linear equation $Ax = 0$ and define $n_0 := max\{\, |y_i(j)| \,|\, 1 \leq i \leq h, 1 \leq j \leq r - h\}$. Now, if $x_0 \in L$ is a solution of the inhomogeneous linear equation $Ax = B$ that satisfies $x_0(i) > n_0$, then $x_0 + y_1, x_0 + y_2, \ldots x_0 + y_{h-r}$ are linearly independent and non-negative solutions of the equation $Ax = B$. $\quad\square$

Proof of Lemma 6: From Lemma 5 we see that $L(A) = B_{k,r} \subseteq C_k$ implies

$$\{v \in R_A \mid A_\Delta \cdot \psi(v) = \mathbf{0}\} = \{v \in R_A \mid \begin{pmatrix} A_\Delta \\ A_\Gamma \end{pmatrix} \cdot \psi(v) = \mathbf{0}\}.$$ Using $K \subseteq R_A$ from Lemma 7 a) with $\psi(K) = \{C + P \cdot Y \mid Y \in \mathbb{N}^h\}$ and by b) there exists $Y_0 \in \mathbb{N}^h$ for each $n_0 \in \mathbb{N}$ with $Y_0(j) > n_0$ for each $1 \leq j \leq h$ and a string $w \in R_A$ such that $\pi_2(w) \in B_{k,r}$, $A_\Gamma \cdot \psi(w) = 0$, $A_\Delta \cdot \psi(w) = 0$, and $C + P \cdot Y_0 = \psi(w)$. This yields the equation

$$(*): \quad \{Y \in \mathbb{N}^h \mid A_\Delta \cdot P \cdot Y = -(A_\Delta) \cdot C\} =$$
$$\left\{ Y \in \mathbb{N}^h \,\middle|\, \begin{pmatrix} A_\Delta \\ A_\Gamma \end{pmatrix} \cdot P \cdot Y = - \begin{pmatrix} A_\Delta \\ A_\Gamma \end{pmatrix} \cdot C \right\}, \text{ and}$$

$Y_0 \in \{Y \in \mathbb{N}^h \mid A_\Delta \cdot P \cdot Y = -(A_\Delta) \cdot C\}$. But by Lemma 9 and Lemma 8 we see

$$rank\,\{Y \in \mathbb{N}^h \mid A_\Delta \cdot P \cdot Y = -(A_\Delta) \cdot C\} >$$
$$rank\,\left\{ Y \in \mathbb{N}^h \,\middle|\, \begin{pmatrix} A_\Delta \\ A_\Gamma \end{pmatrix} \cdot P \cdot Y = - \begin{pmatrix} A_\Delta \\ A_\Gamma \end{pmatrix} \cdot C \right\},$$

which means that equation $(*)$ cannot be fulfilled and $B_{k,r} \notin (k, r+1)$–RBC. $\qquad\square$

The above results yield our main Theorem:

Theorem 1. (k_1, r_1)–*RBC* $\subsetneq$ (k_2, r_2)–*RBC* *iff* $(k_1 < k_2)$ *or* $(k_1 = k_2$ *and* $r_1 > r_2)$.

Proof: The mere inclusion (k_1, r_1)–$RBC \subseteq (k_2, r_2)$–RBC if $(k_1 < k_2)$ or $(k_1 = k_2$ and $r_1 > r_2)$, follows from the definition of the family (k, r)–RBC (Definitions 2 to 4). The strictness of (k_1, r_1)–$RBC \subsetneqq (k_2, r_2)$–$RBC$ if $k_1 < k_2$ is verified as follows:

By definition we have (k, r_1)–$RBC \subseteq (k, 0)$–RBC for any $r_1 \leq k$, the strict inclusion $(k, 0)$–$RBC = \mathcal{M}(C_k) \subsetneqq \mathcal{M}(B_{k+1}) = (k+1, k+1)$–$RBC$ is Theorem 2, and again by definition $(k+1, k+1)$–$RBC \subseteq (k+1, r_2)$–RBC for any $r_2 \leq k+1$. Finally, the inclusion $(k, r+1)$–$RBC \neq (k, r)$–RBC for all $k \in I\!N$ and $0 \leq r < k$ has been shown in Lemma 6. □

Acknowledgment

We thank Berndt Farwer and Olaf Kummer for the fruitful discussion on the preliminary version of this work. We also thank the referees for their comments.

References

Bers 80. J. Berstel, Transductions and Context-free Languages, Teubner Stuttgart (1980).

FiMR 68. P.C. Fischer, A.R. Meyer, and A.L. Rosenberg. Counter machines and counter languages, Math. Syst. Theory, **2** 1968 265–283.

Gins 75. S. Ginsburg, Algebraic and Automata Theoretic Properties of Formal Languages, North Holland Publ. Comp. Amsterdam (1975).

GiGr 70. S. Ginsburg and S. Greibach, Principal AFL, J. Comput. Syst. Sci., **4** (1970) 308–338.

Grei 76. S. Greibach. Remarks on the complexity of nondeterministic counter languages, Theoretical Computer Science, **1** (1976) 269–288.

Grei 78. S. Greibach. Remarks on blind and partially blind one-way multicounter machines, Theoretical Computer Science, **7** (1978) 311–324.

Hrom 86. J. Hromkovič. Hierarchy of reversal bounded one-way multicounter machines, Kybernetika, **22** (1986) 200–206.

Jant 79. M. Jantzen. On the hierarchy of Petri net languages, R.A.I.R.O., Informatique Théorique, **13** (1979) 19–30.

Jant 97. M. Jantzen. On twist-closed trios: A new morphic characterization of the r.e. sets. In: Foundations of Computer Science, Lecture Notes in Comput. Sci., vol 1337, Springer-Verlag, Heidelberg (1997) 135 - 142.

Jant 98. M. Jantzen. Hierarchies of principal twist-closed trios, In: Proceedings of 15th Internat. Symp. on Theoretical Aspects of Computer Science, STACS '98, Paris, Lecture Notes in Comput. Sci., vol 1373, Springer-Verlag, Heidelberg (1998) 344 - 355.

A Simple Undecidable Problem:
The Inclusion Problem for Finite Substitutions
on ab^*c *

Juhani Karhumäki[1] and Leonid P. Lisovik[2]

[1] Department of Mathematics and
Turku Centre for Computer Science, University of Turku
FIN-20014 Turku, Finland
karhumak@cs.utu.fi
[2] Department of Cybernetics, Kiev National University
Kiev, 252017, Ukraine
lis@cyber.univ.kiev.ua

Abstract. As an evidence of the power of finite unary substitutions we show that the inclusion problem for finite substitutions on the language $L = ab^*c$ is undecidable, i.e. it is undecidable whether for two finite substitutions φ and ψ the relation $\varphi(w) \subseteq \psi(w)$ holds for all w in L.

1 Introduction

Finite substitutions between free monoids are natural extensions of corresponding morphisms. However, due to their inherent nondeterministic nature, they behave in many aspects very differently. A goal of this paper is to emphasize this difference in a particularly simple setting.

Finite substitutions, as well as their images, i.e. finite languages, have been studied rather intensively during the last few years. Such research has revealed a number of nice, and also surprising, results. In [LII], see also [HH], it was shown that the question whether two finite substitutions are equivalent, word by word, on the language $L = a\{b,c\}^*d$ is undecidable, in other words, the equivalence problem for finite substitutions on the language L, and hence also on regular languages, is undecidable. In [CKO] all finite languages commuting with a given two-element language were characterized, and as a byproduct Conway's Problem for two element sets was solved affirmately. Conway's Problem, see [C], asks whether the maximal set commuting with a given rational X, referred to as its *centralizer*, is rational as well. Very recently Conway's Problem for three-element sets was also solved in [KP], but the problem remains open even for finite sets X. The general problem, as well as some related ones, seems to be very hard.

An intriguing subcase of the problem solved in [LII] is the case when L' is assumed to be ab^*c, i.e. a very special bounded language. This problem was posed, at least implicitly, in [CuK] and has, so far, avoided all attempts to be solved. In [KL] some special cases, as well as related problems, were considered.

* Research supported under the grant 44087 of the Academy of Finland.

A. Ferreira and H. Reichel (Eds.): STACS 2001, LNCS 2010, pp. 388–395, 2001.
© Springer-Verlag Berlin Heidelberg 2001

One result of [KL] shows that the inclusion problem for finite substitutions on regular languages is decidable if the substitutions are (or in fact only the simulating one is) so-called *prefix* substitutions, that is the images of the letters are prefix sets. Here we show that the restriction to prefix substitutions is essential. Indeed, otherwise the problem becomes undecidable even in the case when the language equals to $L' = ab^*c$. The corresponding equivalence problem remains still open.

This paper is organized as follows.

First in Section 2 we fix the needed terminology, and recall the basic tool used here, the notion of a nondeterministic defense system. Section 3 is devoted to our main undecidability result. In Section 4 we consider applications of our result, as well as some related ones.

In this extended abstract the proof of the main result is only partially presented, and some other proofs are omitted.

2 Preliminaries

In this section we fix our terminology, introduce our problems and recall the basic tools needed. For undefined notions in combinatorics of words we refer to [ChK] and in automata theory to [B].

Let Σ be a finite alphabet, and Σ^* (resp. Σ^+) the free monoid (resp. semigroup) generated by Σ. We denote by 1 the unit of Σ^*, so that $\Sigma^* = \Sigma^+ \cup \{1\}$. For two finite alphabets Σ and Δ we consider *finite substitutions* $\varphi : \Sigma^* \to \Delta^*$ which are many-valued mappings and can be defined as morphisms from Σ^* into the monoid of finite subsets of Δ^*, i.e. into 2^{Δ^*}. If φ is single-valued it is an ordinary semigroup morphism $\Sigma^* \to \Delta^*$. By a 1-*free* (or ε-*free*) finite substitution we mean a finite substitution φ for which 1 is not in $\varphi(a)$ for any a in Σ.

Let φ, ψ be finite substitutions $\Sigma^* \to \Delta^*$ and $L \subseteq \Sigma^*$ a language. We say that φ and ψ are *equivalent* on L if and only if

$$\varphi(w) = \psi(w) \quad \text{for all } w \in L.$$

Similarly, we say that φ is *included* in ψ on L if and only if

$$\varphi(w) \subseteq \psi(w) \quad \text{for all } w \in L.$$

We note that the question $\varphi(w) \overset{?}{\subseteq} \psi(w)$ (for a fixed w) can be viewed as a task of finding a winning strategy in a two player game: in any choice for values in $\varphi(a)$ the ψ must be able to respond following the input word.

Now we can state two important decision problems.

Problem 1 (P_1). Given two finite substitutions φ, $\psi : \Sigma^* \to \Delta^*$ and a rational language $L \subseteq \Sigma^*$, decide whether or not φ and ψ are equivalent on L.

Problem 2 (P_2). Given two finite substitutions φ, $\psi : \Sigma^* \to \Delta^*$ and a rational language $L \subseteq \Sigma^*$, decide whether or not φ is included in ψ on L.

There are two obvious remarks. First, using any standard encoding we can assume that Δ is binary, say $\Delta = \{0,1\}$. Second, in the special case of morphisms the problems are equal, and easily seen to be decidable.

For finite substitutions the situation changes drastically. Indeed, even in the case when the language L is chosen to be fixed, the problem seems to be very difficult. From the point of view of this paper interesting subcases are obtained when L is fixed to be ab^*c - a very special bounded language. In this case we restate the problems as follows:

Problem 3 (UP$_1$). Problem P_1 for the fixed language $L = ab^*c$.

Problem 4 (UP$_2$). Problem P_2 for the fixed language $L = ab^*c$.

We use U above as an indication that the problems deal with finite substitutions which are essentially over a unary input alphabet. More precisely, we consider problems on unary finite substitutions augmented with endmarkers.

Problem P_1, and hence also P_2, was shown to be undecidable in [LII] even in the case when L is fixed to be the language $a\{b,c\}^*d$. Actually the undecidability of Problem P_2 is very easy to conclude. On the other hand, these problems are not decidable only in the case when the mappings are morphisms, but also in the case where they are so-called prefix substitutions, i.e. the images of the letters are prefix sets, cf. [KL]. In fact, for the decidability it is enough that ψ is a prefix substitution. Several related problems are considered in [M], [TI], [LIII] and [TII].

So the interesting remaining problems are UP_1 and UP_2. We are not able to solve UP_1 here, but we do solve UP_2. And surprisingly the answer is negative: the problem is undecidable.

The basic tool in our proof is to use so-called nondeterministic defense systems. A *nondeterministic defense system*, ND-system in short, over the alphabet Δ is a triple $V = (Q, P, q_1)$, where Q is a finite set of states, q_1 is the unique initial (or principal) state and P is a finite set of rules of the form (p, a, q, z), where $p, q \in Q$, $a \in \Delta$ and $z \in \{-1, 0, 1\}$, that is $P \subseteq Q \times \Delta \times Q \times \{-1, 0, 1\}$. We say that the ND-system V is *reliable* if and only if, for each $w = a_1 \ldots a_t$, with $a_i \in \Sigma$ for $i = 1, \ldots, t$, there exist states $q_1, \ldots, q_{t+1}$ such that

$$(q_i, a_i, q_{i+1}, z_i) \in P \quad \text{for } i = 1, \ldots, t, \tag{1}$$

and moreover,

$$\sum_{i=1}^{t} z_i = 0. \tag{2}$$

We emphasize that the sequence (1) can be interpreted, in a natural way, as a computation in a finite transducer: w corresponds to the input, q_i's determine the state transitions and the numbers z_i are the outputs produced in each step. The essential condition is the condition (2) which requires that the sum of the outputs equals zero. Such computations are called *defending*. Hence the reliability means that for each input word there exists a defending computation. Now, a crucial result is the following

Theorem 1. *It is undecidable whether a given ND-system is reliable.*

The proof of Theorem 1 can be found in [LI]. It uses the undecidability of the Post Correspondence Problem. Actually the original ND-systems were equipped with probabilities, but those are not needed in the above proof. It is also obvious that Δ can be fixed as long as it contains at least two symbols. We fix $\Delta = \{0, 1\}$.

3   The Main Result

This section is devoted to the main result of this paper and to its proof.

Theorem 2. *The inclusion problem for 1-free finite substitutions on the language $L = ab^*c$ is undecidable.*

Proof. We reduce the undecidability to that of the reliability of ND-systems. Let $V = (Q, P, q_1)$ be an ND-system over $\{0, 1\}$ and with $Q = \{q_1, \ldots, q_s\}$. We associate V with a pair (φ, ψ) of finite substitutions

$$\{a, b, c\}^* \rightarrow \{0, 1, 2, 3, 4, 5, 6\}^*$$

such that

$$V \text{ is reliable} \tag{3}$$

if and only if

$$\varphi(ab^i c) \subseteq \psi(ab^i c) \text{ for all } i \geq 0. \tag{4}$$

Hence, by Theorem 1, the result would follow.

Before defining φ and ψ we have to fix some terminology. We define

$$W = v_1 \ldots v_{s+1} \quad \text{with} \quad v_i = 0^i 1234 \quad \text{for } i = 1, \ldots, s+1. \tag{5}$$

Consequently, $W \in \{0, 1, 2, 3, 4\}^+$. Further we set $w_{kj} = v_k \ldots v_j$ for $1 \leq k \leq j \leq s+1$. Next, for $k, j \in \{1, \ldots, s\}$, $a \in \{0, 1\}$ and $y \in \{-1, 0, 2\}$ we define words

$$F(a, k, j, y) = w_{k+1, s+1}(S(a)S(a)W)^{y+1}S(a)S(a)w_{1,j},$$

and

$$B_a = S(a)S(a)WS(a)S(a)W,$$

where $S(a) = 5 + a$.

Now, using the word F we define three new sets of words:

(i) $I_1(a, k, j, -1) = F(a, k, j, 2)$

(ii) $\begin{cases} I_2(a, k, j, 0) = F(a, k, j, 0)(34)^{-1} \\ T_2(a, j, j, 0) = 34F(a, j, j, 0) \end{cases}$

(iii) $\begin{cases} I_3(a, k, j, 1) = F(a, k, j, -1)(234)^{-1} \\ M_3(a, j, j, 1) = 234F(a, j, j, 0)4^{-1} \\ T_3(a, j, j, 1) = 4F(a, j, j, -1). \end{cases}$

Here we use the notation uv^{-1} for the right quotient of u by v. Note also that the fourth argument of these words inside any group (i)-(iii) is always a constant, either -1, 0 or 1. The abbreviations I, T and M come from the words initial, terminal and middle, respectively. From now on we may talk about I- or T_2-words, for example.

Next, out of the set of all above I-, T- and M-words we select some, based on the rules of the defense system V, to constitute a language L. It consists of exactly the following words:

$$\begin{cases} I_1(a, k, j, -1), & \text{if } (k, a, j, -1) \in P, \\ I_2(a, k, j, 0) \text{ and } T_2(a, j, j, 0) & \text{if } (k, a, j, 0) \in P, \\ I_3(a, k, j, 1), \ M_3(a, j, j, 1) \text{ and } T_3(a, j, j, 1) & \text{if } (k, a, j, 1) \in P. \end{cases} \tag{6}$$

Here, of course, a, k and j range over the sets

$$\{0, 1\}, \{1, \ldots, s\} \text{ and } \{1, \ldots, s\},$$

respectively.

Now, we are ready to define the required finite substitutions. The substitution φ is defined by

$$\varphi(a) = \varphi(c) = W \text{ and } \varphi(b) = \{B_0 B_0, B_1 B_1\}.$$

Consequently, for each $n \geq 0$, we have

$$\varphi(ab^n c) = W\{(55W)^4, (66W)^4\}^n W,$$

where W is defined in (5).

The substitution ψ, in turn, is defined by the formulas

$$\begin{aligned} \psi(a) &= w_{11} \\ \psi(b) &= LL, \\ \psi(c) &= \{\gamma \mid \gamma = w_{k, s+1} W \text{ with } 2 \leq k \leq s + 1\}. \end{aligned}$$

It remains to be proved that the construction works as intended, that is: the conditions (3) and (4) are equivalent. In this extended abstract we prove the implication in only one direction.

Assume that V is reliable. We have to show that, for each $n \geq 0$ and each word z of the form

$$z = u_0 u_1 \ldots u_{n+1} \text{ with } u_0 = u_{n+1} = W \text{ and } u_i \in \{B_0 B_0, B_1 B_1\}$$
$$\text{for } i = 1, \ldots n$$

there exist words $v_1, \ldots, v_n \in \psi(b)$ and $v_{n+1} \in \psi(c)$ such that

$$w_{11}v_1 \ldots v_n v_{n+1} = z.$$

In the case $n = 0$ we can choose $v_1 = w_{2,s+1}W$ so that $w_{11}v_1 = WW$ as required.

Assume now that $n \geq 1$. Next we use the assumption that V is reliable. Consider the word $t = a_1 \ldots a_n \in \{0, 1, \}^n$ defined by

$$a_i = \alpha \text{ if and only if } u_i = B_\alpha B_\alpha, \text{ for } i = 1, \ldots, n.$$

Since V is reliable, there exist states $q_1 = q_{j_1}, q_{j_2}, \ldots, q_{j_{n+1}}$ and numbers $z_1, \ldots, z_n \in \{-1, 0, 1\}$ such that

$$(q_{j_i}, a_i, q_{j_{i+1}}, z_i) \in P, \text{ for } i = 1, \ldots, n. \tag{7}$$

and moreover,

$$\sum_{i=1}^{n} z_i = 0. \tag{8}$$

The numbers z_i in (7) define, for $i = 1, \ldots, n$, via (6) the words of the types I_1, $I_2 T_2$ or $I_3 M_3 T_3$ depending on whether $z_i = -1$, 0, or 1, respectively. Moreover, such a word, say y_i, is of the form

$$y_i = w_{j_i+1,s+1}(S(a_i)S(a_i)W)^3 S(a_i)S(a_i)w_{1,j_{i+1}}.$$

Consequently, by choosing $v_{n+1} = w_{j_{i+1}+1,s+1}W$ we conclude that

$$v_0 y_1 y_2 \ldots y_n v_{n+1} = z = u_0 u_1 \ldots u_n u_{n+1}.$$

Now the crucial observation is that the word $y_1 \ldots y_n$ consists of altogether $2n + \sum_{i=1}^{n} z_i$ factors of types I, M, and T, that is of L. Hence, by (8), this word can be refactorized as

$$y_1 \ldots y_n = v_1 \ldots v_n \text{ with } v_i \in L^2 = \psi(b) \text{ for } i = 1, \ldots, n.$$

Therefore the factorization $z = w_{11}v_1 \ldots v_n v_{n+1}$ is the required one.

The detailed proof of the other implication can be found in the final version of this paper.

4   Applications and Related Problems

In this section we search for some applications of Theorem 2, as well as its strenghtenings. We first show that one of the endmarkers, say c, can be completely eliminated in the formulation of Theorem 2, and that even both can be essentially eliminated. Both these results are obtained straightforwardly from the constructions of the proof of Theorem 2.

Theorem 3. *It is undecidable whether for two finite 1-free substitutions one is included in the other on the language ab^+.*

Of course, the language L can not be further reduced to b^+.

Theorem 4. *It is undecidable whether for two words α and β and two finite sets C and D the following holds true:*

$$\{\alpha, \beta\}^n \subseteq CD^{n-1} \text{ for all } n \geq 1.$$

Next we state a few applications of our main result. Recall that a finite substitution $\tau : \Sigma^* \to \Delta^*$ can be realized by a nondeterministic generalized sequential machine (ngsm for short) without any states, and that a finite substitution $\tau : \{a, b\}^* \to \Delta^*$ restricted to the language ab^+ can be realized by a two-state ngsm with a unary input alphabet. Indeed, the outputs associated to a can be associated to the reading of b combined with a change of the state (Hence the inputs are changed from ab^i into b^{i+1}) Consequently, Theorem 3 now yields

Theorem 5. *The inclusion (resp. the equivalence) problem for relations defined by two-state (resp. three-state) ngsm's with a unary input alphabet is undecidable.*

In fact, in the inclusion problem of Theorem 5 one of the relations (namely the one which is asked to be included into the other) can be taken to be a finite substitution (on a unary alphabet). Therefore, the statement for the equivalence problem follows by considering the two-state ngsm and the union of it and the one-state ngsm. Hence, the equivalence remains undecidable even if one of the ngsm's has only two states. We also recall that Theorem 5 and its proof techniques are essential strenghtenings of those used in [LIII], where the simulating transducer is required to be only a finite transducer.

The other corollary comes from the fact that the language $L = ab^+$ is a D0L language, cf. [RS]. We call a D0L language *binary*, if it is over a two-letter alphabet.

Theorem 6. *The inclusion problem of finite substitutions on binary D0L languages is undecidable.*

As a contrast to the above theorem we recall that the equivalence of morphisms on D0L languages is decidable, cf. eg. [CuK]. However, even in the case of binary D0L languages the problem is not trivial, although computationally easy: it is enough to consider four first words of the language, cf. [K].

We conclude with a few remarks on our problem P_2, which asks for two finite substitutions φ and ψ and a rational language L whether of not $\varphi(w) \subseteq \psi(w)$ for all $w \in L$. Now, if ψ is a morphism then so must be φ (or the inclusion does not hold), and the problem is trivially decidable. If, in turn, φ is a morphism we are in a nontrivial case: In general, the problem is undecidable cf. [M], [TII] and [LIII], while if the language L is assumed to be of the form ab^*c, or more generally bounded, then the problem becomes decidable, as will be shown in a forthcoming note.

References

B.   Berstel, J. *Transductions and Context-Free Languages*, B.G. Teubner, Stuttgart, 1979.

C.   Conway, J. *Regular Algebra and Finite Machines*, Chapman and Hall Ltd, London 1971.

ChK.   Choffrut, C. and Karhumäki, J., *Combinatorics of words*, in : A. Salomaa and G. Rozenberg (eds), Handbook of Formal Languages, Springer 1997, 324–438.

CuK.   Culik II, K. and Karhumäki, J., *Decision problems solved with the help of Ehrenfeucht Conjecture*, EATCS Bulletin 27, 1985, 30–35.

CKO.   Choffrut, C., Karhumäki, J., and Ollinger, N., *The commutation of finite sets: a challenging problem*, TUCS Report 303, 1999; Special Issue of Theoret. Comput. Sci on Words, to appear.

HH.   Halava, V. and Harju, T., *Undecidability of the equivalence of finite substitutions on regular languages*, Theoret. Inform. and Appl. 33, 1999, 117–124.

K.   Karhumäki, J., *On the equivalence problem for binary D0L systems*, Information and Control 50, 276–284, 1981.

KL.   Karhumäki, J. and Lisovik, L. P., *On the equivalence of finite substitutions and transducers*, in : J. Karhumäki, H. Maurer, Gh. Păun and G. Rozenberg (eds), Jewels are Forever, Springer, 1999, 97–108.

KP.   Karhumäki, J. and Petre, I., *On the centralizer of a finite set*, Proceedings of ICALP00, Lecture Notes in Computer Science 1853, Springer, 2000, 536–546.

LI.   Lisovik, L. P. *An undecidable problem for countable Markov chains*, Kibernetika 2, 1991, 1–6.

LII.   Lisovik, L. P. *The equivalence problem for finite substitutions on regular languages*, Doklady of Academy of Sciences of Russia 357, 1997, 299–301.

LIII.   Lisovik, L. P. *The equivalence problems for transducers with restricted number of states*, Kibernetika i Sistemny Analiz 6, 1997, 109–114.

M.   Maon, Y., *On the equivalence of some transductions involving letter to letter morphisms on regular languages*, Acta inform. 23, 1986, 585–596

RS.   Rozenberg, G. and Salomaa, A., *The Mathematical Theory of L Systems*, Academic Press, 1980.

TI.   Turakainen, P., *On some transducer equivalence problems for families of languages*, Intern. J. Computer Math. 23, 1988, 99–124.

TII.   Turakainen, P., *The Undecidability of some equivalence problems concerning ngsm's and finite substitutions*, Theoret. Comput. Sci. 174, 1997, 269–274.

New Results on Alternating and Non-deterministic Two-Dimensional Finite-State Automata

Jarkko Kari[1]* and Cristopher Moore[2]

[1] Department of Computer Science, 15 MLH, University of Iowa
Iowa City, IA, 52242 USA
jjkari@cs.uiowa.edu

[2] Computer Science Department and Department of Physics and Astronomy
University of New Mexico, Albuquerque NM 87131
and
The Santa Fe Institute, 1399 Hyde Park Road, Santa Fe NM 87501
moore@cs.unm.edu

Abstract. We resolve several long-standing open questions regarding the power of various types of finite-state automata to recognize "picture languages," i.e. sets of two-dimensional arrays of symbols. We show that the languages recognized by 4-way alternating finite-state automata (AFAs) are incomparable to the so-called tiling recognizable languages. Specifically, we show that the set of acyclic directed grid graphs with crossover is AFA-recognizable but not tiling recognizable, while its complement is tiling recognizable but not AFA-recognizable. Since we also show that the complement of an AFA-recognizable language is tiling recognizable, it follows that the AFA-recognizable languages are not closed under complementation. In addition, we show that the set of languages recognized by 4-way NFAs is not closed under complementation, and that NFAs are more powerful than DFAs, even for languages over one symbol.

1   Introduction

Two-dimensional words, or "pictures," are rectangular arrays of symbols over a finite alphabet, and sets of such pictures are "picture languages." Pictures can be accepted or rejected by various types of automata, and this gives rise to different language classes; thus picture languages form an interesting extension of the classical theory of one-dimensional languages and automata, and can be viewed as formal models of image recognition.

In particular, let us consider finite-state automata, which recognize two-dimensional generalizations of the regular languages. In one dimension, we can define the regular languages as those recognized by finite-state automata that can move in one direction (1-way) or both directions (2-way) on the input, and which

---

* Research supported by NSF Grant CCR 97-33101.

A. Ferreira and H. Reichel (Eds.): STACS 2001, LNCS 2010, pp. 396–406, 2001.

are deterministic (DFAs), non-deterministic (NFAs) or alternating (AFAs). In one dimension, these are all equivalent in their computational power.

In two dimensions, natural generalizations of finite-state automata are *4-way finite-state automata,* which at each step can read a symbol of the array, change their internal state, and move up, down, left or right to a neighboring symbol. These can be deterministic, non-deterministic or alternating. Automata of this kind were introduced by Blum and Hewitt [BH67].

Another definition of regular language that we can generalize to two dimensions is the following. A *finite complement* language is one defined by forbidding a finite number of subwords. While not every regular language is finite complement, every regular language is the image of a finite complement language under an alphabetic homomorphism, i.e. a function that maps symbols from one alphabet into another (possibly smaller) one. In two dimensions, a picture language is called *local* if it can be defined by forbidding a finite number of local blocks, and the image of such a language under an alphabetic homomorphism is *tiling recognizable.* Without loss of generality we may assume that all forbidden blocks have size 2×2, or are 1×2 or 2×1 dominoes [GR92].

Tiling recognizable languages have also been called *homomorphisms of local lattice languages* or h(LLL)s [LMN98] or the languages recognizable by *non-deterministic on-line tessellation acceptors* [IN77]. We will follow [GR92] and denote this set of languages REC.

While DFAs, NFAs, AFAs and REC are all equivalent to the regular languages in one dimension, in two or more dimensions they become distinct:

$$\text{DFA} \subset \text{NFA} \begin{array}{c} \subset \text{AFA} \\ \subset \text{REC} \end{array}$$

where all of these inclusions are strict. We recommend [LMN98,GR96,IT91], [Ros79] for reviews of these classes. A bibliography of papers in the subject is maintained by Borchert at [BB].

Note that we restrict our automata to move within the picture they are trying to recognize. (For DFAs, it is known that allowing them to move outside the picture into a field of blanks does not increase their computational power [Ros79]. For NFAs this is known only for $1 \times n$ pictures [LMN98], and for AFAs the question is open.) However, we allow them to sense the boundary of the rectangle, and make different transitions accordingly. Similarly, when defining a language in REC we allow its local pre-image to forbid blocks containing blank symbols ♮ outside the rectangle; for instance, forbidding the block $\begin{array}{cc} ♮ & ♮ \\ ♮ & a \end{array}$ prevents the symbol a from appearing in the upper-left corner of the picture.

A fair amount is known about the closure properties of these classes. The DFAs, NFAs, AFAs and REC are all closed under intersection and union using straightforward constructions. DFAs are also closed under complementation by an argument of Sipser [Sip80] which allows us to remove the danger that a DFA might loop forever and never halt. We construct a new DFA that starts in the final halt state, which we can assume without loss of generality is in the lower right-hand corner. Then this DFA does a depth-first search backwards, attempting to reach the initial state of the original DFA, and using the original DFA's transitions to backtrack along the tree. This gives a loop-free DFA which accepts if and only if the original DFA accepts, and since a loop-free DFA always halts, we can then switch accepting and rejecting states to recognize the complement of the original language. Furthermore, it is known that REC is not closed under complementation [Sze92], even for languages over a unary alphabet [Mat00, Thm. 2.26].

In contrast, up to now it has been an open question whether the 4-way NFA and AFA language classes are closed under complementation [GR96,LMN98]. In this paper we resolve both these questions in the negative, thus completing the following matrix of Boolean closure properties:

	∩	∪	co
DFA	yes	yes	yes
NFA	yes	yes	no
AFA	yes	yes	no
REC	yes	yes	no

Furthermore, the relationship between REC and AFA has been open up to now [IT91]. Here we show that the complement of an AFA language is tiling recognizable and that the classes REC and AFA are incomparable, i.e.

$$\text{co-AFA} \subset \text{REC}, \quad \text{AFA} \not\subset \text{REC}, \quad \text{and} \quad \text{REC} \not\subset \text{AFA}.$$

Specifically, the set of acyclic directed grid graphs with crossover is in AFA but not REC, and its complement is in REC but not AFA.

We also explore picture languages over a unary alphabet. Such pictures are unmarked rectangles and they can be identified with pairs of positive integers indicating the width and the height of the rectangle. In the final section of the paper we show that NFAs are not closed under complementation, and that the inclusions $\text{DFA} \subset \text{NFA} \subset \text{AFA}$ are strict even for unary alphabets.

2 Alternation and Tiling Recognizable Picture Languages

Recall that an *alternating finite-state automaton* has existential and universal states. A computation that meets a universal (resp. existential) state accepts if every transition (resp. at least one transition) from that state leads to an accepting computation. Thus an NFA is an AFA with only existential states.

In this section we prove that AFA and REC are incomparable, by first proving that the complement of an AFA-recognizable language is in REC. To illustrate the idea of the proof, consider a 4-way NFA A over an arbitrary alphabet. A *configuration* consists of the state and position of A. By definition, A does not accept a picture if and only if an accepting state cannot be reached from the initial configuration, that is, iff every possible computation either goes on indefinitely or halts in a non-accepting state. This is clearly equivalent to the existence of a set C of configurations with the following properties:

(i) the initial configuration is in C

(ii) all possible immediate successors of all configurations of C are in C

(iii) there is no accepting configuration in C.

Let A's set of internal states be S. For a given cell of the input picture, call a state in S "reachable" if C contains configurations at that cell in that state. Recall that a language in REC is the image of a local language under some alphabetic homomorphism; this local language can be over a larger alphabet, and so it can include "hidden variables" in each cell, including one bit for each $s \in S$ which indicates whether s is reachable at that cell. Since configurations only make transitions between neighboring cells, conditions (i)-(iii) can then be checked locally. If these hidden variables are then erased by the homomorphism, we have a tiling recognizable picture language that contains exactly those pictures that are not accepted by the NFA.

This construction can be generalized to alternating finite automata:

Theorem 1. *The complement of every AFA recognizable picture language is tiling recognizable, so co-AFA $\subset$ REC.*

Proof. Let A be an AFA with internal states S which accepts a picture language L over an alphabet Σ. A picture P is not accepted by A if and only if there exists a set C of configurations such that

(i) the initial configuration is in C

(ii) if $c \in C$ is existential then all possible successors of c are in C

(iii) if $c \in C$ is universal then at least one of its immediate successors is in C

(iv) there is no accepting configuration in C.

(Notice that we have adopted the convention that a universal state with no successors accepts.) We construct a local pre-image for P over an expanded alphabet $\Sigma \times \{0,1\}^S$, so that each cell contains two variables: the symbol of the input, and $|S|$ bits indicating which states are reachable at that cell in C. It is easy to see that we can verify conditions (i)-(iv) by forbidding a finite set of local blocks. We then use an alphabetic homomorphism from this expanded alphabet into Σ that erases the second variable, giving us a tiling system that accepts exactly those pictures that A does not accept. Thus $\overline{L}$ is in REC. $\square$

Now we know that complements of AFA languages are tiling recognizable. What about AFA languages themselves? This construction fails since a tiling system cannot make sure the AFA's computation path contains no loops. In the case of an NFA, loops can be prevented locally by demanding that each configuration has a unique predecessor, and therefore the construction can be saved by storing this predecessor in the hidden variables as well. But in the presence of universal states the same configuration may appear in different branches of a computation tree with different predecessors. Similarly, the complement of a language recognized by an Π^1 AFA, i.e. one with only universal states, is not necessarily recognized by an NFA, since inputs might be rejected by loopy computation paths which never halt.

In fact, in the following we demonstrate that there exist AFA languages that are not tiling recognizable. The basic reason is the inability of tiling systems to recognize directed acyclic graphs. To illustrate the idea, we first consider a picture language L_P consisting of pictures that represent two identical permutations. A permutation of n elements is represented as an $n \times n$ picture of 0s and 1s such that every row and every column of the picture contains exactly one symbol 1. Two identical permutations are concatenated and a single column of 2s is placed between them as a separator. This way pictures of size $(2n+1) \times n$ are obtained. For example,

1	0	0	2	1	0	0
0	0	1	2	0	0	1
0	1	0	2	0	1	0

is a picture in the language L_P.

Lemma 1. L_P *is not tiling recognizable.*

Proof. Assume the contrary, that L_P is the image of a local language L over k symbols under some alphabetic homomorphism. Without loss of generality we may assume that the forbidden blocks of L have size 2×2. Let n be large enough for $n! > k^n$ to hold. There are $n!$ different permutations of size n so L contains $n!$ pictures of size $(2n+1) \times n$ with different images. The pigeonhole principle states that two of the pictures must match in the middle column. By combining the left- and righthand sides of the two pictures we obtain a new element of L whose image is not in L_P: the left- and righthand sides of the image represent two different permutations. $\qquad\square$

We note that similar counting arguments are used in [Sze92,GR96] to show that REC is not closed under complementation.

Lemma 2. L_P *is accepted by a 4-way AFA whose states are all universal.*

Proof. An AFA, or in fact a DFA, can easily verify that the given picture consists of two permutations separated by a column of 2s: by scanning left to right and top to bottom, it verifies that each column except the middle one contains exactly one 1, and each row contains exactly one 1 on each side of the column of 2s.

Using universal states we then verify that the two permutations are identical. To do this, we try all possibilities of moving in the array in the following fashion:

(*) From a 1 move right to another column but stay on the same side of the wall of 2s. Find the 1 on that column. Then move to the other 1 that is on the same row, on the opposite side of the wall, and repeat.

The picture is accepted if the automaton gets stuck, that is, if it is on the rightmost column and is requested to find another column to the right.

If the two permutations are different then there is a non-accepting infinite loop that repeats instruction (*) indefinitely. There are namely two rows whose corresponding columns are in different order on the two sides of the wall, which allows an infinite loop:

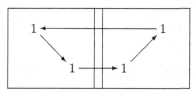

Conversely, if the two permutations are identical then all alternatives lead to a halting accepting computation as two repetitions of (*) always move the automaton at least two positions to the right. Notice that the distance between the two 1s on the same row is constant, so the horizontal movements across the wall cancel each other in two rounds. The remaining instructions move the automaton at least one column to the right on both sides of the wall. This can be continued only until the rightmost column is reached, whereupon the automaton gets stuck and accepts.   □

This gives us the other main results of this section:

Theorem 2. *Language L_P is in AFA but not in REC. The complement of L_P is in REC but not in AFA. Therefore, AFA and REC are incomparable, and AFA is not closed under complementation.*   □

Proof. The first part was proved in Lemma 1 and Lemma 2. The complement of L_P is not recognized by any AFA, because if it were then according to Theorem 1 its complement L_P would be in REC. On the other hand, by Theorem 1 the complement of L_P is in REC since L_P is in AFA.   □

Note that Theorem 2 holds even for Π^1 AFAs, since the AFA in Lemma 2 has only universal states.

More generally, let us consider the picture language of acyclic directed grid graphs, suitably encoded with a finite alphabet. Lemma 2 clearly still holds since an AFA with only universal states can verify acyclicity by following all possible paths in the graph and accepting when they arrive at a sink, while Theorem 1 applies to its complement since a tiling system can guess a cycle and mark it with hidden states.

If crossover is allowed, then Lemma 1 holds as well: the idea is to divide the picture into two halves with n nodes along the boundary between them. The left half (say) induces a relation $\prec$ between these boundary nodes, where we say $a \prec b$ if b is reachable from a by a directed path in that half. If this half has no cycles by itself, then $\prec$ is a partial order.

Now consider the set of $n(n-1)$ right halves consisting of a single directed path from a to b; combining each of these with the left half produces a cycle if and only if $a \prec b$. Therefore, any two partial orders which differ for some pair a, b differ on which right halves will create a cycle, and so each partial order yields a different equivalence class of which right halves are allowed. Since the number of partial orders is at least the number of total orders $n!$, and since all of these are easily achieved by a grid graph with crossover, any local language will run out of states for the interface and we again get a contradiction by the pigeonhole principle. The language L_P above simply restricts to the case where these partial orders are the total orders associated with permutations.

Thus we have

Theorem 3. *The picture language of acyclic directed grid graphs with crossover is in AFA but not in REC, and its complement is in REC but not in AFA.*

However, in the planar case where we disallow crossover, the induced partial orders correspond to outerplanar directed graphs. Since the number of these grows only exponentially in n, this pigeonhole argument fails. We conjecture, in fact, that the set of *planar* acyclic directed grid graphs is in REC.

3   Four-Way Finite Automata Over a Unary Alphabet

In this section, we give an example of a rectangle set S that can be recognized by a 4-way NFA, but not by any DFA. Moreover, we show that the complement of S is recognized by an AFA but not by any NFA. This proves that, unlike the deterministic case, the class of NFA-recognizable picture languages is not closed under complementation, even for pictures over a unary alphabet.

Our main tool in proving these results is to interpret two-dimensional automata as two-way one-dimensional automata by fixing the height of the rectangles and letting the width vary. This approach has become standard, e.g. [Mat00]. Our variant of two-way finite automata can detect when they are reading the first or the last symbol of the input and can make different transitions accordingly. They may move left or right, or remain stationary. They cannot move beyond the ends of the input. A word is accepted iff a final state can be reached.

For a unary alphabet, pictures are just unmarked rectangles, which we can identify with their width and height (w, h). Then we have the following:

Lemma 3. *Let $S \subseteq \mathbb{N}^2$ be the set of rectangles recognized by a k-state 4-way NFA A. Then for every height h there exists a two-way NFA B with kh states recognizing the language*

$$\{1^w \mid (w, h) \in S\}$$

of corresponding widths. Moreover, if A is deterministic then B is deterministic.

Proof. The states of B are pairs (i, s) where i is an integer, $1 \le i \le h$, representing the current vertical position of A in the rectangle, and s is the current state of A. The position of B represents the horizontal position of A inside the rectangle. It is easy to interpret A's transition rules as B's transition rules, so that B will simulate A step-by-step. If A is deterministic then B is also.   □

The following well-known "pumping lemma" allows us to prove that certain languages cannot be accepted by any two-way NFA or DFA with a given number of states:

Lemma 4. *Let A be a two-way NFA with k states over the single-symbol alphabet $\{1\}$, accepting a language $L \subseteq 1^*$. Then, for every $n > k + 2$,*

$$1^n \in L \Longrightarrow 1^{n+k!} \in L$$

Moreover, if A is deterministic then this implication holds in both directions.

Proof. Let $n > k + 2$ and consider an accepting computation C of input 1^n. Let us divide C into segments between consecutive visits of the NFA at the endpoints of the input. The head has to move through all intermediate positions during any segment S of C from one end to the other. There are $n - 2 \geq k + 1$ intermediate positions. Let $s_1, s_2, \ldots, s_{n-2}$ be the states of A when it enters the intermediate positions $1, 2, \ldots, n - 2$, respectively, for the first time within segment S. It follows from the pigeonhole principle that two of the states $s_1, s_2, \ldots, s_{k+1}$ must be the same, say $s_i = s_{i+t}$. The computation between positions i and $i + t$ can then be repeated arbitrarily many times, taking the NFA into position $i + jt$ for any $j \geq 0$, and remaining in state s_i.

Because t divides $k!$ this means that input $1^{n+k!}$ is accepted by a computation that is identical to C except that in any segment of C from one end to the other a loop of length t is repeated $k!/t$ times.

If A is deterministic then each input has a unique computation C. If $n > k+2$ and 1^n is not accepted (either A halts in non-accepting state or loops forever) then the same construction as above yields a non-accepting computation (halting or looping) for input $1^{n+k!}$. In other words, in the deterministic case we have $1^n \in L \Longleftrightarrow 1^{n+k!} \in L$. □

This gives us the immediate corollary:

Corollary 1. *If L is a finite language recognized by a two-way NFA with k states, then its longest word has length at most $k + 2$.*

Now we are ready to prove the main results of this section. Consider the following set of rectangles:

$$S = \{(w, h) \mid w = ih + j(h + 1) \text{ for some non-negative integers } i \text{ and } j \}$$
$$= \{(w, h) \mid w = ih + j \text{ for some } 0 \leq j \leq i \}.$$

For any given height h the set of allowed widths w is the union of contiguous segments

$$ih, ih + 1, \ldots, ih + i$$

for all $i = 0, 1, \ldots$. It is easy to see that the largest width w that is not allowed is $h^2 - h - 1$. (It is the only integer between the segments for $i = h - 2$ and $i = h - 1$. For larger values of i the consecutive segments overlap.)

The set S can be easily recognized by a billiard-ball-like NFA that sets out at a 45° angle from the lower left corner of the rectangle. When the ball hits the upper or lower edge of the rectangle it either bounces back immediately, or moves one cell to the right along the edge before bouncing, as in Figure 1. The rectangle is accepted if the ball is in either corner of the rectangle when it reaches the right edge.

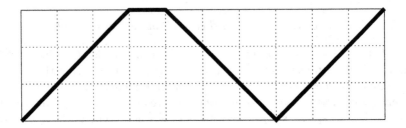

Fig. 1. An Accepting Computation for the 10×3 Rectangle.

However, the complement $\overline{S}$ is not accepted by any NFA. Notice that $\overline{S}$ contains a finite number of rectangles of any given height h, the largest of which has width $w = h^2 - h - 1$. Assume that $\overline{S}$ is accepted by a 4-way NFA with k states. According to Lemma 3 there exists a two-way NFA B with kh states that accepts the finite language

$$L = \{1^w \mid (w, h) \in \overline{S}\}.$$

If we choose the height h such that $h^2 - h - 1 > kh + 2$, then the longest word in L has length greater than the number of states in B plus two. This contradicts the corollary to Lemma 4, so we have proved

Theorem 4. *The NFA-recognizable picture languages are not closed under complementation, even for a unary alphabet.*

Since the DFA-recognizable languages are closed under complementation [Sip80] as discussed above, it follows that S is not in DFA. We can also show this directly using the deterministic variants of Lemma 3 and Lemma 4.

Furthermore, it is easy to see that if a language is recognized by a loop-free NFA, then its complement is accepted by an AFA whose states are all universal. Here, for instance, the AFA moves diagonally at a 45° angle from the lower left corner. When it hits the upper or lower edge of the rectangle a universal state splits the computation into two parts: one in which the AFA bounces from the edge immediately and one in which the bounce is delayed by one step to the right. Both alternatives must lead to accepting computations. A computation is accepting if the AFA is not in the corner when it reaches the right edge of the rectangle. Since S is in NFA but not DFA, and $\overline{S}$ is AFA but not in NFA, we have proved

Theorem 5. *The inclusions DFA ⊂ NFA ⊂ AFA are strict even for picture languages over a unary alphabet.*

4   Conclusions

We have solved several open problems in the field of two-dimensional finite-state automata and picture languages. In particular, we have shown that the NFA- and AFA-recognizable picture languages are not closed under complementation, that the complement of an AFA-recognizable language is in the set REC of tiling recognizable languages, and that AFA and REC are incomparable. In fact, since the AFA in Lemma 2 has only universal states, even this restriction of AFA is incomparable with REC. All these results generalize easily to more dimensions, so they hold for d-dimensional picture languages whenever $d \geq 2$.

Some authors have studied 3-way automata, which are only allowed to move (say) up, down, and right. The NFA for S in Theorem 4 is in fact 3-way, so we can conclude that there exists a 3-way NFA language that is not recognized by any 4-way DFA, and whose complement is not accepted by any 4-way NFA. In addition, the AFA for $\overline{S}$ in Theorem 5 is 3-way so there also exists a 3-way AFA language that is not accepted by any 4-way NFA.

Finally, we leave as an open question our conjecture that the picture language of *planar* acyclic directed graphs is tiling recognizable. Other interesting questions include whether AFA is not closed under complementation for a unary alphabet, and whether REC and co-AFA are distinct.

Acknowledgments

JK thanks the Santa Fe Institute for an enjoyable visit where this work began. We also thank Bernd Borchert, Juraj Hromkovič, Mats Nordahl, Peter Stadler, and the referees for helpful conversations.

References

BH67. M. Blum and C. Hewitt, "Automata on a 2-dimensional tape." *8th IEEE Symp. on Switching and Automata Theory* (1967) 155–160.

BB. B. Borchert, http://math.uni-heidelberg.de/logic/bb/2dpapers.html

GR92. D. Giammarresi and A. Restivo, "Recognizable picture languages." *Int. J. of Pattern Recognition and Artificial Intelligence* **6(2-3)** (1992) 241–256.

Gia93. D. Giammaresi, "Two-dimensional languages and recognizable functions." In G. Rozenberg and A. Salomaa, Eds., *Proc. Developments in Language Theory.* World Scientific (1994) 290–301.

GR96. D. Giammarresi and A. Restivo, "Two-dimensional languages." In G. Rosenberg and A. Salomaa, Eds., *Handbook of Formal Languages,* Volume III. Springer Verlag (1996) 215–267.

IN77.   K. Inoue and A. Nakamura, "Some properties of two-dimensional on-line tessellation acceptors." *Information Sciences* **13** (1977) 95–121.

IN79.   K. Inoue and A. Nakamura, "Two-dimensional finite automata and unacceptable functions." *Int. J. Comput. Math. A* **7** (1979) 207–213.

IT91.   K. Inoue and I. Takanami, "A survey of two-dimensional automata theory." *Information Sciences* **55** (1991) 99–121.

LMN98.   K. Lindgren, C. Moore and M.G. Nordahl, "Complexity of two-dimensional patterns." *Journal of Statistical Physics* **91** (1998) 909–951.

Mat00.   O.Matz, "Dot-depth, monadic quantifier alternation and first-Order closure over grids and pictures." To appear in *Theoretical Computer Science*.

Mil76.   D.L. Milgram, "A region crossing problem for array-bounded automata." *Information and Control* **31** (1976) 147–152.

Min67.   M. Minsky, *Computation: Finite and Infinite Machines*. Prentice-Hall, 1967.

Ros79.   A. Rosenfeld, *Picture Languages: Formal Models for Picture Recognition*. Academic Press, 1979.

Sip80.   M. Sipser, "Halting space-bounded computations." *Theoretical Computer Science* **10** (1980) 335–338.

Sze92.   A. Szepietowski, "Two-dimensional on-line tesselation acceptors are not closed under complement." *Information Sciences*, **64** (1992) 115–120.

The Complexity of Minimal Satisfiability Problems

Lefteris M. Kirousis[1][*] and Phokion G. Kolaitis[2][**]

[1] Department of Computer Engineering and Informatics
University of Patras, University Campus, GR-265 04 Patras, Greece
kirousis@ceid.upatras.gr
[2] Computer Science Department
University of California, Santa Cruz
Santa Cruz, CA 95064, U.S.A.
kolaitis@cse.ucsc.edu

Abstract. A dichotomy theorem for a class of decision problems is a result asserting that certain problems in the class are solvable in polynomial time, while the rest are NP-complete. The first remarkable such dichotomy theorem was proved by T.J. Schaefer in 1978. It concerns the class of generalized satisfiability problems $SAT(S)$, whose input is a $CNF(S)$-formula, i.e., a formula constructed from elements of a fixed set S of generalized connectives using conjunctions and substitutions by variables. Here, we investigate the complexity of minimal satisfiability problems $MIN\ SAT(S)$, where S is a fixed set of generalized connectives. The input to such a problem is a $CNF(S)$-formula and a satisfying truth assignment; the question is to decide whether there is another satisfying truth assignment that is strictly smaller than the given truth assignment with respect to the coordinate-wise partial order on truth assignments. Minimal satisfiability problems were first studied by researchers in artificial intelligence while investigating the computational complexity of propositional circumscription. The question of whether dichotomy theorems can be proved for these problems was raised at that time, but was left open. In this paper, we settle this question affirmatively by establishing a dichotomy theorem for the class of all $MIN\ SAT(S)$-problems.

1 Introduction and Summary of Results

Computational complexity strives to analyze important algorithmic problems by first placing them in suitable complexity classes and then attempting to determine whether they are complete for the class under consideration or they actually belong to a more restricted complexity class. This approach to analyzing algorithmic problems has borne fruit in numerous concrete cases and has led to the successful development of the theory of NP-completeness. In this vein, *dichotomy theorems* for classes of NP-problems are of particular interest, where a dichotomy theorem is a result that concerns an infinite class $\mathcal{F}$ of related decision problems and asserts that certain problems in $\mathcal{F}$ are solvable in polynomial time, while on the contrary all other problems in $\mathcal{F}$ are NP-complete. It

---

[*] Research conducted while on sabbatical at the University of California, Santa Cruz and partially supported by the University of Patras, by NSF grant CCR-9610257 and by the Computer Technology Institute.
[**] Partially supported by NSF grants CCR-9610257 and CCR-9732041.

A. Ferreira and H. Reichel (Eds.): STACS 2001, LNCS 2010, pp. 407–418, 2001.
© Springer-Verlag Berlin Heidelberg 2001

should be pointed out that the a priori existence of dichotomy theorems cannot not be taken for granted. Indeed, Ladner [17] showed that if P $\neq$ NP, then there are problems in NP that are neither NP-complete nor in P. Consequently, a given class $\mathcal{F}$ of NP-problems may contain such problems of intermediate complexity, which rules out the existence of a dichotomy theorem for $\mathcal{F}$.

The first remarkable (and highly non-trivial) dichotomy theorem was established by Schaefer [22], who introduced and studied the class of GENERALIZED SATISFIABILITY problems (see also [8, LO6, page 260]). A *logical relation* (or *generalized connective*) R is a non-empty subset of $\{0, 1\}^k$, for some $k \geq 1$. If $S = \{R_1, \ldots, R_m\}$ is a finite set of logical relations, then a CNF(S)-formula is a conjunction of expressions (called *generalized clauses* or, simply, *clauses*) of the form $R'_i(x_1, \ldots, x_k)$, where each R'_i is a relation symbol representing the logical relation R_i in S and each x_j is a Boolean variable. Each finite set S of logical relations gives rise to the GENERALIZED SATISFIABILITY problem SAT(S): given a CNF(S)-formula φ, is φ satisfiable? Schaefer isolated six efficiently checkable conditions and proved the following dichotomy theorem for the class of all GENERALIZED SATISFIABILITY problems SAT(S): if the set S satisfies at least one of these six conditions, then SAT(S) is solvable in polynomial time; otherwise, SAT(S) is NP-complete. Since that time, only a handful of dichotomy theorems for other classes of decision problems have been established. Two notable ones are the dichotomy theorem for the class of FIXED SUBGRAPH HOMEOMORPHISM problems on directed graphs, obtained by Fortune, Hocroft and Wyllie [6], and the dichotomy theorem for the class of H-COLORING problems on undirected graphs, obtained by Hell and Nešetřil [9]. The latter is a special case of CONSTRAINT SATISFACTION, a rich class of problems that have been the object of systematic study in artificial intelligence. It should be noted that no dichotomy theorem for the entire class of CONSTRAINT SATISFACTION problems has been established thus far, in spite of intensive efforts to this effect (see Feder and Vardi [7], Jeavons, Cooper and Gyssens [11]).

In recent years, researchers have obtained dichotomy theorems for optimization problems, counting problems, and decision problems that are variants of GENERALIZED SATISFIABILITY problems. Creignou [4], Khanna, Sudan and Williamson [14], Khanna, Sudan and Trevisan [13], and Zwick [23] obtained dichotomy theorems for certain classes of optimization problems related to propositional satisfiability and Boolean constraint satisfaction, Creignou and Hermann [3] proved a dichotomy theorem for the class of counting problems that ask for the number of satisfying assignments of a given CNF(S)-formula, and Kavvadias and Sideri [12] established a dichotomy theorem for the class of decision problems INVERSE SAT(S) that ask whether a given set of truth assignments is the set of all satisfying assignments of some CNF(S)-formula, where S is a finite set of logical relations. Even more recently, Reith and Vollmer [21] proved a dichotomy theorem for the class of optimization problems LEXMIN SAT(S) and LEXMAX SAT(S) that ask for the lexicographically minimal (or maximal) truth assignment that satisfies a given CNF(S)-formula. In addition, Istrate [10] investigated the existence of a dichotomy for the restriction of generalized satisfiability problems in which each variable appears at most twice.

Researchers have also investigated the class of decision problems MIN SAT(S), where S is a finite set of logical relations. For a fixed S, the input to the problem is a CNF(S)-formula φ and a satisfying truth assignment α of φ; the question is to decide whether there is another satisfying truth assignment β of φ such that $\beta < \alpha$, where $<$ is the coordinate-wise partial order on truth assignments. These decision problems were introduced and studied by researchers in artificial intelligence while investigating *circumscription*, a well-developed formalism of common-sense reasoning introduced by McCarthy [19] about twenty years ago. The main question left open about MIN SAT(S) was whether a dichotomy theorem holds for the class of all MIN SAT(S) problems, where S is a finite set of logical relations. In the present paper, we settle this question in the affirmative and also provide easily checkable criteria that tell apart the polynomial-time solvable cases of MIN SAT(S) from the NP-complete ones.

In circumscription, properties are specified using formulas of some logic, a natural partial order between models of each formula is considered, and preference is given to models that are minimal with respect to this partial order. McCarthy's key intuition was that minimal models should be preferred because they are the ones that have as few "exceptions" as possible and thus embody common-sense. A fundamental algorithmic problem about every logical formalism is *model checking*, the problem of deciding whether a finite structure satisfies a formula. As regards circumscription, model checking amounts to the problem of deciding whether a finite structure is a minimal model of a formula. The simplest case of circumscription is *propositional circumscription*, where properties are specified using formulas of propositional logic; thus, the model checking problem for propositional circumscription is precisely the problem of deciding whether a satisfying truth assignment of a propositional formula is minimal with respect to the coordinate-wise order. Clearly, this problem is equivalent to the complement of the minimal satisfiability problem; moreover, it is not hard to show that this problem is coNP-complete, when arbitrary propositional formulas are allowed as part of the input. For this reason, researchers in artificial intelligence embarked on the pursuit of tractable cases of the model checking problem for propositional circumscription. In particular, Cadoli [1,2] adopted Schaefer's approach, introduced the class of decision problems MIN SAT(S), identified several tractable cases, and raised the question of the existence of a dichotomy theorem for this class (see [2, page 132]). Moreover, Cadoli pointed out that if a dichotomy theorem for MIN SAT(S) indeed exists, then the dividing line is going to be very different from the dividing line in Schaefer's dichotomy theorem for SAT(S). To see this, consider first the set $S = \{R_{1/3}\}$, where $R_{1/3} = \{(1,0,0),(0,1,0),(0,0,1)\}$. In this case, SAT($S$) is the well-known NP-complete problem POSITIVE-1-IN-3-SAT, while on the contrary MIN SAT(S) is trivial, since it can be easily verified that every satisfying truth assignment of a given CNF(S)-formula is minimal. Thus, an intractable case of SAT(S) becomes a tractable (in fact, a trivial) case of MIN SAT(S). In the opposite direction, Cadoli [1,2] showed that certain tractable (in fact, trivial) cases of SAT(S) become NP-complete cases of MIN SAT(S). Specifically, one of the six tractable cases in Schaefer's dichotomy theorem is the case where S consists entirely of 1-*valid* logical relations, that is, every relation R in S contains the all-ones tuple $(1,\ldots,1)$ (and, hence, every CNF(S)-formula is satisfied by the truth assignment that assigns 1 to every variable). In contrast, Cadoli [1,2] discovered a finite set S of 1-valid relations such that MIN SAT(S) is NP-complete.

As it turns out, the collection of 1-valid relations holds the key to the dichotomy theorem for MIN SAT(S). More precisely, we first establish a dichotomy theorem for the class of MIN SAT(S) problems, where S is a finite set of 1-valid relations. Using this restricted dichotomy theorem as a stepping stone, we then derive the desired dichotomy theorem for the full class of MIN SAT(S) problems, where S is a finite set of arbitrary logical relations. Note that all dichotomy theorems described thus far involve $\mathrm{CNF}(S)$-formulas that do not contain the constant symbols 0 and 1; Schaefer [22], however, also proved a dichotomy theorem for $\mathrm{CNF}(S)$-formulas with constant symbols. Here, we derive dichotomy theorems for minimal satisfiability of $\mathrm{CNF}(S)$ formulas with constant symbols as well. Our results differ from earlier dichotomy theorems for satisfiability problems in two major aspects. First, in all earlier dichotomy theorems the tractable cases arise from conditions that are directly applied to the set S of logical relations under consideration; in our main dichotomy theorem, however, the tractable cases arise from conditions that are applied not to the set S of logical relations at hand, but to a certain set S^* of 1-valid logical relations obtained from S by projecting the relations in S in a particular way. Second, the proofs of essentially all earlier dichotomy theorems for satisfiability problems used Schaefer's dichotomy theorem; furthermore, they often hinged on stronger versions of what has become known as Schaefer's *expressibility theorem* [22, Theorem 3.0, page 219], which asserts that if S does not satisfy at least one of the six conditions that give rise to tractable cases of $\mathrm{SAT}(S)$, then every logical relation is definable from some $\mathrm{CNF}(S)$-formula using existential quantification and substitution by constants. The proof of our dichotomy theorem for $\mathrm{MIN\ SAT}(S)$, however, hinges on new and rather delicate expressibility results that provide precise information about the way particular logical relations, such as the *implication* connective, are definable from $\mathrm{CNF}(S)$-formulas using existential quantification and substitution by constants.

Researchers in artificial intelligence have also investigated various powerful extensions of circumscription in which the partial order among models of a formula is modified, so that some parts of the model are assigned fixed values and some other parts are allowed to vary arbitrarily [18,20]. We have been able to establish dichotomy theorems for the model checking problem for most such extensions of propositional circumscription, thus answering another question left open by Cadoli [1,2]. These results are contained in the full version of the present paper, available in electronic form [15].

2 Preliminaries and Background

This section contains the definitions of the main concepts used in this paper and a minimum amount of the necessary background material from Schaefer's work on the complexity of GENERALIZED SATISFIABILITY problems [22].

Definition 1. Let $S = \{R_1, \ldots, R_m\}$ be a finite set of logical relations of various arities, let $S' = \{R'_1, \ldots, R'_m\}$ be a set of relation symbols whose arities match those of the relations in S, and let V be an infinite set of variables.

A $\mathrm{CNF}(S)$-formula is a finite conjunction $C_1 \wedge \ldots \wedge C_n$ of clauses built using relation symbols from S' and variables from V, that is, each C_i is an atomic formula of the form $R'_j(x_1, \ldots, x_k)$, where R'_j is a relation symbol of arity k in S', and $x_1, \ldots, x_k$ are variables in V. A $\mathrm{CNF}_{\mathrm{C}}(S)$-formula is a formula obtained from a $\mathrm{CNF}(S)$-formula by substituting some of its variables by the constant symbols 0 or 1. The semantics of $\mathrm{CNF}(S)$-formulas and $\mathrm{CNF}_{\mathrm{C}}(S)$-formulas are defined in a standard way by assuming that variables range over the set of bits $\{0, 1\}$, each relation symbol R'_j in S' is interpreted by the corresponding relation R_j in S, and the constant symbols 0 and 1 are interpreted by 0 and 1 respectively.

$SAT(S)$ is the following decision problem: given a $CNF(S)$-formula φ, is it satisfiable? (i.e., is there a truth assignment to the variables of φ that makes every clause of φ true?) The decision problem $SAT_C(S)$ is defined in a similar way. ∎

It is clear that, for each finite set S of logical relations, both $SAT(S)$ and $SAT_C(S)$ are problems in NP. Moreover, several well-known NP-complete problems and several important tractable cases of Boolean satisfiability can easily be cast as $SAT(S)$ problems for particular sets S of logical relations. Indeed, we already saw in the previous section that the NP-complete problem POSITIVE-1-IN-3-SAT ([8, LO4, page 259]) is precisely the problem $SAT(S)$, where S is the singleton consisting of the relation $R_{1/3} = \{(1,0,0),(0,1,0),(0,0,1)\}$. Moreover, the prototypical NP-complete problem 3-SAT coincides with the problem $SAT(S)$, where $S = \{R_0, R_1, R_2, R_3\}$ and $R_0 = \{0,1\}^3 - \{(0,0,0)\}$ (expressing the clause $(x \vee y \vee z)$), $R_1 = \{0,1\}^3 - \{(1,0,0)\}$ (expressing the clause $(\neg x \vee y \vee z)$), $R_2 = \{0,1\}^3 - \{(1,1,0)\}$ (expressing the clause $(\neg x \vee \neg y \vee z)$), and $R_3 = \{0,1\}^3 - \{(1,1,1)\}$ (expressing the clause $(\neg x \vee \neg y \vee \neg z)$). Similarly, but on the side of tractability, 2-SAT is precisely the problem $SAT(S)$, where $S = \{R_0, R_1, R_2\}$ and $R_0 = \{0,1\}^2 - \{(0,0)\}$ (expressing the clause $(x \vee y)$), $R_1 = \{0,1\}^2 - \{(1,0)\}$ (expressing the clause $(\neg x \vee y)$), and $R_2 = \{0,1\}^2 - \{(1,1)\}$ (expressing the clause $(\neg x \vee \neg y)$).

The next two definitions introduce the key concepts needed to formulate Schaefer's dichotomy theorems.

Definition 2. Let φ be a propositional formula.

φ is 1-*valid* if it is satisfied by the truth assignment that assigns 1 to every variable. Similarly, φ is 0-*valid* if it is satisfied by the truth assignment that assigns 0 to every variable.

φ is *bijunctive* if it is a 2CNF-formula, i.e., it is a conjunction of clauses each of which is a disjunction of at most two *literals* (variables or negated variables).

φ is *Horn* if it is the conjunction of clauses each of which is a disjunction of literals such that at most one of them is a variable. Similarly, φ is *dual Horn* if it is the conjunction of clauses each of which is disjunction of literals such that at most one of them is a negated variable.

φ is *affine* if it is the conjunction of subformulas each of which is an *exclusive disjunction* of literals or a negation of an exclusive disjunctions of literals (by definition, an exclusive disjunction of literals is satisfied exactly when an odd number of these literals are true; we will use $\oplus$ as the symbol of the exclusive disjunction). Note that a formula φ is affine precisely when the set of its satisfying assignments is the set of solutions of a system of linear equations over the field $\{0,1\}$. ∎

Definition 3. Let R be a logical relation and S a finite set of logical relations.

R is 1-*valid* if it contains the tuple $(1,1,\ldots,1)$, whereas R is 0-*valid* if it contains the tuple $(0,0,\ldots,0)$. We say that S is 1-*valid* (0-*valid*) if every member of S is 1-valid (0-valid).

R is *bijunctive* (*Horn*, *dual Horn*, or *affine*, respectively) if there is a propositional formula φ which is bijunctive (Horn, dual Horn, or affine, respectively) and such that R coincides with the set of truth assignments satisfying φ.

S is *Schaefer* if at least one of the following four conditions hold: every member of S is bijunctive; every member of S is Horn; every member of S is dual Horn; every member of S is affine. Otherwise, we say that S is *non-Schaefer*. ∎

There are simple criteria to determine whether a logical relation is bijunctive, Horn, dual Horn, or affine. In fact, a set of such criteria was already provided by Schaefer [22]; moreover, Dechter and Pearl [5] gave even simpler criteria for a relation to be Horn or dual Horn. Each of these criteria involves a *closure property* of the logical relations at hand under a certain function. Specifically, a relation R is bijunctive if and only if for all $t_1, t_2, t_3 \in R$, we have that $(t_1 \vee t_2) \wedge (t_2 \vee t_3) \wedge (t_1 \vee t_3) \in R$, where the operators $\vee$ and $\wedge$ are applied coordinate-wise to the bit-tuples. Note that the i-th coordinate of the tuple $(t_1 \vee t_2) \wedge (t_2 \vee t_3) \wedge (t_1 \vee t_3)$ is equal to 1 exactly when the majority of the i-th coordinates of t_1, t_2, t_3 is equal to 1. Thus, this criterion states that R is bijunctive exactly when it is closed under coordinate-wise applications of the ternary *majority* function. R is Horn (respectively, dual Horn) if and only if for all $t_1, t_2 \in R$, we have that $t_1 \wedge t_2 \in R$ (respectively, $t_1 \vee t_2 \in R$). Finally, R is affine if and only if for all $t_1, t_2, t_3 \in R$, we have that $t_1 \oplus t_2 \oplus t_3 \in R$. As an example, it is easy to apply these criteria to the ternary relation $R_{1/3} = \{(1,0,0), (0,1,0), (0,0,1)\}$ and verify that $R_{1/3}$ is neither bijunctive, nor Horn, nor dual Horn, nor affine; moreover, it is obvious that $R_{1/3}$ is neither 1-valid nor 0-valid. Finally, there are polynomial-time algorithms that given a logical relation that is bijunctive (Horn, dual Horn, or affine, respectively), produce a defining propositional formula which is bijunctive (Horn, dual Horn, or affine, respectively). See [5,16].

If S is a 0-valid or a 1-valid set of logical relations, then $\text{SAT}(S)$ is a trivial decision problem (the answer is always "yes"). If S is an affine set of logical relations, then $\text{SAT}(S)$ can easily be solved in polynomial time using Gaussian elimination. Moreover, there are well-known polynomial-time algorithms for the satisfiability problem for the class of all bijunctive formulas (2-SAT), the class of all Horn formulas, and the class of all dual Horn formulas. Schaefer's seminal discovery was that the above six cases are the *only* ones that give rise to tractable cases of $\text{SAT}(S)$; furthermore, the last four are the *only* ones that give rise to tractable cases of $\text{SAT}_C(S)$.

Theorem 4. [Schaefer's Dichotomy Theorems, [22]] *Let S be a finite set of logical relations.*

If S is 0-valid or 1-valid or Schaefer, then $\text{SAT}(S)$ is solvable in polynomial time; otherwise, it is NP-complete.

If S is Schaefer, then $\text{SAT}_C(S)$ is solvable in polynomial time; otherwise, it is NP-complete.

As an application, Theorem 4 immediately implies that POSITIVE-1-IN-3-SAT is NP-complete, since this is the same problem as $\text{SAT}(R_{1/3})$, and $R_{1/3}$ is neither 0-valid, nor 1-valid, nor Schaefer.

To obtain the above dichotomy theorems, Schaefer had to first establish a result concerning the expressive power of $\text{CNF}_C(S)$ formulas. Informally, this result asserts that if S is a non-Schaefer set of logical relations, then $\text{CNF}_C(S)$-formulas have extremely highly expressive power, in the sense that every logical relation can be defined from a $\text{CNF}_C(S)$-formula using existential quantification.

Theorem 5. [Schaefer's Expressibility Theorem, [22]] *Let S be a finite set of logical relations. If S is non-Schaefer, then for every k-ary logical relation R there is a $\mathrm{CNF}_C(S)$-formula $\varphi(x_1, \ldots, x_k, z_1, \ldots, z_m)$ such that R coincides with the set of all truth assignments to the variables $x_1, \ldots, x_k$ that satisfy the formula $(\exists \bar{z})\varphi(\bar{x}, \bar{z})$.*

3 Dichotomy Theorems for Minimal Satisfiability

In this section, we present our main dichotomy theorem for the class of all minimal satisfiability problems MIN SAT(S). We begin with the precise definition of MIN SAT(S), as well as of certain variants of it that will play an important role in the sequel.

Definition 6. Let $\leq$ denote the standard total order on $\{0, 1\}$, which means that $0 \leq 1$.

Let k be a positive integer and let $\alpha = (a_1, \ldots, a_k)$, $\beta = (b_1, \ldots, b_k)$ be two k-tuples in $\{0, 1\}^k$. We write $\beta \leq \alpha$ to denote that $b_i \leq a_i$, for every $i \leq k$. Also, $\beta < \alpha$ denotes that $\beta \leq \alpha$ and $\beta \neq \alpha$.

Let S be a finite set of logical relations. MIN SAT(S) is the following decision problem: given a CNF(S)-formula φ and a satisfying truth assignment α of φ, is there a satisfying truth assignment β of φ such that $\beta < \alpha$? In other words, MIN SAT(S) is the problem to decide whether or not a given truth assignment of a given CNF(S)-formula is minimal. The decision problem MIN SAT$_C(S)$ is defined in a similar way by allowing CNF$_C(S)$-formulas as part of the input.

Let S be a 1-valid set of logical relations. 1-MIN SAT(S) is the following decision problem: given a CNF(S)-formula φ (note that φ is necessarily 1-valid), is there a satisfying truth assignment of φ that is different (and, hence, smaller) from the all-ones truth assignment $(1, \ldots, 1)$?

A CNF$_1(S)$-formula is obtained from a CNF(S)-formula by replacing some of its variable by the constant symbol 1. The decision problem 1-MIN SAT$_1(S)$ is defined the same way as 1-MIN SAT(S), except that CNF$_1(S)$-formulas are allowed as part of the input (arbitrary CNF$_C(S)$-formulas are not allowed, since substituting variables by 0 may destroy 1-validity). ∎

As mentioned earlier, Cadoli [1,2] raised the question of whether a dichotomy theorem for the class of all MIN SAT(S) problems exists. Note that if S is a 0-valid set of logical relations, then MIN SAT(S) is a trivial decision problem. Moreover, Cadoli showed that if S is a Schaefer set, then MIN SAT(S) is solvable in polynomial time. To see this, let φ be a CNF(S)-formula and α be a k-tuple in $\{0, 1\}^k$ that satisfies φ. Assume, without loss of generality, that for some l, $1 \leq l \leq k+1$ the components a_j for $1 \leq j < l$ are all equal to 0 and the rest are all all equal to 1. For each i such that $l \leq i \leq k$, let φ_i be the formula in CNF$_C(S)$ obtained from φ by substituting the variables $x_1, \ldots, x_{l-1}$ *and* the variable x_i with 0. It is easy to see that φ has a satisfying truth assignment strictly less than α if and only if at least one of the formulas φ_i for $l \leq i \leq k$ is satisfied. Therefore MIN SAT(S) is polynomially reducible to SAT$_C(S)$; thus, if S is Schaefer, MIN SAT(S) is polynomially solvable. Actually, this argument also shows that if S is Schaefer, then MIN SAT$_C(S)$ is solvable in polynomial time. On the intractability side, however, Cadoli [1,2] showed that there is a 1-valid set of logical relations such that MIN SAT(S) is NP-complete. Consequently, any dichotomy theorem for MIN SAT(S)

will be substantially different from Schaefer's dichotomy theorem for $\text{SAT}(S)$. Furthermore, such a dichotomy theorem should also yield a dichotomy theorem for the special case of MIN $\text{SAT}(S)$ in which S is restricted to be 1-valid. In what follows, we first establish a dichotomy theorem for this special case of MIN $\text{SAT}(S)$ and then use it to derive the desired dichotomy theorem for MIN $\text{SAT}(S)$, where S is an arbitrary finite set of logical relations.

Theorem 7. [Dichotomy of MIN $\text{SAT}(S)$ for 1-Valid S] *Let S be a 1-valid set of logical relations.*

If S is 0-valid or Schaefer, then MIN $\text{SAT}(S)$ *is solvable in polynomial time; otherwise, it is NP-complete.*

If S is Schaefer, then MIN $\text{SAT}_C(S)$ *is solvable in polynomial time; otherwise, it is NP-complete.*

Proof: Let S be a 1-valid set of logical relations. In view of the remarks preceding the statement of the theorem, it remains to establish the intractable cases of the two dichotomies. The proof involves three main steps; the first step uses Schaefer's Expressibility Theorem 5, whereas the second step requires the development of additional technical machinery concerning the expressibility of the binary logical relation $\{(0,0), (0,1), (1,1)\}$, which represents the *implication* connective $\rightarrow$.

Step 1: If S is 1-valid and non-Schaefer, then $\text{SAT}(R_{1/3})$ is log-space reducible to 1-MIN $\text{SAT}_1(S \cup \{\rightarrow\})$. Consequently, 1-MIN $\text{SAT}_1(S \cup \{\rightarrow\})$ is NP-complete.

Step 2: If S is 1-valid and non-Schaefer, then 1-MIN $\text{SAT}_1(S \cup \{\rightarrow\})$ is log-space reducible to MIN $\text{SAT}_C(S)$. Consequently, MIN $\text{SAT}_C(S)$ is NP-complete.

Step 3: If S is 1-valid but neither 0-valid nor Schaefer, then MIN $\text{SAT}_C(S)$ is log-space reducible to MIN $\text{SAT}(S)$. Consequently, MIN $\text{SAT}(S)$ is NP-complete.

Proof of Step 1: Assuming that S is 1-valid and non-Schaefer, we will exhibit a log-space reduction of $\text{SAT}(R_{1/3})$ to 1-MIN $\text{SAT}_1(S \cup \{\rightarrow\})$. According to Definition 6, the latter problem asks: given a $\text{CNF}_1(S \cup \{\rightarrow\})$-formula, is it satisfied by a truth assignment that is different from the all-ones truth assignment $(1, \ldots, 1)$?

Let $\varphi(\bar{x})$ be a given $\text{CNF}(\{R_{1/3}\})$-formula, where $\bar{x} = (x_1, \ldots, x_n)$ is the list of its variables. By applying Schaefer's Expressibility Theorem 5 to the occurrences of $R_{1/3}$ in $\varphi(\bar{x})$, we can construct in log-space a $\text{CNF}(S)$-formula $\chi(\bar{x}, \bar{z}, w_0, w_1)$, such that $\varphi(\bar{x}) \equiv \exists \bar{z} \chi(\bar{x}, \bar{z}, 0/w_0, 1/w_1)$, where $\bar{z} = (z_1, \ldots, z_m)$, w_0, w_1 are new variables different from $\bar{x}$ (substitutions of different variables by the same constant can be easily consolidated to substitutions of the occurrences of a single variable by that constant). Notice that the formula $\chi(\bar{x}, \bar{z}, w_0, 1/w_1)$, whose variables are $\bar{x}, \bar{z}$, and w_0, is a $\text{CNF}_1(S)$-formula, since it is obtained from a $\text{CNF}(S)$-formula by substitutions by 1 only. Let $\psi(\bar{x}, \bar{z}, w_0)$ be the following formula:

$$\chi(\bar{x}, \bar{z}, w_0, 1/w_1) \wedge \left(\bigwedge_{i=1}^{n} (w_0 \rightarrow x_i) \right) \wedge \left(\bigwedge_{j=1}^{m} (w_0 \rightarrow z_j) \right).$$

It is clear that $\psi(\bar{x}, \bar{z}, w_0)$ is a $\text{CNF}_1(S \cup \{\rightarrow\})$-formula (hence, 1-valid, because S is 1-valid) and that $\varphi(\bar{x}) \equiv \exists \bar{z} \chi(\bar{x}, \bar{z}, 0/w_0, 1/w_1) \equiv \exists \bar{z} \psi(\bar{x}, \bar{z}, 0/w_0)$.

It is now easy to verify that the given $\text{CNF}(\{R_{1/3}\})$-formula $\varphi(\bar{x})$ is satisfiable if and only if the $\text{CNF}_1(S \cup \{\rightarrow\})$-formula $\psi(\bar{x}, \bar{z}, w_0)$ is satisfied by a truth assignment different from the all-ones assignment $(1, \ldots, 1)$. This completes the proof of Step 1. ∎

To motivate the proof of Step 2, let us consider the combined effect of Steps 1 and 2. Once both these steps have been established, it will follow that $\text{SAT}(\{R_{1/3}\})$ is log-space reducible to $\text{MIN SAT}_C(S)$, which means that an NP-complete satisfiability problem will have been reduced to a minimal satisfiability problem. Note that the only information we have about S is that it is a 1-valid, non-Schaefer set of logical relations. Therefore, it is natural to try to use Schaefer's Expressibility Theorem 5 in the desired reduction, since it tells us that $R_{1/3}$ is definable from some $\text{CNF}_C(S)$-formula using existential quantification. The presence of existential quantifiers, however, introduces a new difficulty in our context, because this way we reduce the satisfiability of a $\text{CNF}(\{R_{1/3}\})$-formula $\varphi(\bar{x})$ to the minimal satisfiability of a $\text{CNF}_C(S)$-formula $\psi(\bar{x}, \bar{z})$, where $\bar{z}$ are additional variables. It is the presence of these additional variables that creates a serious difficulty for minimal satisfiability, unlike the case of satisfiability in Schaefer's Dichotomy Theorem 4. Specifically, it is conceivable that, while we toil to preserve the minimality of truth assignments to the variables $\bar{x}$, the witnesses to the existentially quantified variables $\bar{z}$ may very well destroy the minimality of truth assignments to the entire list of variable $\bar{x}, \bar{z}$. Note that this difficulty was bypassed in Step 1 by augmenting S with the implication connective $\rightarrow$, which made it possible to produce formulas in which we control the witnesses to the variables $\bar{z}$. The proof of Step 2, however, hinges on the following crucial technical result that provides precise information about the definability of the implication connective $\rightarrow$ from an arbitrary 1-valid, non-Schaefer set S of logical relations.

Key Lemma 8. *Let S be a 1-valid, non-Schaefer set of logical relations. Then at least one of the following two statements is true about the implication connective.*

1. *There exists a $\text{CNF}_C(S)$-formula $\varepsilon(x, y)$ such that $(x \rightarrow y) \equiv \varepsilon(x, y)$.*
2. *There exists in $\text{CNF}_C(S)$-formula $\eta(x, y, z)$ such that*
   (i) $(x \rightarrow y) \equiv (\exists z)\eta(x, y, z)$; (ii) $\eta(x, y, z)$ is satisfied by the truth assignment $(1, 1, 1)$;
   (iii) if a truth assignment $(1, 1, b)$ satisfies $\eta(x, y, z)$, then $b = 1$.
   In other words, the formula $(\exists z)\eta(x, y, z)$ is logically equivalent to $(x \rightarrow y)$ and has the additional property that 1 is the only witness for the variable z under the truth assignment $(1, 1)$ to the variables (x, y).

The proof of the above Key Lemma 8 and the formal proofs of Steps 1 and 2 can be found in the full version of this paper [15]. This concludes the proof of Theorem 7. ∎

The following three examples illustrate the preceding Theorem 7.

Example 9. Consider the ternary logical relation $K = \{(1, 1, 1), (0, 1, 0), (0, 0, 1)\}$. Since K is 1-valid, the satisfiability problem $\text{SAT}(\{K\})$ is trivial (the answer is always "yes"). In contrast, Theorem 7 implies that the minimal satisfiability problems $\text{MIN SAT}(\{K\})$ and $\text{MIN SAT}_C(\{K\})$ are NP-complete. Indeed, it is obvious that K is not 0-valid. Moreover, using the criteria mentioned after Definition 3, it is easy to verify that K is neither bijunctive, nor Horn, nor dual Horn, nor affine (for instance, K is not Horn because $(0, 1, 0) \wedge (0, 0, 1) = (0, 0, 0) \notin K$).

Note that the logical relation K can also be used to illustrate the Key Lemma 8. Specifically, it is clear that $(x \to y)$ is logically equivalent to the formula $(\exists z)K(x, y, z)$; moreover, 1 is the only witness for the variable z such that $(\exists z)K(1, 1, z)$ holds. As a matter of fact, it was this particular property of K that inspired us to conceive of the Key Lemma 8. ∎

Example 10. Consider the 1-valid set $S = \{R_0, R_1, R_2\}$, where $R_0 = \{0, 1\}^3 - \{(0, 0, 0)\}$ (expressing the clause $(x \lor y \lor z)$), $R_1 = \{0, 1\}^3 - \{(1, 0, 0)\}$ (expressing the clause $(\neg x \lor y \lor z)$), $R_2 = \{0, 1\}^3 - \{(1, 1, 0)\}$ (expressing the clause $(\neg x \lor \neg y \lor z)$). Since S is a 1-valid set, $\text{SAT}(S)$ is trivial. In contrast, Theorem 7 implies that $\text{MIN SAT}(S)$ and $\text{MIN SAT}_C(S)$ are NP-complete, since it is not hard to verify that S is neither 0-valid nor Schaefer. ∎

Theorem 7 yields a dichotomy for $\text{MIN SAT}(S)$, where S is a 1-valid set of logical relations. We will now use this result to establish a dichotomy for $\text{MIN SAT}(S)$, where S is an arbitrary set of logical relations. Before doing so, however, we need to introduce the following crucial concept.

Definition 11. Let R be a k-ary logical relation and R' a relation symbol to be interpreted as R. We say that a logical relation T is a 0-*section* of R if R can be defined from the formula $R'(x_1, \ldots, x_k)$ by replacing some (possibly none), but not all, of the variables $x_1, \ldots, x_k$ by 0. ∎

To illustrate this concept, observe that the 1-valid logical relation $\{(1)\}$ is a 0-section of $R_{1/3} = \{(1, 0, 0), (0, 1, 0), (0, 0, 1)\}$, since it is definable by $R'_{1/3}(x_1, 0, 0)$. Note that the logical relation $\{(1, 0), (0, 1)\}$ is also a 0-section of $R_{1/3}$, since it is definable by the formula $R'_{1/3}(0, x_2, x_3)$, but it is not 1-valid. In fact, it is easy to verify that $\{(1)\}$ is the *only* 0-section of $R_{1/3}$ that is 1-valid.

Theorem 12. [Dichotomy of $\text{MIN SAT}(S)$] *Let S be a set of logical relations and let S^* be the set of all logical relations P such that P is both 1-valid and a 0-section of some relation in S.*

If S^* is 0-valid or Schaefer, then $\text{MIN SAT}(S)$ is solvable in polynomial time; otherwise, it is NP-complete.

If S^* is Schaefer, then $\text{MIN SAT}_C(S)$ is solvable in polynomial time; otherwise, it is NP-complete.

Moreover, each of these two dichotomies can be decided in polynomial time; that is to say, there is a polynomial-time algorithm to decide whether, given a finite set S of logical relations, $\text{MIN SAT}(S)$ is solvable in polynomial time or NP-complete (and similarly for $\text{MIN SAT}_C(S)$).

A complete proof of the above theorem can be found in the full version of this paper [15] (in electronic form). We now present several different examples that illustrate the power of Theorem 12.

Example 13. If m and n are two positive integers with $m < n$, then $R_{m/n}$ is the n-ary logical relation consisting of all n-tuples that have m ones and $n - m$ zeros. Clearly, $R_{m/n}$ is neither 0-valid nor 1-valid. Moreover, it is not hard to verify that $R_{m/n}$ is not

Schaefer. Let S be a set of logical relations each of which is a relation $R_{m/n}$ for some m and n with $m < n$. The preceding remarks and Schaefer's Dichotomy Theorem 4 imply that $\text{SAT}(S)$ is NP-complete. In contrast, the Dichotomy Theorem 12 implies that $\text{MIN SAT}(S)$ and $\text{MIN SAT}_C(S)$ are solvable in polynomial time. Indeed, S^* is easily seen to be Horn (and, hence, Schaefer), since every relation P in S^* is a singleton $P = \{(1, \ldots, 1)\}$ consisting of the m-ary all-ones tuple for some m.

   This family of examples contains POSITIVE-1-IN-3-SAT as the special case where $S = \{R_{1/3}\}$; thus, Theorem 12 provides an explanation for the difference in complexity between the satisfiability problem and the minimal satisfiability problem for POSITIVE-1-IN-3-SAT. ∎

Example 14. Consider the 3-ary logical relation $T = \{0, 1\}^3 - \{(0, 0, 0), (1, 1, 1)\}$. In this case, $\text{SAT}(\{T\})$ is the problem POSITIVE-NOT-ALL-EQUAL-3-SAT: given a 3CNF-formula φ with clauses of the form $(x \vee y \vee z)$, is there a truth assignment such that in each clause of φ at least one variable is assigned value 1 and at least one variable is assigned value 0? This problem is NP-complete. In contrast, the Dichotomy Theorem 12 easily implies that $\text{MIN SAT}(\{T\})$ and $\text{MIN SAT}_C(\{T\})$ are solvable in polynomial time. To see this, observe that $\{T\}^* = \{\{(1)\}, \{(0, 1), (1, 0), (1, 1)\}\}$, where the logical relation $\{(1)\}$ is the 0-section of T obtained from T by setting any two variable to 0 (for instance, it is definable by the formula $T'(x, 0, 0)$) and the logical relation $\{(0, 1), (1, 0), (1, 1)\}$ is the 0-section of T obtained from T by setting any one variable to 0 (for instance, it is definable by the formula $T'(x, y, 0)$). It is clear that each of these two logical relations is bijunctive (actually, each is also dual Horn), hence $\{T\}^*$ is Schaefer. ∎

Example 15. 3-SAT coincides with $\text{SAT}(S)$, where $S = \{R_0, R_1, R_2, R_3\}$ and $R_0 = \{0, 1\}^3 - \{(0, 0, 0)\}$ (expressing the clause $(x \vee y \vee z)$), $R_1 = \{0, 1\}^3 - \{(1, 0, 0)\}$ (expressing the clause $(\neg x \vee y \vee z)$), $R_2 = \{0, 1\}^3 - \{(1, 1, 0)\}$ (expressing the clause $(\neg x \vee \neg y \vee z)$), and $R_3 = \{0, 1\}^3 - \{(1, 1, 1)\}$ (expressing the clause $(\neg x \vee \neg y \vee \neg z)$). Since the logical relations R_0, R_1, R_2 are 1-valid, they are members of S^*. It follows that S^* is not 0-valid, since it contains R_0. Moreover, the logical relation R_1 is not Horn, it is not bijunctive, and it is not affine, whereas the logical relation R_2 is not dual Horn. Consequently, S^* is not Schaefer. We can now apply Theorem 12 and immediately conclude that $\text{MIN SAT}(S)$ (i.e., MIN 3-SAT) is NP-complete. ∎

References

1. M. Cadoli. The complexity of model checking for circumscriptive formulae. *Information Processing Letters*, pages 113–118, 1992.
2. M. Cadoli. *Two Methods for Tractable Reasoning in Artificial Intelligence: Language Restriction and Theory Approximation*. PhD thesis, Univ. Di Roma, 1993.
3. N. Creignou and M. Hermann. Complexity of generalized satisfiability counting problems. *Information and Computation*, 125(1):1–12, 1996.
4. N. Creignou. A dichotomy theorem for maximum generalized satisfiability problems. *Journal of Computer and System Sciences*, 51:511–522, 1995.

5. R. Dechter and J. Pearl. Structure identification in relational data. *Artificial Intelligence*, 48:237–270, 1992.
6. S. Fortune, J. Hopcroft, and J. Wyllie. The directed homeomorphism problem. *Theoretical Computer Science*, 10:111–121, 1980.
7. T.A. Feder and M.Y. Vardi. The computational structure of monotone monadic SNP and constraint satisfaction: a study through Datalog and group theory. *SIAM J. on Computing*, 28:57–104, 1999.
8. M. R. Garey and D. S. Johnson. *Computers and Intractability - A Guide to the Theory of NP-Completeness* W. H. Freeman and Co., 1979.
9. P. Hell and J. Nešetřil. On the complexity of H-coloring. *Journal of Combinatorial Theory, Series B*, 48:92–110, 1990.
10. G. Istrate. Looking for a version of Schaefer's dichotomy theorem when each variable occurs at most twice. T.R. 652, Computer Science Department, The University of Rochester, 1997.
11. P.G. Jeavons, D.A. Cohen, and M. Gyssens. Closure properties of constraints. *Journal of the Association for Computing Machinery*, 44:527–548, 1997.
12. D. Kavvadias and M. Sideri. The inverse satisfiability problem. *SIAM Journal of Computing*, 28(1):152–163, 1998.
13. S. Khanna, M. Sudan, and L. Trevisan. Constraint satisfaction: the approximability of minimization problems. *Proc. 12th IEEE Conf. on Computational Complexity*, 282–296, 1997.
14. S. Khanna, M. Sudan, and D.P. Williamson. A complete classification of the approximability of maximization problems derived from Boolean constraint satisfaction. *Proceedings of the 29th Annual ACM Symposium on Theory of Computing*, 11–20, 1997.
15. L.M. Kirousis and Ph.G. Kolaitis. The complexity of minimal satisfiability problems. Electronic Coll. Computational Complexity (http://www.eccc.uni-trier.de/eccc), No. 82, 2000.
16. Ph.G. Kolaitis and M.Y. Vardi. Conjunctive-query containment and constraint satisfaction. In *Proc. 17th ACM Symp. on Principles of Database Systems*, pages 205–213, 1998.
17. R. Ladner. On the structure of polynomial time reducibility. *Journal of the Association for Computing Machinery*, 22:155–171, 1975.
18. V. Lifschitz. Computing circumscription. In *Proceedings of the 9th International Joint Conference on Artificial Intelligence - AAAI '85*, pages 121–127, 1985.
19. J. McCarthy. Circumscription - a form of nonmonotonic reasoning. *Artificial Intelligence*, 13:27–39, 1980.
20. J. McCarthy. Applications of circumscription in formalizing common sense knowledge. *Artificial Intelligence*, 28:89–116, 1985.
21. S. Reith and H. Vollmer. Optimal satisfiability for propositional calculi and constraint satisfaction problems. In *Proc. 25th Symposium on Mathematical Foundations of Computer Science - MFCS 2000* pages 640–649, Lecture Notes in Computer Science, Springer, 2000.
22. T.J. Schaefer. The complexity of satisfiability problems. In *Proc. 10th ACM Symp. on Theory of Computing*, pages 216–226, 1978.
23. U. Zwick. Finding almost-satisfying assignments. *Proceedings of the 30th Annual ACM Symposium on Theory of Computing*, 551–560, 1998.

On the Minimal Hardware Complexity of Pseudorandom Function Generators

(Extended Abstract)

Matthias Krause and Stefan Lucks*

Theoretische Informatik, Univ. Mannheim, 68131 Mannheim, Germany
{krause,lucks}@informatik.uni-mannheim.de

Abstract. A set F of Boolean functions is called a pseudorandom function generator (PRFG) if communicating with a randomly chosen secret function from
F cannot be efficiently distinguished from communicating with a truly random
function. We ask for the minimal hardware complexity of a PRFG. This question is motivated by design aspects of secure secret key cryptosystems. These
should be efficient in hardware, but often are required to behave like PRFGs. By
constructing efficient distinguishing schemes we show for a wide range of basic
nonuniform complexity classes (including TC_2^0), that they do not contain PRFGs.
On the other hand we show that the PRFG proposed by Naor and Reingold in [24]
consists of TC_4^0-functions. The question if TC_3^0-functions can form PRFGs remains as an interesting open problem. We further discuss relations of our results
to previous work on cryptographic limitations of learning and Natural Proofs.
Keywords: Cryptography, pseudorandomness, Boolean complexity theory,
computational distinguishability.

1 Basic Definitions

A **function generator** F is an efficient (i.e., polynomial time) algorithm which for specific values of plaintext block length n computes for each plaintext block $x \in \{0,1\}^n$
and each key s from a predefined key set $S_n^F \subseteq \{0,1\}^{k(n)}$ a corresponding ciphertext
output block $y = F_n(x,s) \in \{0,1\}^{l(n)}$. $k(n)$ and $l(n)$ are called key length and output
length of F. The efficiency of F implies that $k(n)$ and $l(n)$ are polynomially bounded
in n. Observe that the encryption mechanism of a secret key block cipher can be thought
of as a function generator in a straightforward way. Clearly, cryptographic algorithms
occurring in practice are usually designed for one specific input length n. However, in
many cases the definition can be generalized to infinitely many values of admissible
input length n in a more or less natural way. Correspondingly, we consider function
generators to be sequences $F = (F_n)_{n \in \mathbb{N}}$ of sets of Boolean functions

$$F_n = \left\{ f_{n,s} : \{0,1\}^n \longrightarrow \{0,1\}^{l(n)}; \ s \in S_n^F \right\},$$

where, if n is admissible, we define $f_{n,s}(x) = F_n(x,s)$.

A function generator F is **pseudorandom** if it is infeasible to distinguish between a
(pseudorandom) function, which is randomly chosen from F_n, n admissible, and a truly

---

* Supported by DFG grant Kr 1521/3-1.

A. Ferreira and H. Reichel (Eds.): STACS 2001, LNCS 2010, pp. 419–430, 2001.

random function $f \in B_n^{l(n)}$. (For $l, n \in \mathbb{N}$ let B_n^l denote the set of all $2^{2^{ln}}$ functions $f : \{0,1\}^n \longrightarrow \{0,1\}^l$.) In the sequel, we concentrate on functions $f : \{0,1\}^n \longrightarrow \{0,1\}^1$ and define $B_n = B_n^1$. Note that a truly random function in $B_n^l(n)$ is just a tuple of $l(n)$ independent random functions in B_n.

An **H-oracle** chooses randomly, via the uniform distribution on H, a secret function $h \in H$ and answers membership queries for inputs $x \in \{0,1\}^n$ immediately with $h(x)$. A **distinguishing algorithm** for a function generator $F = F_n$ is a randomized oracle Turing machine D which knows the definition of F, which gets an admissible input parameter n and which communicates via membership queries with an H-oracle, where either $H = B_n^{l(n)}$ (the truly random source) or $H = F_n$ (the pseudorandom source). The aim of D is to find out whether $H = B_n$ (in this case, D outputs 0) or $H = F_n$ (in this case, D outputs 1). Let us denote by $Pr_D(f)$ the probability that D accepts if the unknown oracle function is f. The relevant cost parameters of a distinguishing algorithms D are the **worst case running time** $t_D = t_D(n)$ and the **advantage** $\varepsilon_D = \varepsilon_D(n)$, which is defined as

$$\varepsilon_D(n) = \big| Pr[D \text{ outputs } 1 | H = F_n] - Pr[D \text{ outputs } 1 | H = B_n] \big|$$
$$= \big| \mathbf{E}_{f \in F_n} Pr_D(f) - \mathbf{E}_{f \in B_n^{l(n)}} Pr_D(f) \big|.$$

The **ratio** $r_D = r_D(n)$ of a distinguishing algorithm D is $r_D(n) = t_D(n) \cdot \varepsilon_D^{-1}(n)$.

Observe further that for any function generator F, there are two trivial strategies to distinguish it from a truly random source, which achieve ratio $O(|F_n| \log(|F_n|))$, the trivial upper bound. In both cases the distinguisher fixes a set X of inputs, where $|X|$ is the minimal number satisfying $2^{|X|} \geq 2|F_n|$. The first strategy is to fix a function $f \in F_n$ and to accept if the oracle coincides with f on X. This gives running time $O(|X|) = O(\log |F_n|)$ and advantage $\frac{1}{2}|F_n|^{-1}$. The second strategy is to check via exhaustive search whether there is some $f \in F_n$ which coincides with the oracle function on X. This implies advantage at least $\frac{1}{2}$ but running time $O(|F_n| \log(|F_n|))$.

We call F a **pseudorandom function generator** (for short: **PRFG**) if for all distinguishing algorithms D for F it holds that $r_D \in 2^{n^{\Omega(1)}}$. Observe that this definition is similar to that in [7]. The difference is that in [7] only superpolynomiality is required.

Given a complexity measure M we denote by $P(M)$ the complexity class containing all sequences of (multi-output) Boolean functions which have polynomial size representations with respect to M. We say that a function generator F has M-complexity bounded by a function $c : \mathbb{N} \longrightarrow \mathbb{N}$ if for all n and all keys $s \in S_n^F$ it holds that $M(f_{n,s}) \leq c(n)$, and that F belongs to $P(M)$ if the M-complexity of F is bounded by some $c(n) \in n^{O(1)}$. We will call a complexity class **cryptographically strong** if it contains a PRFG, and **cryptographically weak** otherwise.

It is widely believed that there exist PRFGs (see e.g. section 4), i.e., P/poly is supposed to be cryptographically strong. Pseudorandom function generators are of great interest in cryptography, e.g. as building blocks for block ciphers [20,21], for remotely keyed encryption schemes [22,3], for message authentication [2], and others. As the existence of PRFGs obviously implies $P \neq NP$, recent pseudorandomness proofs refer to unproven cryptographic hardness assumptions. Below we search for cryptographical strength – or weakness – for most of the basic nonuniform complexity classes.

A distinguishing algorithm $D = D(n, m)$, depending on input parameters n (input length) and m (complexity parameter), is a **polynomial distinguishing scheme** with respect to M (resp. P(M)) if there are functions $t(n,m), \varepsilon^{-1}(n,m) \in (n+m)^{O(1)}$ such that for all polynomial bounds $m = m(n) \in n^{O(1)}$ and all (single output) functions $g \in B_n$ with $M(g) \le m(n)$ it holds that $D(n,m)$ runs in time $t(n,m)$ and

$$Pr_D(g) - \mathbf{E}_{f \in B_n} Pr_D(f) \ge \varepsilon(n,m).$$

The definition of a **quasipolynomial distinguishing scheme** with respect to M can be obtained by replacing $t(n,m), \varepsilon^{-1}(n,m) \in (n+m)^{O(1)}$ by $t(n,m), \varepsilon^{-1}(n,m) \in (n+m)^{\log O(1)(n+m)}$. We call a distinguishing scheme **efficient** if it is quasipolynomial or polynomial.

If there is an efficient distinguishing scheme D w.r.t. such a complexity measure M then, obviously, P(M) is cryptographically weak as each output bit of a function generator in P(M) can be efficiently distinguished via D. Consequently, as the **efficiency of key length** is a central design criterion for modern secret key encryption algorithms, these algorithms should have nearly maximal complexity w.r.t. to such complexity measures M. As cryptographers are searching for encryption mechanisms having hardware implementations which are very efficient with respect to time and energy consumption, there is a **low complexity danger** to get into the sphere of influence of one of the distinguishing schemes presented in this paper.

We consider several types of constant depth circuits over unbounded fan-in MOD_m, AND-, OR-, as well as bounded and unbounded weight threshold gates. The gate function MOD_m is defined by $\text{MOD}_m(x_1, \ldots, x_n) = 1$ if and only if $x_1 + \ldots + x_n \not\equiv 0 \bmod m$. Unweighted threshold gates $T^n_{\ge r}$, resp. $T^n_{\le r}$, are defined by the relations

$$T^n_{\ge r}(x_1, \ldots, x_n) = 1 \iff x_1 + \ldots + x_n \ge r$$

and $T^n_{\le r}(x_1, \ldots, x_n) = 1 \iff x_1 + \ldots + x_n \le r$. A weighted threshold gate $T^{\vec{a}}_{\ge r}$, where $\vec{a} \in \mathbb{Z}^n$, is defined by the relation

$$T^{\vec{a}}_{\ge r}(x_1, \ldots, x_n) = 1 \iff a_1 x_1 + \ldots + a_n x_n \ge r.$$

The inputs are the constants 0 and 1 and literals from $\{x_1, \ldots, x_n, \bar{x}_1, \ldots, \bar{x}_n\}$. The definition of the mode of computation as well as the definition of AND- and OR-gates should be known. As usual, by AC^0_k, $AC^0_k[m]$, TC^0_k we denote the complexity classes consisting of all problems having polynomial size depth k circuits over AND-,OR-, resp. AND-, OR-, MOD_m-, resp. unweighted threshold gates.

We further consider branching programs, alternatively called binary decision diagrams (BDDs). A branching program for a Boolean function $f \in B_n$ is a directed acyclic graph $G = (V, E)$ with l sources. Each sink is labeled by a Boolean constant and each inner node by a Boolean variable. Inner nodes have two outgoing edges, one labeled by 0 and the other by 1. Given an input a, the output $f(a)_j$ is equal to the label of the sink reached by the unique path consistent with a and starting at source j, $1 \le j \le l$. Relevant restricted types of branching programs are

- Ordered binary decision diagrams (OBDDs), where each computational path has to respect the same variable ordering. An OBDD which respects a fixed variable ordering π is called a π-OBDD.
- Read-k-BDDs, for which on each path each variable is forbidden to occur more than k times.

2   Related Work, Our Results

Cryptographic Weakness. In section 3 we present efficient distinguishing schemes for the following complexity measures,

- a quasipolynomial scheme for the size of read-k BDDs (Theorem 3),
- a quasipolynomial scheme for the size of weighted Threshold-MOD_2 circuits, i.e. depth 2 circuits with a layer of MOD_2-gates connected with one output layer consisting of weighted threshold gates (Theorem 1),
- a quasipolynomial scheme for the size of constant depth circuits consisting of AND-, OR-, and MOD_p-gates, p prime (Theorem 2),
- a polynomial scheme for the size of unweighted threshold circuits of depth 2 (Theorem 4),
- a quasipolynomial scheme for the size of constant depth circuits having a constant number of layers of AND-, OR-gates connected with one output layer of weighted threshold gates (Theorem 5).

Observe that the function generator $f_{\vec{a}}(x_1, \ldots, x_n) = \sum_{i=1}^{n} a_i x_i$, where $\vec{a} \in \mathbb{Z}^n$, $(x_1, \ldots, x_n) \in \{0, 1\}^n$, corresponding to the NP-hard **Subset Sum Problem**, belongs to TC_2^0 [28], which emphasizes the cryptographic weakness of this operation.

The complexity measures M handled below represent a "frontline" in the sense that they correspond to the most powerful models for which we know effective lower bound arguments, i.e., methods to show $\Pi \notin P(M)$ for some explicitly defined problem Π. Indeed, all our distinguishing schemes are inspired by the known lower bound arguments for the corresponding models and can be seen as some "algorithmic version" of these arguments. It seems that searching for effective lower bound arguments for a complexity measure M is the same problem as searching for methods to distinguish unknown $P(M)$-functions from truly random functions. Note that a similar observation, but with respect to another mode of distinguishing, was made already by *Razborov* and *Rudich* in [27]. For illustrating the difference of their approach with our paper let us review the results in [27] in some more detail and start with the following definition.

Distinguishing Schemes versus Natural Proofs. Let $\Gamma \subseteq P/poly$ denote a complexity class and $T = (T_n) \in \Gamma$ be a sequence of Boolean functions for which the input length of T_n is $N = 2^n$. T is called an efficient Γ-test against a function generator $F = (F_n)_{n \in \mathbb{N}}$ (consisting of single output functions) if for all n
$$\left| Pr_f[T_n(f) = 1] - Pr_s[T_n(f_{n,s}) = 1] \right| \geq p^{-1}(N)$$
for a polynomially (in N) bounded function $p : \mathbb{N} \longrightarrow \mathbb{N}$. Hereby, functions $f \in B_n$ are considered to be strings of length $N = 2^n$. The probability on the left side is taken w.r.t. the uniform distribution on B_n (the truly random case), the probability on the right side is taken w.r.t. the uniform distribution on F_n (the pseudorandom case). The following observation was made in [27].

(1) It seems that all complexity classes Λ for which we know a method for proving that $F \notin \Lambda$ for some explicitly defined problem F have a so called Γ-Natural Proof for some complexity classes $\Gamma \subseteq P/poly$. (the somewhat technical definition of Natural Proofs is omitted here).

(2) On the other hand (and this is the property of Natural Proofs which is important in our context), if Λ has a Γ-Natural Proof then all function generators $F = (F_n)$ belonging to Λ have efficient Γ-tests.

The main implication of [27] is that a $P/poly$-Natural Proof against $P/poly$ would imply the nonexistence of function generators which are pseudorandom w.r.t. $P/poly$-tests. But this implies the nonexistence of pseudorandom bit generators [27], contradicting widely believed cryptographic hardness assumptions.

In contrast to our concept of pseudorandomness, the existence of an efficient Γ-test for a given PRFG F does not yield any feasible attack against the corresponding cipher, because the whole function table has to be processed, which is of exponential size in n. Thus, informally speaken, the message of [27] is that effective lower bound arguments for M, as a rule, imply low complexity circuits which efficiently distinguish P(M)-functions from truly random functions, where the complexity is measured in the size of the whole function table. Our message is that effective lower bound arguments for M, as a rule, imply even efficient distinguishing attacks against each secret key encryption mechanism which belongs to P(M), where the running time is measured in the input length of the function. Observe that our most complicated distinguishing scheme for the size of constant depth circuits over AND, OR, MOD_p, p prime, (Theorem 2) uses an idea from [27] for constructing an NC^2-Natural Proof for $AC^0[p]$, $p > 2$ prime.

Cryptographic Strength. In section 4 we try to identify the smallest complexity classes which are powerful enough to contain PRFGs. In [7], a general method for constructing PRFGs on the basis of pseudorandom bit generators is given. The construction is inherently sequential, and at first glance it seems hopeless to build PRFGs with small parallel time complexity. *Naor* and *Reingold* [23,24] used a modified construction, based on concrete number-theoretic assumptions instead of generic pseudorandom bit generators. They presented a function generator (which we shortly call NR-generator, the definition will be presented in section 4) which is pseudorandom under the condition that the **Decisional Diffie-Hellman Assumption**, a widely believed cryptographic hardness assumption, is true. Moreover, the NR-generator belongs to TC^0, in [24] it is claimed (without proof) that it consists of TC_5^0-functions.

We show in Theorem 6 that the NR-generator even consists of TC_4^0-functions, i.e. TC_4^0 seems to be cryptographic strong while TC_2^0 is weak. It is an interesting open question if TC_3^0 is strong enough to contain PRFGs.

Learning versus Distinguishing. Clearly, a successful distinguishing attack against a secret key encryption algorithm does not automatically imply that relevant information about the secret key can be efficiently computed. Observe that breaking the cipher corresponds to efficiently learning an unknown function from a known concept class. It is intuitively clear and easy to prove that, with respect to any reasonable model of algorithmically learning Boolean concept classes from examples, any efficient learning algorithm for functions from a given complexity class Λ gives an efficient distinguishing scheme for Λ. (Use the learning algorithm to compute a low complexity hypothesis h of the unknown function f and test if h really approximates f.) But without making membership queries, each efficient distinguishing algorithm (which poses oracle queries only for randomly chosen inputs) can be simulated by an efficient weak learning algorithm, computing a $\frac{1}{2} + \varepsilon$-approximator for the unknown function [4]. I.e., efficient **known plaintext** distinguishing attacks do clearly break a cipher. There is some evidence that in the general case, if **chosen plaintext**, i.e., membership queries are allowed, this is not the case. It is not hard to see that there is a polynomial distinguishing scheme for polynomial size OBDDs.[1] On the other hand, there are several results proved in [17] which strongly support the following conjecture: it is impossible to efficiently learn the optimal variable ordering of a function with small OBDDs from examples.

The results of this paper can be considered as cryptographic limitations of proving lower bounds for complexity classes containing TC_4^0, while the results of [27] can be seen as cryptographic limitations of proving lower bounds against P/poly. Cryptographic limitations of learning were already detected by *Kearns* and *Valiant* in [13]: efficient learnability of TC_3^0-functions would contradict the existence of pseudorandom bit generators in TC_3^0 and thus to widely believed cryptographic hardness assumptions like the security of RSA or *Rabin*'s cryptosystem.

Note that for all complexity classes Λ which are shown in section 3 to be cryptographically weak, it is unknown whether Λ- functions are efficiently learnable.

3   Distinguishing Schemes

We start with basis test $T(p, \delta, N)$, where $\delta, p \in (0, 1)$, which accepts if

$$\frac{1}{N} \sum_{i=1}^{N} X_i \notin [p - \delta, p + \delta].$$

The X_i denote N mutually independent random variables defined by $Pr[X_i = 1] = p$ and $Pr[X_i = 0] = 1 - p$. Höffdings Inequality, see e.g., [1, Appendix A], yields

Lemma 1. *The probability that $T(p, \delta, N)$ accepts is smaller than $2e^{-2\delta^2 N}$.* □

Most of our distinguishing schemes first choose a random seed r from an appropriate set R, and then perform a corresponding test $T(r)$ on the oracle function. Such a scheme is called a (p, q, ρ)-**test for a function** $f^* \in B_n$ if it accepts a random function with probability $\leq \rho$ (i.e., $\mathbf{E}_{r \in R}[Pr_{f \in B_n}[T(r) \text{ accepts } f]] \leq \rho$), but if the probability (taken over r) that $T(r)$ accepts $f^* \in F_n$ with probability at least q, is $\geq p$.

Lemma 2. *If $pq > \rho$ then a (p, q, ρ)-test for f^* distinguishes f^* with advantage at least $pq - \rho$ from a truly random function.* □

Theorem 1. *There is a polynomial distinguishing scheme for polynomial size weighted threshold-MOD_2 circuits.*

---

[1] Take disjoint random subsets of variables Y and Z of appropriate logarithmic size and test if the matrix $(f(y, z, \vec{0}))$, where y and z range over all assignments of Y and Z, resp., has small rank. As in the pseudorandom case with probability $1/poly(n)$, Y and Z are separated by the optimal variable ordering of the oracle function f. This gives an efficient test.

Proof. The algorithm is based on a result from *Bruck* [6]. If m is the minimal number of MOD_2-nodes in a weighted threshold-MOD_2-circuit computing a given $f \in B_n$ then there is a MOD_2-function $p(x) = x_{i_1} \oplus \ldots \oplus x_{i_r}$ in B_n such that $\left| \mathbf{E}_{x \in \{0,1\}^n} [f \oplus p(x)] - \frac{1}{2} \right| \geq \frac{1}{2m}$. Let us fix a polynomial bound $m(n) \in n^{O(1)}$. Let the scheme D work as follows on n and $m = m(n)$. It chooses an approriate number $\tilde{n}$, $\log(m) < \tilde{n} < n$, chooses a random MOD_2-function $\tilde{p}(x)$ over $\{x_1, \ldots, x_{\tilde{n}}\}$ and accepts if $\left| \mathbf{E}_{x \in \{0,1\}^n} [f(x, \vec{0}) \oplus \tilde{p}(x)] - \frac{1}{2} \right| \geq \frac{1}{4m}$. Observe that the running time is linear in $N = 2^{\tilde{n}}$ and that this test is a $(1/N, 1, 2e^{-2\frac{1}{16m^2}N})$-test on each function $f^* \in B_n$ having weighted threshold-MOD_2 circuits of size m. (Observe the above mentioned result [6] and the fact that the subfunction $f(\cdot, \vec{0})$ has size $\leq m$.) It is easy to see that we can find some $\tilde{n} \in O(\log(n))$ yielding advantage $\frac{1}{2N}$ (see Lemma 2). $\square$

Theorem 2. *For all primes p and all constant depth bounds d there is a quasipoly-nomial distinguishing scheme for polynomial size depth d circuits over $\{AND, OR, MOD_p\}$.*

The proof is quite lengthy and can be found in the full paper [14]. As MOD_{p^k} belongs to $AC_2^0[p]$ [29], the proof for prime powers follows immediately.

Theorem 3. *For all $k \geq 1$ there is a quasipolynomial distinguishing scheme for non-deterministic read–k BDDs.*

Proof. The first exponential lower bounds on read k branching programs were independently proved in [5] and [26]; see also [12]. We use these methods for our distinguishing scheme. Let us fix an arbitrary natural constant $k \geq 1$, and a polynomial bound $m = m(n) \in n^{O(1)}$. Let us denote $X_n = \{x_1, \ldots, x_n\}$. *Jukna* [12] shows the existence of a number $s \in m^{O(1)} = n^{O(1)}$ and a constant $\gamma \in (0,1)$ such that each $f \in B_n$ which is computable by a nondeterministic syntactic read–k times branching program of size $m(n)$ can be written as $f = \bigvee_{i=1}^{W} f_i$, where for all i, $1 \leq i \leq W$, it holds that there is a partition $X_n = U_i \cup V_i \cup W_i$ of pairwise disjoint subsets U_i, V_i, W_i of X_n such that $f_i(X_n) = g_i(U_i, V_i) \wedge h_i(V_i, W_i)$, where $|U_i| \geq \gamma n$ and $|W_i| \geq \gamma n$.
  The distinguishing scheme D works on n and $m = m(n)$ as follows.

(0) Fix an appropriate $N \in n^{O(1)}$ and test via $T(\frac{1}{2}, \frac{1}{12}, N)$ if the probability that the oracle function outputs 1 is at least $\frac{1}{3}$. If not accept.

(1) Compute s and parameters $q, r \in \log^{O(1)} n$. Let $Q = 2^q$. Choose randomly disjoint subsets U, W from X_n with $|U| = |W| = q$, and a $\{0,1\}$-assignment b of $X \setminus (U \cup W)$. Finally, choose random $\{0,1\}$-assignments $a^1, \ldots, a^r$ of U.

(2) Accept iff $f(a^1, b, c) \wedge \ldots \wedge f(a^r, b, c) = 1$ for at least $\frac{Q}{6s}$ assignments c of W.

The parameters q, N, and r will be specified later. Observe that the running time is $O(rQ)$. Observe further that the probability that a truly random function will be accepted in Step 2 is bounded by $2e^{-2\delta^2 Q}$ for $\delta = \frac{1}{6s} - 2^{-r}$ (see (1)).
  In the pseudorandom case $U \subseteq U_j$ and $W \subseteq W_j$ holds for some j for which $Pr_x[f_j(x) = 1] \geq \frac{1}{3s}$ with probability $\frac{1}{s}(\gamma/2)^{2q}$. Further, with probability $\frac{1}{2s}(\gamma/2)^{2q}$ we have b fixed in such that $Pr_{a,c}[f_j(a, b, c) = 1] \geq \frac{1}{6s}$, where a and c denote the

assignments of U and W respectively. This implies $Pr_a[g_j(a,b) = 1] \geq \frac{1}{6s}$ and $Pr_c[h_j(b,c) = 1] \geq \frac{1}{6s}$. Thus, with probability $p = \frac{1}{6s}^r \frac{1}{2s}(\gamma/2)^{2q} g_j(a^1,b) = \ldots = g_j(a^r,b) = 1$ holds. Under this condition, it holds for all assignments c to W and $l, 1 \leq l \leq r$, that $f_j(a_l,b,c) = 1$ iff $h_j(b,c) = 1$ iff $f_j(a_i,b,c) = 1$ for all $l, 1 \leq l \leq r$. As $f_j(a_i,b,c) = 1$ implies $f(a_i,b,c) = 1$, the function is accepted in Step 2 with probability 1. We obtain that Step 1 and 2 form a $(p, 1, 2e^{-2\delta^2 Q})$-test for each function f of size at most m. It can be easily verified that for $q = \lfloor \log_2(s^2 n) \rfloor$ and $r = \lfloor \log_2(12s) \rfloor$, we can find some $N \in n^{O(1)}$ such that $D(n,m)$ achieves advantage $\varepsilon(n,m)$ fulfilling $\varepsilon(n,m)^{-1} \in n^{O(\log n)}$. $\square$

Theorem 4. *There is a polynomial distinguishing scheme for polynomial size unweighted depth 2 threshold circuits.*

Proof. For all distributed functions $f : \{0,1\}^n \times \{0,1\}^n \longrightarrow \{0,1\}$ consider the invariants $\gamma(f) = \max\left\{ \left|\mathbf{E}_{x,y}[f(x,y) \oplus g(x) \oplus h(y)] - \frac{1}{2}\right|; \, g, h \in B_n \right\}$ and
$$\alpha(f) = \max\left\{ \left|\mathbf{E}_y[f(x,y) \oplus f(x',y)] - \frac{1}{2}\right|; \, x \neq x' \in \{0,1\}^n \right\}.$$
The first exponential lower bound on the size of unweighted depth 2 threshold circuits was proved in [10]. The following two observations are implicitly contained there. Let us fix an arbitrary polynomial bound $m = m(n) \in n^{O(1)}$.

(I) There is a number $S \in m^{O(1)}$ such that if $f : \{0,1\}^n \times \{0,1\}^n \longrightarrow \{0,1\}$ has unweighted depth 2 threshold circuits of size $m(n)$ then $\gamma(f) \geq \frac{1}{S}$.
(II) For all $f : \{0,1\}^n \times \{0,1\}^n \longrightarrow \{0,1\}$ it holds that $\gamma(f) \leq (\frac{1}{2}(\alpha(f) + 2^{-n}))^{1/2}$.

The distinguishing scheme $D = D(n,m)$ is defined to do the following on n and m. It chooses an appropriate number $q \in O(\log(n))$ such that for $Q = 2^q$ the condition $Q \geq S^2$ is satisfied, and two random assignments $x \neq x'$ of $\{x_1, \ldots, x_q\}$. D accepts if
$$\left|\mathbf{E}_{y \in \{0,1\}^q}[f(x,y,\vec{0}) \oplus f(x',y,\vec{0})] - \frac{1}{2}\right| \geq \frac{1}{2S^2}.$$
Observe that the probability that this test accepts a truly random function is the same as the probability that test $T(\frac{1}{2}, \frac{1}{2S^2}, Q)$ accepts, i.e., at most $2e^{-Q/S^2}$. On the other hand, for all oracle functions of size $\leq m$ the following holds: if in Step 1 the pair x, x' determining $\alpha(f(\cdot, \cdot, \vec{0}))$ is chosen (and this occurs with probability $1/(Q(Q-1))$) then Step 2 will accept with probability 1. In other words, we have a $(1/(Q(Q-1)), 1, 2e^{-Q/S^2})$-test. It is quite easy to verify that we can fix some $q \in O(\log(n))$ which gives advantage $\varepsilon(n,m)$ for $D(n,m)$ fulfilling that $\varepsilon^{-1}(n,m) \in n^{O(1)}$. $\square$

Theorem 5. *For all $k \geq 1$ it holds that there is a distinguishing algorithm of quasipolynomially bounded ratio for depth $k + 1$ circuits consisting of k levels of AND and OR gates connected with one weighted threshold gate as output gate.*

The proof exhibits the "Switching Lemma" [11] and can be found in the full paper [14].

4   Pseudorandom TC_4^0-Functions

The NR-generator F is defined as follows. For all n the keys s for F have the form $s = (P, Q, g, r, a_1, \ldots, a_n)$, where P and Q are primes, Q divides $P - 1$, $g \in \mathbb{Z}_P^*$ has multiplicative order Q, and $a_1, \ldots, a_n$ are from $\mathbb{Z}_Q^*$. Define the corresponding function $f_s : \{0,1\}^n \to \mathbb{Z}_P \subseteq \{0,1\}^n$ by

$$f_s(x) = f_s(x_1, \ldots, x_n) = g^{y(x)} \bmod P,$$

where $y(x) = \prod_{i=1}^n a_i^{x_i}$. For our purpose it is obviously sufficient to show

Theorem 6. *The function $f = f_s$ has polynomial size depth 4 unweighted threshold circuits.*

Proof. We use the following terminology and facts about threshold circuits which are mainly based on results from [8,9,28].

Definition 1. *A Boolean function $g : \{0,1\}^n \longrightarrow \{0,1\}$ is called t-bounded if there are integer weights $w_1, \ldots, w_n$ and t pairwise disjoint intervals $[a_k, b_k]$, $1 \le k \le t$ of the real line such that $(g(x_1, \ldots, x_n) = 1 \iff \exists k \text{ s.t. } \sum_{i=1}^n w_i x_i \in [a_k, b_k])$; g is called polynomially bounded if g is t-bounded for some $t \in n^{O(1)}$. A multi-output function is called t-bounded if each output bit is a t-bounded Boolean function.*

Fact 1: Suppose that a function $f : \{0,1\}^n \longrightarrow \{0,1\}^n$ can be computed by a depth d circuit of polynomial size, where each gate of the circuit performs a function which can be written as a sum of at most $s \in n^{O(1)}$ polynomially bounded operations. Then f can be computed by a polynomial size depth $d+1$ unbounded weight threshold circuit.

Observe the following statements which can be easily proved.

Fact 2: If $g(x_1, \ldots, x_n)$ depends only on a linear combination $\sum_{i=1}^n w_i x_i$, where for all i, $1 \le i \le n$, it holds $|w_i| \in n^{O(1)}$, then g is a polynomially bounded operation.

Fact 3: If a Boolean function $g : \{0,1\}^n \longrightarrow \{0,1\}$ can be written as $g = h(g_1, \ldots, g_c)$, where c is a constant and the Boolean functions $g_1, \ldots, g_c : \{0,1\}^n \longrightarrow \{0,1\}$ are polynomially bounded operations, then g is a polynomially bounded operation.

As for many other efficient threshold circuit constructions, the key idea is to parallelize the computation of $f(x)$ via Chinese remaindering. Let us fix the first r prime numbers $p_1, \ldots, p_r$, where r is the smallest number such that $\Pi := \prod_{1 \le k \le r} p_k \ge \prod_{i=1}^n a_i$. Observe that $r \in O(n^2)$ and that all p_i, $1 \le i \le r$, are polynomially bounded in n, i.e., can be written as m-bit numbers for some $m \in O(\log n)$.

Consider the inverse Chinese remaindering transformation CRT^{-1} which assigns to each r-tupel of m bit numbers $(z^1, \ldots, z^r)$, $z^i = (z_{m-1}^i, \ldots, z_0^i)$ for $i = 1, \ldots, r$, the uniquely defined number $y < \Pi$ for which $y \equiv z^i \bmod p_i$ for all $i = 1, \ldots, r$. Denote by CRT_P^{-1} the function $CRT_P^{-1} : (\{0,1\}^m)^r \longrightarrow \{0,1\}^{n^2}$ defined as $(CRT^{-1}(z^1, \ldots, z^r) \bmod P)$, and observe

Fact 4: CRT_P^{-1} can be written as the sum of polynomially (in n) many polynomially bounded operations.

The proof (see, e.g., [28]) is based on the fact that

$$CRT^{-1}(z^1, \ldots, z^r) = \sum_{i=1}^{r} E_i z^i \bmod \Pi,$$

where for $i = 1 \ldots r$ the number E_i denotes the uniquely determined number smaller than Π for which $(E_i \bmod p_j) = \delta_{i,j}$ for all $i, j = 1, \ldots, r$. This implies

$$CRT^{-1}(z^1, \ldots, z^r) = \sum_{i=1}^{r} E_i \left(\sum_{j=0}^{m-1} z_j^i 2^j \right) \bmod \Pi$$
$$= \sum_{i=1}^{r} \sum_{j=0}^{m-1} e_{i,j} z_j^i \bmod \Pi.$$

for $e_{i,j} = (E_i 2^j \bmod \Pi)$.

The computation of $f(x)$ will be performed on 3 consecutive levels consisting of operations which are polynomially bounded (level 1,2) or which can written as polynomial length sums of polynomially bounded operations.

Level 1: Compute $z(x) = (z^1(x), \ldots, z^r(x))$, where for all $i = 1, \ldots, r$, the m-bit number z^i is defined to be $(y(x) \bmod p_i)$.

Observe that for all $i = 1, \ldots, r$, $z^i(x)$ can be written as

$$z^i(x) = \prod_{j=1}^{n} a_j^{x_j} \bmod p_i = \alpha_i^{\sum_{j=1}^{n} r_j^i x_j} \bmod p_i,$$

where α_i denotes a fixed element of order $p_i - 1$ in $\mathbb{Z}_{p_i}^*$ and r_j^i denotes for $j = 1, \ldots, n$ the discrete logarithm of a_j to the base α_i. Because all r_j^i are polynomially bounded in n, it follows by Fact 2 that $z(x)$ is a polynomially bounded operation.

For all inputs $z = (z^1, \ldots, z^r) \in (\{0,1\}^m)^r$ denote by $Y(z)$ the number

$$Y(z) = \sum_{i=1}^{r} \sum_{k=0}^{m-1} e_k^i z_k^i.$$

Observe that for all x it holds that $y(x) \equiv Y(z(x)) \bmod \Pi$ and $Y(z(x)) \leq mr\Pi$. Moreover, there exists exactly one $k, 1 \leq k \leq mr-1$, such that $y(x) = Y(z(x)) - k\Pi$. This k is characterized by $k\Pi \leq Y(z(x)) \leq (k+1)\Pi - 1$. Hence, $f = f_0 + \ldots + f_{mr-1}$ holds, where for each $k = 0, \ldots, mr - 1$, the function f_k is defined as

$$f_k(x) = \chi_k(z(x))(g^{Y(z(x)) - k\Pi} \bmod P),$$

where $\chi_k(z(x)) \in \{0,1\}$ is defined by $\chi_k(z(x)) = 1$ iff $k\Pi \leq Y(z(x)) \leq (k+1)\Pi - 1$.

Further observe that $g^{Y(z) - k\Pi} \bmod P = G_k(z) \bmod P$, where $G_k(z) = c_k \prod_{i=1}^{r} \prod_{j=0}^{m} (b_{i,j})^{z_j^i}$, and the c_k and $b_{i,j}$ are n-bit numbers defined by $c_k = (g^{-k\Pi} \bmod P)$ and $b_{i,j} = (g^{e_{i,j}} \bmod P)$. In contrast to $g^{Y(z) - k\Pi}$, the number $G_k(z)$ has polynomially many bits, namely $n(mr + 1)$. Fix u to be the smallest number with $\prod_{i=1}^{u} p_i \geq 2^{n(mr+1)}$. By the same arguments as above (Level 1), the operation $(G_k(z) \bmod p_i)$ is for all $i = 1, \ldots, u$ polynomially bounded.

Level 2: For all $k = 0 \ldots mr - 1$ and $i = 1 \ldots u$ compute

$$H_k^i(z) = \chi_k(z)(G_k(z) \bmod p_i).$$

This is a polynomially bounded operation as each output bit depends only on two polynomially bounded operations (Fact 3).

Level 3: Compute $f_k(x) = CRT_P^{-1}(H_k^1(z(x)), \ldots, H_k^u(z(x)))$.

Theorem 6 follows from Fact 4 and Fact 1. $\square$

5   Open Problems

We could like to detect for each basic nonuniform complexity class $\Lambda = P(M)$ whether it has an efficient distinguishing scheme (then cryptodesigners should obey the low complexity danger w.r.t. M) or whether Λ contains a PRFG (then lower bound proofs for this model seem to be a very serious task). Alas, for classes like TC_3^0 and $AC_3^0[m]$, m composite, this is still unknown. Is TC_3^0 is strong enough to contain PRFGs? Observe that TC_3^0 seems to contain pseudorandom bit generators. Operations such as squaring modulo the product of two unknown primes is in TC_3^0 [28].

Another open problem is the design of an efficient distinguishing scheme for polynomial size weighted threshold-MOD_p circuits, p an odd prime power. This is the only example of a complexity measure for which we failed to transform the known effective lower bound method (see [15]) into a distinguishing algorithm.

Also, we would like to determine the minimal hardware complexity of other cryptographic primitives like pseudorandom bit generators, pseudorandom permutation generators, one-way functions and cryptographically secure hash functions. Does TC_2^0 contain pseudorandom bit generators? Luby and Rackoff [20] presented a construction for *pseudorandom permutations* by three sequential applications of a pseudorandom function, each followed by an XOR-operation. They also showed how to construct *super pseudorandom permutations* by four such applications. Thus, as a corollary of our results, efficient pseudorandom permutations can be constructed in TC_{10}^0 and efficient super pseudorandom permutations can be constructed in TC_{13}^0. We conjecture that these results can be further improved, perhaps based on the results from [25].

References

1. N. Alon, J. Spencer, P. Erdös. The probabilistic method. Wiley & Sons 1992.
2. M. Bellare, S. Goldwasser. New paradigms for digital signatures and message authentication based on non-interactive zero knowledge proofs. Crypto '89, Springer LNCS, pp. 194–211.
3. M. Blaze, J. Feigenbaum, M. Naor. A Formal Treatment of Remotely Keyed Encryption. Eurocrypt '98, Springer LNCS, 1998.
4. A. Blum, M. Furst, M. Kearns, R. J. Lipton. Cryptographic primitives based on hard learning problems. Proc. CRYPTO 93, LNCS 773, 278-291.
5. A. Borodin, A. Razborov, R. Smolensky. On lower bounds for read k times branching programs. J. Computational Complexity 3, 1993, 1-13.
6. J. Bruck. Harmonic Analysis of polynomial threshold functions. SIAM Journal of Discrete Mathematics. 3:22, 1990, pp. 168-177.
7. O. Goldreich, S. Goldwasser, S. Micali. How to construct random functions. J. of the ACM, vol 33, pp. 792–807, 1986.
8. M. Goldmann, J. Hastad, A. A. Razborov. Majority gates versus general weighted Threshold gates. J. Computational Complexity 2, 1992, 277-300.
9. M. Goldmann, M. Karpinski. Simulating threshold circuits by majority circuits. Proc. 25th ACM Symp. on Theory of Computing (STOC), 1993, 551-560.
10. A. Hajnal, W. Maass, P. Pudlak, M. Szegedy, G. Turan. Threshold circuits of bounded depth. FOCS'87, pp. 99-110.
11. J. Hastad. Almost optimal lower bounds for small depth circuits. STOC'86, pp. 6-20.

12. S. Jukna. A note on read-k time branching programs. Theoretical Informatics and Applications 29(1), 1995, 75-83.
13. M. Kearns, L. Valiant. Cryptographic limitations on learning Boolean formulae and finite automata. J. of the ACM, vol. 41(1), 1994, pp. 67-95.
14. M. Krause, S. Lucks. On the minimal Hardware Complexity of Pseudorandom Function Generators.
    http://th.informatik.uni-mannheim.de/research/research.html.
15. M. Krause, P. Pudlak. On the computational power of depth-2 circuits with threshold and modulo gates. J. Theoretical Computer Science 174, 1997, pp. 137-156. Prel. version in STOC'94, pp. 49-59.
16. M. Krause, P. Pudlak. Computing Boolean functions by polynomials and threshold circuits. J. Comput. complex. 7 (1998), pp. 346-370. Prel. version in FOCS'95, pp. 682-691.
17. M. Krause, P. Savicky, I. Wegener. Approximation by OBDDs, and the variable ordering problem. Lect. Notes Comp. Science 1644, Proc. of ICALP'99, pp. 493-502.
18. M. Krause, S. Waack. Variation ranks of communication matrices and lower bounds for depth two circuits having symmetric gates with unbounded fan-in. J. Mathematical System Theory 28, 1995, 553–564.
19. N. Linial, Y. Mansour, N. Nisan. Constant depth circuits, Fourier transform, and learnability. J. of the ACM, vol. 40(3), 1993, pp. 607-620. Prel. version in FOCS'89, pp. 574-579.
20. M. Luby, C. Rackoff. How to construct pseudorandom permutations from pseudorandom functions. SIAM J. Computing, Vol. 17, No. 2, pp. 373–386, 1988.
21. S. Lucks. Faster Luby-Rackoff Ciphers. Fast Software Encryption 1996, Springer LNCS 1039, 189–203, 1996.
22. S. Lucks. On the Security of Remotely Keyed Encryption. Fast Software Encryption 1997, Springer LNCS 1267, 219–229, 1997.
23. M. Naor, O. Reingold. Synthesizers and their application to the parallel construction of pseudo-random functions. Proc. 36th IEEE Symp. on Foundations of Computer Science, pp. 170–181, 1995.
24. M. Naor, O. Reingold. Number-theoretic constructions of efficient pseudo-random functions. Preliminary Version. Proc. 38th IEEE Symp. on Foundations of Computer Science, 1997.
25. M. Naor, O. Reingold. On the construction of pseudo-random permutations: Luby-Rackoff revisited. J. of Cryptology, Vol. 12, No 1, 29–66, 1999.
26. E. Okolshnikova. On lower bounds for branching programs. Siberian Advances in Mathematics 3(1), 1993, 152-166.
27. A. Razborov, S. Rudich. Natural Proofs. J. of Computer and System Science, vol. 55(1), 1997, pp. 24-35. Prel. version STOC '94, pp. 204-213.
28. K. Siu, J. Bruck, T. Kailath, T. Hofmeister. Depth efficient neural networks for division and related problems. IEEE Trans. of Inform. Theory, vol. 39, 1993, pp. 946-956
29. R. Smolensky. Algebraic methods in the theory of lower bounds for Boolean circuit complexity. STOC'87, pp. 77-82.
30. I. Wegener. The complexity of Boolean functions. John Wiley & Sons, 1987.

Approximation Algorithms for Minimum Size 2-Connectivity Problems*

Piotr Krysta** and V.S. Anil Kumar

Max-Planck-Institut für Informatik, Stuhlsatzenhausweg 85
D-66123 Saarbrücken, Germany
{krysta,kumar}@mpi-sb.mpg.de

Abstract. We study some versions of the problem of finding the minimum size 2-connected subgraph. This problem is NP-hard (even on cubic planar graphs) and MAX SNP-hard. We show that the minimum 2-edge connected subgraph problem can be approximated to within $\frac{4}{3} - \epsilon$ for general graphs, improving upon the recent result of Vempala and Vetta [14]. Better approximations are obtained for planar graphs and for cubic graphs. We also consider the generalization, where requirements of 1 or 2 edge or vertex disjoint paths are specified between every pair of vertices, and the aim is to find a minimum subgraph satisfying these requirements. We show that this problem can be approximated within $\frac{3}{2}$, generalizing earlier results for 2-connectivity. We also analyze the classical local optimization heuristics. For cubic graphs, our results imply a new upper bound on the integrality gap of the linear programming formulation for the 2-edge connectivity problem.

1 Introduction

Graph connectivity is an important topic in theory and practice. It finds applications in the design of computer and telecommunication networks, and in the design of transportation systems. Networks with certain level of connectivity, which intuitively means that they provide certain number of connections between sites, are able to maintain reliable communication between sites, even when some of the network elements fail. For a survey and further applications, see Grötschel *et al.* [9].

Problem Statement. Given a graph with weights on its edges, and an integral *connectivity requirement* function r_{uv} for each pair of vertices u and v, the *vertex connectivity* (*edge connectivity*, respectively) *survivable network design problem* (SNDP) is to find a minimum weight subgraph containing at least r_{uv} vertex (edge, respectively) disjoint paths between each pair u, v of vertices. If $r_{uv} \in X$ for some set X, for each pair u, v, we denote the problem as X-VC-SNDP (X-EC-SNDP, respectively). The term survivable refers to the fact that the network is tolerant to the failures of sites and links (in case

---

* Partially supported by the IST Program of the EU under contract number IST-1999-14186 (ALCOM-FT).

** The author was supported by Deutsche Forschungsgemeinschaft (DFG) Graduate Scholarship. Part of the work by this author was done while he was visiting the Combinatorics & Optimization Dept., University of Waterloo, Ontario, Canada, during January-March, 2000, and was partially supported by NSERC grant no. OGP0138432 of Joseph Cheriyan.

A. Ferreira and H. Reichel (Eds.): STACS 2001, LNCS 2010, pp. 431–442, 2001.
© Springer-Verlag Berlin Heidelberg 2001

of VC-SNDP) or links (for EC-SNDP). Even the simplest versions of these problems are NP-hard, and so approximation algorithms[1] are of interest.

Previous Results for General Cases. For $\mathcal{N}_{\geq 0}$-EC-SNDP with arbitrary edge weights, there is a $O(\log(r_{max}))$-approximation by Goemans *et al.* [6] ($r_{\max} = \max_{u,v}(r_{uv})$), which was recently improved to 2 by Jain [10]. No algorithm with a non-trivial guarantee is known for the general version of VC-SNDP. For $\{0, 1, 2\}$-VC-SNDP with arbitrary edge weights, Ravi and Williamson [13] gave a 3-approximation algorithm.

Unweighted Low-Connectivity Problems. The case of low connectivity requirements is of particular importance, as in practice networks have rather small connectivities. There has been intense research for problems with low connectivity requirements and all the edge weights being equal to one (*unweighted problems*) [2,5,11,14]. We focus on the special unweighted cases of this problem where each $r_{uv} \in \{1, 2\}$. These are the simplest non-trivial versions of this problem and have been studied for a long time, but tight approximation guarantees and inapproximability results are not fully understood.

For the unweighted $\{2\}$-EC-SNDP (or 2-EC) Khuller and Vishkin [11] gave a $\frac{3}{2}$-approximation, which was improved by Cheriyan *et al.* [2] to $\frac{17}{12}$, and recently to $\frac{4}{3}$ by Vempala and Vetta [14]. For the unweighted $\{2\}$-VC-SNDP (or 2-VC) Khuller and Vishkin [11] gave an algorithm with approximation guarantee of $\frac{5}{3}$, which was improved to $\frac{3}{2}$ by Garg *et al.* [5] and to $\frac{4}{3}$ by Vempala and Vetta [14]. Both unweighted 2-VC and 2-EC problems are NP-hard even on cubic planar graphs, and also MAX SNP-hard [3]. For both unweighted problems $\{1, 2, \ldots, k\}$-VC-SNDP and $\{1, 2, \ldots, k\}$-EC-SNDP, the results of Nagamochi and Ibaraki [12] imply k-approximation algorithms.

The Linear Programming (LP) relaxation for the 2-EC problem and the subtour relaxation for TSP are very closely related [2]. The approximation ratio of $\frac{4}{3}$ obtained by Vempala and Vetta [14] has a special significance, because of the connections with the $\frac{4}{3}$ conjecture for metric TSP [1,2]. In this regard, the issue of whether $\frac{4}{3}$ can be improved for 2-EC is an interesting question. Also, the integrality gap[2] of the LP relaxation for 2-EC is not well understood. Vempala and Vetta [14] say that their result does not imply the same bound on the integrality gap for 2-EC. We prove a bound of better than $\frac{4}{3}$ on the integrality gap for cubic graphs, i.e. graphs with maximum degree at most 3.

Little is known about vertex-connectivity generalizations where arbitrary requirements are allowed, even for unweighted graphs. The simplest such generalization is to allow requirements of 1 or 2, instead of 2 for every pair. It should be noted that if the requirement can also take value of *zero*, the unweighted and weighted problems are essentially identical: an edge with an integer weight w can be replaced by a path of Steiner vertices of length w. For instance, unweighted $\{0, 1, 2\}$-VC-SNDP is equivalent to the weighted $\{0, 1, 2\}$-VC-SNDP considered by Ravi and Williamson [13].

Our Contributions. We give improved approximation algorithms for the 2-edge connectivity (2-EC) and the $\{1, 2\}$-connectivity problems. We show a $(\frac{4}{3}-\epsilon)$-approximation

---

[1] A polynomial time algorithm is called an α-*approximation algorithm*, or is said to achieve an *approximation (or performance) guarantee* of α, if it finds a solution of weight at most α times the weight of an optimal solution. α is also called an *approximation ratio (factor)*.

[2] The LP for the unweighted 2-EC is $\min\{\sum_{e \in E} x_e : \sum_{e \in \delta(S)} x_e \geq 2, \forall S \subset V, S \neq \emptyset; x_e \geq 0, \forall e \in E\}$, where $\delta(S)$ is the set of edges with exactly one end vertex in S. The ratio of the optimum integral solution value of 2-EC to the value of the LP, is called the *integrality gap*.

algorithm for 2-EC on general graphs, where $\epsilon = \frac{1}{1344}$. Our algorithm extends the technique of Vempala and Vetta [14] and removes the bottlenecks mentioned in their paper. The main new ideas are a better charging scheme, a refined lower bound, and the use of a local search heuristic. We show a tight example for their lower bound and their analysis, which leads naturally to the lower bound we use. The improved approximation is obtained by proving that paths can be connected for better charge by a more careful analysis, which is of comparable complexity as theirs, but is more uniform because it only deals with paths. Since, it is unlikely that an approximation scheme exists for the 2-EC problem (because of the MAX SNP-hardness), finding the best possible approximation guarantee for this classical problem is interesting (Section 3).

We achieve better guarantees for special classes of graphs, on which 2-EC is still NP-hard. For planar graphs, we show for 2-EC a $\frac{17}{13}$-approximation, but in quasi-polynomial time, by computing a stronger lower bound (Section 4). For 2-EC on cubic graphs we obtain a $\frac{21}{16}$-approximation using a simple local search heuristic. This implies an integrality gap of at most $\frac{21}{16}$ for the standard LP on cubic graphs (Section 5).

For the $\{1,2\}$-VC-SNDP and $\{1,2\}$-EC-SNDP (henceforth denoted by $\{1,2\}$-VC and $\{1,2\}$-EC), we give a $\frac{3}{2}$ approximation. This improves on straightforward 2-approximation. Our algorithms are generalizations of the algorithms of Garg *et al.* [5] and of Khuller and Vishkin [11]. The lower bounds used in [5,11] do not apply to our problems and we generalize them appropriately. We also analyze the performance of the classical local optimization heuristics (Section 6). Finally, Section 7 has some conclusions. Most of the details and proofs are missing in this abstract, and will appear in the full version.

2 Preliminaries

Graph Theory. We consider only undirected simple graphs. Given a graph $G = (V, E)$, we write $V(G) = V$ (*vertices*) and $E(G) = E$ (*edges*). Sets of vertices or contracted subgraphs will sometimes be called *(super) nodes*. C_l denotes a cycle of length l. A $u-v$ *path* is a path with end vertices u, v. $d_G(v)$ is the degree of vertex v in G. For definitions of the following standard graph theory notions the reader is referred to [7]: *cut vertex, two vertices separated by a cut vertex, bridge* (or *cut edge*), *ear decomposition* $\mathcal{E} = \{Q_0, Q_1, \ldots, Q_k\}$ (Q_0 is just one vertex), *ears* (Q_i's), ℓ-*ear* (an ear with ℓ edges), *open/closed ear* and *ear decomposition*. If a graph is 2-vertex(edge)-connected, then we write that it is 2-VC(EC). It is well known, that a graph is 2-EC -VC) iff it has no bridge (cut vertex, resp.), and iff it has an (open, resp.) ear decomposition. Also, an (open) ear decomposition can be found in polynomial time.

Let $\mathcal{E}$ be an ear decomposition of a 2-connected graph. We call an ear $S \in \mathcal{E}$ of length ≥ 2 *pendant* if none of the internal vertices of S is an end vertex of another ear $T \in \mathcal{E}$ of length ≥ 2. Let $\mathcal{E}' \subseteq \mathcal{E}$ be a subset of ears of the ear decomposition $\mathcal{E}$. We say that set $\mathcal{E}'$ is *terminal* in $\mathcal{E}$ if: (1) every ear in $\mathcal{E}'$ is a pendant ear of $\mathcal{E}$, (2) for every pair of ears $S, T \in \mathcal{E}'$ there is no edge between an internal vertex of S and an internal vertex of T, and (3) every ear in $\mathcal{E}'$ is open.

Given a rooted tree T, $[a, b]$ denotes the $a - b$ path in tree T for some two vertices a, b such that b is an ancestor of a, and path $[a, b]$ contains both vertices a, b. We define $[a, b)$ (and $(a, b]$, resp.) similarly, but the path contains a (b, resp.) and does not b (a, resp.). If a is a proper descendant of b in T, we say that a is *below* or *lower than* b, and b is *above* or *higher than* a. A vertex in T is also descendant and ancestor of itself. $opt(G)$ or opt denotes the size of an optimal solution on G to the problem under consideration.

Preliminary Reductions. Given a graph $G = (V, E)$, the problem of finding a minimum size subgraph G' of G in which every vertex of V has degree at least 2, is called $D2$. We refer to G' as $D2$, or $D2$ solution on G. This problem can be solved exactly in polynomial time [14], and it gives a lower bound for both 2-EC and 2-VC. The graph for 2-EC can be assumed to have no pair of adjacent degree 2 vertices (cf. [14]), and no cut vertices (else, we can solve 2-EC separately on each 2-VC component).

We define now *beta structures* [14]. A vertex u is a *beta vertex* if deleting some two adjacent vertices v_1, v_2 leaves at least 3 components, one of which is just u (Fig. 1(a)). Two adjacent vertices u_1, u_2 are called *beta pair* if there are two other vertices v_1, v_2, whose removal leaves at least 3 components, one of which is just u_1 and u_2 (Fig. 1(b)). Fig. 1(b) shows all the four edges $(v_1, u_1), (v_1, u_2), (v_2, u_1), (v_2, u_2)$, but it may be the case for a beta pair that just three of them are present. A graph with no beta vertex or beta pair is called *beta-free*. Vempala and Vetta [14] show that any α-approximation algorithm for 2-EC on beta-free graphs can be turned into an α-approximation algorithm for 2-EC on general graphs. Thus, we can assume for 2-EC that the graph is beta-free.

Let $C = C_l$ ($l \geq 3$) be a given cycle in G. If any solution to 2-EC on G uses at least l' edges from the subgraph induced on $V(C)$, then we can *contract cycle* C (i.e. identify the vertices in $V(C)$ and delete self loops) into a super node and solve recursively 2-EC on the resulting graph, incurring a factor of $\frac{l}{l'}$ as in [14]. The solution to 2-EC on G will then be the union of $E(C)$ with the edges of the recursive solution. l' here is used as a "local" lower bound. The overall approximation ratio of such an algorithm is $\geq l/l'$.

Let T be a given rooted spanning tree, some of whose vertices might be super nodes corresponding to subgraphs of the input graph. For a given non-root super node N of T the tree edge to the parent of N is called the *upper tree edge* of N. If N is a non-leaf super node, then any of the tree edges to its children is called the *lower tree edge* of N.

Local Optimization Heuristics. We define here a general local optimization heuristic. Let Π be a minimization problem on $G = (V, E)$, where we want to find a spanning subgraph of G with minimum number of edges, which is feasible for (or w.r.t.) problem Π. Given a non-negative integer j and any feasible solution $H \subseteq G$ to problem Π, we define the *j-opt heuristic* as the algorithm which repeats, if possible, the following:

- if there are subsets $E_0 \subseteq E \setminus E(H), E_1 \subseteq E(H)$ ($|E_0| \leq j, |E_1| > |E_0|$) such that $(H \setminus E_1) \cup E_0$ is feasible w.r.t. Π, then set $H \leftarrow (H \setminus E_1) \cup E_0$.

The algorithm outputs H, if it can perform no such operation on H any more. We say that such output solution is *j-opt* (or *j-optimal*) w.r.t. Π. If $|E_0| = j$, then we call the operation above a *j-opt exchange*. The algorithm can be implemented to run in polynomial time when j is a fixed constant.

3 Approximating 2-Edge Connectivity: General Graphs

We start by considering the algorithm of Vempala and Vetta [14], which gives a $\frac{4}{3}$ approximation for 2-edge connectivity problem. They use $D2$ as the lower bound. The beta-free example in Fig. $1(c)$ shows that their lower bound is tight. $T_1, \ldots, T_k$ are 3-cycles connected to vertices v, w of C, and C is a small clique. Clearly, the optimum must use 4 edges for each T_i, and asymptotically, $|OPT| = \frac{4n}{3}$, where n is the number of vertices. This suggests that we should use a modified lower bound of $\max(|D2|, 4l)$, where l is the number of such disjoint triangles.

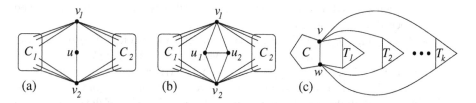

Fig. 1. (a), (b) Beta-Structures. (c) A Tight Example for the Lower Bound in [14].

Our algorithm is an extension and refinement of that of Vempala and Vetta [14]. The main differences in our algorithm are:

- In a necessary preprocessing step, we use a simple local optimization heuristic to eliminate some complicated configurations.
- We observe that paths can be connected up with charge less than $\frac{1}{3}$ by partly charging their child nodes and by using the fact that neither beta-vertices nor cut vertices are present. The analysis of [14] mentions that one of the bottlenecks in their $\frac{4}{3}$ approximation is the charge for 2-connecting paths. The reason that they only get a charge of $\frac{1}{3}$ is because they only use the property that cut vertices are missing.
- We actually obtain a stronger property: paths can not only be connected up with charge less than $\frac{1}{3}$, they can also pay for an extra $\frac{5}{14}$, with charge remaining less than $\frac{1}{3}$ (in some cases, such paths can only pay for an extra $\frac{3}{14}$). The algorithm first computes some DFS spanning tree of the graph, different than the DFS spanning tree in [14]. Then it 2-connects the tree in a top-down manner ([14] does this bottom-up). This is critical, since it allows us to charge the added edges to children.
- Paths whose parent is incident at one end, can be 2-connected with charge better than $\frac{1}{3}$, even if one free edge is not available, by partly charging the children.
- Another bottleneck in [14] are 3-cycles and a $\frac{1}{3}$ charge for these seems inevitable in their analysis. We overcome this by either merging 3-cycles with their children to form longer paths, or charge them with their children. For leaf 3-cycles, we use a stronger lower bound of $\max(|D2|, 4l)$, where l is the number of such leaves.

3.1 The Algorithm

A description of our algorithm is given below.

1. Compute $D2$ solution and partition it into paths and 2-EC components as in [14]. See Section 3.2 below for details.
2. Run a 1-opt heuristic on the $D2$ solution w.r.t. $D2$, and replace that current $D2$ solution with the 1-opt one.

3. Transform each 2-EC component into a cycle of length at least 3 by contracting suitable vertices and edges. See Section 3.2 for details.
4. Shrink all paths in $D2$ into single nodes.
5. Perform DFS on the contracted graph with the following rule for deciding the order of exploring neighbors on entering a node.
   - On entering a path: Suppose that the parent enters this path node at vertex v. Then start DFS from the end of the path that is farther away from v, and proceed till the other end.
   - On entering a cycle $C = v_1, \ldots, v_k$: Traverse the whole C first in the order $v_1, v_2, \ldots, v_{k-6}, v_{\pi(k-5)}, \ldots, v_{\pi(k)}$, where $\pi(k-5), \ldots, \pi(k)$ is a permutation of $k-5, \ldots, k$. The π is chosen such that: (i) one can traverse the cycle in this order, and (ii) if possible, $v_{\pi(k)}$ is a vertex leading to an unvisited vertex. That is, of all the possible traversals of the last 6 vertices of the cycle, choose one that allows the last vertex to have a child. If the child is another cycle, we get a longer path. If no such $v_{\pi(k)}$ exists, any permutation is equally good.

   Uncontract the nodes representing paths in $D2$, and let T be the resulting DFS tree.
6. Decompose T into maximal paths in the following manner: paths in $D2$ remain paths now, and we call them *red paths*. All edges of a cycle are part of one path. If a collection of cycles or paths can be merged, they will form a new path. Paths corresponding only to cycles are called *blue paths*, whereas a path formed by merging cycles and paths of $D2$ is a red path. The only difference between these is that a red path has two free edges (cf. Section 3.2 and [14]). Consider the natural parent child relation between these paths. The parent of a blue path is incident to it at one end.
7. Consider the paths in a top down manner. Each path is connected using extra edges. For a blue path, 2-connecting up just means connecting it within, since one end of it is part of the parent. 2-connecting paths involves forming blocks (Section 3.3).

The next subsection describes the way edges are added and charged for each path. The charge on each edge of each path other than blue paths of length three at leaves is at most $\frac{37}{112} < \frac{1}{3}$. Since our lower bound is $\max(D2, 4l)$, where l is the number of blue paths of length three at leaves, the approximation factor is $1 + \frac{(D2-3l)\alpha+3l\cdot(1/3)}{\max(D2,4l)} \leq 1 + \frac{D2\alpha+l(1-3\alpha)}{\max(D2,4l)} \leq 1 + \alpha + \frac{1-3\alpha}{4} = 1 + \frac{1+\alpha}{4}$. For $\alpha = \frac{37}{112}$, the factor is at most $1 + \frac{149}{448} < \frac{4}{3}$. This gives us the following theorem.

Theorem 1. *The minimum 2-edge connected subgraph problem on unweighted graphs can be approximated within a factor of $\frac{4}{3} - \epsilon$, where $\epsilon = \frac{1}{1344}$, in time $O(n^{2.5})$.*

Remark. The running time, as in [14], is the time required to find the $D2$. If $D2$ has no C_3 or C_4, then our proof shows that the charge on $D2$ is at most $\frac{23}{70}$ (cf. Section 4).

3.2   Preprocessing Details

Partitioning $D2$. We partition the $D2$ solution as described in [14], as follows. Partition $D2$ into connected components. Each such component is formed by 2-EC components, connected by paths. By an easy reduction, we can assume that any 2-EC component with at least 4 $D2$ edges contains no C_3.

Let us fix now some connected component C in $D2$. We will partition any such connected component C as follows. Any maximal 2-EC component of C is called a *supernode*. Contract all the maximal 2-EC components of C into single nodes. Partition the resulting tree (from C) into maximal paths arbitrarily. Then, any of these paths is also called a *supernode*. The key point here is that any such path supernode has two associated "free" edges – the first and last edge of the path, which we call *free edges*.

Contracting into Cycles. We contract parts of each 2-EC component and solve the problem on the resulting graph. The following lemma describes properties of the resulting 2-EC components. The local search in step 2 of our algorithm is used in its proof.

Lemma 1. *The contracted final subgraph resulting from a 2-EC component is a simple cycle of length at least three. Moreover, each contracted super node contains inside at least four edges of the D2 solution.*

3.3   2-Connecting Paths

The paths are 2-connected by adding extra edges. A path is partitioned into blocks, and one extra edge is used for each block, and the cost of this edge is charged to the vertices of the block. This is sufficient if the blocks have length at least 4. For smaller blocks we charge the children also. At a high level, 2-connecting a path node N is as follows. We process the path in the direction opposite to the direction of DFS (the matter of direction is critical). Assume that vertices up to $u_0 \in N$ have been 2-connected already, and the subsequent vertices are $u_1, u_2, \ldots, u_k$, with u_k being the last vertex of N. Contract all the nodes not in the subtree rooted at N, and the vertices of N up to u_0 into u_0. Now consider the farthest going edge e from u_0. We have the following cases.

1. e is incident on $u_i, i \geq 4$: The collection of vertices from u_1 to the other end of e forms a block of size at least 4, and we add e to 2-connect this block.
2. e is incident on u_3: If $k = 3$, this is the last block of N and is connected differently (see below). Else, there must be an edge f from u_1 or u_2 beyond u_3, to prevent it from being a cut vertex. If f is from u_2 to some vertex beyond u_3, add e, f and delete (u_2, u_3). The vertices from u_1 till the other end of f form a block of size at least 4, and a total of 1 edge is used for this block. Now let f be from u_1 to a vertex beyond u_3, and let there be no edge from u_2 to u_0, or beyond u_3. Then u_2 is a beta-vertex, and there must be a child node C from u_2. Vertices u_1, u_2, u_3 form a block of length 3, and we add e, and partly charge C, to reduce the charge.
3. e is incident on u_2: If e goes to an ancestor, and the upper tree edge is incident on u_2, N has to be a red path and we form a block of length 2 (vertices u_1, u_2), which is connected by adding the upper tree edge and e. If $k = 2$, this is the last block, which is handled later. Otherwise, there must be an edge f from u_1, beyond u_2 (else u_2 is a cut vertex). If f goes beyond u_3, form a block of length at least 4 with the vertices from u_1 to the other end of f, and add e, f and delete (u_1, u_2). If $f = (u_1, u_3)$, there must be an edge f' from u_2, beyond u_3. The vertices from u_1 to the other end of f' form a block, and we add e, f, f' and delete $(u_1, u_2), (u_2, u_3)$.

4. e is incident on u_1: If $k = 1$, this is the last block, which is handled later. Otherwise this is possible only if N is a red path, and e goes to an ancestor. In this case, we use part of the free edges available for red paths. The block will be of length 1, and gets connected by installing the upper tree edge and an upper back edge.

Remark. The above cases have been stated in a highly oversimplified manner. The arguments about cut vertices above might not hold if some vertex is a contracted supernode. But in this case, we can charge the $D2$ edges inside this supernode, using Lemma 1. For each extra edge to pay we have at least 4 $D2$ edges available, and $\frac{a+k}{b+4k} \leq \frac{a}{b}$ if $\frac{a}{b} \geq \frac{1}{4}$.

Last Block. The description above assumes that the other end of e is not the last vertex; the blocks formed above are called intermediate blocks. 2-connecting the last block poses problems if its size is at most 3. In the case of red paths, we use the fact that two free edges are available for connecting the path. We use up $\frac{3}{2}$ of these while 2-connecting the intermediate blocks, and $\frac{1}{2}$ is available while dealing with the last block. For the last block of a blue path, we do not have any free edge available and the algorithm adds edges based on the last and the second last block together.

The Charging Strategy. As mentioned before, the edges of all the paths, and two extra (free) edges per each red path are part of the lower bound ($D2$). We add extra edges to 2-connect each path, and the cost of these edges is charged to the edges of the paths. For each block, the cost of adding an edge to connect it is charged to the edges in the block. This works only for blocks of length at least 4, since we want to beat the $\frac{1}{3}$ barrier. Therefore, we will partly charge child nodes of the current node as well, to reduce the charge on the current node. The way the child node is charged depends on whether it is red or blue. This is due to the difference in the structure of red and blue paths: a blue path is connected to its parent at one end, while a red path need not be. For a red path, we shall make sure that it can not only get 2-connected for charge less than $\frac{1}{3}$, but can also pay for some extra charge, which will be used by the parent node.

Claim. Any red path can be 2-connected within and upwards, and also pay for an extra $\frac{3}{14}$, while incurring a charge less than $\frac{37}{112} < \frac{1}{3}$. A red path, such that no back edge from a descendant has already been installed, can pay for an extra $\frac{5}{14}$.

The extra charge for the red path is achieved by making each block pay an extra $\frac{1}{7}$. Finally, we need to prove that the algorithm produces a feasible solution. The path nodes are connected top-down, and each time a path is processed, it is 2-connected to its parent. Thus, if the edges added by the ancestors are not altered, the solution would indeed be 2-connected. This is true, because in most cases (except possibly when considering a red path with 2 internal edges), the upper tree edge from each of the paths is retained. Even in the case of a red path with 2 internal edges, the upper tree edge is not retained only if no back edge from any descendant has been installed. Thus, an ear decomposition of the solution edges can be done, which is a proof of its feasibility.

4   The Algorithm for Planar Graphs

Let $G = (V, E)$ be a given planar graph with $|V| = n$. As before, assume that G is 2-VC and has no beta-structures, as in [14]. We show how to compute the minimum subgraph $CD2$ which is connected and in which each vertex has degree at least two. This is a stronger lower bound than $D2$. We actually compute a $(1 + \epsilon)$ approximation to $CD2$, worsening our overall approximation by an extra $(1 + \epsilon)$ multiplicative factor. By a simple reduction from the Hamilton cycle problem, the $CD2$ problem is NP-hard, even on cubic planar graphs. Modifying the approach of Grigni *et al.* [8], we can obtain:

Lemma 2. *There is a $(1+\varepsilon)$-approximation algorithm for the $CD2$ problem on planar graphs (for any $\varepsilon > 0$), with running time of $n^{O((\log^3 n)/\varepsilon^4)}$.*

If $n \geq 5$, one can show that any $CD2$ solution can be transformed in polynomial time to another $CD2$ solution of no greater size, and without any C_3 or C_4. Thus, the analysis in Section 3 gives a better ratio, since all blue paths have length at least 5 now. This gives us the following.

Theorem 2. *For any given $\varepsilon > 0$, a $\left(\frac{93}{70}+\varepsilon\right)$-approximate solution to the minimum size 2-EC problem on a planar graph can be found in time $n^{O((\log^3 n)/\varepsilon^4)}$.*

Remark. A more careful analysis yields a $\left(\frac{17}{13} + \varepsilon\right)$-approximation for planar graphs with the same running time.

5   The Algorithm for Cubic Graphs

In this section we give a local search based approximation algorithm for the 2-EC problem on cubic graphs. Let $G = (V, E)$ be a given 2-VC cubic graph, with $|V| = n$. Our algorithm has two steps. The first step involves computing an ear decomposition H of G with the minimum number ϕ of even ears (i.e. of even length), using the algorithm of Frank [4,2] (delete all 1-ears, since they are redundant). It is easy to see, that $n + \phi - 1$ is a lower bound on the size of the optimum 2-EC solution in G [2].

Let a j-opt exchange that does not increase the number of even ears in H be called a *parity-preserving j-opt exchange*. In the 2nd step, run parity-preserving 1-opt heuristic on H w.r.t. 2-EC. Let H' be the resulting ear decomposition. Given an ear S in an ear decomposition $\mathcal{E}$, we say that an internal vertex v in S is *free* if $d_{\mathcal{E}}(v) = 2$.

Claim. H' is an open ear decomposition with all 2- and 3-ears being terminal. Any 5-ear in H' has at least two free vertices.

Let p_ℓ be the total number of internal vertices in the ℓ-ears of H'. Then, $p_\ell/(\ell - 1)$ is the number of ℓ-ears. We can estimate the size of the solution as: $|E(H')| \leq \sum_{i=2}^{8} \frac{i}{i-1}p_i + \frac{9}{8}(n - \sum_{i=2}^{8} p_i)$. The first summation in the right denotes the number of edges in ℓ-ears for $\ell = 2, 3, \ldots, 8$. This gives: $|E(H')| \leq (\frac{9}{8}n + \frac{7}{8}p_2 + \frac{5}{24}p_4 + \frac{3}{40}p_6 + \frac{1}{56}p_8) + (\frac{3}{8}p_3 + \frac{1}{8}p_5 + \frac{1}{24}p_7)$.

Since we have used only parity preserving exchanges, the bound of ϕ still applies to the number of even ears: $n + p_2 + \frac{1}{3}p_4 + \frac{1}{5}p_6 + \frac{1}{7}p_8 - 1 \leq n + \phi - 1$, thus $\frac{9}{8}(n + p_2 + \frac{1}{3}p_4 + \frac{1}{5}p_6 + \frac{1}{7}p_8) \leq \frac{9}{8}(n + \phi - 1) + \frac{9}{8} \leq \frac{9}{8}opt + \frac{9}{8}$. So the first term in the brackets in $|E(H')|$ can be upper bounded by $\frac{9}{8}opt + \frac{9}{8}$.

By the claim above, each 3-ear (5-ear, 7-ear, resp.) has at least 4 (4, 2, resp.) vertices associated with it (2, (2, 0,resp.) free vertices, and the two end vertices). Therefore, we can lower bound the number of vertices as follows: $4\frac{p_3}{2} + 4\frac{p_5}{4} + 2\frac{p_7}{6} \le n$, which gives us that $\frac{3}{8}p_3 + \frac{3}{16}p_5 + \frac{1}{16}p_7 \le \frac{3}{16}n$. This we use to upper bound the second term in the brackets in $|E(H')|$. Finally, since $n \le opt$, $|E(H')| \le (\frac{21}{16} + \epsilon)opt$, for any $\epsilon > 0$.

Theorem 3. *There is a polynomial time local search $(\frac{21}{16} + \epsilon)$-approximation algorithm (for any $\epsilon > 0$) for 2-EC on cubic graphs.*

Application: Integrality Gap of the LP. Let LP be the value of the standard linear program for 2-EC. Obviously, $n \le LP$. [2] shows that $n + \phi - 1 \le LP$. Thus, we obtain that the integrality gap of the LP is at most $\frac{21}{16} < \frac{4}{3}$ on unweighted cubic graphs.

6   Approximating $\{1, 2\}$-Connectivity

Due to the lack of space, we will make here a simplifying assumption that G is 2-vertex connected. Define function $\hat{r}$ as: $\hat{r}_u = \max\{r_{uv} : v \in V \setminus u\}$. We can assume that there exists a pair $u, v \in V$ with $r_{uv} = 2$, else the solution is a spanning tree.

6.1   Algorithm for $\{1, 2\}$-VC

We first present our modified lower bound for $\{1, 2\}$-VC. We assume that the input graph has no cut vertices.

Lemma 3. *If there is a pair $u, v \in V$ with $r_{uv} = 2$, then $opt(G) \ge \max(n, 2|I|)$, where $I \subseteq V$ is an independent set of vertices v with $\hat{r}_v = 2$.*

Our algorithm has two phases. The **first phase** is the same as in [5]. It computes a 2-VC spanning subgraph E': very roughly, it first finds a rooted DFS spanning tree, then adds extra back edges to 2-vertex connect the DFS tree, by processing it bottom-up. In the process, *blocks* are formed on addition of a back edge. The blocks form a partition of the vertices, finally. Our second phase is different from that in [5].

The Algorithm – Phase 2. After the 1st phase, set E' consisting of the edges of the DFS tree and of one back edge out of each non-root block has been output. $I = \emptyset$, initially. The 2nd phase will try for each block B to delete a tree edge within B, or if it is impossible to find a vertex s with $\hat{r}_s = 2$, add it to I, such that I remains independent. The 2nd phase will also modify the set of back edges.

Like in the second phase in [5], we traverse the blocks top-down. At each step we fix a block, say B, and B *decides* on the back edges going out of the child blocks of B. The first step is made with the root block: it chooses the farthest going back edge from each of its child blocks. Any child block having its back edge decided, decides on the back edges for its child blocks in a way we will specify. We proceed towards the leaves.

Let B be some non-root and non-leaf block for which the decision about the back edge $e = e(B)$ out of it has been made. Let v be an end vertex of e and $v \in B$, and $u = u(B)$ be the other end vertex of e. Let $p = p(B)$ be the parent vertex of B.

Block Property. We can assume that: (1) the back edge $e(B)$ goes higher than $p(B)$, and (2) there is an $u(B) - p(B)$ path that goes through the ancestor blocks of B.

The above property is obviously true for child blocks of the root block. We show that it is maintained during the algorithm, see Lemma 4.

Let w' be the highest vertex in path $[v, p)$ with $d_T(w') \geq 3$. If there is no such vertex in $[v, p)$, then let $w' = v$. Let w'' be the lowest vertex in $(w', p]$ which has a back edge from some child block of B. It can be shown that both w' and w'' are well defined. We note that the path $[w', w'']$ has length (i.e. the number of tree edges) at least one. Let q be the parent vertex of w'. The algorithm considers the following cases.

1. Assume that $q \neq w''$, and there is a vertex $s \in [q, w'')$ with $\hat{r}_s = 2$. Then, we label block B and vertex s with MARKED, and add s to the set I. B decides to retain the 1st phase choices of the farthest going back edges from all child blocks of B.
2. Assume that either $q \neq w''$, and for any vertex $s \in [q, w'')$, we have $\hat{r}_s = 1$, or $q = w''$. In this case the tree edge (q, w') is deleted from the current solution. We will show later that this step preserves the connectivities. Let e' be the back edge going from some child block B' of B into vertex w''. Then B decides not to take the farthest going back edge from B', but e' instead. For all other child blocks of B, B decides to retain the choice of the back edges made by the 1st phase.

In case 2 of the above algorithm when we delete a tree edge within block B, we pay this way for the back edge going out of B. Thus, in these cases the back edge is for free, and all such blocks are labeled FREE. Finally, the root block has no back edge going out of it and so is labeled FREE. Each leaf block is itself a vertex of requirement two, and so we label it MARKED, and choose all the leaf vertices into I.

Lemma 4. *(1) Each case of the 2nd phase of the algorithm maintains the Block Property, and preserves the feasibility of the solution w.r.t. r. (2) The set I of MARKED vertices is an independent set in G. Also all the vertices in I have the requirement of 2.*

Theorem 4. *The above algorithm is a linear time $\frac{3}{2}$-approximation algorithm for the unweighted $\{1, 2\}$-VC problem.*

6.2   Algorithms for $\{1, 2\}$-EC

Application of the Previous Algorithm. A straightforward application of the algorithm of Section 6.1 gives a $\frac{3}{2}$-approximation algorithm for $\{1, 2\}$-EC.

Simple Algorithm. A simple modification of the algorithm of Khuller and Vishkin [11] leads to a $\frac{3}{2}$-approximation algorithm for $\{1, 2\}$-EC. Let $G = (V, E)$, r be a given instance of $\{1, 2\}$-EC. Find a DFS spanning tree of G and keep all the tree edges in our solution subgraph H. Whenever the DFS backs-up over a tree edge e, check if e is a cut-edge of current H (i.e. none of the back edges in H covers e). If yes, and if the cut $(S, \bar{S})$ given by e separates a vertex pair x, y with $r_{xy} = 2$, add the farthest going back edge covering e into H. Also, "mark" the cut $(S, \bar{S})$, where S is the vertex set of the subtree below e, and $\{x, y\}$ is separated by $(S, \bar{S})$ if S has exactly one of x, y.

The number of tree edges in H is at most $n - 1$ which is at most $opt(G)$. The number of back edges in H is equal to the number of "marked" cuts $(S, \bar{S})$. Because of the DFS tree property and the way back edges were chosen, any two such cuts are edge disjoint. Thus, the optimal solution to $\{1, 2\}$-EC must have at least 2 edges in each of these cuts. So, the number of these cuts is at most $\frac{1}{2}opt(G)$, and $E(H) \leq \frac{3}{2}opt(G)$.

Local Optimization Heuristics. Generalizing the local optimization techniques used in Section 5, we can obtain the following results.

Theorem 5. *There are local search based algorithms which achieve a $\frac{7}{4}$-approximation for the $\{1,2\}$-VC problem and a $\frac{5}{3}$-approximation for the $\{1,2\}$-EC problem.*

Remark. Theorem 5 generalizes a result for the 2-VC communicated by J. Cheriyan.

7 Conclusions

While our improvement for 2-EC is small, it shows that $\frac{4}{3}$ can certainly be improved. With a more careful analysis, it should be possible to improve the approximation factor further. We think that similar methods can also improve the approximation ratio for 2-VC. We have also presented approximation algorithms for $\{1,2\}$-VC and $\{1,2\}$-EC. The algorithms are based on depth-first-search methods and on local search heuristics. It is important to note that almost all our algorithms use local search as the main method or as a subroutine. We think it would be interesting to further develop applications of the local search paradigm in the area of approximating connectivity problems.

References

1. R. Carr and R. Ravi. A new bound for the 2-edge connected subgraph problem. In the *Proc. 6th IPCO*, LNCS **1412**, 1998.
2. J. Cheriyan, A. Sebő and Z. Szigeti. An Improved Approximation Algorithm for Minimum Size 2-Edge Connected Spanning Subgraphs. In the *Proc. 6th IPCO*, LNCS **1412**, 1998.
3. C.G. Fernandes. A better approximation ratio for the minimum size k-edge-connected spanning subgraph problem. *Journal of Algorithms*, **28**, pp. 105–124, 1998.
4. A. Frank. Conservative weightings and ear-decompositions of graphs. *Combinatorica*, **13**, pp. 65–81, 1993.
5. N. Garg, V. Santosh and A. Singla. Improved Approximation Algorithms for Biconnected Subgraphs via Better Lower Bounding Techniques. In the *Proc. 4th ACM-SIAM SODA*, pp. 103–111, 1993.
6. M.X. Goemans, A. Goldberg, S. Plotkin, D.B. Shmoys, È. Tardos and D.P. Williamson. Improved Approximation Algorithms for Network Design Problems. In the *Proc. 5th ACM-SIAM SODA*, pp. 223–232, 1994.
7. R.L. Graham, M. Grötschel and L. Lovász, editors. *Handbook of Combinatorics. Volume I*. North-Holland, 1995.
8. M. Grigni, E. Koutsoupias and C.H. Papadimitriou. An approximation scheme for planar graph TSP. In the *Proc. of the IEEE FOCS*, 1995.
9. M. Grötschel, C. Monma and M. Stoer. Design of survivable networks. In *Handbook in Operations Research and Management Science*, Volume on Networks. North-Holland, 1995.
10. K. Jain. A Factor 2 Approximation Algorithm for the Generalized Steiner Network Problem. In the *Proc. of the IEEE FOCS*, 1998.
11. S. Khuller and U. Vishkin. Biconnectivity Approximations and Graph Carvings. *Journal of the ACM*, **41**(2), pp. 214–235, 1994.
12. H. Nagamochi and T. Ibaraki. A linear-time algorithm for finding a sparse k-connected spanning subgraph of a k-connected graph. *Algorithmica*, **7**, pp. 583–596, 1992.
13. R. Ravi and D.P. Williamson. An Approximation Algorithm for Minimum-Cost Vertex-Connectivity Problems. *Algorithmica*, **18**(1), pp. 21–43, 1997.
14. S. Vempala and A. Vetta. Factor 4/3 Approximations for Minimum 2-Connected Subgraphs. In the *Proc. 3rd International Workshop APPROX*, LNCS **1913**, 2000.

A Model Theoretic Proof of Büchi-Type Theorems and First-Order Logic for N-Free Pomsets

Dietrich Kuske*

Technische Universität Dresden

Abstract. We give a uniform proof for the recognizability of sets of finite words, traces, or N-free pomsets that are axiomatized in monadic second order logic. This proof method uses Shelah's composition theorem for bounded theories. Using this method, we can also show that elementarily axiomatizable sets are aperiodic. In the second part of the paper, it is shown that width-bounded and aperiodic sets of N-free pomsets are elementarily axiomatizable.

1 Introduction

In theoretical computer science, the notion of a recognizable subset of a monoid and more generally of an algebra is of outstanding importance. Here, recognizability means to be recognized by a homomorphism into a finite algebra. Often, this algebraic notion is equivalent to the more combinatorial notion of regularity, i.e. acceptance by a finite automaton. Seen as subset of an algebra, recognizable sets can often be described by certain rational expressions.

On the other hand, often the elements of the algebra in consideration carry an internal structure. For instance, words can be seen as labeled linear orders, terms as labeled ordered trees, Mazurkiewicz traces as dependence graphs etc. If such an internal structure is present, it is natural to consider sets of such structures that share a typical property. Classical results state that properties expressed in monadic second order logic give rise to recognizable sets. This holds for words, terms, Mazurkiewicz traces, computations of stably concurrent automata, local traces, and many others. If one restricts the expressibility of the logic, corresponding restrictions of the set of recognizable sets can be described.

In the first part of this paper, we are interested in the fact that logically expressed properties give rise to recognizable sets. Several very different proofs can be found in the literature: Using closure properties of the set of regular sets of words [17] as well as of traces [18], Thomas shows that any monadically axiomatizable set of words or traces can be accepted by a finite (asynchronous) automaton. An alternative proof by Ladner [12] for words uses Ehrenfeucht-Fraïssé-games. Courcelle [1] interprets the counting monadic second order theory of a graph in the monadic second order theory of the generating term, then he

---

* New address: Department of Mathematics and Computer Science, University of Leicester, Leicester, LE1 7RH, UK, email: D.Kuske@mcs.le.ac.uk.

A. Ferreira and H. Reichel (Eds.): STACS 2001, LNCS 2010, pp. 443–454, 2001.

appeals to Doner's Theorem [3]. Ebinger and Muscholl [7] interpret the monadic theory of a trace in the monadic theory of its linear representations and then use Büchi's theorem for words. This method is extended to concurrency monoids in [5]. In [4,10,11], the respective result is shown using the observation that any monadically axiomatizable set is the projection of a monadically axiomatizable set of traces.

It is the aim of the first part to give a uniform proof method for the fact that monadically axiomatizable sets are recognizable. This is achieved by model theoretic methods, in particular a weak form of Shelah's composition theorem [16]. For words, the idea is the following: Suppose one wants to know whether a monadic sentence φ holds in the concatenation of two words v and w. Then it is not necessary to know the words v and w completely, but it suffices to know which monadic sentences are satisfied by v and by w, respectively. Even more, one can restrict attention to monadic sentences of the quantifier depth of φ. This composition theorem defines a semigroup structure on the set of bounded monadic theories of finite words and a homomorphism from the free semigroup onto the semigroup of bounded theories. As this semigroup is finite, any monadically axiomatizable set is recognizable. This idea is worked out more generally for $\mathcal{D}$-algebras of graphs (see below for a precise definition). As corollary, we obtain one direction of Büchi-type theorems for words, for semitraces, for traces, and for N-free pomsets. The result for N-free pomsets generalizes a result from [11] where we had to impose additional restrictions on the N-free pomsets.

Dealing with first order logic, we again construct an algebra of bounded elementary theories and show that any binary operation in this algebra is an aperiodic semigroup. This implies one direction of McNaughton and Papert's theorem for words [14], for semitraces, for traces [7], and for N-free pomsets where it is new.

The second part of the paper exclusively deals with N-free pomsets, in particular with the question whether aperiodic sets are elementarily axiomatizable. It is shown that width-bounded and aperiodic sets of N-free pomsets are starfree; the restriction to width-bounded sets originates in our proof, but we conjecture that it cannot be droped in general. In particular we believe that the set of trees is not starfree (it is aperiodic since it is elementarily axiomatizable). Finally, we show that starfree sets are elementarily axiomatizable. The only difficulty in this proof is the parallel product of elementarily axiomatizable sets. This problem is solved using Gaifman's Theorem and the observation that connected components in an N-free pomset are elementarily definable. Thus, we obtain the equivalence of starfreeness, aperiodicity, and elementary axiomatizablity for sets of N-free pomsets of bounded width.

2   Basic Definitions

2.1   Graphs and D-Sums

Throughout this paper, let A be a finite set. We will consider finite directed graphs whose vertices are labeled by the elements of A, i.e. structures (V, E, λ) where V is a finite nonempty set of vertices, $E \subseteq V \times V$ is the set of edges, and $\lambda : V \to A$ is a labeling function. For short, we call these structures *graphs*. The set of all graphs is denoted by $\mathbb{G}$. The element a of A is identified with the one-vertex graph labeled by a that does not contain any edge.

Example 2.1. We can, as usual, identify a word $w \in A^+$ with the labeled linear order $(\{1, 2, \ldots, |w|\}, <, \lambda)$ where $\lambda(i)$ is the i-th letter of w. This is a particular graph. Then the concatenation of words $v \cdot w$ corresponds to the disjoint union of the sets of positions. The edge relation of the product is the edge relation of the arguments together with a new edge from any position in v to any position in w. This operation, that can be applied to any two graphs s and t in $\mathbb{G}$, is called *sequential product* and we write $s \cdot t$ for the sequential product of s and t. If s and t are orders, this operation is known as lexicographic sum in the mathematical literature. A graph (V, E, λ) corresponds to some word iff it can be constructed from the letters $a \in A$ by the sequential product.

Example 2.2. The possibly simplest operation on the set of graphs $\mathbb{G}$ is the disjoint union, also known as *parallel product* $s \parallel t$. A graph that can be constructed from the singletons by the sequential and the parallel product is called *series-parallel graph*. One can easily show that the edge relation of any series-parallel graph is actually a (strict) partial order. Furthermore the partial order $(N, <_N)$ cannot be embedded into a series-parallel graph. Since these properties characterize the series-parallel graphs (cf. [8]), we will speak of *N-free pomsets*. The set of all N-free pomsets is denoted by NF.

The poset $(N, <_N)$

Note that the parallel product is not only associative, but also commutative. A structure $(S, \cdot, \parallel)$ where S is a set, $\cdot$ and $\parallel$ are associative operations on S, and $\parallel$ is in addition commutative, is called *sp-algebra* [13]. Thus, $(\mathrm{NF}, \cdot, \parallel)$ is an sp-algebra.

Example 2.3. Let $\mathrm{SD} \subseteq A \times A$ be reflexive. Then one can define the following operation on graphs: For $s = (V_s, E_s, \lambda_s)$ and $t = (V_t, E_t, \lambda_t)$, the *semitrace product of s and t* is the graph (V, E, λ) where V is the disjoint union of V_s and V_t, the labeling is preserved, and $(x, y) \in E$ iff $(x, y) \in E_s \cup E_t$ or $x \in V_s$, $y \in V_t$, and $(\lambda_s(x), \lambda_t(y)) \in \mathrm{SD}$. A *semitrace over* SD is a graph that can be constructed from the singletons using the semitrace product. The set of semitraces together with the semitrace product is a semigroup generated by A.

Example 2.4. Now let $\mathrm{D} \subseteq A \times A$ be reflexive and symmetric. Then the semitrace product from the preceding example becomes the usual trace product as considered in the theory of Mazurkiewicz traces. A semitrace over D is then called Mazurkiewicz trace or simply trace. The set of traces together with the trace product is the well studied trace semigroup $\mathbb{M}(A, \mathrm{D})$.

All the operations in the examples above can be seen as special instances of a *D-sum* that we introduce next:

Definition 2.5. *A sum description of arity n is a family $D = (D_{\ell,m})_{\ell,m \leq n, \ell \neq m}$ of subsets of $A \times A$. For a sum description D of arity n and graphs $s_i = (V_i, E_i, \lambda_i)$ for $1 \leq i \leq n$, the D-sum $\sum_D(s_i) = (V, E, \lambda)$ is defined as follows:*

$$V = \{(\ell, z) \mid \ell \leq n, z \in V_\ell\},$$
$$((\ell, v), (m, w)) \in E \text{ iff } \ell \neq m \text{ and } (\lambda_\ell(v), \lambda_m(w)) \in D_{\ell,m}$$
$$\text{or } \ell = m \text{ and } (v, w) \in E_\ell, \text{ and}$$
$$\lambda(\ell, v) = \lambda_\ell(v).$$

Thus, the D-sum is the disjoint union of the summands. The edge relation within the summands is not altered. Whether two nodes from different summands are connected by an edge depends on their labels and is dictated by the sum description D. This operation is a special case of the "generalized sum" as considered in [16, Def. 2.3].

Example 2.6. All operations we presented above, are binary. Let $D^i_{2,1} = \emptyset$. With $D^1_{1,2} = A \times A$, the D-sum equals the sequential product, with $D^2_{1,2} = \emptyset$ the parallel product, and with $D^3_{1,2} = SD$ the semitrace product. Hence also the trace product is a special instance of our concept of a D-sum.

The construction of terms (seen as trees) seems at first sight not to fit into our setting as the term $f(t_1, t_2, \ldots, t_n)$ contains not only the disjoint union of the terms t_i, but an additional root node that corresponds to the outermost function symbol f. Nevertheless, we can model this construction by a nested application of the disjoint union (of the arguments) and the sequential product with the singleton graph f.

Let $\mathcal{D}$ be a set of sum descriptions. By $\mathbb{G}(\mathcal{D})$, we denote the set of all graphs that can be constructed from the singletons by the D-sums for some $D \in \mathcal{D}$.

Note that $\mathbb{G}(\mathcal{D})$ is closed under the application of the D-sum for $D \in \mathcal{D}$. Hence $(\mathbb{G}(\mathcal{D}), (\sum_D)_{D \in \mathcal{D}})$ is an algebra over the signature $\mathcal{D}$. A $\mathcal{D}$-*algebra* is a structure $(S, (f_D)_{D \in \mathcal{D}})$ where f_D is a operation on S of the arity of D for any $D \in \mathcal{D}$. Hence, $\mathbb{G}(\mathcal{D})$ together with the D-sums for $D \in \mathcal{D}$ is a $\mathcal{D}$-algebra.

Note that the B-sum is associative for any binary sum description B. Furthermore, it is commutative iff $B_{1,2} = B_{2,1}$. Thus, the $\{B\}$-algebra $\mathbb{G}(\{B\})$ is a (commutative) semigroup (if $B_{1,2} = B_{2,1}$).

Example 2.6 (continued) The set $\mathbb{G}(\{D^1\})$ is the set of labeled linear orders, i.e. of those graphs that correspond to words over A. Therefore $(\mathbb{G}(\{D^1\}), \sum_{D^1})$ is the free semigroup generated by A. Similarly, $(\mathbb{G}(\{D^1, D^2\}), (\sum_{D^i})_{i=1,2})$ is isomorphic to the sp-algebra of N-free pomsets $(NF, \cdot, \|)$. Finally, $(\mathbb{G}(\{D^3\}), \sum_{D^3})$ is the semigroup of semitraces and therefore isomorphic to the trace semigroup if SD is symmetric.

A set $X \subseteq \mathbb{G}(\mathcal{D})$ is *recognizable* if there exists a finite $\mathcal{D}$-algebra $(S, (f_D)_{D \in \mathcal{D}})$ and a homomorphism $\eta : \mathbb{G}(\mathcal{D}) \to S$ such that $X = \eta^{-1}\eta(X)$. Note that we could have required the homomorphism η to be surjective without changing the concept of recognizability.

2.2   Logic

Monadic Formulas involve first order variables $x, y, z \ldots$ for vertices and monadic second order variables $X, Y, Z, \ldots$ for sets of vertices. They are built up from the atomic formulas $\lambda(x) = a$ for $a \in A$, $(x, y) \in E$, and $x \in X$ by means of the boolean connectives $\neg, \vee, \wedge, \rightarrow, \leftrightarrow$ and quantifiers $\exists, \forall$ (both for first order and for second order variables). Formulas without free variables are called *sentences*. The satisfaction relation $\models$ between graphs $p = (V, E, \lambda)$ and monadic sentences φ is defined canonically with the understanding that first order variables range over the vertices of V and second order variables over subsets of V.

Let $\mathcal{C}$ be a set of graphs and φ a monadic sentence. Furthermore, let $X = \{p \in \mathcal{C} \mid p \models \varphi\}$ denote the set of graphs from $\mathcal{C}$ that satisfy φ. Then we say that the sentence φ *axiomatizes the set X relative to $\mathcal{C}$* or that X is *monadically axiomatizable relative to $\mathcal{C}$*.

An *elementary formula* is a monadic formula that does not contain any set variable. A set $X \subseteq \mathcal{C}$ is *elementarily axiomatizable relative to $\mathcal{C}$* if there exists an elementary sentence that axiomatizes X relative to $\mathcal{C}$.

The *quantifier depth* of a monadic formula is defined canonically. For a graph s and a positive integer k, let $\mathrm{MTh}_k(s)$ denote the set of all monadic sentences φ of quantifier depth at most k that are satisfied by s. The set $\mathrm{MTh}_k(s)$ is the *k-bounded monadic theory of s*. Analogously, the *k-bounded elementary theory* $\mathrm{Th}_k(s)$ comprises all elementary sentences in $\mathrm{MTh}_k(s)$. By MTH_k and TH_k, we denote the set of all k-bounded monadic and elementary theories of graphs. Up to logical equivalence, there are only finitely many monadic sentences of quantifier depth at most k; hence the set of k-bounded monadic theories is finite (and the same holds for the elementary theories).

3   Axiomatizable Sets of Graphs

3.1   Monadically Axiomatizable Sets Are Recognizable

Theorem 3.1. *Let $k \in \mathbb{N}$ and D be a sum description of arity n. Furthermore, let s_ℓ and t_ℓ be graphs for $\ell \leq n$.*
(1) If $\mathrm{MTh}_k(s_\ell) = \mathrm{MTh}_k(t_\ell)$ for $\ell \leq n$, then $\mathrm{MTh}_k(\sum_D s_\ell) = \mathrm{MTh}_k(\sum_D t_\ell)$.
(2) Similarly, if $\mathrm{Th}_k(s_\ell) = \mathrm{Th}_k(t_\ell)$ for $\ell \leq n$, then $\mathrm{Th}_k(\sum_D s_\ell) = \mathrm{Th}_k(\sum_D t_\ell)$.

We already mentioned that the D-sum of n graphs is a special case of the generalized sum as considered by Shelah. The preceding theorem follows from his result [16, Thm. 2.4]. A condensed proof of the full result can be found in [9], for lexicographic sums of linear orders, [19] contains the full proof. In the terminology of [2], the first statement says that D-sums are Hintikka operations, and it can be derived from their results on quantifier free definable operations. Shelah is also interested in the effective computability of the combined theory from the argument theories. If one is only interested in the result as stated above, an alternative and much simpler proof of both statements can be given using Ehrenfeucht-Fraïssé-games for the respective logics.

Again, let D be a sum description of arity n and $k \in \mathbb{N}$. Then by the above theorem the set MTH_k of k-bounded monadic theories can be equiped with an

n-ary operation f_D such that
$$\mathrm{MTh}_k(\textstyle\sum_D(s_1, s_2, \ldots, s_n)) = f_D(\mathrm{MTh}_k(s_1), \mathrm{MTh}_k(s_2), \ldots, \mathrm{MTh}_k(s_n)).$$
Hence the mapping MTh_k is a homomorphism from the $\mathcal{D}$-algebra of graphs onto the $\mathcal{D}$-algebra $(\mathrm{MTH}_k, (f^D)_{D \in \mathcal{D}})$ for any set of sum descriptions $\mathcal{D}$:

Corollary 3.2. *Let $\mathcal{D}$ be a set of sum descriptions. The set of k-bounded monadic theories MTH_k can be equiped with the structure of a $\mathcal{D}$-algebra such that $\mathrm{MTh}_k : (\mathbb{G}, (\sum_D)_{D \in \mathcal{D}}) \to \mathrm{MTH}_k$ is a homomorphism between $\mathcal{D}$-algebras.*

Now let φ be a sentence of the monadic second order logic of quantifier depth k. Since $t \models \varphi$ iff $\varphi \in \mathrm{MTH}_k(t)$, this homomorphism recognizes the set $\{t \in \mathbb{G}(\mathcal{D}) \mid t \models \varphi\}$, i.e. we showed

Theorem 3.3. *Let φ be a sentence of the monadic second order logic. Then $\{t \in \mathbb{G}(\mathcal{D}) \mid t \models \varphi\}$ is recognizable in $(\mathbb{G}(\mathcal{D}), (\sum_D)_{D \in \mathcal{D}})$.*

For $\mathcal{D}$ finite, this result follows from [2], but their technique of interpreting a graph in a generating term cannot be applied for $\mathcal{D}$ infinite.

Corollary 3.4.   *1. Any monadically axiomatizable set of words, traces, or semi-traces can be recognized by a homomorphism into a finite semigroup.*
  2. Any monadically axiomatizable set of N-free pomsets can be recognized by a homomorphism from $(\mathrm{NF}, \cdot, \|)$ into a finite sp-algebra.

This corollary generalizes Büchi's Theorem for finite words as well as a result by Thomas [18] on traces. The result on semitraces is new. For N-free pomsets, a weaker statement was shown in [11]. There, we had to require that the pomsets $t \in \mathrm{NF}$ that satisfy φ are width bounded.

3.2   Elementarily Axiomatizable Sets Are Aperiodic

Let φ be an elementary sentence. There exists a finite $\mathcal{D}$-algebra and a homomorphism into this algebra recognizing the set of graphs that satisfy φ. In this section, we want to derive additional properties of this finite $\mathcal{D}$-algebra.

Let $B \in \mathcal{D}$ be a binary sum description. To simplify the notions, we will write $s +_B t$ for $\sum_B(s, t)$. A $\mathcal{D}$-algebra $(S, (f_D)_{D \in \mathcal{D}})$ is *aperiodic* iff (S, f_B) is an aperiodic semigroup for any binary $B \in \mathcal{D}$. A set $X \subseteq \mathbb{G}(\mathcal{D})$ is *aperiodic* if it can be recognized by a homomorphism into a finite aperiodic $\mathcal{D}$-algebra. In the following example, we consider some aperiodic sets and their closure properties:

Example 3.5. Obviously, a homomorphism that recognizes $X \subseteq \mathbb{G}(\mathcal{D})$ recognizes $\mathbb{G}(\mathcal{D}) \setminus X$ as well. The set of aperiodic $\mathcal{D}$-algebras is closed under finite direct products. Hence the union as well as the intersection of aperiodic sets in $\mathbb{G}(\mathcal{D})$ are aperiodic, i.e. the set of aperiodic sets is closed under Boolean operations.

Next consider the following sp-algebra: $S = \{1, 2\}$, $s \cdot s' = 1$ and $s \| s' = 2$ for any $s, s' \in S$. Since, for $s \in S$, we have $s \cdot s = 1 = s \cdot s \cdot s$ and $s \| s = 2 = s \| s \| s$, the sp-algebra $(S, \cdot, \|)$ is aperiodic. Then the mapping $\eta : \mathrm{NF} \to \{1, 2\}$ with $\eta(s) = 1$ iff s is connected is a homomorphism. Hence the set of connected N-free pomsets as well as its complement $\mathrm{NF} \| \mathrm{NF}$ are aperiodic.

Adopting the above methods for monadic second order logic, we can define the $\mathcal{D}$-algebra of k-bounded elementary theories $(\mathrm{TH}_k, (f_D)_{D \in \mathcal{D}})$ for any set of sum descriptions $\mathcal{D}$. Then the mapping Th_k from the set of graphs onto this $\mathcal{D}$-algebra is a homomorphism. Similarly to the monadic case, it is immediate that any elementarily axiomatizable set can be recognized by the homomorphism Th_k for some k. Thus, the core of this section is to prove that (TH_k, f_B) is aperiodic for any binary sum description B.

Theorem 3.1 relates two D-sums for some fixed D. Next, we will prove a similarly looking result for different D-sums. To compensate this more general situation, we require that all argument graphs s_ℓ and t_m have the same k-bounded elementary theory T. In addition, we have to relate the different sum descriptions that will appear in the lemma. Loosely speaking, they have to have the same k-bounded elementary theory. To give this intuition a clear meaning, we consider an n-ary sum description D as a relational structure $R(D)$ on the carrier set $\{1, 2, \ldots, n\}$. For any $M \subseteq A \times A$, the relational structure $R(D)$ has a binary relation $\{(i, j) \mid D_{i,j} = M\}$.

Theorem 3.6. *Let $k \in \mathbb{N}$, and let D^1 and D^2 be sum descriptions of arity n_1 and n_2, respectively. Assume that the associated relational structures $R(D^1)$ and $R(D^2)$ have the same k-bounded elementary theory. Furthermore, let s_x and t_y be graphs with $\mathrm{Th}_k(s_x) = \mathrm{Th}_k(t_y)$ for $x \leq n_1$ and $y \leq n_2$. Then $\mathrm{Th}_k(\sum_{D^1} s_x) = \mathrm{Th}_k(\sum_{D^2} t_y)$.*

Proof. The proof uses Ehrenfeucht-Fraïssé-games, see [6, p. 16 ff.] for an introduction. Following the notation in [6], we write $G_k(s, t)$ to denote the game on the structures s and t with k rounds.

By the requirements on D^1, D^2, s_x and t_y, Duplicator has a winning strategy for the games $G_k(R(D^1), R(D^2))$ and $G_k(s_x, t_y)$ for $x \leq n_1$ and $y \leq n_2$. A winning strategy for $G_k(\sum_{D^1} s_x, \sum_{D^2} t_y)$ is described as follows: Suppose in the first $\ell < k$ rounds, Spoiler and Duplicator played $((x_i, a_i), (y_i, b_i))_{1 \leq i \leq \ell}$ with $x_i \leq n_1$, $y_i \leq n_2$, $a_i \in V_{s_{x_i}}$ and $b_i \in V_{t_{y_i}}$. Now, in round $\ell + 1$, Spoiler chooses (x, a) with $x \leq n_1$ and $a \in V_{s_x}$. Note that $(x_i, y_i)_{1 \leq i \leq \ell}$ can be seen as the result of ℓ rounds in the game $G_k(R(D^1), R(D^2))$. The winning strategy for this game tells Duplicator which $y \leq n_2$ to choose. Now let $1 \leq i_1 < i_2 < \cdots < i_n \leq \ell$ be those indices for which $y_{i_j} = y$. Since Duplicator always plays according to the strategy we are describing, we obtain $x_{i_j} = x$ for all these indices. Hence $(a_{i_j}, b_{i_j})_{1 \leq j \leq n}$ is the result of n rounds of the game $G_k(s_x, t_y)$ and Duplicator's winning strategy for this game tells him how to answer to Spoiler's move a. □

A slightly weaker form of the above theorem follows from [16]. Namely, the structures $R(D^1)$ and $R(D^2)$ would have to coincide in their ℓ-bounded theory where ℓ can be computed effectively from k. In our case of D-sums, it suffices to take $\ell = k$.

Corollary 3.7. *Let $k \in \mathbb{N}$ and let $B = (B_{1,2}, B_{2,1})$ be a binary sum description. Then the semigroup of all k-bounded elementary theories (TH_k, f_B) is aperiodic.*

Proof. Let s be a graph.

For $n \in \mathbb{N}$, let D^n be the following n-ary sum description defined by $D_{i,j}^n = B_{1,2}$ and $D_{j,i}^n = B_{2,1}$ for $1 \le i < j \le n$. Since $\sum_B$ is associative, we obtain
$$\sum_{D^{n+1}}(s_1, s_2, \dots, s_n, s) = \sum_{D^n}(s_1, s_2, \dots, s_n) +_B s.$$
A special case of [12, Lemma 4.4] states that
$$\mathrm{Th}_k(\{1, 2, \dots, 2^k - 1\}, <) = \mathrm{Th}_k(\{1, 2, \dots, 2^k\}, <).$$
Note that the associated relational structure $R(D^n)$ has the carrier set $\{1, \dots, n\}$. Furthermore, a pair (i, j) with $i \ne j$ is labeled by $B_{1,2}$ iff $i < j$, and by $B_{2,1}$ otherwise. Hence the k-bounded elementary theory of $R(D^{2^k-1})$ equals that of $R(D^{2^k})$. Now the theorem above implies

$$\mathrm{Th}_k(\underbrace{s +_B s +_B \cdots +_B s}_{(2^k-1)-\text{times}}) = \mathrm{Th}_k(\sum_{D^{2^k-1}}(s, s, s, \dots, s))$$

$$= \mathrm{Th}_k(\sum_{D^{2^k}}(s, s, s, \dots, s)) = \mathrm{Th}_k(\underbrace{s +_B s +_B \cdots +_B s}_{2^k-\text{times}}).$$

Since Th_k is surjective, the result follows. □

Now let $X \subseteq \mathbb{G}(\mathcal{D})$ be an elementarily axiomatizable set. Since Corollary 3.2 holds for elementary theories verbatim, X is recognized by the homomorphism Th_k into the finite $\mathcal{D}$-algebra of k-bounded elementary theories for some k. By Corollary 3.7, this $\mathcal{D}$-algebra is aperiodic. Hence we obtain

Theorem 3.8. *Let φ be an elementary sentence and $\mathcal{D}$ be a set of sum descriptions. Then $\{t \in \mathbb{G}(\mathcal{D}) \mid t \models \varphi\}$ is an aperiodic set in $(\mathbb{G}(\mathcal{D}), (\sum_D)_{D \in \mathcal{D}})$.*

As a corollary, we obtain the known results on finite words [14] as well as on finite traces [18,7]. The corresponding statements for semitraces and for N-free pomsets are new.

Corollary 3.9. *1. Any elementarily axiomatizable set of words, traces, or semitraces can be recognized by a homomorphism into a finite aperiodic semigroup.*
2. Any elementarily axiomatizable set of N-free pomsets can be recognized by a homomorphism from $(\mathrm{NF}, \cdot, \|)$ into a finite aperiodic sp-algebra.

4   Aperiodic Sets of N-Free Pomsets Are Elementarily Axiomatizable

By [14,18,7], the inverse of Corollaries 3.4(1) and 3.9(1) hold for words and for traces. Let $X \subseteq \mathrm{NF}$ be a recognizable set of N-free pomsets of bounded width. Then, according to [11], X is monadically axiomatizable, i.e. in this restricted setting the inverse of Corollary 3.4(2) holds as well. It is the aim of this section to show similarly the inverse of Corollary 3.9(2), i.e. to show that any aperiodic set of N-free pomsets is elementarily axiomatizable. This is shown via starfree sets of N-free pomsets that we consider next.

4.1 Starfree Sets Are Elementarily Axiomatizable

The sequential and the parallel product of N-free pomsets can easily be extended to sets of N-free pomsets. By X^+, we denote the sequential iteration of X, i.e. the set $\{p_1 \cdot p_2 \cdots p_n \mid n > 0, p_i \in X\}$. Recall that we did not allow a poset to be empty. Therefore, in general, $X \cdot \mathrm{NF}$ does not contain X. Since, occasionally it will be convenient to have this, we will use the abbreviations $X \cdot \mathrm{NF}_\varepsilon = X \cdot \mathrm{NF} \cup X$ and similarly $\mathrm{NF}_\varepsilon \cdot X = \mathrm{NF} \cdot X \cup X$ for $X \subseteq \mathrm{NF}$.

The class of *starfree languages* is the least class $\mathcal{C}$ of subsets of NF containing $\{s\} \in \mathcal{C}$ for $s \in \mathrm{NF}$ that is closed under the Boolean operations and the sequential and parallel product.

Example 4.1. The *width* $w(t)$ of a poset t is the maximal size of an antichain in t. For $k \in \mathbb{N}$, let $\mathrm{NF}_k = \{t \in \mathrm{NF} \mid w(t) \leq k\}$. The set $\mathrm{NF}_\varepsilon \cdot (\mathrm{NF} \parallel \mathrm{NF}) \cdot \mathrm{NF}_\varepsilon$ of all N-free pomsets of width at least 2 is starfree. Hence, the set NF_1 of linear pomsets is starfree, too. By Example 2.1, words over A can be identified with linear pomsets, i.e. with the elements of NF_1. Then, of course, $A^+ \setminus L$ corresponds to $\mathrm{NF}_1 \setminus L'$ where $L' \subseteq \mathrm{NF}_1$ corresponds to L. Hence, starfree word languages correspond to starfree subsets of NF.

We saw above that NF_1 is starfree. Note that, for any $k \in \mathbb{N}$, the set $\mathrm{NF}_\varepsilon \cdot ((\mathrm{NF} \setminus \mathrm{NF}_{k-1}) \parallel \mathrm{NF}) \cdot \mathrm{NF}_\varepsilon$ is the set of N-free pomsets of width at least $k + 1$. Hence, its complement NF_k is starfree.

Theorem 4.2. *Let $X \subseteq \mathrm{NF}$ be starfree. Then X is elementarily axiomatizable.*

Proof. As usual, this can be shown by induction on the construction of starfree sets. The base case as well as the Boolean operations and the sequential product are easily dealt with. To handle the parallel product, we have to invest a bit more work: Let $X_i \subseteq \mathrm{NF}$ be axiomatized by φ_i for $i = 1, 2$. The distance between two elements of an N-free pomset is 0, 1, or 2 (if they lie in the same connected component), and ∞ if the two elements belong to different connected components. Hence we can infer from Gaifman's Theorem [6] that any elementary sentence is equivalent to a Boolean combination of statements of the form
"there are $\leq n$ connected components $C \subseteq V$ with $\mathrm{Th}_k(C, \leq, \lambda) = T$" $\qquad (*)$
for some bounded theory $T \in \mathrm{TH}_k$.

Let $t = (V, \leq, \lambda) \in \mathrm{NF}$ and $x, y \in V$. Then x and y lie in the same connected component of t iff they are bounded from above or from below. Hence the statements of the form $(*)$ are expressible by an elementary sentence $\psi(n, T)$.

Thus, for any sentence φ, there exists a Boolean combination ψ of sentences $\psi(n, T)$ such that $t \models \varphi$ iff $t \models \psi$ for any N-free pomset $t \in \mathrm{NF}$. The sentence $\bigwedge \psi(n, T) \wedge \bigwedge \neg\psi(n, T)$ $(**)$ is a particularly simple such Boolean combination where the two conjunctions run over finite subsets of $\mathbb{N} \times \mathrm{TH}_k$. Let δ_1 and δ_2 be two conjunctions of the form $(**)$. Then we define a new such formula δ as the conjunction of the following formulas (where $T \in \mathrm{TH}_k$):

$$\psi(n_1 + n_2, T) \qquad \text{if } \psi(n_1, T) \text{ occurs positively in } \delta_1 \text{ and}$$
$$\psi(n_2, T) \text{ occurs positively in } \delta_2,$$
$$\psi(n_i, T) \qquad \text{if } \psi(n_i, T) \text{ occurs positively in } \delta_i \text{ (for } i = 1, 2), \text{ and}$$
$$\neg\psi(n_1 + n_2 - 1, T) \text{ if } \psi(n_1, T) \text{ occurs negatively in } \delta_1 \text{ and}$$
$$\psi(n_2, T) \text{ occurs negatively in } \delta_2.$$

Now one can check that an N-free pomset t satisfies δ iff it is the parallel product of N-free pomsets t_1 and t_2 that satisfy δ_1 and δ_2, resp.

Coming back to $X_1 \parallel X_2$, we can w.l.o.g. assume that φ_1 and φ_2 are Boolean combinations of sentences $\psi(n, T)$. Even more, one can write φ_i as disjunction of conjunctions of the above form $(**)$. Combining any pair of disjuncts from φ_1 and φ_2 in the above described manner, we obtain a sentence φ that axiomatizes $X_1 \parallel X_2$. $\square$

4.2   Aperiodic Sets of N-Free Pomsets Are Starfree

Let S be a set and let $X_s \subseteq \mathrm{NF}$ for $s \in S$. Then the *set of sets starfree over* $\{X_s \mid s \in S\}$ is the least class $\mathcal{C} \subseteq \mathcal{P}(\mathrm{NF})$ containing X_s for $s \in S$, $\{t\}$ for $t \in \mathrm{NF}$, that is closed under the Boolean operations and the sequential and parallel product. Note that $X \subseteq \mathrm{NF}$ is starfree if it is starfree over the empty family.

Example 4.3. We show that the sequential iteration of a starfree subset of $\mathrm{NF} \parallel \mathrm{NF}$ is starfree: More generally, let $L \subseteq (\mathrm{NF} \parallel \mathrm{NF}) \cup A$. Then L^+ is starfree over $\{L\}$ which, in case L is starfree, implies the starfreeness of L^+: A *minimal sequential factor* of a poset t is a poset $t' \in (\mathrm{NF} \parallel \mathrm{NF}) \cup A$ such that $t \in \mathrm{NF}_\varepsilon \cdot t' \cdot \mathrm{NF}_\varepsilon$. Now one easily observes that L^+ is the set of all pomsets from NF all of whose minimal sequential factors belong to L. Hence
$$L^+ = \mathrm{NF} \setminus (\mathrm{NF}_\varepsilon \cdot (\mathrm{NF} \setminus L \cap ((\mathrm{NF} \parallel \mathrm{NF}) \cup A)) \cdot \mathrm{NF}_\varepsilon),$$
a set starfree over L.

Let S be a set. Then S^+ denotes the set of nonempty words over S. We denote the concatenation operation of words over S by $\odot$. Now let $f : (\mathrm{NF} \parallel \mathrm{NF}) \cup A \to S$ be a function. By [8], the semigroup $(\mathrm{NF}, \cdot)$ is freely generated by $(\mathrm{NF} \parallel \mathrm{NF}) \cup A$. Hence, we can uniquely extend the function f to a semigroup homomorphism (also denoted f) from $(\mathrm{NF}, \cdot)$ to $(S, \odot)$. For a set of words $K \subseteq S^+$, the set $f^{-1}(K) = \{t \in \mathrm{NF} \mid f(t) \in K\}$ is a set of N-free pomsets. The following lemma is shown by induction on the construction of the starfree word language K.

Lemma 4.4. *Let $f : (\mathrm{NF} \parallel \mathrm{NF}) \cup A \to S$ be a function and let K be a starfree word language. Then $K' = f^{-1}(K) \subseteq \mathrm{NF}$ is starfree over $\{f^{-1}(s) \mid s \in S\}$.*

Lemma 4.5. *Let $X \subseteq \mathrm{NF}$ be recognized by a homomorphism $\eta : (\mathrm{NF}, \cdot, \parallel) \to (S, \cdot, \parallel)$ into the finite aperiodic sp-algebra $(S, \cdot, \parallel)$. Then X is starfree over $\{\eta^{-1}(s) \cap (\mathrm{NF} \parallel \mathrm{NF}) \mid s \in S\}$.*

Proof. Let $f : (\mathrm{NF}, \cdot) \to (S^+, \odot)$ be the uniquely determined semigroup homomorphism with $f(t) = \eta(t)$ for $t \in (\mathrm{NF} \parallel \mathrm{NF}) \cup A$. ¿From the free, finitely generated semigroup $(S^+, \odot)$ we have the canonical semigroup homomorphism α onto $(S, \cdot)$. Then $\eta = \alpha \circ f$.

By [15], $K := \alpha^{-1}(\eta(X)) \subseteq S^+$ is a starfree word language since $(S, \cdot)$ is a finite aperiodic semigroup. By Lemma 4.4, $f^{-1}(K) \subseteq \mathrm{NF}$ is starfree over $\{f^{-1}(s) \mid s \in S\}$. Since $f^{-1}(s)$ is the union of $\eta^{-1}(s) \cap (\mathrm{NF} \parallel \mathrm{NF})$ and $\eta^{-1}(s) \cap A$, the set $f^{-1}(K)$ is starfree over $\{\eta^{-1}(s) \cap (\mathrm{NF} \parallel \mathrm{NF}) \mid s \in S\}$. Since $f^{-1}(K) = X$, the result follows. $\qquad\square$

Lemma 4.6. *Let $X \subseteq \mathrm{NF} \parallel \mathrm{NF}$ be aperiodic. Then there exist $n \in \mathbb{N}$ and $K_i, L_i \in \mathrm{NF}$ aperiodic for $1 \le i \le n$ such that $X = \bigcup_{1 \le i \le n} K_i \parallel L_i$.*

Proof. There exist a finite aperiodic sp-algebra $(S, \cdot, \parallel)$ and an sp-homomorphism $\eta : \mathrm{NF} \to S$ such that $X = \eta^{-1}\eta(X)$. Let $G = \{(s_1, s_2) \in S \times S \mid s_1 \parallel s_2 \in \eta(X)\}$. Note that $\eta^{-1}(s) \subseteq \mathrm{NF}$ is aperiodic for any $s \in S$. Now one can show $X = \bigcup\{\eta^{-1}(s_1) \parallel \eta^{-1}(s_2) \mid (s_1, s_2) \in G\}$. $\qquad\square$

Lemma 4.7. *Let $k \in \mathbb{N}$ and let $X \subseteq \mathrm{NF}_k$ be starfree over $\{Y_s \mid s \in S\} = \mathcal{H}$ where $Y_s \subseteq \mathrm{NF}$ for $s \in S$. Then X is starfree over $\{Y_s \cap \mathrm{NF}_\ell \mid s \in S, 1 \le \ell \le k\} = \mathcal{H}_k$*

Proof. One actually shows that $X \cap \mathrm{NF}_k$ is starfree over $\mathcal{H}_k$ whenever X is starfree over $\mathcal{H}$. This is done by induction on the starfree construction of the set X (and *not* by induction on k as one might think). $\qquad\square$

Theorem 4.8. *Let $k \in \mathbb{N}$ and let $X \subseteq \mathrm{NF}_k$ be aperiodic. Then X is starfree.*

Proof. The theorem is shown by induction on k. By Example 4.1, Schützenberger's Theorem [15] implies the theorem for $k = 1$. Now suppose the theorem holds for all $\ell < k$ and let $X \subseteq \mathrm{NF}_k$. Then $X = (X \cap (\mathrm{NF} \parallel \mathrm{NF})) \cup (X \setminus (\mathrm{NF} \parallel \mathrm{NF}))$.

By Lemma 4.6, $X \cap (\mathrm{NF} \parallel \mathrm{NF})$ is a finite union of sets $K \parallel L$ with $K, L \subseteq \mathrm{NF}_{k-1}$ aperiodic. Thus, by the induction hypothesis, K and L and therefore $X \cap (\mathrm{NF} \parallel \mathrm{NF})$ are starfree. In particular, we showed the theorem for aperiodic sets contained in $(\mathrm{NF} \parallel \mathrm{NF}) \cap \mathrm{NF}_k$.

Next, we deal with $X \setminus (\mathrm{NF} \parallel \mathrm{NF})$: It is recognized by an sp-homomorphism $\eta : \mathrm{NF} \to S$ into a finite aperiodic sp-algebra $(S, \cdot, \parallel)$. By Lemma 4.5, $X \setminus (\mathrm{NF} \parallel \mathrm{NF})$ is starfree over $\{\eta^{-1}(s) \cap (\mathrm{NF} \parallel \mathrm{NF}) \mid s \in S\}$. By Lemma 4.7, it is therefore starfree over $\{\eta^{-1}(s) \cap (\mathrm{NF} \parallel \mathrm{NF}) \cap \mathrm{NF}_\ell \mid s \in S, \ell \le k\}$. Note that any of the sets $\eta^{-1}(s) \cap (\mathrm{NF} \parallel \mathrm{NF}) \cap \mathrm{NF}_\ell$ is aperiodic and contained in $(\mathrm{NF} \parallel \mathrm{NF}) \cap \mathrm{NF}_k$. Thus, by what we showed above, they are starfree. $\qquad\square$

References

1. B. Courcelle. The monadic second-order logic of graphs. I: Recognizable sets of finite graphs. *Information and Computation*, 85:12–75, 1990.
2. B. Courcelle and J.A. Makowsky. VR and HR graph grammars: A common algebraic framework compatible with monadic second order logic. In *Graph transformations*, 2000. cf. http://tfs.cs.tu-berlin.de/gratra2000/proceedings.html.
3. J. Doner. Tree acceptors and some of their applications. *J. Comput. Syst. Sci.*, 4:406–451, 1970.
4. M. Droste and P. Gastin. Asynchronous cellular automata for pomsets without autoconcurrency. In *CONCUR'96*, Lecture Notes in Comp. Science vol. 1119, pages 627–638. Springer, 1996.
5. M. Droste and D. Kuske. Logical definability of recognizable and aperiodic languages in concurrency monoids. In *Computer Science Logic*, Lecture Notes in Comp. Science vol. 1092, pages 467–478. Springer, 1996.
6. H.-D. Ebbinghaus and J. Flum. *Finite Model Theory*. Springer, 1991.
7. W. Ebinger and A. Muscholl. Logical definability on infinite traces. *Theoretical Comp. Science*, 154:67–84, 1996.
8. J.L. Gischer. The equational theory of pomsets. *Theoretical Comp. Science*, 61:199–224, 1988.
9. Y. Gurevich. Modest theory of short chains I. *J. of Symb. Logic*, 44:481–490, 1979.
10. D. Kuske. Asynchronous cellular automata and asynchronous automata for pomsets. In *CONCUR'98*, Lecture Notes in Comp. Science vol. 1466, pages 517–532. Springer, 1998.
11. D. Kuske. Infinite series-rational posets: logic and languages. In *ICALP 2000*, Lecture Notes in Comp. Science vol. 1853, pages 648–662. Springer, 2000.
12. R.E. Ladner. Application of model theoretic games to discrete linear orders and finite automata. *Information and Control*, 33:281–303, 1977.
13. K. Lodaya and P. Weil. Series-parallel languages and the bounded-width property. *Theoretical Comp. Science*, 237:347–380, 2000.
14. R. McNaughton and S. Papert. *Counter-Free Automata*. MIT Press, Cambridge, MA, 1971.
15. M.P. Schützenberger. On finite monoids having only trivial subgroups. *Inf. Control*, 8, 1965.
16. S. Shelah. The monadic theory of order. *Annals of Mathematics*, 102:379–419, 1975.
17. W. Thomas. Automata on infinite objects. In J. van Leeuwen, editor, *Handbook of Theoretical Computer Science*, pages 133–191. Elsevier Science Publ. B.V., 1990.
18. W. Thomas. On logical definability of trace languages. In V. Diekert, editor, *Proceedings of a workshop of the ESPRIT BRA No 3166: Algebraic and Syntactic Methods in Computer Science (ASMICS) 1989*, Report TUM-I9002, Technical University of Munich, pages 172–182, 1990.
19. W. Thomas. Ehrenfeucht games, the composition method, and the monadic theory of ordinal words. In J. Mycielski et al., editor, *Structures in Logic and Computer Science, A Selection of Essays in Honor of A. Ehrenfeucht*, Lecture Notes in Comp. Science vol. 1261, pages 118–143. Springer, 1997.

An Ehrenfeucht–Fraïssé Approach to Collapse Results for First-Order Queries over Embedded Databases

Clemens Lautemann and Nicole Schweikardt

Institut für Informatik / FB 17
Johannes Gutenberg-Universität, D–55099 Mainz
{cl,nisch}@informatik.uni-mainz.de

Abstract. We present a new proof technique for *collapse results* for first–order queries on databases which are embedded in $\mathbb{N}$ or $\mathbb{R}_{\geqslant 0}$. Our proofs are by means of an explicitly constructed winning strategy for Duplicator in an Ehrenfeucht–Fraïssé game, and can deal with certain infinite databases where previous, highly involved methods fail. Our main result is that first–order logic has the natural–generic collapse over $\langle \mathbb{N}, \leqslant, + \rangle$ for *arbitrary* (i.e., possibly infinite) databases. Furthermore, a first application of this result shows the natural–generic collapse of first–order logic over $\langle \mathbb{R}_{\geqslant 0}, \leqslant, + \rangle$ for a certain kind of databases over $\mathbb{R}_{\geqslant 0}$ which consist of a possibly *infinite* number of regions.
Classification: Logic in Computer Science, Database Theory.

1 Introduction

One of the issues in database theory that have attracted much interest in recent years is the study of relational databases which are embedded in a fixed, possibly infinite structure. This occurs, e.g., in current applications, such as spatial or temporal databases, where data are represented by (natural or real) numbers (for a recent comprehensive survey see [6]). In many applications, the numerical values only serve as names which are exchangeable. If the underlying structure is linearly ordered, it is often only the relative order of the data which is of interest. For this situation, *locally order generic* queries have been studied, i.e., queries which commute with every *order–preserving* embedding of the active domain into the underlying structure. One central theme here is the question, how much additional power a language, such as first–order logic can gain for locally order generic queries, by using additional (e.g., arithmetical) predicates of the underlying structure. It is well–known that with addition and multiplication first order–logic can, indeed, express more locally order generic queries over $\mathbb{N}$ than with order alone; for other cases, however, these investigations have led to so–called *collapse results* for first–order queries. As an example, take the following theorem (cf. [3], Proposition 3.6.4).

Theorem. *First–order logic has the natural–generic collapse over $\langle \mathbb{N}, \leqslant, + \rangle$ for finite databases.* □

A. Ferreira and H. Reichel (Eds.): STACS 2001, LNCS 2010, pp. 455–466, 2001.
© Springer-Verlag Berlin Heidelberg 2001

In this theorem, databases are considered the active domain of which is a *finite* subset of $\mathbb{N}$. A first–order query has *natural semantics*, if quantifiers are interpreted over all of $\mathbb{N}$. The theorem states that, under natural semantics, every locally order generic first–order query which uses (apart from the database relations) $\leqslant$ and $+$, can be equivalently replaced by one which uses only $\leqslant$ (and the database relations). In other words, under the stated hypotheses, addition does not add to the expressive power of first–order logic.

Theorems like this can be derived from general collapse results in [1,2,4]; for an overview see [3]. The proofs for these results are rather involved; they use non–standard models, and are non–constructive.

Our contribution in this paper is twofold: We extend the natural–generic collapse of first–order logic for the case of $\langle \mathbb{N}, \leqslant, + \rangle$ as underlying structure from *finite* to *arbitrary* databases; and for the case of underlying structure $\langle \mathbb{R}_{\geqslant 0}, \leqslant, + \rangle$, where it was known for *finitely representable* databases, we show that it also holds for *nicely representable* databases (precise formulations are given in Section 3). In both cases, we transcend the finite limitations of the previously used proof techniques. Moreover, our proofs are *constructive* in the sense that they are obtained using an Ehrenfeucht–Fraïssé game with an explicitly constructed winning strategy for Duplicator.

Due to space limitations, most proofs are omitted in this extended abstract. The full paper including complete proofs of all our claims can be obtained from `http://www.informatik.uni-mainz.de/~{}nisch/publications.html`.

2 Preliminaries

We use $\mathbb{Z}$ for the set of all integers, $\mathbb{N}$ for the set of all non–negative integers, $\mathbb{R}$ for the set of all reals, and $\mathbb{R}_{\geqslant 0}$ for the set of all non–negative reals.

Depending on the particular context, we will use $\vec{x}$ as abbreviation for a sequence $x_1, .., x_m$ or a tuple $(x_1, .., x_m)$. Accordingly, if q is a mapping defined on all elements in $\vec{x}$, we will write $q(\vec{x})$ to denote the sequence $q(x_1), .., q(x_m)$ or the tuple $(q(x_1), .., q(x_m))$. If R is an m–ary relation on the domain of q, we write $q(R)$ to denote the relation $\{q(\vec{x}) : \vec{x} \in R\}$; and instead of $\vec{x} \in R$ we often write $R(\vec{x})$. We write $x_1, .., x_m \mapsto y_1, .., y_m$ to denote the mapping q with domain $\{x_1, .., x_m\}$ and range $\{y_1, .., y_m\}$ which satisfies $q(x_i) = y_i$, for all $i \in \{1, .., m\}$. Throughout the rest of this section let $\mathbb{U}$ be $\mathbb{N}$ or $\mathbb{R}_{\geqslant 0}$.

Databases. A *database schema* SC is a finite collection of relation symbols, each of a fixed aritiy. An SC–*database state* $\mathcal{A}$ over $\mathbb{U}$ assigns a concrete l–ary relation $R^{\mathcal{A}} \subseteq \mathbb{U}^l$ to each l–ary relation symbol $R \in SC$. In the literature, attention is often restricted to *finite* databases. Note, however, that we allow database relations to be *infinite*.

The set $adom(\mathcal{A})$ of those elements of $\mathbb{U}$ which occur in one of $\mathcal{A}$'s relations is called the *active domain* of $\mathcal{A}$. If q is a mapping defined on the active domain of $\mathcal{A}$, we write $q(\mathcal{A})$ to denote the SC–database state with $R^{q(\mathcal{A})} = q(R^{\mathcal{A}})$, for every $R \in SC$. Accordingly, we use $q(\mathcal{A}, \vec{a})$ as abbreviation for $(q(\mathcal{A}), q(\vec{a}))$.

An m–ary *SC–query* $\mathcal{Q}$ over $\mathbb{U}$ is a mapping which assigns to every *SC–*database state $\mathcal{A}$ over $\mathbb{U}$ an m–ary relation $\mathcal{Q}(\mathcal{A}) \subseteq \mathbb{U}^m$. For $m = 0$, we have that $\mathcal{Q}(\mathcal{A}) \in \{true, false\}$ and call $\mathcal{Q}$ a *Boolean* query.

Let $\mathcal{K}$ be a class of *SC–*database states over $\mathbb{U}$. We say that $\mathcal{Q}$ is *FO–*definable on $\mathcal{K}$ over $\langle \mathbb{U}, \leqslant \rangle$ (resp., $\langle \mathbb{U}, \leqslant, + \rangle$), if there is an m–ary formula $\varphi(\vec{x}) \in FO(SC, \leqslant)$ (resp., $FO(SC, \leqslant, +)$), such that for every $\mathcal{A} \in \mathcal{K}$ and for every m–tuple $\vec{a} \in \mathbb{U}^m$ it is true that $\vec{a} \in \mathcal{Q}(\mathcal{A})$ if and only if $\langle \mathbb{U}, \leqslant, \mathcal{A}, \vec{a} \rangle \models \varphi(\vec{x})$ (resp., $\langle \mathbb{U}, \leqslant, +, \mathcal{A}, \vec{a} \rangle \models \varphi(\vec{x})$). Here, a formula in $FO(SC, \leqslant)$ is a first–order formula in the language $SC \cup \{=, \leqslant\}$. Similarly, $FO(SC, \leqslant, +)$ consists of all first–order formulas in the language $SC \cup \{=, \leqslant, +\}$.

Ehrenfeucht–Fraïssé Games.   Our main technical tool will be the *Ehrenfeucht–Fraïssé game*. For our purposes, we consider two versions of the game: the $\leqslant$–game and the $+$–game. Both are played by two players, Spoiler and Duplicator, on two structures, $\hat{\mathcal{A}} = \langle \mathbb{U}, \mathcal{A}, \vec{a} \rangle$ and $\hat{\mathcal{B}} = \langle \mathbb{U}, \mathcal{B}, \vec{b} \rangle$, where $\mathcal{A}$ and $\mathcal{B}$ are *SC–*database states over $\mathbb{U}$, and $\vec{a}$ and $\vec{b}$ are sequences of length m in $\mathbb{U}$. There is a fixed number k of rounds. In each round $i \in \{1, \ldots, k\}$

- Spoiler chooses one element, $a^{(i)} \in \mathbb{U}$ in $\hat{\mathcal{A}}$, or an element $b^{(i)} \in \mathbb{U}$ in $\hat{\mathcal{B}}$;
- then Duplicator chooses an element in the other structure, i.e., an element $b^{(i)} \in \mathbb{U}$ in $\hat{\mathcal{B}}$, if Spoiler's move was in $\hat{\mathcal{A}}$, or an element $a^{(i)} \in \mathbb{U}$ in $\hat{\mathcal{A}}$, otherwise.

After k rounds, the game finishes with elements $a^{(1)}, \ldots, a^{(k)}$ chosen in $\hat{\mathcal{A}}$ and $b^{(1)}, \ldots, b^{(k)}$ chosen in $\hat{\mathcal{B}}$. Duplicator has won the $\leqslant$–*game* if, restricted to the sequences $\vec{a}, a^{(1)}, \ldots, a^{(k)}$, and $\vec{b}, b^{(1)}, \ldots, b^{(k)}$, respectively, the structures $\hat{\mathcal{A}}$ and $\hat{\mathcal{B}}$ are indistinguishable with respect to $=$, SC, and $\leqslant$, i.e., if the mapping $a^{(1)}, \ldots, a^{(k)} \mapsto b^{(1)}, \ldots, b^{(k)}$ is a *partial $\leqslant$–isomorphism* (see Definition 1 below) between $\langle \mathbb{U}, \mathcal{A}, \vec{a} \rangle$ and $\langle \mathbb{U}, \mathcal{B}, \vec{b} \rangle$. Duplicator has won the $+$–game, if this mapping even is a *partial $+$–isomorphism*.

Definition 1 (Partial Isomorphism). *Let $\mathcal{A}$ and $\mathcal{B}$ be SC–database states over $\mathbb{U}$, and let $\vec{a}$ and $\vec{b}$ be sequences of length m in $\mathbb{U}$. Furthermore, let $a^{(1)}, \ldots, a^{(k)}$ and $b^{(1)}, \ldots, b^{(k)}$ be elements in $\mathbb{U}$. The mapping $\pi : a^{(1)}, \ldots, a^{(k)} \mapsto b^{(1)}, \ldots, b^{(k)}$ is called a* partial isomorphism *between $\langle \mathbb{U}, \mathcal{A}, \vec{a} \rangle$ and $\langle \mathbb{U}, \mathcal{B}, \vec{b} \rangle$ if the following holds true, where $\vec{a} = a^{(0)}, a^{(-1)}, \ldots, a^{(-m+1)}$ and $\vec{b} = b^{(0)}, b^{(-1)}, \ldots, b^{(-m+1)}$.*

(i) $a^{(i)} = a^{(j)}$ *iff* $b^{(i)} = b^{(j)}$, *for every i, j with $-m < i, j \leqslant k$,*

(ii) $a^{(i)} \in adom(\mathcal{A})$ *iff* $b^{(i)} \in adom(\mathcal{B})$, *for every i with $-m < i \leqslant k$, and* $R^{\mathcal{A}}(a^{(i_1)}, \ldots, a^{(i_l)})$ *iff* $R^{\mathcal{B}}(b^{(i_1)}, \ldots, b^{(i_l)})$, *for every relation symbol $R \in SC$ of arity, say, l, and for all $i_1, \ldots, i_l$ with $-m < i_1, \ldots, i_l \leqslant k$.*

π *is called a* partial $\leqslant$–isomorphism *if additionally we have*

(iii) $a^{(i)} \leqslant a^{(j)}$ *iff* $b^{(i)} \leqslant b^{(j)}$, *for every i, j with $-m < i, j \leqslant k$.*

π *is called a* partial $+$–isomorphism *if, in addition to (i)–(iii), we have*

(iv) $a^{(i)} + a^{(j)} = a^{(l)}$ *iff* $b^{(i)} + b^{(j)} = b^{(l)}$, *for all i, j, l with $-m < i, j, l \leqslant k$.*   □

We write $\langle \mathbb{U}, \mathcal{A}, \vec{a} \rangle \equiv_k^{\leqslant} \langle \mathbb{U}, \mathcal{B}, \vec{b} \rangle$ (resp., $\langle \mathbb{U}, \mathcal{A}, \vec{a} \rangle \equiv_k^{+} \langle \mathbb{U}, \mathcal{B}, \vec{b} \rangle$) to indicate that Duplicator has a *winning strategy* in the k–round $\leqslant$–game (resp., +–game) on $\langle \mathbb{U}, \mathcal{A}, \vec{a} \rangle$ and $\langle \mathbb{U}, \mathcal{B}, \vec{b} \rangle$. The fundamental use of the game comes from the fact that it characterises first–order logic (cf., e.g., [5]). In our context, this can be formulated as follows:

Theorem 1 (Ehrenfeucht, Fraïssé). *Let $\mathcal{K}$ be a class of SC–database states over $\mathbb{U}$. An m–ary SC–query $\mathcal{Q}$ over $\mathbb{U}$ is not FO–definable on $\mathcal{K}$ over $\langle \mathbb{U}, \leqslant \rangle$ (resp., $\langle \mathbb{U}, \leqslant, + \rangle$) if and only if, for every number $k \in \mathbb{N}$, there are $\mathcal{A}, \mathcal{B} \in \mathcal{K}$ and tuples $\vec{a}$ and $\vec{b}$ of length m in $\mathbb{U}$ with $\vec{a} \in \mathcal{Q}(\mathcal{A})$ and $\vec{b} \notin \mathcal{Q}(\mathcal{B})$, such that $\langle \mathbb{U}, \mathcal{A}, \vec{a} \rangle \equiv_k^{\leqslant} \langle \mathbb{U}, \mathcal{B}, \vec{b} \rangle$ (resp., $\langle \mathbb{U}, \mathcal{A}, \vec{a} \rangle \equiv_k^{+} \langle \mathbb{U}, \mathcal{B}, \vec{b} \rangle$).* $\square$

3   Collapse Results

Databases Embedded in $\langle \mathbb{N}, \leqslant, + \rangle$.
We call a mapping $q : \mathbb{N} \to \mathbb{N}$ *order–preserving* if, for every $m, n \in \mathbb{N}$, we have $m \leqslant n$ iff $q(m) \leqslant q(n)$. An SC–query $\mathcal{Q}$ over $\mathbb{N}$ is called *order generic* on an SC–database state $\mathcal{A}$ over $\mathbb{N}$ iff for every order–preserving mapping $q : \mathbb{N} \to \mathbb{N}$, it is true that $q(\mathcal{Q}(\mathcal{A})) = \mathcal{Q}(q(\mathcal{A}))$, i.e., $\vec{a} \in \mathcal{Q}(\mathcal{A})$ iff $q(\vec{a}) \in \mathcal{Q}(q(\mathcal{A}))$, for every $\vec{a}$. Our main theorem states that addition does not add to the expressive power of first–order logic over the natural numbers for defining order generic queries.

Theorem 2. *First–order logic has the natural–generic collapse over $\langle \mathbb{N}, \leqslant, + \rangle$ for arbitrary databases, i.e.: Let $\mathcal{Q}$ be an m–ary SC–query over $\mathbb{N}$, and let $\mathcal{K}$ be a class of SC–database states over $\mathbb{N}$ on which $\mathcal{Q}$ is order generic. If $\mathcal{Q}$ is FO–definable on $\mathcal{K}$ over $\langle \mathbb{N}, \leqslant, + \rangle$, then it already is so over $\langle \mathbb{N}, \leqslant \rangle$.* $\square$

Theorem 2 is a direct consequence of the following result:

Theorem 3. *For every $k \in \mathbb{N}$ there exists a number $r(k) \in \mathbb{N}$ and an order–preserving mapping $q : \mathbb{N} \to \mathbb{N}$ such that, for every database schema SC and every $m \in \mathbb{N}$, the following holds: If $\mathcal{A}$ and $\mathcal{B}$ are SC–database states over $\mathbb{N}$, and if $\vec{a}$ and $\vec{b}$ are sequences of length m in $\mathbb{N}$ with $\langle \mathbb{N}, \mathcal{A}, \vec{a} \rangle \equiv_{r(k)}^{\leqslant} \langle \mathbb{N}, \mathcal{B}, \vec{b} \rangle$, then*

$$\langle \mathbb{N}, q(\mathcal{A}, \vec{a}) \rangle \equiv_k^{+} \langle \mathbb{N}, q(\mathcal{B}, \vec{b}) \rangle.$$ $\square$

The proof of Theorem 3 will be given in Section 4.

Proof of Theorem 2: We assume that $\mathcal{Q}$ is not FO–definable on $\mathcal{K}$ over $\langle \mathbb{N}, \leqslant \rangle$; and we need show that $\mathcal{Q}$ is not FO–definable on $\mathcal{K}$ over $\langle \mathbb{N}, \leqslant, + \rangle$, either. From Theorem 1 we know that, for each number k, there are $\mathcal{A}, \mathcal{B} \in \mathcal{K}$ and tuples $\vec{a}$ and $\vec{b}$ of length m in $\mathbb{N}$ with $\vec{a} \in \mathcal{Q}(\mathcal{A})$ and $\vec{b} \notin \mathcal{Q}(\mathcal{B})$, such that $\langle \mathbb{N}, \mathcal{A}, \vec{a} \rangle \equiv_{r(k)}^{\leqslant} \langle \mathbb{N}, \mathcal{B}, \vec{b} \rangle$. By Theorem 3, then, $\langle \mathbb{N}, q(\mathcal{A}, \vec{a}) \rangle \equiv_k^{+} \langle \mathbb{N}, q(\mathcal{B}, \vec{b}) \rangle$. Furthermore, since q is order–preserving and since $\mathcal{Q}$ is order generic on $\mathcal{A}$ and $\mathcal{B}$, we know that $q(\vec{a}) \in \mathcal{Q}(q(\mathcal{A}))$ and $q(\vec{b}) \notin \mathcal{Q}(q(\mathcal{B}))$. Hence, from Theorem 1 we obtain that $\mathcal{Q}$ is not FO–definable over $\langle \mathbb{N}, \leqslant, + \rangle$. $\square$

Nicely Representable Databases Embedded in $\langle \mathbb{R}_{\geqslant 0}, \leqslant, + \rangle$.
Let $I = (I_n)_{n \in \mathbb{N}}$ be a sequence of real closed intervals $I_n = [l_n, r_n]$. We call I

nice if $r_n < l_{n+1}$, for all $n \in \mathbb{N}$, and the sequence $(r_n)_{n \in \mathbb{N}}$ is unbounded. An SC–database state $\mathcal{C}$ over $\mathbb{R}_{\geqslant 0}$ is called *nicely representable* if

- its active domain is $\bigcup_{n \in \mathbb{N}} I_n$, for a nice sequence $(I_n)_{n \in \mathbb{N}}$ of real closed intervals, and
- all relations of $\mathcal{C}$ are constant on all the intervals, i.e., if R is an m–ary relation in $\mathcal{C}$ and if $x_1, y_1 \in I_{n_1}, \ldots, x_m, y_m \in I_{n_m}$, then we have $R(x_1, .., x_m)$ iff $R(y_1, .., y_m)$.

A mapping $q : \mathbb{R}_{\geqslant 0} \to \mathbb{R}_{\geqslant 0}$ is called an *order–automorphism* of $\mathbb{R}_{\geqslant 0}$ if it is bijective and increasing (note that this, in particular, implies that q is continuous). An SC–query $\mathcal{Q}$ over $\mathbb{R}_{\geqslant 0}$ is called *order generic* on an SC–database state $\mathcal{C}$ over $\mathbb{R}_{\geqslant 0}$ iff for every order–automorphism q of $\mathbb{R}_{\geqslant 0}$, it is true that $q(\mathcal{Q}(\mathcal{C})) = \mathcal{Q}(q(\mathcal{C}))$, i.e., $\vec{c} \in \mathcal{Q}(\mathcal{C})$ iff $q(\vec{c}) \in \mathcal{Q}(q(\mathcal{C}))$, for every $\vec{c}$.

Theorem 4. First–order logic has the natural–generic collapse over $\langle \mathbb{R}_{\geqslant 0}, \leqslant, + \rangle$ for nicely representable databases, *i.e.: Let $\mathcal{Q}$ be an m–ary SC–query over $\mathbb{R}_{\geqslant 0}$, and let $\mathcal{K}$ be a class of nicely representable SC–database states over $\mathbb{R}_{\geqslant 0}$ on which $\mathcal{Q}$ is order generic. If $\mathcal{Q}$ is FO–definable on $\mathcal{K}$ over $\langle \mathbb{R}_{\geqslant 0}, \leqslant, + \rangle$, then it already is so over $\langle \mathbb{R}_{\geqslant 0}, \leqslant \rangle$.* $\qquad\square$

Theorem 4 is a direct consequence of the following result:

Theorem 5. *For every $k \in \mathbb{N}$ there exists a number $r'(k) \in \mathbb{N}$ such that, for every database schema SC and every $m \in \mathbb{N}$, the following holds: If $\mathcal{C}$ and $\mathcal{D}$ are nicely representable SC–database states over $\mathbb{R}_{\geqslant 0}$, and if $\vec{c}$ and $\vec{d}$ are sequences of length m in $\mathbb{R}_{\geqslant 0}$ with $\langle \mathbb{R}_{\geqslant 0}, \mathcal{C}, \vec{c} \rangle \equiv_{r'(k)}^{\leqslant} \langle \mathbb{R}_{\geqslant 0}, \mathcal{D}, \vec{d} \rangle$, then there are order–automorphisms q' and $\tilde{q}'$ of $\mathbb{R}_{\geqslant 0}$ such that $\langle \mathbb{R}_{\geqslant 0}, q'(\mathcal{C}, \vec{c}) \rangle \equiv_k^{+} \langle \mathbb{R}_{\geqslant 0}, \tilde{q}'(\mathcal{D}, \vec{d}) \rangle$.* $\qquad\square$

The proof of Theorem 5 will be given in section 5. We can obtain Theorem 4 from Theorem 5 in exactly the same way as we obtained Theorem 2 from Theorem 3.

4   Proof of Theorem 3

Our proof of Theorem 3 is an adaption of Lynch's proof of his following theorem:

Theorem ([7, Theorem 3.7]). *For every $k \in \mathbb{N}$ there exists a number $d(k) \in \mathbb{N}$ and an infinite set $Q \subseteq \mathbb{N}$ such that, for all subsets $A, B \subseteq Q$, the following holds: If $|A| = |B|$ or $d(k) < |A|, |B| < \infty$, then $\langle \mathbb{N}, A \rangle \equiv_k^{+} \langle \mathbb{N}, B \rangle$.* $\qquad\square$

Unfortunately, neither the statement nor the proof of Lynch's theorem gives us directly what we need. Going through Lynch's proof in detail, we will modify and extend his notation and his reasoning in a way appropriate for obtaining our Theorem 3.

However, to illustrate the overall proof idea, let us first try to explain intuitively Lynch's proof. For simplicity, we concentrate on subsets $A, B \subseteq Q$ of the same size and discuss what Duplicator has to do in order to win the k–round $+$–game on $\langle \mathbb{N}, A \rangle$ and $\langle \mathbb{N}, B \rangle$. Assume that, after $i-1$ rounds $a^{(1)}, \ldots, a^{(i-1)}$ have been played in $\langle \mathbb{N}, A \rangle$, and $b^{(1)}, .., b^{(i-1)}$ in $\langle \mathbb{N}, B \rangle$. Let Spoiler choose some element $a^{(i)}$ in $\langle \mathbb{N}, A \rangle$. When choosing $b^{(i)}$ in $\langle \mathbb{N}, B \rangle$, Duplicator has to make sure that, whatever Spoiler can do in the remaining $k-i$ rounds in one structure, can be matched in the other. In particular, this means that any sum over the $a^{(j)}$ behaves in relation to A exactly as the corresponding sum over the $b^{(j)}$ behaves in relation to B. For instance, for any sets $J, J' \subseteq \{1, .., i\}$, it should hold that there is some $a \in Q$ that lies between $\sum_{j \in J} a^{(j)}$ and $\sum_{j' \in J'} a^{(j')}$ if and only if there is some $b \in Q$ that lies between $\sum_{j \in J} b^{(j)}$ and $\sum_{j' \in J'} b^{(j')}$. But it is not enough to consider simple sums over previously played elements. Since with $O(r)$ additions it is possible to generate $s \cdot a^{(i)}$ from $a^{(i)}$, for any $s \leqslant 2^r$, we also have to consider linear combinations with large coefficients. Furthermore, Spoiler can alternate his moves between the two structures, this results in the necessity of dealing with even more complex linear combinations. One can only handle all these complications because, as the game progresses, the number of rounds left for Spoiler to do all these things decreases. This means, for instance, that the coefficients and the length of the linear combinations we have to consider decrease: after the last round, the only relevant linear combinations are simple additions of chosen elements.

Let us now concentrate on the proof of Theorem 3.

4.1   Notation

We first define a function r and, following Lynch, two functions f and g, which will be used as parameters in the proof. Of course, all we can say at this point to justify this particular choice of functions is that they will make the technicalities of the proof work.

$$r(0) := 1, \quad r(i+1) := r(i) + 2^{i+3},$$
$$f(0) := 1, \quad f(i+1) := 2f(i)^4,$$
$$g(0) := 0, \quad g(i+1) := 2g(i)f(i)^2 + f(i)!.$$

Let $k \in \mathbb{N}$ be fixed. Choose any sequence $p_0, p_1, p_2, \ldots$ in $\mathbb{N}$ with

$$p_0 = 0,$$
$$p_i \geqslant 2^{k+3} f(k)^3 p_{i-1} + 2g(k)f(k)^2, \quad \text{for all } i > 0, \quad \text{and}$$
$$p_i \equiv p_j \pmod{f(k)!}, \quad \text{for all } i, j > 0.$$

Here, $r \equiv s \pmod{n}$ means that that $r-s \in n \cdot \mathbb{Z}$ (for $r, s \in \mathbb{R}$ and $n \in \mathbb{N}$). It is obvious that such a sequence $p_0, p_1, \ldots$ does exist (cf., [7]).

We define the order-preserving mapping $q : \mathbb{N} \to \mathbb{N}$ via $q(i) := p_{i+1}$ for all $i \in \mathbb{N}$; and we define Q to be the range of q, i.e., $Q := q(\mathbb{N}) = \{p_1, p_2, p_3, ..\}$.

Let $\mathcal{A}$ and $\mathcal{B}$ be SC-database states over $\mathbb{N}$, and let $\vec{a}$ and $\vec{b}$ be sequences of length m in $\mathbb{N}$. We define $\mathfrak{A}, \vec{a} := q(\mathcal{A}, \vec{a})$ and $\mathfrak{B}, \vec{b} := q(\mathcal{B}, \vec{b})$. Furthermore, let $\vec{a} = \mathfrak{a}^{(0)}, \mathfrak{a}^{(-1)}, .., \mathfrak{a}^{(-m+1)}$, and $\vec{b} = \mathfrak{b}^{(0)}, \mathfrak{b}^{(-1)}, .., \mathfrak{b}^{(-m+1)}$.

Our assumption is that Duplicator has a winning strategy in the $r(k)$-round $\leqslant$-game on $\langle \mathbb{N}, \mathcal{A}, \vec{a} \rangle$ and $\langle \mathbb{N}, \mathcal{B}, \vec{b} \rangle$ (which, henceforth, will be called the $\leqslant$-game). Our aim is to show that Duplicator has a winning strategy in the k-round $+$-game on $\langle \mathbb{N}, \mathfrak{A}, \vec{a} \rangle$ and $\langle \mathbb{N}, \mathfrak{B}, \vec{b} \rangle$ (which, henceforth, will be called the $+$-game).

For each round $i \in \{1, .., k\}$ of the $+$–game we use $\mathfrak{a}^{(i)}$ and $\mathfrak{b}^{(i)}$, respectively, to denote the element chosen in that round in $\langle \mathbb{N}, \mathfrak{A}, \vec{a} \rangle$ and in $\langle \mathbb{N}, \mathfrak{B}, \vec{b} \rangle$, respectively.

We will translate each move of Spoiler in the $+$–game, say $\mathfrak{a}^{(i)}$ (if Spoiler chooses in $\langle \mathbb{N}, \mathfrak{A}, \vec{a} \rangle$), into a number of moves $a_1^{(i)}, .., a_{n_i}^{(i)}$ for a "virtual Spoiler" in the $\leqslant$–game. Then we can find the answers $b_1^{(i)}, .., b_{n_i}^{(i)}$ of a "virtual Duplicator" playing according to the winning strategy in the $\leqslant$–game, and we can translate these answers into a move $\mathfrak{b}^{(i)}$ for Duplicator in the $+$–game. (The case when Spoiler chooses $\mathfrak{b}^{(i)}$ in $\langle \mathbb{N}, \mathfrak{B}, \vec{b} \rangle$ is symmetric.)

Before we can describe Duplicator's winning strategy in the $+$–game, we have to fix some further notation: As an abbreviation, for $i \in \{0, .., k\}$, we use $\vec{a}^{(i)}$ to denote the sequence $\vec{a}, a_1^{(1)}, .., a_{n_1}^{(1)}, .., a_1^{(i)}, .., a_{n_i}^{(i)}$ of all positions chosen in $\langle \mathbb{N}, \mathcal{A}, \vec{a} \rangle$ until the end of the i–th round of the $+$–game. Analogously, we use $\vec{b}^{(i)}$ to denote the corresponding sequence in $\langle \mathbb{N}, \mathcal{B}, \vec{b} \rangle$. Furthermore, we define $\mathfrak{a}_j^{(i)} := q(a_j^{(i)})$ and $\mathfrak{b}_j^{(i)} := q(b_j^{(i)})$; and we use $\vec{\mathfrak{a}}^{(i)}$ and $\vec{\mathfrak{b}}^{(i)}$ to denote the sequence of the q–images of the elements in the sequence $\vec{a}^{(i)}$ and $\vec{b}^{(i)}$, respectively. Clearly, it holds that $\vec{\mathfrak{a}}^{(i)}, \vec{\mathfrak{b}}^{(i)} \in Q$.

Let $i \in \{0, .., k\}$. A partial mapping c from $\{\mathfrak{a}^{(1)}, .., \mathfrak{a}^{(i)}\} \cup Q$ to $\{\mathfrak{b}^{(1)}, .., \mathfrak{b}^{(i)}\} \cup Q$ is called an i–correspondence if

(i) $\{\mathfrak{a}^{(1)}, .., \mathfrak{a}^{(i)}\} \cup \{\vec{\mathfrak{a}}^{(i)}\}$ is in the domain of c,

(ii) $c(\mathfrak{a}^{(l)}) = \mathfrak{b}^{(l)}$ for all $l \in \{1, .., i\}$,

(iii) $c(\vec{\mathfrak{a}}^{(i)}) = \vec{\mathfrak{b}}^{(i)}$,

(iv) $c(q) \in Q$ if and only if $q \in Q$ (for all q in the domain of c),

(v) c is order–preserving on Q.

An i–correspondence c is called *SC–respecting* if it is a partial isomorphism between $\langle \mathbb{N}, \mathfrak{A}, \vec{a} \rangle$ and $\langle \mathbb{N}, \mathfrak{B}, \vec{b} \rangle$ (note that we do not consider $\leqslant$ or $+$ here). An *i–vector in $\mathfrak{A}$* is a sequence $s := (x_1, .., x_n, \alpha_1, .., \alpha_n, \beta)$, where

(i) $n \leqslant 2^{k-i+1}$,

(ii) $x_1, .., x_n$ are pairwise distinct elements in $\{\mathfrak{a}^{(1)}, .., \mathfrak{a}^{(i)}\} \cup Q$,

(iii) $\alpha_j = \dfrac{u_j}{u_j'}$, with $u_j, u_j' \in \mathbb{Z}$ and $|u_j|, |u_j'| \leqslant f(k-i)$, for each $j \in \{1, .., n\}$,

(iv) $\beta \in \mathbb{R}$ with $|\beta| \leqslant g(k-i)$.

An *i–vector in $\mathfrak{B}$* is defined analogously.

A *minor i–vector* is an i–vector where additionally we have

(iv)' $|\beta| \leqslant g(k-i) - f(k-i-1)! = 2g(k-i-1)f(k-i-1)^2$.

The elements $x_1, .., x_n$ are called the *terms* of the i–vector s; $\alpha_1, .., \alpha_n$ are called the *coefficients* of s; and $\bar{s} := \sum_{j=1}^{n} \alpha_j x_j + \beta$ is called the *evaluation* of s. If c is an i–correspondence and if $s = (x_1, .., x_n, \alpha_1, .., \alpha_n, \beta)$ is an i–vector in $\mathfrak{A}$ whose terms are in the domain of c, then we write $c(s)$ to denote the image of s under c, i.e., $c(s)$ is the i–vector $(c(x_1), .., c(x_n), \alpha_1, .., \alpha_n, \beta)$ in $\mathfrak{B}$.

4.2 Duplicator's Strategy in the +−Game

We will show that Duplicator can play the +−game in such a way that the following four conditions hold at the end of each round i, for $i \in \{0, .., k\}$:

(1) $\langle \mathbb{N}, \mathcal{A}, \vec{a}^{(i)} \rangle \equiv^{\leqslant}_{r(k-i)} \langle \mathbb{N}, \mathcal{B}, \vec{b}^{(i)} \rangle$

(2) $\mathfrak{a}^{(i)} \equiv \mathfrak{b}^{(i)} \pmod{f(k-i)!}$ (if $i \neq 0$)

(3) The mapping $c : \mathfrak{a}^{(1)}, .., \mathfrak{a}^{(i)}, \vec{a}^{(i)} \mapsto \mathfrak{b}^{(1)}, .., \mathfrak{b}^{(i)}, \vec{b}^{(i)}$ is an SC–respecting i–correspondence.

(4) Let d be an i–correspondence and let s_1 and s_2 be i–vectors in $\mathfrak{A}$ whose terms are in the domain of d. Then $\overline{s_1} \leqslant \overline{s_2}$ if and only if $\overline{d(s_1)} \leqslant \overline{d(s_2)}$.

It should be clear that if the conditions (3) and (4) are satisfied for $i{=}k$, then the mapping c defined in condition (3) is a partial +−isomorphism between $\langle \mathbb{N}, \mathfrak{A}, \vec{a} \rangle$ and $\langle \mathbb{N}, \mathfrak{B}, \vec{b} \rangle$, and hence Duplicator has won the game.
We first remark that the four conditions are satisfied at the beginning:

Lemma 1. *If* $\langle \mathbb{N}, \mathcal{A}, \vec{a} \rangle \equiv^{\leqslant}_{r(k)} \langle \mathbb{N}, \mathcal{B}, \vec{b} \rangle$, *then (1)–(4) hold for* $i = 0$.   □

We now assume that (1)–(4) hold for $i{-}1$, where $i \in \{1, .., k\}$; and we show that in the i-th round Duplicator can play in such a way that (1)–(4) hold for i. Let us assume that Spoiler has chosen $\mathfrak{a}^{(i)}$ in $\langle \mathbb{N}, \mathfrak{A}, \vec{a} \rangle$ (the case when Spoiler has chosen $\mathfrak{b}^{(i)}$ in $\langle \mathbb{N}, \mathfrak{B}, \vec{b} \rangle$ is symmetric).

We first determine two $(i{-}1)$–vectors, or minor $(i{-}1)$–vectors, s_m and s_M which approximate $\mathfrak{a}^{(i)}$ from below and from above as closely as possible: If $\mathfrak{a}^{(i)} \in \{\mathfrak{a}^{(1)}, .., \mathfrak{a}^{(i-1)}\} \cup Q$, we take $s_m = s_M = (\mathfrak{a}^{(i)}, 1, 0)$. If $\mathfrak{a}^{(i)} \notin \{\mathfrak{a}^{(1)}, .., \mathfrak{a}^{(i-1)}\} \cup Q$, but there is some $(i{-}1)$–vector s with $\overline{s} = \mathfrak{a}^{(i)}$, we take $s_m = s_M = s$. If there is no $(i{-}1)$–vector s with $\overline{s} = \mathfrak{a}^{(i)}$, then let s_m be a *minor* $(i{-}1)$–vector such that $\overline{s_m}$ is maximal among all minor $(i{-}1)$–vectors s with $\overline{s} < \mathfrak{a}^{(i)}$, and let s_M be a minor $(i{-}1)$–vector such that $\overline{s_M}$ is minimal among all minor $(i{-}1)$–vectors s with $\overline{s} > \mathfrak{a}^{(i)}$. (In particular, $\overline{s_m} \geqslant 0$.)

Let $\mathfrak{a}^{(i)}_1, .., \mathfrak{a}^{(i)}_{n_i}$ be those terms of s_m and s_M that are in Q. ¿From the sequence $\mathfrak{a}^{(i)}_1, .., \mathfrak{a}^{(i)}_{n_i}$ in $Q = q(\mathbb{N})$ we determine the corresponding sequence $a^{(i)}_1, .., a^{(i)}_{n_i}$ in $\langle \mathbb{N}, \mathcal{A}, \vec{a} \rangle$ via $a^{(i)}_j := q^{-1}(\mathfrak{a}^{(i)}_j)$. These are the moves for the "virtual Spoiler" in the $\leqslant$–game. ¿From condition (1) (for $i{-}1$) we know that

$$\langle \mathbb{N}, \mathcal{A}, \vec{a}^{(i-1)} \rangle \equiv^{\leqslant}_{r(k-i+1)} \langle \mathbb{N}, \mathcal{B}, \vec{b}^{(i-1)} \rangle .$$

Thus, the "virtual Duplicator" can find answers $b^{(i)}_1, .., b^{(i)}_{n_i}$ in $\langle \mathbb{N}, \mathcal{B}, \vec{b} \rangle$ such that

$$\langle \mathbb{N}, \mathcal{A}, \vec{a}^{(i-1)}, a^{(i)}_1, .., a^{(i)}_{n_i} \rangle \equiv^{\leqslant}_{r(k-i+1)-n_i} \langle \mathbb{N}, \mathcal{B}, \vec{b}^{(i-1)}, b^{(i)}_1, .., b^{(i)}_{n_i} \rangle .$$

Since $n_i \leqslant 2^{k-i+3}$ and by the choice of r, we have $r(k{-}i{+}1) - n_i \geqslant r(k{-}i)$, and hence condition (1) is satisfied for i.

Let c be the SC–respecting $(i{-}1)$–correspondence obtained from condition (3) (for $i{-}1$). We extend c to $\hat{c}$ by defining it also on $a^{(i)}_1, .., a^{(i)}_{n_i}$ via $\hat{c}(a^{(i)}_j) := b^{(i)}_j :=$

$q(b_j^{(i)})$ (for all $j \in \{1,\ldots,n_i\}$). Since condition (1) is satisfied for i, $\hat{c}$ must be an SC–respecting $(i{-}1)$–correspondence with $\hat{c}(\vec{a}^{(i)}) = \vec{b}^{(i)}$. Furthermore, from $0 \leqslant \overline{s_m} \leqslant \overline{s_M}$ and from condition (4) (for $i{-}1$) we obtain $0 \leqslant \hat{c}(s_m) \leqslant \hat{c}(s_M)$. For her choice of $\mathfrak{b}^{(i)}$ in $\langle \mathbb{N}, \mathfrak{B}, \vec{\mathfrak{b}} \rangle$, Duplicator makes use of the following lemma:

Lemma 2.

(a) $\overline{s_m} \equiv \hat{c}(s_m) \pmod{f(k-i)!}$, and
(b) if $\overline{s_m} < \overline{s_M}$ then $\hat{c}(s_M) - \hat{c}(s_m) > f(k-i)!$. $\qquad \square$

Duplicator chooses $\mathfrak{b}^{(i)}$ in $\langle \mathbb{N}, \mathfrak{B}, \vec{\mathfrak{b}} \rangle$ as follows:
If $\mathfrak{a}^{(i)} = \overline{s_m}$ then $\mathfrak{b}^{(i)} := \hat{c}(s_m) \geqslant 0$, and according to Lemma 2 (a) we have $\mathfrak{a}^{(i)} \equiv \mathfrak{b}^{(i)} \pmod{f(k-i)!}$. In particular, since $\mathfrak{a}^{(i)} \in \mathbb{N}$, this implies that $\mathfrak{b}^{(i)} \in \mathbb{N}$. If $\overline{s_m} < \mathfrak{a}^{(i)} < \overline{s_M}$ then, according to Lemma 2 (b), we have $\hat{c}(s_M) - \hat{c}(s_m) > f(k-i)!$, and hence there exists a $\mathfrak{b}^{(i)} \in \mathbb{N}$ with $0 \leqslant \hat{c}(s_m) < \mathfrak{b}^{(i)} < \hat{c}(s_M)$ and $\mathfrak{a}^{(i)} \equiv \mathfrak{b}^{(i)} \pmod{f(k-i)!}$. In both cases, condition (2) is satisfied for i.

For showing that condition (3) is satisfied for i, we distinguish between the two cases "$\mathfrak{a}^{(i)} \in \{\mathfrak{a}^{(1)},\ldots,\mathfrak{a}^{(i-1)}\} \cup Q$" and "$\mathfrak{a}^{(i)} \notin \{\mathfrak{a}^{(1)},\ldots,\mathfrak{a}^{(i-1)}\} \cup Q$", we make use of the fact that $\hat{c}$ is an SC–respecting $(i{-}1)$–correspondence with $\hat{c}(\vec{\mathfrak{a}}^{(i)}) = \vec{\mathfrak{b}}^{(i)}$, and we make use of the following lemma:

Lemma 3. We have $\mathfrak{a}^{(i)} \in Q$ if and only if $\mathfrak{b}^{(i)} \in Q$. $\qquad \square$

The validity of condition (4) for i is ensured by the following lemma:

Lemma 4. With Duplicator's choice of $\mathfrak{b}^{(i)}$ as described above, condition (4) holds for i. $\qquad \square$

Summing up, we have shown that the conditions (1)–(4) hold for $i{=}0$. Furthermore, we have shown for each $i \in \{1,\ldots,k\}$, that if they hold for $i{-}1$, then Duplicator can play in such a way that they hold for i. In particular, we conclude that Duplicator can play in such a way that the conditions (1)–(4) hold for $i{=}k$, and hence, Duplicator has a winning strategy in the k–round $+$–game on $\langle \mathbb{N}, \mathfrak{A}, \vec{a} \rangle$ and $\langle \mathbb{N}, \mathfrak{B}, \vec{b} \rangle$. This completes our proof of Theorem 3. $\qquad \square$

In fact, our proof shows the following result, which is stronger, but more technical than Theorem 3, and which we will use in the proof of Theorem 5.

Proposition 1. For every $k \in \mathbb{N}$ there exists a number $r(k) \in \mathbb{N}$ and an order–preserving mapping $q : \mathbb{N} \to \mathbb{N}$ such that, for every database schema SC, for every $m \in \mathbb{N}$, for all SC–database states $\mathcal{A}$ and $\mathcal{B}$ over $\mathbb{N}$, and all sequences $\vec{a}$ and $\vec{b}$ of length m in $\mathbb{N}$ with $\langle \mathbb{N}, \mathcal{A}, \vec{a} \rangle \equiv_{r(k)}^{\leqslant} \langle \mathbb{N}, \mathcal{B}, \vec{b} \rangle$, the following holds:

Duplicator can play the $+$–game on $\langle \mathbb{N}, q(\mathcal{A}, \vec{a}) \rangle$ and $\langle \mathbb{N}, q(\mathcal{B}, \vec{b}) \rangle$ in such a way that, for each $i \leqslant k$, after the i–th round the situation is as follows: Let $q(\vec{a}) = \mathfrak{a}^{(0)}, \mathfrak{a}^{(-1)}, \ldots, \mathfrak{a}^{(-m+1)}$, and $q(\vec{b}) = \mathfrak{b}^{(0)}, \mathfrak{b}^{(-1)}, \ldots, \mathfrak{b}^{(-m+1)}$. Furthermore let, for $j \in \{1,\ldots,i\}$, $\mathfrak{a}^{(j)}$ and $\mathfrak{b}^{(j)}$, respectively, be the elements chosen in the j–th round in $\langle \mathbb{N}, q(\mathcal{A}, \vec{a}) \rangle$ and $\langle \mathbb{N}, q(\mathcal{B}, \vec{b}) \rangle$, respectively. The following holds true:

- *The mapping* $\mathfrak{a}^{(1)}, \ldots, \mathfrak{a}^{(i)} \mapsto \mathfrak{b}^{(1)}, \ldots, \mathfrak{b}^{(i)}$ *is a partial* $+$*–isomorphism between* $\langle \mathbb{N}, \mathfrak{A}, \vec{a} \rangle$ *and* $\langle \mathbb{N}, \mathfrak{B}, \vec{b} \rangle$,

- $\mathfrak{a}^{(j)} \equiv \mathfrak{b}^{(j)} \pmod{f(k-i)!}$, *for each* j *with* $-m < j \leqslant i$, *and*

- $\displaystyle\sum_{j=-m+1}^{i} \frac{u_j}{u'_j} \mathfrak{a}^{(j)} + \beta \ \leqslant\ \sum_{j=-m+1}^{i} \frac{v_j}{v'_j} \mathfrak{a}^{(j)} + \delta \quad \text{iff} \quad \sum_{j=-m+1}^{i} \frac{u_j}{u'_j} \mathfrak{b}^{(j)} + \beta \ \leqslant\ \sum_{j=-m+1}^{i} \frac{v_j}{v'_j} \mathfrak{b}^{(j)} + \delta ,$

   for all $u_j, u'_j, v_j, v'_j \in \mathbb{Z}$ *with* $|u_j|, |u'_j|, |v_j|, |v'_j| \leqslant f(k-i)$ *(for* $-m < j \leqslant i$*), and* $\beta, \delta \in \mathbb{R}$ *with* $|\beta|, |\delta| \leqslant g(k-i)$. $\qquad\qquad\square$

5   Proof of Theorem 5

Let the functions r and q be chosen according to Theorem 3. Let $k \in \mathbb{N}$ be fixed and let SC be a database schema. We define $r'(k) := 1 + r(k+2)$.

Let $m \in \mathbb{N}$, let $\mathcal{C}$ and $\mathcal{D}$ be nicely representable SC–database states over $\mathbb{R}_{\geqslant 0}$, and let $\vec{c}$ and $\vec{d}$ be sequences of length m in $\mathbb{R}_{\geqslant 0}$ with $\langle \mathbb{R}_{\geqslant 0}, \mathcal{C}, \vec{c} \rangle \equiv^{\leqslant}_{r'(k)} \langle \mathbb{R}_{\geqslant 0}, \mathcal{D}, \vec{d} \rangle$. We need to find order–automorphisms q' and $\tilde{q}'$ of $\mathbb{R}_{\geqslant 0}$ such that $\langle \mathbb{R}_{\geqslant 0}, q'(\mathcal{C}, \vec{c}) \rangle \equiv^{+}_{k} \langle \mathbb{R}_{\geqslant 0}, \tilde{q}'(\mathcal{D}, \vec{d}) \rangle$.

Our proof makes use of Theorem 3. It is structured as illustrated in Figure 1, i.e.: In the first step, from $\mathcal{C}$, $\vec{c}$ and $\mathcal{D}$, $\vec{d}$, we define $\mathcal{A}$, $\vec{a}$ and $\mathcal{B}$, $\vec{b}$, respectively, such that $\langle \mathbb{N}, \mathcal{A}, \vec{a} \rangle \equiv^{\leqslant}_{r'(k)-1} \langle \mathbb{N}, \mathcal{B}, \vec{b} \rangle$. Since $r'(k) - 1 = r(k+2)$, we obtain from Theorem 3 that $\langle \mathbb{N}, q(\mathcal{A}, \vec{a}) \rangle \equiv^{+}_{k+2} \langle \mathbb{N}, q(\mathcal{B}, \vec{b}) \rangle$. In the second step, we modify $q : \mathbb{N} \to \mathbb{N}$ to order–automorphisms q' and $\tilde{q}'$ of $\mathbb{R}_{\geqslant 0}$, and we translate Duplicator's winning strategy in the $+$–game on $\langle \mathbb{N}, q(\mathcal{A}, \vec{a}) \rangle$ and $\langle \mathbb{N}, q(\mathcal{B}, \vec{b}) \rangle$ to a winning strategy in the $+$–game on $\langle \mathbb{R}_{\geqslant 0}, q'(\mathcal{C}, \vec{c}) \rangle$ and $\langle \mathbb{R}_{\geqslant 0}, \tilde{q}'(\mathcal{D}, \vec{d}) \rangle$.

$$\boxed{\begin{array}{cc} \langle \mathbb{R}_{\geqslant 0}, \mathcal{C}, \vec{c} \rangle \equiv^{\leqslant}_{r'(k)} \langle \mathbb{R}_{\geqslant 0}, \mathcal{D}, \vec{d} \rangle & \langle \mathbb{R}_{\geqslant 0}, q'(\mathcal{C}, \vec{c}) \rangle \equiv^{+}_{k} \langle \mathbb{R}_{\geqslant 0}, \tilde{q}'(\mathcal{D}, \vec{d}) \rangle \\[2mm] \Downarrow \text{ Step 1} & \Uparrow \text{ Step 2} \\[2mm] \langle \mathbb{N}, \mathcal{A}, \vec{a} \rangle \equiv^{\leqslant}_{r'(k)-1} \langle \mathbb{N}, \mathcal{B}, \vec{b} \rangle \ \overset{\text{Thm. 3}}{\Longrightarrow} \ \langle \mathbb{N}, q(\mathcal{A}, \vec{a}) \rangle \equiv^{+}_{k+2} \langle \mathbb{N}, q(\mathcal{B}, \vec{b}) \rangle \end{array}}$$

Fig. 1. The Structure of Our Proof.

We start with Step 1 of Figure 1. Let the elements in the sequence $\vec{c}$ be named by $c^{(0)}, c^{(-1)}, \ldots, c^{(-m+1)}$. Since $\mathcal{C}$ is nicely representable, its active domain is determined by a nice sequence $I = (I_n)_{n \in \mathbb{N}}$ of real closed intervals. We define $I^{\vec{c}} = (I^{\vec{c}}_n)_{n \in \mathbb{N}}$ to be the nice sequence of real closed intervals such that $\{I^{\vec{c}}_n : n \in \mathbb{N}\} = \{I_n : n \in \mathbb{N}\} \cup \{[c^{(i)}, c^{(i)}] : -m < i \leqslant 0, \text{ and there is no } n \in \mathbb{N} \text{ with } c^{(i)} \in I_n\}$. Let e be the partial mapping from $\mathbb{R}_{\geqslant 0}$ to $\mathbb{N}$ which maps, for each $n \in \mathbb{N}$, every element in $I^{\vec{c}}_n$ to n. We define $\mathcal{A}, \vec{a} := e(\mathcal{C}, \vec{c})$. Analogously, we define $\mathcal{B}, \vec{b} := \tilde{e}(\mathcal{D}, \vec{d})$, where $\tilde{e}$ is defined in a similar way as e (where, instead of I and $I^{\vec{c}}$, we use the nice sequence $\tilde{I}$ determined by the active domain of

$\mathcal{D}$, and the corresponding sequence $\tilde{I}^{\vec{d}}$). From our assumption we know that $\langle \mathbb{R}_{\geqslant 0}, \mathcal{C}, \vec{c} \rangle \equiv_{r'(k)}^{\leqslant} \langle \mathbb{R}_{\geqslant 0}, \mathcal{D}, \vec{d} \rangle$. We can use this winning strategy of Duplicator to obtain

Claim 1: $\langle \mathbb{N}, \mathcal{A}, \vec{a} \rangle \equiv_{r'(k)-1}^{\leqslant} \langle \mathbb{N}, \mathcal{B}, \vec{b} \rangle$. $\qquad\qquad\qquad\qquad\qquad\qquad$ □

From Theorem 3 we obtain that $\langle \mathbb{N}, q(\mathcal{A}, \vec{a}) \rangle \equiv_{k+2}^{+} \langle \mathbb{N}, q(\mathcal{B}, \vec{b}) \rangle$.
We now proceed with Step 2 of Figure 1. Choose some $\varepsilon \in \mathbb{R}$ with $0 < \varepsilon < 1$. For each $n \in \mathbb{N}$ let $l_n, r_n \in \mathbb{R}_{\geqslant 0}$ such that $I_n^{\vec{c}} = [l_n, r_n]$. Let q' be an order–automorphism of $\mathbb{R}_{\geqslant 0}$ which satisfies the following, for each $n \in \mathbb{N}$: $q'(l_n) = q(n)$; if $l_n < r_n$, then $q'(r_n) = q(n)+\varepsilon$; and if $c_{i_1} < \cdots < c_{i_s}$ are exactly those elements in the sequence $\vec{c}$ which lie strictly between l_n and r_n, then $q'(c_{i_j}) = q(n) + \frac{j \cdot \varepsilon}{s+1}$, for each $j \in \{1, .., s\}$. The mapping $\tilde{q}'$ is defined analogously (where, instead of $I_n^{\vec{c}}$, we use $\tilde{I}_n^{\vec{d}}$). We already know that $\langle \mathbb{N}, q(\mathcal{A}, \vec{a}) \rangle \equiv_{k+2}^{+} \langle \mathbb{N}, q(\mathcal{B}, \vec{b}) \rangle$. Using Proposition 1, we can translate this winning strategy of Duplicator to obtain

Claim 2: $\langle \mathbb{R}_{\geqslant 0}, q'(\mathcal{C}, \vec{c}) \rangle \equiv_{k}^{+} \langle \mathbb{R}_{\geqslant 0}, \tilde{q}'(\mathcal{D}, \vec{d}) \rangle$. $\qquad\qquad\qquad\qquad$ □

Hence, our proof of Theorem 5 is complete. $\qquad\qquad\qquad\qquad\qquad\qquad\qquad$ □

6   Discussion and Open Questions

Genericity.   Intuitively, order generic queries are those which essentially only depend on the order of the underlying structure. The formalization used in the present paper looks slightly different than the notion of *local order genericity* used, e.g., in [3] and [2]. However, it is not difficult to see that both notions are equivalent.

Natural-Active Collapse.   Another form of collapse which is of some interest concerns the difference between *natural* and *active* semantics. In natural semantics, quantifiers range over all of $\mathbb{U}$. In active semantics they range only over $adom(\mathcal{A})$, implying that only the active domain is relevant for query evaluation. Let us mention that Theorem 2 gives us a collapse from natural–semantics order generic queries over $\langle \mathbb{N}, \leqslant, + \rangle$ to active–semantics queries over $\langle \mathbb{N}, \leqslant \rangle$ for arbitrary databases.

Possible Extensions.   Several extensions are conceivable. One obvious question is whether our results can be extended from $\mathbb{N}$ to $\mathbb{Z}$ and from $\mathbb{R}_{\geqslant 0}$ to $\mathbb{R}$, where in both cases, the active domain is allowed to extend infinitely in *both* directions. Other potential extensions concern the restriction on the database over the real numbers. In [2], *finitely representable* databases are considered. Let us mention that meanwhile one of us has found a *natural generalization* of the notion of finitely representable databases, where an *arbitrary* (i.e., possibly infinite) number of regions is allowed (see [9]).

Other Arithmetical Predicates.   One interesting question is: How much arithmetic is needed to defeat the natural–generic collapse? It is well–known that with full arithmetic, i.e., with underlying structure $\langle \mathbb{N}, \leqslant, +, * \rangle$, no collapse is possible, even for finite databases. Consequently, since $+$ is definable from $*$ and $\leqslant$ over $\mathbb{N}$ (cf., [8]), *FO* does not have the natural–generic collapse over $\langle \mathbb{N}, \leqslant, * \rangle$. In [2], Belegradek, Stolboushkin and Taitslin discuss this question in the context of finite databases embedded in $\mathbb{Z}$ and conjecture that the collapse holds over any extension of $\langle \mathbb{Z}, \leqslant, + \rangle$ which has a decidable first–order theory. We can support this conjecture with a result which, although of less interest in database theory, might be interesting in its own right: Since $\langle \mathbb{N}, * \rangle$ is the weak direct product of ω copies of $\langle \mathbb{N}, + \rangle$, we can translate the result of Theorem 3 to obtain

Corollary 1. *FO has the natural–generic collapse over* $\langle \mathbb{N}, * \rangle$ *for arbitrary databases.* □

Effectivity. Finally, there is the question of *effective* collapse: it would be desirable to have an algorithm which, given an order generic $FO(SC, \leqslant, +)$–formula, constructs an equivalent $FO(SC, \leqslant)$–formula.

Acknowledgements

We thank James Baldwin, Michael Benedikt, Leonid Libkin, James Lynch, Thomas Schwentick, and Luc Ségoufin for helpful discussions on the topics of this paper.

References

1. J.T. Baldwin and M. Benedikt. Embedded finite models, stability theory and the impact of order. In *Thirteenth Annual IEEE Symposium on Logic in Computer Science*, pages 490–500. IEEE, 1998.
2. O.V. Belegradek, A.P. Stolboushkin, and M.A. Taitslin. Extended order-generic queries. *Annals of Pure and Applied Logic*, 97:85–125, 1999.
3. M. Benedikt and L. Libkin. Expressive power: The finite case. In G. Kuper, L. Libkin, and J. Paredaens, editors, *Constraint Databases*, pages 55–87. Springer, 2000.
4. M. Benedikt and L. Libkin. Relational queries over interpreted structures. *Journal of the ACM*, 47, 2000. To appear.
5. H.-D. Ebbinghaus, J. Flum, and W. Thomas. *Mathematical Logic*. Springer-Verlag, New York, 2nd edition, 1994.
6. G. Kuper, L. Libkin, and J. Paredaens, editors. *Constraint Databases*. Springer, 2000.
7. James F. Lynch. On sets of relations definable by addition. *Journal of Symbolic Logic*, 47:659–668, 1982.
8. J. Robinson. Definability and decision problems in arithmetic. *Journal of Symbolic Logic*, 14:98–114, 1949.
9. Nicole Schweikardt. The natural generic collapse for ω-representable databases over the real ordered group. In preparation, 2000.

A New Logical Characterization of Büchi Automata

Giacomo Lenzi*

LaBRI**, Université Bordeaux I
351, Cours de la Libération
F-33405 Talence Cedex, France
lenzi@labri.u-bordeaux.fr

Abstract. We consider the monadic second order logic with two successor functions and equality, interpreted on the binary tree. We show that a set of assignments is definable in the fragment Σ_2 of this logic if and only if it is definable by a Büchi automaton. Moreover we show that every set of second order assignments definable in Σ_2 with equality is definable in Σ_2 without equality as well. The present paper is sketchy due to space constraints; for more details and proofs see [7].

1 Introduction

This paper lies in the framework of *descriptive complexity*, an important and rapidly growing research area in theoretical computer science. Descriptive complexity was proposed by Fagin in [3] as an approach to fundamental problems of complexity theory such as whether $\mathcal{NP}$ equals $co - \mathcal{NP}$. While ordinary computational complexity theory is concerned with the amount of resources (such as time or space) necessary to solve a given problem, the idea of descriptive complexity is of studying the expressibility of problems in some logical formalism. For instance, in [3] Fagin shows that $\mathcal{NP}$ problems coincide (over finite structures) with the problems expressible in existential second order logic. Since then, there has been a large number of results in descriptive complexity. We note that most of these results concern finite structures (which are those interesting for the applications in computational complexity theory), but studying descriptive complexity also over infinite structures makes sense and may lead to a better understanding of the expressiveness of various logical syst! ems.

In particular, in this paper we are interested in the *monadic second order logic (MSOL) over the binary tree* and *Büchi automata*.

Monadic second order logic on the binary tree has been long studied since the seminal paper [8], where Rabin shows that the set of all true sentences in monadic second order logic with two successor functions (let us call *Rabin logic* this logic), interpreted over the binary tree, is decidable. The tool used by Rabin for this result are Rabin automata, a kind of finite automata over infinite trees;

---

* Supported by a CNR grant. The author thanks also the LaBRI for support.
** Laboratoire Bordelais de Recherche en Informatique.

A. Ferreira and H. Reichel (Eds.): STACS 2001, LNCS 2010, pp. 467–477, 2001.
© Springer-Verlag Berlin Heidelberg 2001

in particular, Rabin shows that a property is definable in Rabin logic if and only if it is definable by a Rabin automaton.

Büchi automata are also a kind of finite automata on infinite trees; they were introduced earlier than Rabin automata, in [2], again as a technique for solving decision problems in second order logic, and they have been studied by Rabin in [9] (where they are called special automata). It turns out that Büchi automata are indeed a special case of Rabin automata, and Rabin in [9] gives an example of property definable by a Rabin automaton but not by any Büchi automaton. In particular, Büchi automata correspond to a proper subset of Rabin logic.

Rabin in [9] gives also a logical characterization of Büchi automata by means of the weak monadic second order logic ($WMSOL$), that is monadic second order logic where the second order quantifiers range over finite sets. Rabin's result is that a set is definable by a Büchi automaton if and only if it is definable by a formula formed by a sequence of existential monadic second order quantifiers followed by a weak monadic second order formula.

In this paper we give another logical characterization of Büchi automata: we show that a property is definable by a Büchi automaton if and only if it is definable by a formula in the fragment Σ_2 of Rabin logic. We note that, in the original definition of Rabin logic, the prefix ordering of the binary tree is considered to be primitive; here instead, our result holds only if the prefix ordering is not primitive, since Σ_2 with prefix ordering is equivalent to the entire Rabin logic, as follows from [8].

Finally, as a side issue we note that, in the original definition of Rabin logic, also the equality relation is considered to be primitive, so one can wonder whether Σ_2 with equality is equiexpressive to Σ_2 without equality; here we prove that the answer is affirmative (at least on second order assignments), by showing that every set of second order assignments definable in Σ_2 with equality is definable in Σ_2 without equality as well.

The rest of the paper is organized as follows. In Sects. 2 and 3 we recall some basic notions about logic and Büchi automata. In Sect. 4 we state the main result (that is Büchi equals Σ_2) and the key lemma (that is Π_1 is included in $WMSOL$), and we prove that the key lemma implies the main result. After introducing some definition in Sect. 5, in Sect. 6 we sketch the proof of the key lemma. In Sect. 7 we prove that equality can be eliminated in Σ_2 on second order assignments. Section 8 is devoted to some concluding remarks.

2   Monadic Second Order Logic on the Binary Tree

The kind of logic we are interested in is monadic second order logic. Given a set F of function symbols, each of fixed arity, and a set R of first order relation symbols, each of fixed arity, we denote by $MSOL(F, R)$ the monadic second order logic based on F and R. We recall briefly the syntax of $MSOL(F, R)$ terms and formulas.

First of all, $MSOL(F, R)$ has a countable set of first order variables $x, y, \ldots$ (ranging over individuals) and a countable set of second order variables $X, Y, \ldots$

(ranging over sets of individuals). The (first order) terms of $MSOL(F, R)$ are obtained starting from the first order variables by iteratedly applying the function symbols belonging to F. The atomic formulas of $MSOL(F, R)$ have the form $t \in X$ or $r(t_1, \ldots, t_n)$, where $t, t_1, \ldots, t_n$ are first order terms, X is a second order variable, and r is a relation symbol of arity n belonging to R. The formulas of $MSOL(F, R)$ are then obtained starting from the atomic formulas by applying the boolean connectives $\neg, \wedge, \vee$, the first order, existential or universal, quantifiers $\exists x, \forall x$ and the second order, existential or universal, quantifiers $\exists X, \forall X$.

We recall also that the *weak* monadic second order logic over F and R, denoted by $WMSOL(F, R)$, is the same as $MSOL(F, R)$ with the exception that the second order quantifiers range over *finite* sets rather than on arbitrary sets.

In particular, the logic which we called Rabin logic in the Introduction, and which was studied by Rabin in [8], is $MSOL(s_0, s_1, =, \leq)$, where s_0, s_1 are two unary function symbols (to be interpreted as the successor functions in the binary tree) and $=, \leq$ are two binary relation symbols (to be interpreted as the equality relation and the prefix ordering on the binary tree); and the logic we are interested in here is the fragment $MSOL(s_0, s_1, =)$ of Rabin logic.

We recall the definition of the fragments Σ_n and Π_n of $MSOL$. Σ_0 and Π_0 denote the set of first order formulas of $MSOL$, that is the formulas without second order quantifiers; and inductively we define a Σ_{n+1} formula to be a sequence of existential second order quantifiers followed by a Π_n formula; and dually we define a Π_{n+1} formula to be a sequence of universal second order quantifiers followed by a Σ_n formula.

We note that the logics $MSOL(s_0, s_1, =)$ and $MSOL(s_0, s_1, =, \leq)$ are equivalent, as the prefix ordering is definable in the former; however this is not true in general for the corresponding levels Σ_n: in particular the levels Σ_2 differ, as they correspond to Büchi automata and Rabin automata respectively (on the other hand all levels $\Sigma_m(s_0, s_1, =)$ with $m \geq 3$ and $\Sigma_n(s_0, s_1, =, \leq)$ with $n \geq 2$ are equivalent, as follows from [8]).

The formulas of $MSOL(F, R)$ can be given a semantics by defining an interpretation, that is a set U of individuals, and for each function symbol f in F a function on U of the same arity as f, and for each relation symbol r in R a relation on U of the same arity as r.

Given a set U and two natural numbers m, n, let us call *assignment of type* (m, n) (over U) a $m + n$–tuple whose first m components are elements of U and whose last n components are subsets of U. According to the standard Tarskian semantic rules, each formula with m free first order variables and n free second order variables defines a set of assignments of type (m, n), that is the set of all assignments which make it true. In particular, let us call *second order assignments* the assignments of type $(0, n)$; then a formula without free first order variables and with n free second order variables defines a set of second order assignments.

In this paper we fix the interpretations of our logical symbols as follows: the set of individuals will always be the *binary tree* $\{0, 1\}^*$, that is the set of all finite words over the alphabet $\{0,1\}$, including the empty word ϵ (the root of the tree); the function symbols s_0 and s_1 will be interpreted as the *successor functions*, sending each word $w \in \{0,1\}^*$ to $w0$ and $w1$ respectively; and the symbol $=$ will be interpreted as the *equality* relation.

Sometimes we will adopt the usual, convenient genealogic terminology about the binary tree, for instance we can say that the words $w0$ and $w1$ are the left son and the right son of the word w respectively, that w is the father of $w0$ and $w1$, etc.

Finally, given a word w, we denote by $|w|$ the length of w.

3   Büchi Automata on Trees and Assignments

In this section we review the basic notions about Büchi automata. Recall that, given a set E, the powerset of E is the set $P(E)$ of all the subsets of E.

A *Büchi automaton* is a system

$$A = (\Sigma, Q, M, Q_0, F),$$

where Σ is a finite set (the alphabet of the automaton), Q is a finite set of states, M (the move function) is a function from $Q \times \Sigma$ to $P(Q \times Q)$, $Q_0 \subseteq Q$ is the set of the initial states and $F \subseteq Q$ is the set of the accepting states.

Büchi automata work as follows. A Σ–*tree* is a function from $\{0,1\}^*$ to Σ. A *run* of the automaton $A = (\Sigma, Q, M, Q_0, F)$ on a Σ–tree t is a mapping r from $\{0,1\}^*$ to Q such that, for every $w \in \{0,1\}^*$, we have

$$(r(w0), r(w1)) \in M(r(w), t(w)). \tag{1}$$

A run r is called *accepting* if $r(\epsilon) \in Q_0$ and for every infinite path π through $\{0,1\}^*$, and for infinitely many points $w \in \pi$, we have $r(w) \in F$. We say that the automaton A *accepts* a tree t when it has an accepting run on t. The set of trees *defined* by a Büchi automaton A is the set of all trees accepted by A.

In particular, if Σ is the powerset of a set $\{V_1, \ldots, V_n\}$ of second order variables, then the set of all Σ–trees is isomorphic to the set of all second order assignments of type $(0, n)$; the isomorphism is the map ι which sends the tree t to $X_1, \ldots, X_n$, where $X_i = \{w | V_i \in t(w)\}$. So, for this particular kind of Σ, we say that an automaton A on Σ *accepts* the second order assignment $a = X_1, \ldots, X_n$ when it accepts the corresponding Σ–tree $\iota^{-1}(X_1, \ldots, X_n)$.

The terminology above can be slightly extended in order to take into account first order variables. That is, let m, n be natural numbers, let $V_1, \ldots, V_{m+n}$ be second order variables and let $\Sigma = P(\{V_1, \ldots, V_{m+n}\})$. Given an assignment of type (m, n), say $a = x_1, \ldots, x_m, X_1, \ldots, X_n$, and an automaton A over Σ, we say that A *accepts* a if it accepts the tree $\iota^{-1}(\{x_1\}, \ldots, \{x_m\}, X_1, \ldots, X_n)$.

So both $MSOL$ formulas and Büchi automata (on suitable alphabets) define essentially the same kind of objects, that is sets of assignments, and this allows us to compare Büchi definable sets with $MSOL$ definable sets, as we will do in the sequel.

4   Statement of the Main Result

Theorem 1. *Let m, n be natural numbers and let E be a set of assignments of type (m, n) over $\{0, 1\}^*$.*
The following are equivalent:

1. *E is definable by a Büchi automaton;*
2. *E is definable by a formula in $\Sigma_2(s_0, s_1, =)$.*

The following lemma implies the theorem:

Lemma 1. *(Key Lemma) Let m, n be natural numbers and let E be a set of assignments of type (m, n) over $\{0, 1\}^*$.*
If E is definable in $\Pi_1(s_0, s_1, =)$, then E is definable in $WMSOL(s_0, s_1, =)$ as well.

The proof of this lemma will be given in Sect. 6. Here is the proof of the theorem from the lemma.

To prove that 1) implies 2), we write down in Σ_2 the condition

$$\text{``the automaton } A \text{ accepts the assignment } x_1, \ldots, x_m, X_1, \ldots, X_n \text{''.} \quad (2)$$

Let us consider the tree

$$t = \iota^{-1}(\{x_1\}, \ldots, \{x_m\}, X_1, \ldots, X_n),$$

and let us enumerate the states of the automaton A in an arbitrary way:

$$Q = \{q_1, \ldots, q_k\}.$$

We begin by rewriting the condition (2) as follows:

"there is a k-tuple $p = (Z_{q_1}, \ldots, Z_{q_k})$ of subsets of $\{0, 1\}^*$ such that:

$-$ *p is a partition of $\{0, 1\}^*$;*

$-$ *$\epsilon \in \bigcup_{q \in Q_0} Z_q$;*

$-$ *p is built according to the labeling of t and the move function M of A;*

$-$ *and the run associated to p is accepting".*

Now we rewrite each property of the tuple p above as follows:

$-$ the property

$$\text{``}p \text{ is a partition of } \{0, 1\}^* \text{''} \quad (3)$$

means that the sets Z_q are pairwise disjoint and their union is the entire $\{0, 1\}^*$, and all this is expressible in first order logic (over the empty signature), hence a fortiori in $\Pi_1(s_0, s_1)$;

– the property

$$\text{“}\epsilon \in \bigcup_{q \in Q_0} Z_q\text{”} \tag{4}$$

is equivalent to

$$\text{“}\textit{every nonempty set closed under father meets } \bigcup_{q \in Q_0} Z_q\text{”}, \tag{5}$$

which is $\Pi_1(s_0, s_1)$ (this formalization of (4) a kind of overkilling, but has the advantage of doing without equality, which will be useful in Sect. 7);

– the property

$$\text{“}\textit{p is built according to t and the move function M of A}\text{”} \tag{6}$$

is equivalent to

"for every $w \in \{0,1\}^*$ and for every triple (q, q', q'') of states and for any $\sigma \in P(\{V_1, \ldots, V_{m+n}\})$, if $w \in Z_q$, $w0 \in Z_{q'}$, $w1 \in Z_{q''}$ and $t(w) = \sigma$, then $(q', q'') \in M(q, \sigma)$",

which is $\Sigma_0(s_0, s_1, =)$, hence $\Pi_1(s_0, s_1, =)$ (note that the condition "$t(w) = \sigma$" amounts to a conjunction of $w = x_i$ or negations, with $1 \le i \le m$, and $w \in X_j$ or negations, with $1 \le j \le n$);

– and the property

$$\text{“}\textit{the run associated to p is accepting}\text{”} \tag{7}$$

is equivalent to

$$\text{“}\textit{every infinite path of } \{0,1\}^* \textit{ meets } \bigcup_{q \in F} Z_q \textit{ infinitely often}\text{”}, \tag{8}$$

which turns out to be equivalent to:

"for every set $Y \subseteq \{0,1\}^*$, if Y satisfies the following conditions:

- *for every $q \in F$, $Z_q \subseteq Y$;*
- *for every $w \in \{0,1\}^*$, if $w0 \in Y$ and $w1 \in Y$, then $w \in Y$,*

then $Y = \{0,1\}^*$",

which is $\Pi_1(s_0, s_1)$.

Summing up, the condition (2) is equivalent to a sequence of existential second order quantifiers followed by a $\Pi_1(s_0, s_1, =)$ formula, that is a $\Sigma_2(s_0, s_1, =)$ formula, as desired.

Conversely, assuming the lemma, we have that $\Sigma_2(s_0, s_1, =)$, which is the existential second order closure of $\Pi_1(s_0, s_1, =)$, is included into the existential second order closure of $WMSOL(s_0, s_1, =)$, which coincides with the class of all Büchi definable sets by the classical result of [9] already mentioned; this proves that 2) implies 1) and hence concludes the proof of the theorem from the lemma.

5   Auxiliary Notations

5.1   Bimodalities

Given two words v, w and a second order variable X, denoting a set of words, we introduce the *second order bimodality* $(Xw^{-1})v$, whose semantics is the set of words $\{yv|yw \in X\}$. We can view bimodalities as a kind of (second order) terms in $MSOL$, and we can use them in formulas. For instance we can write $z \in (X1^{-1})0$, which means that the word z is the left brother of an element of the set X.

In the same vein, we allow ourselves to take also *first order bimodalities*, that is we introduce the terms $(xw^{-1})v$, where v and w are binary words and x is a first order variable. This term will denote the set $\{yv|yw = x\}$; note that this set has always at most one element.

For convenience we include among the bimodalities also "constant" expressions of the kind $(Ew^{-1})v$, where E is any fixed set of binary words such as $\{0, 1\}^*$, $\emptyset$, etc., with the obvious semantics.

Finally we define the *length* of a bimodality $(Aw^{-1})v$ to be $|v| + |w|$.

For more comments on bimodalities see the journal version [7].

5.2   Quasiinclusions

In the next section we will build a normal form for first order logic over the binary tree. An ad hoc notion which we need to build our normal form is the notion of *quasiinclusion*.

Given two sets A, B and a natural number k, we say that A is *k–included* in B, written $A \subseteq_k B$, if the difference $A \setminus B$ has at most k elements. That is, A is included in B up to at most k exceptions. We call *quasiinclusion* any expression of the form $A \subseteq_k B$.

In particular, we have $A \subseteq B$ if and only if $A \subseteq_0 B$, so every inclusion is a quasiinclusion.

For more on quasiinclusions see the journal version [7].

6   Proof of the Key Lemma (Sketch)

We are left to prove Lemma 1. To begin the proof, we observe that every formula in $\Pi_1(s_0, s_1, =)$ is a sequence of universal monadic second order quantifiers followed by a formula in $\Sigma_0(s_0, s_1, =)$, that is a first order formula. Now we put every formula of $\Sigma_0(s_0, s_1, =)$ into a normal form as follows.

6.1   A Normal Form Lemma for First Order Formulas

Lemma 2. *Every formula in $\Sigma_0(s_0, s_1, =)$ is equivalent to a boolean combination of quasiinclusions $A \subseteq_k B$, where k is a natural number, A is a finite, nonempty intersection of bimodalities, and B is a finite, nonempty union of bimodalities.*

The proof is omitted for lack of space, see the journal version [7].

By using the conjunctive normal form of propositional logic we have the following corollary:

Corollary 1. *Every formula in $\Sigma_0(s_0, s_1, =)$ is equivalent to a finite conjunction of formulas of the form*

$$(\bigwedge_i A_i \subseteq_{k_i} B_i) \Rightarrow (\bigvee_j C_j \subseteq_{l_j} D_j), \tag{9}$$

where k_i, l_j are natural numbers, A_i, C_j are finite, nonempty intersections of bimodalities and B_i, D_j are finite, nonempty unions of bimodalities.

Since universal quantification distributes over conjunctions, from the previous corollary we obtain:

Corollary 2. *Every formula in $\Pi_1(s_0, s_1, =)$ is equivalent to a finite conjunction of formulas of the form*

$$\forall X.(\bigwedge_i A_i \subseteq_{k_i} B_i) \Rightarrow (\bigvee_j C_j \subseteq_{l_j} D_j), \tag{10}$$

where X is a tuple of second order variables and $k_i, l_j, A_i, C_j, B_i, D_j$ are as above.

6.2   The Main Argument

By Cor. 2, and since $WMSOL$ is closed under conjunction, the proof of Lemma 1 amounts to show that any formula of the form

$$\forall X.(\bigwedge_i A_i(X, Y) \subseteq_{k_i} B_i(X, Y)) \Rightarrow (\bigvee_j C_j(X, Y) \subseteq_{l_j} D_j(X, Y)), \tag{11}$$

where X is a tuple of second order variables, Y is a tuple of parameters (that is, subsets of $\{0, 1\}^*$) and $k_i, l_j, A_i, C_j, B_i, D_j$ are as in the previous subsection, is equivalent to a $WMSOL$ formula.

For notational convenience, we may abbreviate the formula (11) as $\forall X.A(X) \subseteq_k B(X) \Rightarrow C(X) \subseteq_l D(X)$; that is, we may suppress indexes and parameters from the notation.

In order to express the formula (11) in the weak logic, we begin with the following lemma:

Lemma 3. *Let $A(X) \subseteq_k B(X)$ be a finite conjunction of quasiinclusions and let $C(X) \subseteq_l D(X)$ be a finite disjunction of quasiinclusions, where X is a tuple of second order variables, k, l are tuples of natural numbers, A, C are tuples of finite, nonempty intersections of bimodalities, and B, D are tuples of finite, nonempty unions of bimodalities.*

The following are equivalent:

1. $\forall X.A(X) \subseteq_k B(X) \Rightarrow C(X) \subseteq_l D(X)$;
2. *For any tuple F of finite sets and for any tuple G of cofinite sets, both of the same length as the tuple X, if there is X with $F \subseteq X \subseteq G$ (where the inclusion is intended to hold componentwise) and $A(X) \subseteq_k B(X)$, then $C(F) \subseteq_l D(G)$.*

The proof is omitted for lack of space, see the journal version [7].

So, in order to show that the formula (11) is expressible in $WMSOL$, it is enough to show that the property

$$\text{``there is } X \text{ with } F \subseteq X \subseteq G \text{ and } A(X) \subseteq_k B(X)\text{''} \tag{12}$$

is expressible in $WMSOL$. To this aim, we note that the inclusions $F \subseteq X \subseteq G$ together with $A(X) \subseteq_k B(X)$ form a finite system of quasiinclusions between a tuple of finite, nonempty intersections of bimodalities and a tuple of finite, nonempty unions of bimodalities. Hence it is enough to show that the existence of a solution of any such system is expressible in $WMSOL$. The corresponding lemma is:

Lemma 4. *Let*

$$A_1(X,Y) \subseteq_{n_1} A_1'(X,Y) \wedge \ldots \wedge A_r(X,Y) \subseteq_{n_r} A_r'(X,Y) \tag{13}$$

be a finite system of quasiinclusions, where X is a tuple of second order variables (the unknowns), Y is a tuple of subsets of $\{0,1\}^*$ (the parameters), $n_1, \ldots, n_r$ is a tuple of natural numbers, $A_1, \ldots, A_r$ is a tuple of finite, nonempty intersections of bimodalities and $A_1', \ldots, A_r'$ is a tuple of finite, nonempty unions of bimodalities. Let us abbreviate the system by $A(X,Y) \subseteq_n A'(X,Y)$.

Let N be twice the maximum length of a bimodality of the system.

The following are equivalent:

1. *The system (13) has a solution, i.e. there is a tuple X of subsets of $\{0,1\}^*$ such that $A(X,Y) \subseteq_n A'(X,Y)$;*
2. *For any finite subset F of $\{0,1\}^*$ there is a tuple X_F of subsets of $Ball(F,N)$ such that, for any $F' \subseteq F$, we have:*

$$A(X_F \cap F', Y \cap F') \subseteq_n A'(X_F \cap Ball(F',N), Y) \tag{14}$$

(where $Ball(F,N)$ is the set of all elements of $\{0,1\}^*$ which are reachable from some element of F through at most N father–son or son–father steps).

The proof is omitted for lack of space, see the journal version [7].

Since the point 2) of the lemma above is expressible in $WMSOL$, Lemma 1 is proved. Hence, by Sect. 4, Thm. 1 is proved as well.

7   About Second Order Assignments

In the previous sections we stated and proved a theorem which characterizes Büchi definable sets of assignments by means of $\Sigma_2(s_0, s_1, =)$. It turns out that if we consider only second order assignments, we can say a little more: the equality symbol can be eliminated. That is:

Corollary 3. *Let E be a set of second order assignments. If E is definable in $\Sigma_2(s_0, s_1, =)$, then it is definable in $\Sigma_2(s_0, s_1)$ as well.*

Proof. By Thm. 1 it is enough to show that, if E is Büchi definable, then E is definable in $\Sigma_2(s_0, s_1)$. To this aim we look at the proof of Thm. 1 from Lemma 1 in Sect. 4, and we observe that, while writing down in Σ_2 the condition (2), the only place where we need equality is when we write down $w = x_i$ for some i with $1 \leq i \leq m$: so if the elements of E are second order assignments, then $m = 0$ and there is no need of equality. This concludes the proof.   □

8   Conclusion

In this paper we have seen that, on the binary tree, Büchi automata are equivalent to the fragment Σ_2 of monadic second order logic (without prefix ordering).

This has some pleasant consequences. Rabin in [9] exhibits a property which is Büchi definable but whose complement is not Büchi definable; since Büchi is equal to Σ_2, this implies that there is a property which is Σ_2 definable but whose complement is not Σ_2 definable, which is the first theorem of [5]. Likewise, Hafer in [4] exhibits a property which is Rabin definable but which is not definable by any boolean combination of Büchi properties; since Büchi equals Σ_2 and Rabin equals Σ_3, this implies that there is a property which is Σ_3 but it is not defined by any boolean combination of Σ_2 formulas, which is the second theorem of [5]. Thus we found proofs of both theorems of [5] which are considerably more simple than the original proofs.

Another consequence of our result concerns Kozen's mu–calculus (see [6]); that is, on the binary tree, the fragment Π_2^μ of the mu–calculus fixpoint alternation depth hierarchy, which is equal to Büchi by [1], is equal to the level Σ_2 of monadic second order logic. More generally, it would be interesting to compare the various levels of the fixpoint alternation hierarchy of the mu–calculus with the levels Σ_n of the quantifier alternation hierarchy in monadic second order logic, in some interesting class of graphs such as all graphs or all finite graphs: we note that trees are not very interesting in this respect, since the monadic hierarchy on trees collapses to Σ_3 (this is a kind of folklore result, see for instance [11]). This kind of problems will be the subject of subsequent papers.

Note added in proof: an alternate, simple proof that Σ_2 equals Büchi automata has been recently given in [10].

Acknowledgment

The author thanks André Arnold and David Janin for many useful discussions.

References

1. Arnold, A., Niwiński, D.: Fixed point characterization of Büchi automata on infinite trees, J. Inf. Process. Cybern. EIK **26** (1990) 451–459
2. Büchi, J. R.: On a decision method in restricted second order arithmetic, in: E. Nagel et al., eds., Proc. Internat. Congr. on Logic, Methodology and Philosophy of Science (Stanford Univ. Press, Stanford, CA, 1960), 1–11
3. Fagin, R.: Generalized first order spectra and polynomial time recognizable sets, in: Complexity of computation, volume 7, SIAM–AMS, 1974
4. Hafer, T.: On the boolean closure of Büchi tree automaton definable sets of ω–trees, Technical report, Aachener Infor. Ber. Nr. 87–16, RWTH Aachen, 1987
5. Janin, D., Lenzi, G.: On the structure of the monadic logic of the binary tree, in: Proceedings of the conference MFCS'99, Lecture Notes in Computer Science n. 1672, 310–320
6. Kozen, D.: Results on the propositional μ-calculus, Theoretical Computer Science **27** (1983) 333–354
7. Lenzi, G.: A second order characterization of Büchi automata, submitted to the Annals of Pure and Applied Logic (a draft version is available via email, please contact the author)
8. Rabin, M.: Decidability of second order theories and automata on infinite trees, Trans. Amer. Math. Soc. **141** (1969) 1–35
9. Rabin, M.: Weakly definable relations and special automata, in: Y. Bar-Hillel, ed., Mathematical Logic and Foundations of Set theory (North-Holland, Amsterdam 1970), 1–23
10. Skurczyński, J.: A characterization of Büchi tree automata, unpublished manuscript, University of Gdańsk, 2000
11. Walukiewicz, I.: Monadic second order logic on tree-like structures, in: Proceedings of the conference STACS'96, Lecture Notes in Computer Science n. 1046, 401–414

A Primal-Dual Approximation Algorithm for the Survivable Network Design Problem in Hypergraph

Liang Zhao[1], Hiroshi Nagamochi[2], and Toshihide Ibaraki[1]

[1] Kyoto University, Kyoto, Japan
{zhao,ibaraki}@amp.i.kyoto-u.ac.jp
[2] Toyohashi University of Technology, Toyohashi, Aichi, Japan
naga@ics.tut.ac.jp

Abstract. Given a hypergraph with nonnegative costs on hyperedge and a requirement function $r : 2^V \rightarrow \mathbf{Z}^+$, where V is the vertex set, we consider the problem of finding a minimum cost hyperedge set F such that for all $S \subseteq V$, F contains at least $r(S)$ hyperedges incident to S. In the case that r is weakly supermodular (i.e., $r(V) = 0$ and $r(A) + r(B) \leq \max\{r(A \cap B) + r(A \cup B), r(A - B) + r(B - A)\}$ for any $A, B \subseteq V$), and the so-called *minimum violated sets* can be computed in polynomial time, we present a primal-dual approximation algorithm with performance guarantee $d_{\max}\mathcal{H}(r_{\max})$, where $d_{\max}$ is the maximum degree of the hyperedges with positive cost, $r_{\max}$ is the maximum requirement, and $\mathcal{H}(i) = \sum_{j=1}^{i} \frac{1}{j}$ is the harmonic function. In particular, our algorithm can be applied to the survivable network design problem in which the requirement is that there should be at least r_{st} hyperedge-disjoint paths between each pair of distinct vertices s and t, for which r_{st} is prescribed.

1 Introduction

Given an undirected graph with nonnegative edge costs, the network design problem is to find a minimum cost subgraph satisfying certain requirements. In the *survivable network design problem* (SNDP), the requirement is that there should be at least r_{st} *edge-disjoint* paths between each pair of distinct vertices s and t, for which r_{st} is prescribed. It arises from problems of designing a minimum cost network such that certain vertices remain connected after some edges fail. In the Steiner tree problem which is an important special case, we are given a subset T of the vertex set V, and the objective is to find a minimum cost edge set to connect all the vertices in T. Clearly this is an SNDP, in which $r_{st} = 1$ if $s, t \in T$ and $r_{st} = 0$ otherwise. It is known that the Steiner tree problem is NP-hard even for unit cost ([9]). Thus the general SNDP is NP-hard, too.

We focus on developing approximation algorithms. An α-approximation algorithm is a polynomial time algorithm which always outputs a solution of cost at most α times the optimum. The first approximation algorithm for the SNDP is given by Williamson *et al.* [11]. They formalize a basic mechanism for using the primal-dual method. It picks edge sets in $r_{\max} \triangleq \max\{r_{st}\}$ phases, and

A. Ferreira and H. Reichel (Eds.): STACS 2001, LNCS 2010, pp. 478–489, 2001.

each phase tries to augment the size of cuts with deficiency by using an integer program, which is solved within factor 2 by a primal-dual approach. Their algorithm has a performance guarantee of $2r_{\max}$. In [4] Goemans *et al.* show that by augmenting the size of only those cuts with maximum deficiency, a $2\mathcal{H}(r_{\max})$-approximation algorithm can be obtained, where $\mathcal{H}(i) = \sum_{j=1}^{i} \frac{1}{j}$ is the harmonic function. For a detailed overview of these primal-dual algorithms, we refer the readers to the well-written survey [6]. Recently, Jain [7] shows that there is an edge e with $x_e^* \geq \frac{1}{2}$ in any *basic* solution x^* of the LP relaxation of the SNDP (where the constraint $x_e \in \{0,1\}$ is relaxed to $0 \leq x_e \leq 1$ for all edge e). Then it is shown that an *iterative rounding* process yields a 2-approximation algorithm.

In a very recent paper [8], Jain *et al.* considered the *element connectivity problem* (ECP). In that problem, vertices are partitioned into two sets: terminals and non-terminals. Only edges and non-terminals, called the *elements*, can fail and only pairs of terminals have connectivity requirements, specifying the least number of element-disjoint paths to be realized. (Note that only the edges have costs.) The SNDP is a special case of the ECP with empty non-terminal set. Following the basic algorithmic outline established in [11] and [4], they show that a $2\mathcal{H}(r_{\max})$-approximation algorithm can be obtained.

In this paper we consider the SNDP in hypergraphs (SNDPHG). The difference between hypergraph and graph is that edges in hypergraph, called the *hyperedges*, may contain more than two vertices as their endpoints. The *degree* of a hyperedge is defined as the number of endpoints contained in it. By replacing *edges* in the definition of SNDP with *hyperedges*, we get the definition of SND-PHG. Thus the SNDP is a special case of the SNDPHG in which the degrees of all the hyperedges are 2. We note that the ECP is also a special case of the SNDPHG. To see this, consider a non-terminal w. Let $\{v_1, w\}, \ldots, \{v_k, w\}$ be the edges that are incident to w. For each $i = 1, \ldots, k$, replace edge $\{v_i, w\}$ with two edges $\{v_i, w_i\}$ and $\{w_i, w\}$, introducing a new terminal w_i. Let the cost of edge $\{v_i, w_i\}$ be the same as $\{v_i, w\}$. Let $r_{st} = 0$ if at least one of s and t is a new terminal. Then replace w and all the edges $\{\{w_i, w\} | i = 1, \ldots, k\}$ with a hyperedge of zero cost $e_w = \{w_1, \ldots, w_k\}$. Repeat this process until there is no non-terminal left. Clearly in this way we can reduce the ECP to the SNDPHG in linear time. In fact, let $d_{\max}$ denote the maximum degree of the hyperedges with *positive* cost, we have shown that the ECP is a special case of the SND-PHG in which $d_{\max} = 2$. Furthermore, we notice that the SNDPHG can model more general network design problems, e.g., it is easy to see that the problem of multicasting in a network involving router cost can be modeled by an SNDPHG in which routers are modeled by hyperedges.

Clearly, the general SNDPHG is also NP-hard even for unit cost and $r_{\max} = 1$. In [10] Takeshita *et al.* extend the primal-dual approximation algorithm of [5] to the SNDPHG in which $r_{\max} = 1$. They show a k-approximation algorithm, where k is the maximum degree of hyperedges. In this paper we design an approximation algorithm to the SNDPHG based on the primal-dual schema established in [11], [4]. As a result, we can get a performance guarantee of $d_{\max}\mathcal{H}(r_{\max})$. Our result includes (or improves) the former results of [11], [4], [8] in which $d_{\max} = 2$, and [10] in which $r_{\max} = 1$. Like the algorithms for the SNDP in graphs, our algorithm is also applicable to more general problems, provided that the input satisfies two conditions (see Conditions 1 and 2 in the next two sections).

We present the algorithm for problems satisfying Conditions 1 and 2 in Sect. 3. In Sect. 4 we give a proof of the performance guarantee. We then show in Sect. 5 that the SNDPHG satisfies the two conditions.

2   Preliminaries

All (hyper)graphs treated in this paper are undirected unless stated otherwise. Directed graphs are noted as *digraphs*. Let G be a (hyper)graph, and $V(G)$ and $E(G)$ denote the vertex set and (hyper)edge set of G, respectively. A *(hyper)edge* e with end points $v_1, \ldots, v_k$ is denoted by $e = \{v_1, \ldots, v_k\}$ and it may be treated as the set $\{v_1, \ldots, v_k\}$ of the endpoints. For an $S \subseteq V(G)$, the subgraph of G induced by S is denoted by $G[S]$ (i.e., $G[S] = (S, E(G) \cap 2^S)$). The neighbors of S is denoted by $\Gamma(S)$, i.e., $\Gamma(S) \triangleq \{v \in V(G) - S \mid \exists e \in E(G), v \in e, e \cap S \neq \emptyset\}$. The set of (hyper)edges incident to S is denoted by $\delta(S)$, i.e., $\delta(S) \triangleq \{e \in E(G) \mid \emptyset \neq e \cap S \neq e\}$. Let $\delta_A(S) \triangleq A \cap \delta(S)$ for an $A \subseteq E(G)$ (in particular $\delta_{E(G)} = \delta$). It is well known that $|\delta_A|$ for a fixed A is submodular, i.e.,

$$|\delta_A(X)| + |\delta_A(Y)| \geq |\delta_A(X \cap Y)| + |\delta_A(X \cup Y)| \text{ for any } X, Y \subseteq V(G). \quad (1)$$

Since $|\delta_A|$ is symmetric (i.e., $|\delta_A(X)| = |\delta_A(V(G) - X)|$ for any $X \subseteq V(G)$),

$$|\delta_A(X)| + |\delta_A(Y)| \geq |\delta_A(X - Y)| + |\delta_A(Y - X)| \text{ for any } X, Y \subseteq V(G). \quad (2)$$

In this paper we treat the following problem. Given a hypergraph H with nonnegative hyperedge costs, and a requirement function $r : 2^{V(H)} \to \mathbf{Z}^+$ (as an oracle to evaluate $r(X)$ for any given $X \subseteq V(H)$), find a minimum cost hyperedge set $E^* \subseteq E(H)$ such that $|\delta_{E^*}(S)| \geq r(S)$ for all $S \subseteq V(H)$. The problem can be converted into the next equivalent problem.

Definition 1 (Problem $\mathcal{P}$). *Given a bipartite graph $G = (T, W, E)$, where T and W are two disjoint vertex sets and E is a set of edges between T and W, where vertices in T and W are called* terminals *and* non-terminals, *respectively. Let $c : W \to \mathbf{R}^+$ be a cost function, and $r : 2^T \to \mathbf{Z}^+$ be a requirement function. Find a minimum cost $W^* \subseteq W$ such that $|\Gamma(S) \cap \Gamma(T - S) \cap W^*| \geq r(S)$ for all $S \subseteq T$. (Without loss of generality we assume that $r(\emptyset) = r(T) = 0$ and $r_{\max} = \max\{r(S) \mid S \subseteq T\} \leq |W|$, otherwise there is no feasible solution.)*

The equivalence can be seen easily as following. Let $T = V(H)$. Replace each hyperedge $e = \{v_1, \ldots, v_k\}$ with a new non-terminal vertex w_e and k edges $\{v_1, w_e\}, \ldots, \{v_k, w_e\}$. Assign w_e the same cost as the hyperedge e. Notice that $e \in \delta(S)$ in H if and only if $w_e \in \Gamma(S) \cap \Gamma(T - S)$ in G.

In what follows, we will consider the problem $\mathcal{P}$ instead of the original form of the problem. Define $\Delta(S) \triangleq \Gamma(S) \cap \Gamma(T - S) \subseteq W$ for $S \subseteq T$ (here and in what follows, notations Γ and Δ are defined in the input bipartite graph G

unless otherwise stated). Problem $\mathcal{P}$ can be written as the next integer program.

$$
\text{(IP)} \qquad \min \sum c_w x_w
$$
$$
\text{s.t.} \quad x(\Delta(S)) \geq r(S) \quad \text{for all } S \subseteq T,
$$
$$
x_w \in \{0,1\} \quad \text{for all } w \in W,
$$

where $x(\Delta(S)) \triangleq \sum_{w \in \Delta(S)} x_w$. We assume that r satisfies two conditions.

Condition 1. r is weakly supermodular, i.e., $r(T) = 0$, and for any $X, Y \subseteq T$

$$
r(X) + r(Y) \leq \max\{r(X \cap Y) + r(X \cup Y), r(X - Y) + r(Y - X)\}. \tag{3}
$$

(The second condition is stated in Sect. 3.) Let $\Delta_A(S) \triangleq A \cap \Delta(S)$. Notice that $|\Delta|$ defined in G equals to $|\delta|$ in H. Thus $|\Delta_A|$ (for a fixed $A \subseteq W$) is also a symmetric submodular function, from which for any $A \subseteq W$ and $X, Y \subseteq T$,

$$
|\Delta_A(X)| + |\Delta_A(Y)| \geq |\Delta_A(X \cap Y)| + |\Delta_A(X \cup Y)|, \tag{4}
$$
$$
|\Delta_A(X)| + |\Delta_A(Y)| \geq |\Delta_A(X - Y)| + |\Delta_A(Y - X)|. \tag{5}
$$

3   The Primal-Dual Approximation Algorithm for (IP)

For an $S \subseteq T$ and $A \subseteq W$, the *deficiency* of S with respect to A is defined as $r(S) - |\Delta_A(S)|$. Notice that A is feasible if and only if the maximum deficiency over all $S \subseteq T$ is non-positive. Similarly to [4] and [11], our algorithm contains $r_{\max}$ phases. Let $W_0 = \emptyset$ and $W_{i-1} \subseteq W$ be the non-terminal set picked after phase $i - 1$. At the beginning of phase i, the maximum deficiency (with respect to W_{i-1}) is $r_{\max} - i + 1$. We decrease the maximum deficiency by 1 in phase i, by solving an augmenting problem (IP)$_i$. An $A_i \subseteq W - W_{i-1}$ that is feasible to the augmenting problem is found by a primal-dual approach. We then set $W_i = W_{i-1} \cup A_i$ and go to the next phase until $i = r_{\max}$.

The augmenting problem we want to solve in phase i is

$$
\text{(IP)}_i \qquad \min \sum_{w \in W - W_{i-1}} c_w x_w
$$
$$
\text{s.t.} \quad x(\Delta_{W - W_{i-1}}(S)) \geq h_i(S) \quad \text{for all } S \subseteq T,
$$
$$
x_w \in \{0,1\} \quad \text{for all } w \in W - W_{i-1},
$$

where $h_i(\cdot)$ is defined as

$$
h_i(S) = \begin{cases} 1 & \text{if } r(S) - |\Delta_{W_{i-1}}(S)| = r_{\max} - i + 1, \\ 0 & \text{otherwise } (r(S) - |\Delta_{W_{i-1}}(S)| \leq r_{\max} - i) \end{cases} \tag{6}
$$

(hence we have an oracle to evaluate h_i). Clearly the union of W_{i-1} and any feasible solution to (IP)$_i$ decreases the maximum deficiency by at least 1. (We will see that $W_{i-1} \cup A_i$, where A_i is the set found by the primal-dual approach, decreases the maximum deficiency exactly by 1.) Thus at the end of phase $r_{\max}$, a feasible solution to (IP) will be found.

The notation of *violated sets* is needed by the primal-dual approach for (IP)$_i$.

Definition 2 (Violated Set). *Let $A \subseteq W - W_{i-1}$ be a non-terminal set. Set $S \subseteq T$ is said to be* violated *with respect to A if $h_i(S) = 1$ and $\Delta_A(S) = \emptyset$. It is a* minimal violated set *if it is a violated set and minimal under set inclusion. Let $\mathcal{V}(A) \triangleq \{ S \subseteq T \mid S$ is a minimal violated set of $A\}$.*

It is clear that A is feasible to (IP)$_i$ if and only if $\mathcal{V}(A) = \emptyset$. Under the assumption of Condition 1, the violated sets have a nice property as shown in the next lemma.

Lemma 1. *For any $A \subseteq W - W_{i-1}$, if $X, Y \subseteq T$ are two violated sets of A, then either $X \cap Y, X \cup Y$ or $X - Y, Y - X$ are violated sets of A, too.*

Proof. By the definition of violated set and h_i, we see that $r(X) - |\Delta_{W_{i-1}}(X)| = r(Y) - |\Delta_{W_{i-1}}(Y)| = r_{\max} - i + 1$ and $\Delta_A(X) = \Delta_A(Y) = \emptyset$. By (4) and (5),

$$|\Delta_A(X \cap Y)| + |\Delta_A(X \cup Y)| \leq |\Delta_A(X)| + |\Delta_A(Y)| = 0,$$
$$|\Delta_A(X - Y)| + |\Delta_A(Y - X)| \leq |\Delta_A(X)| + |\Delta_A(Y)| = 0.$$

Hence $\Delta_A(X \cap Y) = \Delta_A(X \cup Y) = \Delta_A(X - Y) = \Delta_A(Y - X) = \emptyset$. On the other hand, r is weakly supermodular by Condition 1, $|\Delta_{W_{i-1}}|$ is symmetric and submodular. Thus $r - |\Delta_{W_{i-1}}|$ is also weakly supermodular. Hence

$$(r(X \cap Y) - |\Delta_{W_{i-1}}(X \cap Y)|) + (r(X \cup Y) - |\Delta_{W_{i-1}}(X \cup Y)|)$$
$$\geq (r(X) - |\Delta_{W_{i-1}}(X)|) + (r(Y) - |\Delta_{W_{i-1}}(Y)|) = 2(r_{\max} - i + 1)$$

or

$$(r(X - Y) - |\Delta_{W_{i-1}}(X - Y)|) + (r(Y - X) - |\Delta_{W_{i-1}}(Y - X)|)$$
$$\geq (r(X) - |\Delta_{W_{i-1}}(X)|) + (r(Y) - |\Delta_{W_{i-1}}(Y)|) = 2(r_{\max} - i + 1)$$

holds. Since $r_{\max} - i + 1$ is the maximum deficiency at the beginning of phase i, it holds that $r(S) - |\Delta_{W_{i-1}}(S)| \leq r_{\max} - i + 1$ for all $S \subseteq T$. Thus we have $r(S) - |\Delta_{W_{i-1}}(S)| = r_{\max} - i + 1$ for $S \in \{X \cap Y, X \cup Y\}$ or $S \in \{X - Y, Y - X\}$. Hence either $X \cap Y, X \cup Y$ or $X - Y, Y - X$ are violated sets of A. □

Two sets X, Y are said to *intersect* if $X \cap Y \neq \emptyset$, $X - Y \neq \emptyset$ and $Y - X \neq \emptyset$. An immediate corollary from Lemma 1 is

Corollary 1. *Let $A \subseteq W - W_{i-1}$. Let X and Y be a minimal violated set and a violated set of A, respectively. Then X and Y do not intersect, either $X \subseteq Y$ or $X \cap Y = \emptyset$. Especially, if Y is also a minimal violated set then $X \cap Y = \emptyset$.* □

We now state another condition which needs to be satisfied by r.

Condition 2. *$\mathcal{V}(A)$ can be computed in polynomial time for any $A \subseteq W - W_{i-1}$.*

Relax $x_w \in \{0, 1\}$ to $x_w \geq 0$. The dual of this relaxation of (IP)$_i$ is

$$(D)_i \qquad \max \sum_{S \subseteq T} h_i(S) y_S$$

$$\text{s.t.} \qquad \sum_{S \subseteq T : w \in \Delta(S)} y_S \leq c_w \qquad \text{for all } w \in W - W_{i-1},$$

$$y \geq 0 \ .$$

We consider the next algorithm to (IP)$_i$ according to the primal-dual schema outlined in [4] and [11]. We use $\bar{c}$, A, y and j to denote the reduced cost, primal solution, dual variable and number of iterations, respectively.

$$
\begin{array}{ll}
1 & \bar{c} \leftarrow c, \quad A \leftarrow \emptyset, \quad y \leftarrow 0, \quad j \leftarrow 0 \\
2 & \text{WHILE } A \text{ is not feasible} \\
3 & \quad j \leftarrow j+1 \\
4 & \quad \mathcal{V}_j \leftarrow \text{the minimal violated sets } \mathcal{V}(A) \\
5 & \quad \text{IF exists } S \in \mathcal{V}_j \text{ such that } \Delta_{W-W_{i-1}-A}(S) = \emptyset \text{ THEN} \\
  & \qquad \text{Halt. (IP) has no feasible solution.} \\
6 & \quad w_j \leftarrow \operatorname{argmin}\{\frac{\bar{c}_w}{|\{S\in\mathcal{V}_j|w\in\Delta(S)\}|} \mid w \in W - W_{i-1} - A\} \\
7 & \quad \epsilon_j \leftarrow \frac{\bar{c}_{w_j}}{|\{S\in\mathcal{V}_j|w_j\in\Delta(S)\}|}, \quad y_S \leftarrow y_S + \epsilon_j \text{ for all } S \in \mathcal{V}_j \\
8 & \quad \bar{c}_w \leftarrow \bar{c}_w - |\{S \in \mathcal{V}_j|w \in \Delta(S)\}|\epsilon_j \text{ for all } w \in W - W_{i-1} - A \\
9 & \quad A \leftarrow A \cup \{w_j\} \\
10 & \text{FOR } l = j \text{ DOWN TO } 1 \\
11 & \quad \text{IF } A - \{w_l\} \text{ is feasible THEN } A \leftarrow A - \{w_l\} \\
12 & \text{Output } A \text{ (as } A_i).
\end{array}
$$

Clearly A_i and y are feasible to (IP)$_i$ and (D)$_i$, respectively. Step 1 takes $O(|W|)$ time (only those positive y_S are stored). There are at most $|W - W_{i-1}| \leq |W|$ WHILE iterations since $|A|$ increases by 1 after one iteration. Let θ denote the time complexity to compute $\mathcal{V}(A)$. Notice that steps 4 and 10–11 can be done in θ time since A is feasible if and only if $\mathcal{V}(A) = \emptyset$. By Corollary 1, we see $|\mathcal{V}(A)| \leq |T|$. Thus step 6 can be done in $O(|T||W|)$ time and this dominates other steps. Hence the algorithm for (IP)$_i$ takes $O(|W|(\theta + |T||W|))$ time. Therefore the algorithm for (IP) can be done in $O(r_{\max}|W|(\theta + |T||W|))$ time. Since $r_{\max} \leq |W|$, this is polynomial.

4   Proof of Performance Guarantee

Lemma 2. *Let A_i and y be the output and the corresponding dual variable obtained at the end of the primal-dual algorithm for (IP)$_i$, respectively. Then*

$$
\sum_{w\in A_i} c_w \leq d_{\max} \sum_{S\subseteq T} h_i(S)y_S. \quad \square
$$

Before proving Lemma 2, we show that it implies the claimed guarantee.

Theorem 1. *Let opt$_{IP}$ denote the optimal value of (IP). Let $W_{r_{\max}} = \bigcup_{i=1}^{r_{\max}} A_i$ be the output of our $r_{\max}$-phases algorithm for (IP). Then*

$$
\sum_{w\in W_{r_{\max}}} c_w \leq d_{\max}\mathcal{H}(r_{\max})opt_{IP}. \tag{7}
$$

Proof. Relax $x_w \in \{0,1\}$ to $0 \leq x_w \leq 1$. The dual of this relaxation of (IP) is

$$
\text{(D)} \qquad \max \sum_{S\subseteq T} r(S)y_S - \sum_{w\in W} z_w
$$

$$\text{s.t.} \quad \sum_{S \subseteq T: w \in \Delta(S)} y_S \le c_w + z_w \quad \text{for all } w \in W$$

$$y \ge 0, \quad z \ge 0.$$

Let opt_D denote the optimal value of (D). Notice that $opt_D \le opt_{IP}$ by the weak duality. Fix i. Let y be the dual variable of $(D)_i$ obtained in phase i. Let

$$z_w = \begin{cases} \sum_{S \subseteq T: w \in \Delta(S)} y_S & \text{if } w \in W_{i-1}, \\ 0 & \text{otherwise } (w \in W - W_{i-1}). \end{cases}$$

It is easy to see that (y, z) is a feasible solution to (D). Thus we have

$$opt_D \ge \sum r(S) y_S - \sum z_w = \sum r(S) y_S - \sum_{w \in W_{i-1}} \sum_{S: w \in \Delta(S)} y_S$$

$$= \sum r(S) y_S - \sum |\Delta_{W_{i-1}}(S)| y_S = \sum (r(S) - |\Delta_{W_{i-1}}(S)|) y_S$$

$$= (r_{\max} - i + 1) \sum h_i(S) y_S.$$

The last equality holds because y_S remains to be 0 for all S with $h_i(S) = 0$, and $h_i(S) = 1$ if and only if $r(S) - |\Delta_{W_{i-1}}(S)| = r_{\max} - i + 1$. By Lemma 2 we have

$$\sum_{w \in A_i} c_w \le d_{\max} \sum h_i(S) y_S \le \frac{d_{\max}}{r_{\max} - i + 1} opt_D$$

$$\Rightarrow \sum_{w \in W_{r_{\max}}} c_w = \sum_{i=1}^{r_{\max}} \sum_{w \in A_i} c_w \le d_{\max} \mathcal{H}(r_{\max}) opt_D \le d_{\max} \mathcal{H}(r_{\max}) opt_{IP}. \quad \square$$

Thus we only need to prove Lemma 2 to show the performance guarantee.

Proof of Lemma 2. First suppose that $c_w > 0$ for all $w \in W$. Then $d_{\max}$ is the maximum degree of non-terminals. Let L be the number of WHILE iterations. Notice that $c_{w_l} = \sum_{j=1}^{L} |\{S \in \mathcal{V}_j | w_l \in \Delta(S)\}| \epsilon_j$ for $l = 1, 2, \ldots, L$. Thus

$$\sum_{w \in A_i} c_w = \sum_{w \in A_i} \sum_j |\{S \in \mathcal{V}_j | w \in \Delta(S)\}| \epsilon_j = \sum_j \sum_{S \in \mathcal{V}_j} |\Delta_{A_i}(S)| \epsilon_j.$$

On the other hand, since $y_S = \sum_{j: S \in \mathcal{V}_j} \epsilon_j$, we have

$$\sum_{S \subseteq T} h_i(S) y_S = \sum_{S \subseteq T} y_S = \sum_{S \subseteq T} \sum_{j: S \in \mathcal{V}_j} \epsilon_j = \sum_j |\mathcal{V}_j| \epsilon_j.$$

Thus we only need to show that for all $j \in \{1, \ldots, L\}$, it holds

$$\sum_{S \in \mathcal{V}_j} |\Delta_{A_i}(S)| \le d_{\max} |\mathcal{V}_j|. \tag{8}$$

Fix j, consider $A = \{w_1, \ldots, w_{j-1}\}$. By the reverse delete step 10–11, we see that $B = A \cup A_i$ is a minimal augmentation of A, i.e., $A \subseteq B \subseteq W - W_{i-1}$ and B

is feasible to (IP)$_i$ but the removal of any $w \in B - A$ will violate the feasibility. Thus if we can show that for any infeasible non-terminal set $A \subseteq W - W_{i-1}$ and any minimal augmentation B of A, it holds

$$\sum_{S \in \mathcal{V}(A)} |\Delta_B(S)| \le d_{\max}|\mathcal{V}(A)|, \tag{9}$$

then (8) holds (by letting $A = \{w_1, \ldots, w_{j-1}\}$, $B = A \cup A_i$, notice $|\Delta_{A_i}(S)| \le |\Delta_B(S)|)$, implying Lemma 2. In the following we show (9). For this, a notation of *witness sets* is used. Let $U \triangleq \bigcup_{S \in \mathcal{V}(A)} \Delta_B(S) \subseteq B - A$.

Definition 3 (Witness Set). $C_w \subseteq T$ *is a witness set of $w \in U$ if it satisfies* (i) $h_i(C_w) = 1$, *and* (ii) $\Delta_B(C_w) = \{w\}$.

By (i) and (ii), we see that C_w is a violated set of A (notice that $w \notin A$). For any $w \in U$, there must exist a witness set of w since the removal of w violates the feasibility of B (B is a minimal augmentation of A). Call $\{C_w | w \in U\}$ a witness set family, in which only one C_w is included for each $w \in U$.

Lemma 3. *There exists a laminar (i.e., intersect-free) witness set family.*

Proof. Given a witness set family we construct a laminar one.

Suppose that two witness sets C_v and C_w intersect. Since C_v and C_w are violated sets of A we see that either $C_v \cap C_w$, $C_v \cup C_w$ or $C_v - C_w$, $C_w - C_v$ are violated sets of A (Lemma 1).

Suppose that $C_v \cap C_w$ and $C_v \cup C_w$ are violated sets. By the definition of violated set they must satisfy (i). B is feasible implies that $|\Delta_B(C_v \cap C_w)| \ge 1$ and $|\Delta_B(C_v \cup C_w)| \ge 1$. However, by (4) we see that $|\Delta_B(C_v \cap C_w)| + |\Delta_B(C_v \cup C_w)| \le |\Delta_B(C_v)| + |\Delta_B(C_w)| = 2$. Therefore $|\Delta_B(C_v \cap C_w)| = |\Delta_B(C_v \cup C_w)| = 1$ holds. It is easy to see that $\{v, w\} \subseteq \Delta_B(C_v \cap C_w) \cup \Delta_B(C_v \cup C_w)$, which shows that $\Delta_B(C_v \cap C_w) = \{a\}$ and $\Delta_B(C_v \cup C_w) = \{b\}$ hold for $\{a, b\} = \{v, w\}$. Thus we can replace C_v and C_w by $C_v \cap C_w$ and $C_v \cup C_w$ in the witness family.

Similarly, if $C_v - C_w$ and $C_w - C_v$ are violated sets of A, then we can replace C_v, C_w by $C_v - C_w$ and $C_w - C_v$. In both cases this un-intersecting process will decrease the total number of pairs of intersected sets in the witness family. Thus after a finite number of this process, a laminar witness set family is obtained. □

Let $\mathcal{F}$ be the union of $\{T\}$ and the laminar witness set family. Construct a rooted tree $\mathcal{T}$ by set inclusion relationship as follows. $\mathcal{T}$ contains $|\mathcal{F}|$ nodes: u_C for $C \in \mathcal{F}$, the root is u_T, and for each $C \in \mathcal{F}$, the parent of u_C is the node $u_{C'}$ for the minimum $C' \in \mathcal{F}$ such that $C \subset C'$. For each $S \in \mathcal{V}(A)$, let $u(S) \triangleq u_C$ for the minimum $C \in \mathcal{F}$ such that $S \subseteq C$. Let $\alpha(u_C) = |\{S \in \mathcal{V}(A) | u(S) = u_C\}|$ for $u_C \in \mathcal{T}$. Let $\mathcal{T}_a = \{u_C \in \mathcal{T} | \alpha(u_C) \ge 1\}$. Clearly we have

$$|\mathcal{V}(A)| = \sum_{u_C \in \mathcal{T}} \alpha(u_C) = \sum_{u_C \in \mathcal{T}_a} \alpha(u_C). \tag{10}$$

Let $d_{\mathcal{T}}(u_C)$ denote the degree of node u_C in tree $\mathcal{T}$. For a node u_{C_w} in tree $\mathcal{T}$, C_w is a witness set and is a violated set, implying that it must include some

$S \in \mathcal{V}(A)$. Thus if u_{C_w} is a leaf, then $C_w = u(S)$. Hence all the degree 1 nodes except for the root u_T (if its degree is 1) must be in $\mathcal{T}_a$. This observation shows that $\sum_{u_C \in \mathcal{T} - \mathcal{T}_a} d_T(u_C) \geq 2(|\mathcal{T}| - |\mathcal{T}_a|) - 1$. Since $\mathcal{T}$ is a tree, we have that $\sum_{u_C \in \mathcal{T}} d_T(u_C) = 2(|\mathcal{T}| - 1)$. Thus

$$
\sum_{u_C \in \mathcal{T}_a} d_T(u_C) = \sum_{u_C \in \mathcal{T}} d_T(u_C) - \sum_{u_C \in \mathcal{T} - \mathcal{T}_a} d_T(u_C)
$$
$$
\leq 2(|\mathcal{T}| - 1) - (2(|\mathcal{T}| - |\mathcal{T}_a|) - 1) = 2|\mathcal{T}_a| - 1. \tag{11}
$$

We will show that for each $u_C \in \mathcal{T}_a$,

$$
\sum_{S \in \mathcal{V}(A): u(S) = u_C} |\Delta_B(S)| \leq \min\{d_{\max} - 1, \alpha(u_C)\} d_T(u_C). \tag{12}
$$

Consider an $S \in \mathcal{V}(A)$ and a $w \in \Delta_B(S)$. Let $C_w \in \mathcal{F}$ be the witness set of w. Since C_w is a violated set, either $S \subseteq C_w$ or $S \cap C_w = \emptyset$ holds by Corollary 1.

Case 1: $S \subseteq C_w$. Since $w \in \Delta_B(S)$, there exists an $s \in \Gamma(w) \cap S$. By (ii), there exists a $t \in \Gamma(w) \cap (T - C_w)$. Thus $u(S) = C_w$. Let u_C be the parent of u_{C_w} in $\mathcal{T}$ (it exists since $c_w \neq T$). Then we see that $\Gamma(w) \subseteq C$ (otherwise $w \in \Delta_B(C)$, which implies that $C \neq T$ and C must be a witness set such that $\Delta_B(C) = \{w\}$ contradicting that $\Delta_B(C_w) = \{w\}$ for $C_w \neq C$). We use a *directed* edge (u_{C_w}, u_C) to represent this case. The directed edge (u_{C_w}, u_C) may not be unique since there may exists some other $S' \in \mathcal{V}(A)$ such that $w \in \Delta_B(S')$ and $u(S') = u_{C_w}$. In such cases multiple directed edges (u_{C_w}, u_C) are allowed, but for each S' of such sets ($w \in \Delta_B(S')$ and $u(S') = u_{C_w}$) only one edge (u_{C_w}, u_C) is used. Notice that such sets are disjoint (Corollary 1). It is then easy to see that the total number of such directed edges (u_{C_w}, u_C) is at most $\min\{|\Gamma(w)| - 1, \alpha(u_{C_w})\} \leq \min\{d_{\max} - 1, \alpha(u_{C_w})\}$ (notice that $t \in \Gamma(w) - C_w$).

Case 2: $S \cap C_w = \emptyset$. There must exist an $s \in \Gamma(w) \cap S$ and a $t \in \Gamma(w) \cap C_w$. Let u_C be the parent of u_{C_w}. We see that $\Gamma(w) \subseteq C$, hence $S \subseteq C$ and $u(S) = C$. We use a directed edge (u_C, u_{C_w}) to represent this case. Similarly as in the previous case, the total number of these (u_C, u_{C_w}) edges is at most $\min\{d_{\max} - 1, \alpha(u_C)\}$.

For each $u_C \in \mathcal{T}_a$, the two cases may happen simultaneously. But we have seen that for one edge $\{u_C, u_{C'}\}$ in $\mathcal{T}$, there are at most $\min\{d_{\max} - 1, \alpha(u_C)\}$ directed edges $(u_C, u_{C'})$ that are produced in Case 1 or 2. Thus the total number of the directed edges $(u_C, \cdot)$ is at most $\min\{d_{\max} - 1, \alpha(u_C)\} d_T(u_C)$. On the other hand, the way that the directed edges are produced ensures that the total number of the directed edges $(u_C, \cdot)$ (over all $S \in \mathcal{V}(A)$ and $w \in \Delta_B(S)$) equals to $\sum_{S \in \mathcal{V}(A): u(S) = u_C} |\Delta_B(S)|$. Hence (12) has been shown. Thus

$$
\sum_{S \in \mathcal{V}(A)} |\Delta_B(S)| = \sum_{u_C \in \mathcal{T}} \sum_{S \in \mathcal{V}(A): u(S) = u_C} |\Delta_B(S)|
$$
$$
\leq \sum_{u_C \in \mathcal{T}_a} \min\{d_{\max} - 1, \alpha(u_C)\} d_T(u_C).
$$

To show (9), it suffices to show by (10) that

$$
\sum_{u_C \in \mathcal{T}_a} \min\{d_{\max} - 1, \alpha(u_C)\} d_T(u_C) \leq d_{\max} \sum_{u_C \in \mathcal{T}_a} \alpha(u_C). \tag{13}
$$

Let $X = \{u_C \in \mathcal{T}_a | \alpha(u_C) \geq d_{\max} - 1\}$, $Y = \{u_C \in \mathcal{T}_a | \alpha(u_C) = 1\}$ and $Z = \mathcal{T}_a - X - Y$. The left hand side of (13) is at most

$$(d_{\max} - 1) \sum_{u_C \in X} d_T(u_C) + \sum_{u_C \in Y} d_T(u_C) + (d_{\max} - 2) \sum_{u_C \in Z} d_T(u_C).$$

The right hand side of (13) is at least $d_{\max}((d_{\max} - 1)|X| + |Y| + 2|Z|)$. Then by (11) and $|\mathcal{T}_a| = |X| + |Y| + |Z|$, it is not difficult to verify (13).

Thus we have proved Lemma 2 if $c_w > 0$ for all $w \in W$. It is not difficult to see that it is also true in the general cases. To see this, notice that we only need to show (8) for j with $\epsilon_j > 0$, which implies that $c_w > 0$ for all $w \in \bigcup_{S \in \mathcal{V}_j} \Delta_{A_i}(S)$. Thus $|\Gamma(w)| \leq d_{\max}$ for all $w \in \bigcup_{S \in \mathcal{V}_j} \Delta_{A_i}(S)$ and the proof goes. $\square$

5   Survivable Network Design Problem in Hypergraph

It is equivalent to the next problem defined in a bipartite graph $G = (T, W, E)$. Given $c : W \to \mathbf{R}^+$ and $r_{st} \in \mathbf{Z}^+$ for each pair of distinct terminals $s, t \in T$, find a minimum cost $W^* \subseteq W$ such that, for each pair of s and t, $G[T \cup W^*]$ has at least r_{st} paths which are W-disjoint (i.e., no $w \in W$ belongs to two or more paths). We show that it is equivalent to problem $\mathcal{P}$ (Sect. 2) with r defined as $r(S) = \max\{r_{st} \mid s \in S, t \in T - S\}$ for all $S \subseteq T$ ($r(\emptyset) = r(T) = 0$).

A useful idea when considering W-disjoint paths in G is a transformation from G to a digraph $\overrightarrow{G}$ in the following way.

Definition 4 (Transformation $\mathcal{D}$). *Replace each undirected edge $e = \{v, w\}$ by two directed edges (v, w) and (w, v). Then for each non-terminal w make a copy named w^c, and change the tails of all directed edges (w, v) from w to w^c, then add a new directed edge (w, w^c). Let the capacity of directed edge (w, w^c) be 1 for all non-terminal w, and of others directed edges be $+\infty$.*

Notice that for any pair of terminals s and t, any k W-disjoint s,t-paths in G are transformed to an integer s,t-flow of value k in $\overrightarrow{G}$, and vice versa. Thus a non-terminal set $W' \subseteq W$ is feasible if and only if the maximum s,t-flow in $\overrightarrow{G[T \cup W']}$ has value at least r_{st} for each pair of terminals s and t. By the well known maxflow-mincut theorem [1], this equals that in $\overrightarrow{G[T \cup W']}$ any s,t-cut has capacity at least r_{st} for each pair of terminals s and t. It is not difficult to see that this is equivalent to $|\Delta_{W'}(S)| \geq r(S)$ for all $S \subseteq T$. Thus the SNDPHG is equivalent to problem $\mathcal{P}$ with r defined as $r(S) = \max\{r_{st} \mid s \in S, t \in T - S\}$.

It is easy to see that r is a weakly supermodular function, satisfying Condition 1. We show that the minimum violated sets can be computed in polynomial time (Condition 2), which means that our algorithm in Sect. 3 works for the SNDPHG.

Lemma 4. *Denote $W_{i-1} \cup A$ by $\tilde{A}$. If S is a minimal violated set of A, then there exist vertices $s \in S$ and $t \in T - S$ such that in digraph $\overrightarrow{G[T \cup \tilde{A}]}$, $S = C_{st} \cap T$ for any minimum s,t-cut C_{st} that is minimal under set inclusion.*

Proof. By definition of $r(S)$, there exist two terminals $s \in S$ and $t \in T - S$ such that $r_{st} = r(S)$. Let $C = S \cup \Gamma_{\tilde{A}}(S) \cup \{w^c \mid w \in \Gamma_{\tilde{A}}(S) - \Delta_{\tilde{A}}(S)\}$. It is clear that $S = C \cap T$. We show that $C = C_{st}$, which implies the claimed statement.

We first show that C is a minimum s,t-cut.

The capacity of C is exactly $|\Delta_{\tilde{A}}(S)|$. Let $S' = C_{st} \cap T$. It is easy to see that the capacity of C_{st} is at least $|\Delta_{\tilde{A}}(S')|$. We show that $|\Delta_{\tilde{A}}(S)| \leq |\Delta_{\tilde{A}}(S')|$, which implies that C is a minimum s,t-cut. Notice that S is a violated set of A, implying $h_i(S) = 1$ and $\Delta_A(S) = \emptyset$. Thus $r(S) - |\Delta_{W_{i-1}}(S)| = r_{\max} - i + 1$ and $\Delta_A(S) = \emptyset$. Hence

$$|\Delta_{\tilde{A}}(S)| = |\Delta_{W_{i-1}}(S)| + |\Delta_A(S)| = r_{st} - r_{\max} + i - 1. \tag{14}$$

On the other hand, if S' is also a violated set of A, then similarly

$$|\Delta_{\tilde{A}}(S')| = r(S') - r_{\max} + i - 1 \geq r_{st} - r_{\max} + i - 1 = |\Delta_{\tilde{A}}(S)|. \tag{15}$$

(Notice that $s \in S'$ and $t \in T - S'$.) Otherwise S' is not a violated set of A, and hence either $h_i(S') = 1$, $\Delta_A(S') \neq \emptyset$ or $h_i(S') = 0$, which implies

$$|\Delta_{\tilde{A}}(S')| \geq r(S') - r_{\max} + i \geq r_{st} - r_{\max} + i > |\Delta_{\tilde{A}}(S)|. \tag{16}$$

Thus C is a minimum s,t-cut. Since C_{st} is also a minimum s,t-cut, we have $|\Delta_{\tilde{A}}(S)| = |\Delta_{\tilde{A}}(S')|$. The proof above shows that $S' = C_{st} \cap T$ must be a violated set of A. Thus by Corollary 1 we have $S \subseteq S' = C_{st} \cap T$. It is then easy to see that $C \subseteq C_{st}$. Thus $C = C_{st}$ by the assumption that C_{st} is minimal. $\square$

By Lemma 4, we can identify the minimal violated sets by computing a minimal minimum s,t-cut in $\overrightarrow{G[T \cup \tilde{A}]}$ for all pairs of terminals s and t and checking if they are violated and minimal among these cuts. It is well known that for each pair of s and t, there is only one minimal minimum s,t-cut that can be found by one maxflow computation in $O(n^3)$ time for a digraph with n vertices ([3]). Thus the total running time to find minimal violated sets is dominated by $O(|T|^2)$ maxflow computations. Thus our algorithm for the SNDPHG can be implemented to run in $O(r_{\max}|W||T|^2(|T| + 2|W|)^3)$ time. We summary this as the next theorem.

Theorem 2. *Let $d_{\max}$ be the maximum degree of hyperedges with positive cost, $r_{\max}$ be the maximum requirement, then the SNDPHG can be approximated within factor $d_{\max}\mathcal{H}(r_{\max})$ in $O(r_{\max}mn^2(n + 2m)^3)$ time, where m, n is the number of hyperedges, vertices respectively.* $\square$

6   Conclusion

In this paper, we have shown that the SNDPHG can be approximated by a factor $d_{\max}\mathcal{H}(r_{\max})$ in polynomial time. We note that the performance guarantee $d_{\max}$ to the primal-dual algorithm for $(IP)_i$ (Lemma 2) is tight (a tight example will be given in the full version of this paper). Notice that in [4] Goemans *et al.* have shown that for the SNDP in graphs the performance guarantee $2\mathcal{H}(r_{\max})$ is tight up to a constant factor. It is thus interesting to know whether an algorithm with improved (e.g., constant) performance guarantee can be developed via an iterative rounding process as used in [7] for the SNDP.

Acknowledgment

This research was partially supported by the Scientific Grant-in-Aid from Ministry of Education, Science, Sports and Culture of Japan. The authors would like to thank the anonymous reviewers for helpful suggestions.

References

1. L. R. Ford and D. R. Fulkson. Maximal Flow through a Network. Canad. J. Math. **8** (1956) 399–404
2. H. N. Gabow, M. X. Goemans, and D. P. Williamson. An Efficient Approximation Algorithm for the Survivable Network Design Problem. In Proc. 3rd MPS Conf. on Integer Programming and Combinatorial Optimization, (1993) 57–74
3. A. V. Goldberg and R. E. Tarjan. A New Approach to the Maximum Flow Problem. In Proc. STOC'86 (1986) 136–146. Full paper in J. ACM **35** (1988) 921–940
4. M. X. Goemans, A. V. Goldberg, S. Plotkin, D. Shmoys, E. Tardos, and D. P. Williamson. Improved Approximation Algorithms for Network Design Problems. In Proc. SODA'94 (1994) 223–232
5. M. X. Goemans and D. P. Williamson. A General Approximation Technique for Constrained Forest Problems. In Proc. SODA'92 (1992) 307–315
6. M. X. Goemans and D. P. Williamson. The Primal-Dual Method for Approximation Algorithms and its Application to Network Design Problems, in Approximation Algorithms for NP-hard Problems, (D. Hochbaum, eds.), PWS, (1997) 144–191
7. K. Jain. A Factor 2 Approximation Algorithm for the Generalized Steiner Network Problem. In Proc. FOCS'98 (1998) 448–457
8. K. Jain, I. Măndoiu, V. V. Vazirani, and D. P. Williamson. A Primal-dual Schema Based Approximation Algorithm for the Element Connectivity Problem. In Proc. SODA'99 (1999) 484–489
9. R. M. Karp. Reducibility among combinatorial problems, in Complexity of Computer Computations, (R. E. Miller, J. W. Thatcher, eds.), Plenum Press, New York (1972) 85–103
10. K. Takeshita, T. Fujito, and T. Watanabe. On Primal-Dual Approximation Algorithms for Several Hypergraph Problems (in Japanese). IPSJ Mathematical Modeling and Problem Solving **23-3** (1999) 13–18
11. D. P. Williamson, M. X. Goemans, M. Mihail, and V. V. Vazirani. A Primal-dual Approximation Algorithm for Generalized Steiner Network Problems. In Proc. STOC'93 (1993) 708–717. Full paper in Combinatorica **15** (1995) 435–454

The Complexity of
Copy Constant Detection in Parallel Programs

Markus Müller-Olm

Universität Dortmund, FB Informatik, LS V
44221 Dortmund, Germany
mmo@ls5.cs.uni-dortmund.de

Abstract. Despite of the well-known state-explosion problem, certain simple but important data-flow analysis problems known as gen/kill problems can be solved efficiently and completely for parallel programs with a shared state [7,6,2,3,13]. This paper shows that, in all probability, these surprising results cannot be generalized to significantly larger classes of data-flow analysis problems.

More specifically, we study the complexity of detecting copy constants in parallel programs, a problem that may be seen as representing the next level of difficulty of data-flow problems beyond gen/kill problems. We show that already the intraprocedural problem for loop-free parallel programs is co-NP-complete and that the interprocedural problem is even PSPACE-hard.

1 Introduction

A well-known obstacle for the automatic analysis of parallel programs is the so-called *state-explosion problem*: the number of (control) states of a parallel program grows exponentially with the number of parallel components. It comes as a surprise that certain basic but important data-flow analysis problems can nevertheless be solved completely and efficiently for programs with a fork/join kind of parallelism.

Knoop, Steffen, and Vollmer [7] show that *bitvector analyses*, which comprise, e.g., live/dead variable analysis, available expression analysis, and reaching definition analysis [8], can efficiently be performed on such programs. Knoop shows in [6] that a simple variant of constant detection, that of so-called *strong constants*, is tractable as well. These articles restrict attention to the *intraprocedural* problem, in which each procedure body is analyzed separately with worst-case assumption on called procedures. Seidl and Steffen [13] generalize these results to the *interprocedural* case in which the interplay between procedures is taken into account and to a slightly more extensive class of data-flow problems called *gen/kill problems*[1]. All these papers extend the fixpoint computation technique

---

[1] Gen/kill problems are characterized by the fact that all transfer functions are of the form $\lambda x.(x \wedge a) \vee b$, where a, b are constants from the underlying lattice of data-flow facts.

A. Ferreira and H. Reichel (Eds.): STACS 2001, LNCS 2010, pp. 490–501, 2001.

common in data-flow analysis to parallel programs. Another line of research applies automata-theoretic techniques that were originally developed for the verification of PA-processes, a certain class of infinite-state processes combining sequentiality and parallelism. Specifically, Esparza, Knoop, and Podelski [2,3] demonstrate how live variables analysis can be done and indicate that other bitvector analyses can be approached in a similar fashion.

Can these results be generalized further to considerably richer classes of data-flow problems? The current paper shows that this is very unlikely. We investigate the complexity of detection of copy constants, a problem that may be seen as a canonic representative of the next level of difficulty of data-flow problems beyond gen/kill problems. In the sequential setting the problem gives rise to a *distributive* data-flow framework on a lattice with small chain height and can thus – by the classic result of Kildall [5,8] – completely and efficiently be solved by a fixpoint computation. We show in this paper that copy constant detection is co-NP-complete already for loop-free parallel programs without procedures and becomes even PSPACE-hard if one allows loops and non-recursive procedures. This renders the possibility of complete and efficient data-flow analysis algorithms for parallel programs for more extensive classes of analyses unlikely, as it is generally believed that the inclusions $P \subseteq$ co-NP $\subseteq$ PSPACE are proper.

Our theorems should be contrasted with complexity and undecidability results of Taylor [14] and Ramalingam [11] who consider *synchronization-dependent* data-flow analyses of parallel programs, i.e. analyses that are precise with respect to the synchronization structure of programs. Taylor and Ramalingam largely exploit the strength of rendezvous style synchronization, while we exploit only interference and no kind of synchronization. Our results thus point to a more fundamental limitation in data-flow analysis of parallel programs.

This paper is organized as follows: In Sect. 2 we give some background information on data-flow analysis in general and the constant detection problem in particular. In Sect. 3 we introduce loop-free parallel programs. This sets the stage for the co-NP-completeness result for the loop-free intraprocedural parallel case which is proved afterwards. We proceed by enriching the considered programming language with loops and procedures in Sect. 4. We then show that the interprocedural parallel problem is PSPACE-hard even if we allow only non-recursive procedures. In the Conclusions, Sect. 5, we indicate that the presented results apply also to some other data-flow analysis problems, detection of may-constants and detection of faint code, and discuss directions for future research. Throughout this paper we assume that the reader is familiar with the basic notions and methods of the theory of computational complexity (see, e.g., [10]).

2 Copy Constants

The goal of *data-flow analysis* is to gather information about certain aspects of the behavior of programs by a static analysis. Such information is valuable e.g. in optimizing compilers and in CASE tools. However, most questions about programs are undecidable. This holds in particular for the question whether a condition in a program may be satisfied or not. In order to come to grips with undecidability, it is common in data-flow analysis to abstract from the conditions in the programs and to interpret conditional branching as non-deterministic branching, a point of view adopted in this paper. Of course, an analysis based on this abstraction considers more program executions than actually possible at run-time. One is careful to take this into account when exploiting the results of data-flow analysis.

An expression e is a *constant* at a given point p in a program, if e evaluates to one and the same value whenever control reaches p, i.e. after every run from the start of the program to p. If an expression is detected to be a constant at compile time it can be replaced by its value, a standard transformation in optimizing compilers known as *constant propagation* or *constant folding* [8]. Constant folding is profitable as it decreases both code size and execution time. Constancy information is sometimes also useful for eliminating branches of conditionals that cannot be taken at run-time and for improving the precision of other data-flow analyses.

Reif and Lewis [12] show by a reduction of Hilbert's tenth problem that the general constant detection problem in sequential programs is undecidable, even if branching is interpreted non-deterministically. However, if one restricts the kind of expressions allowed on the right hand side of assignment statements appropriately, the problem becomes decidable. (In practice assignments of a different form are treated by approximating or worst-case assumptions.) A problem that is particularly simple for sequential programs are so-called *copy constants*. In this problem assignment statements take only the simple forms $x := c$ (constant assignment) and $x := y$ (copying assignment), where c is a constant and x, y are variables. In the remainder of this paper we study the complexity of detecting copy constants in parallel programs.

3   Loop-Free Parallel Programs

Let us, first of all, set the stage for the parallel loop-free intraprocedural copy constant detection problem. We consider *loop-free parallel programs* given by the following abstract grammar,

$$\pi ::= x := e \mid \textbf{write}(x) \mid \textbf{skip} \mid \pi_1 \; ; \; \pi_2 \mid \pi_1 \parallel \pi_2 \mid \pi_1 \sqcap \pi_2$$
$$e ::= c \mid x \, ,$$

where x ranges over some set of variables and c over some set of basic constants. As usual we use parenthesis to disambiguate programs. Note that this language has only constant and copying assignments. The specific nature of basic constants and the value domain in which they are interpreted is immaterial; we only need that 0 and 1 are two constants representing different values, which – by abuse of notation – are also denoted by 0 and 1. The atomic statements of the language are assignment statements $x := e$ that assign the current value of e to variable x, 'do-nothing'-statements **skip**, and write-statements. The purpose of write-statements in this paper is to mark prominently the program points at which we are interested in constancy of a certain variable. The operator ; represents sequential composition, $\|$ represents parallel composition, and $\sqcap$ nondeterministic branching.

Parallelism is understood in an interleaving fashion; assignments and write-statements are assumed to be atomic. A run of a program is a maximal sequence of atomic statements that may be executed in this order in an execution of the program. The program $(x := 1 ; x := y) \| y := x$ for example, has the three runs $\langle x := 1, x := y, y := x \rangle$, $\langle x := 1, y := x, x := y \rangle$, and $\langle y := x, x := 1, x := y \rangle$.

In order to allow a formal definition of runs, we need some notation. We denote the empty sequence by ε and the concatenation operator by an infix dot. The concatenation operator is lifted to sets of sequences in the obvious way: If S, T are two sets of sequences then $S \cdot T = \{s \cdot t \mid s \in S, t \in T\}$. Let $r = \langle e_1, \ldots, e_n \rangle$ be a sequence and $I = \{i_1, \ldots, i_k\}$ a subset of positions in r such that $i_1 < i_2 < \cdots < i_k$. Then $r|I$ is the sequence $\langle e_{i_1}, \ldots, e_{i_k} \rangle$. We write $|r|$ for the length of r, viz. n. The *interleaving* of S and T is

$$S \| T \stackrel{\text{def}}{=} \{r \mid \exists I_S, I_T : I_S \cup I_T = \{1, \ldots, |w|\}, I_S \cap I_T = \emptyset, r|I_S \in S, r|I_T \in T\} .$$

The set of runs of a program can now inductively be defined:

$$
\begin{aligned}
\mathsf{Runs}(x := e) &= \{\langle x := e \rangle\} & \mathsf{Runs}(\pi_1 ; \pi_2) &= \mathsf{Runs}(\pi_1) \cdot \mathsf{Runs}(\pi_2) \\
\mathsf{Runs}(\mathbf{write}(x)) &= \{\langle \mathbf{write}(x) \rangle\} & \mathsf{Runs}(\pi_1 \| \pi_2) &= \mathsf{Runs}(\pi_1) \| \mathsf{Runs}(\pi_2) \\
\mathsf{Runs}(\mathbf{skip}) &= \{\varepsilon\} & \mathsf{Runs}(\pi_1 \sqcap \pi_2) &= \mathsf{Runs}(\pi_1) \cup \mathsf{Runs}(\pi_2) .
\end{aligned}
$$

3.1   NP-Completeness of the Loop-Free Intraprocedural Problem

The remainder of this section is devoted to the proof of the following theorem, which shows that complete detection of copy constants is intractable in parallel programs, unless $P = NP$.

Theorem 1. *The problem of detecting copy constants in loop-free parallel programs is co-NP-complete.*

Certainly, the problem lies in co-NP: if a variable x is *not* constant at a certain point in the program we can guess two runs that witness two different values. As the program has no loops, the length of these runs (and thus the time needed to guess them) is at most linear in the size of the program.

For showing co-NP-hardness we reduce SAT, the most widely known NP-complete problem [1,10], to the negation of a copy constant detection problem. An instance of SAT is a conjunction $c_1 \wedge \ldots \wedge c_k$ of *clauses* $c_1, \ldots, c_k$. Each clause is a disjunction of *literals*; a literal l is either a variable x or a negated variable $\neg x$, where x ranges over some set of variables X. It is straightforward to define when a *truth assignment* $T : X \to \mathbb{B}$, where $\mathbb{B} = \{\mathsf{tt}, \mathsf{ff}\}$ is the set of truth values, satisfies $c_1 \wedge \ldots \wedge c_k$. The SAT problem asks us to decide for each instance $c_1 \wedge \ldots \wedge c_k$ whether there is a satisfying truth assignment or not.

Now suppose given a SAT instance $c_1 \wedge \ldots \wedge c_k$ with k clauses over n variables $X = \{x_1, \ldots, x_n\}$. We write $\bar{X} = \{\neg x_1, \ldots, \neg x_n\}$ for the set of negated variables. From this SAT instance we construct a loop-free parallel program. In the program we use $k+1$ variables $z_0, z_1, \ldots, z_k$. Intuitively, z_i is, for $1 \le i \le k$, related to clause c_i; z_0 is an extra variable.

For each literal $l \in X \cup \bar{X}$ we define a program π_l. Program π_l consists of a sequential composition of assignments of the form $z_i := z_{i-1}$ in increasing order of i. The assignment $z_i := z_{i-1}$ is in π_l if and only if the literal l makes clause i true. Formally, $\pi_l = \pi_l^k$, where

$$\pi_l^0 \stackrel{\text{def}}{=} \textbf{skip} \quad \text{and} \quad \pi_l^i \stackrel{\text{def}}{=} \begin{cases} \pi_l^{i-1} \; ; z_i := z_{i-1}, & \text{if clause } c_i \text{ contains } l \\ \pi_l^{i-1}, & \text{if clause } c_i \text{ does not contain } l \end{cases}$$

for $i = 1, \ldots, k$. Now, consider the following program π:

$$z_0 := 1 \; ; z_1 := 0 \; ; \ldots ; z_k := 0 \; ;$$
$$[(\pi_{x_1} \sqcap \pi_{\neg x_1}) \parallel \cdots \parallel (\pi_{x_n} \sqcap \pi_{\neg x_n})] \; ;$$
$$(z_k := 0 \sqcap \textbf{skip}) \; ; \textbf{write}(z_k) \, .$$

Clearly, π can be constructed from the given SAT instance $c_1 \wedge \ldots \wedge c_k$ in polynomial time or logarithmic space. We show that the variable z_k at the write-statement is not a constant if and only if $c_1 \wedge \ldots \wedge c_k$ is satisfiable. This proves the co-NP-hardness claim.

First observe that 0 and 1 are the only values z_k can hold at the write-statement because all variables are initialized by 0 or 1 and the other assignments only copy these values. Clearly, due to the non-deterministic choice just before the write-statement, z_k may hold 0 finally. Thus, z_k is a constant at the write-statement iff it cannot hold 1 there. Hence, our goal reduces to proving that z_k can hold 1 finally if and only if $c_1 \wedge \ldots \wedge c_k$ is satisfiable.

"If": Suppose $T : X \to \mathbb{B}$ is a satisfying truth assignment for $c_1 \wedge \ldots \wedge c_k$. Consider the following run of π: in each parallel component $\pi_{x_i} \sqcap \pi_{\neg x_i}$ we choose the left branch π_{x_i} if $T(x_i) = \text{tt}$ and the right branch $\pi_{\neg x_i}$ otherwise. As T is a satisfying truth assignment, there will be, for any $i \in \{1, \ldots, k\}$, at least one assignment $z_i := z_{i-1}$ in one of the chosen branches. We interleave the branches now in such a way that the assignment(s) to z_1 are executed first, followed by the assignment(s) to z_2 etc. This results in a run that copies the initialization value 1 of z_0 to z_k.

"Only if": Suppose z_k may hold 1 at the write-statement. As the initialization $z_0 := 1$ is the only statement in which the constant 1 occurs, there must be a run in which this value is copied from z_0 to z_k via a sequence of copy instructions. As all copying assignments in π have the form $z_i := z_{i-1}$, the value must be copied from z_0 to z_1, from z_1 to z_2 etc. Consequently, the non-deterministic choices in the parallel components can be resolved in such a way that the chosen branches contain all the assignments $z_i := z_{i-1}$ for $i = 1, \ldots, k$. From such a choice a satisfying truth assignment can easily be constructed.

4   Adding Loops and Procedures

Let us now consider a richer program class: programs with procedures, parallelism and loops. A *procedural parallel program* comprises a finite set Proc of *procedure names* containing a distinguished name *Main*. Each procedure name P is associated with a statement π_P, the corresponding *procedure body*, constructed according to the grammar

$$e \ ::= \ c \mid x$$
$$\pi \ ::= \ x := e \mid \mathbf{write}(x) \mid \mathbf{skip} \mid Q \mid \pi_1 \ ; \ \pi_2 \mid \pi_1 \parallel \pi_2 \mid \pi_1 \sqcap \pi_2 \mid \pi^* \, ,$$

where Q ranges over Proc. A statement of the form Q represents a call to procedure Q and π^* stands for a loop that iterates π an indefinite number of times. Such an indefinite looping construct is consistent with the abstraction that branching is non-deterministic. A program is *non-recursive* if there is an order on the procedure names such that in the body of each procedure only procedures with a strictly smaller name are called.

The definition of runs from the previous section can easily be extended to the enriched language by the following two clauses:[2]

$$\mathsf{Runs}(\pi^*) \ = \ \mathsf{Runs}(\pi)^* \qquad\qquad \mathsf{Runs}(P) \ = \ \mathsf{Runs}(\pi_P).$$

As usual, we define $X^* = \bigcup_{i \geq 0} X^i$, where $X^0 = \{\varepsilon\}$ and $X^{i+1} = X \cdot X^i$ for a set X of sequences. The runs of the program are the runs of *Main*.

4.1   PSPACE-Hardness of Interprocedural Copy Constant Detection

The goal of this section is to prove the following result.

Theorem 2. *The problem of detecting copy constants in non-recursive procedural parallel programs is PSPACE-hard.*

The proof is by means of a reduction of the QBF (quantified Boolean formulas) problem to copy constant detection. QBF (called QSAT in [10]) is a well-known PSPACE-complete problem.

Quantified Boolean Formulas. Let us first recall QBF. A QBF instance is a quantified Boolean formula,

$$\phi \ \equiv \ Q_n x_n : \cdots \forall x_2 : \exists x_1 : c_1 \wedge \ldots \wedge c_k \, ,$$

where Q_n is the quantifier $\exists$ if n is odd and $\forall$ if n is even, i.e. quantifiers are strictly alternating.

---

[2] If the program has recursive procedures, the definition of runs is no longer inductive. Then the clauses are meant to specify the smallest sets obeying the given equations, which exist by the well-known Knaster-Tarski fixpoint theorem. However, only non-recursive programs occur in this paper.

As in SAT, each clause c_i is a disjunction of *literals*, where a literal l is either a variable from $X = \{x_1, \ldots, x_n\}$ or a negated variable from $\bar{X} = \{\neg x_1, \ldots, \neg x_n\}$. The set of indices of clauses made true by literal l is $\mathsf{Cl}(l) \stackrel{\text{def}}{=} \{i \in \{1, \ldots, k\} \mid c_i \text{ contains } l\}$. For later reference the following names are introduced for the sub-formulas of ϕ:

$$\phi_0 \equiv c_1 \wedge \ldots \wedge c_k \qquad \text{and} \qquad \phi_i \equiv Q_i x_i : \phi_{i-1} \quad \text{for } 1 \le i \le n,$$

where again Q_i is $\exists$ if i is odd and $\forall$ if i is even. Clearly, ϕ is just ϕ_n.

Formula ϕ_i is assigned a truth value with respect to a *truth assignment* $T \in \mathsf{TA}_i \stackrel{\text{def}}{=} \{T \mid T : \{x_{i+1}, \ldots, x_n\} \to \mathbb{B}\}$. We write $T[x \mapsto b]$ for the truth assignment that maps x to $b \in \mathbb{B}$ and behaves otherwise like T. We use this notation only if x is not already in the domain of T. For a truth assignment T we denote by $\mathsf{Cl}(T)$ the set of indices of clauses that are made true by T:

$$\mathsf{Cl}(T) \stackrel{\text{def}}{=} \bigcup_{x : T(x)=\mathsf{tt}} \mathsf{Cl}(x) \cup \bigcup_{x : T(x)=\mathsf{ff}} \mathsf{Cl}(\neg x).$$

Note that $\mathsf{Cl}(T[x \mapsto \mathsf{tt}]) = \mathsf{Cl}(T) \cup \mathsf{Cl}(x)$ and $\mathsf{Cl}(T[x \mapsto \mathsf{ff}]) = \mathsf{Cl}(T) \cup \mathsf{Cl}(\neg x)$ (recall that x is not in the domain of T). Note also that TA_n contains only the trivial truth assignment $\emptyset$ for which $\mathsf{Cl}(\emptyset) = \emptyset$.

Using this notation, the truth value of a formula with respect to a truth assignment can be defined as follows:

$$T \models \phi_0 \quad \text{iff} \quad \mathsf{Cl}(T) = \{1, \ldots, k\}$$

$$T \models \phi_i \quad \text{iff} \quad \begin{cases} T[x_i \mapsto \mathsf{tt}] \models \phi_{i-1} \text{ or } T[x_i \mapsto \mathsf{ff}] \models \phi_{i-1}, & \text{if } i \text{ is odd } (Q_i = \exists) \\ T[x_i \mapsto \mathsf{tt}] \models \phi_{i-1} \text{ and } T[x_i \mapsto \mathsf{ff}] \models \phi_{i-1}, & \text{if } i \text{ is even } (Q_i = \forall) \end{cases}$$

The Reduction. From a QBF instance as above, we construct a program, in which we again use $k + 1$ variables $z_0, z_1, \ldots, z_k$ in a similar way as in Sect. 3. Let the programs π_l be defined as in that section.

Let $\mathsf{Proc} = \{Main, P_0, P_1, \ldots, P_n\}$ be the set of procedures. The associated statements are defined as follows:

$$\pi_{Main} \stackrel{\text{def}}{=} z_0 := 1 \; ; \; z_1 := 0 \; ; \; \ldots \; ; \; z_k := 0 \; ; \; P_n \; ; \; (z_0 := 0 \sqcap \mathbf{skip}) \; ; \; \mathbf{write}(z_0)$$

$$\pi_{P_0} \stackrel{\text{def}}{=} z_1 := 0 \; ; \; \ldots \; ; \; z_k := 0 \; ; \; z_0 := z_k$$

$$\pi_{P_i} \stackrel{\text{def}}{=} \begin{cases} (\pi_{x_i}^* \parallel P_{i-1}) \sqcap (\pi_{\neg x_i}^* \parallel P_{i-1}), & \text{if } i \text{ is odd} \\ (\pi_{x_i}^* \parallel P_{i-1}) \; ; \; (\pi_{\neg x_i}^* \parallel P_{i-1}), & \text{if } i \text{ is even} \end{cases} \qquad \text{for } 1 \le i \le n.$$

Clearly, this program can be constructed from the QBF instance in polynomial time or logarithmic space. Note that the introduction of procedures is essential for this to be the case. While we could easily construct an equivalent program without procedures by inlining the procedures, i.e. by successively replacing each call to procedure P_j by its body, for $j = 0, \ldots, n$, the size of the resulting program would in general be exponential in n, as each procedure P_j is called twice in P_{j+1}.

Therefore, we need the procedures to write this program succinctly and to obtain a logspace-reduction.

We show in the following, that the variable z_0 is not a constant at the write-statement in procedure *Main* if and only if the QBF instance is true. This establishes the PSPACE-hardness claim.[3]

Observe again that z_0 can hold only the values 0 and 1 at the write-statement because all variables are initialized by these values and the other assignments only copy them. Clearly, due to the non-deterministic choice just before the write-statement, it can hold 0. Thus, z_0 is a constant at the write-statement iff it cannot hold 1 there. Hence we can rephrase our proof goal as follows:

$$z_0 \text{ can hold the value 1 at the write-statement in } \pi_{Main} \qquad \text{(PG)}$$
$$\text{if and only if } \phi \text{ is true.}$$

In the remainder of this section we separately prove the 'if' and the 'only if' direction.

The "If" Direction. For the 'if' claim, we show that procedure P_n has a run of a special form called a copy chain, if ϕ is true.

Definition 3. *A (total) segment is a sequence of assignment statements of the form* $\langle z_1 := 0, \ldots, z_k := 0, (z_1 := z_0)^{n_1}, \ldots, (z_k := z_{k-1})^{n_k}, z_0 := z_k \rangle$, *where* $n_i \geq 1$ *for* $i = 1, \ldots, n$. *A (total) copy chain is a concatenation of segments.*

Every segment copies the initial value of z_0 back to z_0 via the sub-chain of assignments $z_1 := z_0, z_2 := z_1, \ldots, z_k := z_{k-1}, z_0 := z_k$, where each $z_i := z_{i-1}$ is the last assignment in the block $(z_i := z_{i-1})^{n_i}$. Note that the other statements in a segment do not kill this value; in particular the assignments $\langle z_1 := 0, \ldots, z_k := 0 \rangle$ do not affect z_0. By induction on the number of segments, a total copy chain copies the initial value of z_0 back to z_0 too. Thus, if P_n has a run that is a total copy chain, then z_0 can, at the write-statement in π_{Main}, hold the value 1 by which it was initialized. As a consequence the following lemma implies the 'if'-direction of (PG).

Lemma 4. *If ϕ is true, then P_n has a run that is a total copy chain.*

In order to enable an inductive proof of this lemma we consider *partial copy chains* in which some of the blocks $(z_i := z_{i-1})^{n_i}$ may be missing (i.e. n_i may be zero).

Definition 5. *A partial segment is a sequence of assignment statments of the form* $s = \langle z_1 := 0, \ldots, z_k := 0, (z_1 := z_0)^{n_1}, \ldots, (z_k := z_{k-1})^{n_k}, z_0 := z_k \rangle$, *where now* $n_i \geq 0$ *for* $i = 1, \ldots, n$. *For* $H \subseteq \{1, \ldots, k\}$ *we say that s is a partial segment with holes in H if $H \supseteq \{i \mid n_i = 0\}$. A partial copy chain with holes in H is a concatenation of partial segments with holes in H.*

---

[3] Recall that PSPACE coincides with co-PSPACE because PSPACE is closed under complement.

Intuitively, the holes in a partial copy chain may be filled by programs running in parallel to form a total copy chain. Note that a partial copy chain with holes in $H = \emptyset$ is a total copy chain.

Lemma 6. *For all $i = 0, \dots, n$ and all truth assignments $T \in \mathsf{TA}_i$ the following holds: if $T \models \phi_i$ then P_i has a partial copy chain with holes in $\mathsf{Cl}(T)$.*

Note that Lemma 6 indeed implies Lemma 4: ϕ is true iff the (unique) truth assignment $T \in \mathsf{TA}_n$, viz. $T = \emptyset$, satisfies ϕ_n. By Lemma 6, P_i has then a partial copy chain with holes in $\mathsf{Cl}(\emptyset) = \emptyset$, i.e. a total copy chain.

We show Lemma 6 by induction on i.

Base case ($i = 0$). Suppose given $T \in \mathsf{TA}_0$ with $T \models \phi_0$, i.e. $\mathsf{Cl}(T) = \{1, \dots, k\}$. By definition, P_0 has the run $\langle z_1 := 0, \dots, z_k := 0, z_0 := z_k \rangle$, which may be written as $\langle z_1 := 0, \dots, z_k := 0, (z_1 := z_0)^0, \dots, (z_k := z_{k-1})^0, z_0 := z_k \rangle$, i.e. it is a partial copy chain with holes in $\{1, \dots, k\} = \mathsf{Cl}(T)$.

Induction step ($i \to i+1$). Assume that for a given i, $0 \le i \le k-1$, the claim of Lemma 6 holds for all $T \in \mathsf{TA}_i$ (induction hypothesis). Suppose given $T \in \mathsf{TA}_{i+1}$ with $T \models \phi_{i+1}$.

If $i + 1$ is even, we have, by definition of ϕ_{i+1}, $T \models \forall x_i : \phi_i$, i.e. $T[x_{i+1} \mapsto \mathsf{tt}] \models \phi_i$ and $T[x_{i+1} \mapsto \mathsf{ff}] \models \phi_i$. By the induction hypothesis, there are thus two partial copy chains r_{tt} and r_{ff} with holes in $\mathsf{Cl}(T[x_{i+1} \mapsto \mathsf{tt}]) = \mathsf{Cl}(T) \cup \mathsf{Cl}(x_{i+1})$ and $\mathsf{Cl}(T[x_{i+1} \mapsto \mathsf{ff}]) = \mathsf{Cl}(T) \cup \mathsf{Cl}(\neg x_{i+1})$, respectively.

By interleaving each segment of r_{tt} with a single iteration of $\pi^*_{x_{i+1}}$ appropriately we can fill the holes from $\mathsf{Cl}(x_{i+1})$; this gives us a run r_1 of $\pi^*_{x_{i+1}} \parallel P_i$ that is a partial copy chain with holes in $\mathsf{Cl}(T)$. Similarly, we can fill the holes from $\mathsf{Cl}(\neg x_{i+1})$ in r_{ff} by interleaving each segment with an iteration from $\pi_{\neg x_{i+1}}$; this gives us a run r_2 of $\pi^*_{\neg x_{i+1}} \parallel P_i$ that is a partial copy chain with holes in $\mathsf{Cl}(T)$ too. By concatenating r_1 and r_2 we get a run of P_{i+1} that is a partial copy chain with holes in $\mathsf{Cl}(T)$.

The argumentation for the case that $i + 1$ is odd is similar.

The 'Only If' Direction. As the constant 1 appears only in the initialization to z_0, z_0 can hold the value 1 finally in π_{Main} only if P_n has a run that copies z_0 (perhaps via other variables) back to z_0. We call such a run a *copying run*. Thus, the 'only if' direction of (PG) follows from the following lemma.

Lemma 7. *If P_n has a copying run then ϕ is true.*

Note that, while we could restrict attention to runs of a special form in the 'if'-proof, viz. total and partial copy *chains*, we have to consider arbitrary runs here, as any of them may copy z_0's initial value back to z_0.

In order to enable an inductive proof, we will be concerned with runs that are not (necessarily) yet copying runs but may become so if assignments from a set A are added at appropriate places. Each assignment from A may be added zero, one or many times. The assignment sets A considered are induced by truth assignments T: $A = \mathsf{Asg}(T) \overset{\text{def}}{=} \{z_i := z_{i-1} \mid i \in \mathsf{Cl}(T)\}$. We call such a run a *potentially copying run with holes in* $\mathsf{Asg}(T)$.

Lemma 8. *For all $i = 0, \ldots, n$ and for all $T \in \mathsf{TA}_i$ the following is valid: If there is a potentially copying run of P_i with holes in $\mathsf{Asg}(T)$ then $T \models \phi_i$.*

Note that the case $i = n$ establishes Lemma 7: For the empty truth assignment $\emptyset \in \mathsf{TA}_n$, we have $\mathsf{Asg}(\emptyset) = \emptyset$ and a potentially copying run with holes in $\emptyset$ is just a copying run. Moreover, $\emptyset \models \phi_n$ iff ϕ is true.

We show Lemma 8 by induction on i.

Base case ($i = 0$). Suppose given $T \in \mathsf{TA}_0$. The only run of P_0 is

$$r = \langle z_1 := 0, \ldots, z_k := 0, z_0 := z_k \rangle.$$

If r is a potentially copying run with holes in $\mathsf{Asg}(T)$, assignments from $\mathsf{Asg}(T)$ can be added to r in such a way that the initial value of z_0 influences its final value. As we have only assignments of the form $z_i := z_{i-1}$ available, this can only happen via a sub-chain of assignments of the form $z_1 := z_0, z_2 := z_1, \ldots, z_k := z_{k-1}$, where each assignment $z_i := z_{i-1}$ has to take place after $z_i := 0$ and $z_k := z_{k-1}$ must happen before the final $z_0 := z_k$. Therefore, all assignment $z_1 := z_0, \ldots, z_k := z_{k-1}$ are needed. This means that $\mathsf{Asg}(T)$ must contain all of them, i.e. $\mathsf{Cl}(T)$ must be $\{1, \ldots, k\}$. But then $T \models \phi_0$.

Induction step ($i \rightarrow i + 1$). Suppose given i, $0 \leq i \leq k - 1$, and $T \in \mathsf{TA}_{i+1}$. Assume that there is a potentially copying run r of P_{i+1} with holes in $\mathsf{Asg}(T)$.

If $i + 1$ is odd, r is either a run of $\pi^*_{x_{i+1}} \parallel P_i$ or of $\pi^*_{\neg x_{i+1}} \parallel P_i$. We discuss the case $\pi^*_{x_{i+1}} \parallel P_i$ in detail; the case $\pi^*_{\neg x_{i+1}} \parallel P_i$ is analogous. So let r be an interleaving of a run s of $\pi^*_{x_{i+1}}$ and t of P_i. By definition of $\pi_{x_{i+1}}$, s consists only of assignments from $\mathsf{Asg}(x_{i+1}) \overset{\text{def}}{=} \{z_j := z_{j-1} \mid j \in \mathsf{Cl}(x_{i+1})\}$. As r can be interleaved with the assignments in $\mathsf{Asg}(T)$ to form a copying run, t can be interleaved with assignments from $\mathsf{Asg}(T) \cup \mathsf{Asg}(x_{i+1})$ to form a copying run. Therefore, t is a potentially copying run with holes in $\mathsf{Asg}(T) \cup \mathsf{Asg}(x_{i+1}) = \mathsf{Asg}(T[x_{i+1} \mapsto \mathsf{tt}])$. By the induction hypothesis thus $T[x_{i+1} \mapsto \mathsf{tt}] \models \phi_i$. Consequently, $T \models \exists x_{i+1} : \phi_i$, i.e. $T \models \phi_{i+1}$.

If $i+1$ is even, there are runs s and t of $\pi^*_{x_{i+1}} \parallel P_i$ and $\pi^*_{\neg x_{i+1}} \parallel P_i$ respectively, such that $r = s \cdot t$. It suffices to show that s and t are potentially copying runs with holes in $\mathsf{Asg}(T)$. An argumentation like in the case '$i + 1$ odd' then yields that $T[x_{i+1} \mapsto \mathsf{tt}] \models \phi_i$ and $T[x_{i+1} \mapsto \mathsf{ff}] \models \phi_i$ and thus $T \models \forall x_{i+1} : \phi_i$, i.e. $T \models \phi_{i+1}$.

As $r = s \cdot t$ is a potentially copying run with holes in $\mathsf{Asg}(T)$ it may be interleaved with assignments from $\mathsf{Asg}(T)$ to form a copying run r'. Clearly, we can interleave its two parts s and t separately by assignments from $\mathsf{Asg}(T)$ to sequences s' and t' such that $r' = s' \cdot t'$. It is, however, not obvious that s' and t' really copy from z_0 to z_0 – if they do so, we are done because then s and t

are potentially copying runs with holes in $\mathsf{Asg}(T)$. Of course, there must be a variable z_j such that the value of z_0 is copied by s' to z_j and the value of z_j is copied by t' to z_0; otherwise z_0 cannot be copied to z_0 by r'. But, at first glance, z_j may be different from z_0. It follows from the below lemma, that z_j indeed must be z_0, which completes the proof of Lemma 8.

Lemma 9. *Let r be some interleaving of a run of P_i, $i = 0, \ldots, n$, with assignments of the form $z_l := z_{l-1}$, $l = 1, \ldots, k$. Then r copies none of the variables $z_1, \ldots, z_k$ to some variable.*

This last lemma is proved by induction on i. The interesting argument is in the base case; the induction step is almost trivial.

Base case. Let $i = 0$ and assume given a variable z_j, $j \in \{1, \ldots, k\}$. Then r is an interleaving of $\langle z_1 := 0, \ldots, z_k := 0, z_0 := z_k \rangle$ with assignments of the form $z_l := z_{l-1}$. Assignments of this form can copy only to variables with a higher index. Thus, just before the assignment $z_j := 0$ at most the variables $z_j, z_{j+1}, \ldots, z_k$ can contain the value copied from z_j. The contents of z_j is overwritten by the assignment $z_j := 0$. So immediately after $z_j := 0$ at most $z_{j+1}, \ldots, z_k$ can contain the value copied from z_j. This also holds just before the assignment z_{j+1} which overwrites z_{j+1}; and so on. Just after $z_k := 0$, no variable can still contain the value copied from z_j.

Induction step. Let $i > 0$ and assume that the claim is valid for $i - 1$. Any run of P_i either starts with (if i is even) or is (if i is odd) an interleaving of a run of P_{i-1} with assignments of the described form. Therefore, r starts with or is an interleaving of a run of P_{i-1} with such assignments. The property follows thus immediately from the induction hypothesis.

5 Conclusion

In this paper we have presented two complexity results with detailed proofs. They indicate that the accounts of [7,6,2,3,13] on efficient and complete data-flow analysis of parallel programs cannot be generalized significantly beyond gen/kill problems.

The reductions in this paper apply without change also to the *may-constant* detection problem in parallel programs. In the may-constant problem [9] we ask whether a given variable x can hold a given value k at a certain program point p or not, i.e. whether there is a run from the start of the program to p after which x holds k. In the NP-hardness proof in Sect. 3 we showed that z_k may hold the value 1 at the write-statement iff the given SAT instance is satisfiable and, similarly, in Sect 4 that z_0 may hold 1 at the write-statement iff the given QBF instance is true. This proves that the may constant problem is NP-complete for loop-free parallel programs and PSPACE-hard for programs with procedures and loops. Also the complexity of another data-flow problem, that of detecting faint variables [4] which is related to program slicing [16,15], can be attacked with essentially the same reductions.

For the interprocedural parallel problem the current paper only establishes a lower bound, viz. PSPACE-hardness. It is left for future work to study the precise complexity of this problem. Another interesting question is the complexity of the general intraprocedural problem for parallel programs where we have loops but no procedures.

Acknowledgments. I thank Helmut Seidl for a number of discussions that stimulated the research reported here and Jens Knoop, Oliver Rüthing, and Bernhard Steffen for sharing their knowledge of data-flow analysis with me. I am also grateful to the anonymous referees for their valuable comments.

References

1. S. A. Cook. The complexity of theorem-proving procedures. In *ACM STOC'71*, pages 151–158, 1971.
2. J. Esparza and J. Knoop. An automata-theoretic approach to interprocedural data-flow analysis. In *FOSSACS '99*, LNCS 1578, pages 14–30. Springer, 1999.
3. J. Esparza and A. Podelski. Efficient algorithms for pre* and post* on interprocedural parallel flow graphs. In *ACM POPL'2000*, pages 1–11, 2000.
4. R. Giegerich, U. Möncke, and R. Wilhelm. Invariance of approximative semantics with respect to program transformations. In *GI 11. Jahrestagung*, Informatik Fachberichte 50, pages 1–10. Springer, 1981.
5. G. A. Kildall. A unified approach to global program optimization. In *ACM POPL'73*, pages 194–206, 1973.
6. J. Knoop. Parallel constant propagation. In *Euro-Par'98*, LNCS 1470, pages 445–455. Springer, 1998.
7. J. Knoop, B. Steffen, and J. Vollmer. Parallelism for free: Efficient and optimal bitvector analyses for parallel programs. *ACM Transactions on Programming Languages and Systems*, 18(3):268–299, 1996.
8. S. S. Muchnick. *Advanced compiler design and implementation*. Morgan Kaufmann Publishers, San Francisco, California, 1997.
9. R. Muth and S. Debray. On the complexity of flow-sensitive dataflow analysis. In *ACM POPL'2000*, pages 67–81, 2000.
10. C. H. Papadimitriou. *Computational Complexity*. Addison-Wesley, 1994.
11. G. Ramalingam. Context-sensitive synchronization-sensitive analysis is undecidable. Technical Report RC 21493, IBM T. J. Watson Research Center, 1999. To appear in TOPLAS.
12. J. R. Reif and H. R. Lewis. Symbolic evaluation and the global value graph. In *ACM POPL'77*, pages 104–118, 1977.
13. H. Seidl and B. Steffen. Constraint-based interprocedural analysis of parallel programs. In *ESOP'2000*, LNCS 1782, pages 351–365. Springer, 2000.
14. R. N. Taylor. Complexity of analyzing the synchronization structure of concurrent programs. *Acta Informatica*, 19:57–84, 1983.
15. F. Tip. A survey of program slicing techniques. *Journal of Programming Languages*, 3:121–181, 1995.
16. M. Weiser. Program slicing. *IEEE Transactions on Software Engineering*, SE-10(4):352–357, 1984.

Approximation Algorithms for the Bottleneck Stretch Factor Problem

Giri Narasimhan[1] and Michiel Smid[2]

[1] Department of Mathematical Sciences, The University of Memphis, Memphis TN 38152
giri@msci.memphis.edu
[2] Department of Computer Science, University of Magdeburg, Magdeburg, Germany
michiel@isg.cs.uni-magdeburg.de

1   Introduction

Assume that we are given the coordinates of n airports. Given an airplane that can fly a distance of b miles without refueling, a typical query is to determine the smallest value of t such that the airplane can travel between any pair of airports using flight segments of length at most b miles, such that the sum of the lengths of the flight segments is not longer than t times the direct "as-the-crow-flies" distance between the airports. This problem falls under the general category of *bottleneck problems*. In our case, the *stretch factor*, i.e., the value of t, is a measure of the maximum increase in fuel costs caused by choosing a path other than the direct path between any source and any destination. (Clearly, this direct path cannot be taken if its length is larger than b miles.)

Let us formalize this problem. For simplicity, we take the Euclidean metric for the distance between two airports. In practice, one needs to take into account the curvature of the earth and the wind conditions.

Let $d \geq 2$ be a small constant. For any two points p and q in $\mathbb{R}^d$, we denote their Euclidean distance by $|pq|$. Let S be a set of n points in $\mathbb{R}^d$, and let G be an undirected graph having S as its vertex set. The length of any edge (p, q) of G is defined as $|pq|$. Furthermore, the length of any path in G between two vertices p and q is defined as the sum of the lengths of the edges on this path. We call such a graph G a *Euclidean graph*. For any two vertices p and q of G, we denote by $|pq|_G$ their distance in G, i.e., the length of a shortest path connecting p and q. If there is no path between p and q, then $|pq|_G = \infty$. The *stretch factor* t^* of G is defined as

$$t^* := \max\left\{ \frac{|pq|_G}{|pq|} : p \in S, q \in S, p \neq q \right\}.$$

Note that $t^* = \infty$, if the graph G is not connected.

The *bottleneck stretch factor problem* is to preprocess the points of S into a data structure, such that for any real number $b > 0$, we can efficiently compute the stretch factor of the subgraph of the complete graph on S containing all edges of length at most b.

Let $G = (S, E)$ denote the Euclidean graph on S containing all edges having length at most b. The time complexity of solving the *all-pairs-shortest-path* problem for G is an upper bound on the time complexity of computing the stretch factor of G. Hence, running Dijkstra's algorithm—implemented with Fibonacci heaps—from each vertex

A. Ferreira and H. Reichel (Eds.): STACS 2001, LNCS 2010, pp. 502–513, 2001.

of G, gives the stretch factor of G, in $O(n^2 \log n + n|E|)$ time (c.f., [9]). Note that $|E|$ can be as large as $\binom{n}{2}$. Hence, without any preprocessing, we can answer queries in $O(n^3)$ time. It may be possible to improve the query time, but we are not aware of any algorithm that computes the stretch factor in subquadratic time. (For example, we do not even know if the stretch factor of a Euclidean path can be computed in $o(n^2)$ time.)

A second solution for the bottleneck stretch factor problem is obtained from the observation that there are only $\binom{n}{2}$ "different" query values b. Hence, if we store all $\binom{n}{2}$ different stretch factors, then a query can be solved in $O(\log n)$ time by searching with the query value b in the sorted sequence of all $\binom{n}{2}$ Euclidean distances between the pairs of points of S. Clearly, in this case, the preprocessing time and the amount of space used are at least quadratic in n.

This leads to the question whether more efficient solutions exist, if we are satisfied with an approximation to the stretch factor of the graph G.

Let $c_1 \geq 1$ and $c_2 \geq 1$ be real numbers, let G be an arbitrary Euclidean graph on the point set S, and let t^* be the stretch factor of G. We say that the real number t is a (c_1, c_2)-*approximate stretch factor* of G, if $t/c_1 \leq t^* \leq c_2 t$. The current paper considers the following problem:

Problem 1. The (c_1, c_2)-*approximate bottleneck stretch factor problem* is to preprocess the points of S into a data structure, such that for any real number $b > 0$, we can efficiently compute a (c_1, c_2)-approximate stretch factor of the subgraph of the complete graph on S containing all edges of length at most b.

1.1 Our Results

In this paper, we will present a data structure that solves Problem 1. The general approach, which is given in Section 3, is as follows. We partition the sequence of $\binom{n}{2}$ exact stretch factors into $O(\log n)$ subsequences, such that any two stretch factors in the same subsequence are approximately equal. Our data structure contains a sequence of $O(\log n)$ stretch factors, one from each subsequence. We also store a corresponding sequence of $O(\log n)$ distances between pairs of points. The latter sequence is used to search in $O(\log \log n)$ time in the sequence of $O(\log n)$ stretch factors. The result is a data structure of size $O(\log n)$ that can be used to solve the queries of Problem 1 in $O(\log \log n)$ time. The time to build this data structure, however, is at least quadratic in n.

In Section 4, we show that it suffices to use a sequence of $O(\log n)$ approximate stretch factors instead of the sequence of $O(\log n)$ exact stretch factors. Since the graphs whose stretch factors we have to approximate may have a quadratic number of edges, however, we need to make one more approximation step. That is, in Section 5, we use Callahan and Kosaraju's well-separated pair decomposition [7] to approximate the graph G containing all edges of length at most b by a graph H having $O(n \log n)$ edges and having approximately the same stretch factor. Then we use the algorithm of Narasimhan and Smid [13] to compute an approximate stretch factor of the graph H. In this way, we obtain the main result of this paper: a data structure of size $O(\log n)$, query time $O(\log \log n)$, and that can be built in subquadratic time.

1.2 Related Work

There has been substantial work on the problem of constructing a Euclidean graph on a given set of points whose stretch factor is bounded by a given constant $t > 1$. A good overview of results in this direction can be found in the surveys by Eppstein [11] and Smid [15].

The problem of approximating the stretch factor of any given Euclidean graph has been considered by the authors in [13]. There, we prove the following result, which will be used in the current paper.

Theorem 1 ([13]). *Let S be a set of n points in $\mathbb{R}^d$, let $G = (S, E)$ be an arbitrary connected Euclidean graph, let $\beta \geq 1$ be an integer constant, and let ϵ be a real constant, such that $0 < \epsilon \leq 1/2$. In $O(|E|n^{1/\beta} \log^2 n)$ expected time, we can compute a $(2\beta(1 + \epsilon), 1 + \epsilon)$-approximate stretch factor of G.*

The proof of this theorem uses the well-separated pair decomposition (WSPD) of Callahan and Kosaraju [7]. We use this WSPD in Section 5 to approximate the graph containing all edges of length at most b by a graph having $O(n \log n)$ edges and having approximately the same stretch factor. For other applications of the WSPD, see [2,5,6,7].

To the best of our knowledge, the exact and approximate bottleneck stretch factor problems have not been considered before.

2   Some Preliminary Results

We start by introducing some notation and terminology. Let S be a set of n points in $\mathbb{R}^d$, and let m be the number of distinct distances defined by any two distinct points of S. Let $\delta_1 < \delta_2 < \ldots < \delta_m$ be the sorted sequence of these distances. Note that $m \leq \binom{n}{2}$.

Let G_0 be the graph on S having no edges. Furthermore, for any i, $1 \leq i \leq m$, let G_i be the i-th *bottleneck graph*, i.e., the subgraph of the complete graph on S containing all edges of length at most δ_i. Clearly, for any i, $0 \leq i < m$, G_i is a subgraph of G_{i+1}, and G_m is the complete graph on S. For any i, $0 \leq i \leq m$, we denote by t_i^* the (exact) stretch factor of the graph G_i. The sequence $\mathcal{T} = \langle t_0^*, t_1^*, t_2^*, \ldots, t_m^* \rangle$ will be referred to as the *stretch factor spectrum* of S.

It is clear that determining the stretch factor spectrum of S solves the exact version of the bottleneck stretch factor problem. However, this involves determining the stretch factor of $\Theta(n^2)$ distinct graphs, which is likely to be prohibitively expensive.

First, we observe that $t_0^* = \infty$, $t_m^* = 1$, and $t_{i+1}^* \leq t_i^*$ for all i, $0 \leq i < m$. Also, the graph G_0 is not connected, whereas the graph G_m is connected. Let k be the smallest index such that the graph G_k is connected. Then $t_0^* = t_1^* = \ldots = t_{k-1}^* = \infty$, t_k^* is finite, and $1 = t_m^* \leq t_{m-1}^* \leq \ldots \leq t_{k+1}^* \leq t_k^*$. We will henceforth refer to the distance δ_k (corresponding to index k) as the *connectivity threshold*.

The following lemma characterizes the connectivity threshold. It is a restatement of the well-known folklore theorem that states that the minimum spanning tree is also a bottleneck minimum spanning tree.

Lemma 1. *Let T be a minimum spanning tree of S. Then the longest edge in T has length δ_k.*

Using Lemma 1, we can prove an upper bound on the stretch factor t_k^* of the bottle-neck graph G_k. The bound is useful because it suggests that binary search on the stretch factor spectrum can be performed efficiently.

Lemma 2. *We have $t_k^* \leq n - 1$.*

3   A First Solution

We start by describing the general idea of our solution to the approximate bottleneck stretch factor problem. Let $c > 1$ be an arbitrary constant. For the preprocessing phase, we partition the index set $\{k, k + 1, \ldots, m\}$ into $O(\log n)$ subsets of consecutive integers, such that for any two indices i and i' of the same subset, the stretch factors t_i^* and $t_{i'}^*$ are within a factor of c of each other. This partition induces partitions of the two sequences δ_i, $k \leq i \leq m$, and t_i^*, $k \leq i \leq m$, into $O(\log n)$ subsequences. For each j, we let a_j denote the smallest index in the j-th subset of the partition of $\{k, k+1, \ldots, m\}$.

Our data structure consists of the $O(\log n)$ values δ_{a_j} and $t_{a_j}^*$. For the query phase, given a value $b > 0$, we search for the largest index j, such that $\delta_{a_j} \leq b$, and report the value of $t_{a_j}^*$. We will prove later that $t_{a_j}^*$ approximates the stretch factor of the subgraph of the complete graph on S containing all edges of length at most b. In the rest of this section, we will formalize this approach.

As mentioned above, we fix a constant $c > 1$. For any integer $j \geq 0$, we define

$$X_j := \{i : k \leq i \leq m \text{ and } c^j \leq t_i^* < c^{j+1}\}.$$

Since all stretch factors t_i^* are greater than or equal to one, these sets X_j partition the set $\{k, k + 1, \ldots, m\}$. Also, if $X_j \neq \emptyset$, then there is an index i such that $c^j \leq t_i^*$. Since $t_i^* \leq t_k^*$ and, by Lemma 2, $t_k^* \leq n - 1$, we have $c^j \leq n - 1$, which implies that $j \leq \lfloor \log_c(n - 1) \rfloor$.

Let ℓ be the number of non-empty sets X_j. Then $\ell \leq 1 + \lfloor \log_c(n - 1) \rfloor$. Each non-empty set X_j is a set of consecutive integers. We denote these non-empty sets by $I_1, I_2, \ldots, I_\ell$, and write them as $I_j = \{a_j, a_j + 1, \ldots, a_{j+1} - 1\}$, $1 \leq j \leq \ell$, where $k = a_1 < a_2 < \ldots < a_{\ell+1} = m + 1$.

Lemma 3. *Let j be any integer such that $1 \leq j \leq \ell$, and let i and i' be any two elements of the set I_j. Then $1/c < t_i^*/t_{i'}^* < c$.*

Now we are ready to give the data structure for solving the approximate bottleneck stretch factor problem. This data structure consists of the connectivity threshold δ_k, and two arrays $\Delta[1 \ldots \ell]$ and $SF[1 \ldots \ell]$, where $\Delta[j] = \delta_{a_j}$ and $SF[j] = t_{a_j}^*$. Note that the array Δ is sorted in increasing order, whereas the array SF is sorted in non-increasing order.

Recall that in a query, we get a real number $b > 0$, and have to compute an approx-imate stretch factor t of the graph containing all edges having length at most b. Such a query is answered as follows. If $b < \delta_k$, then the subgraph of the complete graph on S containing all edges of length at most b is not connected. Hence, we report $t := \infty$. If $b \geq \delta_k$, then we search in Δ for the largest index j such that $\Delta[j] \leq b$, and report the value of t defined as $t := SF[j]$.

Lemma 4. *Assume that $b \geq \delta_k$. Let t^* be the exact stretch factor of the subgraph of the complete graph on S containing all edges of length at most b. The value of t reported by the query algorithm satisfies $t/c < t^* < ct$.*

Proof. Consider the index j that was found by the query algorithm. Hence, $t = SF[j] = t^*_{a_j}$. Note that $a_j \in I_j$. Let i be the largest index such that $\delta_i \leq b$. Then $t^* = t^*_i$, and i is also an element of I_j. The claim now follows from Lemma 3.   □

Let us analyze the complexity of our solution. We need $O(\ell) = O(\log n)$ space to store the data structure. If we implement the query algorithm using binary search, then the query time is bounded by $O(\log \ell) = O(\log \log n)$.

It remains to describe and analyze the preprocessing algorithm. First, we compute the sorted sequence of $m \leq \binom{n}{2}$ distances. This takes $O(n^2 \log n)$ time. Then we compute a minimum spanning tree of S. The length of a longest edge in this tree gives us the distance δ_k, and its index k. (See Lemma 1.) This step also takes $O(n^2 \log n)$ time. (Note that a minimum spanning tree of a set of n points in $\mathbb{R}^d$ can be computed faster. The $O(n^2 \log n)$–time bound, however, is good enough for the moment.) Now consider the sequence

$$1 = t^*_m \leq t^*_{m-1} \leq \cdots \leq t^*_{k+1} \leq t^*_k \leq n - 1 \tag{1}$$

of stretch factors. The index sets $I_1, I_2, \ldots, I_\ell$ are obtained by locating the real numbers c^j, $0 \leq j \leq \lfloor \log_c(n - 1) \rfloor$, in the sequence (1). Let $T_{SF}(n)$ denote the worst-case time to compute the exact stretch factor of any Euclidean graph on n points. Then, using binary search, we locate c^j in the sequence (1) in time $O(T_{SF}(n) \log(m - k + 1)) = O(T_{SF}(n) \log n)$. Hence, we can compute all index sets I_j, $1 \leq j \leq \ell$, in $O(T_{SF}(n) \log^2 n)$ total time. Given these index sets, we can compute the two arrays Δ and SF, in $O(T_{SF}(n) \log n)$ time. If we write the constant c as $1 + \epsilon$, then we have proved the following result.

Theorem 2. *Let S be a set of n points in $\mathbb{R}^d$, and let $\epsilon > 0$ be a constant. For the $(1 + \epsilon, 1 + \epsilon)$-approximate bottleneck stretch factor problem, there is a data structure that can be built in $O\left(n^2 \log n + T_{SF}(n) \log^2 n\right)$ time, that has size $O(\log n)$, and whose query time is bounded by $O(\log \log n)$.*

As mentioned in Section 1, the time complexity for computing the stretch factor of an arbitrary Euclidean graph is bounded by $O(n^3)$. Even though it may be possible to improve this upper bound, it is probably very hard to get a subquadratic time bound. Therefore, in the next section, we show that the preprocessing time can be reduced, at the cost of an increase in the approximation factor. The main idea is to store *approximate* stretch factors in the array SF.

4   An Improved Solution

Here we exploit the fact that approximate stretch factors can be computed more efficiently than exact stretch factors. In the previous section, we fixed a constant $c > 1$, and partitioned the sequence (1) of exact stretch factors into $O(\log n)$ subsets, such that any two stretch factors in the same subset are within a factor of c of each other. We obtained this partition, by locating the values c^j, $0 \leq j \leq \lfloor \log_c(n - 1) \rfloor$, in the sorted sequence

(1). In this section, we fix two additional constants c_1 and c_2 that are both greater than or equal to one. For any i, $k \leq i \leq m$, let t_i be a (c_1, c_2)-approximate stretch factor of the bottleneck graph G_i. Hence, we have $t_i/c_1 \leq t_i^* \leq c_2 t_i$. We will show how to use the sequence $t_m, t_{m-1}, \ldots, t_k$ of approximate stretch factors to partition the index set $\{k, k+1, \ldots, m\}$ into $O(\log n)$ subsets, such that for any two indices i and i' within the same subset, the exact stretch factors t_i^* and $t_{i'}^*$ are approximately equal. (The approximation factor depends on c, c_1, and c_2.) This partition is obtained by locating the values c^j in the sequence $t_m, t_{m-1}, \ldots, t_k$. Here, we have to be careful, because the values t_i are not sorted. They are, however, "approximately" sorted, and we will see that this suffices for our purpose.

Let $x > 0$ be a real number. We want to use binary search to "approximately" locate x in the "approximately" sorted sequence $t_m, t_{m-1}, \ldots, t_k$. We specify this algorithm by its decision tree[1]. This tree is a balanced binary tree that enables us to search in a sequence of numbers that have indices $k, k+1, \ldots, m$. More precisely, the leaves of the tree store the indices $k, k+1, \ldots, m$, in this order, from left to right, and each internal node u of the tree stores the smallest index that is contained in the right subtree of u. Given the real number $x > 0$, we search as follows:

Algorithm *search*(x)
    $u :=$ root of the decision tree;
    while $u \neq$ leaf **do**
        $j :=$ index stored in u;
        if $x \leq t_j$ **then** $u :=$ right child of u **else** $u :=$ left child of u **endif**
    endwhile;
    return the index stored in u

Lemma 5. *Let $x > 0$ be a real number, and let z be the index that is returned by algorithm* search(x). *For each i, $k \leq i < z$, we have $t_i^* \geq x/c_1$, whereas for each i, $z < i \leq m$, we have $t_i^* < c_2 x$.*

Hence, running algorithm *search*(x) implicitly partitions the sequence $t_k^*, t_{k+1}^*, \ldots, t_m^*$ of exact stretch factors into the following three subsequences: (i) $t_k^*, t_{k+1}^*, \ldots, t_{z-1}^*$; these are all greater than or equal to x/c_1, (ii) t_z^*, and (iii) $t_{z+1}^*, t_{z+2}^*, \ldots, t_m^*$; these are all less than $c_2 x$.

We are now ready to give the algorithm that partitions the sequence $t_k^*, t_{k+1}^*, \ldots, t_m^*$ of exact stretch factors into $O(\log n)$ subsets, such that any two stretch factors in the same subset are approximately equal. First, we run algorithm *search*(c). Let z be the index returned. Then we report the two sets $\{z\}$ and $\{z+1, z+2, \ldots, m\}$ of indices. Next, we run algorithm *search*(c^2) on the index set $\{k, k+1, \ldots, z-1\}$. This results in a partition of the latter set into three subsets. The "last" two subsets are reported, whereas the "first" subset is partitioned further by running algorithm *search*(c^3). After $O(\log n)$ iterations, we obtain the partition we are looking for.

Let ℓ be the number of non-empty index sets that are computed by this algorithm. As in Section 3, we denote these by $I_1, I_2, \ldots, I_\ell$, and write them as $I_j = \{a_j, a_j + $

---

[1] Note that this decision tree is not constructed (its size is quadratic in n), it is just a convenient way to describe the algorithm. The decision tree represents all possible computations of the algorithm on any input x.

$1, \ldots, a_{j+1} - 1\}, 1 \leq j \leq \ell$, where $k = a_1 < a_2 < \ldots < a_{\ell+1} = m + 1$. It is easy to see that $\ell = O(\log n)$.

Lemma 6. *Let y be any integer such that $1 \leq y \leq \ell$, and let i and i' be any two elements of the set I_y. Then $1/(cc_1 c_2) < t_i^*/t_{i'}^* < cc_1 c_2$.*

The data structure for solving the approximate bottleneck stretch factor problem consists of the connectivity threshold δ_k, and two arrays $\Delta[1 \ldots \ell]$ and $SF_{approx}[1 \ldots \ell]$, where $\Delta[j] = \delta_{a_j}$ and $SF_{approx}[j] = t_{a_j}$.

The query algorithm is basically the same as before. Given any real number $b > 0$, we do the following. If $b < \delta_k$, then the subgraph of the complete graph on S containing all edges of length at most b is not connected. Hence, we report $t := \infty$. If $b \geq \delta_k$, then we search in Δ for the largest index j such that $\Delta[j] \leq b$, and report the value of t defined as $t := SF_{approx}[j]$.

Lemma 7. *Assume that $b \geq \delta_k$. Let t^* be the exact stretch factor of the subgraph of the complete graph on S containing all edges of length at most b. The value of t reported by the query algorithm satisfies $t/(cc_1^2 c_2) < t^* < cc_1 c_2^2 t$.*

Proof. Let j be the largest index such that $\Delta[j] \leq b$. Then $t = SF_{approx}[j] = t_{a_j}$. Let i be the largest index such that $\delta_i \leq b$. Then $t^* = t_i^*$. Since i and a_j both belong to the index set I_j, Lemma 6 implies that $1/(cc_1 c_2) < t^*/t_{a_j}^* < cc_1 c_2$. The lemma now follows from the fact that $1/c_1 \leq t_{a_j}^*/t_{a_j} \leq c_2$. $\qquad\square$

It is clear that the data structure has size $O(\log n)$, and that the query time is bounded by $O(\log \log n)$. In the rest of this section, we analyze the time that is needed to construct the data structure. We will use the following notation.

- $T_{MST}(n)$: the time needed to compute a minimum spanning tree of a set of n points in $\mathbb{R}^d$.
- $T_{rank}(n)$: the time needed to compute the rank of any positive real number δ in the set of distances in a set of n points in $\mathbb{R}^d$. (The *rank* of δ is the number of distances that are less than or equal to δ.)
- $T_{approxSF}(n)$: the time needed to compute a (c_1, c_2)-approximate stretch factor of any bottleneck graph on a set of n points in $\mathbb{R}^d$.
- $T_{sel}(n)$: the time needed to compute the i-th smallest distance in a set of n points in $\mathbb{R}^d$, for any i, $1 \leq i \leq \binom{n}{2}$.

The preprocessing algorithm starts by computing a minimum spanning tree of the point set S. Let δ be the length of a longest edge in this tree. Note that the rank of δ is equal to k. Hence, we can compute the distance $\delta_k = \delta$, and the corresponding index k, in $O(T_{MST}(n) + T_{rank}(n))$ time. Given k and δ_k, we can compute the partition of $\{k, k+1, \ldots, m\}$ into non-empty index sets I_j, in $O(T_{approxSF}(n) \log^2 n)$ time. Given this partition, we can compute the array $SF_{approx}[1 \ldots \ell]$ in $O(T_{approxSF}(n) \log n)$ time. To compute the array $\Delta[1 \ldots \ell]$, we have to solve $O(\log n)$ selection queries of the form "given an index j, compute the a_j-th smallest distance δ_{a_j} in the point set S". One such query takes $T_{sel}(n)$ time. Hence, we can compute the entire array Δ in $O(T_{sel}(n) \log n)$ time.

We observe that $T_{rank}(n) = O(T_{sel}(n) \log n)$: We can compute the rank of a positive real number δ, by performing a binary search in the index set $\{1, 2, \ldots, \binom{n}{2}\}$. During this search, comparisons are resolved in $T_{sel}(n)$ time.

If we write the constant c as $1 + \epsilon$, then we obtain the following result.

Theorem 3. *Let S be a set of n points in $\mathbb{R}^d$, and let $\epsilon > 0$, $c_1 > 1$, and $c_2 > 1$ be constants. For the $((1+\epsilon)c_1^2 c_2, (1+\epsilon)c_1 c_2^2)$-approximate bottleneck stretch factor problem, there is a data structure that can be built in $O(T_{MST}(n) + T_{approxSF}(n) \log^2 n + T_{sel}(n) \log n)$ time, that has size $O(\log n)$, and whose query time is $O(\log \log n)$.*

5   A Fast Implementation of the Improved Algorithm

In order to apply Theorem 3, we need good upper bounds on the functions $T_{MST}(n)$, $T_{sel}(n)$, and $T_{approxSF}(n)$. For the first two functions, subquadratic bounds are known. Theorem 1 implies an upper bound on $T_{approxSF}(n)$: We run the algorithm of [13] on the bottleneck graph. Since such a graph can have a quadratic number of edges, however, this gives a bound that is at least quadratic in n. In Section 5.1, we will show that the bottleneck graph G_i can be approximated by a graph H_i having fewer edges. That is, H_i has $O(n \log n)$ edges, and its stretch factor is approximately equal to that of G_i. This will allow us to approximate the stretch factor of G_i in subquadratic time.

The computation of the graph H_i is based on the *well-separated pair decomposition*, devised by Callahan and Kosaraju [7]. We briefly review well-separated pairs and some of their relevant properties.

Definition 1. *Let $s > 0$ be a real number, and let A and B be two finite sets of points in $\mathbb{R}^d$. We say that A and B are* well-separated *w.r.t. s, if there are two disjoint d-dimensional balls C_A and C_B, having the same radius, such that (i) C_A contains all points of A, (ii) C_B contains all points of B, and (iii) the distance between C_A and C_B is at least equal to s times the radius of C_A.*

We will assume that s is a constant, called the *separation constant*.

Lemma 8. *Let A and B be two finite sets of points that are well-separated w.r.t. s, let x and p be points of A, and let y and q be points of B. Then (i) $|xy| \le (1 + 4/s) \cdot |pq|$, and (ii) $|px| \le (2/s) \cdot |pq|$.*

Definition 2 ([7]). *Let S be a set of n points in $\mathbb{R}^d$, and $s > 0$ a real number. A* well-separated pair decomposition *(WSPD) for S (w.r.t. s) is a sequence of pairs of non-empty subsets of S, $\{A_1, B_1\}, \{A_2, B_2\}, \ldots, \{A_\ell, B_\ell\}$, such that*

1. $A_i \cap B_i = \emptyset$, *for all* $i = 1, 2, \ldots, \ell$,
2. *for any two distinct points p and q of S, there is exactly one pair $\{A_i, B_i\}$ in the sequence, such that (i) $p \in A_i$ and $q \in B_i$, or (ii) $p \in B_i$ and $q \in A_i$,*
3. A_i *and* B_i *are well-separated w.r.t. s, for all* $i = 1, 2, \ldots, \ell$.

The integer ℓ is called the size *of the WSPD.*

In [5], Callahan shows that a WSPD of size $\ell = O(n \log n)$ can be computed, such that each pair $\{A_i, B_i\}$ contains at least one singleton set. This WSPD is computed using a binary tree T, called the *split tree*. We briefly describe the main idea. The split tree is similar to a kd-tree. Callahan starts by computing the bounding box of the points of S, which is successively split by d-dimensional hyperplanes, each of which is orthogonal to one of the axes. If a box is split, he takes care that each of the two resulting boxes contains at least one point of S. As soon as a box contains exactly one point, the process stops (for this box).

The resulting binary tree T stores the points of S at its leaves; one leaf per point. Also, each node u of T is associated with a subset of S. We denote this subset by S_u; it is the set of all points of S that are stored in the subtree of u.

The split tree T can be computed in $O(n \log n)$ time. Callahan shows that, given T, a WSPD of size $\ell = O(n \log n)$ can be computed in $O(n \log n)$ time. Each pair $\{A_i, B_i\}$ in this WSPD is represented by two nodes u_i and v_i of T, i.e., we have $A_i = S_{u_i}$ and $B_i = S_{v_i}$. Since at least one of A_i and B_i is a singleton set, at least one of u_i and v_i is a leaf of T.

Theorem 4 ([5]). *Let S be a set of n points in $\mathbb{R}^d$, and $s > 0$ a separation constant. In $O(n \log n)$ time, we can compute a WSPD for S of size $O(n \log n)$ such that each pair $\{A_i, B_i\}$ contains at least one singleton set.*

5.1 Approximating the Bottleneck Graph

Let $b > 0$ be a fixed real number, and let G be the Euclidean graph on the point set S containing all edges of length at most b. In this section, we show that we can use well-separated pairs to define a graph H whose stretch factor approximates that of G. In Section 5.2, we will give an algorithm that computes such a graph H having only $O(n \log n)$ edges.

Let $s > 4$ be a separation constant, and consider an arbitrary well-separated pair decomposition $\{A_1, B_1\}, \{A_2, B_2\}, \ldots, \{A_\ell, B_\ell\}$ for the point set S. For any index i, $1 \le i \le \ell$, let $x_i \in A_i$ and $y_i \in B_i$ be two points for which $|x_i y_i|$ is minimum.

The graph H has the points of S as its vertices, and contains all edges (x_i, y_i) whose length is less than or equal to b.

Lemma 9. *Let p and q be any two points of S such that $|pq| \le b$. Then $|pq|_H \le (s+4)/(s-4) \cdot |pq|$.*

Proof. The proof is basically the same as Callahan and Kosaraju's proof in [6] that the WSPD yields a spanner for S.   □

Lemma 10. *Let t_G^* and t_H^* denote the exact stretch factors of the graphs G and H, respectively. We have $(s-4)/(s+4) \cdot t_H^* \le t_G^* \le t_H^*$.*

5.2 Computing the Approximation Graph H

We saw in the previous subsection that the graph H approximates the bottleneck graph G. In this section, we show how this graph H can be computed if we use an appropriate WSPD. Consider a WSPD $\{A_1, B_1\}, \{A_2, B_2\}, \ldots, \{A_\ell, B_\ell\}$ in which each pair

$\{A_i, B_i\}$ contains at least one singleton set. By Theorem 4, such a WSPD of size $\ell = O(n \log n)$ can be computed in $O(n \log n)$ time.

The main problem is that we have to compute for each pair $\{A_i, B_i\}$ in this WSPD, the points $x_i \in A_i$ and $y_i \in B_i$ for which $|x_i y_i|$ is minimum. Hence, if A_i is a singleton set, i.e., $A_i = \{x_i\}$, then we have to compute a nearest-neighbor y_i of x_i in the set B_i. We will show that by traversing the split tree T that gives rise to this WSPD, all these pairs (x_i, y_i), $1 \le i \le \ell$, can be computed efficiently.

Recall that for any node u of the split tree T, we denote by S_u the subset of S that is stored in the subtree of u. Also, each pair $\{A_i, B_i\}$ in the WSPD is defined by two nodes u_i and v_i of T. That is, $A_i = S_{u_i}$ and $B_i = S_{v_i}$.

We store with each node u of T, a list of all leaves v such that the two nodes u and v define a pair in the WSPD. (Hence, v defines a singleton set in this pair.)

Let DS be a data structure that stores a set of points in $\mathbb{R}^d$, that supports nearest-neighbor queries of the form "given a query point $q \in \mathbb{R}^d$, find a point in the set that is nearest to q", and that supports insertions of points.

The algorithm that computes the required closest pair of points in each well-separated pair of point sets, traverses the nodes of T in postorder. To be more precise, let u be an internal node of T, and let u' and u'' be the two children of u. At the moment when node u is visited, the nodes u' and u'' store nearest-neighbor data structures $DS(u')$ and $DS(u'')$ storing the point sets $S_{u'}$ and $S_{u''}$, respectively. If $|S_{u'}| \le |S_{u''}|$, then we insert all points of $S_{u'}$ into $DS(u'')$. Otherwise, all points of $S_{u''}$ are inserted into $DS(u')$. Hence, after these insertions, we have a nearest-neighbor data structure $DS(u)$ storing the point set S_u. For each leaf v of T such that u and v define a pair in the WSPD, we query $DS(u)$ to find a point of S_u that is nearest to the point stored at leaf v.

During this postorder traversal of the split tree T, we get all pairs (x_i, y_i), $1 \le i \le \ell$. Clearly, the approximation graph H can be computed from these pairs, in time $O(\ell) = O(n \log n)$.

We analyze the running time of this algorithm. The number of nearest-neighbor queries is equal to the number ℓ of pairs in the WSPD. For any internal node u of T, the data structure $DS(u)$ is obtained by inserting the points from the child's structure whose subtree is smaller, into the structure of the other child of u. It is easy to prove that in this way, each point of S is inserted at most $\log n$ times. The total number of insertions is therefore bounded by $O(n \log n)$.

Let $Q_{NN}(n_0)$ and $I_{NN}(n_0)$ denote the query and insertion times of the data structure DS, respectively, if it stores a set of n_0 points. Since $n_0 \le n$ at any moment during the algorithm, we have proved the following result.

Lemma 11. *Let S be a set of n points in $\mathbb{R}^d$. After $O(n(Q_{NN}(n) + I_{NN}(n)) \log n)$ pre-processing time, we can compute the approximation graph H of any bottleneck graph G, in $O(n \log n)$ time.*

In order to apply Lemma 11, we need to specify the data structure DS. This data structure stores a set of points in $\mathbb{R}^d$, and supports nearest-neighbor queries and insertions of points. We can obtain such a semi-dynamic data structure by applying Bentley's logarithmic method, see [3,4]. This technique transforms an arbitrary static data structure for nearest-neighbor queries into one that also supports insertions of points. To be more specific, let DS^s be a static data structure storing a set of n points in

$\mathbb{R}^d$, that supports nearest-neighbor queries in $Q_{NN}^s(n)$ time, and that can be built in $P_{NN}^s(n)$ time. The logarithmic method transforms DS^s into a semi-dynamic structure DS, in which nearest-neighbor queries can be answered in $O(Q_{NN}^s(n) \log n)$ time, and in which points can be inserted in $O((P_{NN}^s(n)/n) \log n)$ amortized time.

Corollary 1. *Let S be a set of n points in $\mathbb{R}^d$, let $\beta \geq 1$ be an integer constant, and let ϵ be a real constant, such that $0 < \epsilon \leq 1/2$. After $O(n Q_{NN}^s(n) \log^2 n + P_{NN}^s(n) \log^2 n)$ preprocessing time, we can compute a (c_1, c_2)-approximate stretch factor, where $c_1 = 2\beta(1 + \epsilon)^2$ and $c_2 = 1 + \epsilon$, of any bottleneck graph in $O(n^{1+1/\beta} \log^3 n)$ expected time.*

If we combine Theorem 3 and Corollary 1, then we obtain the main result of this paper.

Theorem 5. *Let S be a set of n points in $\mathbb{R}^d$, let $\beta \geq 1$ be an integer constant, and let ϵ be a real constant, such that $0 < \epsilon \leq 1/2$. In*

$$O\left(n Q_{NN}^s(n) \log^2 n + P_{NN}^s(n) \log^2 n + n^{1+1/\beta} \log^5 n + T_{sel}(n) \log n\right)$$

expected time, we can compute a data structure of size $O(\log n)$, such that for any real number $b > 0$, we can compute, in $O(\log \log n)$ time, a real number t, such that

$$\frac{1}{4\beta^2(1 + \epsilon)^6} t \leq t^* \leq 2\beta(1 + \epsilon)^5 t,$$

where t^* is the exact stretch factor of the Euclidean graph containing all edges of length at most b.

We conclude this section by giving concrete bounds on the preprocessing time. We start with the case when the dimension d is equal to two. The static nearest-neighbor problem can be solved using Voronoi diagrams, and a data structure for point location queries, see Preparata and Shamos [14]. For this data structure, we have $Q_{NN}^s(n) = O(\log n)$, and $P_{NN}^s(n) = O(n \log n)$. Chan [8] gives a randomized distance selection algorithm, whose expected running time $T_{sel}(n)$ is bounded by $O(n^{4/3} \log^{5/3} n)$. Hence, if $d = 2$, the expected time needed to build the data structure of Theorem 5 is bounded by $O(n^{1+1/\beta} \log^5 n + n^{4/3} \log^{8/3} n)$. If $\beta = 2$, then the expected preprocessing time is roughly $n^{3/2}$. For $\beta = 3$, it is roughly $n^{4/3}$. For larger values of β, the time bound remains roughly $n^{4/3}$, but then the approximation ratio increases.

Assume that $d \geq 3$. Agarwal, in a personal communication to Dickerson and Eppstein [10], has shown that

$$T_{sel}(n) = O(n^{2(1-1/(d+1))+\eta}), \tag{2}$$

where η is an arbitrarily small positive real constant. Agarwal and Matoušek [1], and Matoušek and Schwarzkopf [12] have given a static nearest-neighbor data structure for which $n Q_{NN}^s(n) \log^2 n + P_{NN}^s(n) \log^2 n$ is asymptotically smaller than the quantity on the right-hand side of (2). Hence, the expected time needed to build the data structure of Theorem 5 is bounded from above by $O(n^{1+1/\beta} \log^5 n + n^{2(1-1/(d+1))+\eta})$. This becomes $O(n^{2(1-1/(d+1))+\eta})$, i.e., subquadratic, if we take $\beta = 2$. Again, for larger values of β, we get the same time bound, but a larger approximation ratio.

6   Concluding Remarks

We have given a subquadratic algorithm for preprocessing a set S of n points in $\mathbb{R}^d$ into a data structure of size $O(\log n)$ such that for an arbitrary query value $b > 0$, we can, in $O(\log \log n)$ time, compute an approximate stretch factor of the bottleneck graph on S containing all edges of length at most b. This result was obtained by (i) approximating the sequence of $\binom{n}{2}$ different stretch factors of all possible bottleneck graphs, and (ii) approximating bottleneck graphs by graphs containing only $O(n \log n)$ edges.

Our algorithms need exact solutions for computing minimum spanning trees, and nearest-neighbor queries, distance selection queries, and distance ranking queries. It would be interesting to know if approximation algorithms for these problems can be used to speed up the preprocessing time.

References

1. P. K. Agarwal and J. Matoušek. Ray shooting and parametric search. *SIAM J. Comput.*, 22:794–806, 1993.
2. S. Arya, G. Das, D. M. Mount, J. S. Salowe, and M. Smid. Euclidean spanners: short, thin, and lanky. In *Proc. 27th Annu. ACM Sympos. Theory Comput.*, pages 489–498, 1995.
3. J. L. Bentley. Decomposable searching problems. *Inform. Process. Lett.*, 8:244–251, 1979.
4. J. L. Bentley and J. B. Saxe. Decomposable searching problems I: Static-to-dynamic transformations. *J. Algorithms*, 1:301–358, 1980.
5. P. B. Callahan. *Dealing with higher dimensions: the well-separated pair decomposition and its applications*. Ph.D. thesis, Dept. Comput. Sci., Johns Hopkins University, Baltimore, Maryland, 1995.
6. P. B. Callahan and S. R. Kosaraju. Faster algorithms for some geometric graph problems in higher dimensions. In *Proc. 4th ACM-SIAM Sympos. Discrete Algorithms*, pages 291–300, 1993.
7. P. B. Callahan and S. R. Kosaraju. A decomposition of multidimensional point sets with applications to k-nearest-neighbors and n-body potential fields. *J. ACM*, 42:67–90, 1995.
8. T. M. Chan. On enumerating and selecting distances. In *Proc. 14th Annu. ACM Sympos. Comput. Geom.*, pages 279–286, 1998.
9. T. H. Cormen, C. E. Leiserson, and R. L. Rivest. *Introduction to Algorithms*. MIT Press, Cambridge, MA, 1990.
10. M. T. Dickerson and D. Eppstein. Algorithms for proximity problems in higher dimensions. *Comput. Geom. Theory Appl.*, 5:277–291, 1996.
11. D. Eppstein. Spanning trees and spanners. In J.-R. Sack and J. Urrutia, editors, *Handbook of Computational Geometry*, pages 425–461. Elsevier Science, Amsterdam, 1999.
12. J. Matoušek and O. Schwarzkopf. On ray shooting in convex polytopes. *Discrete Comput. Geom.*, 10:215–232, 1993.
13. G. Narasimhan and M. Smid. Approximating the stretch factor of Euclidean graphs. *SIAM J. Comput.*, 30:978–989, 2000.
14. F. P. Preparata and M. I. Shamos. *Computational Geometry: An Introduction*. Springer-Verlag, Berlin, 1988.
15. M. Smid. Closest-point problems in computational geometry. In J.-R. Sack and J. Urrutia, editors, *Handbook of Computational Geometry*, pages 877–935. Elsevier Science, Amsterdam, 1999.

Semantical Principles in the Modal Logic of Coalgebras

Dirk Pattinson*

Institut für Informatik, Ludwig-Maximilians-Universität München
pattinso@informatik.uni-muenchen.de

Abstract Coalgebras for a functor on the category of sets subsume many formulations of the notion of transition system, including labelled transition systems, Kripke models, Kripke frames and many types of automata. This paper presents a multimodal language which is bisimulation invariant and (under a natural completeness condition) expressive enough to characterise elements of the underlying state space up to bisimulation. Like Moss' coalgebraic logic, the theory can be applied to an arbitrary signature functor on the category of sets. Also, an upper bound for the size of conjunctions and disjunctions needed to obtain characteristic formulas is given.

1   Introduction

Rutten [17] demonstrates that coalgebras for a functor generalise many notions of transition systems. It was then probably Moss [13] who first realised that modal logic constitutes a natural way to formulate bisimulation-invariant properties on the state spaces of coalgebras. Given an arbitrary signature functor on the category of sets, the syntax of his coalgebraic logic is obtained via an initial algebra construction, where the application of the signature functor is used to construct formulas. This has the advantage of being very general (few restrictions on the signature functor), but the language is abstract in the sense that it lacks the usual modal operators □ and ◇.

Other approaches [8,9,11,15,16] devise multimodal languages, given by a set of modal operators and a set of atomic propositions, which are based on the syntactic analysis of the signature functor (and therefore only work for a restricted class of transition signatures).

This paper aims at combining both methods by exhibiting the underlying semantical structures which give rise to (the interpretation of) modal operators with respect to coalgebras for arbitrary signature functors. After a brief introduction to the general theory of coalgebras (Section 2), we look at examples of modal logics for two different signature functors in Section 3. The analysis of the semantical structures, which permit to use modalities to formulate properties on the state space of coalgebras, reveals that modal operators arise through a special type of natural transformation, which we chose to call "natural relation". Abstracting away from the examples, Section 4 presents a concrete multimodal language which arises through a set of natural relations and can be used to formulate predicates on the state space of coalgebras for arbitrary signature functors. We then prove in Section 5 that the interpretation of the language is indeed invariant under (coalgebraic) bisimulation. In the last section we characterise the expressive power

---

* Research supported by the DFG Graduiertenkolleg "Logik in der Informatik".

A. Ferreira and H. Reichel (Eds.): STACS 2001, LNCS 2010, pp. 514–526, 2001.
© Springer-Verlag Berlin Heidelberg 2001

of the language, and prove that under a natural completeness condition, every point of the state space can be characterised up to bisimulation. We also give an upper bound for the size of conjunctions and disjunctions needed to obtain characteristic formulas.

The present approach is elaborated in more detail in [14], which also contains the proofs of the theorems which are stated in this exposition.

2 Transition Systems and Coalgebras

Given an endofunctor $T : \text{Set} \rightarrow \text{Set}$ on the category of sets and functions, a *T-coalgebra* is a pair (C, γ) where C is a set (the state space or carrier set of the coalgebra) and $\gamma : C \rightarrow TC$ is a function. Using this definition, which dualises the categorical formulation of algebras, many notions of automata and transition systems can be treated in a uniform framework. We only sketch the fundamental definitions and refer the reader to [7,17] for a more detailed account.

Example 1 (Labelled Transition Systems). Suppose L is a set of labels. Labelled transition systems, commonly used to formulate operational semantics of process calculi such as CCS, arise as coalgebras for the functor $TX = \mathcal{P}(L \times X)$. Indeed, given a set C of states and a transition relation R_l for each label $l \in L$, we obtain a T-coalgebra (C, γ) where $\gamma(c) = \{(l, \hat{c}) \in L \times C \mid c \ R_l \ \hat{c}\}$. Conversely, every coalgebra structure $\gamma : C \rightarrow TC$ gives rise to a family of transition relations $(R_l)_{l \in L}$ via $c \ R_l \ c'$ iff $(l, c') \in \gamma(c)$.

Many types of automata can also be viewed as coalgebras for an appropriate type of signature functor on the category of sets:

Example 2 (Deterministic Automata). Let $TX = (O \times X)^I + E$ and $(C, \gamma : C \rightarrow TC)$ be a T-coalgebra. Given an element of the state space $c \in C$, the result $\gamma(c)$ of applying the transition function is either an error condition $e \in E$ or a function $f : I \rightarrow O \times C \in (O \times C)^I$. Supplying an input token $i \in I$, the result $f(i)$ of evaluating f gives us an output token $o \in O$ and a new state $c' \in C$.

Morphisms of coalgebras are functions between the corresponding state spaces, which are compatible with the respective transition structures. Dualising the categorical formulation of algebra morphisms, a *coalgebra morphism* between two T-coalgebras (C, γ) and (D, δ) is a function $f : C \rightarrow D$ such that $Tf \circ \gamma = \delta \circ f$. Diagrammatically, f must make the diagram

$$
\begin{array}{ccc}
C & \xrightarrow{\;\;f\;\;} & D \\
\downarrow{\scriptstyle\gamma} & & \downarrow{\scriptstyle\delta} \\
TC & \xrightarrow{\;\;Tf\;\;} & TD
\end{array}
$$

commutative. The reader may wish to convince himself that in the case of labelled transition systems above, a coalgebra morphism is a functional bisimulation in the sense of Milner [12]. It is an easy exercise to show that coalgebras for a functor T, together with their morphisms, constitute a category.

One important feature of the functional (ie. coalgebraic) formulation of transition systems is that every signature functor comes with a built in notion of bisimulation. Following Aczel and Mendler [1], a *bisimulation* between two coalgebras (C, γ) and (D, δ) is a relation $B \subseteq C \times D$, that can be equipped with a transition structure $\beta : B \to TB$, which is compatible with the projections $\pi_C : B \to C$ and $\pi_D : B \to D$. More precisely, $B \subseteq C \times D$ is a bisimulation, if there exists $\beta : B \to TB$ such that

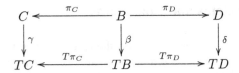

commutes. Again, the reader may wish to convince himself that in the case of labelled transition systems, coalgebraic bisimulations, as just defined, are indeed bisimulations of labelled transition systems.

3 Modal Logic for Coalgebras: Examples

We exemplify the connection between modal logics and coalgebras for a functor by means of the examples given in the previous section. In both examples we observe that the modalities and atomic propositions of the respective languages arise via special types of natural transformation, the "natural relations" already mentioned in the introduction. The general theory developed in the subsequent sections is based on this observation in that it shows, that every set of natural relations induces a multimodal language which allows to formulate bisimulation invariant properties on the state spaces of coalgebras for an arbitrary signature functor.

3.1 Labelled Transition Systems

Consider the functor $TX = \mathcal{P}(L \times X)$ on the category of sets and functions. We have demonstrated in Example 1, that T-coalgebras are labelled transition systems over the set L of labels. It is well known that Hennessy-Milner logic [6] (also discussed in [20]) is an expressive, bisimulation invariant language, which allows to formulate predicate on the state space of labelled transition systems.

Consider the set $\mathcal{L}$ of formulas built up from the atomic propositions tt, ff, conjunctions, disjunctions and a pair of modal operators $\Box_l$ and $\Diamond_l$ for every $l \in L$. Given a T-coalgebra (labelled transition system) (C, γ) and a formula $\phi \in \mathcal{L}$, we write $[\![\phi]\!]_{(C,\gamma)}$ for the set $\{c \in C \mid (c, \gamma) \models \phi\}$ of points $c \in C$, which satisfy the formula ϕ with respect to the transition structure γ, and drop the subscript (C, γ) if the transition structure is clear from the context. Omitting the straightforward interpretation of atomic propositions, conjunctions and disjunctions, the interpretation of the formula $\Box_l \phi$ is given by

$$[\![\Box_l \phi]\!]_{(C,\gamma)} = \{c \in C \mid \forall c' \in C.(l, c') \in \gamma(c) \implies c' \in [\![\phi]\!]_{(C,\gamma)}\} \tag{1}$$

for any $l \in L$. Note that the same definition can be used for any carrier set and transition structure. This leads us to define, given $l \in L$, a parameterised relation $\mu_l(A) \subseteq TA \times$

A, given by

$$\mathfrak{a} \, \mu_l(A) \, a \quad \Longleftrightarrow \quad (l, a) \in \mathfrak{a} \qquad (2)$$

for an arbitrary set A, $\mathfrak{a} \in TA$ and $a \in A$. Using this definition, we can now reformulate (1) as

$$[\![\Box_l \phi]\!]_{(C,\gamma)} = \{c \in C \mid \forall c' \in C.\gamma(c) \, \mu_l(C) \, c' \implies c' \in [\![\phi]\!]_{(C,\gamma)}\} \qquad (3)$$

and obtain the interpretation of the existential modality via

$$[\![\Diamond_l \phi]\!]_{(C,\gamma)} = \{c \in C \mid \exists c' \in C.\gamma(c) \, \mu_l(C) \, c' \land c' \in [\![\phi]\!]_{(C,\gamma)}\}. \qquad (4)$$

The fact that (2) is a canonical definition, which works for any set A, is witnessed by the following universal property: For any function $f : A \to B$, the diagram of sets and relations

$$
\begin{array}{ccc}
TA & \xrightarrow{\;\mu_l(A)\;} & A \\
{\scriptstyle G(Tf)}\big\downarrow & & \big\downarrow{\scriptstyle G(f)} \\
TB & \xrightarrow{\;\mu_l(B)\;} & B
\end{array}
\qquad (5)
$$

commutes (where we write $R : A \nleftrightarrow B$ for a relation $R \subseteq A \times B$ and $G(f)$ for the graph of a function; composition of the arrows in the diagram is relational composition). Parameterised relations, which satisfy condition (5) will be called natural relations in the sequel. Thus summing up, one can say that *natural relations give rise to the interpretation of modalities.*

3.2 Input/Output Automata

In Example 2 we have seen that deterministic input/output automata are coalgebras for the functor $TX = (O \times X)^I + E$. We now go on to demonstrate that the modalities needed to describe properties of these automata also arise via parameterised relations, that is, relations which satisfy the naturality condition (5).

Given a T coalgebra $(C, \gamma : C \to TC)$ and a state $c \in C$, the modality of interest here describes the behaviour of a (possible) successor state, which arises after supplying an input token, if the result $\gamma(c)$ of applying the transition function does not yield an error condition $e \in E$. For $i \in I$ and an arbitrary set A, we consider the relation $\mu_i(A) \subseteq TA \times A$, given by

$$\mathfrak{a} \, \mu_i(A) \, a \quad \text{iff} \quad \exists f : I \to (O \times A) \in (O \times A)^I.\mathfrak{a} = \mathrm{inl}(f) \land \pi_A \circ f(i) = a,$$

where $\mathrm{inl} : (O \times A)^I \to (O \times A)^I + E$ is the canonical injection and π_A denotes the projection function $O \times A \to A$. Note that this parameterised relation also satisfies the naturality condition (5) and allows us to define a pair of modalities $\Box_i$ and $\Diamond_i$ using equations (3) and (4).

In order to obtain a language which allows to specify the behaviour of a state $c \in C$, we furthermore need atomic propositions to be able to formulate that the application $\gamma(c)$ of the transition function yields an error condition $e \in E$ and that – in case $\gamma(c) \in$

$(O \times C)^I$ – supplying an input token $i \in I$ yields an output token $o \in O$. This is taken care of by a set of atomic propositions $\{p_e \mid e \in E\} \cup \{p_{(i,o)} \mid (i,o) \in I \times O\}$. The interpretation of the atomic propositions in this example is straightforward:

$$[\![p_e]\!]_{(C,\gamma)} = \{c \in C \mid \gamma(c) = \mathrm{inr}(e)\} \tag{6}$$

$$[\![p_{(i,o)}]\!]_{(C,\gamma)} = \{c \in C \mid \exists f \in (O \times C)^I . \gamma(c) = \mathrm{inl}(f) \wedge \pi_O \circ f(i) = o\}, \tag{7}$$

where $\mathrm{inr} : E \to (O \times C)^I + E$ is again the canonical injection and $\pi_O : O \times C \to O$ denotes the projection function. In both cases it deserves to be mentioned that the atomic propositions arise as subsets of the set $T1$ (where we write $1 = \{*\}$ for the terminal object in the category of sets and $!_C : C \to 1$ for the unique morphism). To be more precise, consider the sets

$$p_e|_{T1} = \{\mathrm{inr}(e) \mid e \in E\} \tag{8}$$

$$p_{(i,o)}|_{T1} = \{\mathrm{inl}(f) \mid f \in (O \times 1)^I \wedge \pi_O \circ f(i) = o\}, \tag{9}$$

where in this case $\mathrm{inr} : E \to (O \times 1)^I + E$ and $\mathrm{inl} : (O \times 1)^I \to (O \times 1)^I + E$. Using the subsets defined by (8) and (9), we now recover the interpretation of the atomic propositions, originally given by (6) and (7) as

$$[\![p_e]\!]_{(C,\gamma)} = (T!_C \circ \gamma)^{-1}(p_e|_{T1})$$

$$[\![p_{(i,o)}]\!]_{(C,\gamma)} = (T!_C \circ \gamma)^{-1}(p_{(i,o)}|_{T1}),$$

respectively. Thus one can say that *atomic propositions in modal logics for T-coalgebras arise as subsets of the set $T1$.*

4   From Natural Relation to Modal Logics

If $T : \mathrm{Set} \to \mathrm{Set}$ is an endofunctor, the examples in the previous section suggest, that modal logics for coalgebras of a functor are induced by a set of natural relations for T and a set of predicates on $T1$. The remainder of the paper is devoted to showing that this is indeed the case. We start by exhibiting the modal language which arises from a set of natural relations and a set of atomic propositions and show in the subsequent sections, that the language presented is bisimulation invariant and (under a completeness condition on the set of relations) strong enough to distinguish non-bisimilar points.

4.1   Natural Relations

Categorically speaking, natural relations are natural transformations between functors mapping from the category Set of sets and functions to the category Rel of sets and relations. This is captured in

Definition 1 (Natural Relations). *Suppose T is an endofunctor on the category of sets. A* natural relation *for T is a natural transformation*

$$\mathcal{I} \circ T \to \mathcal{I},$$

where $\mathcal{I} : \text{Set} \to \text{Rel}$ is the identity on sets and sends every function to the relation given by its graph.

Unravelling the definition of natural transformations, the reader might wish to convince himself that this definition captures the naturality requirement present in the examples. Note that by moving from a relation $R : A \nrightarrow B$ to a function $\mathbb{S}_R : A \to \mathcal{P}(B)$ given by $\mathbb{S}_R(a) = \{b \in B \mid a \; R \; b\}$, we can also view natural relations $\mathcal{I} \circ T \to \mathcal{I}$ as natural transformations $T \to \mathcal{P}$ (where $\mathcal{P}$ is the covariant powerset functor). This is essentially due to the fact that the category Rel of sets and relations appears as the Kleisli category of the powerset monad[1]. Also every set $\mathcal{A}$ of subsets of $T1$ gives rise to a natural transformation $\mathbb{P}_{\mathcal{A}} : T \to \mathcal{P}(\mathcal{A})$, where $\mathcal{P}(\mathcal{A})$ is the constant functor which sends every set to $\mathcal{P}(\mathcal{A})$. This is elaborated in [14].

For the remainder of this section we assume, that $T : \text{Set} \to \text{Set}$ is an endofunctor on the category of sets and functions, $\mathcal{M}$ is a set of natural relations for T, $\mathcal{A}$ is a set of subsets of $T1$ and κ is a cardinal number.

4.2   Syntax and Semantics of $\mathcal{L}(\mathcal{M}, \mathcal{A}, \kappa)$

As it is often the case with modal languages, we sometimes need infinitary constructs in the language to obtain enough expressive power. In order to be able to deal with the general case later, we fix a cardinal number κ, which serves as upper bound for the size of conjunctions and disjunctions. The language $\mathcal{L}(\mathcal{M}, \mathcal{A}, \kappa)$ induced by the set $\mathcal{M}$ of natural relations and $\mathcal{A}$ of atomic propositions is given by the least set of formulas containing

- An atomic proposition p_a for every $a \in \mathcal{A}$
- The formulas $\bigwedge \Phi$ and $\bigvee \Phi$, if Φ is a set of formulas of cardinality less than or equal to κ, and
- The formulas $\Diamond_\mu \phi$ and $\Box_\mu \phi$ for every $\mu \in \mathcal{M}$ and every formula ϕ of $\mathcal{L}(\mathcal{M}, \mathcal{A}, \kappa)$.

Note that $\mathcal{L}(\mathcal{M}, \mathcal{A}, \kappa)$ contains as a special case the formulas $\bigwedge \emptyset$ and $\bigvee \emptyset$, which we shall abbreviate to $\mathtt{tt}$ and $\mathtt{ff}$, respectively. In order to simplify the exposition of the semantics of $\mathcal{L}(\mathcal{M}, \mathcal{A}, \kappa)$, we introduce an easy bit of notation.

Definition 2. *Suppose $R \subseteq A \times B$ is a relation. Then R induces two operations, which we denote by $\Box_R$ and $\Diamond_R$, both mapping $\mathcal{P}(B) \to \mathcal{P}(A)$, given by*

$$\Diamond_R(\mathfrak{b} \subseteq B) = \{a \in A \mid \exists b \in B . a \; R \; b \wedge b \in \mathfrak{b}\}$$

$$\Box_R(\mathfrak{b} \subseteq B) = \{a \in A \mid \forall b \in B . a \; R \; b \implies b \in \mathfrak{b}\}.$$

---

[1] We would like to thank one of the anonymous referees for pointing this out.

Following Moss [13], we introduce a further operator $\triangle_R : \mathcal{P}(B) \to \mathcal{P}(A)$ *defined by*

$$\triangle_R(\mathfrak{b} \subseteq B) = \Box_R(\mathfrak{b}) \cap \bigcap_{b \in \mathfrak{b}} \Diamond_R(\{b\}),$$

which we will use later.

The semantics of $[\![\phi]\!]_{(C,\gamma)}$ of a formula $\phi \in \mathcal{L}(\mathcal{M}, \mathcal{A}, \kappa)$ can now be inductively defined:

- $[\![p_a]\!]_{(C,\gamma)} = (T!_C \circ \gamma)^{-1}(a)$ for atomic propositions p_a given by $a \in \mathcal{A}$
- $[\![\bigwedge \Phi]\!]_{(C,\gamma)} = \bigcap_{\phi \in \Phi}[\![\phi]\!]_{(C,\gamma)}$ and $[\![\bigvee \Phi]\!]_{(C,\gamma)} = \bigcup_{\phi \in \Phi}[\![\phi]\!]_{(C,\gamma)}$ for conjunctions and disjunctions (following standard conventions, we set $[\![tt]\!]_{(C,\gamma)} = C$ and $[\![ff]\!]_{(C,\gamma)} = \emptyset$), and
- $[\![\Box_\mu \phi]\!]_{(C,\gamma)} = \Box_{\mu(C) \circ G(\gamma)}([\![\phi]\!]_{(C,\gamma)})$ and $[\![\Diamond_\mu \phi]\!]_{(C,\gamma)} = \Diamond_{\mu(C) \circ G(\gamma)}([\![\phi]\!]_{(C,\gamma)})$ for the modal operators.

If the transition structure is clear from the context, we sometimes abbreviate $[\![\phi]\!]_{(C,\gamma)}$ to $[\![\phi]\!]_C$ (and sometimes even to $[\![\phi]\!]$). In case we want to emphasise that a formula ϕ holds at a specific point $c \in C$ of the underlying set, we also write $c \models_\gamma \phi$ for $c \in [\![\phi]\!]_{(C,\gamma)}$.

5 Invariance Properties of $\mathcal{L}(\mathcal{M}, \mathcal{A}, \kappa)$

In this section, we demonstrate that $\mathcal{L}(\mathcal{M}, \mathcal{A}, \kappa)$ is an adequate logic for T-coalgebras. We do this by proving that the semantics of formulas is invariant under coalgebra morphisms and that bisimilar elements of the state space of coalgebras satisfy the same set of formulas.

For the whole section assume that T is an endofunctor on Set, $\mathcal{M}$ is a set of natural relations for T, $\mathcal{A}$ is a set of subsets of $T1$ and κ is a cardinal number.

Theorem 1 (Morphisms Preserve Semantics). *Suppose* $f : (C, \gamma) \to (D, \delta)$ *is a morphism of coalgebras. Then*

$$[\![\phi]\!]_C = f^{-1}([\![\phi]\!]_D)$$

for all formulas ϕ of $\mathcal{L}(\mathcal{M}, \mathcal{A}, \kappa)$.

When proving the theorem, naturality of the relations is essential. We have an easy and immediate

Corollary 1. *Suppose* $f : (C, \gamma) \to (D, \delta)$ *is a morphism of coalgebras and $c \in C$. Then*

$$c \models_\gamma \phi \quad iff \quad f(c) \models_\delta \phi$$

for all formulas $\phi \in \mathcal{L}(\mathcal{M}, \mathcal{A}, \kappa)$.

We now turn to the second invariance property mentioned at the beginning of this chapter and show that bisimilar points satisfy the same sets of formulas. Although this essentially follows from Theorem 1, its importance warrants to state it as

Theorem 2 (Bisimilarity Implies Logical Equivalence). *Suppose* (C, γ) *and* (D, δ) *are T coalgebras and the points $c \in C$ and $d \in D$ are related by a bisimulation. Then*

$$c \models_\gamma \phi \quad iff \quad d \models_\delta \phi$$

for all formulas $\phi \in \mathcal{L}(\mathcal{M}, \mathcal{A}, \kappa)$.

6   Expressivity

This section shows, that the language $\mathcal{L}(\mathcal{M}, \mathcal{A}, \kappa)$ also satisfies an abstractness condition in the sense, that under a natural completeness condition on the pair $(\mathcal{M}, \mathcal{A})$, non-bisimilar points of the carrier set of coalgebras can be distinguished by formulas of $\mathcal{L}(\mathcal{M}, \mathcal{A}, \kappa)$.

For the proof we assume the existence of a terminal coalgebra, that is, of a greatest fixed point for the signature functor T. We represent the greatest fixed point of the signature functor T as limit of the so-called terminal sequence, which makes the succession of state transitions explicit. The categorical dual of terminal sequences is commonly used to construct initial algebras, see [2,19]. We use Theorem 2 of Adámek and Koubek [3], which states that in presence of a terminal coalgebra, the latter can be represented as a fixed point of the terminal sequence. Suppose for the remainder of this section, that T is an endofunctor on the category of sets, $\mathcal{M}$ is a set of natural relations for T and $\mathcal{A}$ is a set of subsets of $T1$.

6.1   Complete Pairs

It is obvious that we cannot in general guarantee that the language $\mathcal{L}(\mathcal{M}, \mathcal{A}, \kappa)$ is strong enough to actually distinguish non-bisimilar points, since the set $\mathcal{M}$ might not contain enough relations or we do not have enough atomic propositions. We start by giving a completeness criterion on the sets $\mathcal{M}$ and $\mathcal{A}$, which ensures that this does not happen.

We write $\mathbb{S}_R(a) = \{b \in B \mid a \, R \, b\}$ if $R : A \nrightarrow B$ is a relation and $a \in A$. We also denote the set of atomic propositions $a \in \mathcal{A}$ satisfied by $x \in TX$ by $\mathbb{P}_{\mathcal{A},X}(x) = \{a \in \mathcal{A} \mid T!_X(x) \in a\}$ if $x \in X$. We shall abbreviate $\mathbb{P}_{\mathcal{A},X}$ to $\mathbb{P}_{\mathcal{A}}$ (or even to $\mathbb{P}$) in the sequel.

Definition 3 (Completeness of $(\mathcal{M}, \mathcal{A})$). *We call the pair $(\mathcal{M}, \mathcal{A})$ complete, if*

$$\{x\} = \bigcap_{\mu \in \mathcal{M}} \{x' \in TX \mid \mathbb{S}_{\mu(X)}(x') = \mathbb{S}_{\mu(X)}(x)\} \cap \bigcap_{a \in \mathbb{P}_{\mathcal{A}}(x)} (T!)^{-1}(a)$$

for all sets X and all elements $x \in TX$.

Intuitively, the pair $(\mathcal{M}, \mathcal{A})$ is complete, if, given any set X, every element $x \in TX$ is determined by its $\mu(X)$-successors and the atomic propositions which are satisfied by x. In case of the powerset functor, this amounts to the axiom of extensionality.

A different way of understanding the completeness condition is by considering complete pairs as natural transformations $T \xrightarrow{.} \mathcal{P}^\kappa \times \mathcal{P}(\mathcal{A})$. Completeness then amounts to the fact that the induced natural transformation is "essentially injective". Details can be found in [14]. We briefly note that the natural relations and atomic propositions defined in Section 3 give rise to complete pairs:

Example 3 (Complete Pairs).

1. Consider the signature functor $TX = \mathcal{P}(L \times X)$. If $\mathcal{M} = \{\mu_l \mid l \in L\}$ is the set of natural relations defined in Section 3.1 and $\mathcal{A} = \emptyset$, then $(\mathcal{M}, \mathcal{A})$ is complete.

2. Suppose $TX = (O \times X)^I + E$ as in Section 3.2 and let $\mathcal{M} = \{\mu_i \mid i \in I\}$ and $\mathcal{A} = \{p_e|_{T1} \mid e \in E\} \cup \{p_{(i,o)}|_{T1} \mid (i,o) \in I \times O\}$ be the set of natural relations and atomic propositions defined there, respectively. Then $(\mathcal{M}, \mathcal{A})$ is complete.

It seems very hard to find a semantical characterisation of functors which admit a complete pair $(\mathcal{M}, \mathcal{A})$ of natural relations and subsets of $T1$. However, it can be proved that the class of these functors contains the identity, constant and powerset functors, and is closed under small limits and small coproducts. For details, we refer to [14]. Note that the class of functors admitting a complete pair is not closed under composition.

6.2   The Expressivity Theorem

This section proves that $\mathcal{L}(\mathcal{M}, \mathcal{A}, \kappa)$ is expressive enough to distinguish non-bisimilar points, subject to the completeness of $(\mathcal{M}, \mathcal{A})$ and the size of κ. The cardinality of conjunctions and disjunctions needed to obtain expressivity is given in terms of the cardinality of the final coalgebra and the convergence of the terminal sequence.

Before we state the expressiveness theorem, we briefly review the construction of greatest fixed points for set functors using terminal sequences. We only give a brief exposition, for details see the original paper by Adámek and Koubek [3] (or Worell [23] for a more categorical treatment). The *terminal sequence* of an endofunctor T on the category of sets is an ordinal-indexed sequence Z_α of sets together with functions $f_{\alpha,\beta} : Z_\alpha \to Z_\beta$ for all ordinals $\beta \leq \alpha$ such that $Z_0 = \{*\}$, $Z_{\alpha+1} = T(Z_\alpha)$ and $Z_\lambda = \mathrm{Lim}_{\alpha<\lambda} Z_\alpha$. It can be seen as the continuation of the sequence

$$1 \xleftarrow{\ !_{T1}\ } T1 \xleftarrow{\ T!_{T1}\ } T^2 1 \xleftarrow{\ T^2(!_{T1})\ } T^3 1 \cdots$$

through the class of all ordinal numbers. Note that the terminal sequence generalises the construction of initial algebras and terminal coalgebras to functors, which do not preserve ω-colimits (resp. ω^{op}-limits). It has been shown in [3], Theorem 2, that in presence of a final T coalgebra, the terminal sequence converges (ie. there exists a (limit) ordinal α such that $f_{\alpha+1,\alpha}$ is an isomorphism) to the terminal coalgebra $(Z_\alpha, f_{\alpha+1,\alpha}^{-1})$. If $f_{\alpha+1,\alpha}$ is an isomorphism, we say that the terminal sequence *stabilises* at α. We are now ready to state the expressiveness theorem:

Theorem 3 $(\mathcal{L}(\mathcal{M}, \mathcal{A}, \kappa)$ **Has Characteristic Formulas).** *Suppose* $(\mathcal{M}, \mathcal{A})$ *is a complete pair,* T *admits a terminal coalgebra* $(Z, \zeta : Z \to TZ)$ *and* κ *is a cardinal such that*

- $\kappa \geq |\mathbb{S}_{\mu(Z)}(z)|$ *for all* $z \in TZ$ *and* $\mu \in \mathcal{M}$
- $\kappa \geq |\mathcal{M}|$ *and* $\kappa \geq |\mathcal{A}|$ *and*
- *The terminal sequence for* T *stabilises at* κ.

Then there is a formula $\phi^z \in \mathcal{L}(\mathcal{M}, \mathcal{A}, \kappa)$ *such that* $[\![\phi^z]\!]_{(Z,\zeta)} = \{z\}$ *for all* $z \in Z$.

Given $z \in Z$, the proof defines a formula $\phi^z(\alpha)$ for each ordinal $\alpha < \kappa$ and $z \in Z$ with the property $[\![\phi^z(\alpha)]\!]_Z = f_{\kappa,\alpha}^{-1}(\{f_{\kappa,\alpha}(z)\})$ by "induction along the terminal sequence" $(Z_\alpha, f_{\alpha,\beta})$ for T. The formula $\phi^z = \phi^z(\kappa)$ then characterises z. The proof can be found in [14].

Some remarks concerning the conditions on the cardinal κ in Theorem 3 are in order. Clearly, we need conjunctions and disjunctions over possibly all atomic propositions and modalities. The third condition is also very natural, since we build the characteristic formula step by step, until we reach the terminal coalgebra, ie. the index, where the terminal sequence stabilises. The only unintuitive condition is the first, giving a lower bound for κ in terms of the final coalgebra. When looking at examples, one however notices that the restriction on the size of successors is very often already implicit in the signature functor T. One can for example show, that all polynomial functors T admit a set of natural relations $\mathcal{M}$, such that for all sets X and all $t \in TX$, the cardinality of the set of successors $\mathbb{S}_{\mu(X)}(x)$ is at most one. Also, since we require T to have a terminal coalgebra, T cannot contain an unbounded powerset construction, hence the signature determines an upper bound of the set of successors in many cases.

As a corollary we conclude that in presence of a terminal coalgebra, any two bisimilar points satisfy the same sets of formulas. Note that for the corollary to work, we need the signature functor T to preserve weak pullbacks, since otherwise also non-bisimilar points are identified in the terminal coalgebra. Since in cases, where the signature functor does not preserve weak pullbacks, bisimulation fails to capture the notion of behavioural equivalence, we do not consider the restriction to weak pullback preserving functors as a defect of our theory.

In cases where the signature functor does not preserve weak pullbacks, Kurz argues in [10], that observable equivalence is not captured by bisimulation as defined by Aczel and Mendler [1], and – in presence of a final coalgebra – one should consider two state bisimilar, when they are identified in the final coalgebra, a notion, which can be equivalently described using co-congruences.

Corollary 2 ($\mathcal{L}(\mathcal{M}, \mathcal{A}, \kappa)$ **Is Adequate**). *Suppose T preserves weak pullbacks and the hypothesis of Theorem 3. If (C, γ) is a T-coalgebra and $c \in C$, there exists a formula $\phi^c \in \mathcal{L}(\mathcal{M}, \mathcal{A}, \kappa)$ such that*

$$[\![\phi^c]\!]_{(C,\gamma)} = \{c' \in C \mid c \leftrightarrows c'\}$$

(where $c \leftrightarrows d$ iff there is a bisimulation $R \subseteq C \times D$ such that $c\,R\,d$).

Theorem 3 also allows us to derive a characterisation of coalgebraic bisimulation in logical terms. To this end, we denote by $\mathrm{Th}(c) = \{\phi \in \mathcal{L}(\mathcal{M}, \mathcal{A}, \kappa) \mid c \models_\gamma \phi\}$ the set of formulas satisfied by a point $c \in C$ for a T-coalgebra (C, γ).

Corollary 3 (**Bisimulation Is Logical Equivalence**). *Suppose T preserves weak pullbacks and the hypothesis of Theorem 3. If (C, γ) and (D, δ) are T-coalgebras and $(c, d) \in C \times D$, then*

$$\mathrm{Th}(c) = \mathrm{Th}(d) \qquad \Longleftrightarrow \qquad c \leftrightarrows d$$

(where again $c \leftrightarrows d$ iff there is a bisimulation $R \subseteq C \times D$ with $c\,R\,d$).

7   Conclusions and Related Work

We have exhibited two semantical principles which allow to use multimodal logics to specify bisimulation invariant properties of coalgebras for an arbitrary signature functor

T. The same issue has been addressed in [4,11,13,15]. We briefly compare the results presented in this paper to the contributions just mentioned.

Regarding the work of Moss [13], it has already been pointed out that the construction of the language used to formulate properties on state spaces of coalgebras is very general, and imposes few restrictions on the signature functor T. Since the construction of the language is carried out in the category of classes and set-continuous functions, T has to be set-based (ie. the action of T on classes has to be defined by its action on sets). In order to obtain a characterisation result, the signature functor T is also assumed to be uniform, a condition, which also appears (in slightly different form) in [21,22]. Note that the defining property of uniformity (taken from [21], section 5.5) is the existence of a natural transformation $\rho : \bar{T} \to \mathcal{P} \circ W$, where $\bar{T}$ is the extension of T to the category of classes, $\mathcal{P}$ is the powerset functor and W maps a class C to the carrier of the $\mathcal{P}$-algebra free over C. Hence T can be embedded into a powerset construction, but it in general this does not seem to imply that T can be embedded into a product $\prod_{\alpha < \kappa} \mathcal{P}$ of the power set functor for a *fixed* cardinal κ. It remains as open question, whether in presence of an accessibility condition on T, such an embedding can be obtained, which would also lead to a better semantical characterisation of the class of functors, which admit complete pairs.

We turn to the work of Baltag [4], where a logical characterisation of simulation is given by extending a set functor T to a relator, that is, to an endofunctor $\mathrm{Rel}(T) :$ $\mathrm{Rel} \to \mathrm{Rel}$ on the category of sets and relations. Baltag argues, that different extensions of T to a relator give rise to different notions of simulation, including bisimulation, which is captured by extending T to a strong relator. The logical language used to obtain a characterisation of (various notions of) simulation is similar to that used in [13]. One of the main goals of the present paper was to obtain languages, which (only) characterise bisimulation. In case the signature functor T preserves weak pullbacks, it is shown in [5] (which is also used in [18] giving – to the authors knowledge – the first characterisation of bisimulation in terms of relators) that T can be uniquely extended to a strong relator $\mathrm{Rel}(T)$. In this case, natural relations can be equivalently described as natural transformations $\mathrm{Rel}(T) \circ \mathcal{I} \dot{\to} \mathcal{I}$, where $\mathcal{I} : \mathrm{Set} \to \mathrm{Rel}$ is the canonical embedding. While this reformulation does not seem to simplify our treatment of coalgebraic modal logic, it would be interesting to see, whether replacing the strong relator $\mathrm{Rel}(T)$ by a different extension of T to a relator, the languages constructed in this paper give also rise to a characterisation of the different forms of simulation as discussed in [4].

The work of [9,11,15] focuses on an inductively defined class of functors, and the languages considered there are built by induction on the structure of the signature functor. We have shown in [14], that most of the functors considered in these approaches admit a complete pair. The notable exception are functors which contain more than one "occurrence" of the powerset functor $\mathcal{P}$, for example $TX = \mathcal{P}(A \times \mathcal{P}(B))$. The logic described in [15] admits a characterisation result even for those functors, but at the expense of a language constructed by an iteration of inductive definitions. That is, at every "occurrence" of the powerset functor, one has to close the language constructed so far under propositional connectives and modalities and uses the set thus obtained as the base case for a new inductive definition. This technique could be mimicked in the framework of natural relations by considering a chain of relations $T = T_k \overset{\mu_k}{\to} T_{k-1} \overset{\mu_{k-1}}{\to} \cdots \overset{\mu_1}{\to} T_0 = \mathrm{Id}$, where each set of relations $T_j \to T_{j-1}$ enjoys a completeness property. Looking at examples, the approach seems promising, but we have not yet worked out the details which then would lead to a more general theory.

Finally, we would like to comment on the predicate liftings used in [9]. By an easy inductive argument, one can see, that the "paths to identity" used in loc. cit. in order to obtain modal operators give rise to natural relations $T \leftrightarrow \mathrm{Id}$. On the other hand, every natural relation μ determines a pair of predicate liftings $\exists_\mu$ and $\forall_\mu$. Here we use the term "predicate lifting" in the general sense, indicating a natural transformation $2 \overset{\cdot}{\to} 2 \circ T$ (2 denotes the contravariant powerset functor) in contrast to [9], where one associates a fixed predicate lifting to each functor T by induction on its syntactical structure. It should also be noted that from a logical perspective, the interpretation of the modal operator associated to the predicate liftings $\exists_\mu$ and $\forall_\mu$ coincides with the interpretation of the existential and universal modality $\Diamond_\mu$ and $\Box_\mu$ induced by a natural relation $\mu : T \leftrightarrow \mathrm{Id}$. It thus seems, that predicate liftings also give rise to logics for coalgebras, but expressiveness results are probably more difficult to obtain, since one can not argue in terms of successors any more (as we did in the proof of Theorem 3).

Acknowledgements

The author would like to thank Alexander Kurz for discussions on the subject of this paper and for useful comments regarding an earlier draft and the people from the Amsterdam Coordination Group at CWI for giving me the opportunity to present and discuss the present approach. Thanks is also due to the anonymous referees for the encouraging (and constructive) remarks, regarding both contents and presentation of the material.

References

1. P. Aczel and N. Mendler. A Final Coalgebra Theorem. In D. H. Pitt et al, editor, *Category Theory and Computer Science*, volume 389 of *LNCS*, pages 357–365. Springer, 1989.
2. J. Adámek. Free algebras and automata realizations in the language of categories. *Comment. Math. Univ. Carolinae*, 15:589–602, 1974.
3. J. Adámek and V. Koubek. On the greatest fixed point of a set functor. *Theor. Comp. Sci.*, 150:57–75, 1995.
4. Alexandru Baltag. A Logic for Coalgebraic Simulation. In Horst Reichel, editor, *Coalgebraic Methods in Computer Science (CMCS'2000)*, volume 33 of *Electr. Notes in Theoret. Comp. Sci.*, 2000.
5. A. Carboni, G. Kelly, and R. Wood. A 2-categorical approach to change of base and geometric morphisms I. Technical Report 90-1, Dept. of Pure Mathematics, Univ. of Sydney, 1990.
6. M. Hennessy and R. Milner. On Observing Nondeterminism and Concurrency. In J. W. de Bakker and J. van Leeuwen, editors, *Automata, Languages and Programming, 7th Colloquium*, volume 85 of *Lecture Notes in Computer Science*, pages 299–309. Springer-Verlag, 1980.
7. B. Jacobs and J. Rutten. A Tutorial on (Co)Algebras and (Co)Induction. *EATCS Bulletin*, 62:222–259, 1997.

8. Bart Jacobs. The temporal logic of coalgebras via Galois algebras. Technical Report CSI-R9906, Computing Science Institute, University of Nijmegen, 1999.

9. Bart Jacobs. Towards a Duality Result in the Modal Logic of Coalgebras. In Horst Reichel, editor, *Coalgebraic Methods in Computer Science (CMCS'2000)*, volume 33 of *Electr. Notes in Theoret. Comp. Sci.*, 2000.

10. Alexander Kurz. *Coalgebras and Applications to Computer Science*. PhD thesis, Universität München, April 2000.

11. Alexander Kurz. Specifying Coalgebras with Modal Logic. *Theor. Comp. Sci.*, 260, 2000.

12. Robin Milner. *Communication and Concurrency*. International series in computer science. Prentice Hall, 1989.

13. Lawrence Moss. Coalgebraic Logic. *Annals of Pure and Applied Logic*, 96:277–317, 1999.

14. Dirk Pattinson. Semantical Principles in the Modal Logic of Coalgebras. Technical report, Institut für Informatik, LMU München, 2000.

15. Martin Rößiger. Coalgebras and Modal Logic. In Horst Reichel, editor, *Coalgebraic Methods in Computer Science (CMCS'2000)*, volume 33 of *Electr. Notes in Theoret. Comp. Sci.*, 2000.

16. Martin Rößiger. From Modal Logic to Terminal Coalgebras. *Theor. Comp. Sci.*, 260, 2000.

17. Jan Rutten. Universal Coalgebra: A theory of systems. Technical Report CS-R 9652, CWI, Amsterdam, 1996.

18. Jan Rutten. Relators and Metric Bisimulations. In B. Jacobs, L. Moss, H. Reichel, and J. Rutten, editors, *Coalgebraic Methods in Computer Science (CMCS'98)*, volume 11 of *Electr. Notes in Theoret. Comp. Sci.*, 1999.

19. M. Smyth and G. Plotkin. The Category-Theoretic Solution of Recursive Domain Equations. *SIAM Journal of Computing*, 11(4):761–783, 1982.

20. Colin Stirling. Modal and temporal logics. In S. Abramsky, D. Gabbay, and T. S. E. Maibaum, editors, *Handbook of Logic in Computer Science*, volume 2. Oxford Science Publications, 1992.

21. D. Turi and J. Rutten. On the foundations of final coalgebra semantics: non-well-founded sets, partial orders, metric spaces. *Mathematical Structures in Computer Science*, 8(5):481–540, 1998.

22. Daniele Turi. *Functorial Operational Semantics and its Denotational Dual*. PhD thesis, Free University, Amsterdam, 1996.

23. James Worell. Terminal Sequences for Accessible Endofunctors. In B. Jacobs and J. Rutten, editors, *Coalgebraic Methods in Computer Science (CMCS'99)*, volume 19 of *Electr. Notes in Theoret. Comp. Sci.*, 1999.

The $\#a = \#b$ Pictures Are Recognizable

Klaus Reinhardt

Wilhelm-Schickhard Institut für Informatik, Universität Tübingen
Sand 13, D-72076 Tübingen, Germany
`reinhard@informatik.uni-tuebingen.de`

Abstract. We show that the language of pictures over $\{a, b\}$ (with a reasonable relation between height and width), where the number of a's is equal to the number of b's, is recognizable using a finite tiling system. This means that counting in rectangular arrays is definable in existential monadic second-order logic.

Classification: Automata and formal languages, logic.

1 Introduction

In [GRST94] pictures are defined as two-dimensional rectangular arrays of symbols of a given alphabet. A set (language) of pictures is called recognizable if it is recognized by a finite tiling system. It was shown in [GRST94] that a picture language is recognizable iff it is definable in existential monadic second-order logic. In [Wil97], it was shown that star-free picture expressions are strictly weaker than first-order logic. The context-sensitive languages are characterized in [LS97a] as frontiers of picture languages. In the same spirit, a link to computational complexity is established in [Bor99], where NP is characterized with the notion of recognizability by padding 1-dimensional words with blanks to form an n-dimensional cube.

A comparison to other regular and context-free formalisms to describe picture languages can be found in [Mat97,Mat98]. Characterizations of the recognizable picture languages by automata can be found in [IN77] and [GR96], where also the subclasses, which are defined by a restriction from nondeterminism to determinism or unambiguity, are considered.

In the one-dimensional case, counting is a kind of a prototype concept for non-recognizability ($\equiv$ non-regularity), but spending one extra dimension easily enables counting (see Section 2) for one line. But it was so far conjectured that counting cannot be done for 2 dimensions without having an extra third dimension available. In [Rei98], the author could only find a nonuniform method for simulating a counter along a picture, which just showed why the attempts, which had been made to disprove Theorem 5, had failed.

1.1 Preliminaries

Definition 1. *[GRST94] A picture* over Σ *is a two-dimensional array of elements of Σ. The set of pictures of size (m, n) is denoted by $\Sigma^{m,n}$. A picture*

A. Ferreira and H. Reichel (Eds.): STACS 2001, LNCS 2010, pp. 527–538, 2001.
© Springer-Verlag Berlin Heidelberg 2001

language *is a subset of* $\Sigma^{*,*} := \bigcup_{m,n \geq 0} \Sigma^{m,n}$.
For a $p \in \Sigma^{m,n}$, *we define* $\hat{p} \in \Sigma^{m+2,n+2}$,
adding a frame of symbols $\# \notin \Sigma$.

Let $T_{m,n}(p)$ *be the set of all sub-pictures of* p
with size (m,n).

$$\hat{p} :=$$

#	#	#	#	#	#
#					#
#		p			#
#					#
#					#
#	#	#	#	#	#

A picture language $L \subseteq \Gamma^{*,*}$ *is called* local *if there is a* Δ *with* $L = \{p \in \Gamma^{*,*}|T_{2,2}(\hat{p}) \subset \Delta\}$. *A picture language* $L \subseteq \Gamma^{*,*}$ *is called* hv-local *if there is a* Δ *with* $L = \{p \in \Gamma^{*,*}|T_{1,2}(\hat{p}) \cup T_{2,1}(\hat{p}) \subset \Delta\}$. *A picture language* $L \subseteq \Sigma^{*,*}$ *is called* recognizable *if there is a mapping* $\pi : \Gamma \to \Sigma$ *and a local language* $L' \subset \Gamma^{*,*}$ *with* $L = \pi(L')$.

This means that in order to recognize a picture, we have to find (non-deterministically) a pre-image in the local pre-image language. According to [LS97b], L is recognizable if and only if there is a mapping $\pi : \Gamma \to \Sigma$ and a hv-local language $L' \subset \Gamma^{*,*}$ with $L = \pi(L')$. This means we can use a hv-local pre-image language, as well.

A necessary condition for recognizability, reflecting that, at most, an exponential amount of information can get from one half of the picture to another, is the following:

Lemma 2. *[Mat98] Let* $L \subseteq \Gamma^{*,*}$ *be recognizable and* $(M_n \subseteq \Gamma^{n,*} \times \Gamma^{n,*})$ *be sets of pairs with* $\forall n, \forall (l,r) \in M_n$ $lr \in L$ *and*
$\forall (l,r) \neq (l',r') \in M_n$ $lr' \notin L$ *or* $l'r \notin L$.
Then $|M_n| \in 2^{O(n)}$.

$$lr = \begin{array}{|c|c|} \hline & \\ l & r \\ & \\ \hline \end{array} \Bigg\} n$$

Let L be the set of pictures over $\{a,b\}$, where the number of a's is equal to the number of b's. Considering pictures where the width is $f(n) \notin 2^{O(n)}$ for the height n, we can find $f(n)$ pairs $(l_1,r_1), (l_2,r_2), ..., (l_{f(n)},r_{f(n)})$, such that l_i has i more a's as b's and r_i has i more b's as a's for all $i \leq f(n)$. Thus $l_i r_i$ is in L but all the $l_i r_j$ with $i \neq j$ have a different number of a's and b's and are, thus, not in L. By contradiction of $f(n) \notin 2^{O(n)}$ with Lemma 2, we get the following:

Corollary 3. *The language of pictures over* $\{a,b\}$, *where the number of* a's *is equal to the number of* b's *(and where sizes* (n,m) *might occur, which do not follow the restriction* $m \leq f(n)$ *or* $n \leq f(m)$ *for a function* $f \in 2^{O(n)}$) *is not recognizable.*

To formulate the main result (Theorem 5) of the paper, we need the following definition:

Definition 4. *The picture language* $L_=$ *(resp.* $L_=^c$) *is the set of pictures over* $\{a,b\}$ *(resp.* $\{a,b,c\}$), *where the number of* a's *is equal to the number of* b's *and having a size* (n,m), *with* $m \leq 2^n$ *and* $n \leq 2^m$.

Remark: We could as well use any other constant base k instead of 2, which means that there is no gap to Corollary 3.

Theorem 5. *The languages $L_=$ and $L_=^c$ are recognizable.*

1.2   Overview

The next section will show how easy it is if we only have to count the difference of a's and b's in the bottom line. The problem which arises in the general case is that the counter is not able to accept simple increments at any local position. Section 3 reduces to the problem of counting only a's and b's in odd columns and rows.

The essential idea of the proof of Theorem 5, in order to overcome this problem, is to construct some 'counting flow', which has small constant capacity at each connection leading from one position to its neighbor. The connections have different orders, which are powers of 4. For example, a flow of 7 in a connection of order 4^i represents a total flow of $7 \cdot 4^i$. Similar to the counter used in [Für82], the number of occurrences of a connection with order 4^i is exponentially decreasing with i. Since the order can not be known on the local level (the alphabet is finite but not the order), some 'skeleton structure' must describe a hierarchy of orders which gives the information at which positions some counted value can be transfered from one order to the next. If, for example, a row having the order 4^i crosses a column having the order 4^{i+1}, the flow in the row may be decreased by 4 simultaneously increasing the flow in the column by 1, which preserves the total flow. This skeleton-language is described in Section 4. Section 5 describes the counting flow for squares of the power of 2 using a variation of the Hilbert-curve. Section 6 shows the generalization to exponentially many appended squares by combining the techniques of Sections 2 and 4 and the generalization to odd cases by folding.

2   Simple Counting for One Line

This section can be viewed as an exercise for Section 5. Here, we consider the language of pictures over $\Sigma = \{a, b, c\}$ with an equal number of a's and b's at the rightmost column and the rest filled with c's (See for example, $\pi(p)$ below).

We use a local pre-image language describing a flow. It is defined over the the alphabet $\Gamma = \{-1, 0, 1\}^4$, where the numbers in (l, r, u, d) describe the flow going out of the position to the left, the right, up and down. In the graphical representation, we describe this by arrows. The sources of the flow correspond to the a's, which is described by $\pi((1, 0, 0, 0)) = \pi(\boxed{\blacktriangleleft}) = a$. Analogously, the b's are the sinks: $\pi((-1, 0, 0, 0)) = \pi(\boxed{\blacktriangleright}) = b$. Everywhere else, the flow has to be continued, which is expressed by $\pi((l, r, u, d)) = c$ if $2l + r + u + d = 0$. The main point is that the flow to the left side has the double order. This means that flows from the rightmost column to the second rightmost column and flows within the second rightmost column have order 1 and, in general, flows between the i-th rightmost column and the $i+1$-th rightmost column and flows within the $i+1$-th rightmost column have the order 2^{i-1}. For example, we may use the pre-image

$\hat{p} =$ (left grid of arrows) for $\pi(\hat{p}) =$ (right grid)

#	#	#	#	#	#	#	#	#	#	#
#	c	c	c	c	c	c	c	a	#	
#	c	c	c	c	c	c	c	a	#	
#	c	c	c	c	c	c	c	a	#	
#	c	c	c	c	c	c	c	a	#	
#	c	c	c	c	c	c	c	b	#	
#	c	c	c	c	c	c	c	a	#	
#	c	c	c	c	c	c	c	b	#	
#	c	c	c	c	c	c	c	b	#	
#	c	c	c	c	c	c	c	b	#	
#	c	c	c	c	c	c	c	b	#	
#	#	#	#	#	#	#	#	#	#	#

Although there might be several possible pre-images, one of them can be obtained in the following way: For a and b, there is only one possible pre-image. The pre-image (l, r, u, d) for c is chosen by taking $r := l'$ for the right neighbor $(l', , ,)$ and $u := d'$ for the upper neighbor $(, , , d')$ ($u := 0$ if the upper neighbor is #). If $r + u = 2$ (resp. -2), let $l := -1$ (resp 1) and $d := 0$. If $r + u = 1$ (resp. -1), let $d := -1$ (resp 1) and $l := 0$; else $d := l := 0$.

In this way the flow from one row down to the next row corresponds to the binary representation of the difference in the number of a's and b's so far. A flow to the left corresponds to a carry-bit.

The formal definition for Δ is

$$\Delta := \{\boxed{\begin{smallmatrix}\#\\\gamma\end{smallmatrix}} \mid \gamma = (l, r, 0, d)\} \cup \{\boxed{\begin{smallmatrix}\gamma\\\#\end{smallmatrix}} \mid \gamma = (l, r, u, 0)\}$$
$$\cup \{\boxed{\#\ \gamma} \mid \gamma = (0, r, u, d)\} \cup \{\boxed{\gamma\ \#} \mid \gamma = (l, 0, 0, 0), l \in \{1, -1\}\}$$
$$\cup \{\boxed{\begin{smallmatrix}\delta\\\gamma\end{smallmatrix}} \mid \delta = (l, r, u, d), \gamma = (l', r', d, d')\}$$
$$\cup \{\boxed{\delta\ \gamma} \mid \delta = (l, r, u, d), \gamma = (r, r', u', d')\}.$$

Remark: We could as well use $\Gamma = \{1 - k, ..., k - 1\}^4$ and $\pi((l, r, u, d)) = c$ for $k \cdot l + r + u + d = 0$ to be able to treat pictures of size (n, m) with $m \leq k^n$.

3   Reduction to Odd Positions

We view a mapping $e : \Sigma \mapsto \Sigma_e^{i,j}$ as lifted to map a picture of size (m, n) over Σ to a picture of size (im, jn) over Σ_e.

Lemma 6. *A picture language L over Σ is recognizable if $e(L)$ is recognizable and e is injective.*

Proof. Let $e(L)$ be recognizable by a mapping $\pi_e : \Gamma_e \mapsto \Sigma_e$ and a tiling $\Delta_e \subseteq \Gamma_e^{2,2}$. Construct $\Gamma := \{g \in \Gamma_e^{i,j} \mid \exists s \in \Sigma \ e(s) = \pi_e(g)\}$, $\pi : \Gamma \mapsto \Sigma$ with $\pi(g) = s$ for $e(s) = \pi_e(g)$ (e injective) and $\Delta := \{p \in (\Gamma_e \cup \{\#\})^{2i,2j} \mid p \in$

$(\Gamma \cup \{\#\})^{2,2}, T_{2,2}(p) \subseteq \Delta_e \cup \{\begin{array}{|c|c|}\hline \# & \# \\ \hline \# & \# \\ \hline\end{array}\}\}$ where, for simplicity, we identify $\#$ with the picture of size (i,j) consisting only of $\#$.

We use $e : \{a,b,c\} \mapsto \{a,b,c,d\}^{2,2}$ with $e(x) = \begin{array}{|c|c|}\hline x & d \\ \hline d & d \\ \hline\end{array}$ for $x \in \{a,b,c\}$. In order to show that $L_{=}^c$ and, thus, $L_{=} = L_{=}^c \cap \{a,b\}^{*,*}$ is recognizable, it remains to show in Lemma 10 that $e(L_{=}^c)$ is recognizable, which means that for L_F in Lemma 8 we only have to care about a's and b's on positions with a odd row and column number by intersecting L_F with the recognizable language $\{\begin{array}{|c|c|}\hline x & d \\ \hline d & d \\ \hline\end{array} \mid x \in \{a,b,c\}\}^{*,*}$.

4 The Skeleton for Squares of the Power of 2

The *skeleton* describes a square, where each corner is surrounded by a square of half the size. The skeleton is described as a hv-local language L_S over the alphabet $\Sigma_S = \{\ulcorner, \urcorner, \llcorner, \lrcorner, \dashv, \vdash, \bullet\!\!-, -\!\!\bullet, \cdot, \cdot\cdot, :, \vee, \wedge, \prec, \succ, \Psi, \Lambda,$ $\prec, \succ, \mid, \vee, \wedge, -, \prec, \succ\}$. Since the last section we are particularly interested in the intersection $L_S \cap L_R := \{p_1, p_2, ...\}$ with the recognizable language

$L_R = \{\begin{array}{|c|c|}\hline x & y \\ \hline z & w \\ \hline\end{array} \mid x \in \{\ulcorner, \urcorner, \llcorner, \lrcorner\}, y,z,w \in \{\dashv, \vdash, \bullet\!\!-, -\!\!\bullet, \cdot, \cdot\cdot, :, \vee, \wedge, \prec,$ $\succ, \Psi, \Lambda, \prec, \succ, \mid, \vee, \wedge, -, \prec, \succ\}\}^{*,*}$. The first 4 examples are

$\hat{p}_1 = $

#	#	#	#
#	⌐⌐-∨-	#	
#	≻	•-	#
#	#	#	#

$\hat{p}_2 = $

#	#	#	#	#	#
#	⌐	-∨-	⌐	·	#
#	≻	•-	→	-∨-	#
#	⌐	-∀-	⌐	:	#
#	·	≻	··	•-	#
#	#	#	#	#	#

$\hat{p}_3 = $

#	#	#	#	#	#	#	#	#	#
#	⌐	-∨-	⌐	·	⌐	-∨-	⌐	·	#
#	≻	•-	→	-∨-	←	-•	≺	·	#
#	⌐	-∀-	⌐	:	⌐	-∀-	⌐	·	#
#	·	≻	··	•———	→	≻	-∨-	#
#	⌐	-∧-	⌐	∣	⌐	-∧-	⌐	:	#
#	≻	•-	→	-∀-	←	-•	≺	:	#
#	⌐	-∧-	⌐	∨	⌐	-∧-	⌐	:	#
#	·	·	·	≻	··	··	··	•-	#
#	#	#	#	#	#	#	#	#	#

and

$$\hat{p}_4 =$$

We define L_S using $\Delta_S := T_{1,2}(\hat{p}_4) \cup T_{2,1}(\hat{p}_4) \cup \{ \ldots \}$.

First we will show by induction that for every i, a picture p_i of size $(2^i, 2^i)$ is in $L_S \cap L_R$:

Consider p_3 as the base of induction and p_4 as an example for a step of induction. Except from the right and lower edge where the picture meets the #'s of the frame, Δ_S has all the symmetries of a square. The upper left quarter of p_{i+1} is exactly p_i. Furthermore, the 3 sub-pictures of size $(2^i-1, 2^i-1)$, starting with $p_{i+1}(1, 2^i+1) = \ulcorner$, $p_{i+1}(2^i+1, 1) = \ulcorner$ and $p_{i+1}(2^i+1, 2^i+1) = \ulcorner$, are rotations of the sub-picture of size $(2^i - 1, 2^i - 1)$, starting with $p_i(1,1) = \ulcorner$ around $p_{i+1}(2^i, 2^i) = \bullet$. Now, Δ_S allows us to continue the 2^i-th row after $\bullet$ with $-$'s until the $\rightarrowtail$ at the column with the only $\vee$ at the lower edge of the upper right sub-picture. ¿From now on, continue with $\succ$'s until the $\vee$ at the last column, which had .'s so far and continues with :'s until the $\bullet$ in the lower right corner. (Column 2^i and last row analogously).

The opposite direction, that for every i, exactly one picture p_i of size $(2^i, 2^i)$ is in $L_S \cap L_R$, follows (considering the only possibility for the right and lower edge) from the following lemma:

Lemma 7. *For every picture $p \in L_S \cap L_R$ and for every $i > 1$, $r, c \geq 0$, the sub-picture q of size $(2^i - 1, 2^i - 1)$ starting at $p(c2^i + 1, r2^i + 1)$ has the following shape at the outside: The upper row is periodically filled by $q(t, 1) = \ulcorner$ (resp. $\curlyvee, \neg, \cdot$) if $t \bmod 4 = 1$ (resp. 2,3,0) and $t < 2^i$. The same holds in a $90°$-rotation symmetric manner. Two exceptions to this are that for even (resp. odd) r we have $q(2^{i-1}, 2^i - 1) = \curlyvee$ (resp. $q(2^{i-1}, 1) = \wedge$) and for even (resp. odd) c we have $q(2^i - 1, 2^{i-1}) = \gg$ (resp. $q(1, 2^{i-1}) = \ll$) if $i > 2$ and that for even (resp. odd) r we have $q(2^{i-1}, 2^i - 1) = \curlyvee\!\!\!-$ (resp. $q(2^{i-1}, 1) = \wedge\!\!\!-$) and for even (resp. odd) c we have $q(2^i - 1, 2^{i-1}) = \gg\!\!\!-$ (resp. $q(1, 2^{i-1}) = \ll\!\!\!-$) if $i = 2$.*

On http://www-fs.informatik.uni-tuebingen.de/~reinhard/picgen.html is a Java program, which demonstrates that in most cases, the symbol is determined by its left and upper neighbors; a wrong choice in the remaining cases will sooner or later result in an unsolvable case at another position. The program is interactive and could help the reader to understand the following proof.

Proof. For each induction step we need the the following additional *Claim C*:

All left neighbors of $q(1, j)$ with $1 \leq j < 2^i$ are in $\{\#, \cdot, :, |, \curlyvee, \wedge\}$ with the only exception that the left neighbor of $q(1, 2^{i-1})$ is in $\{\wedge, \wedge\!\!\!-, \curlyvee, \curlyvee\!\!\!-, \gg\}$ if c is odd. Analogously all upper neighbors of $q(j, 1)$ with $1 \leq j < 2^i$ are in $\{\#, \cdot, \cdots, -, \ll, \gg\}$ with the only exception that the upper neighbor of $q(2^{i-1}, 1)$ is in $\{\ll, \ll\!\!\!-, \gg\!\!\!-, \gg\!\!\!-, \curlyvee\}$ if r is odd.

Base of induction for $i = 2$: Assume furthermore, by induction on r and c, that Claim C holds (The neighbors are $\#$ for $c = 0$ resp. $r = 0$). Because of L_R we have $q(1, 1) \in \{\neg, \ulcorner, \llcorner, \lrcorner\}$. Since $q(1, 1) = \neg$ would, by Δ_S, require the left neighbor to be in $\{\curlyvee, \wedge\}$, and since $q(1, 1) \in \{\llcorner, \lrcorner\}$ would by Δ_S require the upper neighbor in $\{\ll, \ll\!\!\!-, \gg\!\!\!-, \gg\!\!\!-\}$, we have $q(1, 1) = \ulcorner$ which is the only remaining possibility for a position with both odd coordinates. Consequently considering the upper neighbor, $q(2, 1) = \curlyvee$ (resp. $= \wedge\!\!\!-$) if r is even (resp. odd) and $q(1, 2) = \gg\!\!\!-$ (resp. $= \ll\!\!\!-$) if c is even (resp. odd). Each of the 4 combinations forces $q(2, 2) = $ one of four symbols. The two ends of $q(2, 2)$ point to the exceptions of the outside shape. Furthermore, $q(3, 1) = \neg$, $q(1, 3) = \llcorner$. Thus one of the 4 combinations of $q(2, 3) = \wedge$ (resp. $= \curlyvee$) if r is odd (resp. even) and $q(3, 2) = \ll$ (resp. $= \gg\!\!\!-$) if c is odd (resp. even) and $q(3, 3) = \lrcorner$.

Right neighbors of $q(3, 1) = \neg$, $q(3, 2) = \ll$ and $q(3, 3) = \lrcorner$ must be in $\{\cdot, :, |, \curlyvee, \wedge\}$ which proves Claim C for $c + 1$ if c is odd. If c is even, the right neighbor of $q(3, 2) = \gg\!\!\!-$ is in $\{\wedge, \wedge\!\!\!-, \curlyvee, \curlyvee\!\!\!-, \gg\}$ which proves Claim C for $c + 1$. The same holds for $r + 1$ analogously.

Step from i to $i+1$: Assume furthermore, by induction on r and c, that Claim C holds for $i + 1$. (The neighbors are $\#$ for $c = 0$ resp. $r = 0$). By induction on i we have that each of the 4 sub i-pictures of the $i + 1$-sub-picture q has its exceptional side hidden inside q. Since $q(1, 2^i - 1) = \llcorner$, considering the possible left neighbors leads to $q(1, 2^i) = \cdot$ if $i + 1$ is even, resp. $q(1, 2^i) \in \{\gg, \ll\}$ if $i + 1$ is odd. The periodical contents of the rows $2^i - 1$ and $2^i + 1$ only allows us to continue row 2^i with the same symbol until column 2^{i-1}, where

$q(2^{i-1}, 2^i) \in \{ \text{⊁}, \text{⊀}, \text{⊀}, \text{⊀} \}$. This allows us only the combination $q(1, 2^i) = \ldots = q(2^{i-1} - 1, 2^i) = \cdot$ and $q(2^{i-1}, 2^i) = \text{⊁}$ if $i + 1$ is even, resp., $q(1, 2^i) = \ldots = q(2^{i-1} - 1, 2^i) = \text{⊀}$ and $q(2^{i-1}, 2^i) = \text{⊀}$ if $i + 1$ is odd, which has to be continued with $q(2^{i-1} + 1, 2^i) = \ldots = q(2^i - 1, 2^i) = \cdots$, resp., —. Depending on the analogous column 2^i, we get $q(2^i, 2^i) \in \{ \text{⊢}, \text{⊶} \}$, resp., $q(2^i, 2^i) \in \{ \text{⊢}, \text{⊶} \}$ and, further, have to continue with $q(2^i + 1, 2^i) = \ldots = q(2^i + 2^{i-1} - 1, 2^i) = \text{—}$, resp., $\cdots$, $q(2^i + 2^{i-1}, 2^i) = \text{⊁}$, resp., ⊀ and with $q(2^i + 2^{i-1} + 1, 2^i) = \ldots = q(2^{i+1} - 1, 2^i) = \text{⊳}$ if $i + 1$ is even, resp., $\cdot$ if $i + 1$ is odd. Together with the analogous column 2^i, this completes the description of q.

The right neighbor of $q(2^{i+1} - 1, 2^i) = \text{⊳}$ (resp. $\cdot$) must be in in $\{ \text{⅄}, \text{⅄}, \text{⅄}, \text{⅄}, \text{⊳} \}$ if c is even, resp. $\{ \cdot, \cdots, :, |, \text{⅄}, \text{⋏} \}$ if c is odd which proves Claim C for $c + 1$. The same holds for $r + 1$ analogously.

5   The Counting Flow for Squares of the Power of 2

Lemma 8. $e(L_=^c) \cap \bigcup_i \Sigma^{2^i, 2^i}$ *is recognizable.*

Proof. We define a language L_F in the following and show $e(L_=^c) \cap \bigcup_i \Sigma^{2^i, 2^i} = L_F \cap e(\{a, b, c, \}^{*,*})$. We give each flow from one cell to its neighbor a capacity from -9 to 9 by defining $\Sigma_F := \Sigma_S \times \{-9, -8, \ldots, 9\}^4$. Furthermore, we allow only those symbols $(x, l, r, u, d) \in \Sigma$ fulfilling:

$\pi(x, l, r, u, d) := a$ if $x \in \{ \text{⌐}, \text{⌐}, \text{⌐}, \text{⌐} \} \wedge l + r + u + d = 1$,
$\pi(x, l, r, u, d) := b$ if $x \in \{ \text{⌐}, \text{⌐}, \text{⌐}, \text{⌐} \} \wedge l + r + u + d = -1$,
$\pi(x, l, r, u, d) := c$ if $x \in \{ \text{⌐}, \text{⌐}, \text{⌐}, \text{⌐} \} \wedge l + r + u + d = 0$,
$\pi(x, l, r, u, d) := d$ if $x \in \{ \text{⊢}, \text{⊢}, \text{⊶}, \text{⊶} \} \wedge l + r + u + d = 0$,
  or $x \in \{ \cdot, \cdots, :, |, \text{⅄}, \text{⋏}, \text{—}, \text{⋖}, \text{⊳} \} \wedge l = -r \wedge u = -d$,
  or $x \in \{ \text{⅄}, \text{⅄}, \text{⅄}, \text{⅄} \} \wedge l + r + 4u + 4d = 0$,
  or $x \in \{ \text{⊀}, \text{⊁}, \text{⊀}, \text{⊀} \} \wedge 4l + 4r + u + d = 0$. The tiling

$$\Delta_F := \{ \boxed{\begin{smallmatrix} f \\ g \end{smallmatrix}} | \boxed{\begin{smallmatrix} f_1 \\ g_1 \end{smallmatrix}} \in \Delta_S, f = (f_1, l, r, u, d), g = (g_1, l', r', -d, d')\}$$
$$\cup \{ \boxed{f\,g} | \boxed{f_1\,g_1} \in \Delta_S, f = (f_1, l, r, u, d), g = (g_1, -r, r', u', d')\}$$

takes care of the continuation of the flow (additionally, we have the tiles with #

having flow 0). Here the tile $\boxed{f\,g}$ is depicted as $\boxed{\begin{smallmatrix} & u & & u' & \\ l & f_1 & r & -r & g_1 & r' \\ & d & & d' & \end{smallmatrix}}$ illustrating that the flow r going out of f in right direction is the same as going in g from the left. The symbols $\text{⌐}, \text{⌐}, \text{⌐}, \text{⌐}$ allow sources and sinks of the flow; they only occur in odd rows and columns and, therefore, have the order 1; $\text{⊢}, \text{⊢}, \text{⊶}, \text{⊶}$ occur where the flow in the column and the row have the same order; $\cdot, \cdots, :$ $, |, \text{⅄}, \text{⋏}, \text{—}, \text{⋖}, \text{⊳}$ occur where the flow in the column and the row have a completely different order and $\text{⅄}, \text{⅄}, \text{⅄}, \text{⅄}, \text{⊀}, \text{⊁}, \text{⊀}, \text{⊀}$ occur where a rectangle or its elongation meets a rectangle of half the size, which means that a carry of the counter can take place here. Examples are

$$\pi(\boxed{\begin{matrix} -6 \\ -1 \neg \ 3 \\ 5 \end{matrix}}) = a, \ \pi(\boxed{\begin{matrix} -6 \\ -2 \bullet \ 5 \\ 3 \end{matrix}}) = d, \ \pi(\boxed{\begin{matrix} -4 \\ -2 : \ 2 \\ 4 \end{matrix}}) = d, \ \pi(\boxed{\begin{matrix} -1 \\ -4 + 5 \\ -3 \end{matrix}}) = d.$$

In general, for any j and i, the $2^{i-1}+j \cdot 2^i$ -th row, resp., column have the order 4^i (if they are in the picture). The symbols in $\{\neg, \ulcorner, \llcorner, \lrcorner, \uparrow, \upharpoonright, \downharpoonright, \multimap\}$ occur where a $2^{i-1}+j \cdot 2^i$ -th row crosses a $2^{i-1}+k \cdot 2^i$ -th column having the same order. The symbols in $\{\curlyvee, \curlywedge, \curlyveedownarrow, \curlywedge\}$ occur where a $2^{i-1}+j \cdot 2^i$ -th row crosses a $2^i + k \cdot 2^{i+1}$ -th column having the fourfold order ($\prec, \succ, \preceq, \succeq$) vice versa. Thus, from the existence of a flow follows that the number of sources and sinks (a's and b's) must be equal. A picture in $e(\{a, b, c, \}^{*,*})$ has its pre-images in L_R and, thus, Lemma 7 makes sure that the projection to the first component in the 5-tuples has the correct structure, which means that $L_F \cap e(\{a, b, c, \}^{*,*}) \subseteq e(L_\stackrel{c}{=}) \cap \bigcup_i \Sigma^{2^i,2^i}$.

For the other direction, we have to show that for any picture of size $(2^i, 2^i)$ with an equal number of sources and sinks (w.l.o.g. on positions on odd rows and columns), we can construct at least one corresponding flow:

Here we use the Hilbert-curve, where each corner point $(2^{i-1}+j_x \cdot 2^i, 2^{i-1}+j_y \cdot 2^i)$ of the curve having order 4^i is surrounded by 4 corner points $(2^{i-1}+j_x \cdot 2^i \pm 2^{i-2}, 2^{i-1}+j_y \cdot 2^i \pm 2^{i-2})$ of the curve having order 4^{i-1} (see also [NRS97]).

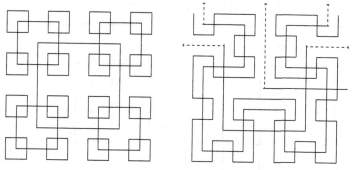

A curve of each order uses 3 lines of a square and then one elongation line to get to the next square. In this way at least one of the 3 lines crosses the curve of the next higher order. (If it crosses it a second time, we ignore the second crossing in the following consideration.) Now we construct the flow along the curve according to the following rules: If a flow of more than 3 (resp. less than -3) crosses a flow of the next higher order or if a flow of more than 0 (resp. less than 0) crosses a negative (resp. positive) flow of the next higher order, it is decreased (resp. increased) by 4 and the flow of the next higher order is increased (resp. decreased) by one. We may assume by induction that a curve has a flow $\in [-6, 6]$ as it enters a square. After at most 3 times crossing the curve of the next lower order (which could bring the flow for example to -9 or 9), it will cross the curve of the next higher order, bringing the flow to $[-5, 5]$. Since we have to consider 4 crossings in the square, the first condition of the rule makes sure that the curve also leaves the square with a flow between -6 and 6 and, thus, never exceeds its capacity. The second condition of the rule makes sure that a small total flow will find itself represented in curves with low order, which is important towards the end of the picture.

Example:

#	b	b	b	b
#	a	a	a	a
#	a	a	a	b
#	a	a	b	b

#	b	d	b	d	b	d	b
#	d	d	d	d	d	d	d
#	a	d	a	d	a	d	a
#	d	d	d	d	d	d	d
#	a	d	a	d	a	d	b
#	d	d	d	d	d	d	d
#	a	d	a	d	b	d	b

In the sub-picture $\pi(p) = e(\ )=$ (last line of d's omitted) the difference of a's and b's is 2. Assume for example that the difference of a's and b's above $\pi(p)$ is $16=4\cdot3+4$ which is represented by a flow of 4 with order 1 and a flow of 3 with order 4 is entering the following pre-image p from above at column 1 and 2. Then the total difference of $18=16+2$ is represented by a flow of 1 with order 16 and a flow of 2 with order 1 leaving p to the right side.

6   The Generalization

Lemma 9. $e(L_{\underline{=}}^c) \cap \bigcup_{i,j} \Sigma^{j2^i,2^i}$ is recognizable.

Proof. Adding the tiles to Δ_S allows skeleton-pictures of size $(j2^i, 2^i)$ for any $j > 0$.

The counting flow described in the previous section can be continued from one square to the next, as the following picture illustrates:

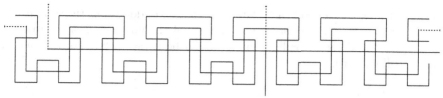

But what can we do with a flow (of order 4^i) reaching the bottom line? The idea is to combine the method with the method of Section 2 by using $\Sigma_e := \Sigma_F \times \{-1, 0, 1\}^4$ as a new alphabet and designing Δ_e in such way that a transfer of the flow from one system to the other is allowed at those symbols ⚲ and ⚲ , which occur at the bottom line. The r-th row (from the bottom) can now have flows of the order $4^i \cdot 2^{r-1}$. This allows us to represent numbers up to $4^i \cdot 2^{2^i}$ and, thus, recognize pictures of size $(j2^i, 2^i)$ for any $0 < j < 2^{2^i}$ (resp. k^{2^i}) with the same number of a's and b's.

Lemma 10. $e(L^c_=)$ *is recognizable.*

Proof. For the general case of pictures of size (m, n) where we assume w.l.o.g. $m > n$, we choose the smallest i with $2^i \geq n$ and the smallest j with $j2^i \geq m$. Then, since $2n > 2^i$ and $2m > j2^i$, a picture of size $(j2^i, 2^i)$, which was recognized with the method described so far, can be folded twice and we get the size (m, n). This folding can be simulated using an alphabet $\Sigma_g := \Sigma_e^4$, where the first layer corresponds to the picture and the other 3 layers may contain the simulated border consisting of # and parts of the flow but no sinks and sources (this means only c's and d's). Δ_g simulates Δ_e on each layer by additionally connecting layer 1 with 4 and 2 with 3 at the top border and 1 with 2 and 3 with 4 at the right border.

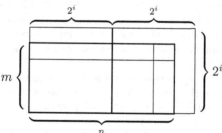

Remark: In the same way, we can further generalize this counting method to n-dimensional recognizable 'picture'-languages by folding $L_=$ into n dimensions like in [Bor99].

7   Outlook

It remains open to find a deterministic counting method on pictures. Obviously, this can not be done using the deterministic version of on-line tessalation acceptors [IN77] as a model, since the automaton can not handle the number occurring in the last line. But a good candidate is the notion of deterministic recognizability in [Rei98]. At least in the case of squares of the power of 2, a construction of the skeleton along the Hilbert-curve should be possible, but working out the details will be hard.

References

Bor99.   B. Borchert. A formal languages characterization of NP. Manuscript at
            http://math.uni-heidelberg.de/logic/bb/papers/NP-char.ps, 1999.
Für82.   Martin Fürer. The tight deterministic time hierarchy. In *Proceedings of the
            Fourteenth Annual ACM Symposium on Theory of Computing*, pages 8–16,
            San Francisco, California, 5–7 May 1982.
GR96.   D. Giammarresi and A. Restivo. Two-dimensional languages. In G. Rozen-
            berg and A. Salomaa, editors, *Handbook of Formal Language Theory*, volume
            III. Springer-Verlag, New York, 1996.
GRST94.   Dora Giammarresi, Antonio Restivo, Sebastian Seibert, and Wolfgang
            Thomas. Monadic second-order logic over pictures and recognizability
            by tiling systems. In P. Enjalbert, E.W. Mayr, and K.W. Wagner, ed-
            itors, *Proceedings of the 11th Annual Symposium on Theoretical Aspects
            of Computer Science, STACS 94 (Caen, France, February 1994)*, LNCS
            775, pages 365–375, Berlin-Heidelberg-New York-London-Paris-Tokyo-Hong
            Kong-Barcelona-Budapest, 1994. Springer-Verlag.
IN77.   K. Inoue and A. Nakamura. Some properties of two-dimensional on-line
            tessellation acceptors. *Information Sciences*, 13:95–121, 1977.
LS97a.   M. Latteux and D. Simplot. Context-sensitive string languages and recog-
            nizable picture languages. *Information and Computation*, 138(2):160–169,
            1 November 1997.
LS97b.   M. Latteux and D. Simplot. Recognizable picture languages and domino
            tiling. *Theoretical Computer Science*, 178(1-2):275–283, 1997. Note.
Mat97.   Oliver Matz. Regular expressions and context-free grammars for picture
            languages. In *14th Annual Symposium on Theoretical Aspects of Computer
            Science*, volume 1200 of *lncs*, pages 283–294, Lübeck, Germany, 27 February–
            March 1 1997. Springer.
Mat98.   Oliver Matz. On piecewise testable, starfree, and recognizable picture lan-
            guages. In Maurice Nivat, editor, *Foundations of Software Science and Com-
            putation Structures*, volume 1378 of *Lecture Notes in Computer Science*,
            pages 203–210. Springer, 1998.
NRS97.   R. Niedermeier, K. Reinhardt, and P. Sanders. Towards optimal locality
            in mesh-indexings. In L. Czaja B.S. Chlebus, editor, *Proceedings of the
            FCT'97*, LNCS 1279, pages 364–375. Springer, sept. 1997.
Rei98.   K. Reinhardt. On some recognizable picture-languages. In L. Brim, editor,
            *Proceedings of the 23th Conference on Mathematical Foundations of Com-
            puter Science*, number 1450 in Lecture Notes in Computer Science, pages
            760–770. Springer-Verlag, August 1998.
Wil97.   Thomas Wilke. Star-free picture expressions are strictly weaker than first-
            order logic. In Pierpaolo Degano, Roberto Gorrieri, and Alberto Marchetti-
            Spaccamela, editors, *Automata, Languages and Programming*, volume 1256
            of *Lect. Notes Comput. Sci.*, pages 347–357, Bologna, Italy, 1997. Springer.

A Logical Approach to Decidability of Hierarchies of Regular Star–Free Languages*

Victor L. Selivanov

A.P. Ershov Institute of Informatics Systems
Siberian Division of Russian Academy of Sciences
vseliv@nspu.ru

Abstract. We propose a new, logical, approach to the decidability problem for the Straubing and Brzozowski hierarchies based on the preservation theorems from model theory, on a theorem of Higman, and on the Rabin tree theorem. In this way, we get purely logical, short proofs for some known facts on decidability, which might be of methodological interest.

Our approach is also applicable to some other similar situations, say to "words" over dense orderings which is relevant to the continuous time and hybrid systems.

Keywords: Star–free regular languages, hierarchies, definability, decidability.

1 Introduction

In automata theory, several natural hierarchies of regular languages were studied. Among the most popular are hierarchies of Brzozowski and Straubing [Pin86], both exhausting the regular star–free languages. A natural question about these hierarchies is formulated as follows: given a level of a hierarchy and a finite automaton, one has to decide effectively whether or not the language of the automaton is in the given level. Till now, this question is solved positively only for lower levels. For higher levels, the problem is still open and seems to be hard (see e.g. [Pin86, Pin94] for more information and references).

In the literature one could identify at least two approaches to the decidability problem, which might be called algebraic and automata–theoretic. The first approach exploits the well known relationship of regular languages to semigroups, the second one tries to find a property of a finite automaton (usually in terms of so called forbidden patterns) equivalent to the property that the language recognized by the automaton is in the given level.

In this paper, we propose another, logical, approach to the problem. From [Th82, PP86] it follows that the problem is similar in formulation to some traditional decidability problems of logic. Our main observation is that one can

---

* Partly supported by the Alexander von Humboldt Foundation, by a grant of the Russian Ministry of Education and by RFBR Grant 00-01-00810.

A. Ferreira and H. Reichel (Eds.): STACS 2001, LNCS 2010, pp. 539–550, 2001.

apply in this situation some old facts known as preservation theorems (see e.g. [Ro63, Ma71]), as well as a theorem of Higman [CKa91]. Observing that the corresponding conditions are interpretable in the Rabin tree theory, we get new, purely logical and short proofs of some known facts on decidability. This might be of methodological interest. Our approach is applicable also to some other similar situations yielding several new results.

The rest of our paper is organized as follows: in Section 2 we consider some versions of the Straubing hierarchy, in Section 3 some versions of the Brzozowski hierarchy, in Section 4 we discuss the role of the empty word and relationships of our versions to the original Straubing and Brzozowski hierarchies, in Section 5 we discuss some relevant results and possible future work.

We close this introduction with reminding notation used throughout the paper. Let A be an alphabet, i.e. a finite nonempty set. Let $A^*(A^+)$ denotes the set of all words (resp., of all nonempty words) over A. As usual, the empty word is denoted by ε, the length of a word u by $|u|$, and the concatenation of words u and v by uv. Concatenation of languages X, Y is denoted XY. For $u = u_0 \ldots u_n \in A^+$ and $i \leq j \leq n$, let $u[i,j]$ denote the segment (or factor) of u bounded by i, j (including the bounds).

2   Straubing-Type Hierarchies

A word $u = u_0 \ldots u_n \in A^+$ may be considered as a structure $\mathbf{u} = (\{0, \ldots, n\}; <, Q_a, \ldots)$, where $<$ has its usual meaning and $Q_a(a \in A)$ are unary predicates on $\{0, \ldots, n\}$ defined by $Q_a(i) \leftrightarrow u_i = a$. As is well known (see e.g. [MP71, Th82, PP86]), there is a close relationship between star-free languages and classes of models $\mathbf{u}$ of sentences of signature $\sigma = \{<, Q_a, \ldots\}$ (in this section, "sentence" means "first order formula of signature σ without free variables"; the only exception is the proof of Theorem 2.1 below where we need also sentences of another kind). Note that the alphabet A is fixed throughout the paper, hence we omit the alphabet from our notation.

Let us consider a first order theory CLO of signature σ that is closely related to the theory of regular languages. The axioms of CLO state that $<$ is a linear ordering and that any element satisfies exactly one of the predicates $Q_a(a \in A)$. Models of CLO are called colored (more precisely, A-colored) linear orderings. We use letters like $\mathbf{u}, \mathbf{v}, \ldots$ (respectively, $\mathbf{U}, \mathbf{V}, \ldots$) to denote finite (respectively, countable) models of CLO. As usual, $\mathbf{U} \subseteq \mathbf{V}$ denotes that $\mathbf{U}$ is a substructure of $\mathbf{V}$. For a sentence ϕ, let M_ϕ be the set of all countable models of CLO satisfying ϕ, in symbols $M_\phi = \{\mathbf{U}|\mathbf{U} \models \phi\}$. Note that any finite model of CLO is isomorphic to a structure $\mathbf{u}$ from the preceding paragraph, for a unique $u \in A^+$. The relation $\subseteq$ induces a partial ordering on A^+ that will be denoted by the same symbol.

For $n > 0$, let Σ_n^0 denote the set of all sentences in prenex normal form starting with the existential quantifier and having $n - 1$ quantifier alternations.

Let S_n be the set of sentences equivalent to a Σ_n^0-sentence (modulo theory CLO). In other words, $S_n = \{\psi | \exists \phi \in \Sigma_n^0(M_\psi = M_\phi)\}$. Let $\check{S}_n$ be the dual set for S_n, i.e. the set of sentences equivalent to negations of S_n-sentences. Let $B(S_n)$ be the set of sentences equivalent to a Boolean combination of Σ_n^0-sentences. Then we have the following assertions.

Lemma 2.1. *(i) For any $n > 0$, $B(S_n) = S_{n+1} \cap \check{S}_{n+1}$.*

(ii) $\phi \in S_1$ iff $\forall \mathbf{U} \models \phi \forall \mathbf{V} \supseteq \mathbf{U}(\mathbf{V} \models \phi)$.

(iii) $\phi \in \check{S}_2$ iff the union of arbitrary chain $\mathbf{U}_0 \subseteq \mathbf{U}_1 \subseteq \cdots$ of models of ϕ is a model of ϕ.

(iv) $\phi \in S_2$ iff $\forall \mathbf{U} \models \phi \exists \mathbf{u} \subseteq \mathbf{U} \forall \mathbf{V}(\mathbf{u} \subseteq \mathbf{V} \subseteq \mathbf{U} \to \mathbf{V} \models \phi)$.

Proof. (i)—(iii) are well known results of logic (see e.g. the "preservation theorems" from [Ro63, Ma71, Sh67, CKe73]), while (iv) easily follows from (iii). Namely, if a sentence $\phi = \exists \bar{x} \forall \bar{y} \psi(\bar{x}, \bar{y})$, where ψ is a quantifier-free formula, is true in $\mathbf{U}$, then let $\mathbf{u}$ be the substructure of $\mathbf{U}$ with the universe $\{x_1, \ldots, x_n\}$, where $\bar{x} = (x_1, \ldots, x_n)$. Then $\mathbf{u}$ clearly satisfies the condition $\forall \mathbf{V}(\mathbf{u} \subseteq \mathbf{V} \subseteq \mathbf{U} \to \mathbf{V} \models \phi)$. Conversely, assume the righthandside condition of (iv) and prove that $\phi \in S_2$. Suppose the contrary; then, by (iii), there is a chain $\mathbf{U}_0 \subseteq \mathbf{U}_1 \subseteq \cdots$ of models of $\neg \phi$ the union $\mathbf{U}$ of which satisfies ϕ. Let $\mathbf{u}$ be a finite substructure of $\mathbf{U}$ satisfying $\forall \mathbf{V}(\mathbf{u} \subseteq \mathbf{V} \subseteq \mathbf{U} \to \mathbf{V} \models \phi)$. Choosing a number i with $\mathbf{u} \subseteq \mathbf{U}_i$, one gets a contradiction (take $\mathbf{U}_i$ in place of $\mathbf{V}$). This completes the proof.

Let $\{D_k\}_{k \geq 0}$ be the difference hierarchy (known also as the Boolean hierarchy) over S_1. Hence, D_0 is the set of false sentences, $D_1 = S_1$, $D_2(D_3, D_4)$ is the set of sentences equivalent to sentences of the form $\phi_0 \wedge \neg \phi_1$ (respectively, $(\phi_0 \wedge \neg \phi_1) \vee \phi_2$, $(\phi_0 \wedge \neg \phi_1) \vee (\phi_2 \wedge \neg \phi_3)$) and so on, where $\phi_i \in \Sigma_1^0$ (for more information on the difference hierarchy see e.g. [Ad65, Se95]). An *alternating chain* for a sentence ϕ is by definition a sequence $\mathbf{U}_0 \subseteq \cdots \subseteq \mathbf{U}_k$ of CLO-models such that $\mathbf{U}_i \models \phi$ iff $\mathbf{U}_{i+1} \models \neg \phi$; k is called the length of such a chain. Such a chain is called a 1-alternating chain, if $\mathbf{U}_0 \models \phi$. One could consider also infinite alternating chains (with order type ω).

The next assertions are also known in a more general form [Ad65, Se91].

Lemma 2.2. *(i) For any k, $D_k \cup \check{D}_k \subseteq D_{k+1}$.*

(ii) $\cup_k D_k = B(S_1)$.

(iii) $\phi \in D_k$ iff there is no 1-alternating chain for ϕ of length k.

We are ready to prove one of our main results on the decidability of some classes of sentences introduced above.

Theorem 2.1. *The classes of sentences $S_1, S_2, B(S_1), D_k (k \geq 0)$ are decidable.*

Proof. Let $T = \{0, 1\}^*$ and let r_0, r_1 be unary functions on T defined by $r_i(u) = ui (i \leq 1)$. According to the celebrated theorem of M. Rabin [Ra69], the monadic second order theory S2S of the structure $(T; r_0, r_1)$ is decidable. We shall use this fact in the following way: for any set $C \in \{S_1, S_2, B(S_1), D_k | k \geq 0\}$ and for any σ-sentence ϕ one can effectively construct a monadic second order sentence $\tilde{\phi}$ of signature $\{r_0, r_1\}$ such that $\phi \in C$ iff $\tilde{\phi} \in$ S2S (the monadic sentence $\tilde{\phi}$ is called the interpretation of the sentence ϕ). This is obviously enough.

We will use some well known facts on definability (by monadic second order formulas) in $(T; r_0, r_1)$ established in [Ra69]. First recall that the lexicographical ordering $\preceq$ on T is definable. Let $B \subseteq T$ be the set of all sequences $x101$ having no subsequence 101 except one at the end. Then B is definable and $(B; \preceq)$ has the order type of rationals. This implies that any countable linear ordering is isomorphic to an ordering of the form $(U; \preceq)$ with $U \subseteq B$. Hence, any countable model of CLO is isomorphic to a structure of the form $\mathbf{U} = (U; \preceq, Q_a, \dots)$ with $U \subseteq B$ and $Q_a \subseteq U$ for $a \in A$ (in this proof, we call such structures *inner structures*). In the monadic logic, one can use variables for subsets of T and even quantify over them. Hence, it is possible to speak about arbitrary inner structures. We can also speak about substructures because for any abstract models $\mathbf{U}$ and $\mathbf{V}$ of CLO, $\mathbf{U}$ is embeddable in $\mathbf{V}$ iff there are inner models $(U; \preceq, Q_a, \dots)$ and $(V; \preceq, Q'_a, \dots)$ isomorphic to $\mathbf{U}$ and $\mathbf{V}$, respectively, and satisfying $U \subseteq V$ and $Q_a \subseteq Q'_a (a \in A)$.

Note also that for any fixed σ-sentence ψ the set of all inner structures $\mathbf{U}$ satisfying ψ is definable (i.e., if ψ is $\forall x \exists y (x \leq y \wedge Q_a(y))$ then $\mathbf{U} \models \psi$ iff $\forall x \in U \exists y \in U (x \preceq y \wedge Q_a(y)))$. In particular, the set of all inner models of CLO is definable.

Now let us return to the proof of the theorem. Let e.g. $C = S_1$ and ϕ be a given σ-sentence. Let $\tilde{\phi}$ be a monadic sentence expressing that for any inner model $\mathbf{U}$ of CLO satisfying ϕ and any inner model $\mathbf{V}$ of CLO extending $\mathbf{U}$, $\mathbf{V}$ satisfies ϕ. By Lemma 2.1 and remarks above, $\phi \in S_1$ iff $\tilde{\phi} \in S2S$. This completes the proof for the set S_1. Remaining cases are treated in the same way (in the case of S_2 one shall note that the class of finite subsets of T is also definable [Ra69]). This completes the proof.

Remark 2.1. The proof implies the known fact that the monadic second order theory of countable models of CLO is decidable.

Theorem 2.1 demonstrates ideas of our approach for a decision problem traditional for logic (though the results seem formally new). It turns out that, due to its abstract nature, the approach is also applicable in the context of automata theory, which we would like now to demonstrate. This application is founded on a close relationship of star–free regular languages to first order definability established in [MP71].

By remarks at the beginning of this section, there is a natural one-one correspondence between subsets of A^+ and classes of finite CLO-models closed under isomorphism. This induces some notions on words corresponding to notions on models introduced above; we will use some of these notions under the same names. Relate to any sentence ϕ the language $L_\phi^+ = \{u \in A^+ | \mathbf{u} \models \phi\}$. By [MP71], such languages are exactly the regular star–free languages. Let $S_n^+, B(S_n^+), D_k^+$ be defined as the corresponding classes above, but with L^+ in place of M; in particular, $S_n^+ = \{\psi | \exists \phi \in \Sigma_n^0 (L_\psi^+ = L_\phi^+)\}$. Then $\{B(S_n^+)\}_{n \geq 1}$ is the version of the Straubing hierarchy mentioned in the introduction.

Note that there is an evident relationship between classes $S_n^+, \ldots$ and corresponding classes without $+$, namely $S_n \subseteq S_n^+$ and so on. But the $+$-classes contain a lot of new sentences. E.g., we have $S_1^+ \not\subseteq B(S_1)$ (the sentence saying that the ordering is dense belongs to S_1^+ but not to S_2).

Recall [CKa96, Theorem 7.2] that a *well partial ordering* is a partial ordering such that for any nonempty subset X the set of all minimal elements of X is nonempty and finite.

Lemma 2.3. *(i)* $(A^+; \subseteq)$ *is a well partial ordering.*

(ii) $\phi \in D_k^+$ *iff there is no 1-alternating chain of words for ϕ of length k.*

(iii) $\phi \in B(S_1^+)$ *iff there is no infinite alternating chain of words for ϕ.*

(iv) $\phi \in B(S_1^+)$ *iff* $\forall \mathbf{U} \exists \mathbf{u} \subseteq \mathbf{U}(\forall \mathbf{v}(\mathbf{u} \subseteq \mathbf{v} \subseteq \mathbf{U} \to \mathbf{v} \models \phi) \lor \forall \mathbf{v}(\mathbf{u} \subseteq \mathbf{v} \subseteq \mathbf{U} \to \mathbf{v} \models \neg\phi)).$

Proof. (i) is a well known result of G. Higman (see e.g. [CKa96, Theorem 7.2]).

(ii) From left to right, the assertion follows from Lemma 2.2.(iii). Now let there is no 1-alternating chain of words for ϕ of length k; we have to show $\phi \in D_k^+$. For simplicity of notation, consider only typical particular case $k = 2$; then there are no words $u_0, u_1, u_2 \in A^+$ with $u_0 \subseteq u_1 \subseteq u_2$ and $\mathbf{u}_0 \models \phi, \mathbf{u}_1 \models \neg\phi, \mathbf{u}_2 \models \phi$. Let $C_0 = \{u \in A^+ | \exists u_0 \in A^+(u_0 \subseteq u \land \mathbf{u}_0 \models \phi)\}$ and $C_1 = \{u \in A^+ | \exists u_0, u_1 \in A^+(u_0 \subseteq u_1 \subseteq u \land \mathbf{u}_0 \models \phi \land \mathbf{u}_1 \models \neg\phi)\}$. One easily checks that $L_\phi^+ = C_0 \setminus C_1$. By (i), any of C_0, C_1 is either empty or of the form $\{v \in A^+ | v_0 \subseteq v \lor \ldots \lor v_m \subseteq v\}$ for some $m \geq 0$ and $v_0, \ldots, v_m \in A^+$. This easily implies that $C_i = L_{\phi_i}^+$ for some $\phi_i \in \Sigma_1^0 (i \leq 1)$. Then $L_\phi^+ = L_{\phi_0 \land \neg\phi_1}^+$. Hence, $\phi \in D_2^+$ completing the proof.

(iii) From left to right, the assertion follows from (ii) and the equality $B(S_1^+) = \cup_k D_k^+$. It remains to show that for any $\phi \notin B(S_1^+)$ there is an infinite alternating chain of words. By (ii), there are alternating chains of words for ϕ of arbitrary finite length.

Let ω^* be the set of all finite sequences of natural numbers, including the empty sequence ε. We construct a partial function $u : \omega^* \to A^*$ as follows. Let $u(\varepsilon) = \varepsilon$ and suppose, by induction on $|\tau|$, that $u(\tau)$ is already defined. If $|\tau|$ is even then find $m \in \omega$ and words $v_0, \ldots, v_m \in A^+$ enumerating without repetitions the $\subseteq$-minimal elements in $X = \{v \in A^+ | u(\tau) \subseteq v \land \mathbf{v} \models \phi\}$. Then we set $u(\tau i) = v_i$ for $i \leq m$ and $u(\tau i)$ is undefined for $i > m$. For $|\tau|$ odd, the definition is similar, but we use the set $X = \{v \in A^+ | u(\tau) \subseteq v \land \mathbf{v} \models \neg\phi\}$.

From (i) and (ii) easily follows that $\{\tau \in \omega^* | u(\tau)$ is defined$\}$ is an infinite finitely branching tree (under the relation of being an initial segment). By König's lemma, there is an infinite path through this tree. The image of this path under u provides the desired infinite alternating chain for ϕ.

(iv) Let $\phi \in B(S_1^+)$, then $L_\phi^+ = L_\psi^+$ for a Boolean combination ψ of Σ_1^0-sentences. Note that $\psi, \neg\psi \in S_2$ and any $\mathbf{U}$ satisfies one of $\psi, \neg\psi$. Hence, the condition on the righthandside of (iv) follows from Lemma 2.1.(iii).

Conversely, suppose that $\phi \notin B(S_1^+)$. By (i), there is an infinite alternating chain $\mathbf{u}_0 \subseteq \mathbf{u}_1 \subseteq \ldots$ for ϕ consisting of finite models of CLO. Then $\mathbf{U} = \cup_k \mathbf{u_k}$ is a countable model of CLO for which the condition on the righthandside of (iv) is false. This completes the proof.

Repeating the proof of Theorem 2.1, one immediately gets

Theorem 2.2. *Classes* $S_1^+, B(S_1^+), D_k^+ (k \geq 0)$ *are decidable.*

Remark 2.2. Till now, we were unable to prove (by purely logical means) the known fact that the class S_2^+ is decidable.

Note that Lemma 2.3 and Theorem 2.2 provide new, shorter proofs for several known facts from automata theory (cf. e.g. [St85,Pin86,SW99]). E.g., decidability of $B(S_1^+)$ is equivalent (using a simple observation of Section 4 below) to the well-known result on decidability of the class of so called piecewise testable languages.

Our method is also applicable to some other similar situations, and now we want to give a couple of examples. There are several natural modifications of the operation $\phi \mapsto L_\phi^+$, among the most popular are ω-languages $L_\phi^\omega = \{\alpha : \omega \to A | \alpha \models \phi\}$ and Z-languages (Z is the set of integers) $L_\phi^Z = \{\alpha : \omega \to A | \{\alpha \models \phi\}$, where α is the structure defined similarly to the case of finite words (one could even consider "words" over more exotic linear orderings, say rationals or ω^2). Such operations induce corresponding classes of sentences $S_n^\omega, B(S_n^\omega), D_k^Z$ and so on. Are such classes of sentences decidable?

Till now, we were unable to answer this question using the methods developed above (the problem is that we do not see an appropriate analog of Lemma 2.3 for the infinite words). But the methods become applicable if we add to infinite words the finite ones, i.e. if we consider "languages" like $L_\phi^{\omega+} = L_\phi^\omega \cup L_\phi^+$, which are also traditional objects of automata theory, and the corresponding classes of sentences $S_n^{\omega+}, \ldots$. Let us formulate the analog of Theorem 2.2 for ω-words (similar results hold also for other kinds of infinite words).

Theorem 2.3. *Classes* $S_1^{\omega+}, B(S_1^{\omega+}), D_k^{\omega+} (k \geq 0)$ *are decidable.*

Proofsketch. From Lemma 2.1.(iii) it follows that if $\phi \in \Sigma_2^0$ then $L_\phi^{\omega+}$ is approximable (i.e., for any ω-word $\alpha \in L_\phi^{\omega+}$ there is a finite word $u \subseteq \alpha$ such that $\mathbf{v} \models \phi$ for any finite word v with $u \subseteq v \subseteq \alpha$). Repeating the proof of Lemma 2.3, one obtains analogs of assertions (ii), (iii) and (iv) for the classes $B(S_1^{\omega+}), D_k^{\omega+} (k \geq 0)$; but one have to add to the righthandsides of these assertions the condition that both $L_\phi^{\omega+}$ and $L_{\neg\phi}^{\omega+}$ are approximable.

With analog of Lemma 2.3 at hand, it is easy to adjust also the proof of Theorem 2.1 to our case. In place of the set B we shall take now the set $B_1 = \{1^k | k < \omega\}$; it is definable and $(B_1; \preceq)$ has order type ω. It remains to modify the notion of inner structures in such a way that their universes are subsets of B_1. This completes the proof.

3   Brzozowski-Type Hierarchies

Here we shall consider some versions of the well known Brzozowski hierarchy. Some results of this Section have some relevance to independent papers [T99, GS00, S00]. I am grateful to an anonymous referee for hints to these papers.

Following [Th82] (with some minor changes), we enrich the signature σ of the preceding section to the signature $\sigma' = \sigma \cup \{\bot, \top, p, s\}$, where $\bot$ and $\top$ are constant symbols while p and s are unary function symbols ($\bot, \top$ are assumed to denote the least and the greatest elements, while p and s are respectively predecessor and successor functions). Let us also add to the axioms of CLO the following axioms:

$$\forall x (\bot \leq x \leq \top),$$
$$\forall x (p(x) \leq x \wedge \neg \exists y (p(x) < y < x)),\ \forall x (x \leq s(x) \wedge \neg \exists y (x < y < s(x))),$$
$$\forall x > \bot (p(x) < x)\ \text{and}\ \forall x < \top (x < s(x)).$$

We denote the resulting theory CLO'. For models $\mathbf{U}, \mathbf{V}$ of this theory, $\mathbf{U} \subseteq' \mathbf{V}$ means that $\mathbf{U}$ is a substructure of $\mathbf{V}$ respecting all symbols from σ'.

There is also a "relational" version of CLO' defined as follows. Let $\sigma'' = \sigma \cup \{\bot, \top, S\}$, where S is a binary predicate symbol ($\bot, \top$ are as above, while S denotes the successor predicate). Let CLO'' be obtained from CLO by adjoining the axioms

$$\forall x (\bot \leq x \leq \top),$$
$$\forall x, y (S(x, y) \leftrightarrow x < y \wedge \neg \exists z (x < z < y)),$$
$$\forall x < \top \exists y S(x, y)\ \text{and}\ \forall x > \bot \exists y S(y, x).$$

Using the standard procedure of extending a theory by definable predicate and function symbols (see e.g. [Sh67]), one easily sees that CLO' and CLO'' are essentially the same theory (e.g., every model of one theory may be in a unique way considered as a model of another, the natural translations respect classes of sentences S_n and analogs of other classes from Section 2, any of these classes modulo one theory is decidable if and only if it is decidable modulo the other theory, and so on). For this reason our notation will not distinguish between these theories.

From the axioms easily follows that countable CLO'-models consist of all finite CLO-models and all countably infinite CLO-models of order type $\omega + Z \cdot L + \omega^-$, where ω, ω^-, Z are respectively order types of positive, negative and all integers, L is a countable (possibly empty) linear ordering, $Z \cdot L$ is the linear ordering obtained by inserting a copy of Z in place of any element of L, and $+$ is the operation of "concatenation" of linear orderings.

For the theory CLO' the analogs of Lemmas 2.1 and 2.2 hold true with some evident changes in formulation (say, the righthandside of 2.1.(iv) now looks like $\forall \mathbf{U} \models \phi \exists \mathbf{u} \subseteq \mathbf{U} \forall \mathbf{V} (\mathbf{u} \subseteq \mathbf{V} \subseteq' \mathbf{U} \rightarrow \mathbf{V} \models \phi)$, where $\subseteq$ has the same meaning as in Section 2 and $\mathbf{u}$ is a finite CLO-model).

Repeating now the proof of Theorem 2.1, we immediately get the following assertion, in which classes of sentences are defined just as in Section 2, but modulo theory CLO'.

Theorem 3.1. *The classes of sentences $S_1, S_2, B(S_1), D_k(k \geq 0)$ modulo theory CLO' are decidable.*

Remark 3.1. As in Section 2, the proof of Theorem 3.1 implies the decidability of the monadic second-order theory of the class of countable CLO'-models.

We see that for the case of all countable "words" the theory CLO' is treated quite similarly to the theory CLO.

Let us now turn to finite words. Classes of sentences $S_n^+, B(S_n^+), D_m^+$ are defined by analogy with Section 2. Again, as in Section 2, these classes include the corresponding classes without +, but the converse inclusions are far from being true. E.g., the sentence $\exists x Q_a(x) \wedge \forall x > \bot(Q_a(x) \rightarrow \exists y(\bot < y < x \wedge Q_a(y)))$ belongs to S_1^+ but not to S_2.

The treatment of the +-classes modulo theory CLO' turns out to be more complicated, as compared with CLO. A reason is that if $\mathbf{U} \subseteq' \mathbf{V}$ and one of the CLO'-models $\mathbf{U}, \mathbf{V}$ is finite then $\mathbf{U} = \mathbf{V}$. Hence, the analog of Lemma 2.3 is false.

In this situation, the following notion from [St85] is of some use. Let $u = u_1 \ldots u_m$ and $v = v_1 \ldots v_n$ be words from A^+, $u_i, v_j \in A$. A k-*embedding* from u to v is an increasing function $\theta : \{1, \ldots, m\} \rightarrow \{1, \ldots, n\}$ such that

(i) $\theta(j) = j, j = 1, \ldots, min(k, m)$,
(ii) $\theta(m - j) = n - j, j = 0, \ldots, min(k - 1, m - 1)$,
(iii) $u_{i+j} = v_{\theta(i)+j}, i = 1, \ldots, m, j = 0, \ldots, k, i + j \leq m$.

This means that u is a subword of v including the first k letters of v and the last k letters and such that any letter used to build u is followed by the same k letters in u and in v.

We write $u \leq^k v$ to denote that there is a k-embedding from u to v. For finite CLO-models $\mathbf{u}$ and $\mathbf{v}$, we write $\mathbf{u} \subseteq^k \mathbf{v}$ to denote that $\mathbf{u} \subseteq \mathbf{v}$ and the identity function is a k-embedding from u to v (u and v are words corresponding to the models as in Section 2). With some evident modifications we may apply the last relation also to countably infinite CLO-models.

We concentrate on formulations and analogies with Section 2, skipping (following a referee suggestion) rather technical proofs.

Lemma 3.1. *(i) If $u \leq^{k+1} v$ then $u \leq^k v$.*

(ii) $\leq^k$ is a partial ordering.

(iii) $\leq^0$ coincides with $\subseteq$.

(iv) If $u \leq^k v$ then $au \leq^k av$ for any $a \in A$.

(v) For all $\mathbf{u}$ and k, there is an existential σ'-sentence ϕ_u^k such that $\mathbf{u} \subseteq^k \mathbf{U}$ iff $\mathbf{U} \models \phi_u^k$.

Let E^k be the set of sentences equivalent in the theory CLO' to a finite conjunction of finite disjunctions of sentences $\phi_u^k (u \in A^+)$. Let $\{D_n^k\}_n$ be the difference hierarchy over E^k. Then we have the following analog of Lemma 2.3.(i)—(iii).

Lemma 3.2. *(i)* $(A^+; \subseteq^k)$ *is a well partial ordering.*

(ii) $\phi \in E^k$ *iff* L_ϕ^+ *is closed upwards under* $\leq^k$.

(iii) $\phi \in D_n^k$ *iff there is no 1-alternating* $\subseteq^k$-*chain of words for* ϕ *of length* n.

(iv) $\phi \in B(E^k)$ *iff there is no infinite alternating* $\subseteq^k$-*chain of words for* ϕ.

The analog of Lemma 2.3.(iv) is more intricate. In the following assertion the boldface letters have the same meaning as in Section 2.

Lemma 3.3. *(i) If* $\mathbf{u} \subseteq \mathbf{v} \subseteq^k \mathbf{U}$ *and* $\mathbf{u} \subseteq^k \mathbf{U}$ *then* $\mathbf{u} \subseteq^k \mathbf{v}$.

(ii) $\phi \in B(E^k)$ *iff* $\forall \mathbf{U} \exists \mathbf{u} \subseteq \mathbf{U} (\forall \mathbf{v} (\mathbf{u} \subseteq \mathbf{v} \subseteq^k \mathbf{U} \to \mathbf{v} \models \phi) \vee \forall \mathbf{v} (\mathbf{u} \subseteq \mathbf{v} \subseteq^k \mathbf{U} \to \mathbf{v} \models \neg \phi))$.

Repeating now the argument from Section 2, we get the following generalization of Theorem 2.2 (by Lemma 3.1.(iii), Theorem 2.2 is obtained if one takes $k = 0$).

Theorem 3.2. *For all k and n, classes D_n^k and $B(E^k)$ are decidable.*

Let us now show that E^{k+1} contains many new sentences as compared with D_n^k.

Lemma 3.4. *If the alphabet A contains at least two letters then $E^{k+1} \not\subseteq B(E^k)$ for any k.*

Now we shall relate the classes D_n^k and D_n^+. Let $n \geq 1, w_1, \ldots, w_n \in A^+, l_i = |w_i|$ and $w_1 \ldots w_n = a_1 \ldots a_m (a_j \in A, m = l_1 + \cdots + l_n)$. Let $\phi(w_1, \ldots, w_n)$ be a Σ_1^0-sentence of signature σ'' saying that there exist $x_1 < \cdots < x_m$ such that $x_1 = \bot, x_m = \top, Q_{a_i}(x_i)$ for $i = 1, \ldots, m$ and $S(x_i, x_{i+1})$ for $i \in \{1, \ldots, m\} \setminus \{l_1, l_1 + l_2, \ldots, l_1 + \cdots + l_{n-1}\}$.

Lemma 3.5. *(i)* $u \models \phi(w_1, \ldots, w_n)$ *iff* $u = w_1 v_1 w_2 v_2 \ldots w_n$ *for some* $v_1, \ldots, v_{n-1} \in A^*$.

(ii) For any $\phi \in \Sigma_1^0, L_\phi^+ \neq \emptyset$, *there is a disjunction* ψ *of sentences of the form* $\phi(w_1, \ldots, w_n)$ *satisfying* $L_\psi^+ = L_\phi^+$.

Now we can state the desired relationship.

Lemma 3.6. *(i)* $S_1^+ = \cup_k E^k$.

(ii) For any n, $D_n^+ = \cup_k D_n^k$.

(iii) $B(S_1^+) = \cup_{n,k} D_n^k$.

Theorem 3.2 together with a result from [St85] implies

Corollary 3.1. *The class $B(S_1^+)$ is decidable.*

Corollary 3.1 is equivalent to the well-known result that the class of so called languages of dot-depth one is decidable.

Remark 3.2. Unfortunately, results of this section are not so complete and elegant as those in Section 2. The proof of the corollary is not completely satisfactory from the point of view of methodology of our paper, because it uses an automata–theoretic argument (in the proof of the cited result from [St85]).

4   The Empty Word

Here we relate the hierarchies considered above to the "real" Straubing and Brzozowski hierarchies which classify subsets of A^* (rather than A^+). We state a simple relationship that aims to avoid annoying discussions (and sometimes even confusions) caused by the role of the empty word ε in this context.

The Straubing hierarchy is defined as follows (see [PP86]): let $\mathcal{B}_0 = \mathcal{A}_0 = \{\emptyset, A^*\}$; let $\mathcal{B}_{n+1}$ be the closure of $\mathcal{A}_n$ under $\cap, \cup$, and the operation relating to languages X, Y and a letter $a \in A$ the concatenation language XaY; finally, let $\mathcal{A}_{n+1} = B(\mathcal{B}_{n+1})$ be the Boolean closure of $\mathcal{B}_{n+1}$. The sequence $\{\mathcal{B}_n\}$ is known as *Straubing hierarchy*.

In [PP86], a natural logical description of the introduced classes of languages was established. Namely, classes of sentences Σ_n and Γ_n were found such that $\mathcal{B}_n = \{L_\phi | \phi \in \Sigma_n\}$ and $\mathcal{A}_n = \{L_\phi | \phi \in \Gamma_n\}$. Here L_ϕ is defined similarly to the the language L_ϕ^+ in Section 2, but now the empty structure is also admitted (with a natural notion of satisfaction).

Let $\mathcal{S}_n = \{L_\phi^+ | \phi \in S_n^+\}$, where S_n^+ is the class from Section 2. For $\mathcal{X} \subseteq P(A^+)$, let $\mathcal{X}^\varepsilon = \{X \cup \{\varepsilon\} | X \in \mathcal{X}\}$. Then the desired relationship between introduced classes looks as follows.

Theorem 4.1. *For any $n > 0$, $\mathcal{B}_n = \mathcal{S}_n \cup \mathcal{S}_n^\varepsilon$ and $\mathcal{A}_n = B(\mathcal{S}_n) \cup B(\mathcal{S}_n)^\varepsilon$.*

Proofsketch. First note that $\mathcal{S}_n \subseteq \mathcal{B}_n$ (if $X \in \mathcal{S}_n$, then $X = L_\phi^+$ for a sentence $\phi \in S_n^+ \subseteq \Sigma_n$ starting with the existential quantifier; hence $\varepsilon \not\models \phi$ and $X = L_\phi \in \mathcal{B}_n$.)

The desired equalities are checked by induction on n. We have already proven that $\mathcal{S}_1 \subseteq \mathcal{B}_1$. Note that $\{\varepsilon\} = L_\phi$, where ϕ is $\forall x (x \neq x)$, hence $\{\varepsilon\} \in \mathcal{B}_1$. But $\mathcal{B}_1$ is closed under $\cup$, so $\mathcal{S}_1^\varepsilon \subseteq \mathcal{B}_1$ and $\mathcal{S}_1 \cup \mathcal{S}_1^\varepsilon \subseteq \mathcal{B}_1$.

For the converse, recall that $\mathcal{B}_1$ is a closure of $\mathcal{A}_0$, hence for proving the inclusion $\mathcal{B}_1 \subseteq \mathcal{S}_1 \cup \mathcal{S}_1^\varepsilon$ it suffices to show that the class $\mathcal{S}_1 \cup \mathcal{S}_1^\varepsilon$ contains $\mathcal{A}_0$ and is closed under $\cup, \cap$ and the operation XaY. Only the last assertion is not evident, so let us deduce $XaY \in \mathcal{S}_1 \cup \mathcal{S}_1^\varepsilon$ from $X, Y \in \mathcal{S}_1 \cup \mathcal{S}_1^\varepsilon$. By the cited result from [PP86], $X = L_\phi$ and $Y = L_\psi$ for some $\phi, \psi \in \Sigma_1$. Let θ be $\exists x (Q_a(x) \wedge \phi^{(<x)} \wedge \psi^{(>x)})$, where $\phi^{(<x)}$ and $\psi^{(>x)})$ are evident relativizations of ϕ and ψ, respectively. By definition of Σ_1 [PP86], $\theta \in S_1^+$, hence $XaY \in \mathcal{S}_1$.

The equality $\mathcal{A}_1 = B(\mathcal{S}_1) \cup B(\mathcal{S}_1)^\varepsilon$ is easy, which completes the induction basis. The argument of induction step is almost the same as for the basis. This completes the proof.

Let $\{\mathcal{D}_{n,k}\}_k$ be the difference hierarchy over $\mathcal{S}_n$ and $\{\mathcal{D}'_{n,k}\}_k$ be the difference hierarchy over $\mathcal{B}_n$. Using Theorem 4.1 and an evident set-theoretic argument, we get

Corollary 4.1. *For all n and k, $\mathcal{D}'_{n,k} = \mathcal{D}_{n,k} \cup \mathcal{D}_{n,k}^\varepsilon$.*

A similar relationship exists between the Brzozowski hierarchy and the corresponding classes from Section 3.

5   Conclusion

We see that some problems of automata theory not only may be formulated in a logical form, they can be even solved by logical means. It is natural to ask a general logical question generalizing problems considered in Sections 2 and 3. For a given theory T, let S_n be the set of sentences equivalent in the theory T to a Σ_n^0-sentence. Let S_n^+ be defined similarly but using the equivalence in finite structures. One can define also classes $D_{n,k}(D_{n,k}^+)$ of the difference hierarchy over S_n (respectively, over S_n^+), and even classes of the fine hierarchy over $\{S_n\}$ (see [Se91, Se95]).

The general question is to determine in what cases the introduced classes of sentences are decidable. Problems considered in Sections 2 and 3 are obtained when one considers the theories CLO and CLO' in place of T.

The question is quite traditional for mathematical logic, hence one could hope to find some relevant information in the logical literature. Indeed, in [Ma71] we find (with the reference to source papers) the following result: if T is undecidable then so are S_n for all $n > 0$. But what about the more interesting case of a decidable theory T (which is the case for CLO and CLO')? It seems that, strangely enough, there is almost nothing known about this natural problem. From results in [Se91a, Se92] (which rely upon Tarski elementary classification of Boolean algebras) one can easily deduce the following result.

Theorem 5.1. *Modulo theory T of Boolean algebras, all classes $D_{n,k}$ (and even all classes of the fine hierarchy) are decidable.*

Proof. In [Se91a, Se92] we have described an effective sequence of sentences $\phi_0, \phi_1, \dots$ such that any sentence ϕ is equivalent (modulo theory of Boolean algebras) to exactly one of ϕ_i, and position of any ϕ_i in the hierarchy $\{D_{n,k}\}$ was completely determined. This evidently implies the desired algorithm completing the proof.

It seems interesting to consider analogs of Theorem 5.1 for other popular decidable theories, say for Abelian groups.

We hope that methods developed in this paper may be used in some other similar situations, say for the case of tree languages.

Acknowledgment

A good deal of this work was done during my stay at RWTH Aachen in spring of 1999. I am grateful to Wolfgang Thomas for hospitality and for many useful bibliographical hints.

References

A65.   J. Addison. The method of alternating chains. In: *The theory of models*, North Holland, Amsterdam, 1965, 1—16.

CKa91.   C. Choffrut and J. Karumäki. Combinatirics of words. In: *Handbook of Formal Languages* (G. Rozenberg and A. Salomaa, ed.), v. 1 Springer, 1996, 329—438.

CKe73.   C.C. Chang, H.J. Keisler. *Model theory*, North Holland, Amsterdam, 1973.

GS00.   C. Glasser, H. Schmitz. The Boolean structure of dot-depth one. *J. Automata, Languages and Combinatorics*, to appear.

Ma71.   A.I. Malcev. *Algebraic Systems*, Springer, Berlin, 1971.

MP71.   R. McNaughton and S. Papert. *Counter–free automata*. MIT Press, Cambridge, Massachusets, 1971.

PP86.   D. Perrin and J.E. Pin. First order logic and star–free sets. *J. Comp. and Syst. Sci.*, 32 (1986), 393—406.

Pin86.   J.E. Pin. *Varieties of Formal Languages*. Plenum, London, 1986.

Pin94.   J.E. Pin. Logic on words. *Bulletin of the EATCS*, 54 (1994), 145—165.

Ra69.   M.O. Rabin. Decidability of second order theories and automata on infinite trees. *Trans. Amer. Math. Soc.*, 141 (1969), 1—35.

Ro63.   A. Robinson. *Introduction to Model Theory and to the Metamathematics of Algebra*, North Holland, Amsterdam, 1963.

S00.   H. Schmitz. Restricted temporal logic and deterministic languages. *J. Automata, Languages and Combinatorics*, 5 (2000), 325—342.

Se91.   V.L. Selivanov. Fine hierarchy of formulas. *Algebra i Logika*, 30 (1991), 568—583 (Russian, English translation: *Algebra and Logic*, 30 (1991), 368—379).

Se91a.   V.L. Selivanov. Fine hierarchies and definable index sets. *Algebra i logika*, 30, No 6 (1991), 705—725 (Russian, English translation: *Algebra and logic*, 30 (1991), 463—475).

Se92.   V.L. Selivanov. Computing degrees of definable classes of sentences. *Contemporary Math.*, 131, part 3 (1992), 657—666.

Se95.   V.L. Selivanov. Fine hierarchies and Boolean terms. *J. Symbolic Logic*, 60 (1995), 289—317.

Sh67.   J.R. Shoenfield. *Mathematical Lodic*, Addison-Wesley, 1967.

St85.   J. Stern. Characterizations of some classes of regular events. *Theor. Comp. Science*, 35 (1985), 17—42.

SW99.   H. Schmitz and K. Wagner. The Boolean hierarchy over level 1/2 of the Sraubing–Therien hierarchy, to appear (currently available at `http://www.informatik.uni-wuerzburg.de`).

Th82.   W. Thomas. Classifying regular events in symbolic logic. *J. Comp. and Syst. Sci.*,25 (1982), 360—376.

T99.   A.N. Trahtman. Computing the order of local testability. In: *Proc. 1st Int. Workshop on Descriptional Complexity of Automata, Grammars and Related Structures*, Magdeburg, 1999, 197—206.

Regular Languages Defined by Generalized First-Order Formulas with a Bounded Number of Bound Variables

Howard Straubing[1] and Denis Thérien[2]

[1] Computer Science Department, Boston College
Chestnut Hill, Massachusetts, USA 02467
[2] School of Computer Science, McGill University
Montréal, Québec, Canada H3A2A7

Abstract. We give an algebraic characterization of the regular languages defined by sentences with both modular and first-order quantifiers that use only two variables.

1 Introduction

One finds in the theory of finite automata a meeting ground between algebra and logic, where difficult questions about expressibility can be classified, and very often effectively decided, by appeal to the theory of semigroups. This line of research began with the work of McNaughton and Papert [7], who showed that the languages definable by first-order sentences over '$<$' are precisely the 'star-free' regular languages, and thus, by a theorem of Schützenberger, the languages whose syntactic monoids are *aperiodic*—that is, contain no nontrivial groups. Let us give an example of the kind of first-order formulas we are considering:

$$\exists x \exists y (Q_\sigma x \wedge Q_\sigma y \wedge x < y \wedge \neg \exists z (x < z \wedge z < y)).$$

This sentence is meant to be interpreted in words over a specified finite alphabet Σ that contains the letter σ. The variables in the sentence denote positions in the word (that is, integers between 1 and the length of the word, inclusive) and the subformula $Q_\sigma x$ means 'the letter in position x is σ'. The subformula $x < y \wedge \neg \exists z (x < z \wedge z < y))$ says that position x is to the left of position y, and that there is no position strictly between them (*i.e.,* that $y = x + 1$), and thus the whole sentence says 'there are two consecutive occurrences of σ'. We say that the sentence defines a language over Σ, namely the set of all strings that contain the factor $\sigma\sigma$.

Since McNaughton and Papert's work, researchers have investigated the expressibility of regular languages in various restrictions and extensions of first-order logic over $<$. For example, we can replace the predicate $x < y$ by the (weaker) predicate $y = x + 1$ (Beauquier and Pin [2]). We can permit the use of *modular* quantifiers, which count, modulo a fixed period, the number of positions of a string satisfying a given condition (Straubing, Thérien and Thomas [16]). For example, the sentence $\exists^{1 \bmod 2} x Q_\sigma x$ is interpreted to mean 'the number of positions containing σ is congruent to 1 modulo 2', and thus defines the set of strings containing an odd number of occurrences of σ.

A. Ferreira and H. Reichel (Eds.): STACS 2001, LNCS 2010, pp. 551–562, 2001.
© Springer-Verlag Berlin Heidelberg 2001

In all the cases considered, the family of regular languages obtained can be characterized in terms of the syntactic monoids or the syntactic morphisms of its members, and in most cases this characterization gives rise to an algebraic algorithm for deciding membership of a given language in the family. The book by Straubing [14] provides an exhaustive catalogue of such results.

Kamp [6] and later, Immerman and Kozen [5] showed that every first-order sentence over $<$ is equivalent to such a sentence in which only three variables are used. The number of bound variables that occur in a formula can be considered as a kind of expressibility resource, along with the kinds and depth of the quantifiers and the set of available atomic formulas.

Thérien and Wilke [17] considered the regular languages defined by sentences in which only two variables are used, and found that these, too, could be characterized in algebraic terms: A language L is definable by a sentence with two variables if and only if its syntactic monoid belongs to a particular family **DA** of finite aperiodic monoids. (We will give the definition of **DA** in the next section.) It was already known that the two-variable definable languages are precisely those definable in the fragment of propositional temporal logic that includes both the past and future versions of the Next and Eventually operators, but excludes the Until operator (Etessami, Vardi and Wilke [4]). Since it is possible to determine from the multiplication table of a finite monoid whether it belongs to **DA**, the Thérien-Wilke result provides an algorithm for determining whether a given regular language is definable in this fragment of temporal logic.

In the present paper we investigate the effect of bounding the number of variables in sentences that include the modular quantifiers $\exists^r \bmod n$ as well as ordinary first-order quantifiers, and we characterize, again in algebraic terms, the regular languages that are thereby defined.

We have shown that the three-variable property continues to hold for formulas that include modular quantifiers. That is,

Theorem 1. *Let ϕ be a sentence over $<$ containing first-order and modular quantifiers. Then ϕ is equivalent to such a sentence with only three variables.*

For formulas that contain *only* modular quantifiers, we have an even stronger result:

Theorem 2. *Let ϕ be a sentence over $<$ in which only modular quantifiers appear. Then ϕ is equivalent to such a sentence with only two variables.*

Our main theorem is that the languages L defined by two-variable sentences are characterized by membership of their syntactic monoids $M(L)$ in the pseudovariety **DA** $* \mathbf{G}_{sol}$, defined in Section 2:

Theorem 3. *Let Σ be a finite alphabet. A regular language $L \subseteq \Sigma^*$ is defined by a two-variable sentence over $<$ containing first-order and modular quantifiers, if and only if $M(L) \in \mathbf{DA} * \mathbf{G}_{sol}$.*

It is important to remark that while our main theorem permits us in many individual cases to show that a language is, or is not, two-variable definable, the general problem of determining membership in **DA** $* \mathbf{G}_{sol}$ is not known to be decidable.

Example. Let $\Sigma = \{\sigma, \tau\}$, and let L be the language defined by the regular expression $(\sigma\tau)^*$. Thus $w \in L$ if and only if w contains no occurrence of the factor $\sigma\sigma$ or $\tau\tau$, and w is either the empty string, or begins with σ and ends with τ. We saw above how to write a first-order sentence that says a string contains no occurrence of $\sigma\sigma$ or $\tau\tau$. We can say that a string is empty or begins with σ with the sentence: $\forall x(\forall y(\neg y < x) \to Q_\sigma x)$, and we similarly say that a string is empty or ends with τ. Thus L is definable by a first-order sentence. We could have obtained the same conclusion by constructing the syntactic monoid of L and verifying that it contains no nontrivial groups. A closer look at the syntactic monoid shows that the image of the word $\sigma\tau$ under the syntactic morphism is idempotent, but that the image of $\sigma\tau\sigma$ is not idempotent. This implies $M(L) \notin \mathbf{DA}$, so by the theorem of Thérien and Wilke cited above, L cannot be defined by a first-order sentence with only two variables. But L *is* definable by a two-variable sentence if we permit modular quantifiers: A string belongs to L if and only if it has even length, and has σ in all the odd-numbered positions and τ in all the even-numbered positions. Thus we can define L by the sentence

$$\exists^{0 \bmod 2} x(x = x) \wedge \forall y(Q_\sigma y \leftrightarrow \exists^{0 \bmod 2} x(x < y)).$$

This example shows that the situation is more complicated, and potentially more interesting, than what one might have supposed, since the modular quantifiers can be used to economically express properties that are not intrinsically periodic (that is, that do not require modular quantifiers for their expression). Incidentally, the same language is definable by another two-variable sentence whose modular quantifiers are all of modulus 3; in fact, *any* modulus greater than 1 will suffice. On the other hand, the set of strings over $\{\sigma, \tau\}$ that do not contain an occurrence of $\sigma\sigma$ has a syntactic monoid that is not in the pseudovariety $\mathbf{DA} * \mathbf{G}_{sol}$—this will follow from our results in Section 5 concerning the ideal structure of the monoids in this family—and thus, by Theorem 3, cannot be defined by a two-variable sentence.

2   Background from Semigroup Theory

A *semigroup* is a set together with an associative multiplication. If the semigroup contains a multiplicative identity element, then we call the semigroup a *monoid,* and we usually denote the identity by 1. If Σ is a finite alphabet, then Σ^*, the set of all strings over Σ, is a monoid with concatenation of strings as the multiplication. Σ^* is the *free monoid* on Σ: This means that if M is any monoid and $f : \Sigma \to M$ any map, then f extends to a unique homomorphism from Σ^* into M.

For a discussion of basic facts about semigroups and monoids, particularly as they pertain to the theory of automata (*i.e.,* recognition of regular languages by finite monoids, syntactic monoid and syntactic morphism, wreath product and semidirect product) the reader is referred to the books by Eilenberg [3] and Pin [9]. Here we briefly discuss the ideal structure of finite semigroups and some facts about pseudovarieties of finite semigroups and monoids.

Ideal Structure and Green's Relations. If S is a semigroup and I is a nonempty subset of S, then we say I is an *ideal* of S if $SI \subseteq I$ and $IS \subseteq I$. Similarly, we say that I is a *right ideal* of S if $IS \subseteq I$, and a *left ideal* if $SI \subseteq I$. If I is an ideal, then the set

$(S - I) \cup \{0\}$ forms a semigroup with multiplication $\times$ given by $s_1 \times s_2 = s_1 s_2$, if $s_1, s_2, s_1 s_2 \in S - I$, and $s_1 \times s_2 = 0$ otherwise. We denote this semigroup by S/I; this is the image of S under the homomorphism that collapses all the elements of I to a single element.

If $s, t \in S$ we write $s \leq_{\mathcal{J}} t$ if s belongs to the ideal $\{t\} \cup St \cup tS \cup StS$ generated by t. If $s \leq_{\mathcal{J}} t$ and $t \leq_{\mathcal{J}} s$, then we say that s and t are $\mathcal{J}$-equivalent and write $s \equiv_{\mathcal{J}} t$. The equivalence classes for this relation are called $\mathcal{J}$-classes. Similarly we define $\leq_{\mathcal{L}}$, $\leq_{\mathcal{R}}$, $\equiv_{\mathcal{L}}$, $\equiv_{\mathcal{R}}$, $\mathcal{L}$-class, and $\mathcal{R}$-class, by considering left and right ideals in place of two-sided ideals.

A $\mathcal{J}$-class is said to be *regular* if it contains an idempotent, that is, an element e such that $e^2 = e$. The *Rees Matrix Theorem*, which we now state, describes the structure of the regular $\mathcal{J}$-classes of a finite semigroup. If A, B are finite sets, G a finite group, and $P : B \times A \to G \cup \{0\}$ a map, then (A, B, G, P) denotes the semigroup $A \times G \times B \cup \{0\}$ with multiplication given by

$$(a, g, b)(a', g', b') = (a, g \cdot P(b, a') \cdot g', b'),$$

if $P(b, a') \neq 0$, and $(a, g, b)(a', g', b') = 0$ otherwise. For each regular $\mathcal{J}$-class J of a finite semigroup, there exist finite sets A and B, a finite group G, and a map $P : B \times A \to G \cup \{0\}$ such that the semigroup $J \cup \{0\}$ is isomorphic to (A, B, G, P). Under this isomorphism, the $\mathcal{R}$-classes contained in J are the sets $\{a\} \times G \times B$, where $a \in A$, the $\mathcal{L}$-classes are the sets $A \times G \times \{b\}$, where $b \in B$, and every $\mathcal{R}$-class contains at least one idempotent, as does every $\mathcal{L}$-class. We call (A, B, G, P) a *Rees matrix representation* of J.

A non-regular $\mathcal{J}$-class is called a *null* $\mathcal{J}$-class. If J is a null $\mathcal{J}$-class of a finite semigroup and $s, t \in J$, then $st \notin J$.

Pseudovarieties. A *pseudovariety* of finite semigroups is a family of finite semigroups that is closed under finite direct products, submonoids and homomorphic images. A pseudovariety of finite monoids is defined analogously. If $\mathbf{V}_1$ and $\mathbf{V}_2$ are pseudovarieties of finite monoids, then $\mathbf{V}_1 * \mathbf{V}_2$ is defined to be the pseudovariety generated by all wreath products $M_1 \circ M_2$, where $M_1 \in \mathbf{V}_1$, $M_2 \in \mathbf{V}_2$. If $\mathbf{V}_1$ is a pseudovariety of finite semigroups, and $\mathbf{V}_2$ a pseudovariety of finite monoids, then $\mathbf{V}_1^{-1}\mathbf{V}_2$ consists of all finite monoids M for which there exist finite monoids K, N and homomorphisms $\phi : K \to N$, $\psi : K \to M$, such that ψ is onto M, $N \in \mathbf{V}_2$, and, for each idempotent $e \in N$, the semigroup $\phi^{-1}(e)$ belongs to $\mathbf{V}_1$. $\mathbf{V}_1^{-1}\mathbf{V}_2$ is a pseudovariety of finite monoids.

In this paper we will be concerned with the following pseudovarieties of finite monoids: $\mathbf{A}$–the finite *aperiodic* monoids; that is, the finite monoids that contain no nontrivial group; $\mathbf{G}$–the finite groups; $\mathbf{G}_{sol}$–the finite solvable groups; $\mathbf{R}$–the finite $\mathcal{R}$-trivial monoids; that is, the finite monoids with one-element $\mathcal{R}$-classes; $\mathbf{DA}$–the finite aperiodic monoids each of whose regular $\mathcal{J}$-classes J is a subsemigroup. (That is, in each Rees matrix representation (A, B, G, P) of J, G is trivial and P never maps to 0.)

We will also consider the pseudovariety $\mathbf{LI}$ of finite semigroups consisting of all semigroups S such that $ese = e$ for all $e, s \in S$ with e idempotent. Such semigroups are called *generalized definite* or *locally trivial* in the literature.

Our main objects of study are pseudovarieties of the form $\mathbf{DA} * \mathbf{H}$, where $\mathbf{H}$ is a pseudovariety of finite groups. There are several alternative characterizations of this pseudovariety:

Lemma 1. *For any pseudovariety $\mathbf{H}$ of finite groups,*

$$\mathbf{DA} * \mathbf{H} = \mathbf{DA}^{-1}\mathbf{H} = \mathbf{LI}^{-1}(\mathbf{J}_1 * \mathbf{H}).$$

For every finite semigroup S there is a *reversed* semigroup S^{rev}, with the same underlying set as S, and with multiplication $\times$ given by $s \times t = ts$ for all $s, t \in S$. If $\mathbf{V}$ is a pseudovariety of semigroups or monoids, then $\mathbf{V}^{rev}$ denotes the pseudovariety consisting of the reversals of members of $\mathbf{V}$. If $\mathbf{H}$ is a pseudovariety of finite groups, then $\mathbf{H} = \mathbf{H}^{rev}$, since the map $h \mapsto h^{-1}$ is an anti-isomorphism of any group. Since obviously $\mathbf{DA} = \mathbf{DA}^{rev}$, we find that $\mathbf{DA}^{-1}\mathbf{H}$ is closed under reversal as well.

3   Formal Languages and Generalized First-Order Logic

Let Σ be a finite alphabet. We build formulas from the unary predicate symbols $\{Q_\sigma : \sigma \in \Sigma\}$, the binary predicate symbol $<$, variable symbols, the boolean connectives $\neg$ and $\wedge$, and two kinds of quantifier symbols: $\exists$ and $\exists^{r \bmod m}$, where $0 \le r < m$. The atomic formulas are those of the form $x < y$, where x and y are variable symbols, and $Q_\sigma x$, where $\sigma \in \Sigma$ and x is a variable symbol.

In our subsequent discussion, we will also use the boolean connectives $\vee$, $\rightarrow$, $\leftrightarrow$, as well as the universal quantifier symbol $\forall$. These are all definable in terms of the original base of symbols.

We have indicated in the introduction the meaning attached to the symbols in our formulas. (For a formal definition of the semantics of formulas, see Straubing [14].) If $w \in \Sigma^*$, and ϕ is a sentence, then we write $w \models \phi$ to mean w satisfies ϕ, and we say that the language $\{w \in \Sigma^* : w \models \phi\}$ is the *language defined by* ϕ. We will also consider *pointed words* (w, i), where $w \in \Sigma^*$ and $1 \le i \le |w|$. A formula ϕ with a single free variable is interpreted in such a pointed word, by substituting i for the free variable. We write $(w, i) \models \phi$ if the pointed word satisfies the formula.

We will define a number of operations on formulas, which we call *relativizations*. Let ϕ be a formula in which the variable x does not appear. The formula $\phi[< x]$ is constructed recursively by beginning with the outermost quantifiers of ϕ, and replacing each subformula $\exists^* y\psi$, where $\exists^*$ is either an ordinary existential quantifier or a modular quantifier, by $\exists^* y(y < x \wedge \psi[< x])$. (If ψ is an atomic formula then $\psi[< x]$ is identical to ψ.) Let ψ be a sentence, and let $w \in \Sigma^*$ with $|w| \ge i$. Let v be the prefix of w of length i. Then $(w, i) \models \psi[< x]$ if and only if $v \models \psi$. We define relativizations $\phi[\le x]$, $\phi[> x]$, and $\phi[\ge x]$ analogously.

Let θ be a *sentence* with the property that $w \models \theta$ if and only if every prefix v of w satisfies θ. Let ϕ be a formula. We define the relativized formula $\phi[\le \theta]$ by recursively replacing each subformula $\exists^* x\psi$ by

$$\exists^* x(\theta[\le x] \wedge \psi[\le \theta]).$$

Let $w \in \Sigma^*$, and let v be the longest prefix of w that satisfies θ. If ϕ is a sentence, then $v \models \phi$ if and only if $w \models \phi[\leq \theta]$.

Let ϕ, θ be as in the preceding paragraph. We define the relativized formula $\phi[> \theta]$ by recursively replacing each quantified subformula $\exists^* x \psi$ by

$$\exists^* x(\neg \theta[\leq x] \wedge \psi[> \theta]).$$

Let $w \in \Sigma^*$, and let $w = vv'$, where v is the longest prefix of w that satisfies θ. If ϕ is a sentence, then $v' \models \phi$ if and only if $w \models \phi[> \theta]$.

In the full paper, we will prove Theorem 1, that every sentence in our language is equivalent to a sentence in which only three variable symbols are used. Our real interest is in what can be expressed with only two variables. To this end, we now sketch the proof of Theorem 2 and a slight generalization (Theorem 4 below), which we will use to prove our main result, Theorem 3. Let $L \subseteq \Sigma^*$ be defined by a sentence in which only modular quantifiers are used. Then $M(L) \in \mathbf{G}_{sol}$ (Straubing, Thérien and Thomas [16]). Thus, by a theorem of Straubing [15], L can be constructed, starting from the empty language, by repeatedly applying boolean operations and the operations $K \mapsto < K, \sigma, r, n >$, where $\sigma \in \Sigma$, $0 \leq r < n$, and $< K, \sigma, r, n >$ denotes the set of strings w such that the number of factorizations $w = u\sigma v$ with $u \in K$ is congruent to r modulo n. We make the claim (stronger than Theorem 2) that L is definable by a *left-relativizable* two-variable sentence; that is, a two-variable sentence ϕ such that the formulas $\phi[< x]$ and $\phi[\leq x]$, which each have one free variable, are themselves equivalent to two-variable formulas. The claim follows by noting, first, that the empty language is defined by the sentence $\exists^{1 \bmod 2} x(x < x)$, which is certainly a left-relativizable two-variable sentence, and second, that if K is definable by a left-relativizable two-variable sentence ϕ, then $< K, \sigma, r, n >$ is defined by the following sentence ψ:

$$\exists^{r \bmod n} x(Q_\sigma x \wedge \phi[< x]).$$

By assumption, $\phi[< x]$ is equivalent to a two-variable formula. Observe now that $\psi[< y]$ is equivalent to

$$\exists^{r \bmod n} x((x < y) \wedge Q_\sigma x \wedge \phi[< x]),$$

which has two variables, and that $\psi[\leq y]$ is equivalent to the same formula with $x < y$ replaced by $x \leq y$. Thus ψ is a left-relativizable two-variable sentence.

Precisely the same reasoning shows that the smallest family of languages closed under boolean operations and the operations $K \mapsto < K, \sigma, r, n >$ and $K \mapsto K\sigma\Sigma^*$ is definable by a left-relativizable two-variable sentence. It follows from results of Stiffler [13] that this is the family of languages whose syntactic monoids belong to the pseudovariety $\mathbf{R} * \mathbf{G}_{sol}$. Thus we have:

Theorem 4. *If $L \subseteq \Sigma^*$ is a regular language with $M(L) \in \mathbf{R} * \mathbf{G}_{sol}$, then L is definable by a left-relativizable two-variable sentence.*

4   Formulas and Games

In this section we sketch the proof that every regular language definable by a two-variable sentence has its syntactic monoid in $\mathbf{DA} * \mathbf{G}_{sol}$. This is one direction of our main result, Theorem 3. We will need the following normal form result, whose proof will be given in the full paper:

Lemma 2. *Let $\theta(x)$ be a two-variable formula with a single free variable x. Then θ is equivalent to a two-variable formula in which an ordinary quantifier never appears within the scope of a modular quantifier.*

Let us fix integers $m > 1$ and $r \geq 0$, and let us treat as atomic formulas all formulas with one free variable using exclusively modular quantifiers of modulus m and quantifier depth no more than r. Observe that there are only finitely many inequivalent formulas of this form. We look at two-variable first-order formulas over this base of atoms. By the *depth* of such a formula we mean the depth of nesting of the ordinary first-order quantifiers. In view of Lemma 2, it is sufficient to prove that the syntactic monoid of any language defined by such a formula is in $\mathbf{DA} * \mathbf{G}_{sol}$.

For each $k \geq 0$ we define two equivalence relations, one on words, and the other on pointed words, both denoted $\equiv_k$: We say $w_1 \equiv_k w_2$ if and only if w_1 and w_2 satisfy the same two-variable sentences of depth k or less, and $(w_1, i) \equiv_k (w_2, j)$ if and only if the two pointed words satisfy the same two-variable formulas $\phi(x)$ (with one free variable) of depth k or less.

Here is an explicit description of $\equiv_0$: Let $\mathbf{H}_m$ be the pseudovariety of finite Abelian groups of exponent m, and let $\mathbf{H}_m^r$ be the pseudovariety consisting of all finite groups that have a normal series of length r or less in which every quotient group belongs to $\mathbf{H}_m$. For every finite alphabet Σ, $\mathbf{H}_m^r$ has a finite Σ-generated free object F. Let π be the canonical homomorphism from Σ^* onto F. It follows from results in [14] that two words are $\equiv_0$-equivalent if and only if they have the same image under π. Furthermore, two pointed words (w_1, i) and (w_2, j) are $\equiv_0$-equivalent if and only if there are factorizations $w_1 = u\sigma v$ and $w_2 = u'\sigma v'$ where $\sigma \in \Sigma$, $|u| = i - 1, |u'| = j - 1, \pi(u) = \pi(u')$ and $\pi(v) = \pi(v')$. From this follows the important fact that not only is $\equiv_0$ a congruence on words, but it is a *congruence on pointed words* in the sense that if $(w_1, i) \equiv_0 (w_2, j)$, $u_1 \equiv_0 u_2$, and $v_1 \equiv_0 v_2$, then $u_1(w_1, i)v_1 \equiv_0 u_2(w_2, j)v_2$. ($u_1(w_1, i)v_1$ is shorthand for $(u_1wv_1, i + |u_1|)$.)

For $k > 0$, we characterize $\equiv_k$ in terms of a variant, due to Wilke [18], of the Ehrenfeucht-Fraïssé game. The game is played on two pointed words (w_1, i) and (w_2, j). If these are not $\equiv_0$-equivalent, then Player I wins at once, in zero rounds. Otherwise, each round proceeds as follows. Think of each pointed word as an ordinary word with a pebble on one position. Player I picks one of the words and moves its pebble one or more positions to the left or right. Player II must now move the pebble in the other word in the same direction (left if Player I moved left, right if Player I moved right). The new pointed words (w_1, i') and (w_2, j') are required to be $\equiv_0$-equivalent—Player II loses if she cannot meet this requirement. If Player II can correctly respond for k successive rounds, then she wins the game. We can also play the game on words. In the first round, Player I places his pebble on a position in one of the words, and Player II pebbles a position in the other word. The resulting structures (w_1, i) (w_2, j) are required to be $\equiv_0$-equivalent, or Player II loses. Play then proceeds as above for $k - 1$ additional rounds.

It's easy to prove that the standard result for model-theoretic games holds for this variant:

Lemma 3. $(w_1, i) \equiv_k (w_2, j)$ *if and only if Player II has a winning strategy in the k-round game on these two pointed words.* $w_1 \equiv_k w_2$ *if and only if Player II has a winning strategy in the k-round game in these two words.*

It follows from this game characterization, and the fact that $\equiv_0$ is a congruence on pointed words, that $\equiv_k$ is a congruence of finite index on Σ^*. By Lemma 2, every language defined by a two-variable sentence is a union of $\equiv_k$-classes for some k, m and r. So it is enough to prove that the quotient monoid $\Sigma^* / \equiv_k$ belongs to $\mathbf{DA} * \mathbf{G}_{sol}$.

We will prove this by induction on k. $\Sigma^* / \equiv_0$ is the free Σ-generated group in $\mathbf{H}_m^r$, and thus is in $\mathbf{G}_{sol}$. The passage from 0 to 1 is a special case: It follows from a result in Straubing [14] that the syntactic monoid of any language defined by a sentence of depth 1 is in $\mathbf{J}_1 * *\mathbf{G}_{sol}$ (where $**$ denotes a symmetric version of the product $*$ of pseudovarieties). From a theorem of Rhodes and Tilson [11], this is the same as $\mathbf{J}_1 * \mathbf{G}_{sol}$. Since each $\equiv_1$-class is such a language, it follows that $\Sigma^* / \equiv_1 \in \mathbf{J}_1 * \mathbf{G}_{sol}$.

We now carry out the inductive step from $\equiv_k$ to $\equiv_{k+1}$, where $k \geq 1$. Since $\equiv_{k+1}$ refines $\equiv_k$, there is a homomorphism from $\Sigma^* / \equiv_{k+1}$ to $\Sigma^* / \equiv_k$. We claim that the preimage of each idempotent under this homomorphism is in $\mathbf{LI}$. Since $\mathbf{DA} * \mathbf{G}_{sol} = \mathbf{LI}^{-1}(\mathbf{J}_1 * \mathbf{G}_{sol})$ (by Lemma 1) and $\mathbf{LI}^{-1}(\mathbf{LI}^{-1}\mathbf{V}) = \mathbf{LI}^{-1}\mathbf{V}$ for every pseudovariety $\mathbf{V}$ of finite monoids, this will complete the proof.

Suppose u and v are $\equiv_k$-equivalent words in Σ^*, and are idempotent in $\Sigma^* / \equiv_k$. Suppose further that u is idempotent in $\Sigma^* / \equiv_{k+1}$. We need to show $uvu \equiv_{k+1} u$ i.e., the inverse image of each idempotent satisfies the identity $ese = e$ whenever e is idempotent. Since u is idempotent in $\Sigma^* / \equiv_{k+1}$, this is equivalent to $uuvuu \equiv_{k+1} uuuuu$. By Lemma 3 it suffices to show that Player II has a winning strategy on this pair of words in the $(k + 1)$-round game. The strategy is this: If Player I moves anywhere but the middle segment of one of the words, Player II will respond on the corresponding position in the other word. If Player I ever moves into the middle segment, Player II will respond according to her strategy for the k-round game in u and v. If Player I moves out of the middle segment and back in again, Player II picks up the middle segment strategy again, starting from the beginning. This strategy will win the game for Player II unless Player I makes *all* his moves in the middle segments. In that case, after k rounds, the two pointed words are $uu(v, i)uu$ and $uu(u, j)uu$. Suppose Player I now moves to the right in the first word, remaining in v, giving $uu(v, i')uu$ with $i < i'$. Player II might not be able to respond in the middle segment of the other word. Instead, she picks a position j' in u such that $(v, i') \equiv_0 (u, j')$ (such a position exists because $u \equiv_k v$ and $k \geq 1$) and moves the pebble to the right to produce $uuu(u, j')u$. Since $\equiv_0$ is a congruence on pointed words, and since u, being idempotent for $\equiv_k$, is idempotent for $\equiv_0$, the resulting pointed words are $\equiv_0$-equivalent. The same strategy works if Player I moves to the left in v, or moves in either direction in the middle segment of $uuuuu$. Thus whatever Player I does, Player II can play safely for $k + 1$ successive rounds, and so wins the game.

5   Ideal Structure of Monoids in DA * G

In this section we state without proof some algebraic properties of pseudovarieties of the form $\mathbf{DA} * \mathbf{H}$, where $\mathbf{H}$ is a pseudovariety of finite groups. Let $\mathcal{F}$ be a set of partial one-to-one functions from a finite set X into itself. We will denote the image of $x \in X$ under $f \in \mathcal{F}$ by xf. We say that $\mathcal{F}$ is **H**-*extendible* if there is a finite set Y with $X \subseteq Y$, and a permutation group G on Y such that $G \in \mathbf{H}$, and for each $f \in \mathcal{F}$ there exists $g \in G$ such that f is equal to the restriction of g to the domain of f.

Lemma 4. *Let* $M \in \mathbf{DA} * \mathbf{H}$, *where* **H** *is a pseudovariety of finite groups. Let* $\psi :$ $\Sigma^* \to M$ *be a homomorphism. Let* J *be a regular* $\mathcal{J}$-*class of* M, *and let* (A, B, G, P) *be a Rees matrix representation of* J.

(a) There exist a partition of A, *a partition of* B, *and a bijection between the sets of blocks of the two partitions such that* $P(b, a) \neq 0$ *if and only if the blocks containing* a *and* b *correspond under the bijection.*

(b) Let $\mathcal{B}$ *denote the set of blocks of the partition of* B. *Let* $s \in M$. *There is a one-to-one partial function* $\pi_s : \mathcal{B} \to \mathcal{B}$ *such that if* $b \in B$, *and* $B(b)$ *is the block containing* b, *then* $B(b)\pi_s$ *is defined if and only if* $(a, g, b)s \in J$, *in which case* $B(b)\pi_s$ *is the block containing the right co-ordinate of* $(a, g, b)s$. *Moreover, the set of partial functions* $\{\pi_s : s \in M\}$ *is* **H**-*extendible.*

(c) Let B_1, B_2 *be two blocks of the partition of* B. *Then the language*

$$\{w \in \Sigma^* : B_1 \pi_{\psi(w)} = B_2\}$$

is recognized by a monoid in $\mathbf{J}_1 * \mathbf{H}$

(d) Suppose $\mathbf{H} * \mathbf{H} = \mathbf{H}$. *Let* $(a, g, b) \in J$, $g' \in G$. *Then the language*

$$\{w \in \Sigma^* : (a, g, b)\psi(w) \in \{a\} \times \{g'\} \times B\}$$

is recognized by a monoid in $\mathbf{R} * \mathbf{H}$.

6   Two-Variable Definability for $\mathbf{DA} * \mathbf{G}_{sol}$

Let Σ be a finite alphabet. We will prove in this section that if $L \subseteq \Sigma^*$ is recognized by a monoid $M \in \mathbf{DA} * \mathbf{G}_{sol}$, then L is definable by a sentence with two variables. This will complete the proof of Theorem 3.

Let $\phi : \Sigma^* \to M$ be a homomorphism. Each $w \in \Sigma^*$ has a unique factorization $w = w_0 \sigma_1 w_1 \cdots \sigma_k w_k$, where each σ_i is in Σ, $\phi(w_0) \equiv_{\mathcal{R}} 1$, and where, for $i = 1, \ldots, k$,

$$\phi(w_0 \sigma_1 \cdots \sigma_i w_i) \equiv_{\mathcal{R}} \phi(w_0 \sigma_1 \cdots w_{i-1}\sigma_i) <_{\mathcal{R}} \phi(w_0\sigma_1 \cdots w_{i-1}).$$

Let $s, t \in M$, with $s \equiv_{\mathcal{R}} t$. We define $L[s, t] = \{w \in \Sigma^* : s \cdot \phi(w) = t\}$. Thus, if $m \in M$, $\phi^{-1}(m)$ is the union of all languages of the form

$$L[1, t_0]\sigma_1 L[t_0 \cdot \phi(\sigma_1), t_1] \cdots \sigma_k L[t_{k-1} \cdot \phi(\sigma_k), t_k], \tag{1}$$

where $t_k = m$, and, for $i = 1, \ldots, k$, $t_i \equiv_{\mathcal{R}} t_{i-1} \cdot \phi(\sigma_i) <_{\mathcal{R}} t_{i-1}$. This union is finite, since k is bounded above by the number of $\mathcal{R}$-classes of M. It is therefore sufficient to show that every language of the form (1) is definable by a two-variable sentence. We prove this by induction on $|M|$: If $|M| = 1$, then the language (1) is Σ^*, which is defined by the one-variable sentence $\neg \exists x(x < x)$. We thus suppose $|M| > 1$. Our inductive hypothesis is that for all $M' \in \mathbf{DA} * \mathbf{G}_{sol}$ with $|M'| < |M|$, languages of the form (1) are two-variable definable.

We prove the assertion for M by a second induction, this time on k. We begin by considering languages of the form $L[s, t]$. First, suppose that the $\mathcal{R}$-class containing s and t is contained in a regular $\mathcal{J}$-class J of M. We identify J with a Rees matrix representation (A, B, G, P). There is a partition of B as specified in Lemma 4; as before, we denote by $B(b)$ the block of this partition containing the element b. Let $s = (a, g, b), t = (a'g', b')$. In order for a word w to belong to $L[s, t]$, we need either:

(a) $\phi(w) \notin J$ and $s\phi(w) = t$, or

(b) $\phi(w) \in J$, $B(b)$ is in the domain of $\pi_{\phi(w)}$, the middle co-ordinate of $(a, g, b)\phi(w)$ is g', and $w = w_1 \sigma w_2$, where $\sigma \in \Sigma$, $\phi(w_2) \notin J$, and $\phi(\sigma w_2) \in J$ with right co-ordinate b'.

The set of strings satisfying condition (a) is recognized by the monoid M/I, where I is the ideal consisting of all elements of M that are not strictly above J in the $\mathcal{J}$-ordering. If $|M/I| = |M|$ then J consists of a single element, which is the zero of M, and $L[s, t] = \Sigma^*$. Thus we may suppose $|M/I| < |M|$. Since $M/I \in \mathbf{DA} * \mathbf{G}_{sol}$, the inductive hypothesis implies that this set of strings is two-variable definable.

By Lemma 4, the set of strings w such that $B(b)$ is in the domain of $\pi_{\phi(w)}$ is recognized by a monoid in $\mathbf{J}_1 * \mathbf{G}_{sol}$, and, since $\mathbf{G}_{sol} * \mathbf{G}_{sol} = \mathbf{G}_{sol}$, the set of strings w such that the middle co-ordinate of $(a, g, b)\phi(w)$ is g' is recognized by a monoid in $\mathbf{R} * \mathbf{G}_{sol}$. By Theorem 4, these are both definable by left-relativizable two-variable sentences. Let $\sigma \in \Sigma$, and let K_σ be the set of strings $w_1 \sigma w_2$, where $\phi(w_2) \notin J$, and $\phi(\sigma w_2) \in J$ with right-co-ordinate b. K_σ is then a union of sets of the form (1), but with the sets $L[s, t]$ replaced by their $\mathcal{L}$-class duals, with respect to the monoid M/I. Since, as we noted in the remarks following Lemma 1, $\mathbf{DA} * \mathbf{G}_{sol}$ is closed under reversal, the inductive hypothesis implies that each K_σ is two-variable definable.

In the case where J is a null $\mathcal{J}$-class, the product of two elements of J is not in J. Thus $L[s, t]$ is recognized by M/I, where I is as defined above, and is thus two-variable definable.

We now suppose that we have a two-variable sentence δ for the language

$$L = L[t_i \phi(\sigma_{i+1}), t_{i+1}] \sigma_{i+2} \cdots L[t_{k-1} \phi(\sigma_k), t_k],$$

and use it to obtain a two-variable definition for $L' = L[t_{i-1} \phi(\sigma_i), t_i] \sigma_{i+1} L$. First we consider the case where the $\mathcal{J}$-class J that contains $t_{i-1} \phi(\sigma_i)$ and t_i is regular. Let $t_{i-1} \phi(\sigma_i) = (a_1, g_1, b_1)$, $t_i = (a_2, g_2, b_2)$. Let θ be a left-relativizable two-variable sentence for the set of strings u such that $B(b_1)$ is in the domain of $\pi_{\phi(u)}$, and η a left-relativizable two-variable sentence for the set of strings u such that the middle co-ordinate of $(a_1, g_1, b_1)\phi(u)$ is g_2. Such sentences exist by Lemma 4 and Theorem 4.

Let ζ be a two-variable sentence for the set of strings u such that $u = u_1 \sigma u_2$, with $\sigma \in \Sigma$, $\phi(u_2) \notin J$, and $\phi(\sigma u_2) \in J$ with right co-ordinate b_2. We showed above that such a sentence exists. Let ζ' be a two-variable sentence for the set of strings u such that $\phi(u) \notin J$, and $(a_1, g_1, b_1)\phi(u) = (a_2, g_2, b_2)$. Again, we showed above that such a sentence exists.

Our sentence defining L' is

$$\exists x(Q_{\sigma_{i+1}} x \wedge \theta[< x] \wedge \neg\theta[\leq x]) \wedge ((\eta \wedge \zeta) \vee \zeta')[\leq \theta] \wedge \delta[> \theta].$$

Observe that because of the left-relativizability of θ, all the relativizations in the above sentence are two-variable formulas.

For the case of a null $\mathcal{J}$-class, we proceed in the identical fashion, except now we do not need formulas analogous to η and ζ.

7   Directions for Further Research

$\mathbf{G}_{sol}$-*Extendibility.* The biggest question left unanswered by our work is whether one can effectively determine if a given regular language is definable by a two-variable sentence. It follows from our arguments that L is two-variable definable if and only if for every regular $\mathcal{J}$-class J of $M(L)$, J admits a block partition of the kind described in Section 5, and the set $\{\pi_s : s \in M\}$ of partial one-to-one transformations on the set of B-blocks of $\mathcal{J}$ is $\mathbf{G}_{sol}$-extendible. In fact, we are able to prove:

Theorem 5. *The following two decision problems are equivalent: (a) To determine whether a given regular language is two-variable definable. (b) To determine whether a given set of partial one-to-one functions on a finite set is $\mathbf{G}_{sol}$-extendible.*

It is not known whether this latter problem is decidable. Margolis, Sapir and Weil [8] show that if $\mathbf{H}$ is a pseudovariety of groups such that $\mathbf{H} * \mathbf{H} = \mathbf{H}$ ($\mathbf{G}_{sol}$ has this property) then the question of $\mathbf{H}$-extendibility is equivalent to the problem of computing the closure of a finitely-generated subgroup of the free group in the profinite topology induced by $\mathbf{H}$. Ribes and Zaleskii [12] showed that this problem is decidable for the pseudovariety $\mathbf{G}_p$ of p-groups for a fixed prime p. As a consequence we have

Theorem 6. *Let p be prime. It is decidable whether a given regular language is definable by a two-variable sentence in which all the modular quantifiers are of modulus p.*

Modular Temporal Logic. First-order logic over $<$ is equivalent in expressive power to linear propositional temporal logic (LPTL), and two-variable first-order logic over $<$ is equivalent to the fragment of LPTL that includes both the past and the future versions of the Next and Eventually operators, but not the Until operator. Baziramwabo, McKenzie and Thérien [1] study an extension of LPTL that includes a modular temporal operator, and show that this has the same expressive power as sentences over $<$ with both modular and ordinary quantifiers. In the full paper we will show that the fragment of modular temporal logic that includes the past and future versions of the Next and Eventually operators, as well as all the modular operators, captures exactly the languages in $\mathbf{DA} * \mathbf{G}_{sol}$.

Related Model-Theoretic Questions. Pin and Weil [10] show that the languages whose syntactic monoids are in $\mathbf{DA}$ are exactly those languages that are simultaneously definable by both a Σ_2 and a Π_2-sentence over $<$, with no restriction on the number of variables. In the full paper we will show that the analogous property holds for languages whose syntactic monoids belong to $\mathbf{DA} * \mathbf{G}_{sol}$—here "$\Sigma_2$-sentence" means a Σ_2-sentence over the base of atoms consisting of all the purely modular formulas.

We would also like to know the effect of bounding the number of variables in sentences that use only the successor relation $y = x + 1$ in place of the more powerful ordering relation.

Acknowledgments

We would like to thank Stuart Margolis for his very helpful comments. The second author's research is supported by grants from NSERC, FCAR and the von Humboldt Foundation.

References

1. A. Baziramwabo, P. McKenzie and D. Thérien, "Modular Temporal Logic", Proc. 1999 IEEE Conference on Logic in Computer Science (LICS) , Trento, Italy, July 1999.
2. D. Beauquier and J. E. Pin, "Factors of Words", *Proc. 16th ICALP*, Springer Lecture Notes in Computer Science **372** (1989) 63–79.
3. S. Eilenberg, *Automata, Languages and Machines*, vol. B, Academic Press, New York, 1976.
4. K. Etessami, M. Vardi, and T. Wilke, "First-Order Logic with Two Variables and Unary Temporal Logic", *Proceedings, 12th IEEE Symposium on Logic in Computer Science*, 228-235 (1996).
5. N. Immerman and D. Kozen, "Definability with a Bounded Number of Bound Variables", *Information and Computation*, **83**, 121-139 (1989).
6. J. Kamp, *Tense Logic and the Theory of Linear Order*, Ph. D. thesis, UCLA (1968).
7. R. McNaughton and S. Papert, *Counter-Free Automata*, MIT Press, Cambridge, Massachusetts, 1971.
8. S. Margolis, M. Sapir, and P. Weil, "Closed Subgroups in Pro-**V** Topologies and the Extension Problem for Inverse Automata", preprint.
9. J. E. Pin, *Varieties of Formal Languages*, Plenum, London, 1986.
10. J. E. Pin and P. Weil, "Polynomial Closure and Unambiguous Product", *Theory Comput. Systems* **30** (1997) 383-422.
11. J. Rhodes and B. Tilson, "The Kernel of Monoid Morphisms", *J. Pure and Applied Algebra* **62** (1989) 227–268.
12. L. Ribes and P. Zaleskii, "The pro-p topology of a free group and algorithmic problems in semigroups, *International Journal of Algebra and Computation* **4** (1994) 359-374.
13. P. Stiffler, "Extensions of the Fundamental Theorem of Finite Semigroups", *Advances in Mathematics*, **11** 159-209 (1973).
14. H. Straubing, *Finite Automata, Formal Languages, and Circuit Complexity*, Birkhäuser, Boston, 1994.
15. H. Straubing, "Families of recognizable sets corresponding to certain varieties of finite monoids", *Journal of Pure and Applied Algebra* **15** (1979), 305-318.
16. H. Straubing, D. Thérien, and W. Thomas, "Regular Languages Defined by Generalized Quantifiers", *Information and Computation* **118** 289-301 (1995).
17. D. Thérien and T. Wilke, "Over Words, Two Variables are as Powerful as One Quantifier Alternation," *Proc. 30th ACM Symposium on the Theory of Computing* 256-263 (1998).
18. T. Wilke, "Classifying Discrete Temporal Properties", Habilitationsschrift, University of Kiel, 1998.

New Bounds on the OBDD-Size of Integer Multiplication via Universal Hashing

Philipp Woelfel

FB Informatik, LS2, Univ. Dortmund, 44221 Dortmund, Germany
woelfel@Ls2.cs.uni-dortmund.de

Abstract. Ordered binary decision diagrams (OBDDs) nowadays belong to the most common representation types for Boolean functions. Although they allow important operations such as satisfiability test and equality test to be performed efficiently, their limitation lies in the fact that they may require exponential size for important functions. Bryant [8] has shown that any OBDD-representation of the function $\mathrm{MUL}_{n-1,n}$, which computes the middle bit of the product of two n-bit numbers, requires at least $2^{n/8}$ nodes. In this paper a stronger bound of $2^{n/2}/61$ is proven by a new technique, using a recently found universal family of hash functions [23]. As a result, one cannot hope anymore to find reasonable small OBDDs even for the multiplication of relatively short integers, since for only a 64-bit multiplication millions of nodes are required. Further, a first non-trivial upper bound of $7/3 \cdot 2^{4n/3}$ for the OBDD size of $\mathrm{MUL}_{n-1,n}$ is provided.

1 Introduction and Results

Binary Decision Diagrams (BDDs), introduced by Lee [15] and Akers [1], are a well established representation type of Boolean functions. While the general model has a large representational power and allows the simulation of any other general model of computation, its generality also has certain severe drawbacks. They lie in the NP-hardness of several important operations which should be available for a representation serving as a dynamic data structure (see also [21]). Among these operations are e.g. the equivalence test (which tests whether two representations describe the same function), the satisfiability test (which tests whether there exists a satisfying input for the represented function), or the minimization problem (which is to minimize the size of the representation of a given function).

In order to overcome these drawbacks, Bryant [7] has introduced restricted BDDs, called *ordered binary decision diagrams* (OBDDs).

Definition 1. Let $X_n = \{x_1, \ldots, x_n\}$ be a set of Boolean variables.

1. A variable ordering π on X_n is a permutation of the indices $\{1, \ldots, n\}$, leading to the ordered list $x_{\pi(1)}, \ldots, x_{\pi(n)}$ of the variables.

A. Ferreira and H. Reichel (Eds.): STACS 2001, LNCS 2010, pp. 563–574, 2001.

2. *A π-OBDD on X_n for a variable ordering π is a directed acyclic graph with the following properties: Each sink is labeled by a constant 0 or 1. Each inner node is labeled by a variable from X_n and has two outgoing edges, one of them labeled by 0, the other by 1. If an edge leads from a node labeled by x_i to a node labeled by x_j, then $x_{\pi^{-1}(i)} < x_{\pi^{-1}(j)}$. This means that any path on the graph passes the nodes in an order respecting the variable ordering π.*

3. *A node v of a π-OBDD is said to compute a Boolean function $f_v : \{0,1\}^n \to \{0,1\}$, if for any $a = (a_1, \ldots, a_n) \in \{0,1\}^n$, the path starting from v and leading from any x_i node over the edge labeled by the value of a_i, finishes at the sink with label $f(a)$. A π-OBDD with a root v is said to compute a Boolean function f, if v computes f.*

4. *The size of a π-OBDD is the number of its nodes. The π-OBDD size of a Boolean function f (short: $\pi - \text{OBDD}(f)$) is the size of the minimum π-OBDD computing f. Finally, the OBDD size of f (short: $\text{OBDD}(f)$) is the minimum of $\pi - \text{OBDD}(f)$ for all variable orderings π.*

Efficient algorithms on π-OBDDs are known generally for all important operations, as e.g. the ones mentioned above (for an in-depth discussion of OBDDs and their operations see [21]). The size of a π-OBDD though, can be quite sensitive to the chosen variable ordering π, and finding a variable ordering leading to small or even minimal π-OBDDs is a hard problem (see [4, 7, 18]). Furthermore, since there exist 2^{2^n} Boolean functions of n variables, it can be seen that almost all functions require an exponential number of elements in any realization by networks using only primitive elements. However, this still is not disturbing as long as for all practical relevant families of functions small representations of a certain kind exist. Therefore, one is interested in finding exponential lower bounds for the size of OBDDs (and other representation types) computing important families of functions, such as integer multiplication.

Definition 2. *The Boolean function $\text{MUL}_{k,n} : \{0,1\}^{2n} \to \{0,1\}$ computes the bit z_k of the product $(z_{2n-1} \ldots z_0)$ of two integers $(y_{n-1} \ldots y_0)$ and $(x_{n-1} \ldots x_0)$.*

The first step towards bounding the size of OBDDs for integer multiplication was done by Bryant in 1986 [7]. He showed that for any variable ordering π, there exists an index k, such that the π-OBDD size for $\text{MUL}_{k,n}$ is at least $2^{n/8}$. This result though, would still allow the possibility that one might obtain polynomial size OBDDs for all functions $\text{MUL}_{k,n}$ by choosing different variable orderings for different output bits. In 1991, Bryant found that computing the middle bit of multiplication (that is $\text{MUL}_{n-1,n}$) requires a π-OBDD containing $2^{n/8}$ nodes for any variable ordering π [8]. More precisely, he showed that for any $1 \leq k \leq n$, $\text{OBDD}(\text{MUL}_{k-1,n})$ and $\text{OBDD}(\text{MUL}_{2n-k-1,n})$ have a value of at least $2^{k/8}$.

Although this proves the exponential size of OBDDs for multiplication, the bound is - as stated e.g. by Bollig and Wegener in [5] - not satisfactory. This is because Bryant's bound would still allow the possibility that one can construct 64-bit multipliers represented by OBDDs containing only 256 nodes, while on the other hand it is widely conjectured that OBDDs computing $\text{MUL}_{n-1,n}$ have a size of at least 2^n. This would mean that such a multiplier could not even be realized with millions of nodes. Since one would like to use OBDDs for realistic applications, such as verification of multipliers, one is interested in either finding such small constructions or a better lower bound. The following result, which is proven in the next two sections, provides the second alternative:

Theorem 1. *Any OBDD computing* $\mathrm{MUL}_{n-1,n}$ *has a size of at least* $2^{\lfloor n/2 \rfloor}/61$.

This bound shows that any OBDD for 64-bit multiplication must be constructed of more than 70 million nodes and thus demonstrates a true weakness of the representation type. The technique leading to this result is new. It highly relies on a recently found universal family of hash functions [23] and makes use of a new lemma showing how such functions distribute two arbitrary sets over the range. Universal hashing is introduced in the next section.

Since it is generally believed though, that the true bound on the OBDD size for $\mathrm{MUL}_{n-1,n}$ is still larger, it is of interest to have an upper bound, too. Note that for any Boolean function on m variables, there exists an OBDD of size $(2 + o(1))2^m/m$ [6], so a trivial upper bound for $\mathrm{OBDD}(\mathrm{MUL}_{n-1,n})$ is roughly $2^{2n}/n$. The following upper bound, proved in section 4, is the first non-trivial one.

Theorem 2. *There exists an OBDD for* $\mathrm{MUL}_{n-1,n}$ *having a size of* $7/3 \cdot 2^{4n/3}$.

The bound shows that the middle bit of a 16-bit multiplication can be computed by an OBDD containing less than 6.2 million nodes. Although the proof is existential, constructions of OBDDs satisfying this bound can be derived from it.

2   Universal Hashing

The concept of universal hashing was introduced by Carter and Wegman in 1979 [9]. While one of its original purposes was to use randomization in hashing schemes instead of relying on the distribution of the inputs, it has found over the years a large variety of applications in areas of all different kinds. They range from algorithms for the various types of hashing based dictionaries [9, 12, 13] over cryptographic aspects as message authentication [2, 17] up to complexity theoretical statements [14, 20].

Universal hash families are usually defined by the following notation: Let $\mathcal{H}$ be a family of hash functions $U \rightarrow R$. U and R are called *universe* and *range*, respectively. For arbitrary $x, x' \in U$ and $h \in \mathcal{H}$, we define

$$\delta_h(x, x') = \begin{cases} 1 & \text{if } x \neq x' \text{ and } h(x) = h(x') \\ 0 & \text{otherwise.} \end{cases}$$

If one or more of h, x and x' are replaced in $\delta_h(x, x')$ by sets, then the sum is taken over the elements from these sets. E.g., for $H \subseteq \mathcal{H}$, $V \subseteq U$ and $x \in U$, $\delta_H(x, V)$ means

$$\sum_{h \in H} \sum_{x' \in V} \delta_h(x, x').$$

Definition 3. *A family $\mathcal{H}$ of hash functions $U \to R$ is universal, if for any two distinct $x, x' \in U$*

$$\delta_{\mathcal{H}}(x, x') \leq \frac{|\mathcal{H}|}{|R|}.$$

A stronger definition of so-called "strongly universal hash families" was given in [22]. Among the many applications, there were also interesting results concerning branching programs (or equivalently BDDs). So, Mansour, Nisan and Tiwari [16] investigated the computational complexity of strongly universal hashing, and gave a lower bound for the time-space tradeoff of branching programs computing the functions of such families. Further, Beame, Tompa and Yan [3] have found results on the communication-space tradeoff of strongly universal hash families in a general model of two communicating branching programs.

For OBDDs, it is not possible to show a general exponential lower bound for universal hash families. E.g., the convolution of two bit strings can be viewed as a strongly universal hash family [16], but it can be easily seen that for any output bit of the convolution, there exists a variable ordering π leading to a linear size π-OBDD.

The property of universal hash families we will use here, can be described in the following way: If there are two large enough subsets of the universe given, then there exists a hash function under which the function values of the elements from each set cover a large fraction of the range. This is in a way a twisted version of the known results, telling that there exists a function under which the number of collisions of elements from a set is small.

For a function $h : U \to R$ and a subset $M \subseteq U$, define $h(M)$ to be the image of M under h, namely

$$h(M) := \{y \in R \mid \exists x \in M : h(x) = y\}.$$

Lemma 1. *Let $\mathcal{H}$ be a universal family of hash functions $U \to R$ and $0 \leq \epsilon < 1$. Then for arbitrary $M, N \subseteq U$ with*

$$|M| = |N| > 2(|R| - 1)\frac{\epsilon}{1 - \epsilon},$$

there exists a hash function $h \in \mathcal{H}$ such that $h(M)$ and $h(N)$ contain more than $\epsilon|R|$ elements each.

Proof. Let $r = |R|$, $m = |M| = |N|$ and for $h \in \mathcal{H}$ let the random variable X_h be the sum of $\delta_h(M, M)$ and $\delta_h(N, N)$. Using the universal property of $\mathcal{H}$, we obtain for a randomly chosen function h an upper bound for the expectation of X_h:

$$\mathop{\mathbf{E}}_{h \in \mathcal{H}}[X_h] = \sum_{\substack{x, x' \in M \\ x \neq x'}} \mathop{\mathbf{E}}_{h \in \mathcal{H}}[\delta_h(x, x')] + \sum_{\substack{y, y' \in N \\ y \neq y'}} \mathop{\mathbf{E}}_{h \in \mathcal{H}}[\delta_h(y, y')]$$

$$\leq |M|(|M| - 1)\frac{1}{r} + |N|(|N| - 1)\frac{1}{r}$$

$$= \frac{2}{r}m(m - 1).$$

This means by the probabilistic method that there exists an $h_0 \in \mathcal{H}$ with

$$X_{h_0} \leq \frac{2}{r} m(m-1). \tag{1}$$

In order to prove that this h_0 fulfills the claim, we assume that $h_0(M)$ contains at most ϵr elements. By summing over the ordered pairs of elements in $h_0^{-1}(y) \cap M$ for each $y \in h_0(M)$, we get

$$\delta_{h_0}(M, M) = \sum_{y \in h_0(M)} \left|h_0^{-1}(y) \cap M\right|\left(\left|h_0^{-1}(y) \cap M\right| - 1\right)$$

$$= \sum_{y \in h_0(M)} \left|h_0^{-1}(y) \cap M\right|^2 - |M|.$$

Clearly, the last sum takes its minimum, if each $h_0^{-1}(y) \cap M$ contains the same number of $|M|/(\epsilon r)$ elements. Therefore,

$$\delta_{h_0}(M, M) \geq \epsilon r \left(\frac{m}{\epsilon r}\right)^2 - m = m\left(\frac{m}{\epsilon r} - 1\right).$$

For N, we obtain with similar arguments that $\delta_{h_0}(N, N) \geq m(m/r - 1)$. So we have the following lower bound on X_{h_0}:

$$X_{h_0} \geq m\left(\frac{m}{\epsilon r} + \frac{m}{r} - 2\right).$$

Together with the upper bound from (1), we obtain

$$\frac{2}{r} m(m-1) \geq m\left(\frac{m}{\epsilon r} + \frac{m}{r} - 2\right).$$

By the assumption that $m > 2(r-1)\epsilon/(1-\epsilon)$, this results into the contradiction

$$2 - \frac{2}{r} \geq m\frac{1-\epsilon}{\epsilon r} > 2\frac{r-1}{r}. \qquad \square$$

We now consider hash functions, which map the n-bit universe $U := \{0, \ldots, 2^n - 1\}$ to the k-bit range $R_k := \{0, \ldots, 2^k - 1\}$. For $a, b \in U$ let

$$h_{a,b}^k : U \to R_k, \qquad x \mapsto \left((ax + b) \bmod 2^n\right) \operatorname{div} 2^{n-k},$$

where "div" is the integer division (i.e. $x \operatorname{div} y = \lfloor x/y \rfloor$). In a bitwise view, the result of the modulo operation $x \bmod 2^n$ is represented by the n least significant bits of x. On the other hand, the division $x \operatorname{div} 2^{n-k}$ can be seen as shifting x by $n - k$ digits to the right. In other words, if the value of the linear function $ax + b$ is represented by $(y_{2n-1} \ldots y_0)$, then $h_{a,b}^k$ is the integer, which is represented by the k bits $(y_{n-1} \ldots y_{n-k})$. The following result has recently been established by the author [23, 24].

Theorem 3. *Let $1 \leq k \leq n$. Then there exist sets $A \subseteq U$ and $B \subseteq \{0, \ldots, 2^{n-k} - 1\}$ such that the family of hash functions $h_{a,b}^k$ with $a \in A$ and $b \in B$ is universal.*

Similar hash classes have been investigated in [10, 11].

3   Lower Bounds

Since the functions $h_{a,b}^k$ are evaluated not only by a multiplication, but also by an addition, we cannot use Lemma 1 for the lower bound proof of OBDD(MUL$_{k,n}$) directly. Let $f_a^k := h_{a,0}$ be the functions that can be evalutated without addition. The following lemma gives a result similar to that of Lemma 1. Note that as stated in [11], the hash functions f_a^k form an "almost" universal hash class (which means that in Definition 3 $|\mathcal{H}|/|R|$ is replaced by $c|\mathcal{H}|/|R|$ for some constant c). This property though, is not sufficient to prove a result as strong as the one given below.

Lemma 2. Let $M, N \subseteq U$ and $1/2 \le \epsilon < 1$. If $|M| = |N| > 2(2^{k+1}-1)\epsilon/(1-\epsilon)$, then there exists an $a \in U$, such that $f_a^k(M)$ and $f_a^k(N)$ contain at least $(2\epsilon-1)2^k$ elements each.

Proof. By Lemma 1 and Theorem 3, there exists an $a \in U$ and an $b \in \{0, \ldots, 2^{n-k-1} - 1\}$ such that $h_{a,b}^{k+1}(M)$ and $h_{a,b}^{k+1}(N)$ contain more than $\epsilon|R_{k+1}| = \epsilon 2^{k+1}$ elements each. Let these a, b be fixed and $f := f_a^k$. We show that $f(M)$ contains at least $(2\epsilon - 1)2^k$ elements; the claim then follows for N with the same argument.

Clearly, there exists a subset $M' \subseteq M$ with $|M'| = \epsilon 2^{k+1}$, such that all $x \in M'$ have distinct function values under $h_{a,b}^{k+1}$. Since R_{k+1} contains exactly 2^k even elements, there are at least $|M'| - 2^k$ elements in M', which have an odd function value under $h_{a,b}^{k+1}$. Let M'' be a subset of M' containing exactly $\epsilon 2^{k+1} - 2^k = 2^k(2\epsilon - 1)$ elements with an odd function value. To prove the claim, it suffices to show that for any two distinct $x, x' \in M''$ we have $f(x) \ne f(x')$. Let $h_{a,b}^{k+1}(x) = z$ and $h_{a,b}^{k+1}(x') = z'$, thus

$$z2^{n-k-1} \le (ax + b) \bmod 2^n < (z+1)2^{n-k-1}.$$

Since by definition $0 \le b < 2^{n-k-1}$, it follows that

$$(z - 1)2^{n-k-1} \le (ax) \bmod 2^n < (z+1)2^{n-k-1}.$$

Further, by z being odd, $(z-1)/2$ equals $\lfloor z/2 \rfloor$ and $(z+1)/2$ equals $\lfloor z/2 \rfloor +1$. Therefore, the above inequalities imply

$$\lfloor z/2 \rfloor 2^{n-k} \le (ax) \bmod 2^n < (\lfloor z/2 \rfloor + 1)2^{n-k}.$$

This means that $f(x) = \lfloor z/2 \rfloor$, and with the same argument also $f(x') = \lfloor z'/2 \rfloor$. But because z and z' are both odd and different, clearly $\lfloor z/2 \rfloor$ and $\lfloor z'/2 \rfloor$ are different, too. So, we obtain the desired result $f(x) \ne f(x')$.   □

We are now ready to prove an intermediate result, from which the lower bound for the OBDD size of MUL$_{n-1,n}$ follows easily. In order to do so, we have to introduce some more notation. Let x be an integer represented in a bitwise notation as $(x_{n-1} \ldots x_0)$. Then we write $[x]_k$ for the $(k + 1)$-th bit x_k. Further, let MUL$_{k,n}^a : \{0, 1\}^n \to \{0, 1\}$ for $a \in \{0, 1\}^n$ be the Boolean function that computes the $(k + 1)$-th bit of the product of a with an n-bit number, i.e. MUL$_{k,n}^a(x) = $ MUL$_{k,n}(a, x)$.

Theorem 4. *Let π be an arbitrary variable ordering on X_n. Then there exists an $a \in \{0, \dots, 2^n - 1\}$ for which any π-OBDD for $\mathrm{MUL}_{n-1,n}^a$ consists of at least $2^{\lfloor n/2 \rfloor}/121 + 1$ nodes.*

Proof. Let the input variables for the π-OBDD be $x_{n-1}, \dots, x_0$ for an n, which is w.l.o.g. even. Consider the top part T of π, which contains the first $n/2$ variables with respect to π and the bottom part B containing the other $n/2$ variables. We construct now two sets M and N of numbers in $\{0, \dots, 2^n - 1\}$ as follows: M contains all numbers which can be represented by $(x_{n-1} \dots x_0)$ if the variables from T are set to 0, and N contains all numbers which can be represented by $(x_{n-1} \dots x_0)$ if the variables from B are set to 0. Note that any number in $\{0, \dots, 2^n - 1\}$ can be uniquely expressed as $p + q$ for $p \in M$ and $q \in N$.

Our goal is to find an appropriate constant a and two subsets $M' \subseteq M$, $N' \subseteq N$ with the following property: For any distinct q, q' in N', there exists such an $p \in M'$ that $a(p + q)$ and $a(p + q')$ differ in the n-th bit. More formally

$$\forall q, q' \in N', q \neq q' \; \exists p \in M' : \quad [a(p + q)]_{n-1} \neq [a(p + q')]_{n-1}. \tag{2}$$

Since q and q' are determined only by the top variables and p is determined by the bottom variables, it follows that among the $2^{n/2}$ subfunctions obtained by replacing the top variables with constants, there are at least $|N'|$ pairwise different ones. So, at level $n/2$, the π-OBDD consists of at least $|N'|$ nodes. Further, a simple inductive argument shows that any OBDD contains in a level i at most one more node than there are nodes in all preceding levels $1, \dots, i-1$ together. Therefore, the total number of nodes in the OBDD for $\mathrm{MUL}_{n-1,n}^a$ is at least $2|N'| + 1$ (including the two sinks).

Let $\epsilon = 16/17$ and $k = n/2 - 6$. Then by an easy calculation one obtains that

$$|M| = |N| = 2^{n/2} > 2(2^{k+1} - 1)\frac{\epsilon}{1 - \epsilon}.$$

By Lemma 2, there exists an a for which $f_a^k(M)$ and $f_a^k(N)$ contain at least $(2\epsilon - 1)2^k = 15/17 \cdot 2^k$ elements each. We fix this a, define $f = f_a^k$ and continue to determine appropriate M' and N'.

As an intermediate step, we choose M^* and N^* to be minimal subsets of M respectively N, such that $f(M^*)$ and $f(N^*)$ contain exactly $13/17 \cdot 2^{k-1}$ even elements. Such sets exist, since at most 2^{k-1} of the 2^k possible function values are odd, and thus at least $15/17 \cdot 2^k - 2^{k-1} = 13/17 \cdot 2^{k-1}$ of the elements in M respectively N have distinct and even function values under f. Note that because we required M^* and N^* to be minimal, no two elements from M^* respectively N^* have the same function value under f.

The following observation is crucial for the rest of the proof: For any $p \in M^*$ and any $q \in N^*$, the k-th bit of $f(p) + f(q)$ has the same value as the n-th bit of $a(p + q)$. Or formally

$$[f(p) + f(q)]_{k-1} = [a(p + q)]_{n-1}. \tag{3}$$

The reason for this is that the rightmost bits of $f(p)$ and $f(q)$ are both zero (since these values are even). Recalling that the division executed by f is in fact a right-shift by $n-k$ bits, we obtain $[ap]_{n-k} = [aq]_{n-k} = 0$. Therefore, the bits of $ap+aq$ with higher index than $n-k$ are not influenced by a carry bit resulting from the addition of the less significant bits $([ap]_{n-k} \ldots [ap]_0) + ([aq]_{n-k} \ldots [aq]_0)$. This means that $f(p) + f(q)$ has in all bits (except possibly the least significant one) the same value as $a(p+q)$ in the bits with indices $n-1, \ldots, n-k$, and equation (3) is true.

In order to satisfy property (2) it is sufficient by the above arguments that the sets M' and N' are subsets of M^* and N^* and that the following holds:

$$\forall q, q' \in N', q \neq q' \; \exists p \in M' : \; [f(p) + f(q)]_{k-1} \neq [f(p) + f(q')]_{k-1}. \quad (4)$$

We set $M' = M^*$ and

$$N' = \{q \in N^* \mid \exists p \in M' : f(q) = 2^k - f(p)\}. \quad (5)$$

In order to prove claim (4), let q and q' be arbitrary distinct elements from N'. Since q and q' are in N^* and therefore have distinct function values under f, we may assume w.l.o.g. that

$$0 < (f(q) - f(q')) \bmod 2^k \leq 2^{k-1} \quad (6)$$

(otherwise we achieve this by exchanging q and q'). By construction, there exists a $p \in M'$ with $f(p) + f(q) = 2^k$. For this p, obviously the k-th bit of $f(p) + f(q)$, that is $[f(p) + f(q)]_{k-1}$, equals 0. But on the other hand, by inequation (6), the value of $(f(p) + f(q')) \bmod 2^k$ is in $\{2^{k-1}, \ldots, 2^k - 1\}$. This means that the k-th bit of $f(p) + f(q')$ equals 1, and thus claim (4) is proven.

So far, we have constructed subsets $M' \subseteq M$ and $N' \subseteq N$, which satisfy claim (2), implying by our arguments a lower bound on the π-OBDD size of $2|N'| + 1$. All that is left to do, is to give an appropriate lower bound on $|N'|$. Recall the definition of N' in (5), and that $f(M')$ and $f(N^*)$ contain $13/17 \cdot 2^{k-1}$ even elements each. Because for any even $f(p)$ also $2^k - f(p)$ is even, the set $\{2^k - f(p) \mid p \in M'\}$ contains $13/17 \cdot 2^{k-1}$ even elements, too. But since there exist only 2^{k-1} even elements in $\{0, \ldots, 2^k\}$, the intersection of $f(N^*)$ and $\{2^k - f(p) \mid p \in M'\}$ - which is $f(N')$ - has a cardinality of at least $d \cdot 2^{k-1}$, where $d = 1 - 2(1 - 13/17) = 9/17$. By the choice of k, $2|N'| + 1$ (and thus also the size of the π-OBDD) is bounded below by

$$2|f(N')| + 1 \geq 2 \cdot \frac{9}{17} \cdot 2^{k-1} + 1 = \frac{9}{17} \cdot 2^{n/2-6} + 1 > \frac{2^{n/2}}{121} + 1 \qquad \square$$

This theorem shows the general result for $\mathrm{MUL}_{n-1,n}$ by the following straight-forward observation: If for some constant B and some variable ordering π there exists an a, for which the π-OBDD size of $\mathrm{MUL}_{n-1,n}^a$ is at least $B + 1$, then the π-OBDD size of $\mathrm{MUL}_{n-1,n}$ is at least $2B$. This is, because in any OBDD computing $\mathrm{MUL}_{n-1,n}(x,y)$ either the input x or the input y may be set to the constant a. In both cases the resulting OBDD contains at at least $B-1$ inner nodes, not counting those for variables fixed to constants (since they may be deleted without changing the function). So, by the last theorem the OBDD for $\mathrm{MUL}_{n-1,n}$ has a size of at least $2 \cdot 2^{\lfloor n/2 \rfloor}/121$, which proves the main result (Theorem 1).

Furthermore, by a straightforward reduction, one can easily obtain a lower bound on computing the other output bits of the multiplication. A simple proof (see [8], Corollary 1) shows that any representation computing $\mathrm{MUL}_{k-1,n}$ or $\mathrm{MUL}_{2n-k-1,n}$ may also compute $\mathrm{MUL}_{k-1,k}$.

Corollary 1. *The size of an OBDD computing* $\mathrm{MUL}_{k-1,n}$ *or* $\mathrm{MUL}_{2n-k-1,n}$ *is at least* $2^{\lfloor k/2 \rfloor}/61$.

Note that our lower bound on $\mathrm{MUL}_{n-1,n}$ relies only on the existence of a constant a for each variable ordering π, for which $\mathrm{MUL}_{n-1,n}^a$ leads to a large π-OBDD representation. If one would want to significantly improve this bound, this would have to be done by a different technique, taking more values for a into consideration. In other words, the result of Theorem 4 is optimal up to a small constant factor:

Theorem 5. *There exists a variable ordering* π *which allows for any* $a \in \{0, \ldots, 2^n - 1\}$ *the construction of a* $\pi - OBDD$ *for* $\mathrm{MUL}_{n-1,n}^a$ *having a size of* $3 \cdot 2^{\lceil n/2 \rceil}$.

The proof will be sketched at the end of the next section.

4 Upper Bounds

In this section, we derive the upper bounds stated in Theorems 2 and 5. Both bounds can be proven by the same technique, which makes use of the fact that the minimal-size π-OBDD for a Boolean function f is unique up to isomorphism [7], and of a theorem by Sieling and Wegener [19], describing the structure of the minimal-size π-OBDD.

Let f be a Boolean function and π be an arbitrary variable ordering on X_n. For $a_1, \ldots, a_i \in \{0,1\}$ $(1 \leq i \leq n)$, denote by $f_{a_1,\ldots,a_i}$ the subfunction of f that computes $f(x_1, \ldots, x_n)$, where for $1 \leq j \leq i$ the j-th input-variable according to π (that is $x_{\pi^{-1}(j)}$) is fixed by the constant a_j. More formally,

$$f_{a_1,\ldots,a_i} := f_{|x_{\pi^{-1}(1)}=a_1,\ldots,x_{\pi^{-1}(i)}=a_i}.$$

Further, we say that a function g essentially depends on an input variable x_i, if $g_{|x_i=0} \neq g_{|x_i=1}$.

Theorem 6 ([19]). *The number of* x_i-*nodes of the minimal-size* π-*OBDD for* f *is the number of different subfunctions* $f_{a_1,\ldots,a_{i-1}}$ *for* $a_1, \ldots, a_{i-1} \in \{0,1\}$, *essentially depending on* x_i.

In order to show the stated upper bounds, let $x = (x_{n-1} \ldots x_0)$ and $y = (y_{n-1} \ldots y_0)$ be the input variables for $\mathrm{MUL}_{n-1,n}$. Further, let $\mathcal{F}_i$ denote the family of subfunctions f_{x,y^*} of $\mathrm{MUL}_{n-1,n}$ that result from replacing the variables $x_0, \ldots, x_{n-1}$ and $y_0, \ldots, y_{i-1}$ with constants. I.e., for $y^* := (y_{i-1} \ldots y_0)$,

$$f_{x,y^*}(y_{n-1} \ldots y_i) = \mathrm{MUL}_{n-1,n}\big(x, \, y_{n-1} \ldots y_i \, y^*\big).$$

Our goal is to bound the number of different subfunctions in $\mathcal{F}_i$. We define for any subfunction $f_{x,y^*} \in \mathcal{F}_i$ its index $ind(f_{x,y^*})$ to be the number represented by $(z_{n-1} \ldots z_i)$, where $z = x \cdot y^*$. Consider arbitrary x and $y = (y_{n-1} \ldots y_i \, y^*)$. By the school-method of multiplication we have

$$x \cdot y = x \cdot y^* + 2^i x \cdot (y_{n-1} \ldots y_i).$$

Since the second term of the sum is a value shifted by i bits to the left (and thus has its i least significant bits set to 0), the addition of xy^* and $2^i x \cdot (y_{n-1} \ldots y_i)$ has no carry at position i. Hence, replacing $x \cdot y^*$ by $2^i \cdot ind(f_{x,y^*})$ in the above sum does not change the result for the output bits with indices $i, \ldots, n-1$. Furthermore, writing $2^i x \cdot (y_{n-1} \ldots y_i)$ as

$$\sum_{j=0}^{n-1} 2^{i+j} x_j \cdot (y_{n-1} \ldots y_i),$$

implies that the bits x_j with $j \geq n - i$ have no influence on the output bit with index $n - 1$. Thus, $\mathrm{MUL}_{n-1,n}(x,y)$ is uniquely determined by $ind(f_{x,y^*})$ and $2^i(x_{n-i-1} \ldots x_0) \cdot (y_{n-1} \ldots y_i)$. We summarize this result in the following claim:

Claim 1. *Each subfunction $f_{x,y^*} \in \mathcal{F}_i$ is uniquely determined by $(x_{n-i-1} \ldots x_0)$ and its index $ind(f_{x,y^*})$.*   □

We are now ready to prove the upper bounds.

Proof (of Theorem 2). Let G be the minimal-size π-OBDD, which reads first all x-bits and then the bits $y_0, \ldots, y_{n-1}$ in this order. Further, let $k = \lceil n/3 \rceil$. Denote the upper part of G to the subgraph, in which the x-variables and the variables $y_0, \ldots, y_{k-1}$ are read. Obviously, this part contains at most as many nodes as a balanced binary tree with $n + k$ levels, thus has a size of at most $2^{n+k} - 1$.

We bound now the number of y_i-nodes, for $i \geq k$. By Theorem 6, this is at most the number of different subfunctions f_{x,y^*} in $\mathcal{F}_i$. But since there are only 2^{n-i} different values for $ind(f_{x,y^*})$ and as many values for $(x_{n-i-1} \ldots x_0)$, it follows from Claim 1 that there are at most $2^{2(n-i)}$ different subfunctions in $\mathcal{F}_i$. So, the bottom part of G consists of at most $2^{2(n-i)}$ inner nodes for each $i \in \{k, \ldots, n-1\}$. An easy calculation shows that both parts contain together at most

$$2^{n+k} - 1 + \sum_{i=k}^{n-1} 2^{2(n-i)} = 2^{n+k} + \frac{4}{3} \cdot 2^{2n-2k} - \frac{7}{3}$$

inner nodes. Since $k = \lceil n/3 \rceil$, we may write $n = 3k - \tau$ for some $0 \leq \tau < 3$. Thus, also counting the two sinks, the π-OBDD-size is bounded above by

$$2^{4k-\tau} + \frac{4}{3} \cdot 2^{4k-2\tau} - \frac{7}{3} + 2 \leq 2^{4k-4\tau/3}\left(2^{\tau/3} + \frac{4}{3} \cdot 2^{-2\tau/3}\right).$$

By case distinction ($\tau = 0, 1, 2$) it can be easily verified that the factor in parenthesis has a value of at most $7/3$. Since further the exponent of the first factor ($4k - 4\tau/3$) equals $4n/3$, the proof is complete. $\qquad\square$

The proof of Theorem 5 for $\mathrm{MUL}^a_{n-1,n}$ uses almost the same line of argument, so that we only sketch the differences. The vector x of variables is replaced with the constant a, and the variables $y_0, \ldots, y_{n-1}$ are read again in this order. But now, the upper part of the OBDD consists of the first $n/2$ variables of y, that is $y_0, \ldots, y_{n/2-1}$, and its size is again bounded by that of a binary tree ($2^{n/2} - 1$). Using the index of the functions f_{a,y^*}, the number of different subfunctions in $\mathcal{F}_i$ is then bounded for $n/2 \leq i \leq n - 1$ similarly to the above proof. In this way, we conclude that the lower part of the OBDD consists of at most $\sum_{i=n/2}^{n-1} 2^{n-i} = 2(2^{n/2} - 1)$ inner nodes, which shows the claim.

Note that it is possible to specify the subfunctions f_{x,y^*} explicitly. This means that the above proofs do not only show the existence of OBDDs with the properties stated in Theorems 2 and 5, but can in fact be used to construct them.

Acknowledgments

I thank Ingo Wegener for valuable hints leading to several improvements, as well as Beate Bollig and Detlef Sieling for their helpful comments.

References

1. S. B. Akers. Binary decision diagrams. *IEEE Transactions on Computers*, C-27:509–516, 1978.
2. M. Atici and D. R. Stinson. Universal hashing and multiple authentication. In *Advances in Cryptology – CRYPTO '96*, pp. 16–30. 1996.
3. P. Beame, M. Tompa, and P. Yan. Communication-space tradeoffs for unrestricted protocols. *SIAM Journal on Computing*, 23:652–661, 1994.
4. B. Bollig and I. Wegener. Improving the variable ordering of OBDDs is NP-complete. *IEEE Transactions on Computers*, 45:993–1002, 1996.
5. B. Bollig and I. Wegener. Asymptotically optimal bounds for OBDDs and the solution of some basic OBDD problems. In *Proceedings of the 25th International Colloquium on Automata, Languages, and Programming*, pp. 187–198. 2000.
6. Y. Breibart, H. B. Hunt III, and D. Rosenkrantz. On the size of binary decision diagrams representing Boolean functions. *Theoretical Computer Science*, 145:45–69, 1995.
7. R. E. Bryant. Graph-based algorithms for boolean function manipulation. *IEEE Transactions on Computers*, C-35:677–691, 1986.

8. R. E. Bryant. On the complexity of VLSI implementations and graph representations of boolean functions with applications to integer multiplication. *IEEE Transactions on Computers*, 40:205–213, 1991.

9. J. L. Carter and M. N. Wegman. Universal classes of hash functions. *Journal of Computer and System Sciences*, 18:143–154, 1979.

10. M. Dietzfelbinger. Universal hashing and k-wise independent random variables via integer arithmetic without primes. In *Proceedings of the 13th Annual Symposium on Theoretical Aspects of Computer Science*, pp. 569–580. 1996.

11. M. Dietzfelbinger, T. Hagerup, J. Katajainen, and M. Penttonen. A reliable randomized algorithm for the closest-pair problem. *Journal of Algorithms*, 25:19–51, 1997.

12. M. Dietzfelbinger, A. Karlin, K. Mehlhorn, F. Meyer auf der Heide, H. Rohnert, and R. E. Tarjan. Dynamic perfect hashing: Upper and lower bounds. *SIAM Journal on Computing*, 23:738–761, 1994.

13. M. L. Fredman, J. Komlós, and E. Szemerédi. Storing a sparse table with $O(1)$ worst case access time. *Journal of the Association for Computing Machinery*, 31:538–544, 1984.

14. R. Impagliazzo and D. Zuckerman. How to recycle random bits. In *Proceedings of the 30th Annual IEEE Symposium on Fountations of Computer Science*, pp. 248–253. 1989.

15. C. Y. Lee. Representation of switching circuits by binary-decision programs. *The Bell Systems Technical Journal*, 38:985–999, 1959.

16. Y. Mansour, N. Nisan, and P. Tiwari. The computational complexity of universal hashing. *Theoretical Computer Science*, 107:121–133, 1993.

17. P. Rogaway. Bucket hashing and its application to fast message authentication. In *Advances in Cryptology – CRYPTO '95*, pp. 29–42. 1995.

18. D. Sieling. On the existence of polynomial time approximation schemes for OBDD minimization. In *Proceedings of the 15th Annual Symposium on Theoretical Aspects of Computer Science*, pp. 205–215. 1998.

19. D. Sieling and I. Wegener. NC-algorithms for operations on binary decision diagrams. *Parallel Processing Letters*, 48:139–144, 1993.

20. M. Sipser. A complexity theoretic approach to randomness. In *Proceedings of the 15th Annual ACM Symposium on Theory of Computing*, pp. 330–335. 1983.

21. I. Wegener. *Branching Programs and Binary Decision Diagrams - Theory and Applications*. Siam, first edition, 2000.

22. M. N. Wegman and J. L. Carter. New classes and applications of hash functions. In *Proceedings of the 20th Annual IEEE Symposium on Fountations of Computer Science*, pp. 175–182. 1979.

23. P. Woelfel. Efficient strongly universal and optimally universal hashing. In *Mathematical Foundations of Computer Science: 24th International Symposium*, pp. 262–272. 1999.

24. P. Woelfel. Klassen universeller Hashfunktionen mit ganzzahliger Arithmetik. Diploma thesis, Univ. Dortmund, 2000.

Author Index

Lecture Notes in Computer Science

For information about Vols. 1–1914
please contact your bookseller or Springer-Verlag

Vol. 1945: W. Grieskamp, T. Santen, B. Stoddart (Eds.), Integrated Formal Methods. Proceedings, 2000. X, 441 pages. 2000.

Vol. 1946: P. Palanque, F. Paternò (Eds.), Interactive Systems. Proceedings, 2000. X, 251 pages. 2001.

Vol. 1948: T. Tan, Y. Shi, W. Gao (Eds.), Advances in Multimodal Interfaces – ICMI 2000. Proceedings, 2000. XVI, 678 pages. 2000.

Vol. 1949: R. Connor, A. Mendelzon (Eds.), Research Issues in Structured and Semistructured Database Programming. Proceedings, 1999. XII, 325 pages. 2000.

Vol. 1950: D. van Melkebeek, Randomness and Completeness in Computational Complexity. XV, 196 pages. 2000.

Vol. 1951: F. van der Linden (Ed.), Software Architectures for Product Families. Proceedings, 2000. VIII, 255 pages. 2000.

Vol. 1952: M.C. Monard, J. Simão Sichman (Eds.), Advances in Artificial Intelligence. Proceedings, 2000. XV, 498 pages. 2000. (Subseries LNAI).

Vol. 1953: G. Borgefors, I. Nyström, G. Sanniti di Baja (Eds.), Discrete Geometry for Computer Imagery. Proceedings, 2000. XI, 544 pages. 2000.

Vol. 1954: W.A. Hunt, Jr., S.D. Johnson (Eds.), Formal Methods in Computer-Aided Design. Proceedings, 2000. XI, 539 pages. 2000.

Vol. 1955: M. Parigot, A. Voronkov (Eds.), Logic for Programming and Automated Reasoning. Proceedings, 2000. XIII, 487 pages. 2000. (Subseries LNAI).

Vol. 1956: T. Coquand, P. Dybjer, B. Nordström, J. Smith (Eds.), Types for Proofs and Programs. Proceedings, 1999. VII, 195 pages. 2000.

Vol. 1957: P. Ciancarini, M. Wooldridge (Eds.), Agent-Oriented Software Engineering. Proceedings, 2000. X, 323 pages. 2001.

Vol. 1960: A. Ambler, S.B. Calo, G. Kar (Eds.), Services Management in Intelligent Networks. Proceedings, 2000. X, 259 pages. 2000.

Vol. 1961: J. He, M. Sato (Eds.), Advances in Computing Science – ASIAN 2000. Proceedings, 2000. X, 299 pages. 2000.

Vol. 1963: V. Hlaváč, K.G. Jeffery, J. Wiedermann (Eds.), SOFSEM 2000: Theory and Practice of Informatics. Proceedings, 2000. XI, 460 pages. 2000.

Vol. 1964: J. Malenfant, S. Moisan, A. Moreira (Eds.), Object-Oriented Technology. Proceedings, 2000. XI, 309 pages. 2000.

Vol. 1965: Ç. K. Koç, C. Paar (Eds.), Cryptographic Hardware and Embedded Systems – CHES 2000. Proceedings, 2000. XI, 355 pages. 2000.

Vol. 1966: S. Bhalla (Ed.), Databases in Networked Information Systems. Proceedings, 2000. VIII, 247 pages. 2000.

Vol. 1967: S. Arikawa, S. Morishita (Eds.), Discovery Science. Proceedings, 2000. XII, 332 pages. 2000. (Subseries LNAI).

Vol. 1968: H. Arimura, S. Jain, A. Sharma (Eds.), Algorithmic Learning Theory. Proceedings, 2000. XI, 335 pages. 2000. (Subseries LNAI).

Vol. 1969: D.T. Lee, S.-H. Teng (Eds.), Algorithms and Computation. Proceedings, 2000. XIV, 578 pages. 2000.

Vol. 1970: M. Valero, V.K. Prasanna, S. Vajapeyam (Eds.), High Performance Computing – HiPC 2000. Proceedings, 2000. XVIII, 568 pages. 2000.

Vol. 1971: R. Buyya, M. Baker (Eds.), Grid Computing – GRID 2000. Proceedings, 2000. XIV, 229 pages. 2000.

Vol. 1972: A. Omicini, R. Tolksdorf, F. Zambonelli (Eds.), Engineering Societies in the Agents World. Proceedings, 2000. IX, 143 pages. 2000. (Subseries LNAI).

Vol. 1973: J. Van den Bussche, V. Vianu (Eds.), Database Theory – ICDT 2001. Proceedings, 2001. X, 451 pages. 2001.

Vol. 1974: S. Kapoor, S. Prasad (Eds.), FST TCS 2000: Foundations of Software Technology and Theoretical Computer Science. Proceedings, 2000. XIII, 532 pages. 2000.

Vol. 1975: J. Pieprzyk, E. Okamoto, J. Seberry (Eds.), Information Security. Proceedings, 2000. X, 323 pages. 2000.

Vol. 1976: T. Okamoto (Ed.), Advances in Cryptology – ASIACRYPT 2000. Proceedings, 2000. XII, 630 pages. 2000.

Vol. 1977: B. Roy, E. Okamoto (Eds.), Progress in Cryptology – INDOCRYPT 2000. Proceedings, 2000. X, 295 pages. 2000.

Vol. 1979: S. Moss, P. Davidsson (Eds.), Multi-Agent-Based Simulation. Proceedings, 2000. VIII, 267 pages. 2001. (Subseries LNAI).

Vol. 1983: K.S. Leung, L.-W. Chan, H. Meng (Eds.), Intelligent Data Engineering and Automated Learning – IDEAL 2000. Proceedings, 2000. XVI, 573 pages. 2000.

Vol. 1984: J. Marks (Ed.), Graph Drawing. Proceedings, 2001. XII, 419 pages. 2001.

Vol. 1987: K.-L. Tan, M.J. Franklin, J. C.-S. Lui (Eds.), Mobile Data Management. Proceedings, 2001. XIII, 289 pages. 2001.

Vol. 1989: M. Ajmone Marsan, A. Bianco (Eds.), Quality of Service in Multiservice IP Networks. Proceedings, 2001. XII, 440 pages. 2001.

Vol. 1991: F. Dignum, C. Sierra (Eds.), Agent Mediated Electronic Commerce. VIII, 241 pages. 2001. (Subseries LNAI).

Vol. 1992: K. Kim (Ed.), Public Key Cryptography. Proceedings, 2001. XI, 423 pages. 2001.

Vol. 1995: M. Sloman, J. Lobo, E.C. Lupu (Eds.), Policies for Distributed Systems and Networks. Proceedings, 2001. X, 263 pages. 2001.

Vol. 1998: R. Klette, S. Peleg, G. Sommer (Eds.), Robot Vision. Proceedings, 2001. IX, 285 pages. 2001.

Vol. 2000: R. Wilhelm (Ed.), Informatics: 10 Years Back, 10 Years Ahead. IX, 369 pages. 2001.

Vol. 2004: A. Gelbukh (Ed.), Computational Linguistics and Intelligent Text Processing. Proceedings, 2001. XII, 528 pages. 2001.

Vol. 2010: A. Ferreira, H. Reichel (Eds.), STACS 2001. Proceedings, 2001. XV, 576 pages. 2001.